Multivariate Time Series Analysis

A complete list of the titles in this series appears at the end of this volume.

Multivariate Time Series Analysis

With R and Financial Applications

RUEY S. TSAY

Booth School of Business
University of Chicago
Chicago, IL

WILEY

Published by John Wiley & Sons, Inc., Hoboken, New Jersey
Published simultaneously in Canada

Library of Congress Cataloging-in-Publication Data:

Tsay, Ruey S., 1951–
Multivariate time series analysis: with R and financial applications / Ruey S. Tsay, Booth School of Business, University of Chicago, Chicago, IL.
pages cm
Includes bibliographical references and index.
ISBN 978-1-118-61790-8 (hardback)
1. Time-series analysis. 2. R (Computer program language) 3. Econometric models. I. Title.
QA280.T73 2014
519.5′5–dc23

2014009453

Printed in the United States of America

10 9 8 7 6 5 4 3 2

To my teacher and friend
George

Contents

Preface

This book is based on my experience in teaching and research on multivariate time series analysis over the past 30 years. It summarizes the basic concepts and ideas of analyzing multivariate dependent data, provides econometric and statistical models useful for describing the dynamic dependence between variables, discusses the identifiability problem when the models become too flexible, introduces ways to search for simplifying structure hidden in high-dimensional time series, addresses the applicabilities and limitations of multivariate time series methods, and, equally important, develops a software package for readers to apply the methods and models discussed in the book.

Multivariate time series analysis provides useful tools and methods for processing information embedded in multiple measurements that have temporal and cross-sectional dependence. The goal of the analysis is to provide a better understanding of the dynamic relationship between variables and to improve the accuracy in forecasting. The models built can also be used in policy simulation or in making inference. The book focuses mainly on linear models as they are easier to comprehend and widely applicable. I tried to draw a balance between theory and applications and kept the notation as consistent as possible. I also tried to make the book self-contained. However, given the complexity of the subject, the level of coverage on selected topics may vary throughout the book. This reflects in part my own preference and understanding of the topics and in part my desire to keep the book at a reasonable length.

The field of high-dimensional data analysis is still under rapid developments, especially for dependent data. Omission of some important topics or methods is not avoidable for a book like this one. For instance, nonlinear models are not discussed, nor the categorical time series. Readers are advised to consult recent articles or journals for further development.

The book starts with some general concepts of multivariate time series in Chapter 1, including assessing and quantifying temporal and cross-sectional dependence. As the dimension increases, the difficulty in presenting multivariate data quickly becomes evident. I tried to keep the presentation in a compact form if possible. In some cases, scalar summary statistics are given. Chapter 2 focuses on vector autoregressive (VAR) models as they are, arguably, the most widely used multivariate time series models. My goal is to make the chapter as comprehensive as possible for readers who are interested in VAR models. Both Bayesian and classical analyses of VAR models are included. Chapter 3 studies stationary vector autoregressive moving-average (VARMA) models. It begins with properties and estimation of vector moving-average (VMA) models. The issue of identifiability of VARMA models is investigated and properties of the models are given. Chapter 4 investigates the structural specification of a multivariate time series. Two methods are introduced to seek the simplifying structure hidden in a vector time series. These methods enable users to discover the *skeleton* of a linear multivariate time series. Chapter 5 deals with unit-root nonstationarity and cointegration. It includes the basic theory for understanding unit-root time series and some applications. In Chapter 6, I discuss factor models and some selected topics of multivariate time series. Both the classical and approximate factor models are studied. My goal is to cover all factor models currently available in the literature and to provide the relationship between them. Chapter 7 focuses on multivariate volatility modeling. It covers volatility models that are relatively easy to use and produce positive-definite volatility matrices. The chapter also discusses ways to detect conditional heteroscedasticity in a vector time series and methods for checking a fitted multivariate volatility model. Throughout the book, real examples are used to demonstrate the analysis. Every chapter contains some exercises that analyze empirical vector time series.

Software is an integral part of multivariate time series analysis. Without software packages, multivariate time series becomes a pure theoretical exercise. I have tried my best to write R programs that enable readers to apply all methods and models discussed in the book. These programs are included in the `MTS` package available in R. Readers can duplicate all the analyses shown in the book with the package and some existing R packages. Not a professional programmer, I am certain that many of the codes in `MTS` are not as efficient as they can be and are likely to have bugs. I would appreciate any suggestions and/or corrections to both the package and the book.

RUEY S. TSAY

Chicago, Illinois
September 2014

Acknowledgements

This book would not have been written without the great teachers I have had. In particular, I would like to express my sincere thanks to Professor George C. Tiao who taught me time series analysis and statistical research. His insightful view of empirical time series and his continual encouragements are invaluable. I would like to thank Professor Tea-Yuan Hwang who introduced me to statistics and has remained a close friend over the past four decades. I would also like to thank Mr. Sung-Nan Chen, my junior high school teacher. Without his foresight, I would not have pursued my college education. I would like to thank many other teachers, including late Professor George E. P. Box and late Professor Gregory Reinsel of University of Wisconsin, and friends, including Dr. David F. Findley, Professors Daniel Peña, Manny Parzen, Buddy Gray, and Howell Tong, and late Professor Hirotugu Akaike, for their support of my research in time series analysis. Dr. David Matteson and Mr. Yongning Wang kindly allowed me to use their programs and Yongning has read over the draft carefully. I appreciate their help. I would also like to thank many students who asked informative questions both in and outside the classrooms. I wish to express my sincere thanks to Stephen Quigley and Sari Friedman for their support in preparing this book. I also wish to acknowledge the financial support of Chicago Booth. Finally, I would like to thank my parents who sacrificed so much to support me and for their unconditional love. As always, my children are my inspiration and sources of energy. Finally, I would like to express my sincere thanks to my wife for her love and constant encouragement. In particular, she has always put my career ahead of her own.

The web page of the book is `http://faculty.chicagobooth.edu/ruey.tsay/teaching/mtsbk`.

R. S. T.

CHAPTER 1

Multivariate Linear Time Series

1.1 INTRODUCTION

Multivariate time series analysis considers simultaneously multiple time series. It is a branch of multivariate statistical analysis but deals specifically with dependent data. It is, in general, much more complicated than the univariate time series analysis, especially when the number of series considered is large. We study this more complicated statistical analysis in this book because in real life decisions often involve multiple inter-related factors or variables. Understanding the relationships between those factors and providing accurate predictions of those variables are valuable in decision making. The objectives of multivariate time series analysis thus include

1. To study the dynamic relationships between variables
2. To improve the accuracy of prediction

Let $\boldsymbol{z}_t = (z_{1t}, \cdots, z_{kt})'$ be a k-dimensional time series observed at equally spaced time points. For example, let z_{1t} be the quarterly U.S. real gross domestic product (GDP) and z_{2t} the quarterly U.S. civilian unemployment rate. By studying z_{1t} and z_{2t} jointly, we can assess the temporal and contemporaneous dependence between GDP and unemployment rate. In this particular case, $k = 2$ and the two variables are known to be instantaneously negatively correlated. Figure 1.1 shows the time plots of quarterly U.S. real GDP (in logarithm of billions of chained 2005 dollars) and unemployment rate, obtained via monthly data with averaging, from 1948 to 2011. Both series are seasonally adjusted. Figure 1.2 shows the time plots of the real GDP growth rate and the changes in unemployment rate from the second quarter of 1948 to the fourth quarter of 2011. Figure 1.3 shows the scatter plot of the two time series given in Figure 1.2. From these figures, we can see that the GDP

Multivariate Time Series Analysis: With R and Financial Applications,
First Edition. Ruey S. Tsay.

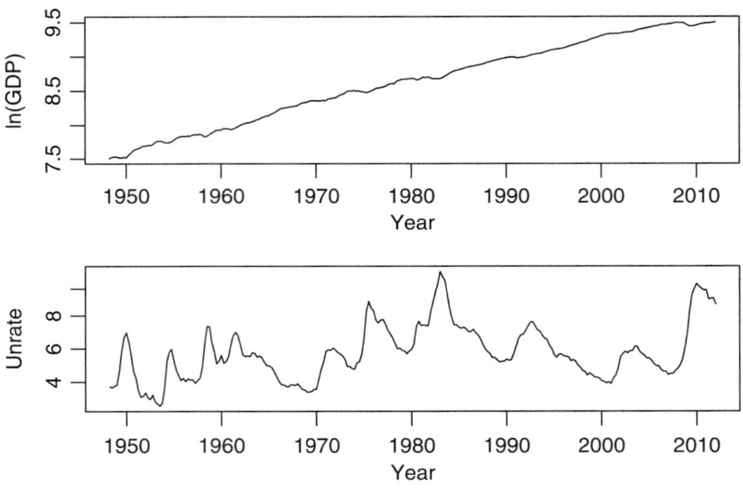

FIGURE 1.1 Time plots of U.S. quarterly real GDP (in logarithm) and unemployment rate from 1948 to 2011. The data are seasonally adjusted.

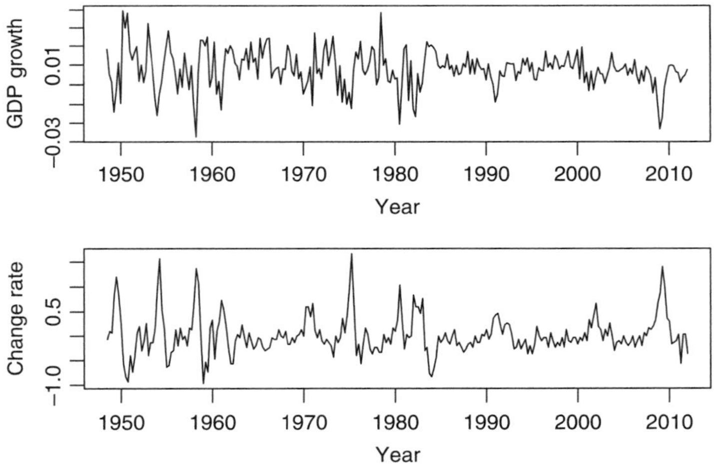

FIGURE 1.2 Time plots of the growth rate of U.S. quarterly real GDP (in logarithm) and the change series of unemployment rate from 1948 to 2011. The data are seasonally adjusted.

and unemployment rate indeed have negative instantaneous correlation. The sample correlation is -0.71.

As another example, consider $k = 3$. Let z_{1t} be the monthly housing starts of the New England division in the United States, and z_{2t} and z_{3t} be the monthly housing starts of the Middle Atlantic division and the Pacific division, respectively. By considering the three series jointly, we can investigate the relationships between the housing markets of the three geographical divisions in the United States. Figure 1.4

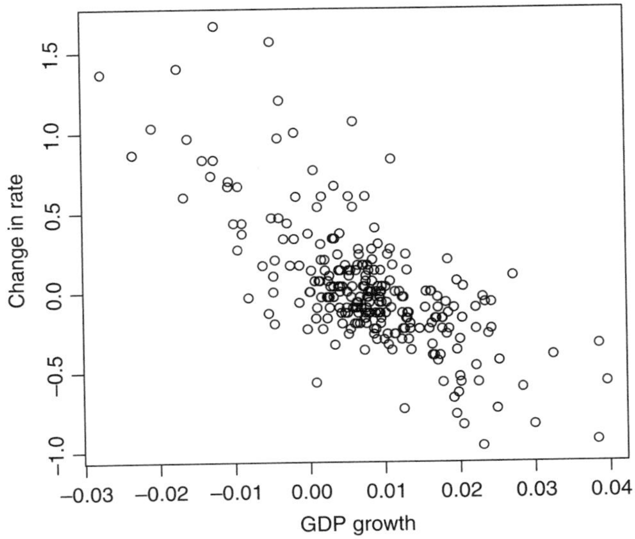

FIGURE 1.3 Scatter plot of the changes in quarterly U.S. unemployment rate versus the growth rate of quarterly real GDP (in logarithm) from the second quarter of 1948 to the last quarter of 2011. The data are seasonally adjusted.

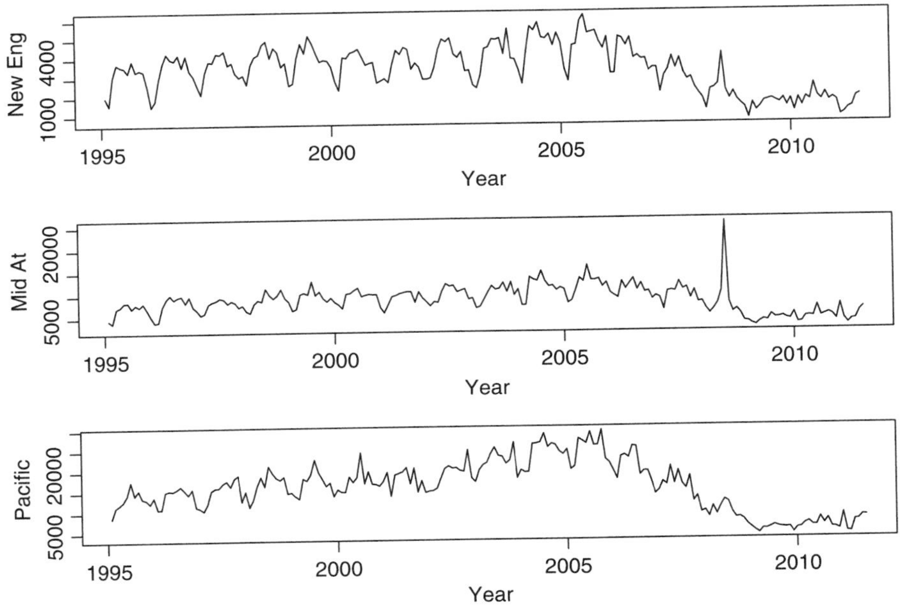

FIGURE 1.4 Time plots of the monthly housing starts for the New England, Middle Atlantic, and Pacific divisions of the United States from January 1995 to June 2011. The data are not seasonally adjusted.

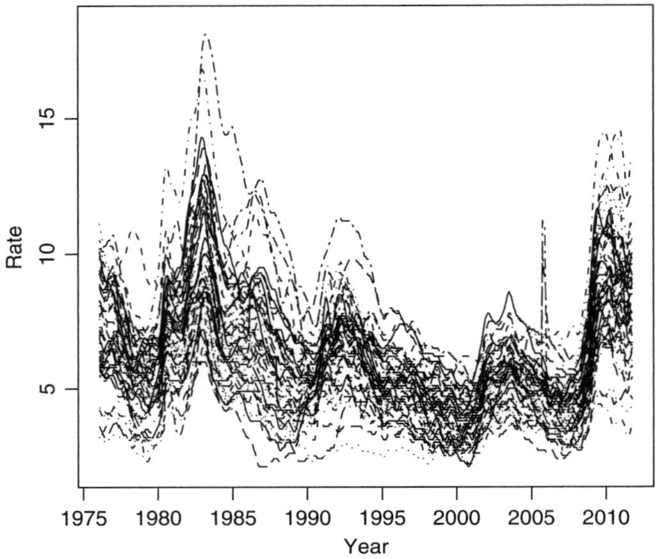

FIGURE 1.5 Time plots of the monthly unemployment rates of the 50 states in the United States from January 1976 to September 2011. The data are seasonally adjusted.

shows the time plots of the three monthly housing starts from January 1995 to June 2011. The data are not seasonally adjusted so that there exists a clear seasonal cycle in the series. From the plots, the three series show certain similarities as well as some marked differences. In some applications, we consider large k. For instance, let $\boldsymbol{z}_t$ be the monthly unemployment rates of the 50 states in the United States. Figure 1.5 shows the time plots of the monthly unemployment rates of the 50 states from January 1976 to September 2011. The data are seasonally adjusted. Here, $k = 50$ and plots are not particularly informative except that the series have certain common behavior. The objective of considering these series simultaneously may be to obtain predictions for the state unemployment rates. Such forecasts are important to state and local governments. In this particular instance, pooling information across states may be helpful in prediction because states may have similar social and economic characteristics.

In this book, we refer to $\{z_{it}\}$ as the ith component of the multivariate time series $\boldsymbol{z}_t$. The objectives of the analysis discussed in this book include (a) to investigate the dynamic relationships between the components of $\boldsymbol{z}_t$ and (b) to improve the prediction of z_{it} using information in all components of $\boldsymbol{z}_t$.

Suppose we are interested in predicting $\boldsymbol{z}_{T+1}$ based on the data $\{\boldsymbol{z}_1, \ldots, \boldsymbol{z}_T\}$. To this end, we may entertain the model

$$\hat{\boldsymbol{z}}_{T+1} = g(\boldsymbol{z}_T, \boldsymbol{z}_{T-1}, \ldots, \boldsymbol{z}_1),$$

where $\hat{\boldsymbol{z}}_{T+1}$ denotes a prediction of $\boldsymbol{z}_{T+1}$ and $g(.)$ is some suitable function. The goal of multivariate time series analysis is to specify the function $g(.)$ based on the available data. In many applications, $g(.)$ is a smooth, differentiable function and can be well approximated by a linear function, say,

$$\hat{\boldsymbol{z}}_{T+1} \approx \boldsymbol{\pi}_0 + \boldsymbol{\pi}_1 \boldsymbol{z}_T + \boldsymbol{\pi}_2 \boldsymbol{z}_{T-1} + \cdots + \boldsymbol{\pi}_T \boldsymbol{z}_1,$$

where $\boldsymbol{\pi}_0$ is a k-dimensional vector, and $\boldsymbol{\pi}_i$ are $k \times k$ constant real-valued matrices (for $i = 1, \ldots, T$). Let $\boldsymbol{a}_{T+1} = \boldsymbol{z}_{T+1} - \hat{\boldsymbol{z}}_{T+1}$ be the forecast error. The prior equation states that

$$\boldsymbol{z}_{T+1} = \boldsymbol{\pi}_0 + \boldsymbol{\pi}_1 \boldsymbol{z}_T + \boldsymbol{\pi}_2 \boldsymbol{z}_{T-1} + \cdots + \boldsymbol{\pi}_T \boldsymbol{z}_1 + \boldsymbol{a}_{T+1}$$

under the linearity assumption.

To build a solid foundation for making prediction described in the previous paragraph, we need sound statistical theories and methods. The goal of this book is to provide some useful statistical models and methods for analyzing multivariate time series. To begin with, we start with some basic concepts of multivariate time series.

1.2 SOME BASIC CONCEPTS

Statistically speaking, a k-dimensional time series $\boldsymbol{z}_t = (z_{1t}, \ldots, z_{kt})'$ is a random vector consisting of k random variables. As such, there exists an underlying probability space on which the random variables are defined. What we observe in practice is a *realization* of this random vector. For simplicity, we use the same notation $\boldsymbol{z}_t$ for the random vector and its realization. When we discuss properties of $\boldsymbol{z}_t$, we treat it as a random vector. On the other hand, when we consider an application, we treat $\boldsymbol{z}_t$ as a realization. In this book, we assume that $\boldsymbol{z}_t$ follows a continuous multivariate probability distribution. In other words, the discrete-valued (or categorical) multivariate time series are not considered. Because we are dealing with random vectors, vector and matrix are used extensively in the book. If necessary, readers can consult Appendix A for a brief review of mathematics and statistics.

1.2.1 Stationarity

A k-dimensional time series $\boldsymbol{z}_t$ is said to be weakly stationary if (a) $E(\boldsymbol{z}_t) = \boldsymbol{\mu}$, a k-dimensional constant vector, and (b) $\text{Cov}(\boldsymbol{z}_t) = E[(\boldsymbol{z}_t - \boldsymbol{\mu})(\boldsymbol{z}_t - \boldsymbol{\mu})'] = \boldsymbol{\Sigma}_z$, a constant $k \times k$ positive-definite matrix. Here, $E(\boldsymbol{z})$ and $\text{Cov}(\boldsymbol{z})$ denote the expectation and covariance matrices of the random vector $\boldsymbol{z}$, respectively. Thus, the mean and covariance matrices of a weakly stationary time series $\boldsymbol{z}_t$ do not depend on time, that is, the first two moments of $\boldsymbol{z}_t$ are time invariant. Implicit in the definition, we require that the mean and covariance matrices of a weakly stationary time series exist.

A k-dimensional time series $\boldsymbol{z}_t$ is strictly stationary if the joint distribution of the m collection, $(\boldsymbol{z}_{t_1}, \ldots, \boldsymbol{z}_{t_m})$, is the same as that of $(\boldsymbol{z}_{t_1+j}, \ldots, \boldsymbol{z}_{t_m+j})'$, where m, j, and $(t_1, \ldots, t_m)$ are arbitrary positive integers. In statistical terms, strict stationarity requires that the probability distribution of an arbitrary collection of $\boldsymbol{z}_t$ is time invariant. An example of strictly stationary time series is the sequence of independent and identically distributed random vectors of standard multivariate normal distribution. From the definitions, a strictly stationary time series $\boldsymbol{z}_t$ is weakly stationary provided that its first two moments exist.

In this chapter, we focus mainly on the weakly stationary series because strict stationarity is hard to verify in practice. We shall consider nonstationary time series later. In what follows, stationarity means weak stationarity.

1.2.2 Linearity

We focus on multivariate linear time series in this book. Strictly speaking, real multivariate time series are nonlinear, but linear models can often provide accurate approximations for making inference. A k-dimensional time series $\boldsymbol{z}_t$ is linear if

$$\boldsymbol{z}_t = \boldsymbol{\mu} + \sum_{i=0}^{\infty} \boldsymbol{\psi}_i \boldsymbol{a}_{t-i}, \tag{1.1}$$

where $\boldsymbol{\mu}$ is a k-dimensional constant vector, $\boldsymbol{\psi}_0 = \boldsymbol{I}_k$, the $k \times k$ identity matrix, $\boldsymbol{\psi}_i$ $(i > 0)$ are $k \times k$ constant matrices, and $\{\boldsymbol{a}_t\}$ is a sequence of independent and identically distributed random vectors with mean zero and a positive-definite covariance matrix $\boldsymbol{\Sigma}_a$.

We require $\boldsymbol{\Sigma}_a$ to be positive-definite; otherwise, the dimension k can be reduced—see the principal component analysis discussed in Chapter 2. The condition that $\boldsymbol{\psi}_0 = \boldsymbol{I}_k$ is satisfied because we allow $\boldsymbol{\Sigma}_a$ to be a general positive-definite matrix. An alternative approach to express a linear time series is to require $\boldsymbol{\psi}_0$ to be a lower triangular matrix with diagonal elements being 1 and $\boldsymbol{\Sigma}_a$ a diagonal matrix. This is achieved by using the Cholesky decomposition of $\boldsymbol{\Sigma}_a$; see Appendix A. Specifically, decomposite the covariance matrix as $\boldsymbol{\Sigma}_a = \boldsymbol{L}\boldsymbol{G}\boldsymbol{L}'$, where $\boldsymbol{G}$ is a diagonal matrix and $\boldsymbol{L}$ is a $k \times k$ lower triangular matrix with 1 being its diagonal elements. Let $\boldsymbol{b}_t = \boldsymbol{L}^{-1}\boldsymbol{a}_t$. Then, $\boldsymbol{a}_t = \boldsymbol{L}\boldsymbol{b}_t$, and

$$\text{Cov}(\boldsymbol{b}_t) = \text{Cov}(\boldsymbol{L}^{-1}\boldsymbol{a}_t) = \boldsymbol{L}^{-1}\boldsymbol{\Sigma}_a(\boldsymbol{L}^{-1})' = \boldsymbol{L}^{-1}(\boldsymbol{L}\boldsymbol{G}\boldsymbol{L}')(\boldsymbol{L}')^{-1} = \boldsymbol{G}.$$

With the sequence $\{\boldsymbol{b}_t\}$, Equation (1.1) can be written as

$$\boldsymbol{z}_t = \boldsymbol{\mu} + \sum_{i=0}^{\infty}(\boldsymbol{\psi}_i\boldsymbol{L})\boldsymbol{b}_{t-i} = \boldsymbol{\mu} + \sum_{i=0}^{\infty} \boldsymbol{\psi}_i^* \boldsymbol{b}_{t-i}, \tag{1.2}$$

where $\boldsymbol{\psi}_0^* = \boldsymbol{L}$, which is a lower triangular matrix, $\boldsymbol{\psi}_i^* = \boldsymbol{\psi}_i\boldsymbol{L}$ for $i > 0$, and the covariance matrix of $\boldsymbol{b}_t$ is a diagonal matrix.

For a stationary, purely stochastic process $\boldsymbol{z}_t$, Wold decomposition states that it can be written as a linear combination of a sequence of serially uncorrelated process $\boldsymbol{e}_t$. This is close, but not identical, to Equation (1.1) because $\{\boldsymbol{e}_t\}$ do not necessarily have the same distribution. An example of $\boldsymbol{z}_t$ that satisfies the Wold decomposition, but not a linear time series, is the multivariate autoregressive conditional heteroscedastic process. We discuss multivariate volatility modeling in Chapter 7. The Wold decomposition, however, shows that the conditional mean of $\boldsymbol{z}_t$ can be written as a linear combination of the lagged values $\boldsymbol{z}_{t-i}$ for $i > 0$ if $\boldsymbol{z}_t$ is stationary and purely stochastic. This provides a justification for starting with linear time series because the conditional mean of $\boldsymbol{z}_t$ plays an important role in forecasting.

Consider Equation (1.1). We see that $\boldsymbol{z}_{t-1}$ is a function of $\{\boldsymbol{a}_{t-1}, \boldsymbol{a}_{t-2}, \cdots\}$. Therefore, at time index $t-1$, the only *unknown* quantity of $\boldsymbol{z}_t$ is $\boldsymbol{a}_t$. For this reason, we call $\boldsymbol{a}_t$ the *innovation* of the time series $\boldsymbol{z}_t$ at time t. One can think of $\boldsymbol{a}_t$ as the *new* information about the time series obtained at time t. We shall make the concept of innovation more precisely later when we discuss forecasting. The innovation $\boldsymbol{a}_t$ is also known as the *shock* to the time series at time t.

For the linear series $\boldsymbol{z}_t$ in Equation (1.1) to be stationary, the coefficient matrices must satisfy

$$\sum_{i=1}^{\infty} \|\boldsymbol{\psi}_i\| < \infty,$$

where $\|\boldsymbol{A}\|$ denotes a norm of the matrix $\boldsymbol{A}$, for example, the Frobenius norm $\|\boldsymbol{A}\| = \sqrt{tr(\boldsymbol{A}\boldsymbol{A}')}$. Based on the properties of a convergence series, this implies that $\|\boldsymbol{\psi}_i\| \to 0$ as $i \to \infty$. Thus, for a stationary linear time series $\boldsymbol{z}_t$ in Equation (1.1), we have $\boldsymbol{\psi}_i \to \boldsymbol{0}$ as $i \to \infty$. Furthermore, we have

$$E(\boldsymbol{z}_t) = \boldsymbol{\mu}, \quad \text{and} \quad \text{Cov}(\boldsymbol{z}_t) = \sum_{i=0}^{\infty} \boldsymbol{\psi}_i \boldsymbol{\Sigma}_a \boldsymbol{\psi}_i'. \tag{1.3}$$

We shall discuss the stationarity conditions of $\boldsymbol{z}_t$ later for various models.

1.2.3 Invertibility

In many situations, for example, forecasting, we like to express the time series $\boldsymbol{z}_t$ as a function of its lagged values $\boldsymbol{z}_{t-i}$ for $i > 0$ plus new information at time t. A time series $\boldsymbol{z}_t$ is said to be invertible if it can be written as

$$\boldsymbol{z}_t = \boldsymbol{c} + \boldsymbol{a}_t + \sum_{j=1}^{\infty} \boldsymbol{\pi}_j \boldsymbol{z}_{t-j}, \tag{1.4}$$

where $\boldsymbol{c}$ is a k-dimensional constant vector, $\boldsymbol{a}_t$ is defined as before in Equation (1.1), and $\boldsymbol{\pi}_i$ are $k \times k$ constant matrices. An obvious example of an invertible time series is a vector autoregressive (VAR) series of order 1, namely, $\boldsymbol{z}_t = \boldsymbol{c} + \boldsymbol{\pi}_1 \boldsymbol{z}_{t-1} + \boldsymbol{a}_t$. Again, we shall discuss the invertibility conditions later. Here, it suffices to say that, for an invertible series $\boldsymbol{z}_t$, $\boldsymbol{\pi}_i \to \boldsymbol{0}$ as $i \to \infty$.

1.3 CROSS-COVARIANCE AND CORRELATION MATRICES

To measure the linear dynamic dependence of a stationary time series $\boldsymbol{z}_t$, we define its lag ℓ cross-covariance matrix as

$$\begin{aligned}\boldsymbol{\Gamma}_\ell &= \text{Cov}(\boldsymbol{z}_t, \boldsymbol{z}_{t-\ell}) = E[(\boldsymbol{z}_t - \boldsymbol{\mu})(\boldsymbol{z}_{t-\ell} - \boldsymbol{\mu})'] \\ &= \begin{bmatrix} E(\tilde{z}_{1t}\tilde{z}_{1,t-\ell}) & E(\tilde{z}_{1t}\tilde{z}_{2,t-\ell}) & \cdots & E(\tilde{z}_{1t}\tilde{z}_{k,t-\ell}) \\ \vdots & \vdots & & \vdots \\ E(\tilde{z}_{kt}\tilde{z}_{1,t-\ell}) & E(\tilde{z}_{kt}\tilde{z}_{2,t-\ell}) & \cdots & E(\tilde{z}_{kt}\tilde{z}_{k,t-\ell}) \end{bmatrix},\end{aligned} \tag{1.5}$$

where $\boldsymbol{\mu} = E(\boldsymbol{z}_t)$ is the mean vector of $\boldsymbol{z}_t$ and $\tilde{\boldsymbol{z}}_t = (\tilde{z}_{1t}, \ldots, \tilde{z}_{kt})' \equiv \boldsymbol{z}_t - \boldsymbol{\mu}$ is the mean-adjusted time series. This cross-covariance matrix is a function of ℓ, not the time index t, because $\boldsymbol{z}_t$ is stationary. For $\ell = 0$, we have the covariance matrix $\boldsymbol{\Gamma}_0$ of $\boldsymbol{z}_t$. In some cases, we use the notation $\boldsymbol{\Sigma}_z$ to denote the covariance matrix of $\boldsymbol{z}_t$, that is, $\boldsymbol{\Sigma}_z = \boldsymbol{\Gamma}_0$.

Denote the (i, j)th element of $\boldsymbol{\Gamma}_\ell$ as $\gamma_{\ell,ij}$, that is, $\boldsymbol{\Gamma}_\ell = [\gamma_{\ell,ij}]$. From the definition in Equation (1.5), we see that $\gamma_{\ell,ij}$ is the covariance between $z_{i,t}$ and $z_{j,t-\ell}$. Therefore, for a positive lag ℓ, $\gamma_{\ell,ij}$ can be regarded as a measure of the linear dependence of the ith component z_{it} on the ℓth lagged value of the jth component z_{jt}. This interpretation is important because we use matrix in the book and one must understand the meaning of each element in a matrix.

From the definition in Equation (1.5), for negative lag ℓ, we have

$$\begin{aligned}\boldsymbol{\Gamma}_\ell &= E[(\boldsymbol{z}_t - \boldsymbol{\mu})(\boldsymbol{z}_{t-\ell} - \boldsymbol{\mu})'] \\ &= E[(\boldsymbol{z}_{t+\ell} - \boldsymbol{\mu})(\boldsymbol{z}_t - \boldsymbol{\mu})'], && \text{(because of stationarity)} \\ &= \{E[(\boldsymbol{z}_t - \boldsymbol{\mu})(\boldsymbol{z}_{t+\ell} - \boldsymbol{\mu})']\}', && \text{(because } \boldsymbol{C} = (\boldsymbol{C}')'\text{)} \\ &= \left\{E[(\boldsymbol{z}_t - \boldsymbol{\mu})(\boldsymbol{z}_{t-(-\ell)} - \boldsymbol{\mu})']\right\}' \\ &= \{\boldsymbol{\Gamma}_{-\ell}\}', && \text{(by definition)} \\ &= \boldsymbol{\Gamma}'_{-\ell}.\end{aligned}$$

Therefore, unlike the case of univariate stationary time series for which the autocovariances of lag ℓ and lag $-\ell$ are identical, one must take the transpose of a positive-lag cross-covariance matrix to obtain the negative-lag cross-covariance matrix.

Remark: Some researchers define the cross-covariance matrix of $\boldsymbol{z}_t$ as $\boldsymbol{G}_\ell = E[(\boldsymbol{z}_{t-\ell} - \boldsymbol{\mu})(\boldsymbol{z}_t - \boldsymbol{\mu})']$, which is the transpose matrix of Equation (1.5). This is also a valid definition; see the property $\boldsymbol{\Gamma}_{-\ell} = \boldsymbol{\Gamma}'_\ell$. However, the meanings of the off-diagonal elements of $\boldsymbol{G}_\ell$ are different from those defined in Equation (1.5) for $\ell > 0$. As a matter of fact, the (i, j)th element $g_{\ell,ij}$ of $\boldsymbol{G}_\ell$ measures the linear dependence of z_{jt} on the lagged value $z_{i,t-\ell}$ of z_{it}. So long as readers understand the meanings of elements of a cross-covariance matrix, either definition can be used. □

For a stationary multivariate linear time series $\boldsymbol{z}_t$ in Equation (1.1), we have, for $\ell \geq 0$,

$$\begin{aligned}
\boldsymbol{\Gamma}_\ell &= E[(\boldsymbol{z}_t - \boldsymbol{\mu})(\boldsymbol{z}_{t-\ell} - \boldsymbol{\mu})'] \\
&= E[(\boldsymbol{a}_t + \boldsymbol{\psi}_1 \boldsymbol{a}_{t-1} + \cdots)(\boldsymbol{a}_{t-\ell} + \boldsymbol{\psi}_1 \boldsymbol{a}_{t-\ell-1} + \cdots)'] \\
&= E[(\boldsymbol{a}_t + \boldsymbol{\psi}_1 \boldsymbol{a}_{t-1} + \cdots)(\boldsymbol{a}'_{t-\ell} + \boldsymbol{a}'_{t-\ell-1} \boldsymbol{\psi}'_1 + \cdots)] \\
&= \sum_{i=\ell}^{\infty} \boldsymbol{\psi}_i \boldsymbol{\Sigma}_a \boldsymbol{\psi}'_{i-\ell}, \qquad (1.6)
\end{aligned}$$

where the last equality holds because $\boldsymbol{a}_t$ has no serial covariances and $\boldsymbol{\psi}_0 = \boldsymbol{I}_k$.

For a stationary series $\boldsymbol{z}_t$, the lag ℓ cross-correlation matrix (CCM) $\boldsymbol{\rho}_\ell$ is defined as

$$\boldsymbol{\rho}_\ell = \boldsymbol{D}^{-1} \boldsymbol{\Gamma}_\ell \boldsymbol{D}^{-1} = [\rho_{\ell,ij}], \qquad (1.7)$$

where $\boldsymbol{D} = \text{diag}\{\sigma_1, \ldots, \sigma_k\}$ is the diagonal matrix of the standard deviations of the components of $\boldsymbol{z}_t$. Specifically, $\sigma_i^2 = \text{Var}(z_{it}) = \gamma_{0,ii}$, that is, the (i, i)th element of $\boldsymbol{\Gamma}_0$. Obviously, $\boldsymbol{\rho}_0$ is symmetric with diagonal elements being 1. The off-diagonal elements of $\boldsymbol{\rho}_0$ are the instantaneous correlations between the components of $\boldsymbol{z}_t$. For $\ell > 0$, $\boldsymbol{\rho}_\ell$ is not symmetric in general because $\rho_{\ell,ij}$ is the correlation coefficient between z_{it} and $z_{j,t-\ell}$, whereas $\rho_{\ell,ji}$ is the correlation coefficient between z_{jt} and $z_{i,t-\ell}$. Using properties of $\boldsymbol{\Gamma}_\ell$, we have $\boldsymbol{\rho}_\ell = \boldsymbol{\rho}'_{-\ell}$.

To study the linear dynamic dependence between the components of $\boldsymbol{z}_t$, it suffices to consider $\boldsymbol{\rho}_\ell$ for $\ell \geq 0$, because for negative ℓ we can use the property $\boldsymbol{\rho}_\ell = \boldsymbol{\rho}'_{-\ell}$. For a k-dimensional series $\boldsymbol{z}_t$, each matrix $\boldsymbol{\rho}_\ell$ is a $k \times k$ matrix. When k is large, it is hard to decipher $\boldsymbol{\rho}_\ell$ simultaneously for several values of ℓ. To summarize the information, one can consider k^2 plots of the elements of $\boldsymbol{\rho}_\ell$ for $\ell = 0, \ldots, m$, where m is a prespecified positive integer. Specifically, for each (i, j)th position, we plot $\rho_{\ell,ij}$ versus ℓ. This plot shows the linear dynamic dependence of z_{it} on $z_{j,t-\ell}$ for $\ell = 0, 1, \ldots, m$. We refer to these k^2 plots as the cross-correlation plots of $\boldsymbol{z}_t$.

1.4 SAMPLE CCM

Given the sample $\{\boldsymbol{z}_t\}_{t=1}^T$, we obtain the sample mean vector and covariance matrix as

$$\hat{\boldsymbol{\mu}}_z = \frac{1}{T} \sum_{t=1}^{T} \boldsymbol{z}_t, \quad \hat{\boldsymbol{\Gamma}}_0 = \frac{1}{T-1} \sum_{t=1}^{T} (\boldsymbol{z}_t - \hat{\boldsymbol{\mu}}_z)(\boldsymbol{z}_t - \hat{\boldsymbol{\mu}}_z)'. \qquad (1.8)$$

These sample quantities are estimates of $\boldsymbol{\mu}$ and $\boldsymbol{\Gamma}_0$, respectively. The lag ℓ sample cross-covariance matrix is defined as

$$\hat{\boldsymbol{\Gamma}}_\ell = \frac{1}{T-1} \sum_{t=\ell+1}^{T} (\boldsymbol{z}_t - \hat{\boldsymbol{\mu}}_z)(\boldsymbol{z}_{t-\ell} - \hat{\boldsymbol{\mu}}_z)'.$$

The lag ℓ sample CCM is then

$$\hat{\boldsymbol{\rho}}_\ell = \hat{\boldsymbol{D}}^{-1} \hat{\boldsymbol{\Gamma}}_\ell \hat{\boldsymbol{D}}^{-1},$$

where $\hat{\boldsymbol{D}} = \text{diag}\{\hat{\gamma}_{0,11}^{1/2}, \cdots, \hat{\gamma}_{0,kk}^{1/2}\}$, in which $\hat{\gamma}_{0,ii}$ is the (i,i)th element of $\hat{\boldsymbol{\Gamma}}_0$. If $\boldsymbol{z}_t$ is a stationary linear process and $\boldsymbol{a}_t$ follows a multivariate normal distribution, then $\hat{\boldsymbol{\rho}}_\ell$ is a consistent estimate of $\boldsymbol{\rho}_\ell$. The normality condition can be relaxed by assuming the existence of finite fourth-order moments of $\boldsymbol{z}_t$. The asymptotic covariance matrix between elements of $\hat{\boldsymbol{\rho}}_\ell$ is complicated in general. An approximate formula has been obtained in the literature when $\boldsymbol{z}_t$ has zero fourth-order cumulants (see Bartlett 1955, Box, Jenkins, and Reinsel 1994, Chapter 11, and Reinsel 1993, Section 4.1.2). However, the formula can be simplified for some special cases. For instance, if $\boldsymbol{z}_t$ is a white noise series with positive-definite covariance matrix $\boldsymbol{\Sigma}_z$, then we have

$$\begin{aligned}
\text{Var}(\hat{\rho}_{\ell,ij}) &\approx \frac{1}{T} \quad \text{for} \quad \ell > 0, \\
\text{Var}(\hat{\rho}_{0,ij}) &\approx \frac{(1-\rho_{0,ij}^2)^2}{T} \quad \text{for} \quad i \neq j, \\
\text{Cov}(\hat{\rho}_{\ell,ij}, \hat{\rho}_{-\ell,ij}) &\approx \frac{\rho_{0,ij}^2}{T}, \\
\text{Cov}(\hat{\rho}_{\ell,ij}, \hat{\rho}_{h,uv}) &\approx 0, \quad \ell \neq h.
\end{aligned}$$

Another special case of interest is that $\boldsymbol{z}_t$ follows a vector moving-average (VMA) model, which will be discussed in Chapter 3. For instance, if $\boldsymbol{z}_t$ is a VMA(1) process, then

$$\text{Var}(\hat{\rho}_{\ell,ii}) \approx \frac{1 - 3\rho_{1,ii}^2 + 4\rho_{1,ii}^4}{T}, \quad \text{Var}(\hat{\rho}_{\ell,ij}) \approx \frac{1 + 2\rho_{1,ii}\rho_{1,jj}}{T},$$

for $\ell = \pm 2, \pm 3, \ldots$. If $\boldsymbol{z}_t$ is a VMA(q) process with $q > 0$, then

$$\text{Var}(\hat{\rho}_{\ell,ij}) \approx \frac{1}{T}\left(1 + 2\sum_{v=1}^{q} \rho_{v,ii}\rho_{v,jj}\right), \quad \text{for} \quad |\ell| > q. \tag{1.9}$$

In data analysis, we often examine the sample CCM $\hat{\boldsymbol{\rho}}_\ell$ to study the linear dynamic dependence in the data. As mentioned before, when the dimension k is large, it is hard to comprehend the k^2 cross-correlations simultaneously. To aid our ability to decipher the dependence structure of the data, we adopt the *simplified matrix* of Tiao and Box (1981). For each sample CCM $\hat{\boldsymbol{\rho}}_\ell$, we define a simplified matrix $\boldsymbol{s}_\ell = [s_{\ell,ij}]$ as

$$s_{\ell,ij} = \begin{cases} + & \text{if } \hat{\rho}_{\ell,ij} \geq 2/\sqrt{T}, \\ - & \text{if } \hat{\rho}_{\ell,ij} \leq -2/\sqrt{T}, \\ \cdot & \text{if } |\hat{\rho}_{\ell,ij}| < 2/\sqrt{T}. \end{cases} \tag{1.10}$$

This simplified matrix provides a summary of the sample CCM $\hat{\boldsymbol{\rho}}_\ell$ by applying the approximate 5% significance test to individual elements of $\boldsymbol{\rho}_\ell$ under the white noise assumption.

Another approach to check the linear dynamic dependence of $\boldsymbol{z}_t$ is to consider the sample counterpart of the cross-correlation plot. For each (i, j)th position of the sample CCM, we plot $\hat{\rho}_{\ell,ij}$ versus ℓ for $\ell = 0, 1, \ldots, m$, where m is a positive integer. This is a generalization of the sample autocorrelation function (ACF) of the univariate time series. For a k-dimensional series $\boldsymbol{z}_t$, we have k^2 plots. To simplify further the reading of these k^2 plots, an approximate 95% pointwise confidence interval is often imposed on the plot. Here, the 95% interval is often computed using $0 \pm 2/\sqrt{T}$. In other words, we use $1/\sqrt{T}$ as the standard error for the sample cross-correlations. This is justified in the sense that we are checking whether the observed time series is a white noise series. As mentioned before, if $\boldsymbol{z}_t$ is a white noise series with a positive-definite covariance matrix, then $\boldsymbol{\rho}_\ell = \boldsymbol{0}$ and the asymptotic variance of the sample cross-correlation $\hat{\rho}_{\ell,ij}$ is $1/T$ for $\ell > 0$.

To demonstrate, we use the command `ccm` of the `MTS` package in R to obtain the cross-correlation plots for a dataset consisting of 300 independent and identically distributed (iid) random draws from the two-dimensional standard Gaussian distribution. In this particular case, we have $\boldsymbol{\Sigma}_z = \boldsymbol{I}_2$ and $\boldsymbol{\rho}_\ell = \boldsymbol{0}$ for $\ell > 0$ so that we expect $\hat{\boldsymbol{\rho}}_\ell$ to be small for $\ell > 0$ and most of the sample cross-correlations to be within the 95% confidence intervals. Figure 1.6 shows the sample cross-correlation plots. As expected, these plots confirm that $\boldsymbol{z}_t$ has zero cross-correlations for all positive lags.

R Demonstration: Output edited.

```
> sig=diag(2) % create the 2-by-2 identity matrix
> x=rmvnorm(300,rep(0,2),sig) % generate random draws
> MTSplot(x) % Obtain time series plots (output not shown)
> ccm(x)
[1] "Covariance matrix:"
        [,1]   [,2]
[1,]  1.006 -0.101
[2,] -0.101  0.994
CCM at lag:  0
        [,1]   [,2]
```

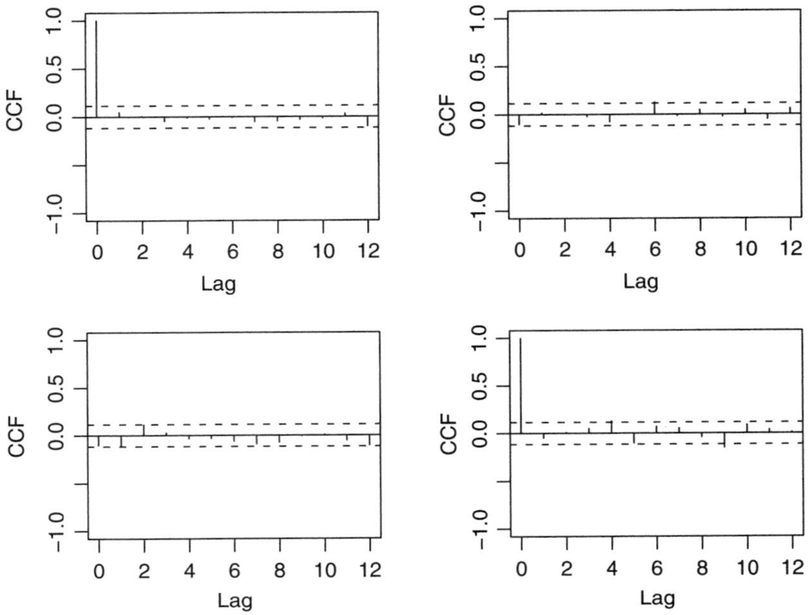

FIGURE 1.6 Sample cross-correlation plots for 300 observations drawn independently from the bivariate standard normal distribution. The dashed lines indicate pointwise 95% confidence intervals.

```
[1,]  1.000 -0.101
[2,] -0.101  1.000
Simplified matrix:
CCM at lag:  1
. .
. .
CCM at lag:  2
. .
. .
CCM at lag:  3
. .
. .
```

1.5 TESTING ZERO CROSS-CORRELATIONS

A basic test in multivariate time series analysis is to detect the existence of linear dynamic dependence in the data. This amounts to testing the null hypothesis $H_0 : \boldsymbol{\rho}_1 = \cdots = \boldsymbol{\rho}_m = \mathbf{0}$ versus the alternative hypothesis $H_a : \boldsymbol{\rho}_i \neq \mathbf{0}$ for some i satisfying $1 \leq i \leq m$, where m is a positive integer. The *Portmanteau test* of univariate time series has been generalized to the multivariate case by several authors. See, for instance, Hosking (1980, 1981), Li and McLeod (1981), and Li (2004). In particular, the multivariate Ljung–Box test statistic is defined as

$$Q_k(m) = T^2 \sum_{\ell=1}^{m} \frac{1}{T-\ell} tr\left(\hat{\boldsymbol{\Gamma}}_\ell' \hat{\boldsymbol{\Gamma}}_0^{-1} \hat{\boldsymbol{\Gamma}}_\ell \hat{\boldsymbol{\Gamma}}_0^{-1}\right), \tag{1.11}$$

where $tr(\boldsymbol{A})$ is the trace of the matrix $\boldsymbol{A}$ and T is the sample size. This is referred to as the *multivariate Portmanteau test*. It can be rewritten as

$$Q_k(m) = T^2 \sum_{\ell=1}^{m} \frac{1}{T-\ell} \hat{\boldsymbol{b}}_\ell' \left(\hat{\boldsymbol{\rho}}_0^{-1} \otimes \hat{\boldsymbol{\rho}}_0^{-1}\right) \hat{\boldsymbol{b}}_\ell,$$

where $\hat{\boldsymbol{b}}_\ell = \text{vec}(\hat{\boldsymbol{\rho}}_\ell')$ and $\otimes$ is the Kronecker product of two matrices. Here, vec($\boldsymbol{A}$) denotes the column-stacking vector of matrix $\boldsymbol{A}$. Readers are referred to Appendix A for the definitions of vectors and the Kronecker product of two matrices.

Under the null hypothesis that $\boldsymbol{\Gamma}_\ell = \boldsymbol{0}$ for $\ell > 0$ and the condition that $\boldsymbol{z}_t$ is normally distributed, $Q_k(m)$ is asymptotically distributed as $\chi^2_{mk^2}$, that is, a chi-square distribution with mk^2 degrees of freedom. Roughly speaking, assume that $E(\boldsymbol{z}_t) = \boldsymbol{0}$ because covariance matrices do not depend on the mean vectors. Under the assumption $\boldsymbol{\Gamma}_\ell = \boldsymbol{0}$ for $\ell > 0$, we have $\boldsymbol{z}_t = \boldsymbol{a}_t$, a white noise series. Then, the lag ℓ sample autocovariance matrix of $\boldsymbol{a}_t$ is

$$\hat{\boldsymbol{\Gamma}}_\ell = \frac{1}{T} \sum_{t=\ell+1}^{T} \boldsymbol{a}_t \boldsymbol{a}_{t-\ell}'.$$

Using vec($\boldsymbol{AB}$) $= (\boldsymbol{B}' \otimes \boldsymbol{I})$vec($\boldsymbol{A}$), and letting $\hat{\boldsymbol{\gamma}}_\ell = \text{vec}(\hat{\boldsymbol{\Gamma}}_\ell)$, we have

$$\hat{\boldsymbol{\gamma}}_\ell = \frac{1}{T} \sum_{t=\ell+1}^{T} (\boldsymbol{a}_{t-\ell} \otimes \boldsymbol{I}_k) \boldsymbol{a}_t.$$

Therefore, we have $E(\hat{\boldsymbol{\gamma}}_\ell) = \boldsymbol{0}$ and

$$\text{Cov}(\hat{\boldsymbol{\gamma}}_\ell) = E(\hat{\boldsymbol{\gamma}}_\ell \hat{\boldsymbol{\gamma}}_\ell') = \frac{T-\ell}{T^2} \boldsymbol{\Sigma}_a \otimes \boldsymbol{\Sigma}_a.$$

In the aforementioned equation, we have used

$$E[(\boldsymbol{a}_{t-\ell} \otimes \boldsymbol{I}_k) \boldsymbol{a}_t \boldsymbol{a}_t' (\boldsymbol{a}_{t-\ell}' \otimes \boldsymbol{I}_k)] = E(\boldsymbol{a}_{t-\ell} \boldsymbol{a}_{t-\ell}') \otimes E(\boldsymbol{a}_t \otimes \boldsymbol{a}_t) = \boldsymbol{\Sigma}_a \otimes \boldsymbol{\Sigma}_a.$$

Moreover, by iterated expectation, we have $\text{Cov}(\hat{\boldsymbol{\gamma}}_\ell, \hat{\boldsymbol{\gamma}}_v) = \boldsymbol{0}$ for $\ell \neq v$. In fact, the vectors $T^{1/2}\hat{\boldsymbol{\gamma}}_\ell$, $\ell = 1, \ldots, m$, are jointly asymptotically normal by application of the martingale central limit theorem; see Hannan (1970, p. 228). Therefore,

$$\frac{T^2}{T-\ell}\hat{\gamma}'_\ell\left(\Sigma_a^{-1}\otimes\Sigma_a^{-1}\right)\hat{\gamma}_\ell = \frac{T^2}{T-\ell}tr\left(\Sigma_a^{-1}\hat{\Gamma}'_\ell\Sigma_a^{-1}\hat{\Gamma}_\ell\right) \tag{1.12}$$

is asymptotically distributed as chi-square with k^2 degrees of freedom.

Remark: Strictly speaking, the test statistic of Li and McLeod (1981) is

$$Q_k^*(m) = T\sum_{\ell=1}^{m}\hat{\boldsymbol{b}}'_\ell\left(\hat{\boldsymbol{\rho}}_0^{-1}\otimes\hat{\boldsymbol{\rho}}_0^{-1}\right)\hat{\boldsymbol{b}}_\ell + \frac{k^2m(m+1)}{2T},$$

which is asymptotically equivalent to $Q_k(m)$. □

To demonstrate the $Q_k(m)$ statistic, we consider the bivariate time series $\boldsymbol{z}_t = (z_{1t}, z_{2t})'$ of Figure 1.2, where z_{1t} is the growth rate of U.S. quarterly real GDP and z_{2t} is the change in the U.S. quarterly unemployment rate. Obviously, there exists certain linear dynamic dependence in the data so that we expect the test statistic to reject the null hypothesis of no cross-correlations. This is indeed the case. The p-values of $Q_k(m)$ are also close to 0 for $m > 0$. See the R demonstration given later, where we use the command `mq` of the `MTS` package to perform the test. We also apply the $Q_k(m)$ statistic to a random sample of 200 observations drawn from the three-dimensional standard normal distribution. In this particular case, the statistic does not reject the null hypothesis of zero cross-correlations. Figure 1.7 shows the time

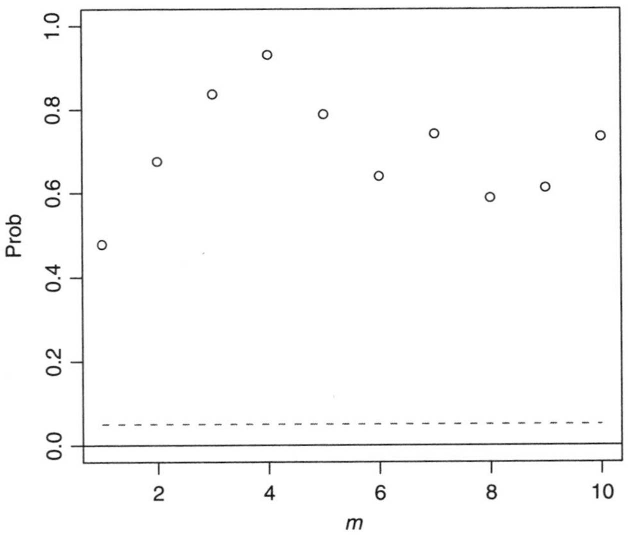

FIGURE 1.7 Plot of p-values for the $Q_k(m)$ statistics for a simulated data consisting of 200 random draws from the three-dimensional standard normal distribution. The dashed line denotes type I error of 5%.

plot of p-values of the $Q_k(m)$ statistic for the simulated three-dimensional white noise series. This is part of the output of the command mq. The dashed line of the plot denotes the type I error of 5%. For this particular simulation, as expected, all p-values are greater than 0.05, confirming that the series has no zero CCMs.

R Demonstration

```
> da=read.table("q-gdpunemp.txt",header=T) % Load the data
> head(da)
  year mon      gdp     rate
1 1948   1 1821.809 3.733333
 ....
6 1949   4 1835.512 5.866667
> x=cbind(diff(da$gdp),diff(da$rate)) % compute differenced
                                       % series
> mq(x,lag=10)  % Compute Q(m) statistics
Ljung-Box Statistics:
        m       Q(m)    p-value
 [1,]     1      140          0
 [2,]     2      196          0
 [3,]     3      213          0
 [4,]     4      232          0
 [5,]     5      241          0
 [6,]     6      246          0
 [7,]     7      250          0
 [8,]     8      261          0
 [9,]     9      281          0
[10,]    10      290          0
>
> sig=diag(3)  %% Simulation study
> z=rmvnorm(200,rep(0,3),sig)
> mq(z,10)
Ljung-Box Statistics:
        m       Q(m)    p-value
 [1,]  1.00     8.56       0.48
 [2,]  2.00    14.80       0.68
 [3,]  3.00    19.86       0.84
 [4,]  4.00    24.36       0.93
 [5,]  5.00    37.22       0.79
 [6,]  6.00    49.73       0.64
 [7,]  7.00    55.39       0.74
 [8,]  8.00    68.72       0.59
 [9,]  9.00    76.79       0.61
[10,] 10.00    81.23       0.73
```

Remark: When the dimension k is large, it becomes cumbersome to plot the CCMs. A possible solution is to summarize the information of $\hat{\mathbf{\Gamma}}_\ell$ by the chi-squared statistic in Equation (1.12). In particular, we can compute the p-value of the chi-squared statistic for testing $H_0 : \boldsymbol{\Gamma}_\ell = \mathbf{0}$ versus $H_a : \boldsymbol{\Gamma}_\ell \neq \mathbf{0}$. By plotting the p-value against the lag, we obtain a multivariate generalization of the ACF plot. □

1.6 FORECASTING

Prediction is one of the objectives of the multivariate time series analysis. Suppose we are interested in predicting $\boldsymbol{z}_{h+\ell}$ based on information available at time $t = h$ (inclusive). Such a prediction is called the ℓ-step ahead forecast of the series at the time index h. Here, h is called the *forecast origin* and ℓ the *forecast horizon*. Let F_t denote the available information at time t, which, in a typical situation, consists of the observations $\boldsymbol{z}_1, \ldots, \boldsymbol{z}_t$. In a time series analysis, the data-generating process is unknown so that we must use the information in F_h to build a statistical model for prediction. As such, the model itself is uncertain. A careful forecaster must consider such uncertainty in making predictions. In practice, it is hard to handle model uncertainty and we make the simplifying assumption that the model used in prediction is the true data-generating process. Keep in mind, therefore, that the forecasts produced by any method that assumes the fitted model as the true model are likely to underestimate the true variability of the time series. In Chapter 2, we discuss the effect of parameter estimates on the mean square of forecast errors for VAR models.

Forecasts produced by an econometric model also depend on the loss function used. In this book, we follow the tradition by using the minimum mean square error (MSE) prediction. Let $\boldsymbol{x}_h$ be an arbitrary forecast of $\boldsymbol{z}_{h+\ell}$ at the forecast origin h. The forecast error is $\boldsymbol{z}_{h+\ell} - \boldsymbol{x}_h$, and the mean square of forecast error is

$$\text{MSE}(\boldsymbol{x}_h) = E[(\boldsymbol{z}_{h+\ell} - \boldsymbol{x}_h)(\boldsymbol{z}_{h+\ell} - \boldsymbol{x}_h)'].$$

Let $\boldsymbol{z}_h(\ell) = E(\boldsymbol{z}_{h+\ell}|F_h)$ be the conditional expectation of $\boldsymbol{z}_{h+\ell}$ given the information F_h, including the model. Then, we can rewrite the MSE of $\boldsymbol{x}_h$ as

$$\begin{aligned}\text{MSE}(\boldsymbol{x}_h) &= E[\{\boldsymbol{z}_{h+\ell} - \boldsymbol{z}_h(\ell) + \boldsymbol{z}_h(\ell) - \boldsymbol{x}_h\}\{\boldsymbol{z}_{h+\ell} - \boldsymbol{z}_h(\ell) + \boldsymbol{z}_h(\ell) - \boldsymbol{x}_h\}'] \\ &= E[\{\boldsymbol{z}_{h+\ell} - \boldsymbol{z}_h(\ell)\}\{\boldsymbol{z}_{h+\ell} - \boldsymbol{z}_h(\ell)\}'] + E[\{\boldsymbol{z}_h(\ell) - \boldsymbol{x}_h\}\{\boldsymbol{z}_h(\ell) - \boldsymbol{x}_h\}'] \\ &= \text{MSE}[\boldsymbol{z}_h(\ell)] + E[\{\boldsymbol{z}_h(\ell) - \boldsymbol{x}_h\}\{\boldsymbol{z}_h(\ell) - \boldsymbol{x}_h\}'], \end{aligned} \tag{1.13}$$

where we have used the property

$$E[\{\boldsymbol{z}_{h+\ell} - \boldsymbol{z}_h(\ell)\}\{\boldsymbol{z}_h(\ell) - \boldsymbol{x}_h\}'] = \mathbf{0}.$$

This equation holds because $\boldsymbol{z}_h(\ell) - \boldsymbol{x}_h$ is a vector of functions of F_h, but $\boldsymbol{z}_{h+\ell} - \boldsymbol{z}_h(\ell)$ is a vector of functions of the innovations $\{\boldsymbol{a}_{h+\ell}, \ldots, \boldsymbol{a}_{h+1}\}$. Consequently, by using the iterative expectation and $E(\boldsymbol{a}_{t+i}) = \mathbf{0}$, the result in Equation (1.13) holds.

Consider Equation (1.13). Since $E[\{\boldsymbol{z}_h(\ell)-\boldsymbol{x}_h\}\{\boldsymbol{z}_h(\ell)-\boldsymbol{x}_h\}']$ is a nonnegative-definite matrix, we conclude that

$$\text{MSE}(\boldsymbol{x}_h) \geq \text{MSE}[\boldsymbol{z}_h(\ell)],$$

and the equality holds if and only if $\boldsymbol{x}_h = \boldsymbol{z}_h(\ell)$. Consequently, the minimum MSE forecast of $\boldsymbol{z}_{h+\ell}$ at the forecast origin $t = h$ is the conditional expectation of $\boldsymbol{z}_{h+\ell}$ given F_h. For the linear model in Equation (1.1), we have

$$\boldsymbol{z}_h(\ell) = \boldsymbol{\mu} + \boldsymbol{\psi}_\ell \boldsymbol{a}_h + \boldsymbol{\psi}_{\ell+1}\boldsymbol{a}_{h-1} + \cdots .$$

Let $\boldsymbol{e}_h(\ell) = \boldsymbol{z}_{h+\ell} - \boldsymbol{z}_h(\ell)$ be the ℓ-step ahead forecast error. Then, we have

$$\boldsymbol{e}_h(\ell) = \boldsymbol{a}_{h+\ell} + \boldsymbol{\psi}_1 \boldsymbol{a}_{h+\ell-1} + \cdots + \boldsymbol{\psi}_{\ell-1}\boldsymbol{a}_{h+1}. \tag{1.14}$$

The covariance matrix of the forecast error is then

$$\text{Cov}[\boldsymbol{e}_h(\ell)] = \boldsymbol{\Sigma}_a + \sum_{i=1}^{\ell-1} \boldsymbol{\psi}_i \boldsymbol{\Sigma}_a \boldsymbol{\psi}_i' = [\sigma_{e,ij}]. \tag{1.15}$$

If we further assume that $\boldsymbol{a}_t$ is multivariate normal, then we can obtain *interval forecasts* for $\boldsymbol{z}_{h+\ell}$. For instance, a 95% interval forecast for the component $z_{i,h+\ell}$ is

$$z_{ih}(\ell) \pm 1.96\sqrt{\sigma_{e,ii}},$$

where $z_{ih}(\ell)$ is the ith component of $\boldsymbol{z}_h(\ell)$ and $\sigma_{e,ii}$ is the (i,i)th diagonal element of $\text{Cov}[\boldsymbol{e}_h(\ell)]$ defined in Equation (1.15). One can also construct confidence regions and simultaneous confidence intervals using the methods available in multivariate statistical analysis; see, for instance, Johnson and Wichern (2007, Section 5.4). An approximate $100(1-\alpha)\%$ confidence region for $\boldsymbol{z}_{t+h}$ is the ellipsoid determined by

$$(\boldsymbol{z}_h(\ell) - \boldsymbol{z}_{h+\ell})'\text{Cov}[\boldsymbol{e}_h(\ell)]^{-1}(\boldsymbol{z}_h(\ell) - \boldsymbol{z}_{h+\ell}) \leq \chi^2_{k,1-\alpha},$$

where $\chi^2_{k,1-\alpha}$ denotes the $100(1-\alpha)$ quantile of a chi-square distribution with k degrees of freedom and $0 < \alpha < 1$. Also, $100(1-\alpha)\%$ simultaneous confidence intervals for all components of $\boldsymbol{z}_t$ are

$$z_{ih}(\ell) \pm \sqrt{\chi^2_{k,1-\alpha} \times \sigma_{e,ii}}, \quad i = 1, \ldots, k.$$

An alternative approach to construct simultaneous confidence intervals for the k components is to use the *Bonferroni's inequality*. Consider a probability space and events $E_1, \ldots, E_k$. The inequality says that

$$Pr(\cup_{i=1}^{k} E_i) \le \sum_{i=1}^{k} Pr(E_i).$$

Therefore,

$$Pr(\cap_{i=1}^{k} E_i) \ge 1 - \sum_{i=1}^{k} Pr(E_i^c),$$

where E_i^c denotes the complement of the event E_i. By choosing a $(100 - (\alpha/k))\%$ forecast interval for each component z_{it}, we apply the inequality to ensure that the probability that the following forecast intervals hold is at least $100(1 - \alpha)$:

$$z_{ih}(\ell) \pm Z_{1-(\alpha/k)} \sqrt{\sigma_{e,ii}},$$

where Z_{1-v} is the $100(1 - v)$ quantile of a standard normal distribution.

From Equation (1.14), we see that the one step ahead forecast error is

$$\boldsymbol{e}_h(1) = \boldsymbol{a}_{h+1}.$$

This says that $\boldsymbol{a}_{h+1}$ is the unknown quantity of $\boldsymbol{z}_{h+1}$ at time h. Therefore, $\boldsymbol{a}_{h+1}$ is called the *innovation* of the series at time index $h + 1$. This provides the justification for using the term innovation in Section 1.2.

1.7 MODEL REPRESENTATIONS

The linear model in Equation (1.1) is commonly referred to as the moving-average (MA) representation of a multivariate time series. This representation is useful in forecasting, such as computing the covariance of a forecast error shown in Equation (1.15). It is also used in studying the *impulse response functions*. Again, details are given in later chapters of the book. For an invertible series, the model in Equation (1.4) is referred to as the autoregressive (AR) representation of the model. This model is useful in understanding how $\boldsymbol{z}_t$ depends on its lag values $\boldsymbol{z}_{t-i}$ for $i > 0$.

If the time series is both stationary and invertible, then these two model presentations are equivalent and one can obtain one representation from the other. To see this, we first consider the mean of $\boldsymbol{z}_t$. Taking expectation on both sides of Equation (1.4), we have

$$\boldsymbol{\mu} = \boldsymbol{c} + \sum_{i=1}^{\infty} \boldsymbol{\pi}_i \boldsymbol{\mu}.$$

Letting $\boldsymbol{\pi}_0 = \boldsymbol{I}_k$, we obtain, from the prior equation,

$$\left(\sum_{i=0}^{\infty} \boldsymbol{\pi}_i\right) \boldsymbol{\mu} = \boldsymbol{c}.$$

Plugging in $\boldsymbol{c}$, we can rewrite Equation (1.4) as

$$\tilde{\boldsymbol{z}}_t = \sum_{i=1}^{\infty} \boldsymbol{\pi}_i \tilde{\boldsymbol{z}}_{t-i} + \boldsymbol{a}_t, \tag{1.16}$$

where, as before, $\tilde{\boldsymbol{z}}_t = \boldsymbol{z}_t - \boldsymbol{\mu}$ is the mean-adjusted time series.

Next, we consider the relationship between the coefficient matrices $\boldsymbol{\psi}_i$ and $\boldsymbol{\pi}_j$, using the mean-adjusted series $\tilde{\boldsymbol{z}}_t$. The MA representation is

$$\tilde{\boldsymbol{z}}_t = \sum_{i=0}^{\infty} \boldsymbol{\psi}_i \boldsymbol{a}_{t-i}.$$

Let B be the back-shift operator defined by $B\boldsymbol{x}_t = \boldsymbol{x}_{t-1}$ for any time series $\boldsymbol{x}_t$. In the econometric literature, the back-shift operator is called the lag operator and the notation L is often used. Using the back-shift operator, the MA representation of $\tilde{\boldsymbol{z}}_t$ becomes

$$\tilde{\boldsymbol{z}}_t = \sum_{i=0}^{\infty} \boldsymbol{\psi}_i \boldsymbol{a}_{t-i} = \sum_{i=0}^{\infty} \boldsymbol{\psi}_i B^i \boldsymbol{a}_t = \boldsymbol{\psi}(B)\boldsymbol{a}_t, \tag{1.17}$$

where $\boldsymbol{\psi}(B) = \boldsymbol{I}_k + \boldsymbol{\psi}_1 B + \boldsymbol{\psi}_2 B^2 + \boldsymbol{\psi}_3 B^3 + \cdots$. On the other hand, we can also rewrite the AR representation in Equation (1.16) using the back-shift operator as

$$\tilde{\boldsymbol{z}}_t - \sum_{i=1}^{\infty} \boldsymbol{\pi}_i \tilde{\boldsymbol{z}}_{t-i} = \boldsymbol{a}_t \quad \text{or} \quad \boldsymbol{\pi}(B)\tilde{\boldsymbol{z}}_t = \boldsymbol{a}_t, \tag{1.18}$$

where $\boldsymbol{\pi}(B) = \boldsymbol{I}_i - \boldsymbol{\pi}_1 B - \boldsymbol{\pi}_2 B^2 - \cdots$. Plugging Equation (1.17) into Equation (1.18), we obtain

$$\boldsymbol{\pi}(B)\boldsymbol{\psi}(B)\boldsymbol{a}_t = \boldsymbol{a}_t.$$

Consequently, we have $\boldsymbol{\pi}(B)\boldsymbol{\psi}(B) = \boldsymbol{I}_k$. That is,

$$(\boldsymbol{I}_k - \boldsymbol{\pi}_1 B - \boldsymbol{\pi}_2 B^2 - \boldsymbol{\pi}_3 B^3 - \cdots)(\boldsymbol{I}_i + \boldsymbol{\psi}_1 B + \boldsymbol{\psi}_2 B^2 + \cdots) = \boldsymbol{I}_k.$$

This equation implies that all coefficient matrices of B^i on the left-hand side, for $i > 0$, must be 0. Consequently, we have

$$\begin{aligned} \boldsymbol{\psi}_1 - \boldsymbol{\pi}_1 &= \mathbf{0}, \quad \text{(coefficient matrix of } B^1\text{)} \\ \boldsymbol{\psi}_2 - \boldsymbol{\pi}_1\boldsymbol{\psi}_1 - \boldsymbol{\pi}_2 &= \mathbf{0}, \quad \text{(coefficient matrix of } B^2\text{)} \\ \boldsymbol{\psi}_3 - \boldsymbol{\pi}_1\boldsymbol{\psi}_2 - \boldsymbol{\pi}_2\boldsymbol{\psi}_1 - \boldsymbol{\pi}_3 &= \mathbf{0}, \quad \text{(coefficient matrix of } B^3\text{)} \\ \vdots &= \vdots \end{aligned}$$

In general, we can obtain $\boldsymbol{\psi}_\ell$ recursively from $\{\boldsymbol{\pi}_i | i = 1, 2, \ldots\}$ via

$$\boldsymbol{\psi}_\ell = \sum_{i=0}^{\ell-1} \boldsymbol{\pi}_{\ell-i}\boldsymbol{\psi}_i, \quad \ell \geq 1, \tag{1.19}$$

where $\boldsymbol{\psi}_0 = \boldsymbol{\pi}_0 = \boldsymbol{I}_k$. Similarly, we can obtain $\boldsymbol{\pi}_\ell$ recursively from $\{\boldsymbol{\psi}_i | i = 1, 2, \ldots\}$ via

$$\boldsymbol{\pi}_1 = \boldsymbol{\psi}_1 \quad \text{and} \quad \boldsymbol{\pi}_\ell = \boldsymbol{\psi}_\ell - \sum_{i=1}^{\ell-1} \boldsymbol{\pi}_i\boldsymbol{\psi}_{\ell-i}, \quad \ell > 1. \tag{1.20}$$

Finally, neither the AR representation in Equation (1.4) nor the MA representation in Equation (1.1) is particularly useful in estimation if they involve too many coefficient matrices. To facilitate model estimation and to gain a deeper understanding of the models used, we postulate that the coefficient matrices $\boldsymbol{\pi}_i$ and $\boldsymbol{\psi}_j$ depend only on a finite number of parameters. This consideration leads to the use of vector autoregressive moving-average (VARMA) models, which are also known as the multivariate autoregressive moving-average (MARMA) models.

A general VARMA(p, q) model can be written as

$$\boldsymbol{z}_t = \boldsymbol{\phi}_0 + \sum_{i=1}^{p} \boldsymbol{\phi}_i \boldsymbol{z}_{t-1} + \boldsymbol{a}_t - \sum_{i=1}^{q} \boldsymbol{\theta}_i \boldsymbol{a}_{t-i}, \tag{1.21}$$

where p and q are nonnegative integers, $\boldsymbol{\phi}_0$ is a k-dimensional constant vector, $\boldsymbol{\phi}_i$ and $\boldsymbol{\theta}_j$ are $k \times k$ constant matrices, and $\{\boldsymbol{a}_t\}$ is a sequence of independent and identically distributed random vectors with mean zero and positive-definite covariance matrix $\boldsymbol{\Sigma}_a$. Using the back-shift operator B, we can write the VARMA model in a compact form as

$$\boldsymbol{\phi}(B)\boldsymbol{z}_t = \boldsymbol{\phi}_0 + \boldsymbol{\theta}(B)\boldsymbol{a}_t, \tag{1.22}$$

where $\boldsymbol{\phi}(B) = \boldsymbol{I}_k - \boldsymbol{\phi}_1 B - \cdots - \boldsymbol{\phi}_p B^p$ and $\boldsymbol{\theta}(B) = \boldsymbol{I}_k - \boldsymbol{\theta}_1 B - \cdots - \boldsymbol{\theta}_q B^q$ are matrix polynomials in B. Certain conditions on $\boldsymbol{\phi}(B)$ and $\boldsymbol{\theta}(B)$ are needed to render the VARMA model stationary, invertible, and identifiable. We shall discuss these conditions in detail in later chapters of the book.

For a stationary series $\boldsymbol{z}_t$, by taking expectation on both sides of Equation (1.21), we have

$$\boldsymbol{\mu} = \boldsymbol{\phi}_0 + \sum_{i=1}^{p} \boldsymbol{\phi}_i \boldsymbol{\mu},$$

where $\boldsymbol{\mu} = E(\boldsymbol{z}_t)$. Consequently, we have

$$\left(\boldsymbol{I}_k - \sum_{i=1}^{p} \boldsymbol{\phi}_i \right) \boldsymbol{\mu} = \boldsymbol{\phi}_0. \tag{1.23}$$

This equation can be conveniently written as $\boldsymbol{\phi}(1)\boldsymbol{\mu} = \boldsymbol{\phi}_0$. Plugging Equation (1.23) into the VARMA model in Equation (1.22), we obtain a mean-adjusted VARMA(p, q) model as

$$\boldsymbol{\phi}(B)\tilde{\boldsymbol{z}}_t = \boldsymbol{\theta}(B)\boldsymbol{a}_t, \tag{1.24}$$

where, as before, $\tilde{\boldsymbol{z}}_t = \boldsymbol{z}_t - \boldsymbol{\mu}$.

The AR and MA representations of $\boldsymbol{z}_t$ can be obtained from the VARMA model by matrix multiplication. Assuming for simplicity that the matrix inversion involved exists, we can rewrite Equation (1.24) as

$$\tilde{\boldsymbol{z}}_t = [\boldsymbol{\phi}(B)]^{-1}\boldsymbol{\theta}(B)\boldsymbol{a}_t.$$

Consequently, comparing with the MA representation in Equation (1.17), we have $\boldsymbol{\psi}(B) = [\boldsymbol{\phi}(B)]^{-1}\boldsymbol{\theta}(B)$, or equivalently

$$\boldsymbol{\phi}(B)\boldsymbol{\psi}(B) = \boldsymbol{\theta}(B).$$

By equating the coefficient matrices of B^i on both sides of the prior equation, we can obtain recursively $\boldsymbol{\psi}_i$ from $\boldsymbol{\phi}_j$ and $\boldsymbol{\theta}_v$ with $\boldsymbol{\psi}_0 = \boldsymbol{I}_k$.

If we rewrite the VARMA model in Equation (1.24) as $[\boldsymbol{\theta}(B)]^{-1}\boldsymbol{\phi}(B)\tilde{\boldsymbol{z}}_t = \boldsymbol{a}_t$ and compare it with the AR representation in Equation (1.18), we see that $[\boldsymbol{\theta}(B)]^{-1}\boldsymbol{\phi}(B) = \boldsymbol{\pi}(B)$. Consequently,

$$\boldsymbol{\psi}(B) = \boldsymbol{\theta}(B)\boldsymbol{\pi}(B).$$

Again, by equating the coefficient matrices of B^i on both sides of the prior equation, we can obtain recursively the coefficient matrix $\boldsymbol{\pi}_i$ from $\boldsymbol{\phi}_j$ and $\boldsymbol{\theta}_v$.

The requirement that both the $\boldsymbol{\phi}(B)$ and $\boldsymbol{\theta}(B)$ matrix polynomials of Equation (1.21) start with the $k \times k$ identity matrix is possible because the covariance matrix of $\boldsymbol{a}_t$ is a general positive-definite matrix. Similar to Equation (1.2), we can have alternative parameterizations for the VARMA(p, q) model. Specifically, consider the Cholesky decomposition $\boldsymbol{\Sigma}_a = \boldsymbol{L}\boldsymbol{\Omega}\boldsymbol{L}'$. Let $\boldsymbol{b}_t = \boldsymbol{L}^{-1}\boldsymbol{a}_t$. We have

$\text{Cov}(\boldsymbol{b}_t) = \boldsymbol{\Omega}$, which is a diagonal matrix, and $\boldsymbol{a}_t = \boldsymbol{L}\boldsymbol{b}_t$. Using the same method as that of Equation (1.2), we can rewrite the VARMA model in Equation (1.21) as

$$\boldsymbol{z}_t = \boldsymbol{\phi}_0 + \sum_{i=1}^{p} \boldsymbol{\phi}_i \boldsymbol{z}_{t-i} + \boldsymbol{L}\boldsymbol{b}_t - \sum_{j=1}^{q} \boldsymbol{\theta}_j^* \boldsymbol{b}_{t-j},$$

where $\boldsymbol{\theta}_j^* = \boldsymbol{\theta}_j \boldsymbol{L}$. In this particular formulation, we have $\theta^*(B) = \boldsymbol{L} - \sum_{j=1}^{q} \boldsymbol{\theta}_j^* B^j$. Also, because $\boldsymbol{L}$ is a lower triangular matrix with 1 being the diagonal elements, $\boldsymbol{L}^{-1}$ is also a lower triangular matrix with 1 being the diagonal elements. Premultiplying Equation (1.21) by $\boldsymbol{L}^{-1}$ and letting $\boldsymbol{\phi}_0^* = \boldsymbol{L}^{-1}\boldsymbol{\phi}_0$, we obtain

$$\boldsymbol{L}^{-1}\boldsymbol{z}_t = \boldsymbol{\phi}_0^* + \sum_{i=1}^{p} \boldsymbol{L}^{-1}\boldsymbol{\phi}_i \boldsymbol{z}_{t-i} + \boldsymbol{b}_t - \sum_{j=1}^{q} \boldsymbol{L}^{-1}\boldsymbol{\theta}_j \boldsymbol{a}_{t-j}.$$

By inserting $\boldsymbol{L}\boldsymbol{L}^{-1}$ in front of $\boldsymbol{a}_{t-j}$, we can rewrite the prior equation as

$$\boldsymbol{L}^{-1}\boldsymbol{z}_t = \boldsymbol{\phi}_0^* + \sum_{i=1}^{p} \boldsymbol{\phi}_i^* \boldsymbol{z}_{t-i} + \boldsymbol{b}_t - \sum_{j=1}^{q} \tilde{\boldsymbol{\theta}}_j \boldsymbol{b}_{t-j},$$

where $\boldsymbol{\phi}_i^* = \boldsymbol{L}^{-1}\boldsymbol{\phi}_i$ and $\tilde{\boldsymbol{\theta}}_j = \boldsymbol{L}^{-1}\boldsymbol{\theta}_j\boldsymbol{L}$. In this particular formulation, we have $\boldsymbol{\phi}^*(B) = \boldsymbol{L}^{-1} - \sum_{i=1}^{p} \boldsymbol{\phi}_i^* B^i$. From the discussion, we see that there are several equivalent ways to write a VARMA(p, q) model. The important issue in studying a VARMA model is not how to write a VARMA model, but what is the dynamic structure embedded in a given model.

1.8 OUTLINE OF THE BOOK

The book comprises seven chapters. Chapter 2 focuses on the VAR models. It considers the properties of VAR models, starting with simple models of orders 1 and 2. It then introduces estimation and model building. Both the least-squares and Bayesian estimation methods are discussed. Estimation with linear parameter constraints is also included. It also discusses forecasting and the decomposition of the forecast-error covariances. The concepts and calculations of impulse response function are given in detail. Chapter 3 studies the stationary and invertible VARMA models. Again, it starts with the properties of simple MA models. For estimation, both the conditional and the exact likelihood methods are introduced. It then investigates the identifiability and implications of VARMA models. Various approaches to study the likelihood function of a VARMA model are given. For model building, the chapter introduces the method of extended CCMs.

Chapter 4 studies the structural specification of VARMA models. Two methods are given that can specify the simplifying structure (or skeleton) of a vector VARMA time series and, hence, overcome the difficulty of identifiability. Chapter 5 focuses on unit-root nonstationarity. The asymptotic properties of unit-root processes are discussed. It then introduces spurious regression, cointegration, and error-correction forms of VARMA models. Finally, the chapter considers cointegration tests and estimation of error-correction models. Applications of cointegration in finance are briefly discussed. Chapter 6 considers factor models and some selected topics in vector time series. Most factor models available in the literature are included and discussed. Both the orthogonal factor models and the approximate factor models are considered. For selected topics, the chapter includes seasonal vector time series, principal component analysis, missing values, regression models with vector time series errors, and model-based clustering. Finally, Chapter 7 studies multivariate volatility models. It discusses various multivariate volatility models that are relatively easy to estimate and produce positive-definite volatility matrices.

1.9 SOFTWARE

Real examples are used throughout the book to demonstrate the concepts and analyses of vector time series. These empirical analyses were carried out via the `MTS` package developed by the author for the book. Not a trained programmer, I am certain that most of the programs in the package are not as efficient as they can be. With high probability, the program may even contain bugs. My goal in preparing the package is to ensure that readers can reproduce the results shown in the book and gain experience in analyzing real-world vector time series. Interested readers and more experienced researchers can certainly improve the package. I sincerely welcome the suggestions for improvements and corrections for any bug.

EXERCISES

1.1 Simulation is helpful in learning vector time series. Define the matrices

$$\boldsymbol{C} = \begin{bmatrix} 0.8 & 0.4 \\ -0.3 & 0.6 \end{bmatrix}, \quad \boldsymbol{S} = \begin{bmatrix} 2.0 & 0.5 \\ 0.5 & 1.0 \end{bmatrix}.$$

Use the command

```
m1 = VARMAsim(300,arlags = c(1),phi = C,sigma = S);zt = m1$series
```

to generate 300 observations from the VAR(1) model

$$\boldsymbol{z}_t = \boldsymbol{C}\boldsymbol{z}_{t-1} + \boldsymbol{a}_t,$$

where $\boldsymbol{a}_t$ are iid bivariate normal random variates with mean zero and $\text{Cov}(\boldsymbol{a}_t) = \boldsymbol{S}$.

- Plot the time series $\boldsymbol{z}_t$.
- Obtain the first five lags of sample CCMs of $\boldsymbol{z}_t$.

- Test $H_0 : \boldsymbol{\rho}_1 = \cdots = \boldsymbol{\rho}_{10} = \mathbf{0}$ versus $H_a : \boldsymbol{\rho}_i \neq \mathbf{0}$ for some i, where $i \in \{1, \ldots, 10\}$. Draw the conclusion using the 5% significance level.

1.2 Use the matrices of Problem 1 and the following command

```
m2 = VARMAsim(200,malags = c(1),theta = C,sigma = S);zt = m2$series
```

to generate 200 observations from the VMA(1) model, $\boldsymbol{z}_t = \boldsymbol{a}_t - \boldsymbol{C}\boldsymbol{a}_{t-1}$, where $\boldsymbol{a}_t$ are iid $N(\mathbf{0}, \boldsymbol{S})$.

- Plot the time series $\boldsymbol{z}_t$.
- Obtain the first two lags of sample CCMs of $\boldsymbol{z}_t$.
- Test $H_0 : \boldsymbol{\rho}_1 = \cdots = \boldsymbol{\rho}_5 = \mathbf{0}$ versus $H_a : \boldsymbol{\rho}_i \neq \mathbf{0}$ for some $i \in \{1, \ldots, 5\}$. Draw the conclusion using the 5% significance level.

1.3 The file `q-fdebt.txt` contains the U.S. quarterly federal debts held by (a) foreign and international investors, (b) federal reserve banks, and (c) the public. The data are from the Federal Reserve Bank of St. Louis, from 1970 to 2012 for 171 observations, and not seasonally adjusted. The debts are in billions of dollars. Take the log transformation and the first difference for each time series. Let $\boldsymbol{z}_t$ be the differenced log series.

- Plot the time series $\boldsymbol{z}_t$.
- Obtain the first five lags of sample CCMs of $\boldsymbol{z}_t$.
- Test $H_0 : \boldsymbol{\rho}_1 = \cdots = \boldsymbol{\rho}_{10} = \mathbf{0}$ versus $H_a : \boldsymbol{\rho}_i \neq \mathbf{0}$ for some i, where $i \in \{1, \ldots, 10\}$. Draw the conclusion using the 5% significance level.

Hint: You may use the following commands of `MTS` to process the data:

```
da=read.table("q-fdebt.txt",header=T)
debt=log(da[,3:5]); tdx=da[,1]+da[,2]/12
MTSplot(debt,tdx); zt=diffM(debt); MTSplot(zt,tdx[-1])
```

1.4 The file `m-pgspabt.txt` consists of monthly simple returns of Procter & Gamble stock, S&P composite index, and Abbott Laboratories from January 1962 to December 2011. The data are from CRSP. Transform the simple returns into log returns. Let $\boldsymbol{z}_t$ be the monthly log returns.

- Plot the time series $\boldsymbol{z}_t$.
- Obtain the first two lags of sample CCMs of $\boldsymbol{z}_t$.
- Test $H_0 : \boldsymbol{\rho}_1 = \cdots = \boldsymbol{\rho}_5 = \mathbf{0}$ versus $H_a : \boldsymbol{\rho}_i \neq \mathbf{0}$ for some $i \in \{1, \ldots, 5\}$. Draw the conclusion using the 5% significance level.

1.5 For a VARMA time series $\boldsymbol{z}_t$, derive the result of Equation (1.20).

REFERENCES

Bartlett, M. S. (1955). *Stochastic Processes*. Cambridge University Press, Cambridge, UK.

Box, G. E. P., Jenkins, G. M., and Reinsel, G. (1994). *Time Series Analysis: Forecasting and Control*. 3rd Edition. Prentice-Hall, Englewood Cliffs, NJ.

Hannan, E. J. (1970). *Multiple Time Series*. John Wiley & Sons, Inc, New York.

Hosking, J. R. M. (1980). The multivariate portmanteau statistic. *Journal of the American Statistical Association*, **75**: 602–607.

Hosking, J. R. M. (1981). Lagrange-multiplier tests of multivariate time series model. *Journal of the Royal Statistical Society, Series B*, **43**: 219–230.

Johnson, R. A. and Wichern, D. W. (2007). *Applied Multivariate Statistical Analysis*. 6th Edition. Pearson Prentice Hall, Upper Saddle River, NJ.

Li, W. K. (2004). *Diagnostic Checks in Time Series*. Chapman & Hall/CRC, Boca Raton, FL.

Li, W. K. and McLeod, A. I. (1981). Distribution of the residual autocorrelations in multivariate time series models. *Journal of the Royal Statistical Society, Series B*, **43**: 231–239.

Reinsel, G. (1993). *Elements of Multivariate Time Series Analysis*. Springer-Verlag, New York.

Tiao, G. C. and Box, G. E. P. (1981). Modeling multiple time series with applications. *Journal of the American Statistical Association*, **76**: 802–816.

CHAPTER 2

Stationary Vector Autoregressive Time Series

2.1 INTRODUCTION

The most commonly used multivariate time series model is the vector autoregressive (VAR) model, particularly so in the econometric literature for good reasons. First, the model is relatively easy to estimate. One can use the least-squares (LS) method, the maximum likelihood (ML) method, or Bayesian method. All three estimation methods have closed-form solutions. For a VAR model, the least-squares estimates are asymptotically equivalent to the ML estimates and the ordinary least-squares (OLS) estimates are the same as the generalized least-squares (GLS) estimates. Second, the properties of VAR models have been studied extensively in the literature. Finally, VAR models are similar to the multivariate multiple linear regressions widely used in multivariate statistical analysis. Many methods for making inference in multivariate multiple linear regression apply to the VAR model.

The multivariate time series $\boldsymbol{z}_t$ follows a VAR model of order p, VAR(p), if

$$\boldsymbol{z}_t = \boldsymbol{\phi}_0 + \sum_{i=1}^{p} \boldsymbol{\phi}_i \boldsymbol{z}_{t-i} + \boldsymbol{a}_t, \tag{2.1}$$

where $\boldsymbol{\phi}_0$ is a k-dimensional constant vector and $\boldsymbol{\phi}_i$ are $k \times k$ matrices for $i > 0$, $\boldsymbol{\phi}_p \neq \boldsymbol{0}$, and $\boldsymbol{a}_t$ is a sequence of independent and identically distributed (iid) random vectors with mean zero and covariance matrix $\boldsymbol{\Sigma}_a$, which is positive-definite. This is a special case of the VARMA(p, q) model of Chapter 1 with $q = 0$. With the back-shift operator, the model becomes $\boldsymbol{\phi}(B)\boldsymbol{z}_t = \boldsymbol{\phi}_0 + \boldsymbol{a}_t$, where $\boldsymbol{\phi}(B) = \boldsymbol{I}_k - \sum_{i=1}^{p} \boldsymbol{\phi}_i B^i$ is a matrix polynomial of degree p. See Equation (1.21). We shall refer to $\boldsymbol{\phi}_\ell = [\phi_{\ell,ij}]$ as the lag ℓ AR coefficient matrix.

Multivariate Time Series Analysis: With R and Financial Applications,
First Edition. Ruey S. Tsay.

To study the properties of VAR(p) models, we start with the simple VAR(1) and VAR(2) models. In many cases, we use bivariate time series in our discussion, but the results continue to hold for the k-dimensional series.

2.2 VAR(1) MODELS

To begin, consider the bivariate VAR(1) model

$$\boldsymbol{z}_t = \boldsymbol{\phi}_0 + \boldsymbol{\phi}_1 \boldsymbol{z}_{t-1} + \boldsymbol{a}_t.$$

This model can be written explicitly as

$$\begin{bmatrix} z_{1t} \\ z_{2t} \end{bmatrix} = \begin{bmatrix} \phi_{10} \\ \phi_{20} \end{bmatrix} + \begin{bmatrix} \phi_{1,11} & \phi_{1,12} \\ \phi_{1,21} & \phi_{1,22} \end{bmatrix} \begin{bmatrix} z_{1,t-1} \\ z_{2,t-1} \end{bmatrix} + \begin{bmatrix} a_{1t} \\ a_{2t} \end{bmatrix}, \tag{2.2}$$

or equivalently,

$$\begin{aligned} z_{1t} &= \phi_{10} + \phi_{1,11} z_{1,t-1} + \phi_{1,12} z_{2,t-1} + a_{1t}, \\ z_{2t} &= \phi_{20} + \phi_{1,21} z_{1,t-1} + \phi_{1,22} z_{2,t-1} + a_{2t}. \end{aligned}$$

Thus, the (1,2)th element of $\boldsymbol{\phi}_1$, that is, $\phi_{1,12}$, shows the linear dependence of z_{1t} on $z_{2,t-1}$ in the presence of $z_{1,t-1}$. The (2,1)th element of $\boldsymbol{\phi}_1$, $\phi_{1,21}$, measures the linear relationship between z_{2t} and $z_{1,t-1}$ in the presence of $z_{2,t-1}$. Other parameters in $\boldsymbol{\phi}_1$ can be interpreted in a similar manner.

2.2.1 Model Structure and Granger Causality

If the off-diagonal elements of $\boldsymbol{\phi}_1$ are 0, that is, $\phi_{1,12} = \phi_{1,21} = 0$, then z_{1t} and z_{2t} are not dynamically correlated. In this particular case, each series follows a univariate AR(1) model and can be handled accordingly. We say that the two series are uncoupled.

If $\phi_{1,12} = 0$, but $\phi_{1,21} \neq 0$, then we have

$$z_{1t} = \phi_{10} + \phi_{1,11} z_{1,t-1} + a_{1t}, \tag{2.3}$$

$$z_{2t} = \phi_{20} + \phi_{1,21} z_{1,t-1} + \phi_{1,22} z_{2,t-1} + a_{2t}. \tag{2.4}$$

This particular model shows that z_{1t} does not depend on the past value of z_{2t}, but z_{2t} depends on the past value of z_{1t}. Consequently, we have a unidirectional relationship with z_{1t} acting as the input variable and z_{2t} as the output variable. In the statistical literature, the two series z_{1t} and z_{2t} are said to have a *transfer function* relationship. Transfer function models, which can be regarded as a special case of the VARMA model, are useful in control engineering as one can adjust the value of z_{1t} to influence the future value of z_{2t}. In the econometric literature, the model implies the existence of Granger causality between the two series with z_{1t} causing z_{2t}, but not being caused by z_{2t}.

Granger (1969) introduces the concept of causality, which is easy to deal with for a VAR model. Consider a bivariate series and the h step ahead forecast. In this case, we can use the VAR model and univariate models for individual components to produce forecasts. We say that z_{1t} causes z_{2t} if the bivariate forecast for z_{2t} is more accurate than its univariate forecast. Here, the accuracy of a forecast is measured by the variance of its forecast error. In other words, under Granger's framework, we say that z_{1t} causes z_{2t} if the past information of z_{1t} improves the forecast of z_{2t}.

We can make the statement more precisely. Let F_t denote the available information at time t (inclusive). Let $F_{-i,t}$ be F_t with all information concerning the ith component z_{it} removed. Consider the bivariate VAR(1) model in Equation (2.2). F_t consists of $\{\boldsymbol{z}_t, \boldsymbol{z}_{t-1}, \ldots\}$, whereas $F_{-2,t}$ consists of the past values $\{z_{1t}, z_{1,t-1}, \ldots\}$ and $F_{-1,t}$ contains $\{z_{2t}, z_{2,t-1}, \ldots\}$. Now, consider the h step ahead forecast $\boldsymbol{z}_t(h)$ based on F_t and the associated forecast error $\boldsymbol{e}_t(h)$. See Section 1.6. Let $z_{j,t+h}|F_{-i,t}$ be the h step ahead prediction of $z_{j,t+h}$ based on $F_{-i,t}$ and $e_{j,t+h}|F_{-i,t}$ be the associated forecast error, where $i \neq j$. Then, z_{1t} causes z_{2t} if Var$[e_{2t}(h)] <$ Var$[e_{2,t+h}|F_{-1,t}]$.

Return to the bivariate VAR(1) model with $\phi_{1,12} = 0$, but $\phi_{1,21} \neq 0$. We see that $z_{2,t+1}$ depends on z_{1t} so that knowing z_{1t} is helpful in forecasting $z_{2,t+1}$. On the other hand, $z_{1,t+1}$ does not depend on any past value of z_{2t} so that knowing the past values of z_{2t} will not help in predicting $z_{1,t+1}$. Thus, z_{1t} causes z_{2t}, but z_{2t} does not cause z_{1t}.

Similarly, if $\phi_{1,21} = 0$, but $\phi_{1,12} \neq 0$, then z_{2t} causes z_{1t}, but z_{1t} does not cause z_{2t}.

Remark: For the bivariate VAR(1) model in Equation (2.2), if $\boldsymbol{\Sigma}_a$ is not a diagonal matrix, then z_{1t} and z_{2t} are instantaneously correlated (or contemporaneously correlated). In this case, z_{1t} and z_{2t} have instantaneous Granger causality. The instantaneous causality goes in both ways. □

Remark: The statement that $\phi_{1,12} = 0$ and $\phi_{1,21} \neq 0$ in a VAR(1) model implies the existence of Granger's causality depends critically on the unique VAR parameterization we use, namely, the matrix polynomial $\boldsymbol{\phi}(B)$ starts with the identity matrix $\boldsymbol{I}_k$. To see this, consider the model in Equations (2.3) and (2.4). For simplicity, assume the constant term being 0, that is, $\phi_{10} = \phi_{20} = 0$. If we multiply Equation (2.4) by a nonzero parameter β and add the result to Equation (2.3), then we have

$$z_{1t} + \beta z_{2t} = (\phi_{1,11} + \beta\phi_{1,21})z_{1,t-1} + \beta\phi_{1,22}z_{2,t-1} + a_{1t} + \beta a_{2t}.$$

Combining this equation with Equation (2.4), we have

$$\begin{bmatrix} 1 & \beta \\ 0 & 1 \end{bmatrix} \begin{bmatrix} z_{1t} \\ z_{2t} \end{bmatrix} = \begin{bmatrix} \phi_{1,11} + \beta\phi_{1,21} & \beta\phi_{1,22} \\ \phi_{1,21} & \phi_{1,22} \end{bmatrix} \begin{bmatrix} z_{1,t-1} \\ z_{2,t-1} \end{bmatrix} + \begin{bmatrix} b_{1t} \\ b_{2t} \end{bmatrix}, \tag{2.5}$$

where $b_{1t} = a_{1t} + \beta a_{2t}$ and $b_{2t} = a_{2t}$. Equation (2.5) remains a VAR(1) model. It has the same underlying structure as that of Equations (2.3) and (2.4). In particular, z_{1t} does not depend on $z_{2,t-1}$. Yet the parameter in the (1,2)th position of the AR coefficient matrix in Equation (2.5) is not zero. This nonzero parameter $\beta\phi_{1,22}$ is induced by the nonzero parameter β in the left side of Equation (2.5). □

The flexibility in the structure of a VAR(1) model increases with the dimension k. For example, consider the three-dimensional VAR(1) model $(\boldsymbol{I}_3 - \boldsymbol{\phi}_1 B)\boldsymbol{z}_t = \boldsymbol{a}_t$ with $\boldsymbol{\phi}_1 = [\phi_{1,ij}]_{3\times 3}$. If $\boldsymbol{\phi}_1$ is a lower triangular matrix, then z_{1t} does not depend on the past values of z_{2t} or z_{3t}, z_{2t} may depend on the past value of z_{1t}, but not on the past value of z_{3t}. In this case, we have a unidirectional relationship from z_{1t} to z_{2t} to z_{3t}. On the other hand, if $\phi_{1,13} = \phi_{1,23} = 0$ and $\phi_{1,ij} \neq 0$, otherwise, then we have a unidirectional relationship from both z_{1t} and z_{2t} to z_{3t}, whereas z_{1t} and z_{2t} are dynamically correlated. There are indeed many other possibilities.

2.2.2 Relation to Transfer Function Model

However, the model representation in Equations (2.3) and (2.4) is in general not a transfer function model because the two innovations a_{1t} and a_{2t} might be correlated. In a transfer function model, which is also known as a distributed-lag model, the input variable should be independent of the disturbance term of the output variable. To obtain a transfer function model, we perform orthogonalization of the two innovations in $\boldsymbol{a}_t$. Specifically, consider the simple linear regression

$$a_{2t} = \beta a_{1t} + \epsilon_t,$$

where $\beta = \text{cov}(a_{1t}, a_{2t})/\text{var}(a_{1t})$ and a_{1t} and ϵ_t are uncorrelated. By plugging a_{2t} into Equation (2.4), we obtain

$$(1 - \phi_{1,22}B)z_{2t} = (\phi_{20} - \beta\phi_{10}) + [\beta + (\phi_{1,21} - \beta\phi_{1,11})B]z_{1t} + \epsilon_t.$$

The prior equation further simplifies to

$$z_{2t} = \frac{\phi_{20} - \beta\phi_{10}}{1 - \phi_{1,22}} + \frac{\beta + (\phi_{1,21} - \beta\phi_{1,11})B}{1 - \phi_{1,22}B} z_{1t} + \frac{1}{1 - \phi_{1,22}B}\epsilon_t,$$

which is a transfer function model. The exogenous variable z_{1t} does not depend on the innovation ϵ_t.

2.2.3 Stationarity Condition

As defined in Chapter 1, a (weakly) stationary time series $\boldsymbol{z}_t$ has time invariant mean and covariance matrix. For these two conditions to hold, the mean of $\boldsymbol{z}_t$ should not depend on when the series started or what was its starting value. A simple way to investigate the stationarity condition for the VAR(1) model is then to exploit this feature of the series. For simplicity in discussion, we assume that the constant term $\boldsymbol{\phi}_0$ is 0, and the model reduces to $\boldsymbol{z}_t = \boldsymbol{\phi}_1 \boldsymbol{z}_{t-1} + \boldsymbol{a}_t$.

Suppose that the time series started at time $t = v$ with initial value $\boldsymbol{z}_v$, where v is a fixed time point. As time advances, the series $\boldsymbol{z}_t$ evolves. Specifically, by repeated substitutions, we have

$$\begin{aligned}
\boldsymbol{z}_t &= \boldsymbol{\phi}_1 \boldsymbol{z}_{t-1} + \boldsymbol{a}_t \\
&= \boldsymbol{\phi}_1(\boldsymbol{\phi}_1 \boldsymbol{z}_{t-2} + \boldsymbol{a}_{t-1}) + \boldsymbol{a}_t \\
&= \boldsymbol{\phi}_1^2 \boldsymbol{z}_{t-2} + \boldsymbol{\phi}_1 \boldsymbol{a}_{t-1} + \boldsymbol{a}_t \\
&= \boldsymbol{\phi}_1^3 \boldsymbol{z}_{t-3} + \boldsymbol{\phi}_1^2 \boldsymbol{a}_{t-2} + \boldsymbol{\phi}_1 \boldsymbol{a}_{t-1} + \boldsymbol{a}_t \\
&= \vdots \\
&= \boldsymbol{\phi}_1^{t-v} \boldsymbol{z}_v + \sum_{i=0}^{t-1} \boldsymbol{\phi}_1^i \boldsymbol{a}_{t-i}.
\end{aligned}$$

Consequently, for $\boldsymbol{z}_t$ to be independent of $\boldsymbol{z}_v$, we need $\boldsymbol{\phi}_1^{t-v}$ goes to 0 as $v \to -\infty$. Here, $v \to -\infty$ means that the series started a long time ago. Recall that if $\{\lambda_1, \ldots, \lambda_k\}$ are the eigenvalues of $\boldsymbol{\phi}_1$, then $\{\lambda_1^n, \ldots, \lambda_k^n\}$ are the eigenvalues of $\boldsymbol{\phi}_1^n$; see Appendix A. Also, if all eigenvalues of a matrix are 0, then the matrix must be 0. Consequently, the condition for $\boldsymbol{\phi}_1^{t-v} \to \boldsymbol{0}$ as $v \to -\infty$ is that all eigenvalues λ_j of $\boldsymbol{\phi}_1$ must satisfy $\lambda_j^{t-v} \to 0$ as $v \to -\infty$. This implies that the absolute values of all eigenvalues λ_j of $\boldsymbol{\phi}_1$ must be less than 1.

Consequently, a necessary condition for the VAR(1) series $\boldsymbol{z}_t$ to be stationary is that all eigenvalues of $\boldsymbol{\phi}_1$ must be less than 1 in absolute value. It can also be shown that if all eigenvalues of $\boldsymbol{\phi}_1$ are less than 1 in absolute value, then the VAR(1) series $\boldsymbol{z}_t$ is stationary.

The eigenvalues of $\boldsymbol{\phi}_1$ are solutions of the determinant equation

$$|\lambda \boldsymbol{I}_k - \boldsymbol{\phi}_1| = 0,$$

which can be written as

$$\lambda^k \left| \boldsymbol{I}_k - \boldsymbol{\phi}_b \frac{1}{\lambda} \right| = 0.$$

Therefore, we can consider the determinant equation $|\boldsymbol{I}_k - \boldsymbol{\phi}_1 x| = 0$, where $x = 1/\lambda$. Eigenvalues of $\boldsymbol{\phi}_1$ are the inverses of the solutions of this new equation.

Consequently, the necessary and sufficient condition for the stationarity of a VAR(1) model is that the solutions of the determinant equation $|\boldsymbol{I}_k - \boldsymbol{\phi}_1 B| = 0$ are greater than 1 in absolute value. That is, the solutions of the determinant equation $|\boldsymbol{\phi}(B)| = 0$ are outside the unit circle.

Example 2.1 Consider the bivariate VAR(1) model

$$\begin{bmatrix} z_{1t} \\ z_{2t} \end{bmatrix} = \begin{bmatrix} 5 \\ 3 \end{bmatrix} + \begin{bmatrix} 0.2 & 0.3 \\ -0.6 & 1.1 \end{bmatrix} \begin{bmatrix} z_{1,t-1} \\ z_{2,t-1} \end{bmatrix} + \begin{bmatrix} a_{1t} \\ a_{2t} \end{bmatrix}, \tag{2.6}$$

where the covariance matrix of $\boldsymbol{a}_t$ is

$$\boldsymbol{\Sigma}_a = \begin{bmatrix} 1.0 & 0.8 \\ 0.8 & 2.0 \end{bmatrix}.$$

Simple calculation shows that the eigenvalues of $\boldsymbol{\phi}_1$ are 0.5 and 0.8, which are less than 1. Therefore, the VAR(1) model in Equation (2.6) is stationary. Note that $\phi_{1,22} = 1.1$, which is greater than 1, but the series is stationary. This simple example demonstrates that the eigenvalues of $\boldsymbol{\phi}_1$ determine the stationarity of $\boldsymbol{z}_t$, not the individual elements of $\boldsymbol{\phi}_1$. □

2.2.4 Invertibility

By definition, a VAR(p) time series is a linear combination of its lagged values. Therefore, the VAR(1) model is always invertible; see the definition of invertibility in Chapter 1.

2.2.5 Moment Equations

Assume that the VAR(1) series in Equation (2.2) is stationary. Taking expectation on both sides of the equation, we have

$$\boldsymbol{\mu} = \boldsymbol{\phi}_0 + \boldsymbol{\phi}_1 \boldsymbol{\mu},$$

where $\boldsymbol{\mu} = E(\boldsymbol{z}_t)$. Consequently, we have $(\boldsymbol{I}_k - \boldsymbol{\phi}_1)\boldsymbol{\mu} = \boldsymbol{\phi}_0$, or equivalently, $\boldsymbol{\mu} = (\boldsymbol{I}_k - \boldsymbol{\phi}_1)^{-1}\boldsymbol{\phi}_0$. We can write this equation in a compact form as $\boldsymbol{\mu} = [\boldsymbol{\phi}(1)]^{-1}\boldsymbol{\phi}_0$. Plugging $\boldsymbol{\phi}_0 = (\boldsymbol{I}_k - \boldsymbol{\phi}_1)\boldsymbol{\mu}$ into the VAR(1) model, we obtain the mean-adjusted model

$$\tilde{\boldsymbol{z}}_t = \boldsymbol{\phi}_1 \tilde{\boldsymbol{z}}_{t-1} + \boldsymbol{a}_t, \tag{2.7}$$

where $\tilde{\boldsymbol{z}}_t = \boldsymbol{z}_t - \boldsymbol{\mu}$. The covariance matrix of $\boldsymbol{z}_t$ is then

$$\begin{aligned}\boldsymbol{\Gamma}_0 &= E\left(\tilde{\boldsymbol{z}}_t\tilde{\boldsymbol{z}}_t'\right) = E\left(\boldsymbol{\phi}_1\tilde{\boldsymbol{z}}_{t-1}\tilde{\boldsymbol{z}}_{t-1}'\boldsymbol{\phi}_1'\right) + E\left(\boldsymbol{a}_t\boldsymbol{a}_t'\right)\\ &= \boldsymbol{\phi}_1\boldsymbol{\Gamma}_0\boldsymbol{\phi}_1' + \boldsymbol{\Sigma}_a,\end{aligned}$$

where we use the fact that $\boldsymbol{a}_t$ is uncorrelated with $\tilde{\boldsymbol{z}}_{t-1}$. This equation can be rewritten as

$$\text{vec}(\boldsymbol{\Gamma}_0) = (\boldsymbol{\phi}_1 \otimes \boldsymbol{\phi}_1)\text{vec}(\boldsymbol{\Gamma}_0) + \text{vec}(\boldsymbol{\Sigma}_a).$$

Consequently, we have

$$(\boldsymbol{I}_{k^2} - \boldsymbol{\phi}_1 \otimes \boldsymbol{\phi}_1)\text{vec}(\boldsymbol{\Gamma}_0) = \text{vec}(\boldsymbol{\Sigma}_a). \tag{2.8}$$

The prior equation can be used to obtain $\boldsymbol{\Gamma}_0$ from the VAR(1) model for which $\boldsymbol{\phi}_1$ and $\boldsymbol{\Sigma}_a$ are known.

Next, for any positive integer ℓ, postmultiplying $\boldsymbol{z}_{t-\ell}'$ to Equation (2.7) and taking expectation, we have

$$\boldsymbol{\Gamma}_\ell = \boldsymbol{\phi}_1\boldsymbol{\Gamma}_{\ell-1}, \quad \ell > 0, \tag{2.9}$$

where we use the property that $\boldsymbol{z}_{t-\ell}$ is uncorrelated with $\boldsymbol{a}_t$. The prior equation is referred to as the multivariate Yule–Walker equation for the VAR(1) model. It can be used in two ways. First, in conjunction with Equation (2.8), the equation can be used recursively to obtain the cross-covariance matrices of $\boldsymbol{z}_t$ and, hence, the cross-correlation matrices. Second, it can be used to obtain $\boldsymbol{\phi}_1$ from the cross-covariance matrices. For instance, $\boldsymbol{\phi}_1 = \boldsymbol{\Gamma}_1\boldsymbol{\Gamma}_0^{-1}$.

To demonstrate, consider the stationary VAR(1) model in Equation (2.6). The mean of the series is $\boldsymbol{\mu} = (4, -6)'$. Using Equations (2.8) and (2.9), we obtain

$$\boldsymbol{\Gamma}_0 = \begin{bmatrix} 2.29 & 3.51 \\ 3.51 & 8.62 \end{bmatrix}, \quad \boldsymbol{\Gamma}_1 = \begin{bmatrix} 1.51 & 3.29 \\ 2.49 & 7.38 \end{bmatrix}, \quad \boldsymbol{\Gamma}_2 = \begin{bmatrix} 1.05 & 2.87 \\ 1.83 & 6.14 \end{bmatrix}.$$

The corresponding cross-correlation matrices of $\boldsymbol{z}_t$ are

$$\boldsymbol{\rho}_0 = \begin{bmatrix} 1.00 & 0.79 \\ 0.79 & 1.0 \end{bmatrix}, \quad \boldsymbol{\rho}_1 = \begin{bmatrix} 0.66 & 0.74 \\ 0.56 & 0.86 \end{bmatrix}, \quad \boldsymbol{\rho}_2 = \begin{bmatrix} 0.46 & 0.65 \\ 0.41 & 0.71 \end{bmatrix}.$$

R Demonstration

```
> phi1=matrix(c(.2,-.6,.3,1.1),2,2) % Input phi_1
> phi1
     [,1] [,2]
[1,]  0.2  0.3
[2,] -0.6  1.1
```

```
> sig=matrix(c(1,0.8,0.8,2),2,2) % Input sigma_a
> sig
     [,1] [,2]
[1,]  1.0  0.8
[2,]  0.8  2.0
> m1=eigen(phi1) % Obtain eigenvalues & vectors
> m1
$values
[1] 0.8 0.5
$vectors
           [,1]       [,2]
[1,] -0.4472136 -0.7071068
[2,] -0.8944272 -0.7071068
> I4=diag(4) ## Create the 4-by-4 identity matrix
> pp=kronecker(phi1,phi1) # Kronecker product
> pp
      [,1]  [,2]  [,3] [,4]
[1,]  0.04  0.06  0.06 0.09
[2,] -0.12  0.22 -0.18 0.33
[3,] -0.12 -0.18  0.22 0.33
[4,]  0.36 -0.66 -0.66 1.21
> c1=c(sig)
> c1
[1] 1.0 0.8 0.8 2.0
> dd=I4-pp
> ddinv=solve(dd)   ## Obtain inverse
> gam0=ddinv%*%matrix(c1,4,1)  # Obtain Gamma_0
> gam0
         [,1]
[1,] 2.288889
[2,] 3.511111
[3,] 3.511111
[4,] 8.622222
> g0=matrix(gam0,2,2)
> g1=phi1%*%g0     ## Obtain Gamma_1
> g1
         [,1]     [,2]
[1,] 1.511111 3.288889
[2,] 2.488889 7.377778
> g2=phi1%*%g1
> g2
         [,1]     [,2]
[1,] 1.048889 2.871111
[2,] 1.831111 6.142222
> D=diag(sqrt(diag(g0))) # To compute cross-correlation
matrices
> D
         [,1]     [,2]
[1,] 1.512907 0.000000
```

```
[2,] 0.000000 2.936362
> Di=solve(D)
> Di%*%g0%*%Di
          [,1]      [,2]
[1,] 1.0000000 0.7903557
[2,] 0.7903557 1.0000000
> Di%*%g1%*%Di
          [,1]      [,2]
[1,] 0.6601942 0.7403332
[2,] 0.5602522 0.8556701
> Di%*%g2%*%Di
          [,1]      [,2]
[1,] 0.4582524 0.6462909
[2,] 0.4121855 0.7123711
```

2.2.6 Implied Models for the Components

In this section, we discuss the implied marginal univariate models for each component z_{it} of a VAR(1) model. Again, for simplicity, we shall employ the mean-adjusted VAR(1) model in Equation (2.7). The AR matrix polynomial of the model is $\boldsymbol{I}_k - \boldsymbol{\phi}_1 B$. This is a $k \times k$ matrix. As such, we can consider its *adjoint* matrix; see Appendix A. For instance, consider the bivariate VAR(1) model of Example 2.1. In this case, we have

$$\boldsymbol{\phi}(B) = \begin{bmatrix} 1 - 0.2B & -0.3B \\ 0.6B & 1 - 1.1B \end{bmatrix}.$$

The adjoint matrix of $\boldsymbol{\phi}(B)$ is

$$\text{adj}[\boldsymbol{\phi}(B)] = \begin{bmatrix} 1 - 1.1B & 0.3B \\ -0.6B & 1 - 0.2B \end{bmatrix}.$$

The product of these two matrices gives

$$\text{adj}[\boldsymbol{\phi}(B)]\boldsymbol{\phi}(B) = |\boldsymbol{\phi}(B)|\boldsymbol{I}_2,$$

where $|\boldsymbol{\phi}(B)| = (1 - 0.2B)(1 - 1.1B) + 0.18B^2 = 1 - 1.3B + 0.4B^2$. The key feature of the product matrix is that it is a diagonal matrix with the determinant being its diagonal elements. This property continues to hold for a general VAR(1) model. Consequently, if we premultiply the VAR(1) model in Equation (2.7) by the adjoint matrix of $\boldsymbol{\phi}(B)$, then we have

$$|\boldsymbol{\phi}(B)|\boldsymbol{z}_t = \text{adj}[\boldsymbol{\phi}(B)]\boldsymbol{a}_t. \tag{2.10}$$

For a k-dimensional VAR(1) model, $|\boldsymbol{\phi}(B)|$ is a polynomial of degree k, and elements of adj$[\boldsymbol{\phi}(B)]$ are polynomials of degree $k-1$, because they are the determinant of a $(k-1) \times (k-1)$ matrix polynomial of order 1. Next, we use the result that any nonzero linear combination of $\{\boldsymbol{a}_t, \boldsymbol{a}_{t-1}, \ldots, \boldsymbol{a}_{t-k+1}\}$ is a univariate MA($k-1$) series. Consequently, Equation (2.10) shows that each component z_{it} follows a univariate ARMA($k, k-1$) model. The orders k and $k-1$ are the maximum orders. The actual ARMA order for the individual z_{it} can be smaller.

To demonstrate, consider again the bivariate VAR(1) model in Example 2.1. For this particular instance, Equation (2.10) becomes

$$(1 - 1.3B + 0.4B^2) \begin{bmatrix} z_{1t} \\ z_{2t} \end{bmatrix} = \begin{bmatrix} 1 - 1.1B & 0.3B \\ -0.6B & 1 - 0.2B \end{bmatrix} \begin{bmatrix} a_{1t} \\ a_{2t} \end{bmatrix}.$$

The marginal model for z_{1t} is then

$$\begin{aligned} \left(1 - 1.3B + 0.4B^2\right) z_{1t} &= (1 - 1.1B)a_{1t} + 0.3Ba_{2t} \\ &= a_{1t} - 1.1a_{1,t-1} + 0.3a_{2,t-1}, \end{aligned}$$

which is an ARMA(2,1) model because the MA part has serial correlation at lag 1 only and can be rewritten as $e_t - \theta e_{t-1}$. The parameters θ and Var(e_t) can be obtained from those of the VAR(1) model. The same result holds for z_{2t}.

2.2.7 Moving-Average Representation

In some applications, for example, computing variances of forecast errors, it is more convenient to consider the moving-average (MA) representation than the AR representation of a VAR model. For the VAR(1) model, one can easily obtain the MA representation using Equation (2.7). Specifically, by repeated substitutions, we have

$$\begin{aligned} \tilde{\boldsymbol{z}}_t &= \boldsymbol{a}_t + \boldsymbol{\phi}_1 \tilde{\boldsymbol{z}}_{t-1} = \boldsymbol{a}_t + \boldsymbol{\phi}_1(\boldsymbol{a}_{t-1} + \boldsymbol{\phi}_1 \tilde{\boldsymbol{z}}_{t-2}) \\ &= \boldsymbol{a}_t + \boldsymbol{\phi}_1 \boldsymbol{a}_{t-2} + \boldsymbol{\phi}_1^2(\boldsymbol{a}_{t-2} + \boldsymbol{\phi} \tilde{\boldsymbol{z}}_{t-3}) \\ &= \cdots \\ &= \boldsymbol{a}_t + \boldsymbol{\phi}_1 \boldsymbol{a}_{t-1} + \boldsymbol{\phi}_1^2 \boldsymbol{a}_{t-2} + \boldsymbol{\phi}_1^3 \boldsymbol{a}_{t-3} + \cdots \end{aligned}$$

Consequently, we have

$$\boldsymbol{z}_t = \boldsymbol{\mu} + \boldsymbol{a}_t + \boldsymbol{\psi}_1 \boldsymbol{a}_{t-1} + \boldsymbol{\psi}_2 \boldsymbol{a}_{t-2} + \cdots, \tag{2.11}$$

where $\boldsymbol{\psi}_i = \boldsymbol{\phi}_1^i$ for $i \geq 0$. For a stationary VAR(1) model, all eigenvalues of $\boldsymbol{\phi}_1$ are less than 1 in absolute value, thus $\boldsymbol{\psi}_i \to \boldsymbol{0}$ as $i \to \infty$. This implies that, as expected, the effect of the remote innovation $\boldsymbol{a}_{t-i}$ on $\boldsymbol{z}_t$ is diminishing as i increases. Eventually the effect vanishes, confirming that the initial condition of a stationary VAR(1) model has no impact on the series as time increases.

2.3 VAR(2) MODELS

A VAR(2) model assumes the form

$$\boldsymbol{z}_t = \boldsymbol{\phi}_0 + \boldsymbol{\phi}_1 \boldsymbol{z}_{t-1} + \boldsymbol{\phi}_2 \boldsymbol{z}_{t-2} + \boldsymbol{a}_t. \tag{2.12}$$

It says that each component z_{it} depends on the lagged values $\boldsymbol{z}_{t-1}$ and $\boldsymbol{z}_{t-2}$. The AR coefficient can be interpreted in a way similar to that of a VAR(1) model. For instance, the $(1, 2)$th element of $\boldsymbol{\phi}_1$, that is, $\phi_{1,12}$, denotes the linear dependence of z_{1t} on $z_{2,t-1}$ in the presence of other lagged values in $\boldsymbol{z}_{t-1}$ and $\boldsymbol{z}_{t-2}$.

The structure and relationship to Granger causality of a VAR(2) model can be generalized straightforwardly from that of the VAR(1) model. For instance, consider a bivariate VAR(2) model. If both $\boldsymbol{\phi}_1$ and $\boldsymbol{\phi}_2$ are diagonal matrices, then z_{1t} and z_{2t} follow a univariate AR(2) model and can be handled accordingly. If $\phi_{1,12} = \phi_{2,12} = 0$, but at least one of $\phi_{1,21}$ and $\phi_{2,21}$ is not zero, then we have a unidirectional relationship from z_{1t} to z_{2t}, because in this case z_{1t} does not depend on any past value of z_{2t}, but z_{2t} depends on some past value of z_{1t}. We can also derive a transfer function model for z_{2t} with z_{1t} as the input variable. In general, for the existence of a unidirectional relationship from component z_{it} to component z_{jt} of a VAR model, the parameters at the (i, j)th position of each coefficient matrix $\boldsymbol{\phi}_v$ must be 0 simultaneously.

2.3.1 Stationarity Condition

To study the stationarity condition for the VAR(2) model in Equation (2.12), we can make use of the result obtained for VAR(1) models. To see this, we consider an expanded $(2k)$-dimensional time series $\boldsymbol{Z}_t = (\boldsymbol{z}_t', \boldsymbol{z}_{t-1}')'$. Using Equation (2.12) and the identity $\boldsymbol{z}_{t-1} = \boldsymbol{z}_{t-1}$, we obtain a model for $\boldsymbol{Z}_t$. Specifically, we have

$$\begin{bmatrix} \boldsymbol{z}_t \\ \boldsymbol{z}_{t-1} \end{bmatrix} = \begin{bmatrix} \boldsymbol{\phi}_0 \\ \boldsymbol{0} \end{bmatrix} + \begin{bmatrix} \boldsymbol{\phi}_1 & \boldsymbol{\phi}_2 \\ \boldsymbol{I}_k & \boldsymbol{0}_k \end{bmatrix} \begin{bmatrix} \boldsymbol{z}_{t-1} \\ \boldsymbol{z}_{t-2} \end{bmatrix} + \begin{bmatrix} \boldsymbol{a}_t \\ \boldsymbol{0} \end{bmatrix}, \tag{2.13}$$

where $\boldsymbol{0}$ is a k-dimensional vector of zero and $\boldsymbol{0}_k$ is a $k \times k$ zero matrix. Consequently, the expanded time series $\boldsymbol{Z}_t$ follows a VAR(1) model, say,

$$\boldsymbol{Z}_t = \boldsymbol{\Phi}_0 + \boldsymbol{\Phi}_1 \boldsymbol{Z}_{t-1} + \boldsymbol{b}_t, \tag{2.14}$$

where the vector $\boldsymbol{\Phi}_0$, the AR coefficient matrix $\boldsymbol{\Phi}_1$, and $\boldsymbol{b}_t$ are defined in Equation (2.13). Using the result of VAR(1) models derived in the previous section, the necessary and sufficient condition for $\boldsymbol{Z}_t$ to be stationary is that all solutions of the determinant equation $|\boldsymbol{I}_{2k} - \boldsymbol{\Phi}_1 B| = 0$ must be greater than 1 in absolute value.

Next, the determinant can be rewritten as

$$\begin{aligned}
|\boldsymbol{I}_{2k} - \boldsymbol{\Phi}_1 B| &= \begin{vmatrix} \boldsymbol{I}_k - \boldsymbol{\phi}_1 B & -\boldsymbol{\phi}_2 B \\ -\boldsymbol{I}_k B & \boldsymbol{I}_k \end{vmatrix} \\
&= \begin{vmatrix} \boldsymbol{I}_k - \boldsymbol{\phi}_1 B - \boldsymbol{\phi}_2 B^2 & -\boldsymbol{\phi}_2 B \\ \boldsymbol{0}_k & \boldsymbol{I}_k \end{vmatrix} \\
&= |\boldsymbol{I}_k - \boldsymbol{\phi}_1 B - \boldsymbol{\phi}_2 B^2| \\
&= |\boldsymbol{\phi}(B)|,
\end{aligned}$$

where the second equality is obtained by multiplying the second column block matrix by B and adding the result to the first column block. Such an operation is valid in calculating the determinant of a matrix. In summary, the necessary and sufficient condition for the stationarity of $\boldsymbol{Z}_t$ (and, hence $\boldsymbol{z}_t$) is that all solutions of the determinant equation $|\boldsymbol{\phi}(B)| = 0$ are greater than 1 in absolute value.

2.3.2 Moment Equations

Assuming that the VAR(2) model in Equation (2.12) is stationary, we derive the moment equations for $\boldsymbol{z}_t$. First, taking expectation of Equation (2.12), we get

$$\boldsymbol{\mu} = \boldsymbol{\phi}_0 + \boldsymbol{\phi}_1 \boldsymbol{\mu} + \boldsymbol{\phi}_2 \boldsymbol{\mu}.$$

Therefore, we have $(\boldsymbol{I}_k - \boldsymbol{\phi}_1 - \boldsymbol{\phi}_2)\boldsymbol{\mu} = \boldsymbol{\phi}_0$. The mean of the series is then $\boldsymbol{\mu} = [\boldsymbol{\phi}(1)]^{-1}\boldsymbol{\phi}_0$. Next, plugging $\boldsymbol{\phi}_0$ into Equation (2.12), we obtain the mean-adjusted model as

$$\tilde{\boldsymbol{z}}_t = \boldsymbol{\phi}_1 \tilde{\boldsymbol{z}}_{t-1} + \boldsymbol{\phi}_2 \tilde{\boldsymbol{z}}_{t-2} + \boldsymbol{a}_t. \tag{2.15}$$

Postmultiplying Equation (2.15) by $\boldsymbol{a}_t'$ and using the zero correlation between $\boldsymbol{a}_t$ and the past values of $\boldsymbol{z}_t$, we obtain

$$E(\tilde{\boldsymbol{z}}_t \boldsymbol{a}_t') = E(\boldsymbol{a}_t \boldsymbol{a}_t') = \boldsymbol{\Sigma}_a.$$

Postmultiplying Equation (2.15) by $\tilde{\boldsymbol{z}}_{t-\ell}'$ and taking expectation, we have

$$\begin{aligned}
\boldsymbol{\Gamma}_0 &= \boldsymbol{\phi}_1 \boldsymbol{\Gamma}_{-1} + \boldsymbol{\phi}_2 \boldsymbol{\Gamma}_{-2} + \boldsymbol{\Sigma}_a, \\
\boldsymbol{\Gamma}_\ell &= \boldsymbol{\phi}_1 \boldsymbol{\Gamma}_{\ell-1} + \boldsymbol{\phi}_2 \boldsymbol{\Gamma}_{\ell-2}, \quad \text{if } \ell > 0.
\end{aligned} \tag{2.16}$$

Using $\ell = 0, 1, 2$, we have a set of matrix equations that relate $\{\boldsymbol{\Gamma}_0, \boldsymbol{\Gamma}_1, \boldsymbol{\Gamma}_2\}$ to $\{\boldsymbol{\phi}_1, \boldsymbol{\phi}_2, \boldsymbol{\Sigma}_a\}$. In particular, using $\ell = 1$ and 2, we have

$$[\boldsymbol{\Gamma}_1, \boldsymbol{\Gamma}_2] = [\boldsymbol{\phi}_1, \boldsymbol{\phi}_2] \begin{bmatrix} \boldsymbol{\Gamma}_0 & \boldsymbol{\Gamma}_1 \\ \boldsymbol{\Gamma}_1' & \boldsymbol{\Gamma}_0 \end{bmatrix}, \tag{2.17}$$

where we use $\mathbf{\Gamma}_{-1} = \mathbf{\Gamma}'_1$. This system of matrix equations is called the multivariate Yule–Walker equation for the VAR(2) model in Equation (2.12). For a stationary series $\boldsymbol{z}_t$, the $2k \times 2k$ matrix on the right side of Equation (2.17) is invertible so that we have

$$[\boldsymbol{\phi}_1, \boldsymbol{\phi}_2] = [\mathbf{\Gamma}_1, \mathbf{\Gamma}_2] \begin{bmatrix} \mathbf{\Gamma}_0 & \mathbf{\Gamma}_1 \\ \mathbf{\Gamma}'_1 & \mathbf{\Gamma}_0 \end{bmatrix}^{-1}.$$

This equation can be used to obtain the AR coefficients from the cross-covariance matrices.

In practice, one may use the expanded series $\boldsymbol{Z}_t$ in Equation (2.14) and the result of VAR(1) model in the previous section to obtain the cross-covariance matrices $\mathbf{\Gamma}_\ell$ of $\boldsymbol{z}_t$ from $\boldsymbol{\phi}_1$, $\boldsymbol{\phi}_2$, and $\mathbf{\Sigma}_a$. Specifically, for Equation (2.14), we have

$$\mathbf{\Phi}_1 = \begin{bmatrix} \boldsymbol{\phi}_1 & \boldsymbol{\phi}_2 \\ \boldsymbol{I}_k & \mathbf{0}_k \end{bmatrix}, \quad \mathbf{\Sigma}_b = \begin{bmatrix} \mathbf{\Sigma}_a & \mathbf{0}_k \\ \mathbf{0}_k & \mathbf{0}_k \end{bmatrix}.$$

In addition, we have

$$\text{Cov}(\boldsymbol{Z}_t) = \mathbf{\Gamma}_0^* = \begin{bmatrix} \mathbf{\Gamma}_0 & \mathbf{\Gamma}_1 \\ \mathbf{\Gamma}'_1 & \mathbf{\Gamma}_0 \end{bmatrix}.$$

For the expanded series $\boldsymbol{Z}_t$, Equation (2.8) becomes

$$[\boldsymbol{I}_{(2k)^2} - \mathbf{\Phi}_1 \otimes \mathbf{\Phi}_1]\text{vec}(\mathbf{\Gamma}_0^*) = \text{vec}(\mathbf{\Sigma}_b). \tag{2.18}$$

Consequently, we can obtain $\mathbf{\Gamma}_0^*$, which contains $\mathbf{\Gamma}_0$ and $\mathbf{\Gamma}_1$. Higher-order cross-covariances $\mathbf{\Gamma}_\ell$ are then obtained recursively using the moment equation in Equation (2.16).

Example 2.2 Consider a three-dimensional VAR(2) model in Equation (2.12) with parameters $\boldsymbol{\phi}_0 = \mathbf{0}$ and

$$\boldsymbol{\phi}_1 = \begin{bmatrix} 0.47 & 0.21 & 0 \\ 0.35 & 0.34 & 0.47 \\ 0.47 & 0.23 & 0.23 \end{bmatrix}, \quad \boldsymbol{\phi}_2 = \begin{bmatrix} 0 & 0 & 0 \\ -0.19 & -0.18 & 0 \\ -0.30 & 0 & 0 \end{bmatrix},$$

$$\mathbf{\Sigma}_a = \begin{bmatrix} 0.285 & 0.026 & 0.069 \\ 0.026 & 0.287 & 0.137 \\ 0.069 & 0.137 & 0.357 \end{bmatrix}.$$

This VAR(2) model was employed later in analyzing the quarterly growth rates of gross domestic product of United Kingdom, Canada, and United States from 1980.II to 2011.II. For this particular VAR(2) model, we can calculate its cross-covariance matrices via Equation (2.18). They are

$$\boldsymbol{\Gamma}_0 = \begin{bmatrix} 0.46 & 0.22 & 0.24 \\ 0.22 & 0.61 & 0.38 \\ 0.24 & 0.38 & 0.56 \end{bmatrix}, \quad \boldsymbol{\Gamma}_1 = \begin{bmatrix} 0.26 & 0.23 & 0.19 \\ 0.25 & 0.35 & 0.38 \\ 0.24 & 0.25 & 0.25 \end{bmatrix},$$

$$\boldsymbol{\Gamma}_2 = \begin{bmatrix} 0.18 & 0.18 & 0.17 \\ 0.16 & 0.17 & 0.20 \\ 0.10 & 0.18 & 0.16 \end{bmatrix}. \qquad \square$$

2.3.3 Implied Marginal Component Models

For the VAR(2) model in Equation (2.12), we can use the same technique as that for VAR(1) models to obtain the univariate ARMA model for the component series z_{it}. The general solution is that z_{it} follows an ARMA$(2k, 2(k-1))$ model. Again, the order $(2k, 2(k-1))$ is the maximum order for each component z_{it}.

2.3.4 Moving-Average Representation

The MA representation of a VAR(2) model can be obtained in several ways. One can use repeated substitutions similar to that of the VAR(1) model. Here, we adopt an alternative approach. Consider the mean-adjusted VAR(2) model in Equation (2.15). The model can be written as

$$\left(\boldsymbol{I}_k - \boldsymbol{\phi}_1 B - \boldsymbol{\phi}_2 B^2\right) \tilde{\boldsymbol{z}}_t = \boldsymbol{a}_t,$$

which is equivalent to

$$\tilde{\boldsymbol{z}}_t = \left(\boldsymbol{I}_k - \boldsymbol{\phi}_1 B - \boldsymbol{\phi}_2 B^2\right)^{-1} \boldsymbol{a}_t.$$

On the other hand, the MA representation of the series is $\tilde{\boldsymbol{z}}_t = \boldsymbol{\psi}(B)\boldsymbol{a}_t$. Consequently, we have $(\boldsymbol{I}_k - \boldsymbol{\phi}_1 B - \boldsymbol{\phi}_2 B^2)^{-1} = \boldsymbol{\psi}(B)$, which is

$$\boldsymbol{I}_k = \left(\boldsymbol{I}_k - \boldsymbol{\phi}_1 B - \boldsymbol{\phi}_2 B^2\right)\left(\boldsymbol{\psi}_0 + \boldsymbol{\psi}_1 B + \boldsymbol{\psi}_2 B^2 + \boldsymbol{\psi}_3 B^3 + \cdots\right), \tag{2.19}$$

where $\boldsymbol{\psi}_0 = \boldsymbol{I}_k$. Since the left-hand side of Equation (2.19) is a constant matrix, all coefficient matrices of B^i for $i > 0$ on the right-hand side of the equation must be 0. Therefore, we obtain

$$\begin{aligned} \mathbf{0} &= \boldsymbol{\psi}_1 - \boldsymbol{\phi}_1 \boldsymbol{\psi}_0, && \text{(coefficient of } B\text{)} \\ \mathbf{0} &= \boldsymbol{\psi}_2 - \boldsymbol{\phi}_1 \boldsymbol{\psi}_1 - \boldsymbol{\phi}_2 \boldsymbol{\psi}_0, && \text{(coefficient of } B^2\text{)} \\ \mathbf{0} &= \boldsymbol{\psi}_v - \boldsymbol{\phi}_1 \boldsymbol{\psi}_{v-1} - \boldsymbol{\phi}_2 \boldsymbol{\psi}_{v-2}, && \text{for } v \geq 3. \end{aligned}$$

Consequently, we have

$$\begin{aligned}\boldsymbol{\psi}_1 &= \boldsymbol{\phi}_1 \\ \boldsymbol{\psi}_v &= \boldsymbol{\phi}_1\boldsymbol{\psi}_{v-1} + \boldsymbol{\phi}_2\boldsymbol{\psi}_{v-2}, \quad \text{for } v \geq 2,\end{aligned} \tag{2.20}$$

where $\boldsymbol{\psi}_0 = \boldsymbol{I}_k$. In other words, we can compute the coefficient matrix $\boldsymbol{\psi}_i$ recursively starting with $\boldsymbol{\psi}_0 = \boldsymbol{I}_k$ and $\boldsymbol{\psi}_1 = \boldsymbol{\phi}_1$.

2.4 VAR(p) MODELS

Consider next the general k-dimensional VAR(p) model, which assumes the form

$$\boldsymbol{\phi}(B)\boldsymbol{z}_t = \boldsymbol{\phi}_0 + \boldsymbol{a}_t, \tag{2.21}$$

where $\boldsymbol{\phi}(B) = \boldsymbol{I}_k - \sum_{i=1}^{p} \boldsymbol{\phi}_i B^i$ with $\boldsymbol{\phi}_p \neq \boldsymbol{0}$. The results for VAR(1) and VAR(2) models discussed in the previous sections continue to hold for the VAR(p) model. For example, a VAR(p) model is invertible and its structure is sufficiently flexible to encompass the transfer function model. In this section, we consider the generalizations of other properties of simple VAR models to the general VAR(p) model.

Assume that the series in Equation (2.21) is stationary. Taking the expectation, we have

$$(\boldsymbol{I}_k - \boldsymbol{\phi}_1 - \cdots - \boldsymbol{\phi}_p)\boldsymbol{\mu} = [\boldsymbol{\phi}(1)]\boldsymbol{\mu} = \boldsymbol{\phi}_0,$$

where, as before, $\boldsymbol{\mu} = E(\boldsymbol{z}_t)$. Therefore, $\boldsymbol{\mu} = [\boldsymbol{\phi}(1)]^{-1}\boldsymbol{\phi}_0$, and the model can be rewritten as

$$\boldsymbol{\phi}(B)\tilde{\boldsymbol{z}}_t = \boldsymbol{a}_t. \tag{2.22}$$

We use this mean-adjusted representation to derive other properties of a stationary VAR(p) model. The mean has no impact on those properties and can be assumed to be 0.

2.4.1 A VAR(1) Representation

Similar to the VAR(2) model, we can express a VAR(p) model in a VAR(1) form by using an expanded series. Define $\boldsymbol{Z}_t = (\tilde{\boldsymbol{z}}_t', \tilde{\boldsymbol{z}}_{t-1}', \ldots, \tilde{\boldsymbol{z}}_{t-p+1}')'$, which is a pk-dimensional time series. The VAR(p) model in Equation (2.22) can be written as

$$\boldsymbol{Z}_t = \boldsymbol{\Phi}\boldsymbol{Z}_{t-1} + \boldsymbol{b}_t, \tag{2.23}$$

where $\boldsymbol{b}_t = (\boldsymbol{a}_t', \boldsymbol{0}')'$ with $\boldsymbol{0}$ being a $k(p-1)$-dimensional zero vector, and

$$\boldsymbol{\Phi} = \begin{bmatrix} \boldsymbol{\phi}_1 & \boldsymbol{\phi}_2 & \cdots & \boldsymbol{\phi}_{p-1} & \boldsymbol{\phi}_p \\ \boldsymbol{I} & \boldsymbol{0} & \cdots & \boldsymbol{0} & \boldsymbol{0} \\ \boldsymbol{0} & \boldsymbol{I} & \cdots & \boldsymbol{0} & \boldsymbol{0} \\ \vdots & \vdots & \ddots & \vdots & \vdots \\ \boldsymbol{0} & \boldsymbol{0} & \cdots & \boldsymbol{I} & \boldsymbol{0} \end{bmatrix},$$

where it is understood that $\boldsymbol{I}$ and $\boldsymbol{0}$ are the $k \times k$ identity and zero matrix, respectively. The matrix $\boldsymbol{\Phi}$ is called the companion matrix of the matrix polynomial $\boldsymbol{\phi}(B) = \boldsymbol{I}_k - \boldsymbol{\phi}_1 B - \cdots - \boldsymbol{\phi}_p B^p$. The covariance matrix of $\boldsymbol{b}_t$ has a special structure; all of its elements are zero except those in the upper-left corner that is $\boldsymbol{\Sigma}_a$.

2.4.2 Stationarity Condition

The sufficient and necessary condition for the weak stationarity of the VAR(p) series $\boldsymbol{z}_t$ can be easily obtained using the VAR(1) representation in Equation (2.23). Since $\boldsymbol{Z}_t$ follows a VAR(1) model, the condition for its stationarity is that all solutions of the determinant equation $|\boldsymbol{I}_{kp} - \boldsymbol{\Phi} B| = 0$ must be greater than 1 in absolute value. Sometimes, we say that the solutions must be greater than 1 in modulus or they are outside the unit circle. By Lemma 2.1, we have $|\boldsymbol{I}_{kp} - \boldsymbol{\Phi} B| = |\boldsymbol{\phi}(B)|$ for the VAR(p) time series. Therefore, the necessary and sufficient condition for the weak stationarity of the VAR(p) series is that all solutions of the determinant equation $|\boldsymbol{\phi}(B)| = 0$ must be greater than 1 in modulus.

Lemma 2.1 For the $k \times k$ matrix polynomial $\boldsymbol{\phi}(B) = \boldsymbol{I}_k - \sum_{i=1}^{p} \boldsymbol{\phi}_i B^i$, $|\boldsymbol{I}_{kp} - \boldsymbol{\Phi} B| = |\boldsymbol{I}_k - \boldsymbol{\phi}_1 B - \cdots - \boldsymbol{\phi}_p B^p|$ holds, where $\boldsymbol{\Phi}$ is defined in Equation (2.23).

A proof of Lemma 2.1 is given in Section 2.12.

2.4.3 Moment Equations

Postmultiplying Equation (2.22) by $\tilde{\boldsymbol{z}}_{t-\ell}$ and taking expectation, we obtain

$$\boldsymbol{\Gamma}_\ell - \boldsymbol{\phi}_1 \boldsymbol{\Gamma}_{\ell-1} - \cdots - \boldsymbol{\phi}_p \boldsymbol{\Gamma}_{\ell-p} = \begin{cases} \boldsymbol{\Sigma}_a & \text{if } \ell = 0, \\ \boldsymbol{0} & \text{if } \ell > 0. \end{cases} \tag{2.24}$$

Consider jointly the matrix equations for $\ell = 1, \ldots, p$. We have a system of matrix equations

$$[\boldsymbol{\Gamma}_1, \boldsymbol{\Gamma}_2, \ldots, \boldsymbol{\Gamma}_p] = [\boldsymbol{\phi}_1, \boldsymbol{\phi}_2, \ldots, \boldsymbol{\phi}_p] \begin{bmatrix} \boldsymbol{\Gamma}_0 & \boldsymbol{\Gamma}_1 & \cdots & \boldsymbol{\Gamma}_{p-1} \\ \boldsymbol{\Gamma}_1' & \boldsymbol{\Gamma}_0 & \cdots & \boldsymbol{\Gamma}_{p-2} \\ \vdots & \vdots & & \vdots \\ \boldsymbol{\Gamma}_{p-1}' & \boldsymbol{\Gamma}_{p-2}' & \cdots & \boldsymbol{\Gamma}_0 \end{bmatrix}, \tag{2.25}$$

where we use $\mathbf{\Gamma}_{-\ell} = \mathbf{\Gamma}'_\ell$. This system of matrix equations is called the multivariate Yule–Walker equation for VAR(p) models. It can be used to obtain the AR coefficient matrices $\boldsymbol{\phi}_j$ from the cross-covariance matrices $\mathbf{\Gamma}_\ell$ for $\ell = 0, \ldots, p$. For a stationary VAR(p) model, the square matrix of Equation (2.25) is nonsingular. On the other hand, to obtain the cross-covariance matrices, hence the cross-correlation matrices, of a stationary VAR(p) model, it is convenient to use the expanded VAR(1) representation in Equation (2.23). For the expanded (kp)-dimensional series $\boldsymbol{Z}_t$, we have

$$\text{Cov}(\boldsymbol{Z}_t) = \mathbf{\Gamma}_0^* = \begin{bmatrix} \mathbf{\Gamma}_0 & \mathbf{\Gamma}_1 & \cdots & \mathbf{\Gamma}_{p-1} \\ \mathbf{\Gamma}'_1 & \mathbf{\Gamma}_0 & \cdots & \mathbf{\Gamma}_{p-2} \\ \vdots & \vdots & & \vdots \\ \mathbf{\Gamma}'_{p-1} & \mathbf{\Gamma}'_{p-2} & \cdots & \mathbf{\Gamma}_0 \end{bmatrix}, \tag{2.26}$$

which is precisely the square matrix in Equation (2.25). Consequently, similar to the VAR(2) case, we can apply Equation (2.8) to obtain

$$[\boldsymbol{I}_{(kp)^2} - \mathbf{\Phi} \otimes \mathbf{\Phi}]\text{vec}(\mathbf{\Gamma}_0^*) = \text{vec}(\mathbf{\Sigma}_b).$$

Thus, given the AR coefficient matrices $\boldsymbol{\phi}_i$ and the covariance matrix $\mathbf{\Sigma}_a$, we can obtain $\mathbf{\Phi}$ and $\mathbf{\Sigma}_b$. The prior equation can be then used to obtain $\mathbf{\Gamma}_0^*$, which contains $\mathbf{\Gamma}_\ell$ for $\ell = 0, \ldots, p-1$. Other higher-order cross-covariance matrices, $\mathbf{\Gamma}_\ell$, can use computed recursively via the moment equation in Equation (2.24).

2.4.4 Implied Component Models

Using the same technique as those used in the VAR(2) model, we see that the component series z_{it} of a VAR(p) model follows a univariate ARMA(kp, $(k-1)p$) model. This ARMA order (kp, $(k-1)p$) is high for a large k or p, but it denotes the maximum order allowed. The actual order of the marginal ARMA model for z_{it} can be substantially lower. Except for seasonal time series, our experience of analyzing univariate time series shows that the order used for real-world series is typically low. Since a VAR(p) model encompasses a wide-rage of component models, one would expect that the VAR orders used for most real-world multivariate time series will not be high.

2.4.5 Moving-Average Representation

Using the same techniques as that of the VAR(2) model, we can obtain the MA representation for a VAR(p) model via a recursive method. The coefficient matrices of the MA representation are

$$\boldsymbol{\psi}_i = \sum_{j=1}^{\min(i,p)} \boldsymbol{\phi}_j \boldsymbol{\psi}_{i-j}, \quad i = 1, 2, \cdots, \tag{2.27}$$

where $\boldsymbol{\psi}_0 = \boldsymbol{I}_k$. The matrices $\boldsymbol{\psi}_i$ are referred to as the ψ-weights of the VAR(p) model. We discuss the meanings of these $\boldsymbol{\psi}_i$ matrices later. Here, it suffices to say that, based on the MA representation, we can easily show the following result.

Lemma 2.2 For a VAR(p) model in Equation (2.21) with $\boldsymbol{a}_t$ being a serially uncorrelated innovation process with mean zero and positive-definite covariance $\boldsymbol{\Sigma}_a$, $\text{Cov}(\boldsymbol{z}_t, \boldsymbol{a}_{t-j}) = \boldsymbol{\psi}_j \boldsymbol{\Sigma}_a$ for $j \geq 0$, where $\boldsymbol{\psi}_j$ denotes the ψ-weight matrix.

2.5 ESTIMATION

A VAR(p) model can be estimated by the LS or ML method or Bayesian method. For the LS methods, we show that the GLS and the OLS methods produce the same estimates; see Zellner (1962). Under the multivariate normality assumption, that is, $\boldsymbol{a}_t$ follows a k-dimensional normal distribution, the ML estimates of a VAR(p) model are asymptotically equivalent to the LS estimates. We also briefly discuss the Bayesian estimation of a VAR(p) model.

Suppose that the sample $\{\boldsymbol{z}_t | t = 1, \ldots, T\}$ is available from a VAR(p) model. The parameters of interest are $\{\boldsymbol{\phi}_0, \boldsymbol{\phi}_1, \ldots, \boldsymbol{\phi}_p\}$ and $\boldsymbol{\Sigma}_a$. In what follows, we discuss various methods for estimating these parameters and properties of the estimates.

2.5.1 Least-Squares Methods

For LS estimation, the available data enable us to consider

$$\boldsymbol{z}_t = \boldsymbol{\phi}_0 + \boldsymbol{\phi}_1 \boldsymbol{z}_{t-1} + \cdots + \boldsymbol{\phi}_p \boldsymbol{z}_{t-p} + \boldsymbol{a}_t, \quad t = p+1, \cdots, T,$$

where the covariance matrix of $\boldsymbol{a}_t$ is $\boldsymbol{\Sigma}_a$. Here, we have $T - p$ data points for effective estimation. To facilitate the estimation, we rewrite the VAR(p) model as

$$\boldsymbol{z}_t' = \boldsymbol{x}_t' \boldsymbol{\beta} + \boldsymbol{a}_t',$$

where $\boldsymbol{x}_t = (1, \boldsymbol{z}_{t-1}', \cdots, \boldsymbol{z}_{t-p}')'$ is a $(kp+1)$-dimensional vector and $\boldsymbol{\beta}' = [\boldsymbol{\phi}_0, \boldsymbol{\phi}_1, \cdots, \boldsymbol{\phi}_p]$ is a $k \times (kp+1)$ matrix. With this new format, we can write the data as

$$\boldsymbol{Z} = \boldsymbol{X}\boldsymbol{\beta} + \boldsymbol{A}, \tag{2.28}$$

where $\boldsymbol{Z}$ is a $(T-p) \times k$ matrix with ith row being $\boldsymbol{z}_{p+i}'$, $\boldsymbol{X}$ is a $(T-p) \times (kp+1)$ design matrix with ith row being $\boldsymbol{x}_{p+i}'$, and $\boldsymbol{A}$ is a $(T-p) \times k$ matrix with ith row being $\boldsymbol{a}'_{p+i}$. The matrix representation in Equation (2.28) is particularly convenient for the VAR(p) model. For example, column j of $\boldsymbol{\beta}$ contains parameters associated with z_{jt}. Taking the vectorization of Equation (2.28), and using the properties of Kronecker product given in Appendix A, we obtain

$$\text{vec}(\boldsymbol{Z}) = (\boldsymbol{I}_k \otimes \boldsymbol{X})\text{vec}(\boldsymbol{\beta}) + \text{vec}(\boldsymbol{A}). \tag{2.29}$$

Note that the covariance matrix of vec($\boldsymbol{A}$) is $\boldsymbol{\Sigma}_a \otimes \boldsymbol{I}_{T-p}$.

2.5.1.1 Generalized Least Squares Estimate

The GLS estimate of $\boldsymbol{\beta}$ is obtained by minimizing

$$\begin{aligned} S(\boldsymbol{\beta}) &= [\text{vec}(\boldsymbol{A})]'(\boldsymbol{\Sigma}_a \otimes \boldsymbol{I}_{T-p})^{-1}\text{vec}(\boldsymbol{A}) \\ &= [\text{vec}(\boldsymbol{Z} - \boldsymbol{X}\boldsymbol{\beta})]' \left(\boldsymbol{\Sigma}_a^{-1} \otimes \boldsymbol{I}_{T-p}\right) \text{vec}(\boldsymbol{Z} - \boldsymbol{X}\boldsymbol{\beta}) \qquad (2.30) \\ &= tr[(\boldsymbol{Z} - \boldsymbol{X}\boldsymbol{\beta})\boldsymbol{\Sigma}_a^{-1}(\boldsymbol{Z} - \boldsymbol{X}\boldsymbol{\beta})']. \qquad (2.31) \end{aligned}$$

The last equality holds because $\boldsymbol{\Sigma}_a$ is a symmetric matrix and we use $tr(\boldsymbol{DBC}) = \text{vec}(\boldsymbol{C}')'(\boldsymbol{B}' \otimes \boldsymbol{I})\text{vec}(\boldsymbol{D})$. From Equation (2.30), we have

$$\begin{aligned} S(\boldsymbol{\beta}) &= [\text{vec}(\boldsymbol{Z}) - (\boldsymbol{I}_k \otimes \boldsymbol{X})\text{vec}(\boldsymbol{\beta})]' \left(\boldsymbol{\Sigma}_a^{-1} \otimes \boldsymbol{I}_{T-p}\right) [\text{vec}(\boldsymbol{Z}) - (\boldsymbol{I}_k \otimes \boldsymbol{X})\text{vec}(\boldsymbol{\beta})] \\ &= \left[\text{vec}(\boldsymbol{Z})' - \text{vec}(\boldsymbol{\beta})'(\boldsymbol{I}_k \otimes \boldsymbol{X}')\right] \left(\boldsymbol{\Sigma}_a^{-1} \otimes \boldsymbol{I}_{T-p}\right) \\ &\quad \times [\text{vec}(\boldsymbol{Z}) - (\boldsymbol{I}_k \otimes \boldsymbol{X})\text{vec}(\boldsymbol{\beta})] \\ &= \text{vec}(\boldsymbol{Z})' \left(\boldsymbol{\Sigma}_a^{-1} \otimes \boldsymbol{I}_{T-p}\right) \text{vec}(\boldsymbol{Z}) - 2\text{vec}(\boldsymbol{\beta})' \left(\boldsymbol{\Sigma}_a^{-1} \otimes \boldsymbol{X}'\right) \text{vec}(\boldsymbol{Z}) \\ &\quad + \text{vec}(\boldsymbol{\beta})' \left(\boldsymbol{\Sigma}_a^{-1} \otimes \boldsymbol{X}'\boldsymbol{X}\right) \text{vec}(\boldsymbol{\beta}). \qquad (2.32) \end{aligned}$$

Taking partial derivatives of $S(\boldsymbol{\beta})$ with respect to vec($\boldsymbol{\beta}$), we obtain

$$\frac{\partial S(\boldsymbol{\beta})}{\partial \text{vec}(\boldsymbol{\beta})} = -2\left(\boldsymbol{\Sigma}_a^{-1} \otimes \boldsymbol{X}'\right) \text{vec}(\boldsymbol{Z}) + 2(\boldsymbol{\Sigma}_a^{-1} \otimes \boldsymbol{X}'\boldsymbol{X})\text{vec}(\boldsymbol{\beta}). \tag{2.33}$$

Equating to 0 gives the normal equations

$$(\boldsymbol{\Sigma}_a^{-1} \otimes \boldsymbol{X}'\boldsymbol{X})\text{vec}(\hat{\boldsymbol{\beta}}) = (\boldsymbol{\Sigma}_a^{-1} \otimes \boldsymbol{X}')\text{vec}(\boldsymbol{Z}).$$

Consequently, the GLS estimate of a VAR(p) model is

$$\begin{aligned} \text{vec}(\hat{\boldsymbol{\beta}}) &= (\boldsymbol{\Sigma}_a^{-1} \otimes \boldsymbol{X}'\boldsymbol{X})^{-1}(\boldsymbol{\Sigma}_a^{-1} \otimes \boldsymbol{X}')\text{vec}(\boldsymbol{Z}) \\ &= [\boldsymbol{\Sigma}_a \otimes (\boldsymbol{X}'\boldsymbol{X})^{-1}](\boldsymbol{\Sigma}_a^{-1} \otimes \boldsymbol{X}')\text{vec}(\boldsymbol{Z}) \\ &= [\boldsymbol{I}_k \otimes (\boldsymbol{X}'\boldsymbol{X})^{-1}\boldsymbol{X}']\text{vec}(\boldsymbol{Z}) \\ &= \text{vec}[(\boldsymbol{X}'\boldsymbol{X})^{-1}(\boldsymbol{X}'\boldsymbol{Z})], \qquad (2.34) \end{aligned}$$

where the last equality holds because vec($\boldsymbol{DB}$) = ($\boldsymbol{I} \otimes \boldsymbol{D}$)vec($\boldsymbol{B}$). In other words, we obtain

$$\hat{\boldsymbol{\beta}} = (\boldsymbol{X}'\boldsymbol{X})^{-1}(\boldsymbol{X}'\boldsymbol{Z}) = \left[\sum_{t=p+1}^{T} \boldsymbol{x}_t\boldsymbol{x}_t'\right]^{-1} \sum_{t=p+1}^{T} \boldsymbol{x}_t\boldsymbol{z}_t', \tag{2.35}$$

which interestingly does not depend on $\boldsymbol{\Sigma}_a$.

Remark: The result in Equation (2.35) shows that one can obtain the GLS estimate of a VAR(p) model equation-by-equation. That is, one can consider the k multiple linear regressions of z_{it} on $\boldsymbol{x}_t$ separately, where $i = 1, \ldots, k$. This estimation method is convenient when one considers parameter constraints in a VAR(p) model. □

2.5.1.2 Ordinary Least-Squares Estimate

Readers may notice that the GLS estimate of VAR(p) model in Equation (2.35) is identical to that of the OLS estimate of the multivariate multiple linear regression in Equation (2.28). Replacing $\boldsymbol{\Sigma}_a$ in Equation (2.31) by $\boldsymbol{I}_k$, we have the objective function of the OLS estimation

$$S_o(\boldsymbol{\beta}) = tr[(\boldsymbol{Z} - \boldsymbol{X\beta})(\boldsymbol{Z} - \boldsymbol{X\beta})']. \tag{2.36}$$

The derivations discussed earlier continue to hold step-by-step with $\boldsymbol{\Sigma}_a$ replaced by $\boldsymbol{I}_k$. One thus obtains the same estimate given in Equation (2.35) for $\boldsymbol{\beta}$. The fact that the GLS estimate is the same as the OLS estimate for a VAR(p) model was first shown in Zellner (1962). In what follows, we refer to the estimate in Equation (2.35) simply as the LS estimate.

The residual of the LS estimate is

$$\hat{\boldsymbol{a}}_t = \boldsymbol{z}_t - \hat{\boldsymbol{\phi}}_0 - \sum_{i=1}^{p} \hat{\boldsymbol{\phi}}_i \boldsymbol{z}_{t-i}, \quad t = p+1, \cdots, T$$

and let $\hat{\boldsymbol{A}}$ be the residual matrix, that is, $\hat{\boldsymbol{A}} = \boldsymbol{Z} - \boldsymbol{X}\hat{\boldsymbol{\beta}} = [\boldsymbol{I}_{T-p} - \boldsymbol{X}(\boldsymbol{X}'\boldsymbol{X})^{-1}\boldsymbol{X}']\boldsymbol{Y}$. The LS estimate of the innovational covariance matrix $\boldsymbol{\Sigma}_a$ is

$$\tilde{\boldsymbol{\Sigma}}_a = \frac{1}{T - (k+1)p - 1} \sum_{t=p+1}^{T} \hat{\boldsymbol{a}}_t\hat{\boldsymbol{a}}_t' = \frac{1}{T - (k+1)p - 1} \hat{\boldsymbol{A}}'\hat{\boldsymbol{A}},$$

where the denominator is $[T - p - (kp+1)]$, which is the effective sample size less the number of parameters in the equation for each component z_{it}. By Equation (2.28), we see that

$$\hat{\boldsymbol{\beta}} - \boldsymbol{\beta} = (\boldsymbol{X}'\boldsymbol{X})^{-1}\boldsymbol{X}'\boldsymbol{A}. \tag{2.37}$$

Since $E(\boldsymbol{A}) = \boldsymbol{0}$, we see that the LS estimate is an unbiased estimator. The LS estimate of a VAR(p) model has the following properties.

Theorem 2.1 For the stationary VAR(p) model in Equation (2.21), assume that $\boldsymbol{a}_t$ are independent and identically distributed with mean zero and positive-definite covariance matrix $\boldsymbol{\Sigma}_a$. Then, (i) $E(\hat{\boldsymbol{\beta}}) = \boldsymbol{\beta}$, where $\boldsymbol{\beta}$ is defined in Equation (2.28), (ii) $E(\tilde{\boldsymbol{\Sigma}}_a) = \boldsymbol{\Sigma}_a$, (iii) the residual $\hat{\boldsymbol{A}}$ and the LS estimate $\hat{\boldsymbol{\beta}}$ are uncorrelated, and (iv) the covariance of the parameter estimates is

$$\text{Cov}[\text{vec}(\hat{\boldsymbol{\beta}})] = \tilde{\boldsymbol{\Sigma}}_a \otimes (\boldsymbol{X}'\boldsymbol{X})^{-1} = \tilde{\boldsymbol{\Sigma}}_a \otimes \left(\sum_{t=p+1}^{T} \boldsymbol{x}_t \boldsymbol{x}_t'\right)^{-1}.$$

Theorem 2.1 follows the LS theory for multivariate multiple linear regression. A proof can be found in Lütkepohl (2005) or in Johnson and Wichern (2007, Chapter 7).

2.5.2 Maximum Likelihood Estimate

Assume further that $\boldsymbol{a}_t$ of the VAR(p) model follows a multivariate normal distribution. Let $\boldsymbol{z}_{h:q}$ denote the observations from $t = h$ to $t = q$ (inclusive). The conditional likelihood function of the data can be written as

$$\begin{aligned}
L(\boldsymbol{z}_{(p+1):T}|\boldsymbol{z}_{1:p}, \boldsymbol{\beta}, \boldsymbol{\Sigma}_a) &= \prod_{t=p+1}^{T} p(\boldsymbol{z}_t|\boldsymbol{z}_{1:(t-1)}, \boldsymbol{\beta}, \boldsymbol{\Sigma}_a) \\
&= \prod_{t=p+1}^{T} p(\boldsymbol{a}_t|\boldsymbol{z}_{1:(t-1)}, \boldsymbol{\beta}, \boldsymbol{\Sigma}_a) \\
&= \prod_{t=p+1}^{T} p(\boldsymbol{a}_t|\boldsymbol{\beta}, \boldsymbol{\Sigma}_a) \\
&= \prod_{t=p+1}^{T} \frac{1}{(2\pi)^{k/2}|\boldsymbol{\Sigma}_a|^{1/2}} \exp\left[\frac{-1}{2}\boldsymbol{a}_t'\boldsymbol{\Sigma}_a^{-1}\boldsymbol{a}_t\right] \\
&\propto |\boldsymbol{\Sigma}_a|^{-(T-p)/2} \exp\left[\frac{-1}{2}\sum_{t=p+1}^{T} tr(\boldsymbol{a}_t'\boldsymbol{\Sigma}_a^{-1}\boldsymbol{a}_t)\right].
\end{aligned}$$

The log-likelihood function then becomes

$$\begin{aligned}\ell(\boldsymbol{\beta}, \boldsymbol{\Sigma}_a) &= c - \frac{T-p}{2}\log(|\boldsymbol{\Sigma}_a|) - \frac{1}{2}\sum_{t=p+1}^{T} tr\left(\boldsymbol{a}_t'\boldsymbol{\Sigma}_a^{-1}\boldsymbol{a}_t\right) \\ &= c - \frac{T-p}{2}\log(|\boldsymbol{\Sigma}_a|) - \frac{1}{2}tr\left(\boldsymbol{\Sigma}_a^{-1}\sum_{t=p+1}^{T}\boldsymbol{a}_t\boldsymbol{a}_t'\right),\end{aligned}$$

where c is a constant, and we use the properties that $tr(\boldsymbol{CD}) = tr(\boldsymbol{DC})$ and $tr(\boldsymbol{C}+\boldsymbol{D}) = tr(\boldsymbol{C}) + tr(\boldsymbol{D})$. Noting that $\sum_{t=p+1}^{T}\boldsymbol{a}_t\boldsymbol{a}_t' = \boldsymbol{A}'\boldsymbol{A}$, where $\boldsymbol{A} = \boldsymbol{Z} - \boldsymbol{X}\boldsymbol{\beta}$ is the error matrix in Equation (2.28), we can rewrite the log-likelihood function as

$$\ell(\boldsymbol{\beta}, \boldsymbol{\Sigma}_a) = c - \frac{T-p}{2}\log(|\boldsymbol{\Sigma}_a|) - \frac{1}{2}S(\boldsymbol{\beta}), \tag{2.38}$$

where $S(\boldsymbol{\beta})$ is given in Equation (2.31).

Since the parameter matrix $\boldsymbol{\beta}$ only appears in the last term of $\ell(\boldsymbol{\beta}, \boldsymbol{\Sigma}_a)$, maximizing the log-likelihood function over $\boldsymbol{\beta}$ is equivalent to minimizing $S(\boldsymbol{\beta})$. Consequently, the ML estimate of $\boldsymbol{\beta}$ is the same as its LS estimate. Next, taking the partial derivative of the log-likelihood function with respective to $\boldsymbol{\Sigma}_a$ and using properties (i) and (j) of **Result 3** of Appendix A, we obtain

$$\frac{\partial\ell(\hat{\boldsymbol{\beta}}, \boldsymbol{\Sigma}_a)}{\partial\boldsymbol{\Sigma}_a} = -\frac{T-p}{2}\boldsymbol{\Sigma}_a^{-1} + \frac{1}{2}\boldsymbol{\Sigma}_a^{-1}\hat{\boldsymbol{A}}'\hat{\boldsymbol{A}}\boldsymbol{\Sigma}_a^{-1}. \tag{2.39}$$

Equating the prior normal equation to 0, we obtain the ML estimate of $\boldsymbol{\Sigma}_a$ as

$$\hat{\boldsymbol{\Sigma}}_a = \frac{1}{T-p}\hat{\boldsymbol{A}}'\hat{\boldsymbol{A}} = \frac{1}{T-p}\sum_{t=p+1}^{T}\hat{\boldsymbol{a}}_t\hat{\boldsymbol{a}}_t'. \tag{2.40}$$

This result is the same as that for the multiple linear regression. The ML estimate of $\boldsymbol{\Sigma}_a$ is only asymptotically unbiased. Finally, the Hessian matrix of $\boldsymbol{\beta}$ can be obtained by taking the partial derivative of Equation (2.33), namely,

$$-\frac{\partial^2\ell(\boldsymbol{\beta}, \boldsymbol{\Sigma}_a)}{\partial\text{vec}(\boldsymbol{\beta})\partial\text{vec}(\boldsymbol{\beta})'} = \frac{1}{2}\frac{\partial^2 S(\boldsymbol{\beta})}{\partial\text{vec}(\boldsymbol{\beta})\partial\text{vec}(\boldsymbol{\beta})'} = \boldsymbol{\Sigma}_a^{-1}\otimes\boldsymbol{X}'\boldsymbol{X}.$$

Inverse of the Hessian matrix provides the asymptotic covariance matrix of the ML estimate of vec($\boldsymbol{\beta}$). Next, using property (e) of **Result 3** and the product rule of Appendix A, and taking derivative of Equation (2.39), we obtain

$$\frac{\partial^2 \ell(\hat{\boldsymbol{\beta}}, \boldsymbol{\Sigma}_a)}{\partial \text{vec}(\boldsymbol{\Sigma}_a)\partial \text{vec}(\boldsymbol{\Sigma}_a)'} = \frac{T-p}{2}\left(\boldsymbol{\Sigma}_a^{-1} \otimes \boldsymbol{\Sigma}_a^{-1}\right) - \frac{1}{2}\left[\left(\boldsymbol{\Sigma}_a^{-1} \otimes \boldsymbol{\Sigma}_a^{-1}\right)\hat{\boldsymbol{A}}'\hat{\boldsymbol{A}}\boldsymbol{\Sigma}_a^{-1}\right] - \frac{1}{2}\left[\boldsymbol{\Sigma}_a^{-1}\hat{\boldsymbol{A}}'\hat{\boldsymbol{A}}\left(\boldsymbol{\Sigma}_a^{-1} \otimes \boldsymbol{\Sigma}_a^{-1}\right)\right].$$

Consequently, we have

$$-E\left(\frac{\partial^2 \ell(\hat{\boldsymbol{\beta}}, \boldsymbol{\Sigma}_a)}{\partial \text{vec}(\boldsymbol{\Sigma}_a)\partial \text{vec}(\boldsymbol{\Sigma}_a)'}\right) = \frac{T-p}{2}\left(\boldsymbol{\Sigma}_a^{-1} \otimes \boldsymbol{\Sigma}_a^{-1}\right).$$

This result provides asymptotic covariance matrix for the ML estimates of elements of $\boldsymbol{\Sigma}_a$.

Theorem 2.2 Suppose that the innovation $\boldsymbol{a}_t$ of a stationary VAR(p) model follows a multivariate normal distribution with mean zero and positive-definite covariance matrix $\boldsymbol{\Sigma}_a$. Then, the ML estimates are $\text{vec}(\hat{\boldsymbol{\beta}}) = (\boldsymbol{X}'\boldsymbol{X})^{-1}\boldsymbol{X}'\boldsymbol{Z}$ and $\hat{\boldsymbol{\Sigma}}_a = (1/T-p)\sum_{t=p+1}^{T} \hat{\boldsymbol{a}}_t\hat{\boldsymbol{a}}_t'$. Also, (i) $(T-p)\hat{\boldsymbol{\Sigma}}_a$ is distributed as $W_{k,T-(k+1)p-1}(\boldsymbol{\Sigma}_a)$, a Wishart distribution, and (ii) $\text{vec}(\hat{\boldsymbol{\beta}})$ is normally distributed with mean $\text{vec}(\boldsymbol{\beta})$ and covariance matrix $\boldsymbol{\Sigma}_a \otimes (\boldsymbol{X}'\boldsymbol{X})^{-1}$, and (iii) $\text{vec}(\hat{\boldsymbol{\beta}})$ is independent of $\hat{\boldsymbol{\Sigma}}_a$, where $\boldsymbol{Z}$ and $\boldsymbol{X}$ are defined in Equation (2.28). Furthermore, $\sqrt{T}[\text{vec}(\hat{\boldsymbol{\beta}}) - \boldsymbol{\beta}]$ and $\sqrt{T}[\text{vec}(\hat{\boldsymbol{\Sigma}}_a) - \text{vec}(\boldsymbol{\Sigma}_a)]$ are asymptotically normally distributed with mean zero and covariance matrices $\boldsymbol{\Sigma}_a \otimes \boldsymbol{G}^{-1}$ and $2\boldsymbol{\Sigma}_a \otimes \boldsymbol{\Sigma}_a$, respectively, where $\boldsymbol{G} = E(\boldsymbol{x}_t\boldsymbol{x}_t')$ with $\boldsymbol{x}_t$ defined in Equation (2.28).

Finally, given the data set $\{\boldsymbol{z}_1, \ldots, \boldsymbol{z}_T\}$, the maximized likelihood of a VAR(p) model is

$$L\left(\hat{\boldsymbol{\beta}}, \hat{\boldsymbol{\Sigma}}_a | \boldsymbol{z}_{1:p}\right) = (2\pi)^{-k(T-p)/2}|\hat{\boldsymbol{\Sigma}}_a|^{-(T-p)/2} \exp\left[-\frac{k(T-p)}{2}\right]. \tag{2.41}$$

This value is useful in likelihood ratio tests to be discussed later.

2.5.3 Limiting Properties of LS Estimate

Consider a k-dimensional stationary VAR(p) model in Equation (2.21). To study the asymptotic properties of the LS estimate $\hat{\boldsymbol{\beta}}$ in Equation (2.35), we need the assumption that the innovation series $\{\boldsymbol{a}_t\}$ is a sequence of independent and identically distributed random vector with mean zero and positive-definite covariance $\boldsymbol{\Sigma}_a$. Also, $\boldsymbol{a}_t = (a_{1t}, \ldots, a_{kt})'$ is continuous and satisfies

$$E|a_{it}a_{jt}a_{ut}a_{vt}| < \infty, \quad \text{for } i, j, u, v = 1, \ldots, k \text{ and all } t. \tag{2.42}$$

In other words, the fourth moment of $\boldsymbol{a}_t$ is finite. Under this assumption, we have the following results.

Lemma 2.3 If the VAR(p) process $\boldsymbol{z}_t$ of Equation (2.21) is stationary and satisfies the condition in Equation (2.42), then, as $T \to \infty$, we have

(i) $\boldsymbol{X}'\boldsymbol{X}/(T-p) \to_p \boldsymbol{G}$,
(ii) $(1/\sqrt{T-p})\text{vec}(\boldsymbol{X}'\boldsymbol{A}) = (1/\sqrt{T-p})(\boldsymbol{I}_k \otimes \boldsymbol{X}')\text{vec}(\boldsymbol{A}) \to_d N(\boldsymbol{0}, \boldsymbol{\Sigma}_a \otimes \boldsymbol{G})$,

where $\to_p$ and $\to_d$ denote convergence in probability and distribution, respectively, $\boldsymbol{X}$ and $\boldsymbol{A}$ are defined in Equation (2.28), and $\boldsymbol{G}$ is a nonsingular matrix given by

$$\boldsymbol{G} = \begin{bmatrix} 1 & \boldsymbol{0}' \\ \boldsymbol{0} & \boldsymbol{\Gamma}_0^* \end{bmatrix} + \begin{bmatrix} 0 \\ \boldsymbol{u} \end{bmatrix} [0, \boldsymbol{u}'],$$

where $\boldsymbol{0}$ is a kp-dimensional vector of zeros, $\boldsymbol{\Gamma}_0^*$ is defined in Equation (2.26) and $\boldsymbol{u} = \boldsymbol{1}_p \otimes \boldsymbol{\mu}$ with $\boldsymbol{1}_p$ being a p-dimensional vector of 1.

A proof of Lemma 2.3 can be found in Theorem 8.2.3 of Fuller (1976, p. 340) or in Lemma 3.1 of Lütkepohl (2005, p. 73). Using Lemma 2.3, one can establish the asymptotic distribution of the LS estimate $\hat{\boldsymbol{\beta}}$.

Theorem 2.3 Suppose that the VAR(p) time series $\boldsymbol{z}_t$ in Equation (2.21) is stationary and its innovation $\boldsymbol{a}_t$ satisfies the assumption in Equation (2.42). Then, as $T \to \infty$,

(i) $\hat{\boldsymbol{\beta}} \to_p \boldsymbol{\beta}$,
(ii) $\sqrt{T-p}[\text{vec}(\hat{\boldsymbol{\beta}}) - \text{vec}(\boldsymbol{\beta})] = \sqrt{T-p}[\text{vec}(\hat{\boldsymbol{\beta}} - \boldsymbol{\beta})] \to_d N(\boldsymbol{0}, \boldsymbol{\Sigma}_a \otimes \boldsymbol{G}^{-1})$,

where $\boldsymbol{G}$ is defined in Lemma 2.3.

Proof. By Equation (2.37), we have

$$\hat{\boldsymbol{\beta}} - \boldsymbol{\beta} = \left(\frac{\boldsymbol{X}'\boldsymbol{X}}{T-p}\right)^{-1} \left(\frac{\boldsymbol{X}'\boldsymbol{A}}{T-p}\right) \to_p \boldsymbol{0},$$

because the last term approaches $\boldsymbol{0}$. This establishes the consistency of $\hat{\boldsymbol{\beta}}$. For result (ii), we can use Equation (2.34) to obtain

$$\begin{aligned} \sqrt{T-p}\left[\text{vec}(\hat{\boldsymbol{\beta}}) - \text{vec}(\boldsymbol{\beta})\right] &= \sqrt{T-p}\left[\boldsymbol{I}_k \otimes (\boldsymbol{X}'\boldsymbol{X})^{-1}\boldsymbol{X}'\right]\text{vec}(\boldsymbol{A}) \\ &= \sqrt{T-p}\left[\boldsymbol{I}_k \otimes (\boldsymbol{X}'\boldsymbol{X})^{-1}\right]\left[\boldsymbol{I}_k \otimes \boldsymbol{X}'\right]\text{vec}(\boldsymbol{A}) \\ &= \left[\boldsymbol{I}_k \otimes \left(\frac{\boldsymbol{X}'\boldsymbol{X}}{T-p}\right)^{-1}\right] \frac{1}{\sqrt{T-p}}\left[\boldsymbol{I}_k \otimes \boldsymbol{X}'\right]\text{vec}(\boldsymbol{A}). \end{aligned}$$

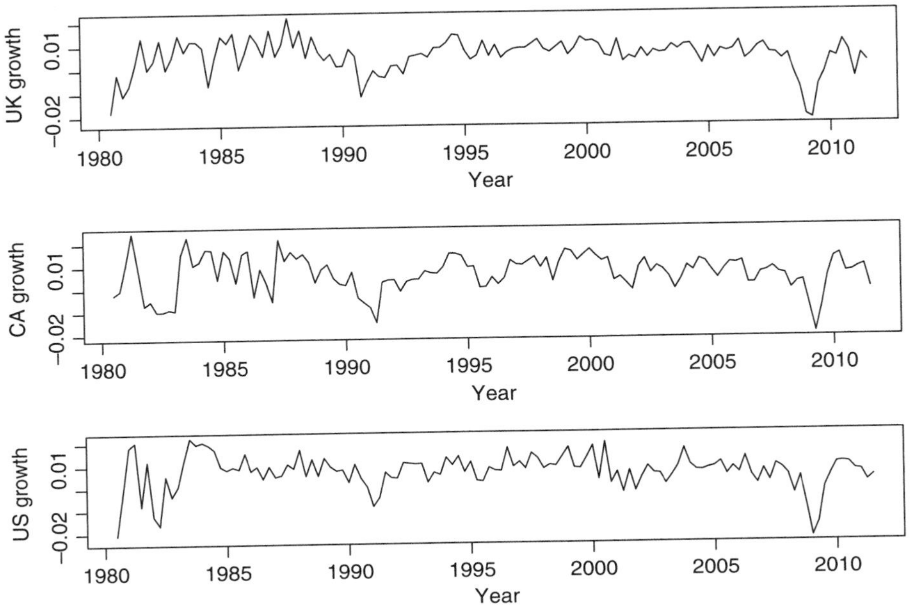

FIGURE 2.1 Time plots of the quarterly growth rates of real gross domestic products of United Kingdom, Canada, and United States from the second quarter of 1980 to the second quarter of 2011.

Therefore, the limiting distribution of $\sqrt{T-p}[\text{vec}(\hat{\boldsymbol{\beta}}) - \text{vec}(\boldsymbol{\beta})]$ is the same as that of

$$\left(\boldsymbol{I}_k \otimes \boldsymbol{G}^{-1}\right) \frac{1}{\sqrt{T-p}}(\boldsymbol{I}_k \otimes \boldsymbol{X}')\text{vec}(\boldsymbol{A}).$$

Hence, by Lemma 2.3, the limiting distribution of $\sqrt{T-p}[\text{vec}(\hat{\boldsymbol{\beta}}) - \text{vec}(\boldsymbol{\beta})]$ is normal and the covariance matrix is

$$\left(\boldsymbol{I}_k \otimes \boldsymbol{G}^{-1}\right)(\boldsymbol{\Sigma}_a \otimes \boldsymbol{G})\left(\boldsymbol{I}_k \otimes \boldsymbol{G}^{-1}\right) = \boldsymbol{\Sigma}_a \otimes \boldsymbol{G}^{-1}.$$

The proof is complete. □

Example 2.3 Consider the quarterly growth rates, in percentages, of real gross domestic product (GDP) of United Kingdom, Canada, and United States from the second quarter of 1980 to the second quarter of 2011. The data were seasonally adjusted and downloaded from the database of Federal Reserve Bank at St. Louis. The GDP were in millions of local currency, and the growth rate denotes the differenced series of log GDP. Figure 2.1 shows the time plots of the three GDP growth rates. In our demonstration, we employ a VAR(2) model. In this particular instance, we have $k=3$, $p=2$, and $T=125$. Using the notation defined in Section 2.5, we have

$$\tilde{\mathbf{\Sigma}}_a = \begin{bmatrix} 0.299 & 0.028 & 0.079 \\ 0.028 & 0.309 & 0.148 \\ 0.079 & 0.148 & 0.379 \end{bmatrix}, \quad \hat{\mathbf{\Sigma}}_a = \begin{bmatrix} 0.282 & 0.027 & 0.074 \\ 0.027 & 0.292 & 0.139 \\ 0.074 & 0.139 & 0.357 \end{bmatrix},$$

and the LS estimates vec($\hat{\boldsymbol{\beta}}$) and their standard errors and t-ratios are given in the following R demonstration. The two estimates of the covariance matrix differ by a factor of $116/123 = 0.943$. From the output, the t-ratios indicate that some of the LS estimates are not statistically significant at the usual 5% level. We shall discuss model checking and refinement later. □

R Demonstration

```
> da=read.table("q-gdp-ukcaus.txt",header=T)
> gdp=log(da[,3:5])
> dim(gdp)
[1] 126   3
> z=gdp[2:126,]-gdp[1:125,]   ## Growth rate
> z=z*100    ## Percentage growth rates
> dim(z)
[1] 125   3
> Z=z[3:125,]
> X=cbind(rep(1,123),z[2:124,],z[1:123,])
> X=as.matrix(X)
> XPX=t(X)%*%X
> XPXinv=solve(XPX)
> Z=as.matrix(Z)
> XPZ=t(X)%*%Z
> bhat=XPXinv%*%XPZ
> bhat
                    uk           ca          us
rep(1, 123) 0.12581630  0.123158083  0.28955814
uk          0.39306691  0.351313628  0.49069776
ca          0.10310572  0.338141505  0.24000097
us          0.05213660  0.469093555  0.23564221
uk          0.05660120 -0.191350134 -0.31195550
ca          0.10552241 -0.174833458 -0.13117863
us          0.01889462 -0.008677767  0.08531363
> A=Z-X%*%bhat
> Sig=t(A)%*%A/(125-(3+1)*2-1)
> Sig
           uk         ca         us
uk 0.29948825 0.02814252 0.07883967
ca 0.02814252 0.30917711 0.14790523
us 0.07883967 0.14790523 0.37850674
```

```
> COV=kronecker(Sig,XPXinv)
> se=sqrt(diag(COV))
> para=cbind(beta,se,beta/se)
> para
             beta         se    t-ratio
 [1,]  0.125816304 0.07266338  1.7314953
 [2,]  0.393066914 0.09341839  4.2075968
 [3,]  0.103105720 0.09838425  1.0479901
 [4,]  0.052136600 0.09112636  0.5721353
 [5,]  0.056601196 0.09237356  0.6127424
 [6,]  0.105522415 0.08755896  1.2051584
 [7,]  0.018894618 0.09382091  0.2013903
 [8,]  0.123158083 0.07382941  1.6681440
 [9,]  0.351313628 0.09491747  3.7012536
[10,]  0.338141505 0.09996302  3.3826660
[11,]  0.469093555 0.09258865  5.0664259
[12,] -0.191350134 0.09385587 -2.0387658
[13,] -0.174833458 0.08896401 -1.9652155
[14,] -0.008677767 0.09532645 -0.0910321
[15,]  0.289558145 0.08168880  3.5446492
[16,]  0.490697759 0.10502176  4.6723437
[17,]  0.240000969 0.11060443  2.1699038
[18,]  0.235642214 0.10244504  2.3001819
[19,] -0.311955500 0.10384715 -3.0039871
[20,] -0.131178630 0.09843454 -1.3326484
[21,]  0.085313633 0.10547428  0.8088572
> Sig1=t(A)%*%A/(125-2)   ## MLE of Sigma_a
> Sig1
           uk         ca         us
uk 0.28244420 0.02654091 0.07435286
ca 0.02654091 0.29158166 0.13948786
us 0.07435286 0.13948786 0.35696571
```

The aforementioned demonstration is to provide details about LS and ML estimation of a VAR model. In practice, we use some available packages in R to perform estimation. For example, we can use the command `VAR` in the `MTS` package to estimate a VAR model. The command and the associated output, which is in the matrix form, are shown next:

R Demonstration: Estimation of VAR models.

```
> da=read.table("q-gdp-ukcaus.txt",header=T)
> gdp=log(da[,3:5])
> z=gdp[2:126,]-gdp[1:125,]
> z=z*100
> m1=VAR(z,2)
Constant term:
Estimates:  0.1258163 0.1231581 0.2895581
```

```
Std.Error:   0.07266338 0.07382941 0.0816888
AR coefficient matrix
AR( 1 )-matrix
      [,1]  [,2]   [,3]
[1,] 0.393 0.103 0.0521
[2,] 0.351 0.338 0.4691
[3,] 0.491 0.240 0.2356
standard error
       [,1]   [,2]   [,3]
[1,] 0.0934 0.0984 0.0911
[2,] 0.0949 0.1000 0.0926
[3,] 0.1050 0.1106 0.1024
AR( 2 )-matrix
        [,1]   [,2]     [,3]
[1,]  0.0566  0.106  0.01889
[2,] -0.1914 -0.175 -0.00868
[3,] -0.3120 -0.131  0.08531
standard error
       [,1]   [,2]   [,3]
[1,] 0.0924 0.0876 0.0938
[2,] 0.0939 0.0890 0.0953
[3,] 0.1038 0.0984 0.1055

Residuals cov-mtx:
           [,1]       [,2]       [,3]
[1,] 0.28244420 0.02654091 0.07435286
[2,] 0.02654091 0.29158166 0.13948786
[3,] 0.07435286 0.13948786 0.35696571

det(SSE) =  0.02258974
AIC =  -3.502259
BIC =  -3.094982
HQ  =  -3.336804
```

From the output, the fitted VAR(2) model for the percentage growth rates of quarterly GDP of United Kingdom, Canada, and United States is

$$\boldsymbol{z}_t = \begin{bmatrix} 0.13 \\ 0.12 \\ 0.29 \end{bmatrix} + \begin{bmatrix} 0.38 & 0.10 & 0.05 \\ 0.35 & 0.34 & 0.47 \\ 0.49 & 0.24 & 0.24 \end{bmatrix} \boldsymbol{z}_{t-1} + \begin{bmatrix} 0.06 & 0.11 & 0.02 \\ -0.19 & -0.18 & -0.01 \\ -0.31 & -0.13 & 0.09 \end{bmatrix} \boldsymbol{z}_{t-2} + \boldsymbol{a}_t,$$

where the residual covariance matrix is

$$\hat{\boldsymbol{\Sigma}}_a = \begin{bmatrix} 0.28 & 0.03 & 0.07 \\ 0.03 & 0.29 & 0.14 \\ 0.07 & 0.14 & 0.36 \end{bmatrix}.$$

Standard errors of the coefficient estimates are given in the output. Again, some of the estimates are not statistically significant at the usual 5% level.

Remark: In our derivation, we used Equation (2.28). An alternative approach is to use

$$\boldsymbol{Y} = \boldsymbol{\varpi}\boldsymbol{W} + \boldsymbol{U},$$

where $\boldsymbol{Y} = [\boldsymbol{z}_{p+1}, \boldsymbol{z}_{p+2}, \ldots, \boldsymbol{z}_T]$, a $k \times (T-p)$ matrix, $\boldsymbol{\varpi} = [\boldsymbol{\phi}_0, \boldsymbol{\phi}_1, \ldots, \boldsymbol{\phi}_p]$, a $k \times (kp+1)$ parameter matrix, $\boldsymbol{W} = [\boldsymbol{x}_{p+1}, \ldots, \boldsymbol{x}_T]$ with $\boldsymbol{x}_t$ defined in Equation (2.28), and $\boldsymbol{U} = [\boldsymbol{a}_{p+1}, \ldots, \boldsymbol{a}_T]$. Obviously, we have $\boldsymbol{Y} = \boldsymbol{Z}'$, $\boldsymbol{\varpi} = \boldsymbol{\beta}'$, $\boldsymbol{W} = \boldsymbol{X}'$, and $\boldsymbol{U} = \boldsymbol{A}'$. One can then derive the limiting distribution of the LS estimate $\hat{\boldsymbol{\varpi}}$ of $\boldsymbol{\varpi}$ via using exactly the same argument as we used. Under this approach, Lemma 2.3 (ii) becomes

$$\frac{1}{\sqrt{T_p}}\text{vec}(\boldsymbol{U}\boldsymbol{W}') = \frac{1}{\sqrt{T_p}}(\boldsymbol{W} \otimes \boldsymbol{I}_k)\text{vec}(\boldsymbol{U}) \to_d N(\boldsymbol{0}, \boldsymbol{G} \otimes \boldsymbol{\Sigma}_a),$$

where $T_p = T - p$, and $\boldsymbol{G} = \lim(\boldsymbol{W}\boldsymbol{W}')/T_p$, which is the same as defined in Lemma 2.3. Furthermore, Theorem 2.3 becomes

$$\sqrt{T_p}\text{vec}\left(\hat{\boldsymbol{\varpi}} - \boldsymbol{\varpi}\right) = \sqrt{T_p}\text{vec}\left(\hat{\boldsymbol{\beta}}' - \boldsymbol{\beta}'\right) \to_d N\left(\boldsymbol{0}, \boldsymbol{G}^{-1} \otimes \boldsymbol{\Sigma}_a\right). \qquad (2.43)$$

□

2.5.4 Bayesian Estimation

We consider Bayesian estimation of a stationary VAR(p) model in this section. The basic framework used is the multivariate multiple linear regression in Equation (2.28). We begin with a brief review of Bayesian inference.

2.5.4.1 Review of Bayesian Paradigm

Consider a statistical inference problem. Denote the set of unknown parameters by $\boldsymbol{\Theta}$. Our prior beliefs of $\boldsymbol{\Theta}$ are often expressed via a probability distribution with density function $f(\boldsymbol{\Theta})$. Let $\boldsymbol{D}$ denote the observed data. The information provided by $\boldsymbol{D}$ is the likelihood function $f(\boldsymbol{D}|\boldsymbol{\Theta})$. By Bayes' theorem, we can combine the prior and likelihood to produce the distribution of the parameters conditional on the data and prior:

$$f(\boldsymbol{\Theta}|\boldsymbol{D}) = \frac{f(\boldsymbol{D}, \boldsymbol{\Theta})}{f(\boldsymbol{D})} = \frac{f(\boldsymbol{D}|\boldsymbol{\Theta})f(\boldsymbol{\Theta})}{f(\boldsymbol{D})}, \tag{2.44}$$

where $f(\boldsymbol{D}) = \int f(\boldsymbol{D}, \boldsymbol{\Theta})d\boldsymbol{\Theta} = \int f(\boldsymbol{D}|\boldsymbol{\Theta})f(\boldsymbol{\Theta})d\boldsymbol{\Theta}$ is the marginal distribution of $\boldsymbol{D}$, obtained by integrating out $\boldsymbol{\Theta}$. The density function $f(\boldsymbol{\Theta}|\boldsymbol{D})$ is called the *posterior* distribution. Bayesian inference on $\boldsymbol{\Theta}$ is drawn from this posterior distribution.

The marginal distribution $f(\boldsymbol{D})$ in Equation (2.44) serves as the normalization constant, referred to as the constant of proportionality, and its actual value is not critical in many applications provided that it exists. Therefore, we may rewrite the equation as

$$f(\boldsymbol{\Theta}|\boldsymbol{D}) \propto f(\boldsymbol{D}|\boldsymbol{\Theta})f(\boldsymbol{\Theta}). \tag{2.45}$$

If the prior distribution $f(\boldsymbol{\Theta})$ and the posterior distribution $f(\boldsymbol{\Theta}|\boldsymbol{D})$ belong to the same family of distributions, then the prior is called a *conjugate prior*. Conjugate priors are often used in Bayesian inference because they enable us to obtain analytic expressions for the posterior distribution.

2.5.4.2 VAR Estimation

To derive the Bayesian estimation of a stationary VAR(p) model, we employ the model in Equation (2.28), namely,

$$\boldsymbol{Z} = \boldsymbol{X}\boldsymbol{\beta} + \boldsymbol{A}, \tag{2.46}$$

where $\boldsymbol{Z}$ and $\boldsymbol{A}$ are $(T-p) \times k$ matrices, $\boldsymbol{\beta}' = [\boldsymbol{\phi}_0, \boldsymbol{\phi}_1, \ldots, \boldsymbol{\phi}_p]$ is a $k \times (kp+1)$ matrix of coefficient parameters, and the ith rows of $\boldsymbol{Z}$ and $\boldsymbol{A}$ are $\boldsymbol{z}'_{p+i}$ and $\boldsymbol{a}'_{p+i}$, respectively. The matrix $\boldsymbol{X}$ is a $(T-p) \times (kp+1)$ design matrix with ith row being $(1, \boldsymbol{z}'_{p+i-1}, \ldots, \boldsymbol{z}'_i)$. The unknown parameters of the VAR(p) model are $\boldsymbol{\Theta} = [\boldsymbol{\beta}', \boldsymbol{\Sigma}_a]$. Rossi, Allenby, and McCulloch (2005) provide a good treatment of Bayesian estimation for the model in Equation (2.46). We adopt their approach.

For ease in notation, let $n = T - p$ be the effective sample size for estimation. Also, we shall omit the condition $\boldsymbol{z}_{1:p}$ from all equations. As stated in Section 2.5.2, the likelihood function of the data is

$$f(\boldsymbol{Z}|\boldsymbol{\beta}, \boldsymbol{\Sigma}_a) \propto |\boldsymbol{\Sigma}_a|^{-n/2} \exp\left[-\tfrac{1}{2}tr\left\{(\boldsymbol{Z} - \boldsymbol{X}\boldsymbol{\beta})'(\boldsymbol{Z} - \boldsymbol{X}\boldsymbol{\beta})\boldsymbol{\Sigma}_a^{-1}\right\}\right].$$

Using the LS properties in Appendix A, we have

$$(\boldsymbol{Z}-\boldsymbol{X}\boldsymbol{\beta})'(\boldsymbol{Z}-\boldsymbol{X}\boldsymbol{\beta}) = \hat{\boldsymbol{A}}'\hat{\boldsymbol{A}} + (\boldsymbol{\beta}-\hat{\boldsymbol{\beta}})'\boldsymbol{X}'\boldsymbol{X}(\boldsymbol{\beta}-\hat{\boldsymbol{\beta}}).$$

where $\hat{\boldsymbol{\beta}} = (\boldsymbol{X}'\boldsymbol{X})^{-1}\boldsymbol{X}'\boldsymbol{Z}$ is the LS estimate of $\boldsymbol{\beta}$ and $\hat{\boldsymbol{A}} = \boldsymbol{Z} - \boldsymbol{X}\hat{\boldsymbol{\beta}}$ is the residual matrix. The likelihood function can be written as

$$\begin{aligned} f(\boldsymbol{Z}|\boldsymbol{\beta}, \boldsymbol{\Sigma}_a) &\propto |\boldsymbol{\Sigma}_a|^{-(n-k)/2} \exp\left[-\tfrac{1}{2}tr(\boldsymbol{S}\boldsymbol{\Sigma}_a^{-1})\right] \\ &\times |\boldsymbol{\Sigma}_a|^{-k/2} \exp\left[-\tfrac{1}{2}tr\left\{(\boldsymbol{\beta}-\hat{\boldsymbol{\beta}})'\boldsymbol{X}'\boldsymbol{X}(\boldsymbol{\beta}-\hat{\boldsymbol{\beta}})\boldsymbol{\Sigma}_a^{-1}\right\}\right], \end{aligned} \tag{2.47}$$

where $\boldsymbol{S} = \hat{\boldsymbol{A}}'\hat{\boldsymbol{A}}$. The first term of Equation (2.47) does not depend on $\boldsymbol{\beta}$. This suggests that the natural conjugate prior for $\boldsymbol{\Sigma}_a$ is the inverted Wishart distribution and the prior for $\boldsymbol{\beta}$ can be conditioned on $\boldsymbol{\Sigma}_a$. Let $\boldsymbol{K} = (\boldsymbol{\beta}-\hat{\boldsymbol{\beta}})'\boldsymbol{X}'\boldsymbol{X}(\boldsymbol{\beta}-\hat{\boldsymbol{\beta}})\boldsymbol{\Sigma}_a^{-1}$ be the matrix in the exponent of Equation (2.47). The exponent of the second term in Equation (2.47) can be rewritten as

$$\begin{aligned} tr(\boldsymbol{K}) &= \left[\text{vec}\left(\boldsymbol{\beta}-\hat{\boldsymbol{\beta}}\right)\right]' \text{vec}\left[\boldsymbol{X}'\boldsymbol{X}\left(\boldsymbol{\beta}-\hat{\boldsymbol{\beta}}\right)\boldsymbol{\Sigma}_a^{-1}\right] \\ &= \left[\text{vec}\left(\boldsymbol{\beta}-\hat{\boldsymbol{\beta}}\right)\right]' (\boldsymbol{\Sigma}_a^{-1} \otimes \boldsymbol{X}'\boldsymbol{X})\, \text{vec}\left(\boldsymbol{\beta}-\hat{\boldsymbol{\beta}}\right) \\ &= \left[\text{vec}(\boldsymbol{\beta}) - \text{vec}(\hat{\boldsymbol{\beta}})\right]' (\boldsymbol{\Sigma}_a^{-1} \otimes \boldsymbol{X}'\boldsymbol{X}) \left[\text{vec}(\boldsymbol{\beta}) - \text{vec}(\hat{\boldsymbol{\beta}})\right]. \end{aligned}$$

Consequently, the second term in Equation (2.47) is a multivariate normal kernel. This means that the natural conjugate prior for vec($\boldsymbol{\beta}$) is multivariate normal conditional on $\boldsymbol{\Sigma}_a$.

The conjugate priors for the VAR(p) model in Equation (2.46) are of the form

$$\begin{aligned} f(\boldsymbol{\beta}, \boldsymbol{\Sigma}_a) &= f(\boldsymbol{\Sigma}_a) f(\boldsymbol{\beta}|\boldsymbol{\Sigma}_a) \\ \boldsymbol{\Sigma}_a &\sim W^{-1}(\boldsymbol{V}_o, n_o) \\ \text{vec}(\boldsymbol{\beta}) &\sim N\left[\text{vec}(\boldsymbol{\beta}_o), \boldsymbol{\Sigma}_a \otimes \boldsymbol{C}^{-1}\right], \end{aligned} \tag{2.48}$$

where $\boldsymbol{V}_o$ is $k \times k$ and $\boldsymbol{C}$ is $(kp+1) \times (kp+1)$, both are positive-definite, $\boldsymbol{\beta}_o$ is a $k \times (kp+1)$ matrix and n_o is a real number. These quantities are referred to as the hyper parameters in Bayesian inference and their values are assumed to be known in this section. For information on Wishart and inverted Wishart distributions, see Appendix A.

Using the pdf of inverted Wishart distribution in Equation (A.12), the posterior distribution is

$$\begin{aligned} f(\boldsymbol{\beta}, \boldsymbol{\Sigma}_a|\boldsymbol{Z}, \boldsymbol{X}) \propto & |\boldsymbol{\Sigma}_a|^{-(v_o+k+1)/2} \exp\left[-\tfrac{1}{2}tr(\boldsymbol{V}_o\boldsymbol{\Sigma}_a^{-1})\right] \\ & \times |\boldsymbol{\Sigma}_a|^{-k/2} \exp\left[-\tfrac{1}{2}tr\left\{(\boldsymbol{\beta}-\boldsymbol{\beta}_o)'\boldsymbol{C}(\boldsymbol{\beta}-\boldsymbol{\beta}_o)\boldsymbol{\Sigma}_a^{-1}\right\}\right] \\ & \times |\boldsymbol{\Sigma}_a|^{-n/2} \exp\left[-\tfrac{1}{2}tr\left\{(\boldsymbol{Z}-\boldsymbol{X}\boldsymbol{\beta})'(\boldsymbol{Z}-\boldsymbol{X}\boldsymbol{\beta})\boldsymbol{\Sigma}_a^{-1}\right\}\right]. \quad (2.49) \end{aligned}$$

To simplify, we can combine the two terms in Equation (2.49) involving $\boldsymbol{\beta}$ via the LS properties in Appendix A. Specifically, denote the Cholesky decomposition of $\boldsymbol{C}$ as $\boldsymbol{C} = \boldsymbol{U}'\boldsymbol{U}$, where $\boldsymbol{U}$ is an upper triangular matrix. Define

$$\boldsymbol{W} = \begin{bmatrix} \boldsymbol{X} \\ \boldsymbol{U} \end{bmatrix}, \quad \boldsymbol{Y} = \begin{bmatrix} \boldsymbol{Z} \\ \boldsymbol{U}\boldsymbol{\beta}_o \end{bmatrix}.$$

Then, we have

$$(\boldsymbol{\beta}-\boldsymbol{\beta}_o)'\boldsymbol{C}(\boldsymbol{\beta}-\boldsymbol{\beta}_o) + (\boldsymbol{Z}-\boldsymbol{X}\boldsymbol{\beta})'(\boldsymbol{Z}-\boldsymbol{X}\boldsymbol{\beta}) = (\boldsymbol{Y}-\boldsymbol{W}\boldsymbol{\beta})'(\boldsymbol{Y}-\boldsymbol{W}\boldsymbol{\beta}).$$

Applying Property (ii) of the LS estimation in Appendix A, we have

$$\begin{aligned} (\boldsymbol{Y}-\boldsymbol{W}\boldsymbol{\beta})'(\boldsymbol{Y}-\boldsymbol{W}\boldsymbol{\beta}) &= (\boldsymbol{Y}-\boldsymbol{W}\tilde{\boldsymbol{\beta}})'(\boldsymbol{Y}-\boldsymbol{W}\tilde{\boldsymbol{\beta}}) + (\boldsymbol{\beta}-\tilde{\boldsymbol{\beta}})'\boldsymbol{W}'\boldsymbol{W}(\boldsymbol{\beta}-\tilde{\boldsymbol{\beta}}), \\ &= \tilde{\boldsymbol{S}} + (\boldsymbol{\beta}-\tilde{\boldsymbol{\beta}})'\boldsymbol{W}'\boldsymbol{W}(\boldsymbol{\beta}-\tilde{\boldsymbol{\beta}}), \quad (2.50) \end{aligned}$$

where

$$\tilde{\boldsymbol{\beta}} = (\boldsymbol{W}'\boldsymbol{W})^{-1}\boldsymbol{W}'\boldsymbol{Y} = (\boldsymbol{X}'\boldsymbol{X}+\boldsymbol{C})^{-1}(\boldsymbol{X}'\boldsymbol{X}\hat{\boldsymbol{\beta}} + \boldsymbol{C}\boldsymbol{\beta}_o),$$

and

$$\tilde{\boldsymbol{S}} = (\boldsymbol{Y}-\boldsymbol{W}\tilde{\boldsymbol{\beta}})'(\boldsymbol{Y}-\boldsymbol{W}\tilde{\boldsymbol{\beta}}) = (\boldsymbol{Z}-\boldsymbol{X}\tilde{\boldsymbol{\beta}})'(\boldsymbol{Z}-\boldsymbol{X}\tilde{\boldsymbol{\beta}}) + (\tilde{\boldsymbol{\beta}}-\boldsymbol{\beta}_o)'\boldsymbol{C}(\tilde{\boldsymbol{\beta}}-\boldsymbol{\beta}_o).$$

Using Equation(2.50), we can write the posterior distribution as

$$\begin{aligned} f(\boldsymbol{\beta}, \boldsymbol{\Sigma}_a|\boldsymbol{Z}, \boldsymbol{X}) \propto & |\boldsymbol{\Sigma}_a|^{-k/2} \exp\left[-\tfrac{1}{2}tr\left\{(\boldsymbol{\beta}-\tilde{\boldsymbol{\beta}})'\boldsymbol{W}'\boldsymbol{W}\left(\boldsymbol{\beta}-\tilde{\boldsymbol{\beta}}\right)\boldsymbol{\Sigma}_a^{-1}\right\}\right] \\ & \times |\boldsymbol{\Sigma}_a|^{-(n_o+n+k+1)/2} \exp\left[-\tfrac{1}{2}tr\left\{\left(\boldsymbol{V}_o+\tilde{\boldsymbol{S}}\right)\boldsymbol{\Sigma}_a^{-1}\right\}\right]. \quad (2.51) \end{aligned}$$

The first term of Equation (2.51) is a multivariate normal kernel, whereas the second term is an inverted Wishart kernel. Consequently, the posterior distributions of $\boldsymbol{\beta}$ and $\boldsymbol{\Sigma}_a$ are

$$\begin{aligned} \boldsymbol{\Sigma}_a|\boldsymbol{Z},\boldsymbol{X} &\sim W^{-1}(\boldsymbol{V}_o+\tilde{\boldsymbol{S}}, n_o+n) \\ \text{vec}(\boldsymbol{\beta})|\boldsymbol{Z},\boldsymbol{X},\boldsymbol{\Sigma}_a &\sim N[\text{vec}(\tilde{\boldsymbol{\beta}}), \boldsymbol{\Sigma}_a \otimes (\boldsymbol{X}'\boldsymbol{X}+\boldsymbol{C})^{-1}], \end{aligned} \tag{2.52}$$

where $n = T - p$ and

$$\begin{aligned} \tilde{\boldsymbol{\beta}} &= (\boldsymbol{X}'\boldsymbol{X}+\boldsymbol{C})^{-1}(\boldsymbol{X}'\boldsymbol{X}\hat{\boldsymbol{\beta}}+\boldsymbol{C}\boldsymbol{\beta}_o), \\ \tilde{\boldsymbol{S}} &= (\boldsymbol{Z}-\boldsymbol{X}\tilde{\boldsymbol{\beta}})'(\boldsymbol{Z}-\boldsymbol{X}\tilde{\boldsymbol{\beta}}) + (\tilde{\boldsymbol{\beta}}-\boldsymbol{\beta}_o)'\boldsymbol{C}(\tilde{\boldsymbol{\beta}}-\boldsymbol{\beta}_o). \end{aligned}$$

The inverse of the covariance matrix of a multivariate normal distribution is called the *precision* matrix. We can interpret $\boldsymbol{X}'\boldsymbol{X}$ and $\boldsymbol{C}$ as the precision matrices of the LS estimate $\hat{\boldsymbol{\beta}}$ and the prior distribution of $\boldsymbol{\beta}$, respectively. (They are relative precision matrices as the covariance matrices involve $\boldsymbol{\Sigma}_a$.) The posterior mean $\tilde{\boldsymbol{\beta}}$ is a weighted average between the LS estimates and the prior mean. The precision of the posterior distribution of $\boldsymbol{\beta}$ is simply the sum of the two precision matrices. These results are generalizations of those of the multiple linear regression; see, for instance, Tsay (2010, Chapter 12).

We can use posterior means as point estimates of the VAR(p) parameters. Therefore, the Bayesian estimates for $\boldsymbol{\beta}$ and $\boldsymbol{\Sigma}_a$ are

$$\breve{\boldsymbol{\beta}} = \tilde{\boldsymbol{\beta}} \quad \text{and} \quad \breve{\boldsymbol{\Sigma}}_a = \frac{\boldsymbol{V}_o + \tilde{\boldsymbol{S}}}{n_o + T - p - k - 1}.$$

The covariance matrix of vec($\breve{\boldsymbol{\beta}}$) is $\breve{\boldsymbol{\Sigma}}_a \otimes (\boldsymbol{X}'\boldsymbol{X}+\boldsymbol{C})^{-1}$.

In practice, we may not have concrete prior information about $\boldsymbol{\beta}$ for a stationary VAR(p) model, especially the correlations between the parameters. In this case, we may choose $\boldsymbol{\beta}_o = \boldsymbol{0}$ and a large covariance matrix $\boldsymbol{C}^{-1}$, say $\boldsymbol{C}^{-1} = \delta \boldsymbol{I}_{kp+1}$ with a large δ. These particular prior choices are referred to as *vague priors* and would result in a small $\boldsymbol{C} = \delta^{-1}\boldsymbol{I}_{kp+1}$, and Equation (2.52) shows that the Bayesian estimate $\tilde{\boldsymbol{\beta}}$ is close to the LS estimate $\hat{\boldsymbol{\beta}}$. A small $\boldsymbol{C}$ also leads to $\tilde{\boldsymbol{S}}$ being close to $\boldsymbol{S}$ of the LS estimate given in Equation (2.47). Consequently, if we also choose a small $\boldsymbol{V}_o$ for the prior of $\boldsymbol{\Sigma}_a$, then the Bayesian estimate of $\boldsymbol{\Sigma}_a$ is also close to the LS estimate.

The choices of prior distributions are subjective in a real application. Our use of conjugate priors and $\boldsymbol{C} = \lambda \boldsymbol{I}_{kp+1}$ with a small λ are for convenience. One can apply several priors to study the sensitivity of the analysis to prior specification. For instance, one can specify $\boldsymbol{C}$ as a diagonal matrix with different diagonal elements to reflect the common prior belief that higher-order AR lags are of decreasing importance. For stationary VAR(p) models, Bayesian estimates are typically not too sensitive to any reasonable prior specification, especially when the sample size is large. In the literature, Litterman (1986) and Doan, Litterman, and Sims (1984) describe a specific prior for stationary VAR(p) models. The prior is known as the *Minnesota prior*. For $\boldsymbol{\beta}$, the prior is multivariate normal with mean zero and a diagonal covariance matrix. That is, Minnesota prior replaces $\boldsymbol{\Sigma}_a \otimes \boldsymbol{C}^{-1}$ by a diagonal matrix $\boldsymbol{V}$. For the AR coefficient $\phi_{\ell,ij}$, the prior variance is given by

$$\text{Var}(\phi_{\ell,ij}) = \begin{cases} (\lambda/\ell)^2 & \text{if } i = j \\ (\lambda\theta/\ell)^2 \times (\sigma_{ii}/\sigma_{jj}) & \text{if } i \neq j, \end{cases}$$

where λ is a real number, $0 < \theta < 1$, σ_{ii} is the (i,i)th element of $\boldsymbol{\Sigma}_a$, and $\ell = 1, \ldots, p$. From the specification, the prior states that $\phi_{\ell,ij}$ is close to 0 as ℓ increases. The choices of λ and θ for the Minnesota prior are subjective in an application.

Example 2.4 Consider, again, the percentage growth rates of quarterly real GDP of United Kingdom, Canada, and United States employed in Example 2.3. We specify a VAR(2) model and use the noninformative conjugate priors with

$$\boldsymbol{C} = 0.1 \times \boldsymbol{I}_7, \quad \boldsymbol{V}_o = \boldsymbol{I}_3, \quad n_o = 5, \quad \boldsymbol{\beta}_o = \boldsymbol{0}.$$

The estimation results are shown in the following R demonstration. As expected, the Bayesian estimates of $\boldsymbol{\phi}_i$ and $\boldsymbol{\Sigma}_a$ are close to the LS estimates. The command for Bayesian VAR estimation in the `MTS` package is `BVAR`. □

R Demonstration: Bayesian estimation.

```
> da=read.table("q-gdp-ukcaus.txt",header=T)
> x=log(da[,3:5])
> dim(x)
[1] 126   3
> dx=x[2:126,]-x[1:125,]
> dx=dx*100
> C=0.1*diag(7)   ### lambda = 0.1
> V0=diag(3) ### Vo = I_3
> mm=BVAR(dx,p=2,C,V0)
Bayesian estimate:
                 Est        s.e.     t-ratio
 [1,]  0.125805143 0.07123059  1.76616742
 [2,]  0.392103983 0.09150764  4.28493158
 [3,]  0.102894946 0.09633822  1.06805941
 [4,]  0.052438976 0.08925487  0.58751947
 [5,]  0.056937547 0.09048722  0.62923303
 [6,]  0.105553695 0.08578002  1.23051603
 [7,]  0.019147973 0.09188759  0.20838475
 [8,]  0.123256168 0.07237470  1.70302833
 [9,]  0.350253306 0.09297745  3.76707803
[10,]  0.337525508 0.09788562  3.44816232
```

```
[11,]  0.468440207 0.09068850  5.16537628
[12,] -0.190144541 0.09194064 -2.06812294
[13,] -0.173964344 0.08715783 -1.99596908
[14,] -0.008627966 0.09336351 -0.09241262
[15,]  0.289317667 0.07987129  3.62229886
[16,]  0.489072359 0.10260807  4.76641231
[17,]  0.239456311 0.10802463  2.21668257
[18,]  0.235601116 0.10008202  2.35408023
[19,] -0.310286945 0.10146386 -3.05810301
[20,] -0.130271750 0.09618566 -1.35437813
[21,]  0.085039470 0.10303411  0.82535258
Covariance matrix:
           uk         ca         us
uk 0.28839063 0.02647455 0.07394349
ca 0.02647455 0.29772937 0.13875034
us 0.07394349 0.13875034 0.36260138
```

2.6 ORDER SELECTION

Turn to model building. We follow the iterated procedure of Box and Jenkins consisting of model specification, estimation, and diagnostic checking. See Box, Jenkins, and Reinsel (2008). For VAR models, model specification is to select the order p. Several methods have been proposed in the literature to select the VAR order. We discuss two approaches. The first approach adopts the framework of multivariate multiple linear regression and uses sequential likelihood ratio tests. The second approach employs information criteria.

2.6.1 Sequential Likelihood Ratio Tests

This approach to selecting the VAR order was recommended by Tiao and Box (1981). The basic idea of the approach is to compare a VAR(ℓ) model with a VAR($\ell - 1$) model. In statistical terms, it amounts to consider the hypothesis testing

$$H_0 : \boldsymbol{\phi}_\ell = \mathbf{0} \quad \text{versus} \quad H_a : \boldsymbol{\phi}_\ell \neq \mathbf{0}. \tag{2.53}$$

This is a problem of nested hypotheses, and a natural test statistic to use is the likelihood ratio statistic. As shown in Section 2.5, we can employ the multivariate linear regression framework for a VAR model. Let $\boldsymbol{\beta}'_\ell = [\boldsymbol{\phi}_0, \boldsymbol{\phi}_1, \ldots, \boldsymbol{\phi}_\ell]$ be the matrix of coefficient parameters of a VAR(ℓ) model and $\boldsymbol{\Sigma}_{a,\ell}$ be the corresponding innovation covariance matrix. Under the normality assumption, the likelihood ratio for the testing problem in Equation (2.53) is

$$\Lambda = \frac{\max L(\boldsymbol{\beta}_{\ell-1}, \boldsymbol{\Sigma}_a)}{\max L(\boldsymbol{\beta}_\ell, \boldsymbol{\Sigma}_a)} = \left(\frac{|\hat{\boldsymbol{\Sigma}}_{a,\ell}|}{|\hat{\boldsymbol{\Sigma}}_{a,\ell-1}|} \right)^{(T-\ell)/2}.$$

This equation follows the maximized likelihood function of a VAR model in Equation (2.41). Note that here we estimate the VAR($\ell - 1$) model using the regression setup of the VAR(ℓ) model. In other words, the $\boldsymbol{Z}$ matrix of Equation (2.28) consists of $\boldsymbol{z}_{\ell+1}, \ldots, \boldsymbol{z}_T$. The likelihood ratio test of H_0 is equivalent to rejecting H_0 for large values of

$$-2\ln(\Lambda) = -(T-\ell)\ln\left(\frac{|\hat{\boldsymbol{\Sigma}}_{a,\ell}|}{|\hat{\boldsymbol{\Sigma}}_{a,\ell-1}|}\right).$$

A commonly used test statistic is then

$$M(\ell) = -(T-\ell-1.5-k\ell)\ln\left(\frac{|\hat{\boldsymbol{\Sigma}}_{a,\ell}|}{|\hat{\boldsymbol{\Sigma}}_{a,\ell-1}|}\right),$$

which follows asymptotically a chi-square distribution with k^2 degrees of freedom. This test statistic is widely used in the multivariate statistical analysis. See, for instance, Result 7.11 of Johnson and Wichern (2007).

To simplify the computation, Tiao and Box (1981) suggest the following procedure to compute the $M(\ell)$ statistic and to select the VAR order:

1. Select a positive integer P, which is the maximum VAR order entertained.
2. Setup the multivariate multiple linear regression framework of Equation (2.28) for the VAR(P) model. That is, there are $T - P$ observations in the $\boldsymbol{Z}$ data matrix for estimation.
3. For $\ell = 0, \ldots, P$, compute the LS estimate of the AR coefficient matrix, that is, compute $\hat{\boldsymbol{\beta}}_\ell$. For $\ell = 0$, $\boldsymbol{\beta}'$ is simply the constant vector $\boldsymbol{\phi}_0$. Then, compute the ML estimate of $\boldsymbol{\Sigma}_a$, that is, compute $\hat{\boldsymbol{\Sigma}}_{a,\ell} = (1/T-P)\hat{\boldsymbol{A}}_\ell'\hat{\boldsymbol{A}}_\ell$, where $\hat{\boldsymbol{A}}_\ell = \boldsymbol{Z} - \boldsymbol{X}\hat{\boldsymbol{\beta}}_\ell$ is the residual matrix of the fitted VAR(ℓ) model.
4. For $\ell = 1, \ldots, P$, compute the modified likelihood ratio test statistic

$$M(\ell) = -(T-P-1.5-k\ell)\ln\left(\frac{|\hat{\boldsymbol{\Sigma}}_{a,\ell}|}{|\hat{\boldsymbol{\Sigma}}_{a,\ell-1}|}\right), \tag{2.54}$$

   and its p-value, which is based on the asymptotic $\chi^2_{k^2}$ distribution.
5. Examine the test statistics sequentially starting with $\ell = 1$. If all p-values of the $M(\ell)$ test statistics are greater than the specified type I error for $\ell > p$, then a VAR(p) model is specified. This is so because the test rejects the null hypothesis $\boldsymbol{\phi}_p = \boldsymbol{0}$, but fails to reject $\boldsymbol{\phi}_\ell = \boldsymbol{0}$ for $\ell > p$.

In practice, a simpler model is preferred. Thus, we often start with a small p. This is particularly so when the dimension k is high.

2.6.2 Information Criteria

Information criteria have been shown to be effective in selecting a statistical model. In the time series literature, several criteria have been proposed. All criteria are likelihood based and consist of two components. The first component is concerned with the goodness of fit of the model to the data, whereas the second component penalizes more heavily complicated models. The goodness of fit of a model is often measured by the maximized likelihood. For normal distribution, the maximized likelihood is equivalent to the determinant of the covariance matrix of the innovations; see Equation (2.41). This determinant is known as the *generalized variance* in multivariate analysis. The selection of the penalty, on the other hand, is relatively subjective. Different penalties result in different information criteria.

Three criteria functions are commonly used to determine VAR order. Under the normality assumption, these three criteria for a VAR(ℓ) model are

$$\begin{aligned}
\text{AIC}(\ell) &= \ln|\hat{\boldsymbol{\Sigma}}_{a,\ell}| + \frac{2}{T}\ell k^2, \\
\text{BIC}(\ell) &= \ln|\hat{\boldsymbol{\Sigma}}_{a,\ell}| + \frac{\ln(T)}{T}\ell k^2, \\
\text{HQ}(\ell) &= \ln|\hat{\boldsymbol{\Sigma}}_{a,\ell}| + \frac{2\ln[\ln(T)]}{T}\ell k^2,
\end{aligned}$$

where T is the sample size, $\hat{\boldsymbol{\Sigma}}_{a,\ell}$ is the ML estimate of $\boldsymbol{\Sigma}_a$ discussed in Section 2.5.2, AIC is the Akaike information criterion proposed in Akaike (1973). BIC stands for Bayesian information criterion (see Schwarz 1978), and HQ(ℓ) is proposed by Hannan and Quinn (1979) and Quinn (1980). The AIC penalizes each parameter by a factor of 2. BIC and HQ, on the other hand, employ penalties that depend on the sample size. For large T, BIC penalizes complicated models more heavily; for example, when $\ln(T) > 2$. HQ penalizes each parameter by $2\ln(\ln(T))$, which is greater than 2 when $T > 15$.

If $\boldsymbol{z}_t$ is indeed a Gaussian VAR(p) time series with $p < \infty$, then both BIC and HQ are consistent in the sense that they will select the true VAR(p) model with probability 1 as $T \to \infty$. The AIC, on the other hand, is not consistent as it has positive probabilities to select VAR(ℓ) models for $\ell > p$. The criterion, however, does not select a VAR(ℓ) model with $\ell < p$ when $T \to \infty$. See Quinn (1980). There is discussion in the literature about the validity of using consistency to compare information criteria because consistency requires the existence of a true model, yet there are no true models in real applications. Shibata (1980) derived asymptotic optimality properties of AIC for univariate time series.

Example 2.5 Consider, again, the growth rates of quarterly real GDP of United Kingdom, Canada, and United States from 1980.II to 2011.II. We apply the sequential likelihood ratio tests and all three information criteria to the data. The maximum order entertained is 13. Table 2.1 summarizes these statistics. From the table, we see that the orders selected by AIC, BIC, and HQ are 2, 1, and 1, respectively. The

TABLE 2.1 Order Selection Statistics for the Quarterly Growth Rates of the Real GDP of United Kingdom, Canada, and United States from the Second Quarter of 1980 to the Second Quarter of 2011

p	AIC	BIC	HQ	$M(p)$	p-Value
0	−30.956	−30.956	−30.956	0.000	0.000
1	−31.883	−31.679	−31.800	115.13	0.000
2	−31.964	−31.557	−31.799	23.539	0.005
3	−31.924	−31.313	−31.675	10.486	0.313
4	−31.897	−31.083	−31.566	11.577	0.238
5	−31.782	−30.764	−31.368	2.741	0.974
6	−31.711	−30.489	−31.215	6.782	0.660
7	−31.618	−30.192	−31.039	4.547	0.872
8	−31.757	−30.128	−31.095	24.483	0.004
9	−31.690	−29.857	−30.945	6.401	0.669
10	−31.599	−29.563	−30.772	4.322	0.889
11	−31.604	−29.364	−30.694	11.492	0.243
12	−31.618	−29.175	−30.626	11.817	0.224
13	−31.673	−29.025	−30.596	14.127	0.118

The information criteria used are AIC, BIC, and HQ. The M-statistics are given in Equation (2.54).

sequential M-statistic of Equation (2.54) selects the order $p = 2$, except for a violation at $p = 8$. This example demonstrates that different criteria may select different orders for a multivariate time series. Keep in mind, however, that these statistics are estimates. As such, one cannot take the values too seriously. Figure 2.2 shows the time plots of the three information criteria. The AIC shows a relatively close values for $p \in \{1, 2, 3, 4\}$, BIC shows a clear minimum at $p = 1$, whereas HQ shows a minimum at $p = 1$ with $p = 2$ as a close second. All three criteria show a drop at $p = 8$. In summary, a VAR(1) or VAR(2) model may serve as a starting model for the three-dimensional GDP series. □

R Demonstration: Order selection.

```
> z1=z/100   ### Original growth rates
> m2=VARorder(z1)
selected order: aic =  2
selected order: bic =  1
selected order: hq =  1
M statistic and its p-value
        Mstat       pv
 [1,] 115.133 0.000000
 [2,]  23.539 0.005093
 [3,]  10.486 0.312559
```

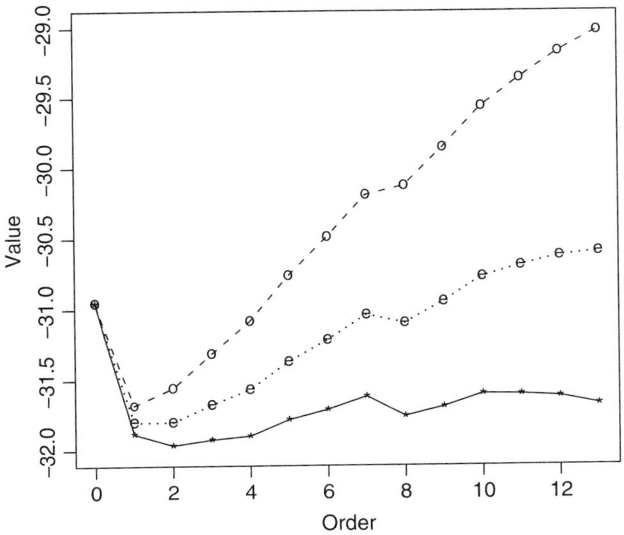

FIGURE 2.2 Information criteria for the quarterly growth rates, in percentages, of real gross domestic products of United Kingdom, Canada, and United States from the second quarter of 1980 to the second quarter of 2011. The solid, dashed, and dotted lines are for AIC, BIC, and HQ, respectively.

```
 [4,]  11.577 0.238240
 [5,]   2.741 0.973698
 [6,]   6.782 0.659787
 [7,]   4.547 0.871886
 [8,]  24.483 0.003599
 [9,]   6.401 0.699242
[10,]   4.323 0.888926
[11,]  11.492 0.243470
[12,]  11.817 0.223834
[13,]  14.127 0.117891
Summary table:
       p     AIC     BIC      HQ     M(p)   p-value
 [1,]  0 -30.956 -30.956 -30.956   0.0000 0.0000000
 [2,]  1 -31.883 -31.679 -31.800 115.1329 0.0000000
 [3,]  2 -31.964 -31.557 -31.799  23.5389 0.0050930
 [4,]  3 -31.924 -31.313 -31.675  10.4864 0.3125594
 [5,]  4 -31.897 -31.083 -31.566  11.5767 0.2382403
 [6,]  5 -31.782 -30.764 -31.368   2.7406 0.9736977
 [7,]  6 -31.711 -30.489 -31.215   6.7822 0.6597867
 [8,]  7 -31.618 -30.192 -31.039   4.5469 0.8718856
 [9,]  8 -31.757 -30.128 -31.095  24.4833 0.0035992
[10,]  9 -31.690 -29.857 -30.945   6.4007 0.6992417
[11,] 10 -31.599 -29.563 -30.772   4.3226 0.8889256
[12,] 11 -31.604 -29.364 -30.694  11.4922 0.2434698
```

```
[13,] 12 -31.618 -29.175 -30.626  11.8168 0.2238337
[14,] 13 -31.672 -29.025 -30.596  14.1266 0.1178914

> names(m2)
[1] "aic"   "aicor" "bic"   "bicor" "hq"    "hqor"   "Mstat" "Mpv"
```

Remark: There are different ways to compute the information criteria for a given time series realization $\{\boldsymbol{z}_1, \ldots, \boldsymbol{z}_T\}$. The first approach is to use the same number of observations as discussed in the calculation of the $M(\ell)$ statistics in Equation (2.54). Here, one uses the data from $t = P + 1$ to T to evaluate the likelihood functions, where P is the maximum AR order. In the `MTS` package, the command `VARorder` uses this approach. The second approach is to use the data from $t = \ell + 1$ to T to fit a VAR(ℓ) model. In this case, different VAR models use different numbers of observations in estimation. For a large T, the two approaches should give similar results. However, when the sample size T is moderate compared with the dimension k, the two approaches may give different order selections even for the same criterion function. In the `MTS` package, the command `VARorderI` uses the second approach. □

2.7 MODEL CHECKING

Model checking, also known as diagnostic check or residual analysis, plays an important role in model building. Its main objectives include (i) to ensure that the fitted model is adequate and (ii) to suggest directions for further improvements if needed. The adequacy of a fitted model is judged according to some selected criteria, which may depend on the objective of the analysis. Typically a fitted model is said to be adequate if (a) all fitted parameters are statistically significant (at a specified level), (b) the residuals have no significant serial or cross-sectional correlations, (c) there exist no structural changes or outlying observations, and (d) the residuals do not violate the distributional assumption, for example, multivariate normality. We discuss some methods for model checking in this section.

2.7.1 Residual Cross-Correlations

The residuals of an adequate model should behave like a white noise series. Checking the serial and cross-correlations of the residuals thus becomes an integral part of model checking. Let $\hat{\boldsymbol{A}} = \boldsymbol{Z} - \boldsymbol{X}\hat{\boldsymbol{\beta}}$ be the residual matrix of a fitted VAR(p) model, using the notation in Equation (2.28). The ith row of $\hat{\boldsymbol{A}}$ contains $\hat{\boldsymbol{a}}_{p+i} = \boldsymbol{z}_{p+i} - \hat{\boldsymbol{\phi}}_0 - \sum_{i=1}^{p} \hat{\boldsymbol{\phi}}_i \boldsymbol{z}_{t-i}$. The lag ℓ cross-covariance matrix of the residual series is defined as

$$\hat{\boldsymbol{C}}_\ell = \frac{1}{T-p} \sum_{t=p+\ell+1}^{T} \hat{\boldsymbol{a}}_t \hat{\boldsymbol{a}}'_{t-\ell}.$$

In particular, we have $\hat{C}_0 = \hat{\Sigma}_a$ is the residual covariance matrix. In matrix notation, we can rewrite the lag ℓ residual cross-covariance matrix $\hat{C}_\ell$ as

$$\hat{C}_\ell = \frac{1}{T-p} \hat{A}' B^\ell \hat{A}, \quad \ell \geq 0 \tag{2.55}$$

where B is a $(T-p) \times (T-p)$ back-shift matrix defined as

$$B = \begin{bmatrix} 0 & 0'_{T-p-1} \\ I_{T-p-1} & 0_{T-p-1} \end{bmatrix},$$

where 0_h is the h-dimensional vector of zero. The lag ℓ residual cross-correlation matrix is defined as

$$\hat{R}_\ell = \hat{D}^{-1} \hat{C}_\ell \hat{D}^{-1}, \tag{2.56}$$

where $\hat{D}$ is the diagonal matrix of the standard errors of the residual series, that is, $\hat{D} = \sqrt{\text{diag}(\hat{C}_0)}$. In particular, $\hat{R}_0$ is the residual correlation matrix.

Residual cross-covariance matrices are useful tools for model checking, so we study next their limiting properties. To this end, we consider the asymptotic joint distribution of the residual cross-covariance matrices $\hat{\Xi}_m = [\hat{C}_1, \ldots, \hat{C}_m]$. Using the notation in Equation (2.55), we have

$$\hat{\Xi}_m = \frac{1}{T-p} \hat{A}' \left[B\hat{A}, B^2\hat{A}, \ldots, B^m\hat{A} \right] = \frac{1}{T-p} \hat{A}' B_m (I_m \otimes \hat{A}), \tag{2.57}$$

where $B_m = [B, B^2, \ldots, B^m]$ is a $(T-p) \times m(T-p)$ matrix. Next, from Equation (2.28), we have

$$\hat{A} = Z - X\hat{\beta} = Z - X\beta + X\beta - X\hat{\beta} = A - X(\hat{\beta} - \beta).$$

Therefore, via Equation (2.57) and letting $T_p = T - p$, we have

$$\begin{aligned} T_p \hat{\Xi}_m &= A' B_m (I_m \otimes A) - A' B_m \left[I_m \otimes X \left(\hat{\beta} - \beta \right) \right] \\ &\quad - \left(\hat{\beta} - \beta \right)' X' B_m (I_m \otimes A) + \left(\hat{\beta} - \beta \right)' X' B_m \left[I_m \otimes X \left(\hat{\beta} - \beta \right) \right]. \end{aligned} \tag{2.58}$$

We can use Equation (2.58) to study the limiting distribution of $\hat{\Xi}_m$. Adopting an approach similar to that of Lütkepohl (2005), we divide the derivations into several steps.

Lemma 2.4 Suppose that z_t follows a stationary VAR(p) model of Equation (2.21) with a_t being a white noise process with mean zero and positive covariance matrix $\boldsymbol{\Sigma}_a$. Also, assume that the assumption in Equation (2.42) holds and the parameter matrix $\boldsymbol{\beta}$ of the model in Equation (2.28) is consistently estimated by a method discussed in Section 2.5 and the residual cross-covariance matrix is defined in Equation (2.55). Then, $\sqrt{T_p}\text{vec}(\hat{\boldsymbol{\Xi}}_m)$ has the same limiting distribution as $\sqrt{T_p}\text{vec}(\boldsymbol{\Xi}_m) - \sqrt{T_p}\boldsymbol{H}\text{vec}[(\hat{\boldsymbol{\beta}} - \boldsymbol{\beta})']$, where $T_p = T - p$, $\boldsymbol{\Xi}_m$ is the theoretical counterpart of $\hat{\boldsymbol{\Xi}}_m$ obtained by dividing the first term of Equation (2.58) by T_p, and $\boldsymbol{H} = \boldsymbol{H}'_* \otimes \boldsymbol{I}_k$ with

$$\boldsymbol{H}_* = \begin{bmatrix} \mathbf{0}' & \mathbf{0}' & \cdots & \mathbf{0}' \\ \boldsymbol{\Sigma}_a & \boldsymbol{\psi}_1\boldsymbol{\Sigma}_a & \cdots & \boldsymbol{\psi}_{m-1}\boldsymbol{\Sigma}_a \\ \mathbf{0}_k & \boldsymbol{\Sigma}_a & \cdots & \boldsymbol{\psi}_{m-2}\boldsymbol{\Sigma}_a \\ \vdots & \vdots & & \vdots \\ \mathbf{0}_k & \mathbf{0}_k & \cdots & \boldsymbol{\psi}_{m-p}\boldsymbol{\Sigma}_a \end{bmatrix}_{(kp+1)\times km},$$

where $\mathbf{0}$ is a k-dimensional vector of zero, $\mathbf{0}_k$ is a $k \times k$ matrix of zero, and $\boldsymbol{\psi}_i$ are the coefficient matrices of the MA representation of the VAR(p) model in Equation (2.27).

Proof of Lemma 2.4 is given in Section 2.12. The first term of Equation (2.58) is

$$\boldsymbol{A}'\boldsymbol{B}_m(\boldsymbol{I}_m \otimes \boldsymbol{A}) = \left[\boldsymbol{A}'\boldsymbol{B}\boldsymbol{A}, \boldsymbol{A}'\boldsymbol{B}^2\boldsymbol{A}, \ldots, \boldsymbol{A}'\boldsymbol{B}^m\boldsymbol{A}\right] \equiv T_p\boldsymbol{\Xi}_m.$$

It is then easy to see that

$$\sqrt{T_p}\text{vec}(\boldsymbol{\Xi}_m) \to_d N(\mathbf{0}, \boldsymbol{I}_m \otimes \boldsymbol{\Sigma}_a \otimes \boldsymbol{\Sigma}_a).$$

In fact, by using part (ii) of Lemma 2.3 and direct calculation, we have the following result. Details can be found in Ahn (1988).

Lemma 2.5 Assume that $\boldsymbol{z}_t$ is a stationary VAR(p) series satisfying the conditions of Lemma 2.4, then

$$\begin{bmatrix} \frac{1}{T_p}\text{vec}(\boldsymbol{A}'\boldsymbol{X}) \\ \sqrt{T_p}\text{vec}(\boldsymbol{\Xi}_m) \end{bmatrix} \to_d N\left(\mathbf{0}, \begin{bmatrix} \boldsymbol{G} & \boldsymbol{H}_* \\ \boldsymbol{H}'_* & \boldsymbol{I}_m \otimes \boldsymbol{\Sigma}_a \end{bmatrix} \otimes \boldsymbol{\Sigma}_a\right),$$

where $\boldsymbol{G}$ is defined in Lemma 2.3 and $\boldsymbol{H}_*$ is defined in Lemma 2.4.

Using Lemmas 2.4 and 2.5, we can obtain the limiting distribution of the cross-covariance matrices of a stationary VAR(p) time series.

Theorem 2.4 Suppose that $\boldsymbol{z}_t$ follows a stationary VAR(p) model of Equation (2.21) with $\boldsymbol{a}_t$ being a white noise process with mean zero and positive covariance matrix $\boldsymbol{\Sigma}_a$. Also, assume that the assumption in Equation (2.42) holds and the parameter matrix $\boldsymbol{\beta}$ of the model in Equation (2.28) is consistently estimated by a method discussed in Section 2.5 and the residual cross-covariance matrix is defined in Equation (2.55). Then,

$$\sqrt{T_p}\text{vec}(\hat{\boldsymbol{\Xi}}_m) \to_d N(\mathbf{0}, \boldsymbol{\Sigma}_{c,m}),$$

where

$$\begin{aligned}\boldsymbol{\Sigma}_{c,m} &= (\boldsymbol{I}_m \otimes \boldsymbol{\Sigma}_a - \boldsymbol{H}_*'\boldsymbol{G}^{-1}\boldsymbol{H}_*) \otimes \boldsymbol{\Sigma}_a \\ &= \boldsymbol{I}_m \otimes \boldsymbol{\Sigma}_a \otimes \boldsymbol{\Sigma}_a - \tilde{\boldsymbol{H}}[(\boldsymbol{\Gamma}_0^*)^{-1} \otimes \boldsymbol{\Sigma}_a]\tilde{\boldsymbol{H}}',\end{aligned}$$

where $\boldsymbol{H}_*$ and $\boldsymbol{G}$ are defined in Lemma 2.5, $\boldsymbol{\Gamma}_0^*$ is the expanded covariance matrix defined in Equation (2.26), and $\tilde{\boldsymbol{H}} = \tilde{\boldsymbol{H}}_* \otimes \boldsymbol{I}_k$ with $\tilde{\boldsymbol{H}}_*$ being a submatrix of $\boldsymbol{H}_*$ with the first row of zeros removed.

Proof. Using Lemma 2.4, the limiting distribution of $\sqrt{T_p}\text{vec}(\hat{\boldsymbol{\Xi}}_m)$ can be obtained by considering

$$\begin{aligned}&\sqrt{T_p}\text{vec}(\boldsymbol{\Xi}_m) - \sqrt{T_p}\text{vec}[(\hat{\boldsymbol{\beta}} - \boldsymbol{\beta})'] \\ &= [-\boldsymbol{H}, \boldsymbol{I}_{mk^2}]\begin{bmatrix}\sqrt{T_p}\text{vec}[(\hat{\boldsymbol{\beta}} - \boldsymbol{\beta})'] \\ \sqrt{T_p}\text{vec}(\boldsymbol{\Xi}_m)\end{bmatrix} \\ &= [-\boldsymbol{H}_*' \otimes \boldsymbol{I}_k, \boldsymbol{I}_{mk^2}]\begin{bmatrix}\left(\dfrac{\boldsymbol{X}'\boldsymbol{X}}{\boldsymbol{T_p}}\right)^{-1} \otimes \boldsymbol{I}_k & \mathbf{0} \\ \mathbf{0}' & \boldsymbol{I}_{mk^2}\end{bmatrix}\begin{bmatrix}\dfrac{1}{T_p}\text{vec}(\boldsymbol{A}'\boldsymbol{X}) \\ \sqrt{T_p}\text{vec}(\boldsymbol{\Xi}_m)\end{bmatrix},\end{aligned}$$

where $\mathbf{0}$ is a $k(kp+1) \times mk^2$ matrix of zero, and we have used $\hat{\boldsymbol{\beta}} - \boldsymbol{\beta} = (\boldsymbol{X}'\boldsymbol{X})^{-1}\boldsymbol{X}'\boldsymbol{A}$ and properties of vec operator. Since $\boldsymbol{X}'\boldsymbol{X}/T_p$ converges to the nonsingular matrix $\boldsymbol{G}$ defined in Lemma 2.3, we can apply Lemma 2.5 and properties of multivariate normal distribution to complete the proof. Specifically, the first two factors of the prior equation converge to

$$[-\boldsymbol{H}_*' \otimes \boldsymbol{I}_k, \boldsymbol{I}_{mk^2}]\begin{bmatrix}\boldsymbol{G}^{-1} \otimes \boldsymbol{I}_k & \mathbf{0} \\ \mathbf{0}' & \boldsymbol{I}_{mk^2}\end{bmatrix} = [-\boldsymbol{H}_*'\boldsymbol{G}^{-1} \otimes \boldsymbol{I}_k, \boldsymbol{I}_{mk^2}],$$

and we have

$$\begin{aligned}&[-\boldsymbol{H}_*'\boldsymbol{G}^{-1} \otimes \boldsymbol{I}_k, \boldsymbol{I}_{mk^2}]\left\{\begin{bmatrix}\boldsymbol{G} & \boldsymbol{H}_* \\ \boldsymbol{H}_*' & \boldsymbol{I}_m \otimes \boldsymbol{\Sigma}_a\end{bmatrix} \otimes \boldsymbol{\Sigma}_a\right\}\begin{bmatrix}-\boldsymbol{G}^{-1}\boldsymbol{H}_* \otimes \boldsymbol{I}_k \\ \boldsymbol{I}_{mk^2}\end{bmatrix} \\ &= (\boldsymbol{I}_m \otimes \boldsymbol{\Sigma}_a - \boldsymbol{H}_*'\boldsymbol{G}^{-1}\boldsymbol{H}_*) \otimes \boldsymbol{\Sigma}_a \\ &= \boldsymbol{I}_m \otimes \boldsymbol{\Sigma}_a \otimes \boldsymbol{\Sigma}_a - (\boldsymbol{H}_*' \otimes \boldsymbol{I}_k)(\boldsymbol{G}^{-1} \otimes \boldsymbol{\Sigma}_a)(\boldsymbol{H}_* \otimes \boldsymbol{I}_k) \\ &= \boldsymbol{I}_m \otimes \boldsymbol{\Sigma}_a \otimes \boldsymbol{\Sigma}_a - \tilde{\boldsymbol{H}}[(\boldsymbol{\Gamma}_0^*)^{-1} \otimes \boldsymbol{\Sigma}_a]\tilde{\boldsymbol{H}}',\end{aligned}$$

where the last equality holds because the first row of $\boldsymbol{H}_*$ is 0. □

Comparing results of Lemma 2.5 and Theorem 2.4, we see that the asymptotic variances of elements of the cross-covariance matrices of residuals $\hat{\boldsymbol{a}}_t$ are less than or equal to those of elements of the cross-covariance matrices of the white noise series $\boldsymbol{a}_t$. This seems counter-intuitive, but the residuals are, strictly speaking, not independent.

Let $\boldsymbol{D}$ be the diagonal matrix of the standard errors of the components of $\boldsymbol{a}_t$, that is, $\boldsymbol{D} = \text{diag}\{\sqrt{\sigma_{11,a}}, \ldots, \sqrt{\sigma_{kk,a}}\}$, where $\boldsymbol{\Sigma}_a = [\sigma_{ij,a}]$. We can apply Theorem 2.4 to obtain the limiting distribution of the cross-correlation matrices $\hat{\boldsymbol{\xi}}_m = [\hat{\boldsymbol{R}}_1, \ldots, \hat{\boldsymbol{R}}_m]$, where the cross-correlation matrix $\hat{\boldsymbol{R}}_j$ is defined in Equation (2.56).

Theorem 2.5 Assume that the conditions of Theorem 2.4 hold. Then,

$$\sqrt{T_p}\text{vec}(\hat{\boldsymbol{\xi}}_m) \to_d N(\boldsymbol{0}, \boldsymbol{\Sigma}_{r,m}),$$

where $\boldsymbol{\Sigma}_{r,m} = [(\boldsymbol{I}_m \otimes \boldsymbol{R}_0) - \boldsymbol{H}_0'\boldsymbol{G}^{-1}\boldsymbol{H}_0] \otimes \boldsymbol{R}_0$, where $\boldsymbol{R}_0$ is the lag-0 cross-correlation matrix of $\boldsymbol{a}_t$, $\boldsymbol{H}_0 = \boldsymbol{H}_*(\boldsymbol{I}_m \otimes \boldsymbol{D}^{-1})$, and $\boldsymbol{G}$, as before, is defined in Lemma 2.3.

Proof. Let $\hat{\boldsymbol{D}}^{-1}$ be the diagonal matrix defined in Equation (2.56). It is easy to see that $\hat{\boldsymbol{D}}^{-1}$ is a consistent estimate of $\boldsymbol{D}^{-1}$. We can express the statistic of interest as

$$\begin{aligned}\text{vec}(\hat{\boldsymbol{\xi}}_m) &= \text{vec}\left[\hat{\boldsymbol{D}}^{-1}\hat{\boldsymbol{\Xi}}_m\left(\boldsymbol{I}_m \otimes \hat{\boldsymbol{D}}^{-1}\right)\right] \\ &= \left(\boldsymbol{I}_m \otimes \hat{\boldsymbol{D}}^{-1} \otimes \hat{\boldsymbol{D}}^{-1}\right)\text{vec}(\hat{\boldsymbol{\Xi}}_m).\end{aligned}$$

Applying the result of Theorem 2.4, we obtain that $\sqrt{T_p}\text{vec}(\hat{\boldsymbol{\xi}}_m)$ follows asymptotically a multivariate normal distribution with mean zero and covariance matrix

$$\begin{aligned}&\left(\boldsymbol{I}_m \otimes \boldsymbol{D}^{-1} \otimes \boldsymbol{D}^{-1}\right)\left[\left(\boldsymbol{I}_m \otimes \boldsymbol{\Sigma}_a - \boldsymbol{H}_*'\boldsymbol{G}^{-1}\boldsymbol{H}_*\right) \otimes \boldsymbol{\Sigma}_a\right]\left(\boldsymbol{I}_m \otimes \boldsymbol{D}^{-1} \otimes \boldsymbol{D}^{-1}\right) \\ &\quad = \left[(\boldsymbol{I}_m \otimes \boldsymbol{R}_0) - \boldsymbol{H}_0'\boldsymbol{G}^{-1}\boldsymbol{H}_0\right] \otimes \boldsymbol{R}_0,\end{aligned}$$

where we have used $\boldsymbol{R}_0 = \boldsymbol{D}^{-1}\boldsymbol{\Sigma}_a\boldsymbol{D}^{-1}$. □

Based on Theorem 2.5, we can obtain the limiting distribution of the lag j cross-correlation matrix $\hat{\boldsymbol{R}}_j$. Specifically,

$$\sqrt{T_p}\text{vec}(\hat{\boldsymbol{R}}_j) \to_d N(\boldsymbol{0}, \boldsymbol{\Sigma}_{r,(j)}),$$

where

$$\boldsymbol{\Sigma}_{r,(j)} = \left[\boldsymbol{R}_0 - \boldsymbol{D}^{-1}\boldsymbol{\Sigma}_a\boldsymbol{\Psi}_p\boldsymbol{G}^{-1}\boldsymbol{\Psi}_p'\boldsymbol{\Sigma}_a\boldsymbol{D}^{-1}\right] \otimes \boldsymbol{R}_0,$$

where $\boldsymbol{\Psi}_p = [\mathbf{0}, \boldsymbol{\psi}'_{j-1}, \ldots, \boldsymbol{\psi}'_{j-p}]$ with $\boldsymbol{\psi}_\ell$ being the coefficient matrices of the MA representation of $\boldsymbol{z}_t$ so that $\boldsymbol{\psi}_\ell = \mathbf{0}$ for $\ell < 0$ and $\mathbf{0}$ is a k-dimensional vector of zero.

2.7.2 Multivariate Portmanteau Statistics

Let $\boldsymbol{R}_\ell$ be the theoretical lag ℓ cross-correlation matrix of innovation $\boldsymbol{a}_t$. The hypothesis of interest in model checking is

$$H_0 : \boldsymbol{R}_1 = \cdots = \boldsymbol{R}_m = \mathbf{0} \quad \text{versus} \quad H_a : \boldsymbol{R}_j \neq \mathbf{0} \quad \text{for some } 1 \leq j \leq m, \tag{2.59}$$

where m is a prespecified positive integer. The Portmanteau statistic of Equation (1.11) is often used to perform the test. For residual series, the statistic becomes

$$\begin{aligned}
Q_k(m) &= T^2 \sum_{\ell=1}^{m} \frac{1}{T-\ell} tr\left(\hat{\boldsymbol{R}}'_\ell \hat{\boldsymbol{R}}_0^{-1} \hat{\boldsymbol{R}}_\ell \hat{\boldsymbol{R}}_0^{-1}\right) \\
&= T^2 \sum_{\ell=1}^{m} \frac{1}{T-\ell} tr\left(\hat{\boldsymbol{R}}'_\ell \hat{\boldsymbol{R}}_0^{-1} \hat{\boldsymbol{R}}_\ell \hat{\boldsymbol{R}}_0^{-1} \hat{\boldsymbol{D}}^{-1} \hat{\boldsymbol{D}}\right) \\
&= T^2 \sum_{\ell=1}^{m} \frac{1}{T-\ell} tr\left(\hat{\boldsymbol{D}} \hat{\boldsymbol{R}}'_\ell \hat{\boldsymbol{D}} \hat{\boldsymbol{D}}^{-1} \hat{\boldsymbol{R}}_0^{-1} \hat{\boldsymbol{D}}^{-1} \hat{\boldsymbol{D}} \hat{\boldsymbol{R}}_\ell \hat{\boldsymbol{D}} \hat{\boldsymbol{D}}^{-1} \hat{\boldsymbol{R}}_0^{-1} \hat{\boldsymbol{D}}^{-1}\right) \\
&= T^2 \sum_{\ell=1}^{m} \frac{1}{T-\ell} tr\left(\hat{\boldsymbol{C}}'_\ell \hat{\boldsymbol{C}}_0^{-1} \hat{\boldsymbol{C}}_\ell \hat{\boldsymbol{C}}_0^{-1}\right).
\end{aligned} \tag{2.60}$$

Theorem 2.6 Suppose that $\boldsymbol{z}_t$ follows a stationary VAR(p) model of Equation (2.21) with $\boldsymbol{a}_t$ being a white noise process with mean zero and positive covariance matrix $\boldsymbol{\Sigma}_a$. Also, assume that the assumption in Equation (2.42) holds and the parameter matrix $\boldsymbol{\beta}$ of the model in Equation (2.28) is consistently estimated by a method discussed in Section 2.5 and the residual cross-covariance matrix is defined in Equation (2.55). Then, the test statistic $Q_k(m)$ is asymptotically distributed as a chi-square distribution with $(m-p)k^2$ degrees of freedom.

The proof of Theorem 2.6 is relatively complicated. Readers may consult Li and McLeod (1981), Hosking (1981), and Lütkepohl (2005) for further information. Compared with the Portmanteau test of Chapter 1, the degrees of freedom of the chi-square distribution in Theorem 2.6 is adjusted by pk^2, which is the number of AR parameters in a VAR(p) model. In practice, some of the AR parameters in a VAR(p) model are fixed to 0. In this case, the adjustment in the degrees of freedom of the chi-square distribution is set to the number of estimated AR parameters.

In the literature, some Lagrange multiplier tests have also been suggested for checking a fitted VAR model. However, the asymptotic chi-square distribution of the Lagrange multiplier test is found to be a poor approximation. See, for instance,

Edgerton and Shukur (1999). For this reason, we shall not discuss the Lagrange multiplier tests.

Example 2.6 Consider again the quarterly GDP growth rates, in percentages, of United Kingdom, Canada, and United States employed in Example 2.3. We apply the multivariate Portmanteau test statistics to the residuals of the fitted VAR(2) model. The results are given in the following R demonstration. Since we used $p=2$, the $Q_2(m)$ statistics requires $m > 2$ to have positive degrees of freedom for the asymptotic chi-square distribution. For this reason, the p-values are set to 1 for $m=1$ and 2. From the demonstration, it is seen that the fitted VAR(2) model has largely removed the dynamic dependence in the GDP growth rates, except for some minor violation at $m=4$. We shall return to this point later as there exist some insignificant parameter estimates in the fitted VAR(2) model. □

R Demonstration: Multivariate Portmanteau statistics.

```
> names(m1)
 [1] "data"       "cnst"      "order"     "coef"      "aic"      "bic"
 [7] "residuals" "secoef"  "Sigma"     "Phi"       "Ph0"
> resi=m1$residuals  ### Obtain the residuals of VAR(2) fit.
> mq(resi,adj=18) ## adj is used to adjust the degrees of
freedom. Ljung-Box Statistics:
          m       Q(m)   p-value
 [1,]   1.000    0.816      1.00
 [2,]   2.000    3.978      1.00
 [3,]   3.000   16.665      0.05
 [4,]   4.000   35.122      0.01
 [5,]   5.000   38.189      0.07
 [6,]   6.000   41.239      0.25
 [7,]   7.000   47.621      0.37
 [8,]   8.000   61.677      0.22
 [9,]   9.000   67.366      0.33
[10,]  10.000   76.930      0.32
[11,]  11.000   81.567      0.46
[12,]  12.000   93.112      0.39
```

2.7.3 Model Simplification

Multivariate time series models may contain many parameters if the dimension k is moderate or large. In practice, we often observe that some of the parameters are not statistically significant at a given significant level. It is then advantageous to simplify the model by removing the insignificant parameters. This is particularly so when no prior knowledge is available to support those parameters. However, there exists no optimal method to simplify a fitted model. We discuss some methods commonly used in practice.

2.7.3.1 *Testing Zero Parameters*

An obvious approach to simplify a fitted VAR(p) model is to remove insignificant parameters. Given a specified significant level, for example, $\alpha = 0.05$, we can identify the *target parameters* for removal. By target parameters we meant those parameters whose individual t-ratio is less than the critical value of the normal distribution with type I error α. They are the target for removal because parameter estimates are correlated and marginal statistics could be misleading. To confirm that those parameters can indeed be removed, we consider a test procedure using the limiting distribution of Theorem 2.3.

Let $\hat{\boldsymbol{\omega}}$ be a v-dimensional vector consisting of the target parameters. In other words, v is the number of parameters to be fixed to 0. Let $\boldsymbol{\omega}$ be the counterpart of $\hat{\boldsymbol{\omega}}$ in the parameter matrix $\boldsymbol{\beta}$ in Equation (2.28). The hypothesis of interest is

$$H_0 : \boldsymbol{\omega} = \mathbf{0} \quad \text{versus} \quad H_a : \boldsymbol{\omega} \neq \mathbf{0}.$$

Clearly, there exists a $v \times k(kp+1)$ locating matrix $\boldsymbol{K}$ such that

$$\boldsymbol{K}\text{vec}(\boldsymbol{\beta}) = \boldsymbol{\omega}, \quad \text{and} \quad \boldsymbol{K}\text{vec}(\hat{\boldsymbol{\beta}}) = \hat{\boldsymbol{\omega}}. \tag{2.61}$$

By Theorem 2.3 and properties of multivariate normal distribution, we have

$$\sqrt{T_p}\,(\hat{\boldsymbol{\omega}} - \boldsymbol{\omega}) \rightarrow_d N\left[\mathbf{0}, \boldsymbol{K}\left(\boldsymbol{\Sigma}_a \otimes \boldsymbol{G}^{-1}\right)\boldsymbol{K}'\right], \tag{2.62}$$

where $T_p = T - p$ is the effective sample size. Consequently, under H_0, we have

$$T_p \hat{\boldsymbol{\omega}}'\left[\boldsymbol{K}\left(\boldsymbol{\Sigma}_a \otimes \boldsymbol{G}^{-1}\right)\boldsymbol{K}'\right]^{-1}\hat{\boldsymbol{\omega}} \rightarrow_d \chi^2_v, \tag{2.63}$$

where $v = \dim(\boldsymbol{\omega})$. This chi-square test can also be interpreted as a likelihood-ratio test under the normality assumption of $\boldsymbol{a}_t$. The null hypothesis H_0: $\boldsymbol{\omega} = \mathbf{0}$ denotes a reduced VAR(p) model. Therefore, one can use likelihood-ratio statistic, which is asymptotically equivalent to the chi-square test of Equation (2.63).

If the hypothesis of interest is

$$H_0 : \boldsymbol{\omega} = \boldsymbol{\omega}_o \quad \text{versus} \quad H_a : \boldsymbol{\omega} \neq \boldsymbol{\omega}_o,$$

where $\boldsymbol{\omega}_o$ is a prespecified v-dimensional vector, and we replace $\boldsymbol{\Sigma}_a$ and $\boldsymbol{G}$ by their estimates, then the test statistic in Equation (2.63) becomes

$$\begin{aligned} \lambda_W &= T_p(\hat{\boldsymbol{\omega}} - \boldsymbol{\omega}_o)'\left[\boldsymbol{K}\left(\hat{\boldsymbol{\Sigma}}_a \otimes \hat{\boldsymbol{G}}^{-1}\right)\boldsymbol{K}'\right]^{-1}(\hat{\boldsymbol{\omega}} - \boldsymbol{\omega}_o) \\ &= (\hat{\boldsymbol{\omega}} - \boldsymbol{\omega}_o)'\left[\boldsymbol{K}\left\{\hat{\boldsymbol{\Sigma}}_a \otimes (\boldsymbol{X}'\boldsymbol{X})^{-1}\right\}\boldsymbol{K}'\right]^{-1}(\hat{\boldsymbol{\omega}} - \boldsymbol{\omega}_o). \end{aligned} \tag{2.64}$$

This is a *Wald statistic* and is asymptotically distributed as χ^2_v, provided that the assumptions of Theorem 2.3 hold so that $\hat{\boldsymbol{\Sigma}}_a$ and $\hat{\boldsymbol{G}} = \boldsymbol{X}'\boldsymbol{X}/T_p$ are consistent. This Wald test can be used to make inference such as testing for Granger causality.

Example 2.7 Consider again the quarterly GDP growth rates, in percentages, of United Kingdom, Canada, and United States employed in Example 2.3. Based on the R demonstration of Example 2.3, there are insignificant parameters in a fitted VAR(2) model. To simplify the model, we apply the chi-square test of Equation (2.63). We used type I error $\alpha = 0.05$ and $\alpha = 0.1$, respectively, to identify target parameters. The corresponding critical values are 1.96 and 1.645, respectively. It turns out that there are 10 and 8 possible zero parameters for these two choices of α. The chi-square test for all 10 targeted parameters being 0 is 31.69 with p-value 0.0005. Thus, we cannot simultaneously set all 10 parameters with the smallest absolute t-ratios to 0. On the other hand, the chi-square test is 15.16 with p-value 0.056 when we test that all eight parameters with the smallest t-ratios, in absolute value, are 0. Consequently, we can simplify the fitted VAR(2) model by letting the eight parameters with smallest t-ratio (in absolute) to 0. □

Remark: To perform the chi-square test for zero parameters, we use the command `VARchi` in the `MTS` package. The subcommand `thres` of `VARchi` can be used to set type I error for selecting the target parameters. The default is `thres = 1.645`. □

R Demonstration: Testing zero parameters.

```
> m3=VARchi(z,p=2)
Number of targeted parameters:  8
Chi-square test and p-value:  15.16379 0.05603778
> m3=VARchi(z,p=2,thres=1.96)
Number of targeted parameters:  10
Chi-square test and p-value:  31.68739 0.000451394
```

2.7.3.2 Information Criteria

An alternative approach to the chi-square test is to use the information criteria discussed in Section 2.6. For instance, we can estimate the unconstrained VAR(p) model (under H_a) and the constrained VAR(p) model (under H_0). If the constrained model has a smaller value for a selected criterion, then H_0 cannot be rejected according to that criterion.

2.7.3.3 Stepwise Regression

Finally, we can make use of the fact that for VAR(p) models the estimation can be performed equation-by-equation; see the discussion in Section 2.5. As such, each equation is a multiple linear regression, and we can apply the traditional idea of stepwise regression to remove insignificant parameters. Readers may consult variable selection in multiple linear regression for further information about the stepwise regression.

Remark: We use the command `refVAR` of the `MTS` package to carry out model simplification of a fitted VAR model. The command uses a threshold to select target parameters for removal and computes information criteria of the simplified model for validation. The default threshold is 1.00. The command also allows the user to specify zero parameters with a subcommand `fixed`. □

To demonstrate, we consider again the quarterly growth rates, in percentages, of United Kingdom, Canada, and United States employed in Example 2.7. The simplified VAR(2) model has 12 parameters, instead of 21, for the unconstrained VAR(2) model. The AIC of the simplified model is -3.53, which is smaller than -3.50 of the unconstrained model. For this particular instance, all three criteria have a smaller value for the constrained model. It is possible in practice that different criteria lead to different conclusions.

R Demonstration: Model simplification.

```
> m1=VAR(zt,2)  # fit a un-constrained VAR(2) model.
> m2=refVAR(m1,thres=1.96)  # Model refinement.
Constant term:
Estimates:  0.1628247 0 0.2827525
Std.Error:  0.06814101 0 0.07972864
AR coefficient matrix
AR( 1 )-matrix
      [,1]  [,2]  [,3]
[1,] 0.467 0.207 0.000
[2,] 0.334 0.270 0.496
[3,] 0.468 0.225 0.232
standard error
       [,1]   [,2]   [,3]
[1,] 0.0790 0.0686 0.0000
[2,] 0.0921 0.0875 0.0913
[3,] 0.1027 0.0963 0.1023
AR( 2 )-matrix
       [,1] [,2] [,3]
[1,]  0.000    0    0
[2,] -0.197    0    0
[3,] -0.301    0    0
standard error
       [,1] [,2] [,3]
[1,] 0.0000    0    0
[2,] 0.0921    0    0
[3,] 0.1008    0    0

Residuals cov-mtx:
           [,1]       [,2]       [,3]
[1,] 0.29003669 0.01803456 0.07055856
[2,] 0.01803456 0.30802503 0.14598345
[3,] 0.07055856 0.14598345 0.36268779
```

```
det(SSE) =  0.02494104
AIC =  -3.531241
BIC =  -3.304976
HQ  =  -3.439321
```

From the output, the simplified VAR(2) model for the percentage growth rates of quarterly GDP of United Kingdom, Canada, and United States is

$$\boldsymbol{z}_t = \begin{bmatrix} 0.16 \\ - \\ 0.28 \end{bmatrix} + \begin{bmatrix} 0.47 & 0.21 & - \\ 0.33 & 0.27 & 0.50 \\ 0.47 & 0.23 & 0.23 \end{bmatrix} \boldsymbol{z}_{t-1} + \begin{bmatrix} - & - & - \\ -0.20 & - & - \\ -0.30 & - & - \end{bmatrix} \boldsymbol{z}_{t-2} + \boldsymbol{a}_t, \tag{2.65}$$

where the residual covariance matrix is

$$\hat{\boldsymbol{\Sigma}}_a = \begin{bmatrix} 0.29 & 0.02 & 0.07 \\ 0.02 & 0.31 & 0.15 \\ 0.07 & 0.15 & 0.36 \end{bmatrix}.$$

All estimates are now significant at the usual 5% level. Limiting properties of the constrained parameter estimates are discussed in the next section.

Finally, we return to model checking for the simplified VAR(2) model in Equation (2.65). Since all estimates are statistically significant at the usual 5% level, we perform a careful residual analysis. Figure 2.3 shows the time plots of the three residual series, whereas Figure 2.4 shows the residual cross-correlation matrices. The dashed lines of the plots in Figure 2.4 indicate the approximate 2 standard-error limits of the cross-correlations, that is $\pm 2/\sqrt{T}$. Strictly speaking, these limits are only valid for higher-order lags; see Theorem 2.5. However, it is relatively complicated to compute the asymptotic standard errors of the cross-correlations so that the simple approximation is often used in model checking. Based on the plots, the residuals of the model in Equation (2.65) do not have any strong serial or cross-correlations. The only possible exception is the $(1,3)$th position of lag-4 cross-correlation matrix.

The multivariate Portmanteau test can be used to check the residual cross-correlations. The results are given in the following R demonstration. Figure 2.5 plots the p-values of the $Q_3(m)$ statistics applied to the residuals of the simplified VAR(2) model in Equation (2.65). Since there are 12 parameters, the degrees of freedom of the chi-square distribution for $Q_3(m)$ is $9m - 12$. Therefore, the asymptotic chi-square distribution works if $m \geq 2$. From the plot and the R demonstration, the $Q_k(m)$ statistics indicate that there are no strong serial or cross-correlations in the residuals of the simplified VAR(2) model. Based on $Q_k(4)$, one may consider some further improvement, such as using a VAR(4) model. We leave it as an exercise.

Remark: Model checking of a fitted VARMA model can be carried out by the command `MTSdiag` of the `MTS` package. The default option checks 24 lags of cross-correlation matrix. □

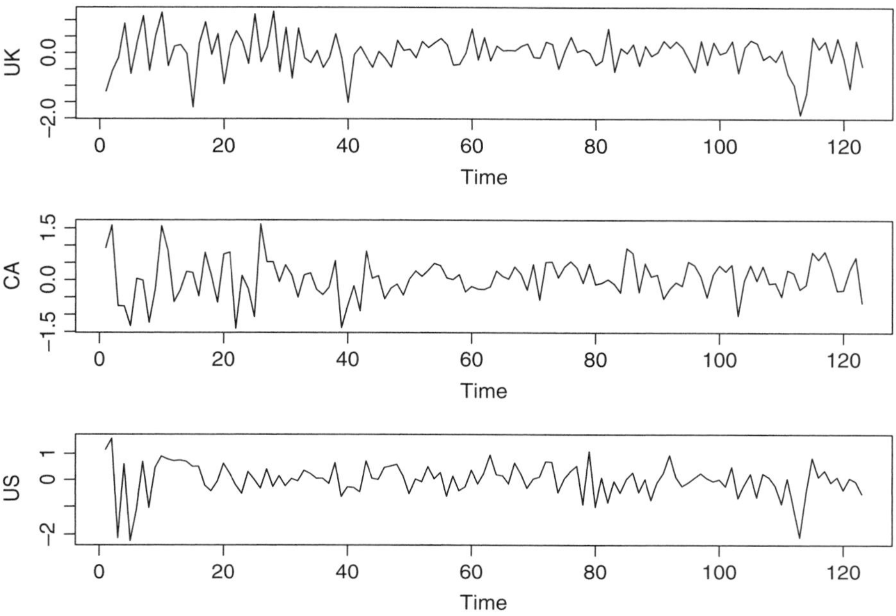

FIGURE 2.3 Residual plots of the simplified VAR(2) model in Equation (2.65) for the quarterly growth rates of real gross domestic products of United Kingdom, Canada, and United States from the second quarter of 1980 to the second quarter of 2011. The growth rates are in percentages.

R Demonstration: Model checking.

```
> MTSdiag(m2,adj=12)
[1] "Covariance matrix:"
       [,1]   [,2]   [,3]
[1,] 0.2924 0.0182 0.0711
[2,] 0.0182 0.3084 0.1472
[3,] 0.0711 0.1472 0.3657
CCM at lag:  0
       [,1]   [,2]  [,3]
[1,] 1.0000 0.0605 0.218
[2,] 0.0605 1.0000 0.438
[3,] 0.2175 0.4382 1.000
Simplified matrix:
CCM at lag:  1
. . .
. . .
. . .
CCM at lag:  2
. . .
. . .
. . .
```

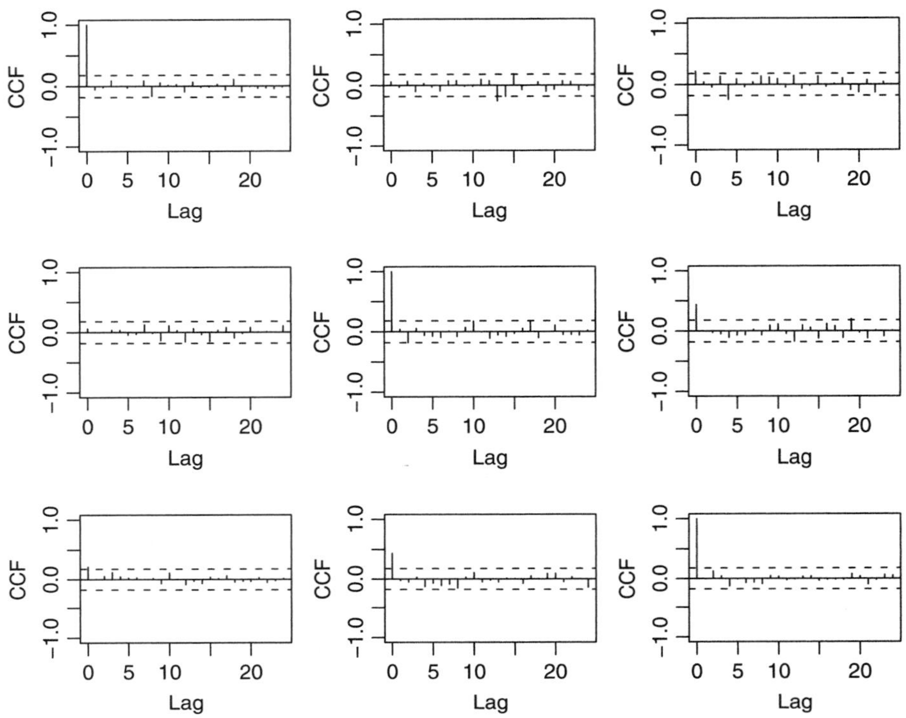

FIGURE 2.4 Residual cross-correlation matrices of the simplified VAR(2) model in Equation (2.65) for the quarterly growth rates of real gross domestic products of United Kingdom, Canada, and United States from the second quarter of 1980 to the second quarter of 2011. The growth rates are in percentages.

```
CCM at lag:  3
. . .
. . .
. . .
CCM at lag:  4
. . -
. . .
. . .
Hit Enter to compute MQ-statistics:

Ljung-Box Statistics:
          m        Q(m)   p-value
 [1,]   1.00        1.78     1.00
 [2,]   2.00       12.41     0.05
 [3,]   3.00       22.60     0.09
 [4,]   4.00       37.71     0.04
 [5,]   5.00       41.65     0.14
 [6,]   6.00       44.95     0.35
 [7,]   7.00       51.50     0.45
```

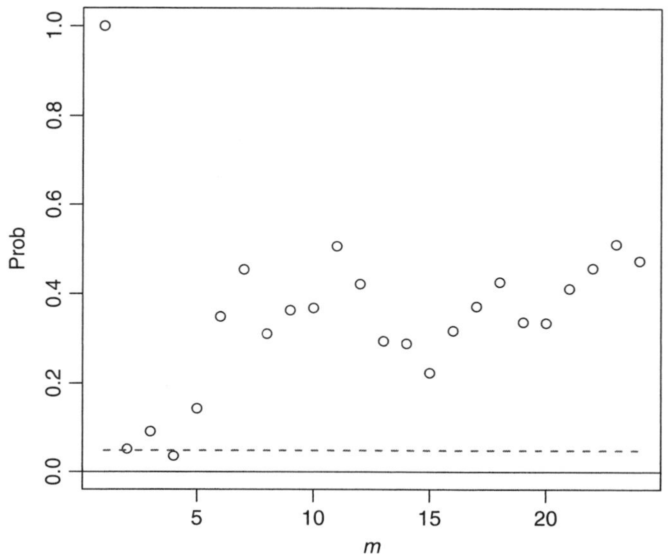

FIGURE 2.5 Plot of p-values of the $Q_k(m)$ statistics applied to the residuals of the simplified VAR(2) model in Equation (2.65) for the quarterly growth rates of real gross domestic products of United Kingdom, Canada, and United States from the second quarter of 1980 to the second quarter of 2011. The growth rates are in percentages.

```
 [8,]    8.00      64.87     0.31
 [9,]    9.00      72.50     0.36
[10,]   10.00      81.58     0.37
[11,]   11.00      86.12     0.51
[12,]   12.00      98.08     0.42
```

In conclusion, the simplified VAR(2) model in Equation (2.65) is adequate for the GDP growth rate series. The model can be written as

$$\begin{aligned}
\text{UK}: \quad & z_{1t} = 0.16 + 0.47z_{1,t-1} + 0.21z_{2,t-1} + a_{1t}, \\
\text{CA}: \quad & z_{2t} = 0.33z_{1,t-1} + 0.27z_{2,t-1} + 0.5z_{3,t-1} - 0.2z_{1,t-2} + a_{2t}, \\
\text{US}: \quad & z_{3t} = 0.28 + 0.47z_{1,t-1} + 0.23z_{2,t-1} + 0.23z_{3,t-1} - 0.3z_{1,t-2} + a_{3t}.
\end{aligned}$$

The correlation matrix of the residuals is

$$\boldsymbol{R}_0 = \begin{bmatrix} 1.00 & 0.06 & 0.22 \\ 0.06 & 1.00 & 0.44 \\ 0.22 & 0.44 & 1.00 \end{bmatrix}.$$

This correlation matrix indicates that the quarterly GDP growth rates of United Kingdom and Canada are not instantaneously correlated. The fitted three-dimensional model shows that the GDP growth rate of United Kingdom does not depend on the lagged growth rates of the United States in the presence of lagged Canadian GDP growth rates, but the United Kingdom growth rate depends on the past growth rate of Canada. On the other hand, the GDP growth rate of Canada is dynamically related to the growth rates of United Kingdom and United States. Similarly, the GDP growth rate of the United States depends on the lagged growth rates of United Kingdom and Canada. If one further considers the dependence of z_{1t} on $z_{3,t-4}$, then all three GDP growth rates are directly dynamically correlated. In summary, the simplified VAR(2) model indicates that the GDP growth rate of United Kingdom is conditionally independent of the growth rate of the United States given the Canadian growth rate.

2.8 LINEAR CONSTRAINTS

Linear parameter constraints can be handled easily in estimation of a VAR(p) model. Consider the matrix representation in Equation (2.28). Any linear parameter constraint can be expressed as

$$\text{vec}(\boldsymbol{\beta}) = \boldsymbol{J\gamma} + \boldsymbol{r}, \tag{2.66}$$

where $\boldsymbol{J}$ is $k(kp+1) \times P$ constant matrix of rank P, $\boldsymbol{r}$ is a $k(kp+1)$-dimensional constant vector, and $\boldsymbol{\gamma}$ denotes a P-dimensional vector of unknown parameters. Here $\boldsymbol{J}$ and $\boldsymbol{r}$ are known. For example, consider the two-dimensional VAR(1) model, $\boldsymbol{z}_t = \boldsymbol{\phi}_0 + \boldsymbol{\phi}_1 \boldsymbol{z}_{t-1} + \boldsymbol{a}_t$. Here we have

$$\boldsymbol{\beta} = \begin{bmatrix} \phi_{0,1} & \phi_{0,2} \\ \phi_{1,11} & \phi_{1,21} \\ \phi_{1,12} & \phi_{1,22} \end{bmatrix}, \quad \text{where} \quad \boldsymbol{\phi}_0 = \begin{bmatrix} \phi_{0,1} \\ \phi_{0,2} \end{bmatrix}, \quad \boldsymbol{\phi}_1 = [\phi_{1,ij}].$$

Suppose the actual model is

$$\boldsymbol{z}_t = \begin{bmatrix} 1 \\ 0 \end{bmatrix} + \begin{bmatrix} 0.6 & 0 \\ 0.2 & 0.8 \end{bmatrix} \boldsymbol{z}_{t-1} + \boldsymbol{a}_t,$$

which has four parameters. In this particular case, we have $\boldsymbol{r} = \boldsymbol{0}_6$ and

$$\text{vec}(\boldsymbol{\beta}) = \begin{bmatrix} \phi_{0,1} \\ \phi_{1,11} \\ \phi_{1,12} \\ \phi_{0,2} \\ \phi_{1,21} \\ \phi_{1,22} \end{bmatrix} = \begin{bmatrix} 1 & 0 & 0 & 0 \\ 0 & 1 & 0 & 0 \\ 0 & 0 & 0 & 0 \\ 0 & 0 & 0 & 0 \\ 0 & 0 & 1 & 0 \\ 0 & 0 & 0 & 1 \end{bmatrix} \begin{bmatrix} 1.0 \\ 0.6 \\ 0.2 \\ 0.8 \end{bmatrix} \equiv \boldsymbol{J\gamma}.$$

Under the linear constraints in Equation (2.66), Equation (2.29) becomes

$$\begin{aligned}\text{vec}(\boldsymbol{Z}) &= (\boldsymbol{I}_k \otimes \boldsymbol{X})(\boldsymbol{J}\boldsymbol{\gamma} + \boldsymbol{r}) + \text{vec}(\boldsymbol{A}) \\ &= (\boldsymbol{I}_k \otimes \boldsymbol{X})\boldsymbol{J}\boldsymbol{\gamma} + (\boldsymbol{I}_k \otimes \boldsymbol{X})\boldsymbol{r} + \text{vec}(\boldsymbol{A}).\end{aligned}$$

Since $\boldsymbol{r}$ is known, $(\boldsymbol{I}_k \otimes \boldsymbol{X})\boldsymbol{r}$ is a known $k(T-p) \times 1$ vector. Therefore, we can rewrite the prior equation as

$$\text{vec}(\boldsymbol{Z}_*) = (\boldsymbol{I}_k \otimes \boldsymbol{X})\boldsymbol{J}\boldsymbol{\gamma} + \text{vec}(\boldsymbol{A}),$$

where $\text{vec}(\boldsymbol{Z}_*) = \text{vec}(\boldsymbol{Z}) - (\boldsymbol{I}_k \otimes \boldsymbol{X})\boldsymbol{r}$. Following the same argument as the VAR(p) estimation in Equation (2.30), the GLS estimate of $\boldsymbol{\gamma}$ is obtained by minimizing

$$\begin{aligned}S(\boldsymbol{\gamma}) &= [\text{vec}(\boldsymbol{A})]'(\boldsymbol{\Sigma}_a \otimes \boldsymbol{I}_{T-p})^{-1}\text{vec}(\boldsymbol{A}) \\ &= [\text{vec}(\boldsymbol{Z}_*) - (\boldsymbol{I}_k \otimes \boldsymbol{X})\boldsymbol{J}\boldsymbol{\gamma}]'(\boldsymbol{\Sigma}_a^{-1} \otimes \boldsymbol{I}_{T-p})[\text{vec}(\boldsymbol{Z}_*) - (\boldsymbol{I}_k \otimes \boldsymbol{X})\boldsymbol{J}\boldsymbol{\gamma}].\end{aligned} \tag{2.67}$$

Using the same method as that in Equation (2.32), we have

$$\begin{aligned}\hat{\boldsymbol{\gamma}} &= \left[\boldsymbol{J}'\left(\boldsymbol{\Sigma}_a^{-1} \otimes \boldsymbol{X}'\boldsymbol{X}\right)\boldsymbol{J}\right]^{-1}\left[\boldsymbol{J}'(\boldsymbol{\Sigma}_a^{-1} \otimes \boldsymbol{X}')\text{vec}(\boldsymbol{Z}_*)\right) \\ &= \left[\boldsymbol{J}'\left(\boldsymbol{\Sigma}_a^{-1} \otimes \boldsymbol{X}'\boldsymbol{X}\right)\boldsymbol{J}\right]^{-1}\boldsymbol{J}'\left(\boldsymbol{\Sigma}_a^{-1} \otimes \boldsymbol{X}'\right)[(\boldsymbol{I}_k \otimes \boldsymbol{X})\boldsymbol{J}\boldsymbol{\gamma} + \text{vec}(\boldsymbol{A})] \\ &= \boldsymbol{\gamma} + \left[\boldsymbol{J}'(\boldsymbol{\Sigma}_a^{-1} \otimes \boldsymbol{X}'\boldsymbol{X})\boldsymbol{J}\right]^{-1}\boldsymbol{J}'\left(\boldsymbol{\Sigma}_a^{-1} \otimes \boldsymbol{X}'\right)\text{vec}(\boldsymbol{A}) \\ &= \boldsymbol{\gamma} + \left[\boldsymbol{J}'(\boldsymbol{\Sigma}_a^{-1} \otimes \boldsymbol{X}'\boldsymbol{X})\boldsymbol{J}\right]^{-1}\boldsymbol{J}'\text{vec}\left(\boldsymbol{X}'\boldsymbol{A}\boldsymbol{\Sigma}_a^{-1}\right) \\ &= \boldsymbol{\gamma} + \left[\boldsymbol{J}'(\boldsymbol{\Sigma}_a^{-1} \otimes \boldsymbol{X}'\boldsymbol{X})\boldsymbol{J}\right]^{-1}\boldsymbol{J}'\left(\boldsymbol{\Sigma}_a^{-1} \otimes \boldsymbol{I}_{kp+1}\right)\text{vec}\left(\boldsymbol{X}'\boldsymbol{A}\right).\end{aligned} \tag{2.68}$$

From Equation (2.68), we have

$$\sqrt{T_p}(\hat{\boldsymbol{\gamma}} - \boldsymbol{\gamma}) = \left[\boldsymbol{J}'\left(\boldsymbol{\Sigma}_a^{-1} \otimes \frac{\boldsymbol{X}'\boldsymbol{X}}{T_p}\right)\boldsymbol{J}\right]^{-1}\boldsymbol{J}'(\boldsymbol{\Sigma}_a^{-1} \otimes \boldsymbol{I}_{kp+1})\frac{1}{\sqrt{T_p}}\text{vec}(\boldsymbol{X}'\boldsymbol{A}), \tag{2.69}$$

where, as before, $T_p = T - p$.

Theorem 2.7 Assume that the stationary VAR(p) process $\boldsymbol{z}_t$ satisfies the conditions of Theorem 2.3. Assume further that the parameters of the model satisfy the linear constraints given in Equation (2.66), where $\boldsymbol{J}$ is of rank P, which is the number of coefficient parameters to be estimated. Then, the GLS estimate $\hat{\boldsymbol{\gamma}}$ of Equation (2.68) is a consistent estimate of $\boldsymbol{\gamma}$ and

$$\sqrt{T_p}\,(\hat{\boldsymbol{\gamma}} - \boldsymbol{\gamma}) \to_d N\left(\boldsymbol{0}, \left[\boldsymbol{J}'(\boldsymbol{\Sigma}_a^{-1} \otimes \boldsymbol{G})\boldsymbol{J}\right]^{-1}\right),$$

where $T_p = T - p$ and $\boldsymbol{G}$ is the limit of $\boldsymbol{X}'\boldsymbol{X}/T_p$ as $T \to \infty$ and is defined in Lemma 2.3.

Proof. By part (ii) of Lemma 2.3,

$$\frac{1}{\sqrt{T_p}}\text{vec}(\boldsymbol{X}'\boldsymbol{A}) \to_d N(\boldsymbol{0}, \boldsymbol{\Sigma}_a^{-1} \otimes \boldsymbol{G}).$$

The theorem then follows from Equation (2.69). □

2.9 FORECASTING

Let h be the forecast origin, $\ell > 0$ be the forecast horizon, and F_h be the information available at time h (inclusive). We discuss forecasts of a VAR(p) model in this section.

2.9.1 Forecasts of a Given Model

To begin, assume that the VAR model is known, that is, we ignore for a moment that the parameters are estimated. Following the discussion in Section 1.6, the minimum mean-squared error forecast of $\boldsymbol{z}_{h+\ell}$ is simply the conditional expectation of $\boldsymbol{z}_{h+\ell}$ given F_h. For the VAR(p) model in Equation (2.21), the one-step ahead prediction is trivial,

$$\boldsymbol{z}_h(1) = E(\boldsymbol{z}_{h+1}|F_h) = \boldsymbol{\phi}_0 + \sum_{i=1}^{p} \boldsymbol{\phi}_i \boldsymbol{z}_{h+1-i}.$$

For two-step ahead prediction, we have

$$\begin{aligned}\boldsymbol{z}_h(2) &= E(\boldsymbol{z}_{h+2}|F_h)\\ &= \boldsymbol{\phi}_0 + \boldsymbol{\phi}_1 E(\boldsymbol{z}_{h+1}|F_h) + \sum_{i=2}^{p} \boldsymbol{\phi}_i \boldsymbol{z}_{h+2-i}\\ &= \boldsymbol{\phi}_0 + \boldsymbol{\phi}_1 \boldsymbol{z}_h(1) + \sum_{i=2}^{p} \boldsymbol{\phi}_i \boldsymbol{z}_{h+2-i}.\end{aligned}$$

In general, for the ℓ-step ahead forecast, we have

$$\boldsymbol{z}_h(\ell) = E(\boldsymbol{z}_{h+\ell}|F_h) = \boldsymbol{\phi}_0 + \sum_{i=1}^{p} \boldsymbol{\phi}_i \boldsymbol{z}_h(\ell - i), \tag{2.70}$$

where it is understood that $\boldsymbol{z}_h(j) = \boldsymbol{z}_{h+j}$ for $j \leq 0$. Thus, the point forecasts of a VAR(p) model can be computed recursively.

Using $\boldsymbol{\phi}_0 = (\boldsymbol{I}_k - \sum_{i=1}^{p} \boldsymbol{\phi}_1)\boldsymbol{\mu}$, Equation (2.70) can be rewritten as

$$\boldsymbol{z}_h(\ell) - \boldsymbol{\mu} = \sum_{i=1}^{p} \boldsymbol{\phi}_i[\boldsymbol{z}_h(\ell - i) - \boldsymbol{\mu}].$$

The prior equation implies that

$$[\boldsymbol{z}_h(\ell) - \boldsymbol{\mu}] - \sum_{i=1}^{p} \boldsymbol{\phi}_i[\boldsymbol{z}_h(\ell - i) - \boldsymbol{\mu}] = \boldsymbol{0},$$

which can be rewritten as

$$\left(\boldsymbol{I}_k - \sum_{i=1}^{p} \boldsymbol{\phi}_i B^i\right)[\boldsymbol{z}_h(\ell) - \boldsymbol{\mu}] = \boldsymbol{0}, \quad \text{or simply} \quad \boldsymbol{\phi}(B)[\boldsymbol{z}_h(\ell) - \boldsymbol{\mu}] = \boldsymbol{0}, \tag{2.71}$$

where it is understood that the back-shift operator B operates on ℓ as the forecast origin h is fixed. Define the expanded forecast vector $\boldsymbol{Z}_h(\ell)$ as

$$\boldsymbol{Z}_h(\ell) = ([\boldsymbol{z}_h(\ell) - \boldsymbol{\mu}]', [\boldsymbol{z}_h(\ell - 1) - \boldsymbol{\mu}]', \ldots, [\boldsymbol{z}_h(\ell - p + 1) - \boldsymbol{\mu}]')'.$$

Then, Equation (2.71) implies that

$$\boldsymbol{Z}_h(\ell) = \boldsymbol{\Phi} \boldsymbol{Z}_h(\ell - 1), \quad \ell > 1, \tag{2.72}$$

where $\boldsymbol{\Phi}$, defined in Equation (2.23), is the companion matrix of the polynomial matrix $\boldsymbol{\phi}(B) = \boldsymbol{I}_k - \sum_{i=1}^{p} \boldsymbol{\phi}_i B^i$. By repeated application of Equation (2.72), we have

$$\boldsymbol{Z}_h(\ell) = \boldsymbol{\Phi}^{\ell-1} \boldsymbol{Z}_h(1), \quad \ell > 1. \tag{2.73}$$

For a stationary VAR(p) model, all eigenvalues of $\boldsymbol{\Phi}$ are less than 1 in absolute value. Therefore, $\boldsymbol{\Phi}^j \to \boldsymbol{0}$ as $j \to \infty$. Consequently, we have

$$\boldsymbol{z}_h(\ell) - \boldsymbol{\mu} \to \boldsymbol{0}, \quad \text{as} \quad \ell \to \infty.$$

This says that the stationary VAR(p) process is mean-reverting because its point forecasts converge to the mean of the process as the forecast horizon increases. The speed of mean-reverting is determined by the magnitude of the largest eigenvalue, in modulus, of $\boldsymbol{\Phi}$.

Turn to forecast errors. For ℓ-step ahead forecast, the forecast error is

$$\boldsymbol{e}_h(\ell) = \boldsymbol{z}_{h+\ell} - \boldsymbol{z}_h(\ell).$$

To study this forecast error, it is most convenient to use the MA representation of the VAR(p) model,

$$\boldsymbol{z}_t = \boldsymbol{\mu} + \sum_{i=0}^{\infty} \boldsymbol{\psi}_i \boldsymbol{a}_{t-i},$$

where $\boldsymbol{\mu} = [\boldsymbol{\phi}(1)]^{-1}\boldsymbol{\phi}_0$, $\boldsymbol{\psi}_0 = \boldsymbol{I}_k$, and $\boldsymbol{\psi}_i$ can be obtained recursively via Equation (2.27). As shown in Equation (1.14), the ℓ-step ahead forecast error is

$$\boldsymbol{e}_h(\ell) = \boldsymbol{a}_{h+\ell} + \boldsymbol{\psi}_1 \boldsymbol{a}_{h+\ell-1} + \cdots + \boldsymbol{\psi}_{\ell-1} \boldsymbol{a}_{h+1}.$$

Consequently, the covariance matrix of the forecast error is

$$\text{Cov}[\boldsymbol{e}_h(\ell)] = \boldsymbol{\Sigma}_a + \sum_{i=1}^{\ell-1} \boldsymbol{\psi}_i \boldsymbol{\Sigma}_a \boldsymbol{\psi}_i'. \tag{2.74}$$

As the forecast horizon increases, we see that

$$\text{Cov}[\boldsymbol{e}_h(\ell)] \to \boldsymbol{\Sigma}_a + \sum_{i=1}^{\infty} \boldsymbol{\psi}_i \boldsymbol{\Sigma}_a \boldsymbol{\psi}_i' = \text{Cov}(\boldsymbol{z}_t).$$

This is consistent with the mean-reverting of $\boldsymbol{z}_t$, as $\boldsymbol{z}_h(\ell)$ approaches $\boldsymbol{\mu}$ the uncertainty in forecasts is the same as that of $\boldsymbol{z}_t$. Also, from Equation (2.74), it is easy to see that

$$\text{Cov}[\boldsymbol{e}_h(\ell)] = \text{Cov}[\boldsymbol{e}_h(\ell-1)] + \boldsymbol{\psi}_{\ell-1} \boldsymbol{\Sigma}_a \boldsymbol{\psi}_{\ell-1}', \quad \ell > 1.$$

The covariances of forecast errors thus can also be computed recursively.

2.9.2 Forecasts of an Estimated Model

In practice, the parameters of a VAR(p) model are unknown, and one would like to take into account the parameter uncertainty in forecasting. For simplicity and similar to real-world applications, we assume that the parameters are estimated using the information available at the forecast origin $t = h$. That is, estimation is carried out based on the available information in F_h. Under this assumption, parameter estimates are functions of F_h and, hence, the ℓ-step ahead minimum mean squared error (MSE) forecast of $\boldsymbol{z}_{h+\ell}$ with estimated parameters is

$$\hat{\boldsymbol{z}}_h(\ell) = \hat{\boldsymbol{\phi}}_0 + \sum_{i=1}^{p} \hat{\boldsymbol{\phi}}_i \hat{\boldsymbol{z}}_h(\ell - i), \tag{2.75}$$

where, as before, $\hat{\boldsymbol{z}}_h(j) = \boldsymbol{z}_{h+j}$ for $j \leq 0$. The point forecasts using estimated parameters thus remain the same as before. This is so because the estimates are unbiased and the forecast is out-of-sample forecast. However, the associated forecast error is

$$\begin{aligned}\hat{\boldsymbol{e}}_h(\ell) &= \boldsymbol{z}_{h+\ell} - \hat{\boldsymbol{z}}_h(\ell) = \boldsymbol{z}_{h+\ell} - \boldsymbol{z}_h(\ell) + \boldsymbol{z}_h(\ell) - \hat{\boldsymbol{z}}_h(\ell) \\ &= \boldsymbol{e}_h(\ell) + [\boldsymbol{z}_h(\ell) - \hat{\boldsymbol{z}}_h(\ell)]. \end{aligned} \tag{2.76}$$

Notice that $\boldsymbol{e}_h(\ell)$ are functions of $\{\boldsymbol{a}_{h+1}, \ldots, \boldsymbol{a}_{h+\ell}\}$ and the second term in the right side of Equation (2.76) is a function of F_h. The two terms of the forecast errors $\hat{\boldsymbol{e}}_h(\ell)$ are therefore uncorrelated and we have

$$\begin{aligned}\text{Cov}[\hat{\boldsymbol{e}}_h(\ell)] &= \text{Cov}[\boldsymbol{e}_h(\ell)] + E\{[\boldsymbol{z}_h(\ell) - \hat{\boldsymbol{z}}_h(\ell)][\boldsymbol{z}_h(\ell) - \hat{\boldsymbol{z}}_h(\ell)]'\} \\ &\equiv \text{Cov}[\boldsymbol{e}_h(\ell)] + \text{MSE}[\boldsymbol{z}_h(\ell) - \hat{\boldsymbol{z}}_h(\ell)], \end{aligned} \tag{2.77}$$

where the notation $\equiv$ is used to denote equivalence. To derive the MSE in Equation (2.77), we follow the approach of Samaranayake and Hasza (1988) and Basu and Sen Roy (1986).

Using the model form in Equation (2.28) for a VAR(p) model, we denote the parameter matrix via $\boldsymbol{\beta}$. Letting $T_p = T - p$ be the effective sample size in estimation, we assume that the parameter estimates satisfy

$$\sqrt{T_p}\text{vec}\left(\hat{\boldsymbol{\beta}}' - \boldsymbol{\beta}'\right) \to_d N(\boldsymbol{0}, \boldsymbol{\Sigma}_{\beta'}).$$

As discussed in Section 2.5, several estimation methods can produce such estimates for a stationary VAR(p) model. Since $\boldsymbol{z}_h(\ell)$ is a differentiable function of $\text{vec}(\boldsymbol{\beta}')$, one can show that

$$\sqrt{T_p}[\hat{\boldsymbol{z}}_h(\ell) - \boldsymbol{z}_h(\ell)|F_h] \to_d N\left(\boldsymbol{0}, \frac{\partial \boldsymbol{z}_h(\ell)}{\partial \text{vec}(\boldsymbol{\beta}')'} \boldsymbol{\Sigma}_\beta \frac{\partial \boldsymbol{z}_h(\ell)'}{\partial \text{vec}(\boldsymbol{\beta}')}\right).$$

This result suggests that we can approximate the MSE in Equation (2.77) by

$$\boldsymbol{\Omega}_\ell = E\left[\frac{\partial \boldsymbol{z}_h(\ell)}{\partial \text{vec}(\boldsymbol{\beta}')'} \boldsymbol{\Sigma}_\beta \frac{\partial \boldsymbol{z}_h(\ell)'}{\partial \text{vec}(\boldsymbol{\beta}')}\right].$$

If we further assume that $\boldsymbol{a}_t$ is multivariate normal, then we have

$$\sqrt{T_p}\,[\hat{\boldsymbol{z}}_h(\ell) - \boldsymbol{z}_h(\ell)] \to_d N(\boldsymbol{0}, \boldsymbol{\Omega}_\ell).$$

Consequently, we have

$$\text{Cov}\left[\hat{\boldsymbol{e}}_h(\ell)\right] = \text{Cov}[\boldsymbol{e}_h(\ell)] + \frac{1}{T_p}\boldsymbol{\Omega}_\ell. \tag{2.78}$$

It remains to derive the quantity $\boldsymbol{\Omega}_\ell$. To this end, we need to obtain the derivatives $\partial \boldsymbol{z}_h(\ell)/\partial \text{vec}(\boldsymbol{\beta}')$. (The reason for using $\boldsymbol{\beta}'$ instead of $\boldsymbol{\beta}$ is to simplify the matrix derivatives later.) As shown in Equation (2.70), $\boldsymbol{z}_h(\ell)$ can be calculated recursively. Moreover, we can further generalize the expanded series to include the constant $\boldsymbol{\phi}_0$. Specifically, let $\boldsymbol{x}_h = (1, \boldsymbol{z}_h', \boldsymbol{z}_{h-1}', \ldots, \boldsymbol{z}_{h-p+1}')'$ be the $(kp+1)$-dimensional vector at the forecast origin $t = h$. Then, by Equation (2.70), we have

$$\boldsymbol{z}_h(\ell) = \boldsymbol{J}\boldsymbol{P}^\ell \boldsymbol{x}_h, \quad \ell \geq 1, \tag{2.79}$$

where

$$\boldsymbol{P} = \begin{bmatrix} 1 & \boldsymbol{0}_{kp}' \\ \boldsymbol{\nu} & \boldsymbol{\Phi} \end{bmatrix}_{(kp+1)\times(kp+1)}, \qquad \boldsymbol{J} = [\boldsymbol{0}_k, \boldsymbol{I}_k, \boldsymbol{0}_{k\times k(p-1)}]_{k\times(kp+1)},$$

and $\boldsymbol{\nu} = [\boldsymbol{\phi}_0', \boldsymbol{0}_{k(p-1)}]'$, where $\boldsymbol{\Phi}$ is the companion matrix of $\boldsymbol{\phi}(B)$ as defined in Equation (2.23), $\boldsymbol{0}_m$ is an m-dimensional vector of zero, and $\boldsymbol{0}_{m\times n}$ is an $m \times n$ matrix of zero. This is a generalized version of Equation (2.73) to include the constant vector $\boldsymbol{\phi}_0$ in the recursion and it can be shown by mathematical induction. Using Equation (2.79) and part (k) of **Result 3** in Appendix A, we have

$$\begin{aligned}
\frac{\partial \boldsymbol{z}_h(\ell)}{\partial \text{vec}(\boldsymbol{\beta}')'} &= \frac{\partial \text{vec}(\boldsymbol{J}\boldsymbol{P}^\ell \boldsymbol{x}_h)}{\partial \text{vec}(\boldsymbol{\beta}')'} = (\boldsymbol{x}_h' \otimes \boldsymbol{J})\frac{\partial \text{vec}(\boldsymbol{P}^\ell)}{\partial \text{vec}(\boldsymbol{\beta}')'} \\
&= (\boldsymbol{x}_h' \otimes \boldsymbol{J})\left[\sum_{i=0}^{\ell-1}(\boldsymbol{P}')^{\ell-1-i} \otimes \boldsymbol{P}^i\right]\frac{\partial \text{vec}(\boldsymbol{P}^\ell)}{\partial \text{vec}(\boldsymbol{\beta}')'} \\
&= (\boldsymbol{x}_h' \otimes \boldsymbol{J})\left[\sum_{i=0}^{\ell-1}(\boldsymbol{P}')^{\ell-1-i} \otimes \boldsymbol{P}^i\right](\boldsymbol{I}_{kp+1} \otimes \boldsymbol{J}') \\
&= \sum_{i=0}^{\ell-1} \boldsymbol{x}_h'(\boldsymbol{P}')^{\ell-1-i} \otimes \boldsymbol{J}\boldsymbol{P}^i\boldsymbol{J}' \\
&= \sum_{i=0}^{\ell-1} \boldsymbol{x}_h'(\boldsymbol{P}')^{\ell-1-i} \otimes \boldsymbol{\psi}_i,
\end{aligned}$$

where we have used the fact that $\boldsymbol{J}\boldsymbol{P}^i\boldsymbol{J} = \boldsymbol{\psi}_i$. Using the LS estimate $\hat{\boldsymbol{\beta}}$, we have, via Equation (2.43), $\boldsymbol{\Sigma}_{\beta'} = \boldsymbol{G}^{-1} \otimes \boldsymbol{\Sigma}_a$. Therefore,

$$
\begin{aligned}
\boldsymbol{\Omega}_\ell &= E\left[\frac{\partial \boldsymbol{z}_h(\ell)}{\partial \text{vec}(\boldsymbol{\beta}')'}\left(\boldsymbol{G}^{-1}\otimes\boldsymbol{\Sigma}_a\right)\frac{\partial \boldsymbol{z}_h(\ell)'}{\partial \text{vec}(\boldsymbol{\beta}')}\right] \\
&= \sum_{i=0}^{\ell-1}\sum_{j=0}^{\ell-1} E\left(\boldsymbol{x}_h'(\boldsymbol{P}')^{\ell-1-i}\boldsymbol{G}^{-1}\boldsymbol{P}^{\ell-1-j}\boldsymbol{x}_h\right)\otimes\boldsymbol{\psi}_i\boldsymbol{\Sigma}_a\boldsymbol{\psi}_j' \\
&= \sum_{i=0}^{\ell-1}\sum_{j=0}^{\ell-1} E\left[tr\left(\boldsymbol{x}_h'(\boldsymbol{P}')^{\ell-1-i}\boldsymbol{G}^{-1}\boldsymbol{P}^{\ell-1-j}\boldsymbol{x}_h\right)\right]\boldsymbol{\psi}_i\boldsymbol{\Sigma}_a\boldsymbol{\psi}_j' \\
&= \sum_{i=0}^{\ell-1}\sum_{j=0}^{\ell-1} tr\left[(\boldsymbol{P}')^{\ell-1-i}\boldsymbol{G}^{-1}\boldsymbol{P}^{\ell-1-j}E\left(\boldsymbol{x}_h\boldsymbol{x}_h'\right)\right]\boldsymbol{\psi}_i\boldsymbol{\Sigma}_a\boldsymbol{\psi}_j' \\
&= \sum_{i=0}^{\ell-1}\sum_{j=0}^{\ell-1} tr\left[(\boldsymbol{P}')^{\ell-1-i}\boldsymbol{G}^{-1}\boldsymbol{P}^{\ell-1-j}\boldsymbol{G}\right]\boldsymbol{\psi}_i\boldsymbol{\Sigma}_a\boldsymbol{\psi}_j'. \qquad (2.80)
\end{aligned}
$$

In particular, if $\ell = 1$, then

$$\boldsymbol{\Omega}_1 = tr(\boldsymbol{I}_{kp+1})\boldsymbol{\Sigma}_a = (kp+1)\boldsymbol{\Sigma}_a,$$

and

$$\text{Cov}[\hat{\boldsymbol{z}}_h(1)] = \boldsymbol{\Sigma}_a + \frac{kp+1}{T_p}\boldsymbol{\Sigma}_a = \frac{T_p+kp+1}{T_p}\boldsymbol{\Sigma}_a.$$

Since $kp+1$ is the number of parameters in the model equation for z_{it}, the prior result can be interpreted as each parameter used increases the MSE of one-step ahead forecasts by a factor of $1/T_p$, where T_p is the effective sample size used in the estimation. When K or p is large, but T_p is small, the impact of using estimated parameters in forecasting could be substantial. The result, therefore, provides support for removing insignificant parameters in a VAR(p) model. In other words, it pays to employ parsimonious models.

In practice, we can replace the quantities in Equation (2.80) by their LS estimates to compute the $\text{Cov}[\hat{\boldsymbol{z}}_h(\ell)]$.

Example 2.8 Consider, again, the quarterly GDP growth rates, in percentages, of United Kingdom, Canada, and United States employed in Example 2.6, where a VAR(2) model is fitted. Using this model, we consider one-step to eight-step ahead forecasts of the GDP growth rates at the forecast origin 2011.II. We also provide the standard errors and root mean-squared errors of the predictions. The root mean-squared errors include the uncertainty due to the use of estimated parameters. The results are given in Table 2.2. From the table, we make the following observations. First, the point forecasts of the three series move closer to the sample means of the data as the forecast horizon increases, showing evidence of mean reverting. Second,

TABLE 2.2 Forecasts of Quarterly GDP Growth Rates, in Percentages, for United Kingdom, Canada, and United States via a VAR(2) Model

	Forecasts			Standard Errors			Root MSE		
Step	United Kingdom	Canada	United States	United Kingdom	Canada	United States	United Kingdom	Canada	United States
1	0.31	0.05	0.17	0.53	0.54	0.60	0.55	0.55	0.61
2	0.26	0.32	0.49	0.58	0.72	0.71	0.60	0.78	0.75
3	0.31	0.48	0.52	0.62	0.77	0.73	0.64	0.79	0.75
4	0.38	0.53	0.60	0.65	0.78	0.74	0.66	0.78	0.75
5	0.44	0.57	0.63	0.66	0.78	0.75	0.67	0.78	0.75
6	0.48	0.59	0.65	0.67	0.78	0.75	0.67	0.78	0.75
7	0.51	0.61	0.66	0.67	0.78	0.75	0.67	0.78	0.75
8	0.52	0.62	0.67	0.67	0.78	0.75	0.67	0.78	0.75
Data	0.52	0.62	0.65	0.71	0.79	0.79	0.71	0.79	0.79

The forecast origin is the second quarter of 2011. The last row of the table gives the sample means and sample standard errors of the series.

as expected, the standard errors and root mean-squared errors of forecasts increase with the forecast horizon. The standard errors should converge to the standard errors of the time series as the forecast horizon increases. Third, the effect of using estimated parameters is evident when the forecast horizon is small. The effect vanishes quickly as the forecast horizon increases. This is reasonable because a stationary VAR model is mean-reverting. The standard errors and mean-squared errors of prediction should converge to the standard errors of the series. The standard errors and root mean-squared errors of Table 2.2 can be used to construct interval predictions. For instance, a two-step ahead 95% interval forecast for U.S. GDP growth rate is $0.49 \pm 1.96 \times 0.71$ and $0.49 \pm 1.96 \times 0.75$, respectively, for predictions without and with parameter uncertainty.

Finally, we provide estimates of the $\mathbf{\Omega}_\ell$ matrix in the following R demonstration. As expected, $\mathbf{\Omega}_\ell$ decreases in magnitude as the forecast horizon increases. The forecasting results of the example are obtained by using the command `VARpred` of the `MTS` package. □

R Demonstration: Prediction.

```
> VARpred(m1,8)
Forecasts at origin:  125
     uk      ca     us
 0.3129 0.05166 0.1660
 0.2647 0.31687 0.4889
 ....
 0.5068 0.60967 0.6630
 0.5247 0.61689 0.6688
```

```
Standard Errors of predictions:
        [,1]   [,2]   [,3]
[1,] 0.5315 0.5400 0.5975
[2,] 0.5804 0.7165 0.7077
 ....
[7,] 0.6719 0.7842 0.7486
[8,] 0.6729 0.7843 0.7487
Root Mean square errors of predictions:
        [,1]   [,2]   [,3]
[1,] 0.5461 0.5549 0.6140
[2,] 0.6001 0.7799 0.7499
 ....
[7,] 0.6730 0.7844 0.7487
[8,] 0.6734 0.7844 0.7487
> colMeans(z)   ## Compute sample means
       uk        ca        us
0.5223092 0.6153672 0.6473996
> sqrt(apply(z,2,var))   ## Sample standard errors
       uk        ca        us
0.7086442 0.7851955 0.7872912
>
Omega matrix at horizon:  1
            [,1]        [,2]       [,3]
[1,] 0.015816875 0.001486291 0.00416376
[2,] 0.001486291 0.016328573 0.00781132
[3,] 0.004163760 0.007811320 0.01999008
Omega matrix at horizon:  2
           [,1]       [,2]       [,3]
[1,] 0.02327855 0.03708068 0.03587541
[2,] 0.03708068 0.09490535 0.07211282
[3,] 0.03587541 0.07211282 0.06154730
Omega matrix at horizon:  3
           [,1]       [,2]       [,3]
[1,] 0.02044253 0.02417433 0.01490480
[2,] 0.02417433 0.03218999 0.02143570
[3,] 0.01490480 0.02143570 0.01652968
Omega matrix at horizon:  4
            [,1]        [,2]        [,3]
[1,] 0.015322037 0.010105520 0.009536199
[2,] 0.010105520 0.007445308 0.006700067
[3,] 0.009536199 0.006700067 0.006181433
```

2.10 IMPULSE RESPONSE FUNCTIONS

In studying the structure of a VAR model, we discussed the Granger causality and the relation to transfer function models. There is another approach to explore the relation between variables. As a matter of fact, we are often interested in knowing the effect of changes in one variable on another variable in multivariate time series analysis. For instance, suppose that the bivariate time series $\boldsymbol{z}_t$ consists of monthly income and expenditure of a household, we might be interested in knowing the effect on expenditure if the monthly income of the household is increased or decreased by a certain amount, for example, 5%. This type of study is referred to as the *impulse response function* in the statistical literature and the *multiplier analysis* in the econometric literature. In this section, we use the MA representation of a VAR(p) model to assess such effects.

In multiplier analysis, we can assume $E(\boldsymbol{z}_t) = \boldsymbol{0}$ because the mean does not affect the pattern of the response of $\boldsymbol{z}_t$ to any shock. To study the effects of changes in z_{1t} on $\boldsymbol{z}_{t+j}$ for $j > 0$ while holding other quantities unchanged, we can assume that $t = 0$, $\boldsymbol{z}_t = \boldsymbol{0}$ for $t \leq 0$, and $\boldsymbol{a}_0 = (1, 0, \ldots)'$. In other words, we would like to study the behavior of $\boldsymbol{z}_t$ for $t > 0$ while z_{10} increases by 1. To this end, we can trace out $\boldsymbol{z}_t$ for $t = 1, 2, \ldots$, assuming $\boldsymbol{a}_t = \boldsymbol{0}$ for $t > 0$. Using the MA representation of a VAR(p) model with coefficient matrix $\boldsymbol{\psi}_\ell = [\psi_{\ell,ij}]$ given in Equation (2.27), we have

$$\boldsymbol{z}_0 = \boldsymbol{a}_0 = \begin{bmatrix} 1 \\ 0 \\ \vdots \\ 0 \end{bmatrix}, \quad \boldsymbol{z}_1 = \boldsymbol{\psi}_1 \boldsymbol{a}_0 = \begin{bmatrix} \psi_{1,11} \\ \psi_{1,21} \\ \vdots \\ \psi_{1,k1} \end{bmatrix}, \quad \boldsymbol{z}_2 = \boldsymbol{\psi}_2 \boldsymbol{a}_0 = \begin{bmatrix} \psi_{2,11} \\ \psi_{2,21} \\ \vdots \\ \psi_{2,k1} \end{bmatrix}, \ldots$$

The results are simply the first columns of the coefficient matrices $\boldsymbol{\psi}_i$. Therefore, in this particular case, we have $\boldsymbol{z}_t = \boldsymbol{\psi}_{t,.1}$ where $\boldsymbol{\psi}_{\ell,.1}$ denotes the first column of $\boldsymbol{\psi}_\ell$. Similarly, to study the effect on $\boldsymbol{z}_{t+j}$ by increasing the ith series z_{it} by 1, we have $\boldsymbol{a}_0 = \boldsymbol{e}_i$, where $\boldsymbol{e}_i$ is the ith unit vector in R^k, the k-dimensional Euclidean space, we have

$$\boldsymbol{z}_0 = \boldsymbol{e}_i, \quad \boldsymbol{z}_1 = \boldsymbol{\psi}_{1,.i}, \quad \boldsymbol{z}_2 = \boldsymbol{\psi}_{2,.i}, \quad \ldots.$$

They are the ith columns of the coefficient matrices $\boldsymbol{\psi}_i$ of the MA representation of $\boldsymbol{z}_t$. For this reason, the coefficient matrix $\boldsymbol{\psi}_i$ of the MA representation of a VAR(p) model is referred to as the coefficients of impulse response functions. The summation

$$\underline{\boldsymbol{\psi}}_n = \sum_{i=0}^{n} \boldsymbol{\psi}_i$$

denotes the *accumulated responses* over n periods to a unit shock to $\boldsymbol{z}_t$. Elements of $\underline{\boldsymbol{\psi}}_n$ are referred to as the nth interim multipliers. The total accumulated responses for all future periods are defined as

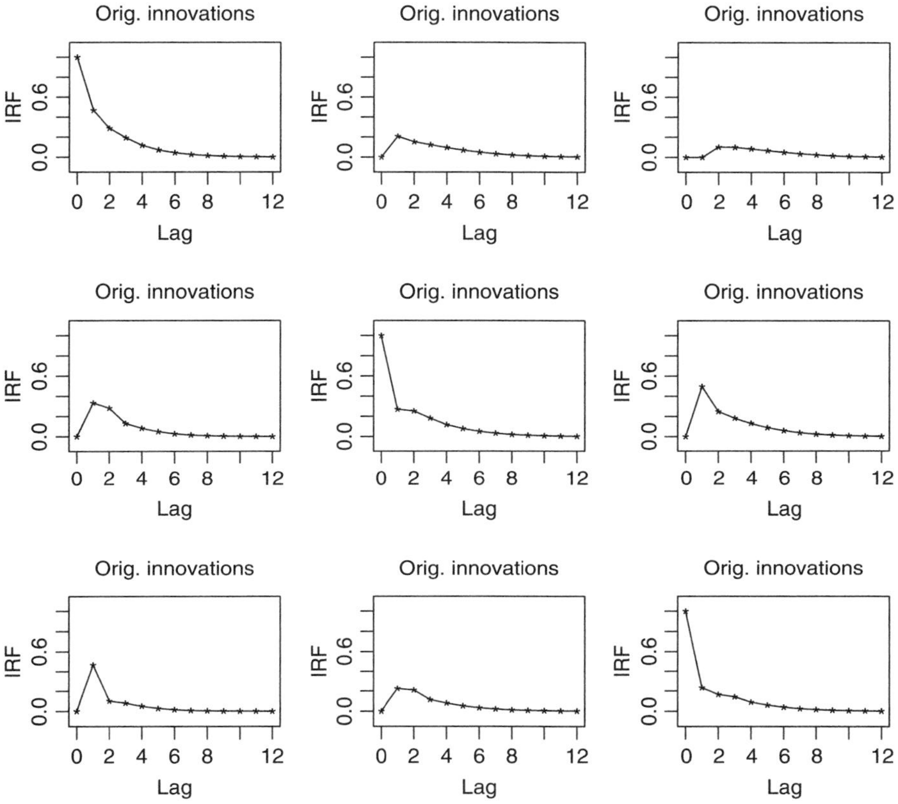

FIGURE 2.6 Impulse response functions of the simplified VAR(2) model in Equation (2.65) for the quarterly growth rates of real gross domestic products of United Kingdom, Canada, and United States from the second quarter of 1980 to the second quarter of 2011. The growth rates are in percentages.

$$\underline{\psi}_{\infty} = \sum_{i=0}^{\infty} \psi_i.$$

Often $\underline{\psi}_{\infty}$ is called the *total multipliers* or *long-run effects*.

To demonstrate, consider again the simplified VAR(2) model in Equation (2.65) for the quarterly growth rates, in percentages, of United Kingdom, Canada, and United States. Figure 2.6 shows the impulse response functions of the fitted three-dimensional model. From the plots, the impulse response functions decay to 0 quickly. This is expected for a stationary series. The upper-right plot shows that there is a delayed effect on the U.K. GDP growth rate if one changed the U.S. growth rate by 1. This delayed effect is due to the fact that a change in U.S. rate at time t affects the Canadian rate at time $t + 1$, which in turn affects the U.K. rate at time $t + 2$. This plot thus shows that the impulse response functions show the marginal

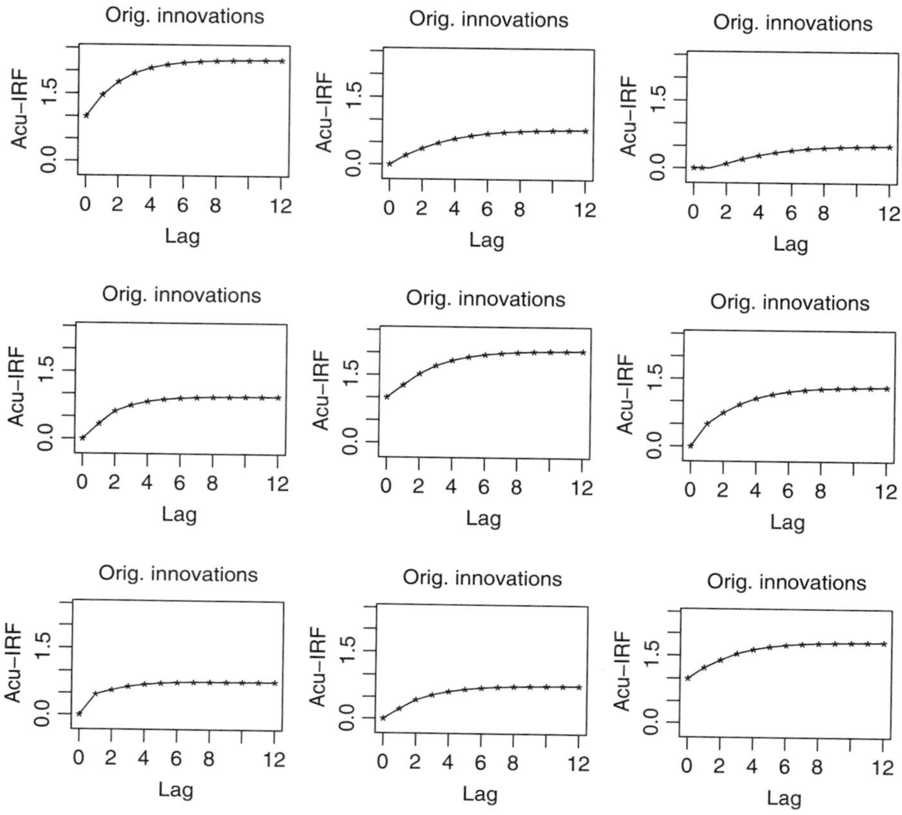

FIGURE 2.7 Accumulated responses of the simplified VAR(2) model in Equation (2.65) for the quarterly growth rates of real gross domestic products of United Kingdom, Canada, and United States from the second quarter of 1980 to the second quarter of 2011. The growth rates are in percentages.

effects, not the conditional effects. Recall that conditional on the lagged Canadian rate, the growth rate of U.K. GDP does not depend on lagged values of U.S. growth rate. Figure 2.7 shows the accumulated responses implied by the simplified VAR(2) model. As expected, the accumulated effects converge to the total multipliers quickly.

2.10.1 Orthogonal Innovations

In practice, elements of $\boldsymbol{a}_t$ tend to be correlated, that is, $\boldsymbol{\Sigma}_a$ is not a diagonal matrix. As such, change in one component of $\boldsymbol{a}_t$ will simultaneously affect other components of $\boldsymbol{a}_t$. Consequently, the impulse response functions introduced in the previous section would encounter difficulties in a real application because one cannot arbitrarily increase z_{1t} by 1 without altering other components z_{it}. Mathematically, the problem can be shown as follows. The effect of change in a_{1t} on the future series $\boldsymbol{z}_{t+j}$ $(j \geq 0)$ can be quantified as $\partial \boldsymbol{z}_{t+j}/\partial a_{1t}$. Since $\{\boldsymbol{a}_t\}$ are serially uncorrelated, we can use the MA representation of $\boldsymbol{z}_{t+j}$ and obtain

$$\frac{\partial \boldsymbol{z}_{t+j}}{\partial a_{1t}} = \boldsymbol{\psi}_j \frac{\partial \boldsymbol{a}_t}{\partial a_{1t}} = \boldsymbol{\psi}_j \begin{bmatrix} \frac{\partial a_{1t}}{\partial a_{1t}} \\ \frac{\partial a_{2t}}{\partial a_{1t}} \\ \vdots \\ \frac{\partial a_{kt}}{\partial a_{1t}} \end{bmatrix} = \boldsymbol{\psi}_j \begin{bmatrix} 1 \\ \frac{\sigma_{a,21}}{\sigma_{a,11}} \\ \vdots \\ \frac{\sigma_{a,k1}}{\sigma_{a,11}} \end{bmatrix} = \boldsymbol{\psi}_j \boldsymbol{\Sigma}_{a,.1} \sigma_{a,11}^{-1},$$

where $\boldsymbol{\Sigma}_{a,.1}$ denotes the first column of $\boldsymbol{\Sigma}_a$, and we have used $\boldsymbol{\Sigma}_a = [\sigma_{a,ij}]$ and the simple linear regression $a_{it} = (\sigma_{a,i1}/\sigma_{a,11})a_{1t} + \epsilon_{it}$ with ϵ_{it} denoting the error term. In general, we have $\partial \boldsymbol{z}_{t+j}/\partial a_{it} = \boldsymbol{\psi}_j \boldsymbol{\Sigma}_{a,.i} \sigma_{a,ii}^{-1}$. Thus, the correlations between components of $\boldsymbol{a}_t$ cannot be ignored.

To overcome this difficulty, one can take a proper transformation of $\boldsymbol{a}_t$ such that components of the innovation become uncorrelated, that is, diagonalize the covariance matrix $\boldsymbol{\Sigma}_a$. A simple way to achieve orthogonalization of the innovation is to consider the Cholesky decomposition of $\boldsymbol{\Sigma}_a$ as discussed in Chapter 1. Specifically, we have

$$\boldsymbol{\Sigma}_a = \boldsymbol{U}'\boldsymbol{U},$$

where $\boldsymbol{U}$ is an upper triangular matrix with positive diagonal elements. Let $\boldsymbol{\eta}_t = (\boldsymbol{U}')^{-1}\boldsymbol{a}_t$. Then,

$$\text{Cov}(\boldsymbol{\eta}_t) = (\boldsymbol{U}')^{-1}\text{Cov}(\boldsymbol{a}_t)\boldsymbol{U}^{-1} = (\boldsymbol{U}')^{-1}(\boldsymbol{U}'\boldsymbol{U})(\boldsymbol{U})^{-1} = \boldsymbol{I}_k.$$

Thus, components of $\boldsymbol{\eta}_t$ are uncorrelated and have unit variance.

From the MA representation of $\boldsymbol{z}_t$, we have

$$\begin{aligned} \boldsymbol{z}_t &= \boldsymbol{\psi}(B)\boldsymbol{a}_t = \boldsymbol{\psi}(B)\boldsymbol{U}'(\boldsymbol{U}')^{-1}\boldsymbol{a}_t, \\ &= [\boldsymbol{\psi}(B)\boldsymbol{U}']\boldsymbol{\eta}_t, \\ &= [\underline{\boldsymbol{\psi}}_0 + \underline{\boldsymbol{\psi}}_1 B + \underline{\boldsymbol{\psi}}_2 B^2 + \cdots]\boldsymbol{\eta}_t, \end{aligned} \quad (2.81)$$

where $\underline{\boldsymbol{\psi}}_\ell = \boldsymbol{\psi}_\ell \boldsymbol{U}'$ for $\ell \geq 0$.

Letting $[\underline{\psi}_{\ell,ij}] = \underline{\boldsymbol{\psi}}_\ell$, we can easily see that $\partial \boldsymbol{z}_{t+\ell}/\partial \eta_{it} = \underline{\boldsymbol{\psi}}_{\ell,.i}$ for $\ell > 0$. We call $\underline{\boldsymbol{\psi}}_\ell$ the impulse response coefficients of $\boldsymbol{z}_t$ with *orthogonal innovations*. The plot of $\underline{\psi}_{\ell,ij}$ against ℓ is called the *impulse response function* of $\boldsymbol{z}_t$ with orthogonal innovations. Specifically, $\underline{\psi}_{\ell,ij}$ denotes the impact of a shock with size being "one standard deviation" of the jth innovation at time t on the future value of $z_{i,t+\ell}$. In particular, the (i, j)th element of the transformation matrix $(\boldsymbol{U}')^{-1}$ (where $i > j$) denotes the instantaneous effect of the shock η_{jt} on z_{it}. We can define accumulated responses by summing over the coefficient matrices $\underline{\boldsymbol{\psi}}_\ell$ in a similar manner as before.

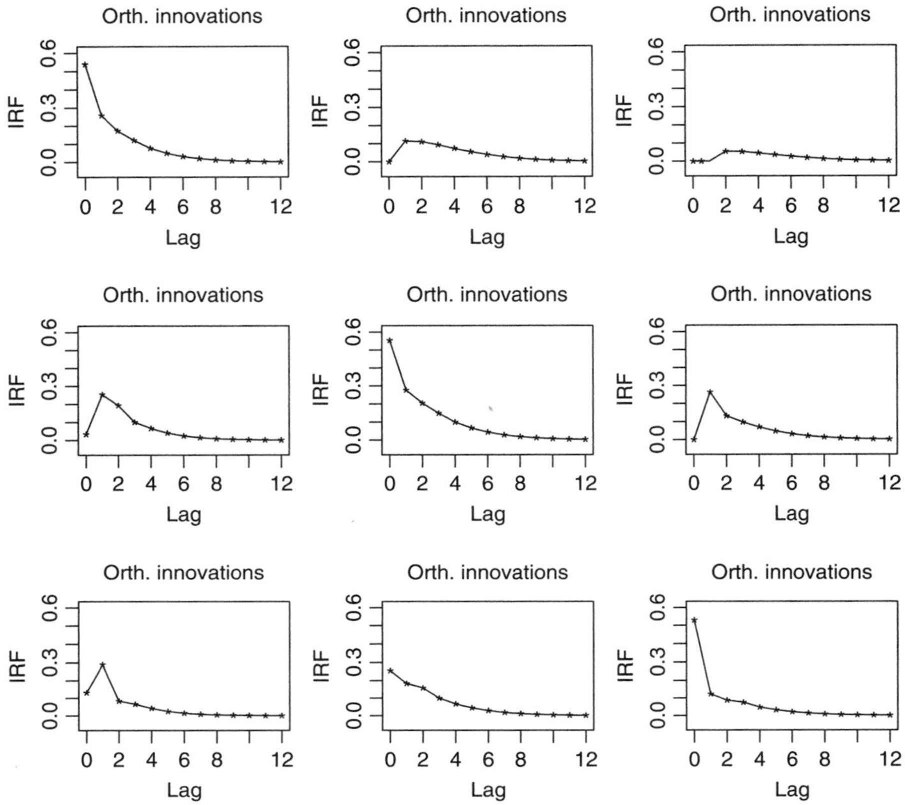

FIGURE 2.8 Impulse response functions of the simplified VAR(2) model in Equation (2.65) for the quarterly growth rates of real gross domestic products of United Kingdom, Canada, and United States from the second quarter of 1980 to the second quarter of 2011. The growth rates are in percentages and innovations are orthogonalized.

Figure 2.8 shows the impulse response functions of the simplified VAR(2) model in Equation (2.65) when the innovations are orthogonalized. For this particular instance, the instantaneous correlations between the components of $\boldsymbol{a}_t$ are not strong so that the orthogonalization of the innovations has only a small impact. Figure 2.9 shows the accumulated responses of the GDP series with orthogonal innovations. Again, the pattern of the impulse response functions is similar to those shown in Figure 2.7.

Remark: The prior definition of impulse response function depends on the *ordering* of the components in $\boldsymbol{z}_t$. The lower triangular structure of $\boldsymbol{U}'$ indicates that the η_{1t} is a function of a_{1t}, η_{2t} is a function of a_{1t} and a_{2t}, and so on. Thus, η_{it} is not affected by a_{jt} for $j > i$. Different orderings of the components of $\boldsymbol{a}_t$ thus lead to different impulse response functions for a given VAR(p) model. However, one

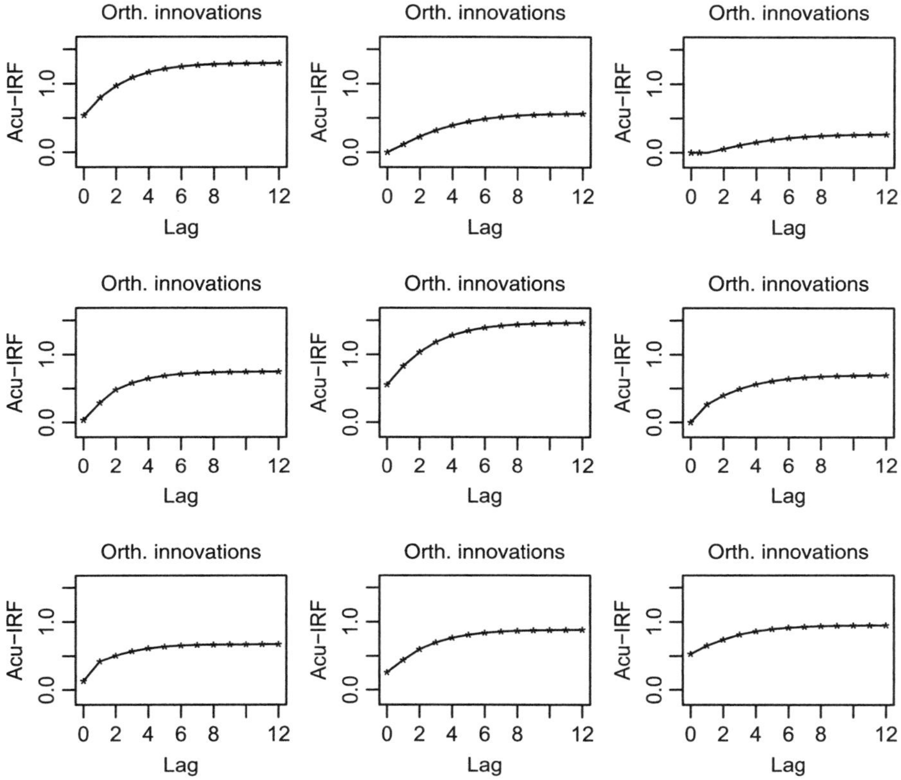

FIGURE 2.9 Accumulated responses of the simplified VAR(2) model in Equation (2.65) for the quarterly growth rates of real gross domestic products of United Kingdom, Canada, and United States from the second quarter of 1980 to the second quarter of 2011. The growth rates are in percentages and innovations are orthogonalized.

should remember that the meanings of the transformed innovations $\boldsymbol{\eta}_t$ also depend on the ordering of $\boldsymbol{a}_t$. □

Remark: The impulse response functions of a VAR(p) with or without the orthogonalization of the innovations can be obtained via the command `VARirf` in the `MTS` package. The input includes the AR coefficient matrix $\boldsymbol{\Phi} = [\boldsymbol{\phi}_1, \ldots, \boldsymbol{\phi}_p]$ and the innovation covariance matrix $\boldsymbol{\Sigma}_a$. The default option uses orthogonal innovations. □

R Demonstration: Impulse response functions of a VAR model.

```
> Phi = m2$Phi   ### m2 is the simplified VAR(2) model
> Sig = m2$Sigma
> VARirf(Phi,Sig)   ### Orthogonal innovations
> VARirf(Phi,Sig,orth=F) ## Original innovations
```

2.11 FORECAST ERROR VARIANCE DECOMPOSITION

Using the MA representation of a VAR(p) model in Equation (2.81) and the fact that $\text{Cov}(\boldsymbol{\eta}_t) = \boldsymbol{I}_k$, we see that the ℓ-step ahead forecast error of $\boldsymbol{z}_{h+\ell}$ at the forecast origin $t = h$ can be written as

$$\boldsymbol{e}_h(\ell) = \underline{\boldsymbol{\psi}}_0 \boldsymbol{\eta}_{h+\ell} + \underline{\boldsymbol{\psi}}_1 \boldsymbol{\eta}_{h+\ell-1} + \cdots + \underline{\boldsymbol{\psi}}_{\ell-1} \boldsymbol{\eta}_{h+1},$$

and the covariance matrix of the forecast error is

$$\text{Cov}[\boldsymbol{e}_h(\ell)] = \sum_{v=0}^{\ell-1} \underline{\boldsymbol{\psi}}_v \underline{\boldsymbol{\psi}}_v'. \tag{2.82}$$

From Equation (2.82), the variance of the forecast error $e_{h,i}(\ell)$, which is the ith component of $\boldsymbol{e}_h(\ell)$, is

$$\text{Var}[e_{h,i}(\ell)] = \sum_{v=0}^{\ell-1} \sum_{j=1}^{k} \underline{\psi}_{v,ij}^2 = \sum_{j=1}^{k} \sum_{v=0}^{\ell-1} \underline{\psi}_{v,ij}^2. \tag{2.83}$$

Using Equation (2.83), we define

$$w_{ij}(\ell) = \sum_{v=0}^{\ell-1} \underline{\psi}_{v,ij}^2,$$

and obtain

$$\text{Var}[e_{h,i}(\ell)] = \sum_{j=1}^{k} w_{ij}(\ell). \tag{2.84}$$

Therefore, the quantity $w_{ij}(\ell)$ can be interpreted as the contribution of the jth shock η_{jt} to the variance of the ℓ-step ahead forecast error of z_{it}. Equation (2.84) is referred to as the forecast error decomposition. In particular, $w_{ij}(\ell)/\text{Var}[e_{h,i}(\ell)]$ is the percentage of contribution from the shock η_{jt}.

To demonstrate, consider the simplified VAR(2) model in Equation (2.65) for the quarterly percentage GDP growth rates of United Kingdom, Canada, and United States. The forecast error variance decompositions for the one-step to five-step ahead predictions at the forecast origin 2011.II are given in Table 2.3. From the table, we see that the decomposition, again, depends on the ordering of the components in $\boldsymbol{z}_t$. However, the results confirm that the three growth rate series are interrelated.

TABLE 2.3 Forecast Error Variance Decomposition for One-step to Five-step Ahead Predictions for the Quarterly GDP Growth Rates, in Percentages, of United Kingdom, Canada, and United States

Variable	Step	United Kingdom	Canada	United States
United Kingdom	1	1.0000	0.0000	0.0000
	2	0.9645	0.0355	0.0000
	3	0.9327	0.0612	0.0071
	4	0.9095	0.0775	0.0130
	5	0.8956	0.0875	0.0170
Canada	1	0.0036	0.9964	0.0000
	2	0.1267	0.7400	0.1333
	3	0.1674	0.6918	0.1407
	4	0.1722	0.6815	0.1462
	5	0.1738	0.6767	0.1495
United States	1	0.0473	0.1801	07726
	2	0.2044	0.1999	0.5956
	3	0.2022	0.2320	0.5658
	4	0.2028	0.2416	0.5556
	5	0.2028	0.2460	0.5512

The forecast origin is the second quarter of 2011. The simplified VAR(2) model in Equation (2.65) is used.

R Demonstration: Forecast error decomposition.

```
> m1=VAR(z,2)
> m2=refVAR(m1)
> names(m2)
 [1] "data"   "order"      "cnst"    "coef"    "aic"   "bic"
 [7] "hq"     "residuals"  "secoef"  "Sigma"   "Phi"   "Ph0"
> Phi=m2$Phi
> Sig=m2$Sigma
> Theta=NULL
> FEVdec(Phi,Theta,Sig,lag=5)
Order of the ARMA mdoel:
[1] 2 0
Standard deviation of forecast error:
          [,1]      [,2]      [,3]      [,4]      [,5]
# Forecat horison
[1,] 0.5385505 0.6082891 0.6444223 0.6644656 0.6745776
[2,] 0.5550000 0.7197955 0.7839243 0.8100046 0.8217975
[3,] 0.6022357 0.7040833 0.7317336 0.7453046 0.7510358
Forecast-Error-Variance Decomposition
Forecast horizon:  1
            [,1]      [,2]      [,3]
[1,] 1.000000000 0.0000000 0.0000000
[2,] 0.003640595 0.9963594 0.0000000
[3,] 0.047327504 0.1801224 0.7725501
```

```
Forecast horizon:   2
          [,1]      [,2]      [,3]
[1,] 0.9645168 0.0354832 0.0000000
[2,] 0.1266584 0.7400392 0.1333023
[3,] 0.2044415 0.1999232 0.5956353
```

2.12 PROOFS

We provide proofs for some lemmas and theorems stated in the chapter.

Proof of Lemma 2.1:

$$|\boldsymbol{I} - \boldsymbol{\Phi} B| = \begin{vmatrix} \boldsymbol{I}_k - \boldsymbol{\phi}_1 B & -\boldsymbol{\phi}_2 B & \cdots & -\boldsymbol{\phi}_{p-1} B & -\boldsymbol{\phi}_p B \\ -\boldsymbol{I}_k B & \boldsymbol{I}_k & \cdots & \mathbf{0} & \mathbf{0} \\ \mathbf{0} & -\boldsymbol{I}_k B & \cdots & \mathbf{0} & \mathbf{0} \\ \vdots & \vdots & \cdots & \vdots & \vdots \\ \mathbf{0} & \mathbf{0} & \cdots & \boldsymbol{I}_k & \mathbf{0} \\ \mathbf{0} & \mathbf{0} & \cdots & -\boldsymbol{I}_k B & \boldsymbol{I}_k \end{vmatrix}.$$

Multiplying the last column block by B and adding the result to the $(p-1)$th column block, we have

$$|\boldsymbol{I} - \boldsymbol{\Phi} B| = \begin{vmatrix} \boldsymbol{I}_k - \boldsymbol{\phi}_1 B & -\boldsymbol{\phi}_2 B & \cdots & -\boldsymbol{\phi}_{p-1} B - \boldsymbol{\phi}_p B^2 & -\boldsymbol{\phi}_p B \\ -\boldsymbol{I}_k B & \boldsymbol{I}_k & \cdots & \mathbf{0} & \mathbf{0} \\ \mathbf{0} & -\boldsymbol{I}_k B & \cdots & \mathbf{0} & \mathbf{0} \\ \vdots & \vdots & \cdots & \vdots & \vdots \\ \mathbf{0} & \mathbf{0} & \cdots & \boldsymbol{I}_k & \mathbf{0} \\ \mathbf{0} & \mathbf{0} & \cdots & \mathbf{0} & \boldsymbol{I}_k \end{vmatrix}.$$

Next, multiplying the $(p-1)$th column block by B and adding the result to the $(p-2)$th column block, we have $-\boldsymbol{\phi}_{p-2} B - \boldsymbol{\phi}_{p-1} B^2 - \boldsymbol{\phi}_p^3$ at the $(1, p-2)$ block and 0 at $(p-1, p-2)$ block. By repeating the procedure, we obtain $|\boldsymbol{I} - \boldsymbol{\Phi} B| =$

$$\begin{vmatrix} \boldsymbol{\phi}(B) & -\sum_{j=2}^{p} \boldsymbol{\phi}_j B^{j-1} & \cdots & -\boldsymbol{\phi}_{p-1} B - \boldsymbol{\phi}_p B^2 & -\boldsymbol{\phi}_p B \\ \mathbf{0} & \boldsymbol{I}_k & \cdots & \mathbf{0} & \mathbf{0} \\ \mathbf{0} & \mathbf{0} & \cdots & \mathbf{0} & \mathbf{0} \\ \vdots & \vdots & \cdots & \vdots & \vdots \\ \mathbf{0} & \mathbf{0} & \cdots & \boldsymbol{I}_k & \mathbf{0} \\ \mathbf{0} & \mathbf{0} & \cdots & \mathbf{0} & \boldsymbol{I}_k \end{vmatrix} = |\boldsymbol{\phi}(B)|.$$

□

Proof of Lemma 2.4: We make use of Equation (2.58), Lemma 2.2, and the limiting distribution of the estimate $\hat{\boldsymbol{\beta}}$. Dividing the second last term of Equation (2.58) by $\sqrt{T_p}$ and applying vec operator, we have

$$\begin{aligned}
&\text{vec}\{\boldsymbol{A}'\boldsymbol{B}_m[\boldsymbol{I}_m \otimes \boldsymbol{X}(\hat{\boldsymbol{\beta}} - \boldsymbol{\beta})]/\sqrt{T_p}\} \\
&= \sqrt{T_p}\text{vec}\left[\frac{1}{T_p}\boldsymbol{A}'\boldsymbol{B}_m(\boldsymbol{I}_m \otimes \boldsymbol{X})[\boldsymbol{I}_m \otimes (\hat{\boldsymbol{\beta}} - \boldsymbol{\beta})]\right] \\
&= \left[\boldsymbol{I}_k \otimes \frac{1}{T_p}\boldsymbol{A}'\boldsymbol{B}_m(\boldsymbol{I}_m \otimes \boldsymbol{X})\right]\sqrt{T_p}\text{vec}[\boldsymbol{I}_m \otimes (\hat{\boldsymbol{\beta}} - \boldsymbol{\beta})].
\end{aligned}$$

The first factor of the prior equation converges to $\mathbf{0}$, whereas the second factor has a limiting normal distribution. Therefore, the second term of Equation (2.58) converges to $\mathbf{0}$ in probability after it is divided by $\sqrt{T_p}$.

Dividing the last term of Equation (2.58) by $\sqrt{T_p}$ and applying the vec operator, we have

$$\begin{aligned}
&\sqrt{T_p}\text{vec}\left\{\left(\hat{\boldsymbol{\beta}} - \boldsymbol{\beta}\right)'\boldsymbol{X}'\boldsymbol{B}_m\left[\boldsymbol{I}_m \otimes \boldsymbol{X}(\hat{\boldsymbol{\beta}} - \boldsymbol{\beta})\right]\right\} \\
&= \left\{\left[\boldsymbol{I}_m \otimes (\hat{\boldsymbol{\beta}} - \boldsymbol{\beta})'\right]\frac{(\boldsymbol{I}_m \otimes \boldsymbol{X}')\,\boldsymbol{B}_m'\boldsymbol{X}}{T_p} \otimes \boldsymbol{I}_k\right\}\sqrt{T_p}\text{vec}\left[\left(\hat{\boldsymbol{\beta}} - \boldsymbol{\beta}\right)'\right].
\end{aligned}$$

The last factor of the prior equation follows a limiting distribution, the second factor converges to a fixed matrix, but the first factor converges to $\mathbf{0}$. Therefore, the last term of Equation (2.58) converges to $\mathbf{0}$ in probability after it is divided by $\sqrt{T_p}$.

It remains to consider the third term of Equation (2.58). To this end, consider

$$\begin{aligned}
\boldsymbol{X}'\boldsymbol{B}_m(\boldsymbol{I}_m \otimes \boldsymbol{A}) &= \boldsymbol{X}'[\boldsymbol{B}\boldsymbol{A}, \boldsymbol{B}^2\boldsymbol{A}, \ldots, \boldsymbol{B}^m\boldsymbol{A}] \\
&= [\boldsymbol{X}'\boldsymbol{B}\boldsymbol{A}, \boldsymbol{X}'\boldsymbol{B}^2\boldsymbol{A}, \ldots, \boldsymbol{X}'\boldsymbol{B}^m\boldsymbol{A}].
\end{aligned}$$

Also, for $i = 1, \ldots, m$,

$$\begin{aligned}
\boldsymbol{X}'\boldsymbol{B}^i\boldsymbol{A} &= \sum_{t=p+1}^{n} \boldsymbol{x}_t\boldsymbol{a}_{t-i}' \\
&= \sum_{t=p+1}^{T} \begin{bmatrix} 1 \\ \boldsymbol{z}_{t-1} \\ \vdots \\ \boldsymbol{z}_{t-p} \end{bmatrix} \boldsymbol{a}_{t-i}'.
\end{aligned}$$

Using the MA representation for $\boldsymbol{z}_{t-j} (j = 1, \ldots, p)$ and Lemma 2.2, we obtain

$$\frac{1}{T_p}\boldsymbol{X}'\boldsymbol{B}^i\boldsymbol{A} \to_p \begin{bmatrix} \boldsymbol{0}' \\ \boldsymbol{\psi}_{i-1}\boldsymbol{\Sigma}_a \\ \boldsymbol{\psi}_{i-2}\boldsymbol{\Sigma}_a \\ \vdots \\ \boldsymbol{\psi}_{i-p}\boldsymbol{\Sigma}_a \end{bmatrix},$$

where it is understood that $\boldsymbol{\psi}_j = \boldsymbol{0}$ for $j < 0$. Therefore,

$$\frac{1}{T_p}\boldsymbol{X}'\boldsymbol{B}_m(\boldsymbol{I}_m \otimes \boldsymbol{A}) \to_p \boldsymbol{H}_*,$$

where $\boldsymbol{H}_*$ is defined in the statement of Lemma 2.4. Dividing Equation (2.58) by $\sqrt{T_p} = \sqrt{T-p}$, taking the vec operator, and using the properties of vec operator, we have

$$\begin{aligned}\sqrt{T_p}\text{vec}(\hat{\boldsymbol{\Xi}}_m) &\approx \sqrt{T_p}\text{vec}\,(\boldsymbol{\Xi}_m) - \frac{1}{\sqrt{T_p}}(\hat{\boldsymbol{\beta}} - \boldsymbol{\beta})'\boldsymbol{X}'\boldsymbol{B}_m\,(\boldsymbol{I}_m \otimes \boldsymbol{A}) \\ &= \sqrt{T_p}\text{vec}(\boldsymbol{\Xi}_m) - \sqrt{T_p}\left[\frac{(\boldsymbol{I}_m \otimes \boldsymbol{A}')\boldsymbol{B}_m'\boldsymbol{X}}{T_p} \otimes \boldsymbol{I}_k\right]\text{vec}[(\hat{\boldsymbol{\beta}} - \boldsymbol{\beta})'],\end{aligned}$$

where $\approx$ is used to denote asymptotic equivalence. This completes the proof of Lemma 2.4. □

EXERCISES

2.1 Prove Lemma 2.2.

2.2 Consider the growth rates, in percentages, of the quarterly real GDP of United Kingdom, Canada, and the United States used in the chapter. Fit a VAR(4) model to the series, simplify the model by removing insignificant parameters with type I error $\alpha = 0.05$, and perform model checking. Finally, compare the simplified VAR(4) model with the simplified VAR(2) model of Section 2.7.3.

2.3 Consider a bivariate time series $\boldsymbol{z}_t$, where z_{1t} is the change in monthly U.S. treasury bills with maturity 3 months and z_{2t} is the inflation rate, in percentage, of the U.S. monthly consumer price index (CPI). The CPI used is the consumer price index for all urban consumers: all items (CPIAUCSL). The original data are downloaded from the Federal Reserve Bank of St. Louis. The CPI rate is 100 times the first difference of the log CPI index. The sample period is from January 1947 to December 2012. The original data are in the file `m-cpitb3m.txt`.

- Construct the z_t series. Obtain the time plots of z_t.
- Select a VAR order for z_t using the BIC criterion.
- Fit the specified VAR model and simplify the fit by the command `refVAR` with threshold 1.65. Write down the fitted model.
- Is the fitted model adequate? Why?
- Compute the impulse response functions of the fitted model using orthogonal innovations. Show the plots and draw conclusion based on the plots.
- Consider the residual covariance matrix. Obtain its Cholesky decomposition and the transformed innovations. Plot the orthogonal innovations.

2.4 Consider the U.S. quarterly gross private saving (GPSAVE) and gross private domestic investment (GPDI) from first quarter of 1947 to the third quarter of 2012. The data are from the Federal Reserve Bank of St. Louis and are in billions of dollars. See the file `m-gpsavedi.txt`.

- Construct the growth series by taking the first difference of the log data. Denote the growth series by z_t. Plot the growth series.
- Build a VAR model for z_t, including simplification and model checking. Write down the fitted model.
- Perform a chi-square test to confirm that one can remove the insignificant parameters in the previous question. You may use 5% significant level.
- Obtain the impulse response functions of the fitted model. What is the relationship between the private investment and saving?
- Obtain one-step to eight-step ahead predictions of z_t at the forecast origin 2012.III (last data point).
- Obtain the forecast error variance decomposition.

2.5 Consider, again, the quarterly growth series z_t of Problem 4. Obtain Bayesian estimation of a VAR(4) model. Write down the fitted model.

2.6 Consider four components of U.S. monthly industrial production index from January 1947 to December 2012 for 792 data points. The four components are durable consumer goods (IPDCONGD), nondurable consumer goods (IPN-CONGD), business equivalent (IPBUSEQ), and materials (IPMAT). The original data are from the Federal Reserve Bank of St. Louis and are seasonally adjusted. See the file `m-ip4comp.txt`.

- Construct the growth rate series z_t of the four industrial production index, that is, take the first difference of the log data. Obtain time plots of z_t. Comment on the time plot.
- Build a VAR model for z_t, including simplification and model checking. Write down the fitted model.
- Compute one-step to six-step ahead predictions of z_t at the forecast origin $h = 791$ (December 2012). Obtain 95% interval forecasts for each component series.

2.7 Consider, again, the z_t series of Problem 6. The time plots show the existence of possible aberrant observations, especially at the beginning of the series. Repeat the analyses of Problem 6, but use the subsample for t from 201 to 791.

2.8 Consider the quarterly U.S. federal government debt from the first quarter of 1970 to the third quarter of 2012. The first series is the federal debt held by foreign and international investors and the second series is federal debt held by Federal Reserve Banks. The data are from Federal Reserve Bank of St Louis, in billions of dollars, and are not seasonally adjusted. See the file `q-fdebt.txt`.

- Construct the bivariate time series z_t of the first difference of log federal debt series. Plot the data.
- Fit a VAR(6) model to the z_t series. Perform model checking. Is the model adequate? Why?
- Perform a chi-square test to verify that all coefficient estimates of the VAR(6) model with t-ratios less than 1.96 can indeed be removed based on an approximate 5% type I error.
- Based on the simplified model, is there any Granger causality between the two time series? Why?

2.9 Consider the quarterly growth rates (percentage change a year ago) of real gross domestic products of Brazil, South Korea, and Israel from 1997.I to 2012.II for 62 observations. The data are from the Federal Reserve Bank of St. Louis and in the file `q-rdgp-brkris.txt`.

- Specify a VAR model for the three-dimensional series.
- Fit the specified model, refine it if necessary, and perform model checking. Is the model adequate? Why?
- Compute the impulse response functions (using the observed innovations) of the fitted model. State the implications of the model.

2.10 Consider the monthly unemployment rates of the States of Illinois, Michigan, and Ohio of the United States from January 1976 to November 2009. The data were seasonally adjusted and are available from FRED of the Federal Reserve Bank of St. Louis. See also the file `m-3state-un.txt`.

- Build a VAR model for the three unemployment rates. Write down the fitted model.
- Use the impulse response functions of the VAR model built to study the relationship between the three unemployment rates. Describe the relationships.
- Use the fitted model to produce point and interval forecasts for the unemployment rates of December 2009 and January 2010 at the forecast origin of November 2009.

REFERENCES

Ahn, S. K. (1988). Distribution for residual autocovariances in multivariate autoregressive models with structured parameterization. *Biometrika*, **75**: 590–593.

Akaike, H. (1973). Information theory and an extension of the maximum likelihood principle. In B. N. Petrov and F. Csaki (eds.). *2nd International Symposium on Information Theory*, pp. 267–281. Akademia Kiado, Budapest.

Basu, A. K. and Sen Roy, S. (1986). On some asymptotic results for multivariate autoregressive models with estimated parameters. *Calcutta Statistical Association Bulletin*, **35**: 123–132.

Box, G. E. P., Jenkins, G. M., and Reinsel, G. (2008). *Time Series Analysis: Forecasting and Control*. 4th Edition. John Wiley & Sons, Inc, Hoboken, NJ.

Doan, T., Litterman, R. B., and Sims, C. A. (1984). Forecasting and conditional projection using realistic prior distributions. *Econometric Reviews*, **3**: 1–144.

Edgerton, D. and Shukur, G. (1999). Testing autocorrelation in a system perspective. *Econometric Reviews*, **18**: 343–386.

Fuller, W. A. (1976). *Introduction to Statistical Time Series*. John Wiley & Sons, Inc, New York.

Granger, C. W. J. (1969). Investigating causal relations by econometric models and cross-spectral methods. *Econometrica*, **37**: 424–438.

Hannan, E. J. and Quinn, B. G. (1979). The determination of the order of an autoregression. *Journal of the Royal Statistical Society, Series B* **41**: 190–195.

Hosking, J. R. M. (1981). Lagrange-multiplier tests of multivariate time series model. Journal of the Royal Statistical Society, Series B, **43**: 219–230.

Johnson, R. A. and Wichern, D. W. (2007). *Applied Multivariate Statistical Analysis*. 6th Edition. Pearson Prentice Hall, Upper Saddle River, NJ.

Li, W. K. and McLeod, A. I. (1981). Distribution of the residual autocorrelations in multivariate time series models. *Journal of the Royal Statistical Society, Series B*, **43**: 231–239.

Litterman, R. B. (1986). Forecasting with Bayesian vector autoregressions – five years of experience. *Journal of Business & Economic Statistics*, **4**: 25–38.

Lütkepohl, H. (2005). *New Introduction to Multiple Time Series Analysis*. Springer, New York.

Quinn, B. G. (1980). Order determination for a multivariate autoregression. *Journal of the Royal Statistical Society, Series B*, **42**: 182–185.

Rossi, P. E., Allenby, G. M., and McCulloch, R. E. (2005). *Bayesian Statistics and Marketing*. John Wiley & Sons, Inc, Hoboken, NJ.

Samaranayake, V. A. and Hasza, D. P. (1988). Properties of predictors for multivariate autoregressive models with estimated parameters. *Journal of Time Series Analysis*, **9**: 361–383.

Schwarz, G. (1978). Estimating the dimension of a model. *Annals of Statistics*, **6**: 461–464.

Shibata, R. (1980). Asymptotically efficient selection of the order of the model for estimating parameters of a linear process. *Annals of Statistics*, **8**: 147–164.

Tiao, G. C. and Box, G. E. P.. (1981). Modeling multiple time series with applications. *Journal of the American Statistical Association*, **76**: 802–816.

Tsay, R. S. (2010). *Analysis of Financial Time Series*. 3rd Edition. John Wiley & Sons, Inc, Hoboken, NJ.

Zellner, A. (1962). An efficient method of estimating seemingly unrelated regressions and tests of aggregation bias. *Journal the American Statistical Association*, **57**: 348–368.

CHAPTER 3

Vector Autoregressive Moving-Average Time Series

For parsimonious parameterization and further simplification in modeling multivariate time series, we consider in this chapter the vector moving-average (VMA) models and the vector autoregressive moving-average (VARMA) models. See Equation (1.21). We study properties of the models and discuss their implications. We also address some new challenges facing the VARMA models that do not occur in studying VAR models of Chapter 2. Similar to the case of VAR models, we start with simple models and provide justifications for using the models. We then provide results for the general stationary and invertible VARMA(p, q) models. Some of the generalizations are obtained by expanding the dimension of the underlying time series in order to reduce the order of a model. Special attention is paid to the exact likelihood function of a VARMA model and the calculation of the likelihood function.

We also study linear transformation and temporal aggregation of a VARMA model. Real examples and some simulations are used to demonstrate the applications and to emphasize the main points of the VARMA model. Finally, we analyze some real time series via the VARMA models.

Heavy matrix notation is used in Sections 3.9 and 3.10 to describe the likelihood function of a VARMA time series. For readers who focus on applications of VARMA models, these two sections can be skipped on the first read.

Multivariate Time Series Analysis: With R and Financial Applications,
First Edition. Ruey S. Tsay.

3.1 VECTOR MA MODELS

A k-dimensional time series $\boldsymbol{z}_t$ follows a VMA model of order q if

$$\boldsymbol{z}_t = \boldsymbol{\mu} + \boldsymbol{a}_t - \sum_{i=1}^{q} \boldsymbol{\theta}_i \boldsymbol{a}_{t-i}, \tag{3.1}$$

where $\boldsymbol{\mu}$ is a constant vector denoting the mean of $\boldsymbol{z}_t$, $\boldsymbol{\theta}_i$ are $k \times k$ matrices with $\boldsymbol{\theta}_q \neq \boldsymbol{0}$, and $\{\boldsymbol{a}_t\}$ is a white noise series defined in Equation (2.1) with $\text{Var}(\boldsymbol{a}_t) = \boldsymbol{\Sigma}_a = [\sigma_{a,ij}]$, which is positive definite. Using the back-shift operator the model becomes $\boldsymbol{z}_t = \boldsymbol{\mu} + \boldsymbol{\theta}(B)\boldsymbol{a}_t$, where $\boldsymbol{\theta}(B) = \boldsymbol{I}_k - \sum_{i=1}^{q} \boldsymbol{\theta}_i B^i$ is a matrix polynomial of degree q.

VMA models exist for many reasons. In finance, it is well known that the bid and ask bounce can introduce negative lag-1 serial correlation in high-frequency returns, for example, minute-by-minute returns, and nonsynchronous trading can lead to serial correlations in asset returns. In economics, survey data are commonly used. If the survey design employed a rotating panel that replaces periodically and systematically part of the sampling units, then the results can exhibit certain serial dependence. The dependence disappears after a finite lag depending on the number of periods for the survey to complete the rotation. As another example, smoothing is commonly used in data processing to reduce the variability. The technique can easily lead to serial correlations that can be handled by MA models if the length of the window used in the smoothing is not long. Finally, in vector time series analysis, over-differencing can lead to MA dependence. This latter issue is related to cointegration discussed in Chapter 5.

3.1.1 VMA(1) Model

To study VMA models, we start with the simplest two-dimensional VMA(1) model,

$$\boldsymbol{z}_t = \boldsymbol{\mu} + \boldsymbol{a}_t - \boldsymbol{\theta}_1 \boldsymbol{a}_{t-1}. \tag{3.2}$$

The model can be written, using $\boldsymbol{\theta}_1 = [\theta_{1,ij}]$, as

$$\begin{bmatrix} z_{1t} \\ z_{2t} \end{bmatrix} = \begin{bmatrix} \mu_1 \\ \mu_2 \end{bmatrix} + \begin{bmatrix} a_{1t} \\ a_{2t} \end{bmatrix} - \begin{bmatrix} \theta_{1,11} & \theta_{1,12} \\ \theta_{1,21} & \theta_{1,22} \end{bmatrix} \begin{bmatrix} a_{1,t-1} \\ a_{2,t-1} \end{bmatrix},$$

or equivalently,

$$\begin{aligned} z_{1t} &= \mu_1 + a_{1t} - \theta_{1,11} a_{1,t-1} - \theta_{1,12} a_{2,t-1} \\ z_{2t} &= \mu_2 + a_{2t} - \theta_{1,21} a_{1,t-1} - \theta_{1,22} a_{2,t-1}. \end{aligned}$$

From the prior equations, the coefficient $\theta_{1,12}$ measures the impact of $a_{2,t-1}$ on z_{1t} in the presence of $a_{1,t-1}$, and $\theta_{1,21}$ denotes the effect of $a_{1,t-1}$ on z_{2t} in the presence of $a_{2,t-1}$. In general, the VMA(1) model is in the form of MA representation of a multivariate time series with the impulse response matrices $\boldsymbol{\psi}_1 = -\boldsymbol{\theta}_1$ and $\boldsymbol{\psi}_i = \mathbf{0}$ for $i > 1$. The elements $\theta_{1,ij}$ can be interpreted in terms of the impulse response function discussed in Chapter 2.

3.1.1.1 Moment Equations

The VMA(1) model in Equation (3.2) is stationary so long as $\{\boldsymbol{a}_t\}$ is a white noise series. Taking the expectation of the model, we have $E(\boldsymbol{z}_t) = \boldsymbol{\mu}$, confirming that $\boldsymbol{\mu}$ is the mean of $\boldsymbol{z}_t$. Also, it is easy to see that

$$\text{Var}(\boldsymbol{z}_t) = \boldsymbol{\Gamma}_0 = \boldsymbol{\Sigma}_a + \boldsymbol{\theta}_1 \boldsymbol{\Sigma}_a \boldsymbol{\theta}_1'.$$

Moreover, we have

$$\boldsymbol{\Gamma}_1 = -\boldsymbol{\theta}_1 \boldsymbol{\Sigma}_a, \quad \boldsymbol{\Gamma}_j = \mathbf{0}, \quad j > 1. \tag{3.3}$$

Consequently, similar to the univariate MA(1) model, the dynamic dependence of a VMA(1) model lasts only one time period. Based on Equation (3.3), the cross-correlation matrices of a VMA(1) model satisfy (i) $\boldsymbol{\rho}_1 \neq \mathbf{0}$ and (ii) $\boldsymbol{\rho}_j = \mathbf{0}$ for $j > 1$.

Example 3.1 Consider the monthly log returns, in percentages, of some capitalization-based portfolios obtained from the Center for Research in Security Prices (CRSP). The data span is from January 1961 to December 2011 with sample size $T = 612$. The portfolios consist of stocks listed on NYSE, AMEX, and NASDAQ. The two portfolios used are decile 5 and decile 8 and the returns are total returns, which include capital appreciation and dividends. Figure 3.1 shows the time plots of the two log return series, whereas Figure 3.2 shows the plots of sample cross-correlation matrices. From the sample cross-correlation plots, the two return series have significant lag-1 dynamic dependence, indicating that a VMA(1) model might be appropriate for the series. Some further details of the portfolio returns are given in the following R demonstration. □

R Demonstration: Bivariate monthly portfolio returns.

```
> da=read.table("m-dec15678-6111.txt",header=T)
> head(da)
      date      dec1      dec5      dec6      dec7      dec8
1 19610131  0.058011  0.081767  0.084824  0.087414  0.099884
2 19610228  0.029241  0.055524  0.067772  0.079544  0.079434
 ...
> x=log(da[,2:6]+1)*100
> rtn=cbind(x$dec5,x$dec8)
> tdx=c(1:612)/12+1961
> par(mfcol=c(2,1))
> plot(tdx,rtn[,1],type='l',xlab='year',ylab='d5')
```

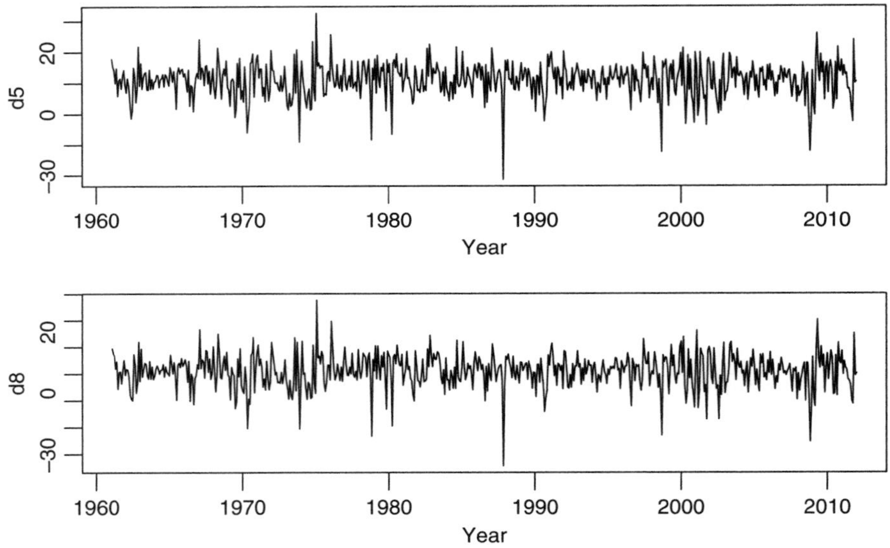

FIGURE 3.1 Time plots of monthly log returns, in percentages, of CRSP decile 5 and decile 8 portfolios from January 1961 to December 2011. The returns are total returns.

```
> plot(tdx,rtn[,2],type='l',xlab='year',ylab='d8')
> ccm(rtn)
[1] "Covariance matrix:"
     [,1] [,2]
[1,] 30.7 34.3
[2,] 34.3 41.2
CCM at lag:  0
      [,1]  [,2]
[1,] 1.000 0.964
[2,] 0.964 1.000
Simplified matrix:
CCM at lag:  1
+ +
+ +
CCM at lag:  2
. .
. .
CCM at lag:  3
. .
. .
```

3.1.1.2 Relation to Transfer Function Models

If $\theta_{1,12} = 0$, but $\theta_{1,21} \neq 0$, then the VMA(1) model in Equation (3.2) indicates the existence of a transfer function relation (or Granger causality) in the system with z_{1t} as the input variable and z_{2t} as the output variable. We can further consider the

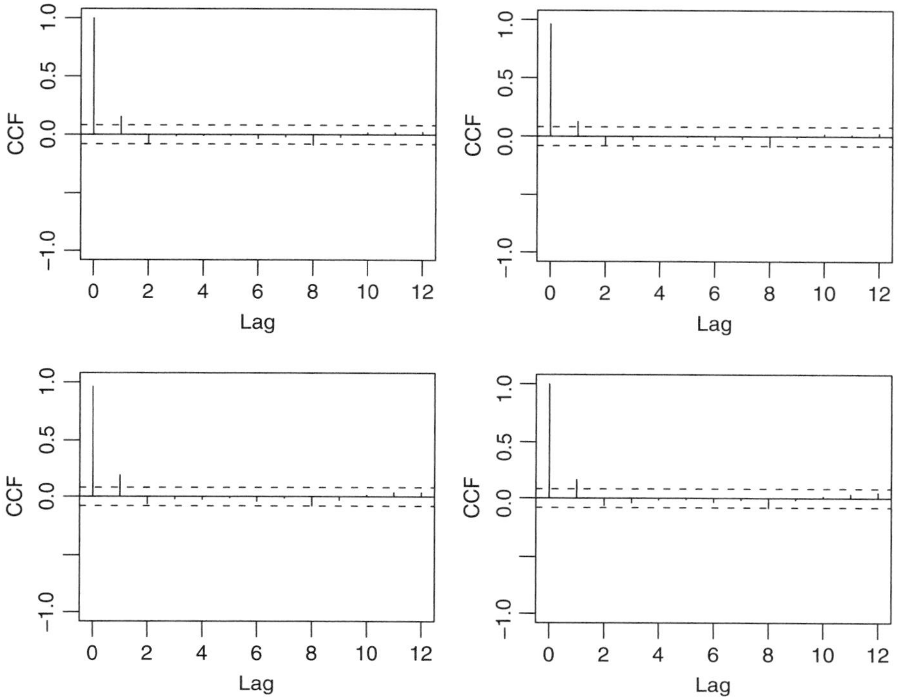

FIGURE 3.2 Plots of sample cross-correlation matrices of monthly log returns of CRSP decile 5 and decile 8 portfolios from January 1961 to December 2011. The returns are total returns.

orthogonalization $a_{2t} = \beta a_{1t} + \epsilon_t$, where $\beta = \sigma_{a,12}/\sigma_{a,11}$ and ϵ_t is uncorrelated with a_{1t}, and obtain a transfer function model

$$z_{2t} = \mu_2 + [\beta - (\theta_{1,22}\beta + \theta_{1,21})B]a_{1t} + (1 - \theta_{1,22}B)\epsilon_t,$$

where $a_{1t} = (1 - \theta_{1,11}B)^{-1}z_{1t}$. If $\theta_{1,21} = 0$, but $\theta_{1,12} \neq 0$, then z_{1t} becomes the output variable and z_{2t} the input variable. If $\theta_{1,12} = \theta_{1,21} = 0$, then z_{1t} and z_{2t} are not dynamically dependent. They may be instantaneously correlated if $\sigma_{a,12} \neq 0$.

3.1.1.3 The Invertibility Condition

Since the mean vector does not affect invertibility, we assume $\boldsymbol{\mu} = \mathbf{0}$. For the VMA(1) model in Equation (3.2), we have $\boldsymbol{a}_t = \boldsymbol{z}_t + \boldsymbol{\theta}_1\boldsymbol{a}_{t-1}$. By repeated substitutions, we have

$$\boldsymbol{a}_t = \boldsymbol{z}_t + \boldsymbol{\theta}_1\boldsymbol{z}_{t-1} + \boldsymbol{\theta}_1^2\boldsymbol{z}_{t-2} + \boldsymbol{\theta}_1^3\boldsymbol{z}_{t-3} + \cdots.$$

Therefore, for the VMA(1) model to have a VAR representation, the prior equation must be a convergent series, and $\boldsymbol{\theta}^j \to \mathbf{0}$ as $j \to \infty$. This is achieved if all eigenvalues of $\boldsymbol{\theta}_1$ are less than 1 in absolute value. Consequently, the invertibility condition of a VMA(1) model is that all eigenvalues of $\boldsymbol{\theta}_1$ must be less than 1 in modulus.

3.1.1.4 *Marginal Component Models*

For a white noise series $\{\boldsymbol{a}_t\}$, consider the expanded $2k$-dimensional random vector $\boldsymbol{y}_t = (\boldsymbol{a}_t', \boldsymbol{a}_{t-1}')'$. It is easy to see that $\text{Cov}(\boldsymbol{y}_t, \boldsymbol{y}_{t-j}) = \boldsymbol{0}$ for $j \geq 2$. Consequently, $x_t = \boldsymbol{c}'\boldsymbol{y}_t$ does not have any serial correlations after lag 1, where $\boldsymbol{c}$ is an arbitrary $2k$-dimensional nonzero constant vector. Using this property, we see that each component z_{it} of the VMA(1) process $\boldsymbol{z}_t$ follows a univariate MA(1) model.

3.1.2 Properties of VMA(q) Models

Properties of the VMA(1) model discussed in the prior subsection can easily be extended to the VMA(q) model, where q is a fixed positive integer. We briefly outline those properties. The VMA(q) series is always weakly stationary provided that $\boldsymbol{a}_t$ is a white noise series. The moment equations of the VMA(q) model can be summarized as follows:

(a) $E(\boldsymbol{z}_t) = \boldsymbol{\mu}$
(b) $\text{Var}(\boldsymbol{z}_t) = \boldsymbol{\Gamma}_0 = \boldsymbol{\Sigma}_a + \sum_{i=1}^{q} \boldsymbol{\theta}_i \boldsymbol{\Sigma}_a \boldsymbol{\theta}_i'$
(c) $\text{Cov}(\boldsymbol{z}_t, \boldsymbol{z}_{t-j}) = \boldsymbol{\Gamma}_j = \sum_{i=j}^{q} \boldsymbol{\theta}_j \boldsymbol{\Sigma}_a \boldsymbol{\theta}_{i-j}'$ where $\boldsymbol{\theta}_0 = -\boldsymbol{I}_k$, for $j = 1, \ldots, q$
(d) $\boldsymbol{\Gamma}_j = \boldsymbol{0}$ for $j > q$.

Therefore, the dynamic dependence of a VMA(q) model disappears after q lags, suggesting that a VMA(q) model has a finite memory. From the moment equations, the cross-correlation matrices $\boldsymbol{\rho}_\ell$ satisfy $\boldsymbol{\rho}_q \neq \boldsymbol{0}$, but $\boldsymbol{\rho}_j = \boldsymbol{0}$ for $j > q$.

Turn to impulse response matrices. It is clear that for a VMA(q) model, (a) $\boldsymbol{\psi}_j = -\boldsymbol{\theta}_j$ for $j = 1, \ldots, q$ and (b) $\boldsymbol{\psi}_j = \boldsymbol{0}$ for $j > q$. Furthermore, one can deduce Granger causality or transfer function relation from a VMA(q) model if the MA coefficient matrices exhibit certain zero block or triangular patterns. For instance, if $\theta_{\ell,ij} = 0$ for all $\ell \in \{1, \ldots, q\}$, then z_{it} does not depend on any lagged values of z_{jt} in the presence of lagged values of other components. This leads naturally to the conditional independence between z_{it} and z_{jt} under the normality assumption.

3.1.2.1 *The Invertibility Condition*

The invertibility condition of a VMA(q) model can be deduced from that of a VMA(1) model. To see this, we can express the VMA(q) model as a kq-dimensional VMA(1) model, where $q > 1$. Specifically, we assume, without loss of generality, that $\boldsymbol{\mu} = \boldsymbol{0}$ and consider the following equations:

$$\boldsymbol{z}_t = \boldsymbol{a}_t - \sum_{i=1}^{q} \boldsymbol{\theta}_i \boldsymbol{a}_{t-i}$$

$$\boldsymbol{0} = \boldsymbol{a}_{t-i} - \boldsymbol{I}_k \boldsymbol{a}_{t-i}, \quad i = 1, \ldots, q-1,$$

where $\mathbf{0}$ is the k-dimensional vector of zero. Putting these equations together, we have

$$\begin{bmatrix} \boldsymbol{z}_t \\ \mathbf{0} \\ \vdots \\ \mathbf{0} \end{bmatrix} = \begin{bmatrix} \boldsymbol{a}_t \\ \boldsymbol{a}_{t-1} \\ \vdots \\ \boldsymbol{a}_{t-q+1} \end{bmatrix} - \begin{bmatrix} \boldsymbol{\theta}_1 & \boldsymbol{\theta}_2 & \cdots & \boldsymbol{\theta}_{q-1} & \boldsymbol{\theta}_q \\ \boldsymbol{I}_k & \mathbf{0}_k & \cdots & \mathbf{0}_k & \mathbf{0}_k \\ \vdots & \vdots & & \vdots & \vdots \\ \mathbf{0}_k & \mathbf{0}_k & \cdots & \boldsymbol{I}_k & \mathbf{0}_k \end{bmatrix} \begin{bmatrix} \boldsymbol{a}_{t-1} \\ \boldsymbol{a}_{t-2} \\ \vdots \\ \boldsymbol{a}_{t-q} \end{bmatrix},$$

where $\mathbf{0}_k$ denotes the $k \times k$ zero matrix. The prior equation is in the form

$$\boldsymbol{y}_t = \boldsymbol{b}_t - \boldsymbol{\Theta} \boldsymbol{b}_{t-1}, \tag{3.4}$$

where $\boldsymbol{y}_t = (\boldsymbol{z}_t', \mathbf{0})'$ with $\mathbf{0}$ being a $k(q-1)$-dimensional row vector of zero, $\boldsymbol{b}_t = (\boldsymbol{a}_t', \ldots, \boldsymbol{a}_{t-q+1}')'$, and $\boldsymbol{\Theta}$ is the $kq \times kq$ companion matrix of the matrix polynomial $\boldsymbol{\theta}(B)$. The mean of $\boldsymbol{b}_t$ is 0 and Cov$(\boldsymbol{b}_t) = \boldsymbol{I}_q \otimes \boldsymbol{\Sigma}_a$. For the kq-dimensional series $\boldsymbol{y}_t$ to be invertible, all eigenvalues of $\boldsymbol{\Theta}$ must be less than 1 in absolute value. By Lemma 2.1, we see that this is equivalent to all solutions of the determinant equation $|\boldsymbol{\theta}(B)| = 0$ are greater than 1 in modulus.

3.1.2.2 *AR Representation*

For an invertible VMA(q) model in Equation (3.1), we can obtain an AR representation for $\boldsymbol{z}_t$. Using $\boldsymbol{z}_t - \boldsymbol{\mu} = \boldsymbol{\theta}(B)\boldsymbol{a}_t$ and $\boldsymbol{\pi}(B)(\boldsymbol{z}_t - \boldsymbol{\mu}) = \boldsymbol{a}_t$, where $\boldsymbol{\pi}(B) = \boldsymbol{I}_k - \sum_{i=1}^{\infty} \boldsymbol{\pi}_i B^i$, we have $[\boldsymbol{\pi}(B)]^{-1} = \boldsymbol{\theta}(B)$. In other words, $\boldsymbol{\theta}(B)\boldsymbol{\pi}(B) = \boldsymbol{I}$. By equating the coefficient matrix of B^i, for $i > 0$, we obtain

$$\boldsymbol{\pi}_i = \sum_{j=1}^{\min\{i,q\}} \boldsymbol{\theta}_i \boldsymbol{\pi}_{i-j}, \quad i > 0, \tag{3.5}$$

where it is understood that $\boldsymbol{\pi}_0 = -\boldsymbol{I}_k$. For instance, for the VMA(1) model in Equation (3.2), we have $\boldsymbol{\pi}_i = -\boldsymbol{\theta}_1^i$.

3.1.2.3 *Marginal Models*

Consider the VMA(q) model in Equation (3.1). Let $\boldsymbol{C}$ be a $k \times g$ matrix of rank $g > 0$ and define $\boldsymbol{x}_t = \boldsymbol{C}'\boldsymbol{z}_t$. From the moment equations of $\boldsymbol{z}_t$ discussed before, we have

(a) $E(\boldsymbol{x}_t) = \boldsymbol{C}'\boldsymbol{\mu}$
(b) Var$(\boldsymbol{x}_t) = \boldsymbol{C}'\boldsymbol{\Sigma}_a\boldsymbol{C} + \sum_{i=1}^{q} \boldsymbol{C}'\boldsymbol{\theta}_i\boldsymbol{\Sigma}_a\boldsymbol{\theta}_i'\boldsymbol{C}$
(c) Cov$(\boldsymbol{x}_t, \boldsymbol{x}_{t-j}) = \sum_{i=j}^{q} \boldsymbol{C}'\boldsymbol{\theta}_j\boldsymbol{\Sigma}_a\boldsymbol{\theta}_{i-j}'\boldsymbol{C}$, where $\boldsymbol{\theta}_0 = -\boldsymbol{I}_k$
(d) Cov$(\boldsymbol{x}_t, \boldsymbol{x}_{t-j}) = \mathbf{0}$ for $j > q$.

Consequently, the linearly transformed series $\boldsymbol{x}_t$ follows a VMA(q) model. In particular, by selecting $\boldsymbol{C}$ to be the ith unit vector in R^k, we see that the component

z_{it} of $\boldsymbol{z}_t$ follows a univariate MA(q) model. As in the VAR case, the order q is the maximum order allowed. We summarize this result into a theorem.

Theorem 3.1 Suppose that $\boldsymbol{z}_t$ follows a VMA(q) model in Equation (3.1) and $\boldsymbol{C}$ is a $k \times g$ constant matrix with rank $g > 0$. Then, $\boldsymbol{x}_t = \boldsymbol{C}'\boldsymbol{z}_t$ is a g-dimensional VMA(q) series.

3.2 SPECIFYING VMA ORDER

The order of a VMA process can be easily identified via the cross-correlation matrices. For a VAM(q) model, the cross-correlation matrices satisfy $\boldsymbol{\rho}_j = \boldsymbol{0}$ for $j > q$. Therefore, for a given j, one can consider the null hypothesis H_0: $\boldsymbol{\rho}_j = \boldsymbol{\rho}_{j+1} = \cdots = \boldsymbol{\rho}_m = \boldsymbol{0}$ versus the alternative H_a:$\boldsymbol{\rho}_\ell \neq 0$ for some ℓ between j and m, where m is a prespecified positive integer. A simple test statistic to use is then

$$Q_k(j,m) = T^2 \sum_{\ell=j}^{m} \frac{1}{T-\ell} tr(\hat{\boldsymbol{\Gamma}}_\ell' \hat{\boldsymbol{\Gamma}}_0^{-1} \hat{\boldsymbol{\Gamma}}_\ell \hat{\boldsymbol{\Gamma}}_0) = T^2 \sum_{\ell=j}^{m} \frac{1}{T-\ell} \boldsymbol{b}_\ell' (\hat{\boldsymbol{\rho}}_o^{-1} \otimes \hat{\boldsymbol{\rho}}_0^{-1}) \boldsymbol{b}_\ell, \tag{3.6}$$

where T is the sample size, $\boldsymbol{b}_\ell = \text{vec}(\hat{\boldsymbol{\rho}}_\ell')$, $\hat{\boldsymbol{\Gamma}}_\ell$ is the lag ℓ sample autocovariance matrix, and $\hat{\boldsymbol{\rho}}_\ell$ is the lag ℓ sample cross-correlation matrix of $\boldsymbol{z}_t$. Under the regularity condition stated in Chapter 1, the test statistic $Q_k(j,m)$ is asymptotically distributed as $\chi^2_{k^2(m-j+1)}$ if $\boldsymbol{z}_t$ follows a VMA(q) model and $j > q$. In practice, we consider $Q_k(j,m)$ for $j = 1, 2, \ldots$. For a VMA(q) model, $Q_k(q,m)$ should be significant, but $Q_k(j,m)$ are insignificant for all $j > q$.

To demonstrate, consider the monthly log returns of CRSP deciles 5 and 8 portfolios from 1961 to 2011 used in Example 3.1. Here, we apply the test statistics in Equation (3.6) to confirm that a VMA(1) model can easily be identified for the series. As expected, in this particular instance, only $Q_k(1,20)$ is significant at the 5% level. See the following R demonstration. Figure 3.3 shows the time plot of the p-values of $Q_k(j,m)$ for the data with $m = 20$. The horizontal dashed line denotes the 5% type I error.

Demonstration: VMA order specification. Output edited.

```
> VMAorder(zt,lag=20)   # Command for identifying MA order
Q(j,m) Statistics:
          j      Q(j,m)    p-value
 [1,]    1.00    109.72      0.02
 [2,]    2.00     71.11      0.64
 [3,]    3.00     63.14      0.76
 [4,]    4.00     58.90      0.78
```

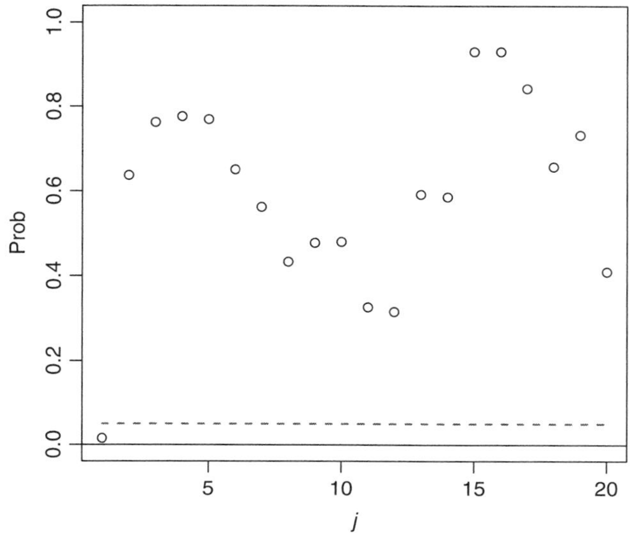

FIGURE 3.3 Time plot of the p-values of the test statistics $Q_k(j, m)$ of Equation (3.6) for the monthly log returns of CRSP deciles 5 and 8 portfolios from 1961 to 2011. The dashed line denotes type-I error.

```
 ....
[19,]  19.00      5.23    0.73
[20,]  20.00      3.97    0.41
```

3.3 ESTIMATION OF VMA MODELS

Unlike the VAR models, there are no closed-form solutions for the estimates of MA parameters. The parameters are commonly estimated by the likelihood method, assuming that the shocks $\boldsymbol{a}_t$ follow a multivariate normal distribution. Two likelihood methods are available. The first method is the conditional likelihood method and the second exact likelihood method. For a given VMA(q) model, the constant term $\boldsymbol{\mu}$ is the mean vector and its maximum likelihood estimate (MLE) is the sample mean. Therefore, we focus our discussion of VMA estimation on zero-mean series. Moreover, via Equation (3.4), a VMA(q) model can always be expressed as a kq-dimensional VMA(1) model so that we entertain the estimation of the Gaussian VMA(1) model

$$\boldsymbol{z}_t = \boldsymbol{a}_t - \boldsymbol{\theta}_1 \boldsymbol{a}_{t-1}.$$

3.3.1 Conditional Likelihood Estimation

The conditional likelihood method assumes that $\boldsymbol{a}_t = \boldsymbol{0}$ for $t \leq 0$. This is equivalent to assume $\boldsymbol{z}_t = \boldsymbol{0}$ for $t \leq 0$. Consequently, using the AR representation, we have

$$
\begin{aligned}
\boldsymbol{a}_1 &= \boldsymbol{z}_1 \\
\boldsymbol{a}_2 &= \boldsymbol{z}_2 + \boldsymbol{\theta}_1 \boldsymbol{z}_1 \\
\vdots &= \vdots \\
\boldsymbol{a}_T &= \boldsymbol{z}_T + \sum_{j=1}^{T-1} \boldsymbol{\theta}_1^{T-j} \boldsymbol{z}_{T-j}.
\end{aligned}
$$

In matrix form, the prior equations become

$$
\begin{bmatrix} \boldsymbol{a}_1 \\ \boldsymbol{a}_2 \\ \boldsymbol{a}_3 \\ \vdots \\ \boldsymbol{a}_T \end{bmatrix} = \begin{bmatrix} \boldsymbol{I}_k & \boldsymbol{0}_k & \boldsymbol{0}_k & \cdots & \boldsymbol{0}_k \\ \boldsymbol{\theta}_1 & \boldsymbol{I}_k & \boldsymbol{0}_k & \cdots & \boldsymbol{0}_k \\ \boldsymbol{\theta}_1^2 & \boldsymbol{\theta}_1 & \boldsymbol{I}_k & \cdots & \boldsymbol{0}_k \\ \vdots & \vdots & \vdots & \ddots & \vdots \\ \boldsymbol{\theta}_1^{T-1} & \boldsymbol{\theta}_1^{T-2} & \boldsymbol{\theta}_1^{T-3} & \cdots & \boldsymbol{I}_k \end{bmatrix} \begin{bmatrix} \boldsymbol{z}_1 \\ \boldsymbol{z}_2 \\ \boldsymbol{z}_3 \\ \vdots \\ \boldsymbol{z}_T \end{bmatrix}. \tag{3.7}
$$

From Equation (3.7), the transformation from $\{\boldsymbol{z}_t\}_{t=1}^T$ to $\{\boldsymbol{a}_t\}_{t=1}^T$ has a unity Jacobian. Consequently, the conditional likelihood function of the data is

$$
p(\boldsymbol{z}_1, \cdots, \boldsymbol{z}_T | \boldsymbol{\theta}_1, \boldsymbol{\Sigma}_a) = p(\boldsymbol{a}_1, \cdots, \boldsymbol{a}_T | \boldsymbol{\theta}_1, \boldsymbol{\Sigma}_a) = \prod_{t=1}^{T} p(\boldsymbol{a}_t | \boldsymbol{\theta}_1, \boldsymbol{\Sigma}_a).
$$

This conditional likelihood function of a VMA(1) model is readily available for estimation as we can evaluate $\boldsymbol{a}_t$ recursively. From the derivation of likelihood function, it is clear that the conditional likelihood function is highly nonlinear of the MA parameters and there exist no closed-form solutions for the parameter estimates. Iterative methods such as the Newton–Raphson method are used to obtain the estimates. Furthermore, it can be shown that the conditional MLEs are consistent and asymptotically normal. The proof follows standard arguments. Readers are referred to Reinsel (1993, Section 5.1.4) for details.

Example 3.2 Consider, again, the monthly log returns, in percentages, of the CRSP decile 5 and decile 8 portfolios used in Example 3.1. The sample cross-correlation matrices and the $Q_k(j, m)$ statistics of Equation (3.6) suggest a VMA(1) model. Letting $\boldsymbol{z}_t = (z_{1t}, z_{2t})'$ with z_{1t} be the log returns of decile 5 and applying the conditional likelihood method, we obtain the fitted model

$$
\boldsymbol{z}_t = \begin{bmatrix} 0.92 \\ 0.98 \end{bmatrix} + \boldsymbol{a}_t - \begin{bmatrix} -0.43 & 0.23 \\ -0.60 & 0.31 \end{bmatrix} \boldsymbol{a}_{t-1}, \quad \hat{\boldsymbol{\Sigma}}_a = \begin{bmatrix} 29.6 & 32.8 \\ 32.8 & 39.1 \end{bmatrix}.
$$

The estimated parameters are significant at the 5% level except $\theta_{1,12}$, which is 0.23 with a marginal t-ratio 1.82. Model checking as discussed in Chapter 2 indicates that the fitted model is adequate. For instance, the Ljung–Box statistics of the residuals give $Q_2(12) = 43.40$ with p-value 0.50. The p-value is based on a chi-square distribution with 44 degrees of freedom because the fitted model uses four MA coefficients. Details of the fitted model are included in the following R demonstration. The eigenvalues of $\hat{\boldsymbol{\theta}}_1$ are -0.089 and -0.031, indicating that the fitted model is invertible, and the dynamic dependence of the decile portfolio returns is weak. The two return series, however, have strong contemporaneous correlation; see $\hat{\boldsymbol{\Sigma}}_a$.

If we treat $\theta_{1,12}$ as 0, then the fitted VMA(1) model indicates that the log return of decile 5 does not depend dynamically on that of decile 8, but the log return of decile 8 depends on the past return of decile 5. This seems plausible as decile 5 consists of relatively larger companies that may affect the return of smaller companies in decile 8. □

R Demonstration: Estimation of VMA models. Output edited. Based on the command `VMA` of the `MTS` package.

```
> m1=VMA(rtn,q=1)
Number of parameters:  6
initial estimates:  0.893 0.946 -0.371 0.185 -0.533 0.266
Final   Estimates:
Coefficient(s):
      Estimate  Std. Error  t value Pr(>|t|)
[1,]    0.9203      0.2595    3.546 0.000392 ***
[2,]    0.9838      0.3026    3.251 0.001151 **
[3,]   -0.4322      0.1448   -2.985 0.002832 **
[4,]    0.2301      0.1255    1.833 0.066764 .
[5,]   -0.5978      0.1676   -3.568 0.000360 ***
[6,]    0.3122      0.1454    2.147 0.031807 *
---
Estimates in matrix form:
Constant term:
Estimates:  0.9202588 0.9838202
MA coefficient matrix
MA( 1 )-matrix
       [,1]  [,2]
[1,] -0.432 0.230
[2,] -0.598 0.312
Residuals cov-matrix:
         [,1]     [,2]
[1,] 29.64753 32.81584
[2,] 32.81584 39.13148
----
aic=  4.44172; bic=  4.422112
> MTSdiag(m1)
[1] "Covariance matrix:"
     [,1] [,2]
[1,] 29.7 32.9
[2,] 32.9 39.2
```

```
CCM at lag:  0
      [,1]  [,2]
[1,] 1.000 0.963
[2,] 0.963 1.000
Simplified matrix:
CCM at lag:  1
. .
. .
CCM at lag:  2
. .
. .
CCM at lag:  3
. .
. .
> r1=m1$residuals
> mq(r1,adj=4) ## Adjust the degrees of freedom.
Ljung-Box Statistics:
          m       Q(m)   p-value
 [1,]   1.000     0.112     1.00
 [2,]   2.000     9.379     0.05
 ...
[12,] 12.000    43.399     0.50
...
[24,] 24.000    90.118     0.54
```

3.3.2 Exact Likelihood Estimation

The exact likelihood method treats initial values of a time series as random variables. They must be handled accordingly in estimation. For the VMA(1) model, the initial value is $\boldsymbol{a}_0$ because $\boldsymbol{z}_1 = \boldsymbol{a}_1 - \boldsymbol{\theta}_1\boldsymbol{a}_0$. The likelihood function of the data depends on this initial value. A proper procedure to obtain the likelihood function of the data then is to integrate out the effect of $\boldsymbol{a}_0$. To this end, we consider the following system of equations:

$$
\begin{aligned}
\boldsymbol{a}_0 &= \boldsymbol{a}_0 \\
\boldsymbol{a}_1 &= \boldsymbol{z}_1 + \boldsymbol{\theta}_1\boldsymbol{a}_0 \\
\boldsymbol{a}_2 &= \boldsymbol{z}_2 + \boldsymbol{\theta}_1\boldsymbol{a}_1 = \boldsymbol{z}_2 + \boldsymbol{\theta}_1\boldsymbol{z}_1 + \boldsymbol{\theta}_1^2\boldsymbol{a}_0 \\
\boldsymbol{a}_3 &= \boldsymbol{z}_3 + \boldsymbol{\theta}_1\boldsymbol{a}_2 = \boldsymbol{z}_3 + \boldsymbol{\theta}_1\boldsymbol{z}_2 + \boldsymbol{\theta}_1^2\boldsymbol{z}_1 + \boldsymbol{\theta}_1^3\boldsymbol{a}_0 \\
\vdots &= \vdots \\
\boldsymbol{a}_T &= \boldsymbol{z}_T + \sum_{j=1}^{T-1}\boldsymbol{\theta}_1^j\boldsymbol{z}_{T-j} + \boldsymbol{\theta}_1^T\boldsymbol{a}_0.
\end{aligned}
$$

Similar to the case of conditional likelihood approach, we can put the prior equations in matrix form as

$$\begin{bmatrix} \boldsymbol{a}_0 \\ \boldsymbol{a}_1 \\ \boldsymbol{a}_2 \\ \vdots \\ \boldsymbol{a}_T \end{bmatrix} = \begin{bmatrix} \boldsymbol{I}_k & \boldsymbol{0}_k & \boldsymbol{0}_k & \cdots & \boldsymbol{0}_k \\ \boldsymbol{\theta}_1 & \boldsymbol{I}_k & \boldsymbol{0}_k & \cdots & \boldsymbol{0}_k \\ \boldsymbol{\theta}_1^2 & \boldsymbol{\theta}_1 & \boldsymbol{I}_k & \cdots & \boldsymbol{0}_k \\ \vdots & \vdots & \vdots & \ddots & \vdots \\ \boldsymbol{\theta}_1^T & \boldsymbol{\theta}_1^{T-1} & \boldsymbol{\theta}_1^{T-2} & \cdots & \boldsymbol{I}_k \end{bmatrix} \begin{bmatrix} \boldsymbol{a}_0 \\ \boldsymbol{z}_1 \\ \boldsymbol{z}_2 \\ \vdots \\ \boldsymbol{z}_T \end{bmatrix}. \tag{3.8}$$

From Equation (3.8), we can transform $\{\boldsymbol{a}_0, \boldsymbol{z}_1, \cdots, \boldsymbol{z}_T\}$ into $\{\boldsymbol{a}_0, \boldsymbol{a}_1, \cdots, \boldsymbol{a}_T\}$ with unit Jacobian. Therefore, we have

$$\begin{aligned} p(\boldsymbol{a}_0, \boldsymbol{z}_1, \cdots, \boldsymbol{z}_T|\boldsymbol{\theta}_1, \boldsymbol{\Sigma}_a) &= p(\boldsymbol{a}_0, \cdots, \boldsymbol{a}_T|\boldsymbol{\theta}_1, \boldsymbol{\Sigma}_a) = \prod_{t=0}^{T} p(\boldsymbol{a}_t|\boldsymbol{\theta}_1, \boldsymbol{\Sigma}_a) \\ &= \frac{1}{(2\pi|\boldsymbol{\Sigma}_a|)^{(T+1)/2}} \exp\left[-\frac{1}{2} tr\left(\sum_{t=0}^{T} \boldsymbol{a}_t \boldsymbol{a}_t' \boldsymbol{\Sigma}_a^{-1}\right)\right]. \end{aligned} \tag{3.9}$$

Next, the likelihood function of the data can be obtained by integrating out $\boldsymbol{a}_0$, namely,

$$p(\boldsymbol{z}_1, \cdots, \boldsymbol{z}_T|\boldsymbol{\theta}_1, \boldsymbol{\Sigma}_a) = \int p(\boldsymbol{a}_0, \boldsymbol{z}_1, \cdots, \boldsymbol{z}_T|\boldsymbol{\theta}_1, \boldsymbol{\Sigma}_a) d\boldsymbol{a}_0. \tag{3.10}$$

The remaining issue is how to integrate out $\boldsymbol{a}_0$. The least squares (LS) properties of Appendix A become helpful. To see this, we rewrite Equation (3.8) by separating the first column block on the right-hand side as

$$\begin{bmatrix} \boldsymbol{a}_0 \\ \boldsymbol{a}_1 \\ \boldsymbol{a}_2 \\ \vdots \\ \boldsymbol{a}_T \end{bmatrix} = \begin{bmatrix} \boldsymbol{0}_k & \boldsymbol{0}_k & \cdots & \boldsymbol{0}_k \\ \boldsymbol{I}_k & \boldsymbol{0}_k & \cdots & \boldsymbol{0}_k \\ \boldsymbol{\theta}_1 & \boldsymbol{I}_k & \cdots & \boldsymbol{0}_k \\ \vdots & \vdots & \ddots & \vdots \\ \boldsymbol{\theta}_1^{T-1} & \boldsymbol{\theta}_1^{T-2} & \cdots & \boldsymbol{I}_k \end{bmatrix} \begin{bmatrix} \boldsymbol{z}_1 \\ \boldsymbol{z}_2 \\ \vdots \\ \boldsymbol{z}_T \end{bmatrix} + \begin{bmatrix} \boldsymbol{I}_k \\ \boldsymbol{\theta}_1 \\ \boldsymbol{\theta}_1^2 \\ \vdots \\ \boldsymbol{\theta}_1^T \end{bmatrix} \boldsymbol{a}_0.$$

The prior equation is in the form of a multiple linear regression:

$$\begin{bmatrix} \boldsymbol{0}_k & \boldsymbol{0}_k & \cdots & \boldsymbol{0}_k \\ \boldsymbol{I}_k & \boldsymbol{0}_k & \cdots & \boldsymbol{0}_k \\ \boldsymbol{\theta}_1 & \boldsymbol{I}_k & \cdots & \boldsymbol{0}_k \\ \vdots & \vdots & \ddots & \vdots \\ \boldsymbol{\theta}_1^{T-1} & \boldsymbol{\theta}_1^{T-2} & \cdots & \boldsymbol{I}_k \end{bmatrix} \begin{bmatrix} \boldsymbol{z}_1 \\ \boldsymbol{z}_2 \\ \vdots \\ \boldsymbol{z}_T \end{bmatrix} = \begin{bmatrix} -\boldsymbol{I}_k \\ -\boldsymbol{\theta}_1 \\ -\boldsymbol{\theta}_1^2 \\ \vdots \\ -\boldsymbol{\theta}_1^T \end{bmatrix} \boldsymbol{a}_0 + \begin{bmatrix} \boldsymbol{a}_0 \\ \boldsymbol{a}_1 \\ \boldsymbol{a}_2 \\ \vdots \\ \boldsymbol{a}_T \end{bmatrix}, \tag{3.11}$$

with $\boldsymbol{a}_0$ as the vector of unknown parameters. More specifically, letting $\boldsymbol{Y}$ denote the $k(T+1)$-dimensional vector in the left side of Equation (3.11), $\boldsymbol{A}$ be the $k(T+1)$-dimensional error vector, and $\boldsymbol{X}$ be the $k(T+1) \times k$ coefficient matrix of $\boldsymbol{a}_0$ in the same equation, we have

$$\boldsymbol{Y} = \boldsymbol{X}\boldsymbol{a}_0 + \boldsymbol{A}. \tag{3.12}$$

This is indeed a multiple linear regression, except that $\text{Cov}(\boldsymbol{A}) = \boldsymbol{I}_{T+1} \otimes \boldsymbol{\Sigma}_a$. The vector $\boldsymbol{Y}$ is readily available from the data and a given $\boldsymbol{\theta}_1$. Let $\boldsymbol{\Sigma}_a^{1/2}$ be the positive-definite square-root matrix of the covariance matrix $\boldsymbol{\Sigma}_a$, and $\boldsymbol{\Sigma}^{-1/2} = \boldsymbol{I}_{T+1} \otimes \boldsymbol{\Sigma}_a^{-1/2}$, which is a $k(T+1) \times k(T+1)$ matrix. Premultiplying Equation (3.12) by $\boldsymbol{\Sigma}^{-1/2}$, we obtain the ordinary multiple linear regression

$$\tilde{\boldsymbol{Y}} = \tilde{\boldsymbol{X}}\boldsymbol{a}_0 + \tilde{\boldsymbol{A}}, \tag{3.13}$$

where $\tilde{\boldsymbol{Y}} = \boldsymbol{\Sigma}^{-1/2}\boldsymbol{Y}$ and $\tilde{\boldsymbol{X}}$ and $\tilde{\boldsymbol{A}}$ are defined similarly. Obviously, $\text{Cov}(\tilde{\boldsymbol{A}}) = \boldsymbol{I}_{k(T+1)}$. The LS estimate of the initial value is

$$\hat{\boldsymbol{a}}_0 = (\tilde{\boldsymbol{X}}'\tilde{\boldsymbol{X}})^{-1}\tilde{\boldsymbol{X}}'\tilde{\boldsymbol{Y}}. \tag{3.14}$$

From Equation (3.13), we have

$$\tilde{\boldsymbol{A}} = \tilde{\boldsymbol{Y}} - \tilde{\boldsymbol{X}}\boldsymbol{a}_0 = \tilde{\boldsymbol{Y}} - \tilde{\boldsymbol{X}}\hat{\boldsymbol{a}}_0 + \tilde{\boldsymbol{X}}(\hat{\boldsymbol{a}}_0 - \boldsymbol{a}_0).$$

Using Property (ii) of the LS estimate in Appendix A, we have

$$\tilde{\boldsymbol{A}}'\tilde{\boldsymbol{A}} = (\tilde{\boldsymbol{Y}} - \tilde{\boldsymbol{X}}\hat{\boldsymbol{a}}_0)'(\tilde{\boldsymbol{Y}} - \tilde{\boldsymbol{X}}\hat{\boldsymbol{a}}_0) + (\hat{\boldsymbol{a}}_0 - \boldsymbol{a}_0)'\tilde{\boldsymbol{X}}'\tilde{\boldsymbol{X}}(\hat{\boldsymbol{a}}_0 - \boldsymbol{a}_0). \tag{3.15}$$

From the definitions of $\boldsymbol{A}$ and $\tilde{\boldsymbol{A}}$ in Equations (3.12) and (3.13), respectively, it is easily seen that

$$\begin{aligned}\tilde{\boldsymbol{A}}'\tilde{\boldsymbol{A}} &= \boldsymbol{A}'(\boldsymbol{I}_{T+1} \otimes \boldsymbol{\Sigma}_a^{-1})\boldsymbol{A} = \sum_{t=0}^{T} \boldsymbol{a}_t'\boldsymbol{\Sigma}_a^{-1}\boldsymbol{a}_t = \sum_{t=0}^{T} tr(\boldsymbol{a}_t'\boldsymbol{\Sigma}_a^{-1}\boldsymbol{a}_t) \\ &= \sum_{t=0}^{T} tr(\boldsymbol{a}_t\boldsymbol{a}_t'\boldsymbol{\Sigma}_a^{-1}) = tr\left(\sum_{t=0}^{T} \boldsymbol{a}_t\boldsymbol{a}_t'\boldsymbol{\Sigma}_a^{-1}\right).\end{aligned} \tag{3.16}$$

Using Equations (3.15) and (3.16), we can write the exponent of the joint density function in Equation (3.9) as

$$
\begin{aligned}
S &= tr\left(\sum_{t=0}^{T} \boldsymbol{a}_t \boldsymbol{a}_t' \boldsymbol{\Sigma}_a^{-1}\right) \\
&= (\tilde{\boldsymbol{Y}} - \tilde{\boldsymbol{X}}\hat{\boldsymbol{a}}_0)'(\tilde{\boldsymbol{Y}} - \tilde{\boldsymbol{X}}\hat{\boldsymbol{a}}_0) + (\hat{\boldsymbol{a}}_0 - \boldsymbol{a}_0)'\tilde{\boldsymbol{X}}'\tilde{\boldsymbol{X}}(\hat{\boldsymbol{a}}_0 - \boldsymbol{a}_0). \qquad (3.17)
\end{aligned}
$$

Ignoring the normalization constant, Equation (3.9) becomes

$$
\begin{aligned}
p(\boldsymbol{a}_0, \boldsymbol{z}_1, \ldots, \boldsymbol{z}_T | \boldsymbol{\theta}_1, \boldsymbol{\Sigma}_a) \propto \exp\left[-\tfrac{1}{2}(\tilde{\boldsymbol{Y}} - \tilde{\boldsymbol{X}}\hat{\boldsymbol{a}}_0)'(\tilde{\boldsymbol{Y}} - \tilde{\boldsymbol{X}}\hat{\boldsymbol{a}}_0)\right] \\
\times \exp\left[-\tfrac{1}{2}(\hat{\boldsymbol{a}}_0 - \boldsymbol{a}_0)'\tilde{\boldsymbol{X}}'\tilde{\boldsymbol{X}}(\hat{\boldsymbol{a}}_0 - \boldsymbol{a}_0)\right]. \qquad (3.18)
\end{aligned}
$$

Notice that the first exponent of Equation (3.18) does not involve $\boldsymbol{a}_0$. Therefore, the integration of Equation (3.10) only involves the second exponent of Equation (3.18). In addition, the second exponent is in the form of a k-dimensional normal density with mean $\hat{\boldsymbol{a}}_0$ and covariance matrix $(\tilde{\boldsymbol{X}}'\tilde{\boldsymbol{X}})^{-1}$. Using properties of multivariate normal distribution, we can carry out the integration in Equation (3.10) and obtain the exact likelihood function of the data $\boldsymbol{Z} = [\boldsymbol{z}_1, \ldots, \boldsymbol{z}_T]$ as

$$
p(\boldsymbol{Z}|\boldsymbol{\theta}_1, \boldsymbol{\Sigma}_a) = \frac{1}{(2\pi|\boldsymbol{\Sigma}_a|)^{T/2}|\tilde{\boldsymbol{X}}'\tilde{\boldsymbol{X}}|^{1/2}} \exp\left[-\frac{1}{2}(\tilde{\boldsymbol{Y}} - \tilde{\boldsymbol{X}}\hat{\boldsymbol{a}}_0)'(\tilde{\boldsymbol{Y}} - \tilde{\boldsymbol{X}}\hat{\boldsymbol{a}}_0)\right]. \qquad (3.19)
$$

This likelihood function can be calculated recursively for given $\boldsymbol{\theta}_1$ and $\boldsymbol{\Sigma}_a$. Specifically, for given $\boldsymbol{\theta}_1$ and $\boldsymbol{\Sigma}_a$, we can formulate the multiple linear regression in Equation (3.13). The exponent of Equation (3.19) is the sum of squares of residuals of the multiple linear regression. The determinant $|\tilde{\boldsymbol{X}}'\tilde{\boldsymbol{X}}|$ can also be obtained easily. As expected, the exact MLFs are also consistent and asymptotically normal.

From the derivation, estimation of VMA models via the exact likelihood method requires more intensive computation than that uses the conditional likelihood method because the former must estimate the initial value $\boldsymbol{a}_0$ to evaluate the likelihood function. Experience indicates that the two likelihood functions provide similar estimates when the sample size T is large, especially when the VMA(q) model is away from the noninvertible region. However, if the VMA(q) model is close to being noninvertible, then exact likelihood estimates are preferred. See, for instance, Hillmer and Tiao (1979).

Example 3.3 Consider, again, the monthly log returns, in percentages, of the decile 5 and decile 8 portfolios of CRSP from January 1961 to December 2011 employed in Example 3.2. Here, we apply the exact likelihood method and obtain the fitted model

$$\boldsymbol{z}_t = \begin{bmatrix} 0.92 \\ 0.98 \end{bmatrix} + \boldsymbol{a}_t - \begin{bmatrix} -0.43 & 0.23 \\ -0.60 & 0.31 \end{bmatrix} \boldsymbol{a}_{t-1}, \quad \hat{\boldsymbol{\Sigma}}_a = \begin{bmatrix} 29.6 & 32.8 \\ 32.8 & 39.1 \end{bmatrix}.$$

In this particular instance, the same size is $T = 612$, which is reasonably large so that the estimates are close to those of the conditional likelihood method, to the two-digit approximation. Again, model checking suggests that the fitted model is adequate. For example, the Ljung–Box statistics of the residuals give $Q_2(12) = 43.40$ with p-value 0.50. Details of the exact likelihood estimates are given in the following R demonstration. □

R Demonstration: Exact likelihood estimation. Output edited. Based on the command `VMAe` of the `MTS` package.

```
> m2=VMAe(rtn,q=1)
Number of parameters:  6
initial estimates:  0.893 0.946 -0.371 0.185 -0.533 0.266
Final   Estimates:
Coefficient(s):
      Estimate  Std. Error  t value Pr(>|t|)
[1,]    0.9196      0.2594    3.544 0.000394 ***
[2,]    0.9829      0.3025    3.249 0.001158 **
[3,]   -0.4332      0.1447   -2.993 0.002760 **
[4,]    0.2306      0.1255    1.838 0.066102 .
[5,]   -0.5992      0.1676   -3.576 0.000349 ***
[6,]    0.3129      0.1454    2.152 0.031423 *
---
Estimates in matrix form:
Constant term:
Estimates:  0.9195531 0.9828963
MA coefficient matrix
MA( 1 )-matrix
       [,1]  [,2]
[1,] -0.433 0.231
[2,] -0.599 0.313
Residuals cov-matrix:
         [,1]     [,2]
[1,] 29.64754 32.81583
[2,] 32.81583 39.13143
----
aic=  4.44172; bic=  4.422112
> MTSdiag(m2)   % Model checking
[1] "Covariance matrix:"
     [,1] [,2]
[1,] 29.7 32.9
```

```
[2,] 32.9 39.2
CCM at lag:  0
      [,1]  [,2]
[1,] 1.000 0.963
[2,] 0.963 1.000
Simplified matrix:
CCM at lag:  1
. .
. .
CCM at lag:  2
. .
. .
> r2=m2$residuals
> mq(r2,adj=4)
Ljung-Box Statistics:
          m        Q(m)    p-value
 [1,]   1.000     0.117      1.00
 [2,]   2.000     9.380      0.05
 ...
[12,]  12.000    43.399      0.50
 ...
[24,]  24.000    90.115      0.54
```

Example 3.4 To demonstrate the difference between conditional and exact likelihood estimations of VMA models, we consider the monthly log returns of IBM and Coca Cola stocks from January 2001 to December 2011 for $T = 132$ observations. Figure 3.4 shows the time plots of the two monthly log returns. Figure 3.5 plots the sample cross-correlation matrices of the data. From the plots, there are no dynamic dependence in the two return series. Furthermore, the Ljung–Box statistics give $Q_2(5) = 25.74$ with p-value 0.17, confirming no significant serial or cross-correlation coefficients. The simple t-tests show that the mean returns of the two stocks are not significantly different from 0 at the 5% level. Therefore, the two return series are essentially a bivariate white noise series. To confirm this observation, we estimate a VMA(1) model to the series using both the conditional and exact likelihood methods. Let z_{1t} and z_{2t} be the percentage log returns of IBM and KO stock, respectively. The fitted model based on the conditional likelihood method is

$$\boldsymbol{z}_t = \boldsymbol{a}_t - \begin{bmatrix} 0.067 & 0.192 \\ -0.015 & 0.013 \end{bmatrix} \boldsymbol{a}_{t-1}, \quad \hat{\boldsymbol{\Sigma}}_a = \begin{bmatrix} 57.4 & 7.0 \\ 7.0 & 26.3 \end{bmatrix},$$

where all estimates are statistically insignificant at the usual 5% level. The fitted model of the exact likelihood method is

$$\boldsymbol{z}_t = \boldsymbol{a}_t - \begin{bmatrix} 0.074 & 0.213 \\ -0.019 & 0.009 \end{bmatrix} \boldsymbol{a}_{t-1}, \quad \hat{\boldsymbol{\Sigma}}_a = \begin{bmatrix} 57.4 & 7.0 \\ 7.0 & 26.3 \end{bmatrix}.$$

Again, as expected, all estimated MA coefficients are statistically insignificant at the 5% level.

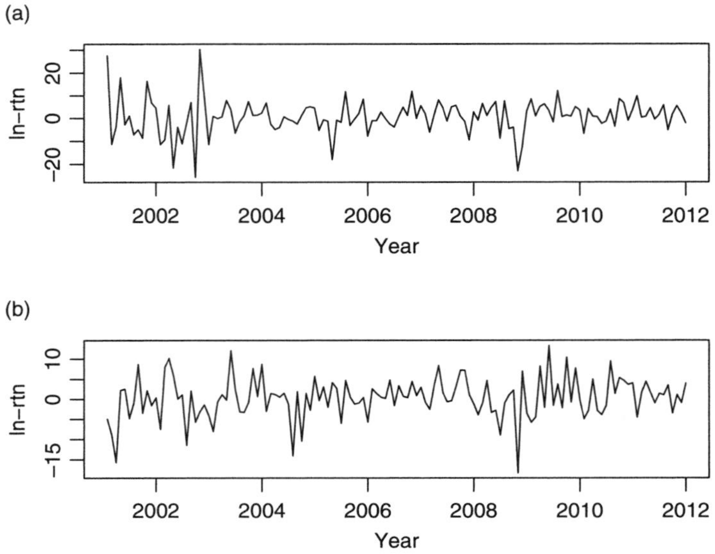

FIGURE 3.4 Time plots of monthly log returns, in percentages, of (a) IBM and (b) Coca Cola stocks from January 2001 to December 2011.

In multivariate time series analysis, noninvertible models can result from over-differencing. While over-differencing would not occur for asset returns, we use it to demonstrate the effect of initial value $\boldsymbol{a}_0$ on the estimation of VMA models. To this end, we purposely difference the two monthly return series. That is, we consider $\boldsymbol{y}_t = (1 - B)\boldsymbol{z}_t$. The sample cross-correlation matrices of $\boldsymbol{y}_t$ suggest a VMA(1) model for the series. The conditional likelihood estimation of the VMA(1) model for $\boldsymbol{y}_t$ is

$$\boldsymbol{y}_t = \boldsymbol{a}_t - \begin{bmatrix} 0.734 & 0.176 \\ 0.053 & 0.965 \end{bmatrix} \boldsymbol{a}_{t-1}, \quad \hat{\boldsymbol{\Sigma}}_a = \begin{bmatrix} 76.1 & 6.2 \\ 6.2 & 26.0 \end{bmatrix}, \tag{3.20}$$

where all MA coefficients are statistically significant at the usual 5% level. On the other hand, the exact likelihood estimation of the VMA(1) model for $\boldsymbol{y}_t$ is

$$\boldsymbol{y}_t = \boldsymbol{a}_t - \begin{bmatrix} 0.877 & 0.087 \\ 0.025 & 0.982 \end{bmatrix} \boldsymbol{a}_{t-1}, \quad \hat{\boldsymbol{\Sigma}}_a = \begin{bmatrix} 83.9 & 5.8 \\ 5.8 & 25.9 \end{bmatrix}, \tag{3.21}$$

where only the two diagonal elements of $\hat{\boldsymbol{\theta}}_1$ are statistically significant at the 5% level.

It is interesting to compare the two models in Equations (3.20) and (3.21). First, the two fitted models are rather different. For instance, all four coefficients of $\hat{\boldsymbol{\theta}}_1$

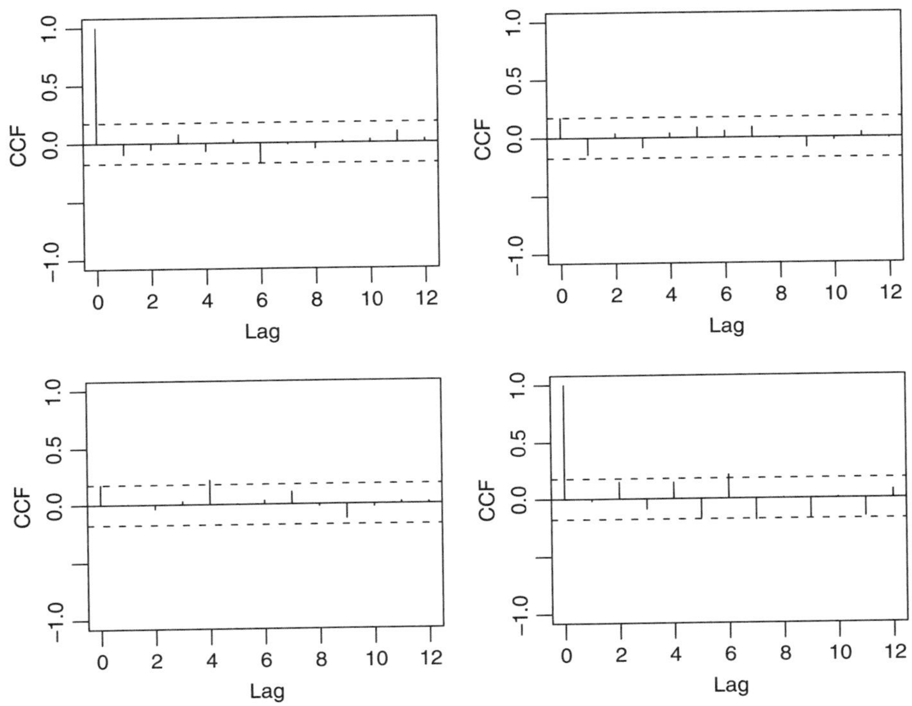

FIGURE 3.5 Cross-correlation matrices of monthly log returns, in percentages, of IBM and Coca Cola stocks from January 2001 to December 2011.

of Model (3.20) are significant, whereas the off-diagonal elements of $\hat{\boldsymbol{\theta}}_1$ in Model (3.21) are not. Second, the eigenvalues of $\hat{\boldsymbol{\theta}}_1$ for Model (3.20) are 1.0 and 0.70, respectively, whereas those of Model (3.21) are 1.0 and 0.86, respectively. Both models are noninvertible as $\hat{\boldsymbol{\theta}}_1$ contains a unit eigenvalue. However, the second eigenvalue of Model (3.21) is much closer to the unit-circle. This indicates that the exact likelihood estimation provides more accurate estimates of the MA parameters because we expect both eigenvalues to be close to one caused by the artificial over-differencing. □

R Demonstration: Comparison between conditional and exact likelihood estimations. Output edited.

```
> rtn=cbind(ibm,ko)
> mq(rtn,10)
Ljung-Box Statistics:
          m       Q(m)   p-value
 [1,]  1.00      3.46      0.48
 ...
 [5,]  5.00     25.74      0.17
> yt=diffM(rtn)
```

```
> mm=ccm(yt)
Simplified matrix:
CCM at lag:  1
- -
. -
> m1=VMA(rtn,q=1,include.mean=F)
Number of parameters:  4
initial estimates:  0.022 0.228 -0.060 0.048
Final   Estimates:
Coefficient(s):
      Estimate  Std. Error  t value Pr(>|t|)
[1,]   0.06672     0.09223    0.723    0.469
[2,]   0.19217     0.13083    1.469    0.142
[3,]  -0.01508     0.05896   -0.256    0.798
[4,]   0.01256     0.07747    0.162    0.871
---
Estimates in matrix form:
MA coefficient matrix
MA( 1 )-matrix
        [,1]   [,2]
[1,]  0.0667 0.1922
[2,] -0.0151 0.0126
  Residuals cov-matrix:
          [,1]      [,2]
[1,] 57.375878  7.024367
[2,]  7.024367 26.326870
----
> m2=VMAe(rtn,q=1,include.mean=F)
Number of parameters:  4
initial estimates:  0.022 0.228 -0.060 0.048
Final   Estimates:
Coefficient(s):
      Estimate  Std. Error  t value Pr(>|t|)
[1,]  0.074385    0.096034    0.775    0.439
[2,]  0.213643    0.136284    1.568    0.117
[3,] -0.019201    0.059104   -0.325    0.745
[4,]  0.009468    0.076993    0.123    0.902
---
Estimates in matrix form:
MA coefficient matrix
MA( 1 )-matrix
        [,1]    [,2]
[1,]  0.0744 0.21364
[2,] -0.0192 0.00947
  Residuals cov-matrix:
          [,1]      [,2]
[1,] 57.403576  7.022986
[2,]  7.022986 26.324071
----
```

```
> yt=diffM(rtn) ### difference the returns
> m1=VMA(yt,q=1,include.mean=F)
Number of parameters:  4
initial estimates:  0.861 0.242 -0.033 0.877
Final   Estimates:
Coefficient(s):
      Estimate  Std. Error  t value Pr(>|t|)
[1,]  0.733690    0.004071   180.23   <2e-16 ***
[2,]  0.175509    0.015932    11.02   <2e-16 ***
[3,]  0.053132    0.004843    10.97   <2e-16 ***
[4,]  0.964984    0.001383   697.74   <2e-16 ***
---
Estimates in matrix form:
MA coefficient matrix
MA( 1 )-matrix
       [,1]  [,2]
[1,] 0.7337 0.176
[2,] 0.0531 0.965
  Residuals cov-matrix:
          [,1]      [,2]
[1,] 76.118900  6.218605
[2,]  6.218605 26.026172
----
> m2=VMAe(yt,q=1,include.mean=F)
Number of parameters:  4
initial estimates:  0.861 0.242 -0.033 0.877
Final   Estimates:
Coefficient(s):
      Estimate  Std. Error  t value Pr(>|t|)
[1,]  0.876916    0.003838  228.455   <2e-16 ***
[2,]  0.087439    0.075615    1.156    0.248
[3,]  0.024876    0.021511    1.156    0.248
[4,]  0.982328    0.001452  676.615   <2e-16 ***
---
Estimates in matrix form:
MA coefficient matrix
MA( 1 )-matrix
       [,1]   [,2]
[1,] 0.8769 0.0874
[2,] 0.0249 0.9823
  Residuals cov-matrix:
          [,1]      [,2]
[1,] 83.868320  5.793222
[2,]  5.793222 25.887593
----
> t1=m1$Theta; t2=m2$Theta
> eigen(t1)
$values
[1] 1.0000000 0.6986745
```

```
> eigen(t2)
$values
[1] 1.0000000 0.8592442
```

3.3.3 Initial Parameter Estimation

Both the conditional and exact likelihood methods for estimating a VMA model require an iterated optimization procedure, such as the Newton–Raphson method, to obtain parameter estimates. Any iterated method in turn requires initial parameter estimates. A good initial estimate can simplify the estimation, especially for constrained estimation. For the VMA models, since $\boldsymbol{a}_t$ are not available before estimation, we use a VAR approximation to obtain a proxy for $\boldsymbol{a}_t$. Specifically, for a given VMA(q) series $\boldsymbol{z}_t$, we use the Akaike information criterion to select a VAR model. Other criterion functions can be used. Suppose that the selected model is VAR(p). We use the ordinary LS method of Chapter 2 to fit the VAR(p) model. Let the residual series of the fitted VAR(p) be $\tilde{\boldsymbol{a}}_t$. We then fit the model

$$\boldsymbol{z}_t = \boldsymbol{c} + \sum_{i=1}^{q} \boldsymbol{\beta}_i \tilde{\boldsymbol{a}}_{t-i} + \boldsymbol{\epsilon}_t, \quad t = q+1, \ldots, T, \tag{3.22}$$

via the ordinary LS method. The initial estimates of the VMA parameters are defined as $\hat{\boldsymbol{\mu}}_o = \hat{\boldsymbol{c}}$ and $\hat{\boldsymbol{\theta}}_i = -\hat{\boldsymbol{\beta}}_i$ for $i = 1, \ldots, q$. If the VMA parameters satisfy some zero restrictions, we fit the model in Equation (3.22) equation-by-equation with the zero restrictions to obtain initial estimates of the VMA parameters. This procedure is used in the `MTS` package.

3.4 FORECASTING OF VMA MODELS

VMA models have finite memory so that their predictions are mean-reverting in a finite number of steps. Suppose t is the forecast origin and the model is known. The one-step ahead prediction of the VMA(q) model in Equation (3.1) is

$$\begin{aligned}
\boldsymbol{z}_t(1) &= E(\boldsymbol{z}_{t+1}|F_t) \\
&= E(\boldsymbol{\mu} + \boldsymbol{a}_{t+1} - \boldsymbol{\theta}_1 \boldsymbol{a}_t - \cdots - \boldsymbol{\theta}_q \boldsymbol{a}_{t+1-q}|F_t) \\
&= \boldsymbol{\mu} - \boldsymbol{\theta}_1 \boldsymbol{a}_t - \cdots - \boldsymbol{\theta}_q \boldsymbol{a}_{t+1-q}.
\end{aligned}$$

The associated forecast error and its covariance matrix are

$$\boldsymbol{e}_t(1) = \boldsymbol{a}_{t+1}, \quad \text{Cov}[\boldsymbol{e}_t(1)] = \boldsymbol{\Sigma}_a.$$

In general, for h-step ahead prediction with $h \le q$, we have

$$\boldsymbol{z}_t(h) = \boldsymbol{\mu} - \sum_{i=h}^{q} \boldsymbol{\theta}_i \boldsymbol{a}_{t+h-i},$$

$$\boldsymbol{e}_t(h) = \boldsymbol{a}_{t+h} - \sum_{i=1}^{h-1} \boldsymbol{\theta}_i \boldsymbol{a}_{t+h-i},$$

$$\text{Cov}[\boldsymbol{e}_t(h)] = \boldsymbol{\Sigma}_a + \sum_{i=1}^{h-1} \boldsymbol{\theta}_i \boldsymbol{\Sigma}_a \boldsymbol{\theta}_i'.$$

For $h > q$, we have

$$\boldsymbol{z}_t(h) = \boldsymbol{\mu},$$

$$\boldsymbol{e}_t(h) = \boldsymbol{a}_{t+h} - \sum_{i=1}^{q} \boldsymbol{\theta}_i \boldsymbol{a}_{t+h-i},$$

$$\text{Cov}[\boldsymbol{e}_t(h)] = \boldsymbol{\Sigma}_a + \sum_{i=1}^{q} \boldsymbol{\theta}_i \boldsymbol{\Sigma}_a \boldsymbol{\theta}_i' = \text{Cov}(\boldsymbol{z}_t).$$

Therefore, the VMA(q) model is mean-reverting in $q+1$ steps.

3.5 VARMA MODELS

A k-dimensional time series $\boldsymbol{z}_t$ is a vector autoregressive moving-average, VARMA(p, q), process if

$$\boldsymbol{\phi}(B)\boldsymbol{z}_t = \boldsymbol{\phi}_0 + \boldsymbol{\theta}(B)\boldsymbol{a}_t \tag{3.23}$$

where $\boldsymbol{\phi}_0$ is a constant vector, $\boldsymbol{\phi}(B) = \boldsymbol{I}_k - \sum_{i=1}^{p} \boldsymbol{\phi}_i B^i$ and $\boldsymbol{\theta}(B) = \boldsymbol{I}_k - \sum_{i=1}^{q} \boldsymbol{\theta}_i B^i$ are two matrix polynomials, and $\boldsymbol{a}_t$ is a sequence of independent and identically distributed random vectors with mean zero and positive-definite covariance matrix $\boldsymbol{\Sigma}_a$. See Equation (1.21) of Chapter 1. In Equation (3.23), we require two additional conditions:

1. $\boldsymbol{\phi}(B)$ and $\boldsymbol{\theta}(B)$ are left co-prime, that is, if $\boldsymbol{u}(B)$ is a left common factor of $\boldsymbol{\phi}(B)$ and $\boldsymbol{\theta}(B)$, then $|\boldsymbol{u}(B)|$ is a nonzero constant. Such a polynomial matrix $\boldsymbol{u}(B)$ is called a *unimodular* matrix. In theory, $\boldsymbol{u}(B)$ is unimodular if and only if $\boldsymbol{u}^{-1}(B)$ exists and is a matrix polynomial (of finite degrees).
2. The MA order q is as small as possible and the AR order p is as small as possible for that q, and the matrices $\boldsymbol{\phi}_p$ (with $p > 0$) and $\boldsymbol{\theta}_q$ (with $q > 0$) satisfy the condition that the rank of the joint matrix $[\boldsymbol{\phi}_p, \boldsymbol{\theta}_q]$ is k, the dimension of $\boldsymbol{z}_t$.

These two conditions are sufficient conditions for VARMA models to be identifiable. In the literature, these conditions are referred to as *block identifiability*. Condition (1) is easily understandable as any nonunimodular left common factor between $\boldsymbol{\phi}(B)$ and $\boldsymbol{\theta}(B)$ can be canceled to reduce the orders p and q. Condition (2) is discussed in Dunsmuir and Hannan (1976). It can be refined by considering column degrees of $\boldsymbol{\phi}(B)$ and $\boldsymbol{\theta}(B)$ instead of the overall degrees p and q; see Hannan and Deistler (1988, Section 2.7). We discuss some of the identifiability issues of VARMA models in the next section.

3.5.1 Identifiability

Unlike the VAR or VMA models, VARMA models encounter the problem of identifiability. Consider the linear vector process,

$$\boldsymbol{z}_t = \boldsymbol{\mu} + \sum_{i=0}^{\infty} \boldsymbol{\psi}_i \boldsymbol{a}_{t-i}, \tag{3.24}$$

where $\boldsymbol{\psi}_0 = \boldsymbol{I}_k$, and $\{\boldsymbol{a}_t\}$ is an iid sequence of random vectors with mean zero and positive-definite covariance matrix $\boldsymbol{\Sigma}_a$. See Equation (1.1). A VARMA model is said to be identified if its matrix polynomials $\boldsymbol{\phi}(B)$ and $\boldsymbol{\theta}(B)$ are uniquely determined by the ψ-weight matrices $\boldsymbol{\psi}_i$ in Equation (3.24). There are cases for which multiple pairs of AR and MA matrix polynomials give rise to identical $\boldsymbol{\psi}_i$ matrices. In this case, care must be exercised to understand the structure of the underlying model. We use simple bivariate models to discuss the issue.

Example 3.5 Consider the VMA(1) model

$$\begin{bmatrix} z_{1t} \\ z_{2t} \end{bmatrix} = \begin{bmatrix} a_{1t} \\ a_{2t} \end{bmatrix} - \begin{bmatrix} 0 & 2 \\ 0 & 0 \end{bmatrix} \begin{bmatrix} a_{1,t-1} \\ a_{2,t-1} \end{bmatrix}, \tag{3.25}$$

which is a well-defined VMA(1) model. However, it can also be written as the VAR(1) model

$$\begin{bmatrix} z_{1t} \\ z_{2t} \end{bmatrix} - \begin{bmatrix} 0 & -2 \\ 0 & 0 \end{bmatrix} \begin{bmatrix} z_{1,t-1} \\ z_{2,t-1} \end{bmatrix} = \begin{bmatrix} a_{1t} \\ a_{2t} \end{bmatrix}. \tag{3.26}$$

To see this, the VMA(1) model in Equation (3.25) implies that

$$z_{1t} = a_{1t} - 2a_{2,t-1} \quad \text{and} \quad z_{2t} = a_{2t}.$$

In other words, z_{2t} is a white noise series. As such, we have

$$z_{1t} = a_{1t} - 2a_{2,t-1} = a_{1t} - 2z_{2,t-1}.$$

Consequently, we have

$$z_{1t} + 2z_{2,t-1} = a_{1t} \quad \text{and} \quad z_{2t} = a_{2t},$$

which is precisely the VAR(1) model in Equation (3.26). Therefore, in this particular example, we have two sets of AR and MA matrix polynomials that produce the same MA representation in Equation (3.24). This type of nonuniqueness in model specification is harmless because either model can be used in a real application.

Note that for the VMA(1) model in Equation (3.25), we have $|\boldsymbol{\theta}(B)| = 1$, implying that $\boldsymbol{\theta}(B)$ is a unimodular matrix. Its inverse is the unimodular matrix $\phi(B)$ of the VAR(1) model in Equation (3.26). Also, for the VMA(1) model in Equation (3.25), we have $p=0$ and $q=1$, but $\boldsymbol{\theta}_1$ is not full rank so that Condition (2) is violated. □

Example 3.6 Consider the VARMA(1,1) model

$$\begin{bmatrix} z_{1t} \\ z_{2t} \end{bmatrix} - \begin{bmatrix} 0.8 & 2 \\ 0 & 0 \end{bmatrix} \begin{bmatrix} z_{1,t-1} \\ z_{2,t-1} \end{bmatrix} = \begin{bmatrix} a_{1t} \\ a_{2t} \end{bmatrix} - \begin{bmatrix} 0.3 & 0 \\ 0 & 0 \end{bmatrix} \begin{bmatrix} a_{1,t-1} \\ a_{2,t-1} \end{bmatrix}. \quad (3.27)$$

It is easy to see that the model is identical to

$$\begin{bmatrix} z_{1t} \\ z_{2t} \end{bmatrix} - \begin{bmatrix} 0.8 & 2+\omega \\ 0 & \beta \end{bmatrix} \begin{bmatrix} z_{1,t-1} \\ z_{2,t-1} \end{bmatrix} = \begin{bmatrix} a_{1t} \\ a_{2t} \end{bmatrix} - \begin{bmatrix} 0.3 & \omega \\ 0 & \beta \end{bmatrix} \begin{bmatrix} a_{1,t-1} \\ a_{2,t-1} \end{bmatrix}, \quad (3.28)$$

for any $\omega \neq 0$ and $\beta \neq 0$. From Equation (3.27), we have

$$z_{1t} = 0.8z_{1,t-1} + 2z_{2,t-1} + a_{1t} - .3a_{1,t-1} \quad (3.29)$$

$$z_{2t} = a_{2t}. \quad (3.30)$$

Thus, z_{2t} is a white noise series. Multiplying Equation (3.30) by ω for $t-1$ and adding Equation (3.29), we obtain the first equation of Equation (3.28). The second equation in Equation (3.28) holds because $z_{2,t-1} = a_{2,t-1}$. This type of identifiability is serious because, without proper constraints, the likelihood function of the VARMA(1,1) model is not uniquely defined. In other words, without proper constraints, one cannot estimate the VARMA(1,1) model. □

For the VARMA(1,1) model in Equation (3.27), we have $\text{Rank}[\phi_1, \boldsymbol{\theta}_1] = 1$, which is smaller than the dimension of $\boldsymbol{z}_t$. This is a clear violation of Condition (2) of Equation (3.23). Also, the two polynomial matrices of Equation (3.28) are not left co-prime, because

$$\begin{bmatrix} 1-0.8B & -(2+\omega)B \\ 0 & 1-\beta B \end{bmatrix} = \begin{bmatrix} 1 & -\omega B \\ 0 & 1-\beta B \end{bmatrix} \begin{bmatrix} 1-0.8B & -2B \\ 0 & 1 \end{bmatrix}$$

$$\begin{bmatrix} 1-0.3B & -\omega B \\ 0 & 1-\beta B \end{bmatrix} = \begin{bmatrix} 1 & -\omega B \\ 0 & 1-\beta B \end{bmatrix} \begin{bmatrix} 1-0.3B & 0 \\ 0 & 1 \end{bmatrix},$$

and the left common factor is not a unimodular matrix. Cancellation of the left common factor reduces Equation (3.28) to Equation (3.27).

Finally, the identifiability problem can occur even if none of the components of $\boldsymbol{z}_t$ is white noise. Therefore, identifiability is an important issue for VARMA models. It implies that model specification of VARMA models involves more than identifying the order (p, q). Indeed, model identification of VARMA models must include structural specification to overcome the problem of identifiability. In the literature, two approaches are available to perform structural specification of VARMA models. The first approach uses state–space formulation with Kronecker indices, and the second approach uses scalar component models. See Tsay (1991). We shall discuss these two approaches in Chapter 4. Here, we assume that the two identifiability conditions associated with the VARMA(p, q) in Equation (3.23) hold.

3.5.2 VARMA(1,1) Models

To gain a deeper understanding of VARMA models, we consider in detail the simplest VARMA(1,1) model,

$$\boldsymbol{z}_t = \boldsymbol{\phi}_0 + \boldsymbol{\phi}_1 \boldsymbol{z}_{t-1} + \boldsymbol{a}_t - \boldsymbol{\theta}_1 \boldsymbol{a}_{t-1}, \tag{3.31}$$

or equivalently, $(\boldsymbol{I}_k - \boldsymbol{\phi}_1 B)\boldsymbol{z}_t = \boldsymbol{\phi}_0 + (\boldsymbol{I}_k - \boldsymbol{\theta}_1 B)\boldsymbol{a}_t$. Under Conditions (1) and (2) of block identifiability, the order of the model cannot be reduced. Assume that $\boldsymbol{z}_t$ is stationary. Taking the expectation of the model in Equation (3.31), we have

$$\boldsymbol{\mu} = \boldsymbol{\phi}_0 + \boldsymbol{\phi}_1 \boldsymbol{\mu}, \quad \text{or} \quad (\boldsymbol{I}_k - \boldsymbol{\phi}_1)\boldsymbol{\mu} = \boldsymbol{\phi}_0.$$

Thus, we can rewrite the model as

$$\tilde{\boldsymbol{z}}_t = \boldsymbol{\phi}_1 \tilde{\boldsymbol{z}}_{t-1} + \boldsymbol{a}_t - \boldsymbol{\theta}_1 \boldsymbol{a}_{t-1}, \tag{3.32}$$

where $\tilde{\boldsymbol{z}}_t = \boldsymbol{z}_t - \boldsymbol{\mu}$ is the mean-adjusted series. Let $\boldsymbol{w}_t = \boldsymbol{a}_t - \boldsymbol{\theta}_1 \boldsymbol{a}_{t-1}$. Clearly, $\boldsymbol{w}_t$ is a VMA(1) process so that it is stationary. Using $\tilde{\boldsymbol{z}}_t = \boldsymbol{\phi}_1 \tilde{\boldsymbol{z}}_{t-1} + \boldsymbol{w}_t$ and by repeated substitutions, we have

$$\tilde{\boldsymbol{z}}_t = \boldsymbol{w}_t + \boldsymbol{\phi}_1 \boldsymbol{w}_{t-1} + \boldsymbol{\phi}_1^2 \boldsymbol{w}_{t-2} + \cdots.$$

Therefore, for $\tilde{\boldsymbol{z}}_t$ to be stationary, $\boldsymbol{\phi}_1^i$ must converge to 0 as i increases. This implies that the stationarity condition of $\boldsymbol{z}_t$ is that all eigenvalues of $\boldsymbol{\phi}_1$ are less than 1 in modulus or equivalently, the solutions of the determinant equation $|\boldsymbol{I}_k - \boldsymbol{\phi}_1 B| = 0$

are greater than 1 in modulus. This is identical to the weak stationarity condition of the VAR(1) model of Chapter 2.

Next, using

$$(\boldsymbol{I}_k - \boldsymbol{\phi}_1 B)(\boldsymbol{I}_k + \boldsymbol{\phi}_1 B + \boldsymbol{\phi}_1^2 B^2 + \cdots) = \boldsymbol{I}_k,$$

we have $(\boldsymbol{I}_k - \boldsymbol{\phi}_1 B)^{-1} = \boldsymbol{I}_k + \boldsymbol{\phi}_1 B + \boldsymbol{\phi}_1^2 B^2 + \cdots$. Consequently, we get

$$\begin{aligned}\tilde{\boldsymbol{z}}_t &= (\boldsymbol{I}_k - \boldsymbol{\phi}_1 B)^{-1}(\boldsymbol{I}_k - \boldsymbol{\theta}_1 B)\boldsymbol{a}_t \\ &= \boldsymbol{a}_t + (\boldsymbol{\phi}_1 - \boldsymbol{\theta}_1)\boldsymbol{a}_{t-1} + \boldsymbol{\phi}_1(\boldsymbol{\phi}_1 - \boldsymbol{\theta}_1)\boldsymbol{a}_{t-2} + \cdots \\ &= \sum_{i=0}^{\infty} \boldsymbol{\psi}_i \boldsymbol{a}_{t-i},\end{aligned}$$

where $\boldsymbol{\psi}_0 = \boldsymbol{I}_k$ and $\boldsymbol{\psi}_i = \boldsymbol{\phi}_1^{i-1}(\boldsymbol{\phi}_1 - \boldsymbol{\theta}_1)$. This is an MA representation for the VARMA(1,1) model. From the aforementioned MA representation, we obtain $\text{Cov}(\boldsymbol{z}_t, \boldsymbol{a}_{t-j}) = \boldsymbol{\psi}_j \boldsymbol{\Sigma}_a$ for $j \geq 0$. Next, postmultiplying Equation (3.32) by $\tilde{\boldsymbol{z}}'_{t-j}$ with $j \geq 0$ and taking expectation, we obtain

$$\boldsymbol{\Gamma}_0 = \boldsymbol{\phi}_1 \boldsymbol{\Gamma}_{-1} + \boldsymbol{\Sigma}_a - \boldsymbol{\theta}_1 \boldsymbol{\Sigma}_a \boldsymbol{\psi}'_1, \tag{3.33}$$

$$\boldsymbol{\Gamma}_1 = \boldsymbol{\phi}_1 \boldsymbol{\Gamma}_0 - \boldsymbol{\theta}_1 \boldsymbol{\Sigma}_a, \tag{3.34}$$

$$\boldsymbol{\Gamma}_j = \boldsymbol{\phi}_1 \boldsymbol{\Gamma}_{j-1}, \quad j > 1. \tag{3.35}$$

From Equation (3.35), the autocovariance matrices of a VARMA(1,1) model satisfy the matrix polynomial equation $(\boldsymbol{I}_k - \boldsymbol{\phi}_1 B)\boldsymbol{\Gamma}_j = \boldsymbol{0}$ for $j > 1$, where the lag operator B applies to j. Consequently, to obtain all autocovariance matrices of a VARMA(1,1) process $\boldsymbol{z}_t$, one needs to find the first two autocovariance matrices $\boldsymbol{\Gamma}_0$ and $\boldsymbol{\Gamma}_1$. To this end, Equations (3.33) and (3.34) are useful. Using $\boldsymbol{\Gamma}_{-1} = \boldsymbol{\Gamma}'_1$ and Equation (3.34), we can rewrite Equation (3.33) as

$$\boldsymbol{\Gamma}_0 - \boldsymbol{\phi}_1 \boldsymbol{\Gamma}_0 \boldsymbol{\phi}'_1 = \boldsymbol{\Sigma}_a - \boldsymbol{\theta}_1 \boldsymbol{\Sigma}_a \boldsymbol{\phi}'_1 - (\boldsymbol{\phi}_1 - \boldsymbol{\theta}_1)\boldsymbol{\Sigma}_a \boldsymbol{\theta}'_1.$$

Let $\boldsymbol{S}$ denote the right side of the aforementioned equation. Clearly, $\boldsymbol{S}$ can be obtained for a given VARMA(1,1) model. The prior equation then becomes

$$\boldsymbol{\Gamma}_0 - \boldsymbol{\phi}_1 \boldsymbol{\Gamma}_0 \boldsymbol{\phi}'_1 = \boldsymbol{S},$$

or equivalently,

$$\text{vec}(\boldsymbol{\Gamma}_0) - (\boldsymbol{\phi}_1 \otimes \boldsymbol{\phi}_1)\text{vec}(\boldsymbol{\Gamma}_0) = \text{vec}(\boldsymbol{S}).$$

Consequently, for a given stationary VARMA(1,1) model, we have

$$\text{vec}(\boldsymbol{\Gamma}_0) = (\boldsymbol{I}_{k^2} - \boldsymbol{\phi}_1 \otimes \boldsymbol{\phi}_1)^{-1}\text{vec}(\boldsymbol{S}),$$

where $\boldsymbol{S} = \boldsymbol{\Sigma}_a - \boldsymbol{\theta}_1\boldsymbol{\Sigma}_a\boldsymbol{\phi}_1' - (\boldsymbol{\phi}_1 - \boldsymbol{\theta}_1)\boldsymbol{\Sigma}_a\boldsymbol{\theta}_1'$. Conversely, assume that $\boldsymbol{\Gamma}_i$ are known for $i = 0, 1$, and 2, and the condition that rank$[\boldsymbol{\phi}_1, \boldsymbol{\theta}_1] = k$ holds. We see that $\boldsymbol{\Gamma}_1 = [\boldsymbol{\phi}_1, -\boldsymbol{\theta}_1](\boldsymbol{\Gamma}_0, \boldsymbol{\Sigma}_a)'$ is of full rank k; otherwise, rank$[\boldsymbol{\Gamma}_0, \boldsymbol{\Sigma}_a]$ is less than k, implying that a linear combination $\boldsymbol{z}_t$ and $\boldsymbol{a}_t$ is degenerated, which is impossible for a well-defined VARMA(1,1) model. Therefore, we have $\boldsymbol{\phi}_1 = \boldsymbol{\Gamma}_2\boldsymbol{\Gamma}_1^{-1}$. With $\boldsymbol{\phi}_1$ known, we can solve for $\boldsymbol{\theta}_1$ and $\boldsymbol{\Sigma}_a$ from $\boldsymbol{\Gamma}_0$ and $\boldsymbol{\Gamma}_1$, although via complicated quadratic matrix equations.

Finally, since $\boldsymbol{w}_t = \tilde{\boldsymbol{z}}_t - \boldsymbol{\phi}_1\tilde{\boldsymbol{z}}_{t-1}$ is a stationary process and follows a VMA(1) model, it is easy to see that the sufficient condition for invertibility of $\boldsymbol{z}_t$ is that all eigenvalues of $\boldsymbol{\theta}_1$ are less than 1 in modulus, or equivalently solutions of the determinant equation $|\boldsymbol{I}_k - \boldsymbol{\theta}_1 B| = 0$ are greater than 1 in modulus. Again, this condition is identical to the invertibility condition of a VMA(1) model of Section 3.1. Using $(\boldsymbol{I}_k - \boldsymbol{\theta}_1 B)^{-1} = \boldsymbol{I}_k + \boldsymbol{\theta}_1 B + \boldsymbol{\theta}_1^2 B^2 + \cdots$, we can obtain an AR representation for $\boldsymbol{z}_t$ as

$$\tilde{\boldsymbol{z}}_t = \boldsymbol{a}_t + (\boldsymbol{\phi}_1 - \boldsymbol{\theta}_1)\tilde{\boldsymbol{z}}_{t-1} + \boldsymbol{\theta}_1(\boldsymbol{\phi}_1 - \boldsymbol{\theta}_1)\tilde{\boldsymbol{z}}_{t-2} + \cdots.$$

In general, we have $\boldsymbol{\pi}_i = \boldsymbol{\theta}_1^{i-1}(\boldsymbol{\phi}_1 - \boldsymbol{\theta}_1)$ for $i \geq 1$.

To demonstrate, consider a two-dimensional VARMA(1,1) model with parameters

$$\boldsymbol{\phi}_1 = \begin{bmatrix} 0.2 & 0.3 \\ -0.6 & 1.1 \end{bmatrix}, \quad \boldsymbol{\theta}_1 = \begin{bmatrix} -0.5 & -0.2 \\ -0.1 & -0.6 \end{bmatrix}, \quad \boldsymbol{\Sigma}_a = \begin{bmatrix} 4 & 1 \\ 1 & 1 \end{bmatrix}. \quad (3.36)$$

Based on the results discussed in the section, we obtain the ψ-weight matrices

$$\boldsymbol{\psi}_1 = \begin{bmatrix} 0.7 & 0.5 \\ -0.5 & 1.7 \end{bmatrix}, \quad \boldsymbol{\psi}_2 = \begin{bmatrix} -0.01 & 0.61 \\ -0.97 & 1.57 \end{bmatrix}, \quad \boldsymbol{\psi}_3 = \begin{bmatrix} -0.29 & 0.59 \\ -1.06 & 1.36 \end{bmatrix},$$

and the π-weight matrices

$$\boldsymbol{\pi}_1 = \begin{bmatrix} 0.7 & 0.5 \\ -0.5 & 1.7 \end{bmatrix}, \quad \boldsymbol{\pi}_2 = \begin{bmatrix} -0.25 & -0.59 \\ 0.23 & -1.07 \end{bmatrix}, \quad \boldsymbol{\pi}_3 = \begin{bmatrix} 0.08 & 0.51 \\ -0.11 & 0.70 \end{bmatrix},$$

and the autocovariance matrices

$$\boldsymbol{\Gamma}_0 = \begin{bmatrix} 9.56 & 7.16 \\ 7.16 & 19.8 \end{bmatrix}, \quad \boldsymbol{\Gamma}_1 = \begin{bmatrix} 6.26 & 8.08 \\ 3.14 & 18.1 \end{bmatrix}, \quad \boldsymbol{\Gamma}_2 = \begin{bmatrix} 2.19 & 7.05 \\ -0.30 & 15.1 \end{bmatrix},$$

and the cross-correlation matrices

$$\rho_0 = \begin{bmatrix} 1.00 & 0.52 \\ 0.52 & 1.00 \end{bmatrix}, \quad \rho_1 = \begin{bmatrix} 0.65 & 0.59 \\ 0.23 & 0.92 \end{bmatrix}, \quad \rho_2 = \begin{bmatrix} 0.23 & 0.51 \\ -0.02 & 0.77 \end{bmatrix}.$$

Remark: The ψ-weights, π-weights, autocovariance matrices, and cross-correlation matrices of a given VARMA model can be obtained via the commands `PSIwgt`, `PIwgt`, and `VARMAcov` of the `MTS` package. The input includes $\boldsymbol{\Phi} = [\boldsymbol{\phi}_1, \ldots, \boldsymbol{\phi}_p]$, $\boldsymbol{\Theta} = [\boldsymbol{\theta}_1, \ldots, \boldsymbol{\theta}_q]$, and $\boldsymbol{\Sigma}_a$ matrix. The autocovariance matrices are computed via the ψ-weight matrices with a default 120 lags. This truncation lag can be specified if needed. See Equation (3.42) of the next section. An alternative approach to compute autocovariance matrices is given in the next section. □

3.5.3 Some Properties of VARMA Models

In this section, we study some properties of a VARMA(p, q) model with $p > 0$ and $q > 0$. The model is assumed to be identifiable and the innovation $\boldsymbol{a}_t$ has mean zero and covariance matrix $\boldsymbol{\Sigma}_a$, which is positive definite. These properties are extensions of those of VARMA(1,1) models.

3.5.3.1 Weak Stationarity

Similar to a VAR(p) model, the necessary and sufficient condition for weak stationarity of the $\boldsymbol{z}_t$ process in Equation (3.23) is that all solutions of the determinant equation $|\boldsymbol{\phi}(B)| = 0$ are outside the unit circle, that is, they are greater than 1 in modulus. Let $\boldsymbol{w}_t = \boldsymbol{\theta}(B)\boldsymbol{a}_t$, which is weakly stationary. The model for $\boldsymbol{z}_t$ then becomes $\boldsymbol{\phi}(B)\boldsymbol{z}_t = \boldsymbol{\phi}_0 + \boldsymbol{w}_t$, from which one can apply the same argument as that of a VAR(p) model to obtain the stationarity condition.

For a stationary VARMA(p, q) model, we have $|\boldsymbol{\phi}(1)| \neq 0$ because 1 is not a solution of $|\boldsymbol{\phi}(B)| = 0$. Taking expectation of Equation (3.23), we have $\boldsymbol{\phi}(1)\boldsymbol{\mu} = \boldsymbol{\phi}_0$. See Equation (1.23) of Chapter 1. Therefore, $\boldsymbol{\mu} = [\boldsymbol{\phi}(1)]^{-1}\boldsymbol{\phi}_0$ and the model can be rewritten as

$$\boldsymbol{\phi}(B)(\boldsymbol{z}_t - \boldsymbol{\mu}) = \boldsymbol{\theta}(B)\boldsymbol{a}_t.$$

Let $\tilde{\boldsymbol{z}}_t = \boldsymbol{z}_t - \boldsymbol{\mu}$ be the mean-adjusted series. The VARMA(p, q) model then becomes

$$\boldsymbol{\phi}(B)\tilde{\boldsymbol{z}}_t = \boldsymbol{\theta}(B)\boldsymbol{a}_t. \tag{3.37}$$

For ease in presentation, we often use Equation (3.37) to derive the properties of the VARMA(p, q) series.

When $\boldsymbol{z}_t$ is stationary, it has an MA representation as

$$\tilde{\boldsymbol{z}}_t = \boldsymbol{a}_t + \boldsymbol{\psi}_1\boldsymbol{a}_{t-1} + \boldsymbol{\psi}_2\boldsymbol{a}_{t-2} + \cdots = \boldsymbol{\psi}(B)\boldsymbol{a}_t, \tag{3.38}$$

where $\boldsymbol{\psi}(B) = \sum_{i=0}^{\infty} \boldsymbol{\psi}_i B^i$ with $\boldsymbol{\psi}_0 = \boldsymbol{I}_k$, the $k \times k$ identity matrix. The coefficient matrices $\boldsymbol{\psi}_i$ can be obtained recursively by equating the coefficients of B^i in

$$\boldsymbol{\phi}(B)\boldsymbol{\psi}(B) = \boldsymbol{\theta}(B).$$

For simplicity, let $\boldsymbol{\theta}_i = \boldsymbol{0}$ if $i > q$, $\boldsymbol{\phi}_i = \boldsymbol{0}$ if $i > p$, and $m = \max\{p, q\}$. Then, the coefficient matrices $\boldsymbol{\psi}_i$ can be obtained recursively as follows:

$$\begin{aligned}
\boldsymbol{\psi}_1 &= \boldsymbol{\phi}_1 - \boldsymbol{\theta}_1 \\
\boldsymbol{\psi}_2 &= \boldsymbol{\phi}_1\boldsymbol{\psi}_1 + \boldsymbol{\phi}_2 - \boldsymbol{\theta}_2 \\
\boldsymbol{\psi}_3 &= \boldsymbol{\phi}_1\boldsymbol{\psi}_2 + \boldsymbol{\phi}_2\boldsymbol{\psi}_1 + \boldsymbol{\phi}_3 - \boldsymbol{\theta}_3 \\
\vdots &= \vdots \\
\boldsymbol{\psi}_m &= \boldsymbol{\phi}_1\boldsymbol{\psi}_{m-1} + \boldsymbol{\phi}_2\boldsymbol{\psi}_{m-2} + \cdots + \boldsymbol{\phi}_{m-1}\boldsymbol{\psi}_1 + \boldsymbol{\phi}_m - \boldsymbol{\theta}_m \\
\boldsymbol{\psi}_\ell &= \boldsymbol{\phi}_1\boldsymbol{\psi}_{\ell-1} + \boldsymbol{\phi}_2\boldsymbol{\psi}_{\ell-2} + \cdots + \boldsymbol{\phi}_m\boldsymbol{\psi}_{\ell-m}, \quad \ell > m.
\end{aligned}$$

A unified equation for $\boldsymbol{\psi}_i$ is

$$\boldsymbol{\psi}_i = \sum_{j=1}^{\min\{p,i\}} \boldsymbol{\phi}_j\boldsymbol{\psi}_{i-j} - \boldsymbol{\theta}_i, \quad i = 1, 2, \ldots, \tag{3.39}$$

where it is understood that $\boldsymbol{\theta}_i = \boldsymbol{0}$ for $i > q$ and $\boldsymbol{\psi}_0 = \boldsymbol{I}_k$. In particular, we have

$$\boldsymbol{\psi}_i = \sum_{j=1}^{\min\{p,i\}} \boldsymbol{\phi}_j\boldsymbol{\psi}_{j-i}, \quad i > q.$$

The MA representation of a VARMA model is useful in many ways. We make extensive use of the representation throughout the book.

3.5.3.2 Invertibility

A sufficient and necessary invertibility condition for the VARMA(p, q) process $\boldsymbol{z}_t$ in Equation (3.23) is that all solutions of determinant equation $|\boldsymbol{\theta}(B)| = 0$ are outside the unit circle. This is so because the invertibility of the model is determined by the MA polynomial matrix $\boldsymbol{\theta}(B)$ and one can apply the result of VMA(q) models. If $\boldsymbol{z}_t$ is invertible, then it has an AR representation, that is,

$$\boldsymbol{\pi}(B)(\boldsymbol{z}_t - \boldsymbol{\mu}) = \boldsymbol{a}_t, \tag{3.40}$$

where $\boldsymbol{\pi}(B) = \boldsymbol{I} - \boldsymbol{\pi}_1 B - \boldsymbol{\pi}_2 B^2 - \cdots$ and the coefficient matrices $\boldsymbol{\pi}_i$ can be obtained by equating the coefficients of B^i in

$$\boldsymbol{\theta}(B)\boldsymbol{\pi}(B) = \boldsymbol{\phi}(B).$$

A unified expression for the $\boldsymbol{\pi}_i$ matrices is then

$$\boldsymbol{\pi}_i = \boldsymbol{\phi}_i + \sum_{j=1}^{\min\{q,i\}} \boldsymbol{\theta}_j \boldsymbol{\pi}_{i-j}, \quad i = 1, 2, \ldots, \tag{3.41}$$

where it is understood that $\boldsymbol{\phi}_i = \mathbf{0}$ for $i > p$ and $\boldsymbol{\pi}_0 = -\boldsymbol{I}_k$. In particular, for $i > p$, we have

$$\boldsymbol{\pi}_i = \sum_{j=1}^{\min\{q,i\}} \boldsymbol{\theta}_j \boldsymbol{\pi}_{i-j} \quad i > p.$$

Note that $\boldsymbol{\pi}(B)\boldsymbol{\psi}(B) = \boldsymbol{I}_k = \boldsymbol{\psi}(B)\boldsymbol{\pi}(B)$.

3.5.3.3 Autocovariance and Cross-Correlation Matrices

For a stationary VARMA(p, q) model, one can use the MA representation to obtain the lag ℓ auto- and cross-covariance matrix as

$$\text{Cov}(\boldsymbol{z}_t, \boldsymbol{z}_{t-\ell}) = \boldsymbol{\Gamma}_\ell = \sum_{i=0}^{\infty} \boldsymbol{\psi}_{\ell+i} \boldsymbol{\Sigma}_a \boldsymbol{\psi}'_i, \tag{3.42}$$

where $\boldsymbol{\psi}_0 = \boldsymbol{I}_k$. The lag ℓ cross-correlation matrix is given by

$$\boldsymbol{\rho}_\ell = \boldsymbol{D}^{-1} \boldsymbol{\Gamma}_\ell \boldsymbol{D}^{-1} \tag{3.43}$$

where $\boldsymbol{D} = \text{diag}\{\sqrt{\gamma_{0,11}}, \ldots, \sqrt{\gamma_{0,kk}}\}$ with $\boldsymbol{\Gamma}_0 = [\gamma_{0,ij}]$.

The autocovariance matrices $\boldsymbol{\Gamma}_\ell$ of Equation (3.42) involve the summation of an infinite series. There exist other approaches to calculating $\boldsymbol{\Gamma}_\ell$ directly for a given stationary VARMA(p, q) model. In what follows, we adopt the approach of Mittnik (1990) to obtain an explicit solution for $\boldsymbol{\Gamma}_\ell$. To this end, we need the following result, which can easily be shown using the MA representation of $\boldsymbol{z}_t$.

Theorem 3.2 For a stationary VARMA(p, q) series $\boldsymbol{z}_t$,

$$E(\boldsymbol{a}_t \boldsymbol{z}'_{t+\ell}) = \begin{cases} \mathbf{0} & \text{if } \ell < 0, \\ \boldsymbol{\Sigma}_a & \text{if } \ell = 0, \\ \boldsymbol{\Sigma}_a \boldsymbol{\psi}'_\ell & \text{if } \ell > 0. \end{cases}$$

Postmultiplying the model in Equation (3.37) by $\tilde{\boldsymbol{z}}'_{t-\ell}$, taking expectation, and using the result of Theorem 3.2, we obtain the moment equations for $\boldsymbol{z}_t$ as

$$\boldsymbol{\Gamma}_\ell - \sum_{i=1}^{p} \boldsymbol{\phi}_i \boldsymbol{\Gamma}_{\ell-i} = \begin{cases} -\sum_{j=\ell}^{q} \boldsymbol{\theta}_j \boldsymbol{\Sigma}_a \boldsymbol{\psi}'_{j-\ell} & \text{if } \ell = 0, 1, \ldots, q, \\ \mathbf{0} & \text{if } \ell > q, \end{cases} \tag{3.44}$$

where, for convenience, $\boldsymbol{\theta}_0 = -\boldsymbol{I}$. The right side of Equation (3.44) is available from the model. Therefore, the key to obtain all autocovariance matrices of $\boldsymbol{z}_t$ is to derive $\boldsymbol{\Gamma}_0, \ldots, \boldsymbol{\Gamma}_p$. To this end, we focus on the system of moment equations in Equation (3.44) for $\ell = 0, \ldots, p$. Let $\boldsymbol{\Gamma} = [\boldsymbol{\Gamma}_0', \boldsymbol{\Gamma}_1', \ldots, \boldsymbol{\Gamma}_p']'$, $\boldsymbol{\Gamma}_* = [\boldsymbol{\Gamma}_0, \boldsymbol{\Gamma}_1, \ldots, \boldsymbol{\Gamma}_p]'$, and $\boldsymbol{\psi} = [\boldsymbol{\psi}_0, \boldsymbol{\psi}_1, \ldots, \boldsymbol{\psi}_q]'$, where $\boldsymbol{\psi}_0 = \boldsymbol{I}_k$. Then, the system of moment equations in Equation (3.44) for $\ell = 0, 1, \ldots, p$ can be written as

$$\boldsymbol{\Gamma} = \boldsymbol{\Phi}_L \boldsymbol{\Gamma} + \boldsymbol{\Phi}_U \boldsymbol{\Gamma}_* - \boldsymbol{\Omega}\boldsymbol{\psi}, \tag{3.45}$$

where

$$\boldsymbol{\Phi}_L = \begin{bmatrix} \mathbf{0}_{k\times kp} & \mathbf{0}_{k\times k} \\ \boldsymbol{L}_{kp\times kp} & \mathbf{0}_{kp\times k} \end{bmatrix}, \quad \boldsymbol{\Phi}_U = \begin{bmatrix} \mathbf{0}_{kp\times k} & \boldsymbol{U}_{kp\times kp} \\ \mathbf{0}_{k\times k} & \mathbf{0}_{k\times kp} \end{bmatrix},$$

with

$$\boldsymbol{L} = \begin{bmatrix} \boldsymbol{\phi}_1 & \mathbf{0}_k & \cdots & \mathbf{0}_k \\ \boldsymbol{\phi}_2 & \boldsymbol{\phi}_1 & \cdots & \mathbf{0}_k \\ \vdots & \vdots & \ddots & \vdots \\ \boldsymbol{\phi}_p & \boldsymbol{\phi}_{p-1} & \cdots & \boldsymbol{\phi}_1 \end{bmatrix}, \quad \boldsymbol{U} = \begin{bmatrix} \boldsymbol{\phi}_1 & \boldsymbol{\phi}_2 & \cdots & \boldsymbol{\phi}_{p-1} & \boldsymbol{\phi}_p \\ \boldsymbol{\phi}_2 & \boldsymbol{\phi}_3 & \cdots & \boldsymbol{\phi}_p & \mathbf{0}_k \\ \vdots & \vdots & & \vdots & \vdots \\ \boldsymbol{\phi}_p & \mathbf{0}_k & \cdots & \mathbf{0}_k & \mathbf{0}_k \end{bmatrix},$$

with $\mathbf{0}_k$ being the $k \times k$ matrix of zeros, and $\boldsymbol{\Omega}$ is a $k(p+1) \times k(q+1)$ matrix defined by $\boldsymbol{\Omega} = \boldsymbol{\Omega}_1(\boldsymbol{I}_{q+1} \otimes \boldsymbol{\Sigma}_a)$, where $\boldsymbol{\Omega}_1$ is given next. If $q > p$, then

$$\boldsymbol{\Omega}_1 = \begin{bmatrix} \boldsymbol{\theta}_0 & \boldsymbol{\theta}_1 & \cdots & \boldsymbol{\theta}_{q-1} & \boldsymbol{\theta}_q \\ \boldsymbol{\theta}_1 & \boldsymbol{\theta}_2 & \cdots & \boldsymbol{\theta}_q & \mathbf{0}_k \\ \boldsymbol{\theta}_2 & \boldsymbol{\theta}_3 & \cdots & \mathbf{0}_k & \mathbf{0}_k \\ \vdots & \vdots & \cdots & \vdots & \vdots \\ \boldsymbol{\theta}_p & \cdots & \boldsymbol{\theta}_q & \cdots & \mathbf{0}_k \end{bmatrix},$$

where $\boldsymbol{\theta}_0 = -\boldsymbol{I}_k$. If $q \le p$, then

$$\boldsymbol{\Omega}_1 = \begin{bmatrix} \boldsymbol{\theta}_0 & \boldsymbol{\theta}_1 & \cdots & \boldsymbol{\theta}_{q-1} & \boldsymbol{\theta}_q \\ \boldsymbol{\theta}_1 & \boldsymbol{\theta}_2 & \cdots & \boldsymbol{\theta}_q & \mathbf{0}_k \\ \boldsymbol{\theta}_2 & \boldsymbol{\theta}_3 & \cdots & \mathbf{0}_k & \mathbf{0}_k \\ \vdots & \vdots & \cdots & \vdots & \vdots \\ \boldsymbol{\theta}_q & \mathbf{0}_k & \cdots & \mathbf{0}_k & \mathbf{0}_k \\ \mathbf{0}_{(p-q)} & \mathbf{0}_{(p-q)} & \cdots & \mathbf{0}_{(p-q)} & \mathbf{0}_{(p-q)} \end{bmatrix},$$

where $\mathbf{0}_{(p-q)}$ is the $k(p-q) \times k$ matrix of zeros. Taking the transposition of Equation (3.45) and applying vectorization, we obtain

$$\text{vec}(\boldsymbol{\Gamma}') = (\boldsymbol{\Phi}_L \otimes \boldsymbol{I}_k)\text{vec}(\boldsymbol{\Gamma}') + (\boldsymbol{\Phi}_U \otimes \boldsymbol{I}_k)\text{vec}(\boldsymbol{\Gamma}_*') - (\boldsymbol{\Omega} \otimes \boldsymbol{I}_k)\text{vec}(\boldsymbol{\psi}'). \tag{3.46}$$

Next, using $\boldsymbol{\Gamma}'_\ell = \boldsymbol{\Gamma}_{-\ell}$ and the properties of vectorization in Appendix A, we have $\text{vec}(\boldsymbol{\Gamma}'_*) = \boldsymbol{I}_{p+1} \otimes \boldsymbol{K}_{kk}\text{vec}(\boldsymbol{\Gamma}')$, where $\boldsymbol{K}_{kk}$ is a commutation matrix. For simplicity, letting $\boldsymbol{K} = \boldsymbol{I}_{p+1} \otimes \boldsymbol{K}_{kk}$, we rewrite Equation (3.46) as

$$[\boldsymbol{I}_{k^2(p+1)} - \boldsymbol{\Phi}_L \otimes \boldsymbol{I}_k - (\boldsymbol{\Phi}_U \otimes \boldsymbol{I}_k)\boldsymbol{K}]\text{vec}(\boldsymbol{\Gamma}') = -(\boldsymbol{\Omega} \otimes \boldsymbol{I}_k)\text{vec}(\boldsymbol{\psi}').$$

Therefore, we have

$$\text{vec}(\boldsymbol{\Gamma}') = -[\boldsymbol{I}_{k^2(p+1)} - \boldsymbol{\Phi}_L \otimes \boldsymbol{I}_k - (\boldsymbol{\Phi}_U \otimes \boldsymbol{I}_k)\boldsymbol{K}]^{-1}(\boldsymbol{\Omega} \otimes \boldsymbol{I}_k)\text{vec}(\boldsymbol{\psi}'). \quad (3.47)$$

This equation can be used to obtain $\boldsymbol{\Gamma}_\ell$ for $\ell = 0, 1, \ldots, p$. All other $\boldsymbol{\Gamma}_\ell$ with $\ell > p$ can be obtained from the moment equations in Equation (3.44).

Finally, for a stationary VARMA(p, q) process $\boldsymbol{z}_t$, the system of equations

$$\boldsymbol{\Gamma}_\ell = \boldsymbol{\phi}_1\boldsymbol{\Gamma}_{\ell-1} + \cdots + \boldsymbol{\phi}_p\boldsymbol{\Gamma}_{\ell-p}, \quad \ell = q+1, \ldots, q+p,$$

is referred to as the multivariate generalized Yule–Walker equations.

Example 3.7 Consider the VARMA(2,1) model

$$\boldsymbol{z}_t = \boldsymbol{\phi}_1\boldsymbol{z}_{t-1} + \boldsymbol{\phi}_2\boldsymbol{z}_{t-2} + \boldsymbol{a}_t - \boldsymbol{\theta}_1\boldsymbol{a}_{t-1},$$

where the parameters are

$$\boldsymbol{\phi}_1 = \begin{bmatrix} 0.816 & -0.623 \\ -1.116 & 1.074 \end{bmatrix}, \quad \boldsymbol{\phi}_2 = \begin{bmatrix} -0.643 & 0.592 \\ 0.615 & -0.133 \end{bmatrix},$$
$$\boldsymbol{\theta}_1 = \begin{bmatrix} 0 & -1.248 \\ -0.801 & 0 \end{bmatrix}, \quad \boldsymbol{\Sigma}_a = \begin{bmatrix} 4 & 2 \\ 2 & 5 \end{bmatrix}.$$

The AR and MA coefficient matrices are obtained from an empirical VARMA(2,1) model in Reinsel (1993, p. 144). We shall compute the theoretical values of the autocovariance and cross-correlation matrices of $\boldsymbol{z}_t$ using Equation (3.47). For this particular instance, we have

$$\boldsymbol{L} = \begin{bmatrix} 0.816 & -0.623 & 0.000 & 0.000 \\ -1.116 & 1.074 & 0.000 & 0.000 \\ -0.643 & 0.592 & 0.816 & -0.623 \\ 0.615 & -0.133 & -1.116 & 1.074 \end{bmatrix}, \quad \boldsymbol{\Phi}_L = \begin{bmatrix} \boldsymbol{0}_{2\times 4} & \boldsymbol{0}_{2\times 2} \\ \boldsymbol{L} & \boldsymbol{0}_{4\times 2} \end{bmatrix},$$

$$\boldsymbol{U} = \begin{bmatrix} 0.816 & -0.623 & -0.643 & 0.592 \\ -1.116 & 1.074 & 0.615 & -0.133 \\ -0.643 & 0.592 & 0.000 & 0.000 \\ 0.615 & -0.133 & 0.000 & 0.000 \end{bmatrix}, \quad \boldsymbol{\Phi}_U = \begin{bmatrix} \boldsymbol{0}_{4\times 2} & \boldsymbol{U} \\ \boldsymbol{0}_{2\times 2} & \boldsymbol{0}_{2\times 4} \end{bmatrix},$$

$$\boldsymbol{\Omega} = \begin{bmatrix} -4.000 & -2.000 & -2.496 & -6.240 \\ -2.000 & -5.000 & -3.204 & -1.602 \\ -2.496 & -6.240 & 0.000 & 0.000 \\ -3.204 & -1.602 & 0.000 & 0.000 \\ 0.000 & 0.000 & 0.000 & 0.000 \\ 0.000 & 0.000 & 0.000 & 0.000 \end{bmatrix}, \quad \boldsymbol{\psi} = \begin{bmatrix} 1 & 0 \\ 0 & 1 \\ 0.816 & -0.315 \\ 0.625 & 1.074 \end{bmatrix}.$$

Applying Equation (3.47), we obtain the autocovariance matrices

$$\boldsymbol{\Gamma}_0 = \begin{bmatrix} 15.71 & 3.20 \\ 3.20 & 29.33 \end{bmatrix}, \ \boldsymbol{\Gamma}_1 = \begin{bmatrix} 10.87 & 7.69 \\ -5.22 & 23.23 \end{bmatrix}, \ \boldsymbol{\Gamma}_2 = \begin{bmatrix} 3.92 & 7.10 \\ -8.51 & 14.44 \end{bmatrix}.$$

and the cross-correlation matrices are

$$\boldsymbol{\rho}_0 = \begin{bmatrix} 1.00 & 0.15 \\ 0.15 & 1.00 \end{bmatrix}, \ \boldsymbol{\rho}_1 = \begin{bmatrix} 0.69 & 0.36 \\ -0.24 & 0.79 \end{bmatrix}, \ \boldsymbol{\rho}_2 = \begin{bmatrix} 0.25 & 0.33 \\ -0.40 & 049 \end{bmatrix}.$$

As expected, these results are very close to those obtained by the command `VARMAcov` of the `MTS` package. □

R Demonstration: Computing autocovariance matrices.

```
> phi=matrix(c(.816,-1.116,-.623,1.074,-.643,.615,.592,
                -.133),2,4)
> phi
       [,1]   [,2]   [,3]   [,4]
[1,]  0.816 -0.623 -0.643  0.592
[2,] -1.116  1.074  0.615 -0.133
> theta=matrix(c(0,-.801,-1.248,0),2,2)
> sig=matrix(c(4,2,2,5),2,2)
> VARMAcov(Phi=phi,Theta=theta,Sigma=sig,lag=2)
Auto-Covariance matrix of lag:  0
         [,1]     [,2]
[1,] 15.70537  3.20314
[2,]  3.20314 29.33396
Auto-Covariance matrix of lag:  1
         [,1]     [,2]
[1,] 10.87468  7.68762
[2,] -5.21755 23.23317
Auto-Covariance matrix of lag:  2
         [,1]     [,2]
[1,]  3.92198  7.10492
[2,] -8.50700 14.44155
cross correlation matrix of lag:  0
       [,1]   [,2]
[1,] 1.0000 0.1492
[2,] 0.1492 1.0000
```

```
cross correlation matrix of lag:  1
          [,1]   [,2]
[1,]  0.6924 0.3582
[2,] -0.2431 0.7920
cross correlation matrix of lag:  2
          [,1]   [,2]
[1,]  0.2497 0.3310
[2,] -0.3963 0.4923
```

3.6 IMPLICATIONS OF VARMA MODELS

VARMA models are flexible in describing the dynamic relation between variables. They encompass many commonly used econometric and statistical models. In this section, we discuss some of the implied relationships between components of a VARMA(p, q) model.

3.6.1 Granger Causality

The MA representation of a VARMA model remains an easy way to infer the dynamic relationships between variables. Without loss of generality, we assume that $E(\boldsymbol{z}_t) = \boldsymbol{0}$ and consider the simple case of partitioning $\boldsymbol{z}_t = (\boldsymbol{x}_t', \boldsymbol{y}_t')'$, where $\boldsymbol{x}_t$ and $\boldsymbol{y}_t$ are, respectively, k_1 and k_2 dimensional subvectors, where $k_1 + k_2 = k$. We partition the ψ-weight, the AR and the MA matrix polynomials, and the innovation $\boldsymbol{a}_t = (\boldsymbol{u}_t', \boldsymbol{v}_t')'$ accordingly. Write the MA representation as

$$\begin{bmatrix} \boldsymbol{x}_t \\ \boldsymbol{y}_t \end{bmatrix} = \begin{bmatrix} \boldsymbol{\psi}_{xx}(B) & \boldsymbol{\psi}_{xy}(B) \\ \boldsymbol{\psi}_{yx}(B) & \boldsymbol{\psi}_{yy}(B) \end{bmatrix} \begin{bmatrix} \boldsymbol{u}_t \\ \boldsymbol{v}_t \end{bmatrix}. \tag{3.48}$$

Similar to the VAR case, if $\boldsymbol{\psi}_{xy}(B) = \boldsymbol{0}$, but $\boldsymbol{\psi}_{yx}(B) \neq \boldsymbol{0}$, then there exists a *uni-directional* relation from $\boldsymbol{x}_t$ to $\boldsymbol{y}_t$. In this particular case, $\boldsymbol{x}_t$ does not depend on any past information of $\boldsymbol{y}_t$, but $\boldsymbol{y}_t$ depends on some lagged values of $\boldsymbol{x}_t$. Consequently, $\boldsymbol{x}_t$ causes $\boldsymbol{y}_t$, in the Granger causality sense, and the VARMA model implies the existence of a linear transfer function model.

The condition that $\boldsymbol{\psi}_{xy}(B) = \boldsymbol{0}$, but $\boldsymbol{\psi}_{yx}(B) \neq \boldsymbol{0}$ can be expressed in terms of the AR and MA matrix polynomials of the VARMA(p, q) model. In the VARMA representation, we have

$$\begin{bmatrix} \boldsymbol{\phi}_{xx}(B) & \boldsymbol{\phi}_{xy}(B) \\ \boldsymbol{\phi}_{yx}(B) & \boldsymbol{\phi}_{yy}(B) \end{bmatrix} \begin{bmatrix} \boldsymbol{x}_t \\ \boldsymbol{y}_t \end{bmatrix} = \begin{bmatrix} \boldsymbol{\theta}_{xx}(B) & \boldsymbol{\theta}_{xy}(B) \\ \boldsymbol{\theta}_{yx}(B) & \boldsymbol{\theta}_{yy}(B) \end{bmatrix} \begin{bmatrix} \boldsymbol{u}_t \\ \boldsymbol{v}_t \end{bmatrix}. \tag{3.49}$$

Making use of the definition of $\boldsymbol{\psi}(B)$, we have

$$\begin{bmatrix} \boldsymbol{\phi}_{xx}(B) & \boldsymbol{\phi}_{xy}(B) \\ \boldsymbol{\phi}_{yx}(B) & \boldsymbol{\phi}_{yy}(B) \end{bmatrix}^{-1} \begin{bmatrix} \boldsymbol{\theta}_{xx}(B) & \boldsymbol{\theta}_{xy}(B) \\ \boldsymbol{\theta}_{yx}(B) & \boldsymbol{\theta}_{yy}(B) \end{bmatrix} = \begin{bmatrix} \boldsymbol{\psi}_{xx}(B) & \boldsymbol{\psi}_{xy}(B) \\ \boldsymbol{\psi}_{yx}(B) & \boldsymbol{\psi}_{yy}(B) \end{bmatrix}.$$

Using properties of matrix partition in Appendix A, we have $[\boldsymbol{\phi}(B)]^{-1} =$

$$\left[\begin{array}{cc} \boldsymbol{D}(B) & -\boldsymbol{D}(B)\boldsymbol{\phi}_{xy}(B)\boldsymbol{\phi}_{yy}^{-1}(B) \\ -\boldsymbol{\phi}_{yy}^{-1}(B)\boldsymbol{\phi}_{yx}(B)\boldsymbol{D}(B) & \boldsymbol{\phi}_{yy}^{-1}(B)[\boldsymbol{I}_{k_2} + \boldsymbol{\phi}_{yx}(B)\boldsymbol{D}(B)\boldsymbol{\phi}_{xy}(B)\boldsymbol{\phi}_{yy}^{-1}(B) \end{array}\right],$$

where $\boldsymbol{D}(B) = [\boldsymbol{\phi}_{xx}(B) - \boldsymbol{\phi}_{xy}(B)\boldsymbol{\phi}_{yy}^{-1}(B)\boldsymbol{\phi}_{yx}(B)]^{-1}$. Consequently, the condition that $\boldsymbol{\psi}_{xy}(B) = \boldsymbol{0}$ is equivalent to

$$\boldsymbol{D}(B)[\boldsymbol{\theta}_{xy}(B) - \boldsymbol{\phi}_{xy}(B)\boldsymbol{\phi}_{yy}^{-1}(B)\boldsymbol{\theta}_{yy}(B)] = \boldsymbol{0},$$

or equivalently,

$$\boldsymbol{\theta}_{xy}(B) - \boldsymbol{\phi}_{xy}(B)\boldsymbol{\phi}_{yy}^{-1}(B)\boldsymbol{\theta}_{yy}(B) = \boldsymbol{0}. \tag{3.50}$$

Using the same techniques but expressing $[\boldsymbol{\phi}(B)]^{-1}$ differently, we obtain that the condition $\boldsymbol{\psi}_{yx}(B) \neq \boldsymbol{0}$ is equivalent to

$$\boldsymbol{\theta}_{yx}(B) - \boldsymbol{\phi}_{yx}(B)\boldsymbol{\phi}_{xx}^{-1}(B)\boldsymbol{\theta}_{xx}(B) \neq \boldsymbol{0}. \tag{3.51}$$

From Equation (3.50), the condition $\boldsymbol{\phi}_{xy}(B) = \boldsymbol{\theta}_{xy}(B) = \boldsymbol{0}$, which results in having simultaneous lower triangular matrices in the AR and MA matrix polynomials of $\boldsymbol{z}_t$, is only a sufficient condition for the existence of Granger causality from $\boldsymbol{x}_t$ to $\boldsymbol{y}_t$. If Equation (3.50) holds and Equation (3.51) changes the inequality to equality, then $\boldsymbol{x}_t$ and $\boldsymbol{y}_t$ are uncoupled in the sense that they are not dynamically correlated. We summarize the prior derivation into a theorem.

Theorem 3.3 Consider the k-dimensional VARMA(p, q) process $\boldsymbol{z}_t$. Suppose that $\boldsymbol{z}_t = (\boldsymbol{x}_t', \boldsymbol{y}_t')'$, where $\boldsymbol{x}_t$ and $\boldsymbol{y}_t$ are k_1 and k_2-dimensional subvectors of $\boldsymbol{z}_t$, respectively, and $k_1 + k_2 = k$. Partition the AR and MA matrix polynomials of $\boldsymbol{z}_t$ according to the dimensions of $\boldsymbol{x}_t$ and $\boldsymbol{y}_t$, then there exists a Granger causality from $\boldsymbol{x}_t$ to $\boldsymbol{y}_t$ if and only if

$$\boldsymbol{\theta}_{xy}(B) = \boldsymbol{\phi}_{xy}(B)\boldsymbol{\phi}_{yy}^{-1}(B)\boldsymbol{\theta}_{yy}(B), \quad \boldsymbol{\theta}_{yx}(B) \neq \boldsymbol{\phi}_{yx}(B)\boldsymbol{\phi}_{xx}^{-1}(B)\boldsymbol{\theta}_{xx}(B).$$

To demonstrate, consider the bivariate VARMA(1,1) process $\boldsymbol{z}_t = (x_t, y_t)'$,

$$\left[\begin{array}{cc} 1 - \phi_{1,11}B & -\phi_{1,12}B \\ -\phi_{1,21}B & 1 - \phi_{1,22}B \end{array}\right]\left[\begin{array}{c} x_t \\ y_t \end{array}\right] = \left[\begin{array}{cc} 1 - \theta_{1,11}B & -\theta_{1,12}B \\ -\theta_{1,21}B & 1 - \theta_{1,22}B \end{array}\right]\left[\begin{array}{c} u_t \\ v_t \end{array}\right].$$

The first condition of Theorem 3.3 implies that

$$\theta_{1,12}B = \phi_{1,12}B(1 - \phi_{1,22}B)^{-1}(1 - \theta_{1,22}B),$$

which, in turn, says that

$$\theta_{1,12} = \phi_{1,12}, \quad \text{and} \quad \phi_{1,22}\theta_{1,12} = \phi_{1,12}\theta_{1,22}.$$

Consequently, if $\theta_{1,12} = 0$, then $\phi_{1,12} = 0$ and the condition holds for infinitely many $\theta_{1,22}$ and $\phi_{1,22}$. On the other hand, if $\theta_{1,12} \neq 0$, then we require $\phi_{1,12} = \theta_{1,12}$ and $\phi_{1,22} = \theta_{1,22}$. The model for the latter case can be written as

$$y_t = v_t + \frac{\phi_{1,21}x_{t-1} - \theta_{1,21}u_{t-1}}{1 - \phi_{1,22}B}$$
$$x_t = \frac{1 - \theta_{1,11}B}{1 - \phi_{1,11}B}u_t + \frac{\phi_{1,12}(\phi_{1,21}x_{t-2} - \theta_{1,21}u_{t-2})}{(1 - \phi_{1,11}B)(1 - \phi_{1,22}B)}.$$

The model for y_t is indeed in the form of a transfer function model and x_t does not depend on any past values of y_t.

Remark: Theorem 3.3 only considers the case of two subvectors $\boldsymbol{x}_t$ and $\boldsymbol{y}_t$ of $\boldsymbol{z}_t$. In practice, the theorem can be applied sequentially to a nested partition of $\boldsymbol{z}_t$ or can be generalized to include further partitions of $\boldsymbol{z}_t$. □

3.6.2 Impulse Response Functions

The MA representation of a VARMA(p, q) model can also be used to obtain the impulse response functions of $\boldsymbol{z}_t$ or to perform the forecast error variance decomposition in predicting $\boldsymbol{z}_t$. The same methods discussed in the VAR case of Chapter 2 continue to apply. Thus, details are omitted.

Remark: The impulse response functions of a fitted VARMA model can be obtained by the command `VARMAirf` of the `MTS` package. Similar to the command `VARirf`, one can perform orthogonal transformation of the innovations in using the command `VARMAirf`. □

3.7 LINEAR TRANSFORMS OF VARMA PROCESSES

First, consider a nonsingular linear transformation of $\boldsymbol{z}_t$. Specifically, let $\boldsymbol{y}_t = \boldsymbol{H}\boldsymbol{z}_t$, where $\boldsymbol{H}$ is a $k \times k$ nonsingular transformation matrix. In this case, $\boldsymbol{y}_t$ follows a VARMA(p, q) model given by

$$\boldsymbol{\phi}^*(B)\boldsymbol{y}_t = \boldsymbol{\phi}_0^* + \boldsymbol{\theta}^*(B)\boldsymbol{b}_t, \tag{3.52}$$

where $\boldsymbol{\phi}_0^* = \boldsymbol{H}\boldsymbol{\phi}_0$, $\boldsymbol{\phi}^*(B) = \boldsymbol{H}\boldsymbol{\phi}(B)\boldsymbol{H}^{-1}$, and $\boldsymbol{\theta}^*(B) = \boldsymbol{H}\boldsymbol{\theta}(B)\boldsymbol{H}^{-1}$, and $\boldsymbol{b}_t = \boldsymbol{H}\boldsymbol{a}_t$ is a sequence of iid random vectors with mean zero and positive-definite covariance matrix $\boldsymbol{\Sigma}_b = \boldsymbol{H}\boldsymbol{\Sigma}_a\boldsymbol{H}'$. In other words, we have $\boldsymbol{\phi}_i^* = \boldsymbol{H}\boldsymbol{\phi}_i\boldsymbol{H}^{-1}$

and $\boldsymbol{\theta}_j^* = \boldsymbol{H}\boldsymbol{\theta}_j\boldsymbol{H}^{-1}$. This result can be easily obtained by premultiplying the VARMA(p, q) model of $\boldsymbol{z}_t$ by $\boldsymbol{H}$, then inserting $\boldsymbol{H}^{-1}\boldsymbol{H}$ in front of $\boldsymbol{z}_{t-i}$ and $\boldsymbol{a}_{t-j}$. Since rank of $[\boldsymbol{\phi}_p, \boldsymbol{\theta}_q]$ is k (under the identifiability condition of $\boldsymbol{z}_t$) and $\boldsymbol{H}$ is of full rank, we see that rank of $[\boldsymbol{\phi}_p^*, \boldsymbol{\theta}_q^*]$ is also k. Therefore, the VARMA(p, q) model for $\boldsymbol{y}_t$ is also identifiable.

Next, we consider the generalization of Theorem 3.1 to VARMA(p, q) processes. Suppose that $\boldsymbol{y}_t = \boldsymbol{H}\boldsymbol{z}_t$, where $\boldsymbol{H}$ is an $m \times k$ matrix of rank m and $m < k$.

Theorem 3.4 Suppose that $\boldsymbol{z}_t$ is a k-dimensional stationary and invertible VARMA(p, q) process. Let $\boldsymbol{x}_t = \boldsymbol{H}\boldsymbol{z}_t$, where $\boldsymbol{H}$ is an $m \times k$ matrix of rank m with $m < k$. Then, the transformed process $\boldsymbol{x}_t$ has a VARMA(p_*, q_*) model representation with $p_* \leq kp$ and $q_* \leq (k-1)p + q$.

Proof. Using the mean-adjusted form, the model for $\boldsymbol{z}_t$ is

$$\boldsymbol{\phi}(B)\tilde{\boldsymbol{z}}_t = \boldsymbol{\theta}(B)\boldsymbol{a}_t.$$

Let $\boldsymbol{\phi}^a(B)$ be the adjoint matrix of $\boldsymbol{\phi}(B)$. Then, we have $\boldsymbol{\phi}^a(B)\boldsymbol{\phi}(B) = |\boldsymbol{\phi}(B)|\boldsymbol{I}_k$. Premultiplying the prior equation by $\boldsymbol{\phi}^a(B)$, we have

$$|\boldsymbol{\phi}(B)|\tilde{\boldsymbol{z}}_t = \boldsymbol{\phi}^a(B)\boldsymbol{\theta}(B)\boldsymbol{a}_t.$$

Since $\boldsymbol{z}_t$ is weakly stationary, $|\boldsymbol{\phi}(x)| \neq 0$ for $|x| \leq 1$. The prior equation is also a valid model representation for $\tilde{\boldsymbol{z}}_t$. Premultiplying the prior equation by $\boldsymbol{H}$, we have

$$|\boldsymbol{\phi}(B)|\tilde{\boldsymbol{x}}_t = \boldsymbol{H}\boldsymbol{\phi}^a(B)\boldsymbol{\theta}(B)\boldsymbol{a}_t. \tag{3.53}$$

Note that the maximum degree of $|\boldsymbol{\phi}(B)|$ is kp, because each term of the determinant is a product of k polynomials of degree p. On the other hand, the degrees of elements of $\boldsymbol{\phi}^a(B)$ and $\boldsymbol{\theta}(B)$ are $(k-1)p$ and q, respectively. By Theorem 3.1, the right side of Equation (3.53) is a VMA model of maximum order $(k-1)p + q$. Consequently, the order (p_*, q_*) for $\tilde{\boldsymbol{x}}_t$ satisfies the conditions that $p_* \leq kp$ and $q_* \leq (k-1)p + q$. □

Letting $\boldsymbol{H}$ be the ith unit vector in R^k and applying Theorem 3.4, we have the following corollary.

Corollary 3.1 Suppose that $\boldsymbol{z}_t$ is a k-dimensional stationary and invertible VARMA(p, q) process. Then, each component z_{it} of $\boldsymbol{z}_t$ follows a univariate ARMA(p_*, q_*) model, where $p_* \leq kp$ and $q_* \leq p(k-1) + q$.

Similar to the VAR models, the orders p_* and q_* of Theorem 3.4 are the maximum orders allowed. The true orders can be lower. In fact, we have the following corollary, which provides lower orders when the subset of $\boldsymbol{z}_t$ is a vector process.

Corollary 3.2 Suppose that $\boldsymbol{z}_t$ is a k-dimensional stationary and invertible VARMA(p, q) process. Let $\boldsymbol{H}$ be an $m \times k$ matrix of rank m, where $m < k$. Then, the process $\boldsymbol{x}_t = \boldsymbol{H}\boldsymbol{z}_t$ follows an m-dimensional VARMA(p_*, q_*) model with $p_* \leq (k - m + 1)p$ and $q_* \leq (k - m)p + q$.

Proof. We consider first $\boldsymbol{H} = [\boldsymbol{I}_m, \boldsymbol{0}_{m,k-m}]$. That is, $\boldsymbol{x}_t$ consists of the first m components of $\boldsymbol{z}_t$. Partition $\boldsymbol{z}_t = (\boldsymbol{x}_t', \boldsymbol{y}_t')'$, where $\boldsymbol{y}_t$ consists of the last $(k - m)$ components of $\boldsymbol{z}_t$. We also partition the AR and MA matrix polynomials and the innovation $\boldsymbol{a}_t = (\boldsymbol{u}_t', \boldsymbol{v}_t')'$ accordingly. Using Equation (3.49) and assuming, without loss of generality, that $E(\boldsymbol{z}_t) = \boldsymbol{0}$, we have

$$\boldsymbol{\phi}_{xx}(B)\boldsymbol{x}_t + \boldsymbol{\phi}_{xy}(B)\boldsymbol{y}_t = \boldsymbol{\theta}_{xx}(B)\boldsymbol{u}_t + \boldsymbol{\theta}_{xy}(B)\boldsymbol{v}_t \tag{3.54}$$

$$\boldsymbol{\phi}_{yx}(B)\boldsymbol{x}_t + \boldsymbol{\phi}_{yy}(B)\boldsymbol{y}_t = \boldsymbol{\theta}_{yx}(B)\boldsymbol{u}_t + \boldsymbol{\theta}_{yy}(B)\boldsymbol{v}_t. \tag{3.55}$$

Premultiplying Equation (3.55) by the adjoint matrix $\boldsymbol{\phi}_{yy}^a(B)$ of $\boldsymbol{\phi}_{yy}(B)$, we obtain

$$|\boldsymbol{\phi}_{yy}(B)|\boldsymbol{y}_t = -\boldsymbol{\phi}_{yy}^a(B)\boldsymbol{\phi}_{yx}(B)\boldsymbol{x}_t + \boldsymbol{\phi}_{yy}^a(B)\boldsymbol{\theta}_{yx}(B)\boldsymbol{u}_t + \boldsymbol{\phi}_{yy}^a(B)\boldsymbol{\theta}_{yy}(B)\boldsymbol{v}_t. \tag{3.56}$$

Next, premultiplying Equation (3.54) by $|\boldsymbol{\phi}_{yy}(B)|$, replacing $|\boldsymbol{\phi}_{yy}(B)|\boldsymbol{y}_t$ by the right side of Equation (3.56), and rearranging terms, we obtain

$$\begin{aligned}
&[|\boldsymbol{\phi}_{yy}(B)|\boldsymbol{\phi}_{xx}(B) - \boldsymbol{\phi}_{xy}(B)\boldsymbol{\phi}_{yy}^a(B)\boldsymbol{\phi}_{yx}(B)]\boldsymbol{x}_t \\
&\quad = [|\boldsymbol{\phi}_{yy}(B)|\boldsymbol{\theta}_{xx}(B) - \boldsymbol{\phi}_{xy}(B)\boldsymbol{\phi}_{yy}^a(B)\boldsymbol{\theta}_{yx}(B)]\boldsymbol{u}_t \\
&\qquad + [|\boldsymbol{\phi}_{yy}(B)|\boldsymbol{\theta}_{xy}(B) - \boldsymbol{\phi}_{xy}(B)\boldsymbol{\phi}_{yy}^a(B)\boldsymbol{\theta}_{yy}(B)]\boldsymbol{v}_t.
\end{aligned} \tag{3.57}$$

The AR matrix polynomial of Equation (3.57) is of order

$$p_* \leq \max\{(k - m)p + p,\ p + (k - m - 1)p + p\} = (k - m + 1)p.$$

The MA matrix polynomial of Equation (3.57) is of order

$$q_* \leq \max\{(k - m)p + q,\ p + (k - m - 1)p + q\} = (k - m)p + q.$$

Therefore, we have shown Corollary 3.2 when $\boldsymbol{H} = [\boldsymbol{I}_m, \boldsymbol{0}_{m,k-m}]$.

For a general $m \times k$ matrix $\boldsymbol{H}$ of rank m, we can construct a nonsingular matrix $\boldsymbol{H}_* = [\boldsymbol{H}', \boldsymbol{H}_\perp']$, where $\boldsymbol{H}_\perp$ is a $(k - m) \times k$ matrix of rank $k - m$ such that $\boldsymbol{H}'\boldsymbol{H}_\perp = \boldsymbol{0}$. In other words, $\boldsymbol{H}_\perp$ is the orthogonal matrix of $\boldsymbol{H}$ in R^k. We then transform $\boldsymbol{z}_t$ to $\boldsymbol{w}_t = \boldsymbol{H}_*\boldsymbol{z}_t$, which remains a k-dimensional stationary and invertible VARMA(p, q) process. Clearly, $\boldsymbol{x}_t$ is the first m components of $\boldsymbol{w}_t$. This completes the proof. □

To demonstrate, consider the simple bivariate VARMA(1,1) model,

$$\boldsymbol{z}_t - \begin{bmatrix} 0.4 & -0.1 \\ 0.4 & 0.9 \end{bmatrix} \boldsymbol{z}_{t-1} = \boldsymbol{a}_t - \begin{bmatrix} -0.4 & -0.5 \\ -0.4 & -0.5 \end{bmatrix} \boldsymbol{a}_{t-1}. \tag{3.58}$$

The eigenvalues of $\boldsymbol{\phi}_1$ are 0.8 and 0.5, whereas those of $\boldsymbol{\theta}_1$ are -0.9 and 0. The rank of $[\boldsymbol{\phi}_1, \boldsymbol{\theta}_1]$ is 2 and the AR and MA matrix polynomials are

$$\boldsymbol{\phi}(B) = \begin{bmatrix} 1-0.4B & 0.1B \\ -0.4B & 1-0.9B \end{bmatrix}, \quad \boldsymbol{\theta}(B) = \begin{bmatrix} 1+0.4B & 0.5B \\ 0.4B & 1+0.5B \end{bmatrix}.$$

These two matrix polynomials are left co-prime so that the model is identifiable. Premultiplying the model by the adjoint matrix of $\boldsymbol{\phi}(B)$, we obtain

$$(1-1.3B+0.4B^2)\boldsymbol{z}_t = \begin{bmatrix} 1-0.5B-0.36B^2 & 0.4B-0.5B^2 \\ 0.8B & 1+0.1B \end{bmatrix} \boldsymbol{a}_t.$$

From the model, it appears that the first component z_{1t} follows a univariate ARMA (2,2) model, whereas the second component z_{2t} follows a univariate ARMA(2,1) model. If we further assume that

$$\text{Cov}(\boldsymbol{a}_t) = \boldsymbol{\Sigma}_a = \begin{bmatrix} 4 & 1 \\ 1 & 2 \end{bmatrix},$$

then we can work out the exact model for z_{2t}. More specifically, let $b_t = 0.8Ba_{1t} + (1+0.1B)a_{2t}$, which is a MA(1) process with $\text{Var}(b_t) = 4.58$ and $\text{Cov}(b_t, b_{t-1}) = 1.0$. Write the model for b_t as $b_t = u_t - \theta u_{t-1}$, where u_t denotes a univariate white noise series with mean zero and variance σ_u^2. We have $(1+\theta^2)\sigma_u^2 = 4.58$ and $-\theta\sigma_u^2 = 1.0$. Simple calculation shows that $\theta \approx -0.23$ and $\sigma_u^2 \approx 4.35$. Therefore, the model for z_{2t} is

$$(1-1.3B+0.4B^2)z_{2t} = (1+0.23B)u_t, \quad \sigma_u^2 = 4.35.$$

Corollary 3.1 states that both z_{1t} and z_{2t} follow a univariate ARMA(2,2) model. In this particular case, the actual model for z_{2t} is simpler. On the other hand, there is no simplification for the z_{1t} process.

3.8 TEMPORAL AGGREGATION OF VARMA PROCESSES

The results of the prior section on linear transformations of a VARMA model can be used to study temporal aggregations of VARMA processes. We demonstrate the application by considering a simple example. The idea and methods used, however, apply to the general temporal aggregations of VARMA processes with orders p and q being finite.

Suppose that the k-dimensional process $\boldsymbol{z}_t$ follows the VARMA(2,1) model

$$(\boldsymbol{I}_k - \boldsymbol{\phi}_1 B - \boldsymbol{\phi}_2 B^2)\boldsymbol{z}_t = (\boldsymbol{I}_k - \boldsymbol{\theta}_1 B)\boldsymbol{a}_t.$$

Suppose also that we wish to aggregate $\boldsymbol{z}_t$ over $h = 3$ consecutive time periods. For instance, we can aggregate monthly time series to obtain quarterly series. Let ℓ be the time index of the aggregated series. To apply the results of the prior section, we construct an hk-dimensional time series $\boldsymbol{y}_\ell = (\boldsymbol{z}'_{h(\ell-1)+1}, \boldsymbol{z}'_{h(\ell-1)+2}, \boldsymbol{z}'_{h\ell})'$, where $h = 3$ and $\ell = 1, 2, \ldots$. From the VARMA(2,1) model of $\boldsymbol{z}_t$, we have

$$\begin{bmatrix} \boldsymbol{I}_k & \boldsymbol{0}_k & \boldsymbol{0}_k \\ -\boldsymbol{\phi}_1 & \boldsymbol{I}_k & \boldsymbol{0}_k \\ -\boldsymbol{\phi}_2 & -\boldsymbol{\phi}_1 & \boldsymbol{I}_k \end{bmatrix} \begin{bmatrix} \boldsymbol{z}_{h(\ell-1)+1} \\ \boldsymbol{z}_{h(\ell-1)+2} \\ \boldsymbol{z}_{h\ell} \end{bmatrix}$$
$$= \begin{bmatrix} \boldsymbol{0}_k & \boldsymbol{\phi}_2 & \boldsymbol{\phi}_1 \\ \boldsymbol{0}_k & \boldsymbol{0}_k & \boldsymbol{\phi}_2 \\ \boldsymbol{0}_k & \boldsymbol{0}_k & \boldsymbol{0}_k \end{bmatrix} \begin{bmatrix} \boldsymbol{z}_{h(\ell-2)+1} \\ \boldsymbol{z}_{h(\ell-2)+2} \\ \boldsymbol{z}_{h(\ell-1)} \end{bmatrix} + \begin{bmatrix} \boldsymbol{I}_k & \boldsymbol{0}_k & \boldsymbol{0}_k \\ -\boldsymbol{\theta}_1 & \boldsymbol{I}_k & \boldsymbol{0}_k \\ \boldsymbol{0}_k & -\boldsymbol{\theta}_1 & \boldsymbol{I}_k \end{bmatrix} \begin{bmatrix} \boldsymbol{a}_{h(\ell-1)+1} \\ \boldsymbol{a}_{h(\ell-1)+2} \\ \boldsymbol{a}_{h\ell} \end{bmatrix}$$
$$- \begin{bmatrix} \boldsymbol{0}_k & \boldsymbol{0}_k & \boldsymbol{\theta}_1 \\ \boldsymbol{0}_k & \boldsymbol{0}_k & \boldsymbol{0}_k \\ \boldsymbol{0}_k & \boldsymbol{0}_k & \boldsymbol{0}_k \end{bmatrix} \begin{bmatrix} \boldsymbol{a}_{h(\ell-2)+1} \\ \boldsymbol{a}_{h(\ell-2)+2} \\ \boldsymbol{a}_{h(\ell-1)} \end{bmatrix},$$

where, as before, $\boldsymbol{0}_k$ is the $k \times k$ matrix of zeros. The prior equation is a VARMA(1,1) model for the hk-dimensional series $\boldsymbol{y}_t$ in the form

$$\boldsymbol{\Phi}_0 \boldsymbol{y}_\ell = \boldsymbol{\Phi}_1 \boldsymbol{y}_{\ell-1} + \boldsymbol{\Theta}_0 \boldsymbol{b}_\ell - \boldsymbol{\Theta}_1 \boldsymbol{b}_{\ell-1},$$

where $\boldsymbol{b}_t = (\boldsymbol{a}'_{h(\ell-1)+1}, \boldsymbol{a}'_{h(\ell-1)+2}, \boldsymbol{a}'_{h\ell})'$. Since $\boldsymbol{\Phi}_0$ and $\boldsymbol{\Theta}_0$ are invertible, we can rewrite the model as

$$\begin{aligned} \boldsymbol{y}_\ell &= \boldsymbol{\Phi}_0^{-1}\boldsymbol{\Phi}_1 \boldsymbol{y}_{\ell-1} + \boldsymbol{\Phi}_0^{-1}\boldsymbol{\Theta}_0 \boldsymbol{b}_\ell - \boldsymbol{\Phi}_0^{-1}\boldsymbol{\Theta}_1\boldsymbol{\Theta}_0^{-1}\boldsymbol{\Phi}_0(\boldsymbol{\Phi}_0^{-1}\boldsymbol{\Theta}_0 \boldsymbol{b}_{\ell-1}) \\ &= (\boldsymbol{\Phi}_0^{-1}\boldsymbol{\Phi}_1)\boldsymbol{y}_{\ell-1} + \boldsymbol{c}_\ell - (\boldsymbol{\Phi}_0^{-1}\boldsymbol{\Theta}_1\boldsymbol{\Theta}_0^{-1}\boldsymbol{\Phi}_0)\boldsymbol{c}_{\ell-1}, \end{aligned} \tag{3.59}$$

where $\boldsymbol{c}_\ell = \boldsymbol{\Phi}_0^{-1}\boldsymbol{\Theta}_0 \boldsymbol{b}_\ell$. We can then apply Corollary 3.2 to the hk-dimensional VARMA(1,1) model in Equation (3.59) to obtain a VARMA model for any h-aggregated series of $\boldsymbol{z}_t$. For instance, for the simple aggregation $\boldsymbol{x}_\ell = \boldsymbol{z}_{h(\ell-1)+1} + \boldsymbol{z}_{h(\ell-1)+2} + \boldsymbol{z}_{h\ell}$, we have $\boldsymbol{x}_\ell = \boldsymbol{H}\boldsymbol{y}_\ell$, where $\boldsymbol{H} = [1, 1, 1] \otimes \boldsymbol{I}_k$, which is a $k \times hk$ matrix. The vector $[1, 1, 1]$ is used because we assumed $h = 3$ in our demonstration. Applying Corollary 3.2, we see that $\boldsymbol{x}_\ell$ follows a k-dimensional VARMA(p_*, q_*) model with $p_* \leq k(h-1) + 1$ and $q_* \leq k(h-1) + 1$.

3.9 LIKELIHOOD FUNCTION OF A VARMA MODEL

Estimation of VARMA models can be carried out using either the conditional or the exact likelihood method. We start with a brief description of the conditional likelihood function, then devote more efforts to the exact likelihood method.

3.9.1 Conditional Likelihood Function

For a stationary and invertible VARMA(p, q) model, the conditional likelihood function of the data can be evaluated recursively by assuming that $\boldsymbol{a}_t = \mathbf{0}$ for $t \leq 0$ and $\boldsymbol{z}_t = \bar{\boldsymbol{z}}$ for $t \leq 0$, where $\bar{\boldsymbol{z}}$ denotes the sample mean of $\boldsymbol{z}_t$. For simplicity, we assume that $E(\boldsymbol{z}_t) = \mathbf{0}$ so that all presample values are 0, that is, $\boldsymbol{a}_t = \mathbf{0} = \boldsymbol{z}_t$ for $t \leq 0$. Given the data $\{\boldsymbol{z}_t | t = 1, \ldots, T\}$, we define two kT-dimensional random variables: $\boldsymbol{Z} = (\boldsymbol{z}_1', \boldsymbol{z}_2', \ldots, \boldsymbol{z}_t')'$ and $\boldsymbol{A} = (\boldsymbol{a}_1', \boldsymbol{a}_2', \ldots, \boldsymbol{a}_T')'$. Under the simplifying conditions that $\boldsymbol{a}_t = \boldsymbol{z}_t = \mathbf{0}$ for $t \leq 0$, we have

$$\boldsymbol{\Phi Z} = \boldsymbol{\Theta A}, \tag{3.60}$$

where the $kT \times kT$ matrices $\boldsymbol{\Phi}$ and $\boldsymbol{\Theta}$ are

$$\boldsymbol{\Phi} = \begin{bmatrix} \boldsymbol{I}_k & \mathbf{0}_k & \mathbf{0}_k & \cdots & \mathbf{0}_k \\ -\boldsymbol{\phi}_1 & \boldsymbol{I}_k & \mathbf{0}_k & \cdots & \mathbf{0}_k \\ -\boldsymbol{\phi}_2 & -\boldsymbol{\phi}_1 & \boldsymbol{I}_k & \cdots & \mathbf{0}_k \\ \vdots & \vdots & \vdots & \ddots & \vdots \\ \mathbf{0}_k & \mathbf{0}_k & \mathbf{0}_k & \cdots & \boldsymbol{I}_k \end{bmatrix}, \quad \boldsymbol{\Theta} = \begin{bmatrix} \boldsymbol{I}_k & \mathbf{0}_k & \mathbf{0}_k & \cdots & \mathbf{0}_k \\ -\boldsymbol{\theta}_1 & \boldsymbol{I}_k & \mathbf{0}_k & \cdots & \mathbf{0}_k \\ -\boldsymbol{\theta}_2 & -\boldsymbol{\theta}_2 & \boldsymbol{I}_k & \cdots & \mathbf{0}_k \\ \vdots & \vdots & \vdots & \ddots & \vdots \\ \mathbf{0}_k & \mathbf{0}_k & \mathbf{0}_k & \cdots & \boldsymbol{I}_k \end{bmatrix},$$

where $\mathbf{0}_k$ denotes the $k \times k$ zero matrix. More precisely, both $\boldsymbol{\Phi}$ and $\boldsymbol{\Theta}$ are low triangular block matrices with the diagonal blocks being $\boldsymbol{I}_k$. For $\boldsymbol{\Phi}$, the first off-diagonal blocks are $-\boldsymbol{\phi}_1$, the second off-diagonal blocks are $-\boldsymbol{\phi}_2$, and so on. All jth off-diagonal blocks are 0 for $j > p$. The $\boldsymbol{\Theta}$ matrix is defined in a similar manner with $\boldsymbol{\phi}_i$ replaced by $\boldsymbol{\theta}_i$ and p replaced by q. Equation (3.60) is nothing but the result of a recursive expression of the model $\boldsymbol{\phi}(B)\boldsymbol{z}_t = \boldsymbol{\theta}(B)\boldsymbol{a}_t$ under the simplifying condition of zero presample quantities. If we define a $T \times T$ lag matrix $\boldsymbol{L} = [L_{ij}]$, where $L_{i,i-1} = 1$ for $i = 2, \ldots, T$ and 0 otherwise, then we have $\boldsymbol{\Phi} = (\boldsymbol{I}_T \otimes \boldsymbol{I}_k) - \sum_{i=1}^{p} (\boldsymbol{L}^i \otimes \boldsymbol{\phi}_i)$ and $\boldsymbol{\Theta} = (\boldsymbol{I}_T \otimes \boldsymbol{I}_k) - \sum_{j=1}^{q} (\boldsymbol{L}^j \otimes \boldsymbol{\theta}_j)$.

Let $\boldsymbol{W} = \boldsymbol{\Phi Z}$. We have $\boldsymbol{W} = \boldsymbol{\Theta A}$ or equivalently $\boldsymbol{A} = \boldsymbol{\Theta}^{-1}\boldsymbol{W}$. Note that $\boldsymbol{\Theta}^{-1}$ exists because $|\boldsymbol{\Theta}| = 1$. Consequently, the transformation from $\boldsymbol{A}$ to $\boldsymbol{W}$ has a unit Jacobian. Under the normality assumption, we have $\boldsymbol{A} \sim N(\mathbf{0}, \boldsymbol{I}_T \otimes \boldsymbol{\Sigma}_a)$. Therefore, the conditional log-likelihood function of the data is

$$\begin{aligned} \ell(\boldsymbol{\beta}, \boldsymbol{\Sigma}_a; \boldsymbol{Z}) &= -\frac{T}{2} \log(|\boldsymbol{\Sigma}_a|) - \frac{1}{2} \sum_{t=1}^{T} \boldsymbol{a}_t' \boldsymbol{\Sigma}_a^{-1} \boldsymbol{a}_t \\ &= -\frac{T}{2} \log(|\boldsymbol{\Sigma}_a|) - \frac{1}{2} \boldsymbol{A}'(\boldsymbol{I}_T \otimes \boldsymbol{\Sigma}_a^{-1})\boldsymbol{A}, \end{aligned} \tag{3.61}$$

where $\boldsymbol{\beta} = \text{vec}[\boldsymbol{\phi}_1, \ldots, \boldsymbol{\phi}_p, \boldsymbol{\theta}_1, \ldots, \boldsymbol{\theta}_q]$ denotes the vector of AR and MA parameters, and the term involving 2π is omitted.

3.9.1.1 Normal Equations for Conditional Likelihood Estimates

Using Properties (i) and (j) of matrix differentiation in Appendix A, we have

$$\frac{\partial \ell(\boldsymbol{\beta}, \boldsymbol{\Sigma}_a; \boldsymbol{Z})}{\partial \boldsymbol{\Sigma}_a} = -\frac{T}{2}\boldsymbol{\Sigma}_a^{-1} + \frac{1}{2}\sum_{t=1}^{n}(\boldsymbol{\Sigma}_a^{-1}\boldsymbol{a}_t\boldsymbol{a}_t'\boldsymbol{\Sigma}_a^{-1}).$$

Consequently, given $\boldsymbol{\beta}$, the conditional MLE of $\boldsymbol{\Sigma}_a$ is $\hat{\boldsymbol{\Sigma}}_a = (1/T)\sum_{t=1}^{T}\boldsymbol{a}_t\boldsymbol{a}_t'$.

Using $\text{vec}(\boldsymbol{AB}) = (\boldsymbol{B}' \otimes \boldsymbol{I})\text{vec}(\boldsymbol{A})$ so that $\boldsymbol{\theta}_i\boldsymbol{a}_{t-i} = (\boldsymbol{a}_{t-i}' \otimes \boldsymbol{I}_k)\text{vec}(\boldsymbol{\theta}_i)$, we can rewrite the innovation $\boldsymbol{a}_t$ of a VARMA(p, q) model as

$$\boldsymbol{a}_t = \boldsymbol{z}_t - \sum_{i=1}^{p}(\boldsymbol{z}_{t-i}' \otimes \boldsymbol{I}_k)\text{vec}(\boldsymbol{\phi}_i) + \sum_{i=1}^{q}\boldsymbol{\theta}_i\boldsymbol{a}_{t-i}. \tag{3.62}$$

Taking transposition, then partial derivatives, we have

$$\frac{\partial \boldsymbol{a}_t'}{\partial \text{vec}(\boldsymbol{\phi}_i)} = \sum_{j=1}^{q}\frac{\partial \boldsymbol{a}_{t-j}'}{\partial \text{vec}(\boldsymbol{\phi}_i)}\boldsymbol{\theta}_j' - (\boldsymbol{z}_{t-i} \otimes \boldsymbol{I}_k), \quad i = 1, \ldots, p \tag{3.63}$$

$$\frac{\partial \boldsymbol{a}_t'}{\partial \text{vec}(\boldsymbol{\theta}_i)} = \sum_{j=1}^{q}\frac{\partial \boldsymbol{a}_{t-j}'}{\partial \text{vec}(\boldsymbol{\theta}_i)}\boldsymbol{\theta}_j' + (\boldsymbol{a}_{t-i} \otimes \boldsymbol{I}_k) \quad i = 1, \ldots, q. \tag{3.64}$$

Define $\boldsymbol{u}_{i,t} = \partial\boldsymbol{a}_t'/\partial\text{vec}(\boldsymbol{\phi}_i')$ for $i = 1, \ldots, p$ and $\boldsymbol{v}_{i,t} = \partial\boldsymbol{a}_t'/\partial\text{vec}(\boldsymbol{\theta}_i')$ for $i = 1, \ldots, q$. Equations (3.63) and (3.64) show that these partial derivatives follow the models

$$\boldsymbol{u}_{i,t} - \sum_{j=1}^{q}\boldsymbol{u}_{i,t-j}\boldsymbol{\theta}_j' = -\boldsymbol{z}_{t-i} \otimes \boldsymbol{I}_k, \quad i = 1, \ldots, p \tag{3.65}$$

$$\boldsymbol{v}_{i,t} - \sum_{j=1}^{q}\boldsymbol{v}_{i,t-j}\boldsymbol{\theta}_j' = \boldsymbol{a}_{t-i} \otimes \boldsymbol{I}_k, \quad i = 1, \ldots, q. \tag{3.66}$$

From the log-likelihood function in Equation (3.61), we see that

$$\partial\ell(\boldsymbol{\beta})/\partial\boldsymbol{\beta} = -\partial\boldsymbol{A}'/\partial\boldsymbol{\beta}(\boldsymbol{I}_T \otimes \boldsymbol{\Sigma}_a^{-1})\boldsymbol{A},$$

where for simplicity we drop the arguments $\boldsymbol{\Sigma}_a$ and $\boldsymbol{Z}$ for the likelihood function. Consequently, we need the partial derivatives of $-\boldsymbol{a}_t'$ with respect to $\boldsymbol{\beta}$, that is,

$$\frac{-\partial \boldsymbol{a}_t'}{\partial \text{vec}(\boldsymbol{\beta})} = - \begin{bmatrix} \boldsymbol{u}_{1,t} \\ \vdots \\ \boldsymbol{u}_{p,t} \\ \boldsymbol{v}_{1,t} \\ \vdots \\ \boldsymbol{v}_{q,t} \end{bmatrix} \equiv \boldsymbol{m}_t,$$

which is a $k^2(p+q) \times k$ matrix. The partial derivative of $-\boldsymbol{A}'$ with respect to $\boldsymbol{\beta}$ is then $\boldsymbol{M} = [\boldsymbol{m}_1, \ldots, \boldsymbol{m}_T]$, which is a $k^2(p+q) \times kT$ matrix.

To obtain an explicit representation for the partial derivative matrix $\boldsymbol{M}$, we can make use of the models in Equations (3.65) and (3.66). Specifically, from the two sets of equations, we define

$$\boldsymbol{n}_t = \begin{bmatrix} \boldsymbol{z}_{t-1} \otimes \boldsymbol{I}_k \\ \vdots \\ \boldsymbol{z}_{t-p} \otimes \boldsymbol{I}_k \\ -\boldsymbol{a}_{t-1} \otimes \boldsymbol{I}_k \\ \vdots \\ -\boldsymbol{a}_{t-q} \otimes \boldsymbol{I}_k \end{bmatrix}_{k^2(p+q)\times k}, \quad t = 1, \ldots, T.$$

We can then put the two sets of equations in Equations (3.65) and (3.66) together as

$$\boldsymbol{M} \times \begin{bmatrix} \boldsymbol{I}_k & -\boldsymbol{\theta}_1' & -\boldsymbol{\theta}_2' & \cdots & \boldsymbol{0}_k \\ \boldsymbol{0}_k & \boldsymbol{I}_k & -\boldsymbol{\theta}_1' & \cdots & \boldsymbol{0}_k \\ \vdots & \vdots & \ddots & \ddots & \vdots \\ \boldsymbol{0}_k & \boldsymbol{0}_k & \cdots & \boldsymbol{I}_k & -\boldsymbol{\theta}_1' \\ \boldsymbol{0}_k & \boldsymbol{0}_k & \cdots & \boldsymbol{0}_k & \boldsymbol{I}_k \end{bmatrix} = \boldsymbol{N}, \tag{3.67}$$

where $\boldsymbol{N} = [\boldsymbol{n}_1, \boldsymbol{n}_2, \ldots, \boldsymbol{n}_T]$. Equation (3.67) says that the partial derivative matrix $\boldsymbol{M}$ of $-\boldsymbol{A}'$ with respect to $\boldsymbol{\beta}$ satisfies

$$\boldsymbol{M}\boldsymbol{\Theta}' = \boldsymbol{N},$$

where $\boldsymbol{\Theta}$ is defined in Equation (3.60). Consequently, we have

$$\frac{-\partial \boldsymbol{A}'}{\partial \boldsymbol{\beta}} = \boldsymbol{M} = \boldsymbol{N}(\boldsymbol{\Theta}')^{-1}. \tag{3.68}$$

Putting the aforementioned derivation together, we have

$$\frac{\partial \ell(\boldsymbol{\beta})}{\partial \boldsymbol{\beta}} = \boldsymbol{N}(\boldsymbol{\Theta}')^{-1}(\boldsymbol{I}_T \otimes \boldsymbol{\Sigma}_a^{-1})\boldsymbol{A}, \tag{3.69}$$

where $\boldsymbol{N}$ is defined in Equation (3.67). Letting the prior equation to 0, we obtain the normal equations of the conditional log-likelihood function of a stationary and invertible VARMA(p, q) model.

To appreciate the result derived earlier, we consider the case of a pure VAR(p) model for which a closed-form solution of the conditional MLE is available; see Chapter 2. When $q = 0$, the $\boldsymbol{n}_t$ matrix of Equation (3.67) is

$$\boldsymbol{n}_t = \begin{bmatrix} \boldsymbol{z}_{t-1} \otimes \boldsymbol{I}_k \\ \vdots \\ \boldsymbol{z}_{t-p} \otimes \boldsymbol{I}_k \end{bmatrix} = \boldsymbol{x}_t \otimes \boldsymbol{I}_k,$$

where $\boldsymbol{x}_t = (\boldsymbol{z}'_{t-1}, \boldsymbol{z}'_{t-2}, \ldots, \boldsymbol{z}'_{t-p})'$ and $\boldsymbol{\Theta} = \boldsymbol{I}_{kT}$. Therefore, we have $\boldsymbol{N} = \boldsymbol{X} \otimes \boldsymbol{I}_k$, where $\boldsymbol{X} = [\boldsymbol{x}_1, \ldots, \boldsymbol{x}_T]$ is a $k \times T$ data matrix. Also, for VAR(p) models, the more presentation of Equation (3.62) can be written as $\boldsymbol{a}_t = \boldsymbol{z}_t - (\boldsymbol{x}'_t \otimes \boldsymbol{I}_k)\boldsymbol{\beta}$, where $\boldsymbol{x}_t = (\boldsymbol{z}'_{t-1}, \ldots, \boldsymbol{x}'_{t-p})'$. In matrix form, we have

$$\boldsymbol{A} = \boldsymbol{Z} - (\boldsymbol{X}' \otimes \boldsymbol{I}_k)\boldsymbol{\beta}.$$

Consequently, the normal equations in Equation(3.69) for the VAR(p) models become

$$(\boldsymbol{X} \otimes \boldsymbol{I}_k)(\boldsymbol{I}_T \otimes \boldsymbol{\Sigma}_a^{-1})[\boldsymbol{Z} - (\boldsymbol{X}' \otimes \boldsymbol{I}_k)\boldsymbol{\beta}] = \boldsymbol{0}.$$

Next, since

$$(\boldsymbol{X} \otimes \boldsymbol{I}_k)(\boldsymbol{I}_T \otimes \boldsymbol{\Sigma}_a^{-1}) = \boldsymbol{X} \otimes \boldsymbol{\Sigma}_a^{-1} = (\boldsymbol{I}_{kp} \otimes \boldsymbol{\Sigma}_a^{-1})(\boldsymbol{X} \otimes \boldsymbol{I}_k),$$

the normal equations reduce to

$$(\boldsymbol{X} \otimes \boldsymbol{I}_k)\boldsymbol{Z} = (\boldsymbol{X} \otimes \boldsymbol{I}_k)(\boldsymbol{X}' \otimes \boldsymbol{I}_k)\boldsymbol{\beta} = (\boldsymbol{X}\boldsymbol{X}' \otimes \boldsymbol{I}_k)\boldsymbol{\beta}.$$

Again, using $(\boldsymbol{B}' \otimes \boldsymbol{I}_k)\text{vec}(\boldsymbol{A}) = \text{vec}(\boldsymbol{A}\boldsymbol{B})$ and $\boldsymbol{\beta} = \text{vec}[\boldsymbol{\phi}_1, \ldots, \boldsymbol{\phi}_p]$, the aforementioned equation is equivalent to

$$\boldsymbol{Y}\boldsymbol{X}' = \boldsymbol{\phi}\boldsymbol{X}\boldsymbol{X}',$$

where $\boldsymbol{Y} = [\boldsymbol{z}_1, \ldots, \boldsymbol{z}_T]$ and $\boldsymbol{\phi} = [\boldsymbol{\phi}_1, \ldots, \boldsymbol{\phi}_p]$. The conditional MLE of the AR coefficient vector is then

$$\hat{\boldsymbol{\phi}} = \boldsymbol{Y}\boldsymbol{X}'[\boldsymbol{X}\boldsymbol{X}']^{-1},$$

which is the OLS estimate of $\boldsymbol{\phi}$ in Chapter 2.

Finally, it can be shown that on ignoring terms that, when divided by T, converge to 0 in probability as $T \to \infty$, the Hessian matrix of the log-likelihood function is

$$-\frac{\partial^2}{\partial \boldsymbol{\beta} \partial \boldsymbol{\beta}'} \approx \frac{\partial \boldsymbol{A}'}{\partial \boldsymbol{\beta}} (\boldsymbol{I}_T \otimes \boldsymbol{\Sigma}_a^{-1}) \frac{\partial \boldsymbol{A}}{\partial \boldsymbol{\beta}'} = \sum_{t=1}^{T} \frac{\partial \boldsymbol{a}_t'}{\partial \boldsymbol{\beta}} \boldsymbol{\Sigma}_a^{-1} \frac{\partial \boldsymbol{a}_t}{\partial \boldsymbol{\beta}'}$$
$$\approx \boldsymbol{N}(\boldsymbol{\Theta}')^{-1}(\boldsymbol{I}_T \otimes \boldsymbol{\Sigma}_a^{-1})\boldsymbol{\Theta}^{-1}\boldsymbol{N}', \tag{3.70}$$

where $\boldsymbol{N}$ is defined in Equation (3.67).

3.9.2 Exact Likelihood Function

The exact likelihood function of a stationary VARMA(p, q) model in Equation (3.23) has been derived by several authors, for example, Hillmer and Tiao (1979) and Nicholls and Hall (1979). In this section, we follow the approach of Reinsel (1993, Section 5.3) to derive the likelihood function. For simplicity, we assume $E(\boldsymbol{z}_t) = \boldsymbol{0}$ or, equivalently, we use the model in Equation (3.37). Suppose that the data set is $\{\boldsymbol{z}_1, \ldots, \boldsymbol{z}_T\}$.

Let $\boldsymbol{Z} = (\boldsymbol{z}_1', \boldsymbol{z}_2', \ldots, \boldsymbol{z}_T')'$ be the $kT \times 1$ vector of the data, and $\boldsymbol{A} = (\boldsymbol{a}_1', \boldsymbol{a}_2', \ldots, \boldsymbol{a}_T')'$. Also, let $\boldsymbol{Z}_* = (\boldsymbol{z}_{1-p}', \boldsymbol{z}_{2-p}', \ldots, \boldsymbol{z}_0')'$ be the $kp \times 1$ vector of presample data and $\boldsymbol{A}_* = (\boldsymbol{a}_{1-q}', \boldsymbol{a}_{2-q}', \ldots, \boldsymbol{a}_0')'$ be the $kq \times 1$ vector of presample innovations. Finally, let $\boldsymbol{U}_0 = (\boldsymbol{Z}_*', \boldsymbol{A}_*')'$ be the vector of presample variables. Using these newly defined variables, the data of the vector time series can be written as

$$\boldsymbol{\Phi Z} = \boldsymbol{\Theta A} + \boldsymbol{P U}_0, \tag{3.71}$$

where $\boldsymbol{\Phi} = (\boldsymbol{I}_T \otimes \boldsymbol{I}_k) - \sum_{i=1}^{p} (\boldsymbol{L}^i \otimes \boldsymbol{\phi}_i)$, $\boldsymbol{\Theta} = (\boldsymbol{I}_T \otimes \boldsymbol{I}_k) - \sum_{j=1}^{q} (\boldsymbol{L}^j \otimes \boldsymbol{\theta}_j)$, and $\boldsymbol{L}$ is the $T \times T$ lag matrix that has 1 on its main subdiagonal elements and 0s elsewhere. More specifically, $\boldsymbol{L} = [L_{ij}]$ such that $L_{i,i-1} = 1$ for $i = 2, \ldots, T$ and $= 0$ otherwise. The matrix $\boldsymbol{P}$ in Equation (3.71) is given by

$$\boldsymbol{P} = \begin{bmatrix} \boldsymbol{G}_p & \boldsymbol{H}_q \\ \boldsymbol{0}_{k(T-p),kp} & \boldsymbol{0}_{k(T-q),kq} \end{bmatrix},$$

where

$$\boldsymbol{G}_p = \begin{bmatrix} \boldsymbol{\phi}_p & \boldsymbol{\phi}_{p-1} & \cdots & \boldsymbol{\phi}_1 \\ \boldsymbol{0}_k & \boldsymbol{\phi}_p & \cdots & \boldsymbol{\phi}_2 \\ \vdots & \vdots & \ddots & \vdots \\ \boldsymbol{0}_k & \boldsymbol{0}_k & \cdots & \boldsymbol{\phi}_p \end{bmatrix}, \quad \boldsymbol{H}_q = -\begin{bmatrix} \boldsymbol{\theta}_q & \boldsymbol{\theta}_{q-1} & \cdots & \boldsymbol{\theta}_1 \\ \boldsymbol{0}_k & \boldsymbol{\theta}_q & \cdots & \boldsymbol{\theta}_2 \\ \vdots & \vdots & \ddots & \vdots \\ \boldsymbol{0}_k & \boldsymbol{0}_k & \cdots & \boldsymbol{\theta}_q \end{bmatrix}.$$

Since $\boldsymbol{A}$ is independent of $\boldsymbol{U}_0$ and $\text{Cov}(\boldsymbol{A}) = \boldsymbol{I}_T \otimes \boldsymbol{\Sigma}_a$, the covariance matrix of $\boldsymbol{Z}$ in Equation (3.71) is

$$\boldsymbol{\Gamma}_Z = \boldsymbol{\Phi}^{-1}[\boldsymbol{\Theta}(\boldsymbol{I}_T \otimes \boldsymbol{\Sigma}_a)\boldsymbol{\Theta}' + \boldsymbol{P}\boldsymbol{\Xi}\boldsymbol{P}'](\boldsymbol{\Phi}')^{-1}, \tag{3.72}$$

where $\boldsymbol{\Xi} = \text{Cov}(\boldsymbol{U}_0)$. For the stationary VARMA(p, q) model $\boldsymbol{z}_t$ in Equation (3.23), the covariance matrix $\boldsymbol{\Xi}$ can be written as

$$\boldsymbol{\Xi} = \begin{bmatrix} \boldsymbol{\Gamma}(p) & \boldsymbol{C}' \\ \boldsymbol{C} & \boldsymbol{I}_q \otimes \boldsymbol{\Sigma}_a \end{bmatrix},$$

where $\boldsymbol{\Gamma}(p)$ is the $kp \times kp$ covariance matrix of $\boldsymbol{Z}_*$ such that the (i, j)th block of which is $\boldsymbol{\Gamma}_{i-j}$ with $\boldsymbol{\Gamma}_\ell$ being the lag ℓ autocovariance matrix of $\boldsymbol{z}_t$, and $\boldsymbol{C} = \text{Cov}(\boldsymbol{A}_*, \boldsymbol{Z}_*)$ that can be obtained by using Theorem 3.2 as $\boldsymbol{C} = (\boldsymbol{I}_q \otimes \boldsymbol{\Sigma}_a)\boldsymbol{\Psi}$ with

$$\boldsymbol{\Psi} = \begin{bmatrix} \boldsymbol{\psi}'_{q-p} & \cdots & \boldsymbol{\psi}'_{q-2} & \boldsymbol{\psi}'_{q-1} \\ \boldsymbol{\psi}'_{q-1-p} & \cdots & \boldsymbol{\psi}'_{q-3} & \boldsymbol{\psi}'_{q-2} \\ \vdots & & \vdots & \vdots \\ \boldsymbol{0}_k & \cdots & \boldsymbol{I}_k & \boldsymbol{\psi}'_1 \\ \boldsymbol{0}_k & \cdots & \boldsymbol{0}_k & \boldsymbol{I}_k \end{bmatrix}$$

being a $kq \times kp$ matrix of the $\boldsymbol{\psi}$-weight matrices of $\boldsymbol{z}_t$, where it is understood that $\boldsymbol{\psi}_j = \boldsymbol{0}$ if $j < 0$ and $\boldsymbol{\psi}_0 = \boldsymbol{I}_k$. Using the matrix inversion formula

$$[\boldsymbol{C} + \boldsymbol{B}\boldsymbol{D}\boldsymbol{B}']^{-1} = \boldsymbol{C}^{-1} - \boldsymbol{C}^{-1}\boldsymbol{B}(\boldsymbol{B}'\boldsymbol{C}^{-1}\boldsymbol{B} + \boldsymbol{D}^{-1})^{-1}\boldsymbol{B}'\boldsymbol{C}^{-1},$$

we obtain the inverse of $\boldsymbol{\Gamma}_Z$ as

$$\boldsymbol{\Gamma}_Z^{-1} = \boldsymbol{\Phi}'(\boldsymbol{\Omega}^{-1} - \boldsymbol{\Omega}^{-1}\boldsymbol{P}\boldsymbol{Q}^{-1}\boldsymbol{P}'\boldsymbol{\Omega}^{-1})\boldsymbol{\Phi}, \tag{3.73}$$

where $\boldsymbol{\Omega} = \boldsymbol{\Theta}(\boldsymbol{I}_T \otimes \boldsymbol{\Sigma}_a)\boldsymbol{\Theta}' = \text{Cov}(\boldsymbol{\Theta}\boldsymbol{A})$ and $\boldsymbol{Q} = \boldsymbol{\Xi}^{-1} + \boldsymbol{P}'\boldsymbol{\Omega}^{-1}\boldsymbol{P}$. Note that we also have $\boldsymbol{\Omega}^{-1} = (\boldsymbol{\Theta}')^{-1}(\boldsymbol{I}_T \otimes \boldsymbol{\Sigma}_a^{-1})\boldsymbol{\Theta}^{-1}$. Next, using the following matrix inversion formula

$$\begin{bmatrix} \boldsymbol{H} & \boldsymbol{C}' \\ \boldsymbol{C} & \boldsymbol{E} \end{bmatrix}^{-1} = \begin{bmatrix} \boldsymbol{D}^{-1} & -\boldsymbol{D}^{-1}\boldsymbol{C}'\boldsymbol{E}^{-1} \\ -\boldsymbol{E}^{-1}\boldsymbol{C}\boldsymbol{D}^{-1} & \boldsymbol{E}^{-1} + \boldsymbol{E}^{-1}\boldsymbol{C}\boldsymbol{D}^{-1}\boldsymbol{C}'\boldsymbol{E}^{-1} \end{bmatrix},$$

where $\boldsymbol{D} = \boldsymbol{H} - \boldsymbol{C}'\boldsymbol{D}^{-1}\boldsymbol{C}$, with $\boldsymbol{C} = (\boldsymbol{I}_q \otimes \boldsymbol{\Sigma}_a)\boldsymbol{\Psi}$ and $\boldsymbol{E} = \boldsymbol{I}_q \otimes \boldsymbol{\Sigma}_a$, we have

$$\boldsymbol{\Xi}^{-1} = \begin{bmatrix} \boldsymbol{\Delta}^{-1} & -\boldsymbol{\Delta}^{-1}\boldsymbol{\Psi}' \\ -\boldsymbol{\Psi}\boldsymbol{\Delta}^{-1} & (\boldsymbol{I}_q \otimes \boldsymbol{\Sigma}_a^{-1}) + \boldsymbol{\Psi}\boldsymbol{\Delta}^{-1}\boldsymbol{\Psi}' \end{bmatrix},$$

where $\boldsymbol{\Delta} = \boldsymbol{\Gamma}(p) - \boldsymbol{\Psi}'(\boldsymbol{I}_q \otimes \boldsymbol{\Sigma}_a)\boldsymbol{\Psi}$.

We are ready to write down the likelihood function of the sample $\boldsymbol{Z}$. Using the property of determinant

$$\begin{vmatrix} \boldsymbol{C} & \boldsymbol{J} \\ \boldsymbol{B} & \boldsymbol{D} \end{vmatrix} = |\boldsymbol{C}|\,|\boldsymbol{D} - \boldsymbol{B}\boldsymbol{C}^{-1}\boldsymbol{J}|,$$

and the definition of Ξ, we obtain

$$|\Xi| = |(\boldsymbol{I}_q \otimes \boldsymbol{\Sigma}_a)|\,|\boldsymbol{\Gamma}(p) - \boldsymbol{\Psi}'(\boldsymbol{I}_q \otimes \boldsymbol{\Sigma}_a)\boldsymbol{\Psi}| = |(\boldsymbol{I}_q \otimes \boldsymbol{\Sigma}_a)|\,|\boldsymbol{\Delta}| = |\boldsymbol{\Sigma}_a|^q|\boldsymbol{\Delta}|.$$

For a given order (p, q), let $\boldsymbol{\beta}$ be a vector consisting of all parameters in a VARMA(p, q) model. Let $\boldsymbol{W} = (\boldsymbol{w}_1', \ldots, \boldsymbol{w}_T')' = \boldsymbol{\Phi}\boldsymbol{Z}$. That is, $\boldsymbol{w}_t = \boldsymbol{z}_t - \sum_{i=1}^{t-1} \boldsymbol{\phi}_i \boldsymbol{z}_{t-i}$ for $t = 1, \ldots, p$ and $\boldsymbol{w}_t = \boldsymbol{z}_t - \sum_{i=1}^{p} \boldsymbol{\phi}_i \boldsymbol{z}_{t-i}$, where it is understood that the summation term disappears if the lower limit exceeds the upper limit. From Equation (3.72) and using the identity

$$|\boldsymbol{A}_{22}|\,|\boldsymbol{A}_{11} - \boldsymbol{A}_{12}\boldsymbol{A}_{22}^{-1}\boldsymbol{A}_{21}| = |\boldsymbol{A}_{11}|\,|\boldsymbol{A}_{22} - \boldsymbol{A}_{21}\boldsymbol{A}_{11}^{-1}\boldsymbol{A}_{12}|,$$

where the inverses involved exist, we obtain the determinant of $\boldsymbol{\Gamma}_Z$ as

$$\begin{aligned} |\boldsymbol{\Gamma}_Z| &= \frac{1}{|\boldsymbol{\Phi}|^2}|\boldsymbol{\Omega} + \boldsymbol{P}\Xi\boldsymbol{P}'| = |\boldsymbol{\Omega}|\,|\Xi|\,|\Xi^{-1} + \boldsymbol{P}'\boldsymbol{\Omega}^{-1}\boldsymbol{P}| = |\boldsymbol{\Omega}|\,|\Xi|\,|\boldsymbol{Q}| \\ &= |\boldsymbol{\Sigma}_a|^T|\Xi|\,|\boldsymbol{Q}|, \end{aligned}$$

where we have used $|\boldsymbol{\Phi}| = 1$ and $|\boldsymbol{\Omega}| = |\boldsymbol{\Sigma}_a|^T$. Consequently, ignoring the normalizing factor involving 2π, the exact likelihood function of $\boldsymbol{Z}$ can be written as

$$\begin{aligned} L(\boldsymbol{\beta}; \boldsymbol{Z}) = {} & |\boldsymbol{\Sigma}|^{-(T+q)/2}|\boldsymbol{Q}|^{-1/2}|\boldsymbol{\Delta}|^{-1/2} \\ & \times \exp\left\{-\tfrac{1}{2}\boldsymbol{W}'\left[\boldsymbol{\Omega}^{-1} - \boldsymbol{\Omega}^{-1}\boldsymbol{P}\boldsymbol{Q}^{-1}\boldsymbol{P}'\boldsymbol{\Omega}^{-1}\right]\boldsymbol{W}\right\}. \end{aligned} \tag{3.74}$$

The quadratic form in the exponent of Equation (3.74) can also be written as

$$\boldsymbol{Z}'\boldsymbol{\Gamma}_Z^{-1}\boldsymbol{Z} = \boldsymbol{W}'\boldsymbol{\Omega}^{-1}\boldsymbol{W} - \hat{\boldsymbol{U}}_0'\boldsymbol{Q}\hat{\boldsymbol{U}}_0, \tag{3.75}$$

where $\hat{\boldsymbol{U}}_0 = \boldsymbol{Q}^{-1}\boldsymbol{P}'\boldsymbol{\Omega}^{-1}\boldsymbol{W}$, which is an estimate of the presample data.

3.9.3 Interpreting the Likelihood Function

To gain insight into the exact likelihood function of the VARMA(p, q) model in Equation (3.74), we provide some interpretations. Considering Equation (3.71) and using the transformed variable $\boldsymbol{W} = \boldsymbol{\Phi}\boldsymbol{Z}$, we have

$$\boldsymbol{W} = \boldsymbol{P}\boldsymbol{U}_0 + \boldsymbol{\Theta}\boldsymbol{A}, \tag{3.76}$$

which is a nonhomogeneous multiple linear regression and the covariance matrix of the error term is $\text{Cov}(\boldsymbol{\Theta}\boldsymbol{A}) = \boldsymbol{\Theta}(\boldsymbol{I}_T \otimes \Sigma_a)\boldsymbol{\Theta}' = \boldsymbol{\Omega}$ defined in Equation (3.73). Therefore, the generalized least squares (GLS) estimate of the presample variable $\boldsymbol{U}_0$ is

$$\hat{\boldsymbol{U}}_0^g = (\boldsymbol{P}'\boldsymbol{\Omega}^{-1}\boldsymbol{P})^{-1}(\boldsymbol{P}'\boldsymbol{\Omega}^{-1}\boldsymbol{W}), \tag{3.77}$$

with covariance matrix $\text{Cov}(\hat{\boldsymbol{U}}_0^g) = (\boldsymbol{P}'\boldsymbol{\Omega}^{-1}\boldsymbol{P})^{-1}$, where the superscript g denotes GLS estimate.

On the other hand, given the VARMA(p, q) model, we have the following *prior distribution* for $\boldsymbol{U}_0$,

$$\boldsymbol{U}_0 \sim N(\boldsymbol{0}, \Xi),$$

where Ξ is given in Equation (3.72). Under the normality assumption, this is a conjugate prior so that the posterior distribution of $\boldsymbol{U}_0$ is multivariate normal with mean $\hat{\boldsymbol{U}}_0$ and covariance $\boldsymbol{Q}^{-1}$ given by

$$\hat{\boldsymbol{U}}_0 = \boldsymbol{Q}^{-1}(\boldsymbol{P}'\boldsymbol{\Omega}^{-1}\boldsymbol{P})\hat{\boldsymbol{U}}_0^q = \boldsymbol{Q}^{-1}\boldsymbol{P}'\boldsymbol{\Omega}^{-1}\boldsymbol{W}, \tag{3.78}$$

$$\boldsymbol{Q} = \Xi^{-1} + \boldsymbol{P}'\boldsymbol{\Omega}^{-1}\boldsymbol{P}, \tag{3.79}$$

where the precision matrix $\boldsymbol{Q}$ is defined in Equation (3.73). These results are standard in Bayesian inference for multivariate normal distribution. See, for instance, Box and Tiao (1973) and Tsay (2010, Chapter 12). Equation (3.78) is precisely the estimate used in the exact likelihood function of $\boldsymbol{z}_t$ in Equation (3.74). Consequently, given the model and the data, we essentially apply a Bayesian estimate of the presample variables.

From Equation (3.76) and the fact that $\boldsymbol{A}$ and $\boldsymbol{U}_0$ are uncorrelated, we can obtain the following identity after some algebra,

$$\begin{aligned}
&\boldsymbol{A}'(\boldsymbol{I}_T \otimes \Sigma_a^{-1})\boldsymbol{A} + \boldsymbol{U}_0'\Xi^{-1}\boldsymbol{U}_0 \\
&= (\boldsymbol{W} - \boldsymbol{P}\boldsymbol{U}_0)'\boldsymbol{\Omega}^{-1}(\boldsymbol{W} - \boldsymbol{P}\boldsymbol{U}_0) + \boldsymbol{U}_0'\Xi^{-1}\boldsymbol{U}_0 \\
&= \boldsymbol{W}'\boldsymbol{\Omega}^{-1}\boldsymbol{W} - \hat{\boldsymbol{U}}_0'\boldsymbol{Q}\hat{\boldsymbol{U}}_0 + (\boldsymbol{U}_0 - \hat{\boldsymbol{U}}_0)'\boldsymbol{Q}(\boldsymbol{U}_0 - \hat{\boldsymbol{U}}_0),
\end{aligned} \tag{3.80}$$

where, again, $\boldsymbol{Q}$ and $\boldsymbol{\Omega}$ are identify in Equation (3.73). This identity provides an interpretation for the determination of the density function of $\boldsymbol{Z}$ given in Equation (3.74). The left side of Equation (3.80) represents the quadratic form in the exponent of the joint density of $(\boldsymbol{A}, \boldsymbol{U}_0)$, the middle expression in Equation (3.80) gives the exponent of the joint density function of $(\boldsymbol{Z}, \boldsymbol{U}_0)$ under the transformation of Equation (3.71), and the right side of Equation (3.80) corresponds to the same density when expressed as the product of the marginal density of $\boldsymbol{Z}$ [as shown in Equation (3.75)] and the conditional density of $\boldsymbol{U}_0$ given $\boldsymbol{Z}$ [as shown by the last term of Equation (3.80)]. Therefore, the marginal density of $\boldsymbol{Z}$ is obtained from the joint density of $(\boldsymbol{Z}, \boldsymbol{U}_0)$ by integrating out the variables in $\boldsymbol{U}_0$. This idea is the same as that of the VMA(1) model discussed earlier. The only difference is that here we use a Bayesian estimate of $\boldsymbol{U}_0$.

3.9.4 Computation of Likelihood Function

To evaluate the likelihood function in Equation (3.74), we note that

$$\boldsymbol{W}'\boldsymbol{\Omega}^{-1}\boldsymbol{W} = \boldsymbol{Z}'\boldsymbol{\Phi}'[\boldsymbol{\Theta}^{-1}(\boldsymbol{I}_T \otimes \boldsymbol{\Sigma}_a^{-1})\boldsymbol{\Theta}^{-1}]\boldsymbol{\Phi}\boldsymbol{Z} = \boldsymbol{A}_0'(\boldsymbol{I}_T \otimes \boldsymbol{\Sigma}_a^{-1})\boldsymbol{A}_0$$
$$= \sum_{t=1}^{T} \boldsymbol{a}_{t,0}'\boldsymbol{\Sigma}_a^{-1}\boldsymbol{a}_{t,0},$$

where $\boldsymbol{A}_0 = \boldsymbol{\Theta}^{-1}\boldsymbol{\Phi}\boldsymbol{Z} = (\boldsymbol{a}_{1,0}', \ldots, \boldsymbol{a}_{T,0}')'$ can be computed recursively from $\boldsymbol{\Theta}\boldsymbol{A}_0 = \boldsymbol{\Phi}\boldsymbol{Z}$ and the element $\boldsymbol{a}_{t,0}$ is given by

$$\boldsymbol{a}_{t,0} = \boldsymbol{z}_t - \sum_{i=1}^{p} \boldsymbol{\phi}_i \boldsymbol{z}_{t-i} + \sum_{j=1}^{q} \boldsymbol{\theta}_j \boldsymbol{a}_{t-j,0}, \quad t = 1, \ldots, T,$$

which is the *conditional residual* obtained by setting the presample variable to 0, that is, $\boldsymbol{U}_0 = \boldsymbol{0}$. Note that assuming $\boldsymbol{W}'\boldsymbol{\Omega}^{-1}\boldsymbol{W}$ alone in the exponent and ignoring $|\boldsymbol{Q}|$ and $|\boldsymbol{\Delta}|$ from Equation (3.74) results in using the conditional likelihood function. See, for instance, Tunnicliffe-Wilson (1973).

Also, note that $\hat{\boldsymbol{U}}_0 = \boldsymbol{Q}^{-1}\boldsymbol{P}'\boldsymbol{\Omega}^{-1}\boldsymbol{W} = \boldsymbol{Q}^{-1}\boldsymbol{P}'(\boldsymbol{\Theta}')^{-1}(\boldsymbol{I}_T \otimes \boldsymbol{\Sigma}_a^{-1})\boldsymbol{\Theta}^{-1}\boldsymbol{\Phi}\boldsymbol{Z} = \boldsymbol{Q}^{-1}\boldsymbol{P}'(\boldsymbol{\Theta}')^{-1}(\boldsymbol{I}_T \otimes \boldsymbol{\Sigma}_a^{-1})\boldsymbol{A}_0$, which can be written as

$$\hat{\boldsymbol{U}}_0 = \boldsymbol{Q}^{-1}\boldsymbol{P}'\boldsymbol{D},$$

where $\boldsymbol{D} = (\boldsymbol{\Theta}')^{-1}(\boldsymbol{I}_T \otimes \boldsymbol{\Sigma}_a^{-1})\boldsymbol{A}_0 \equiv (\boldsymbol{d}_1', \ldots, \boldsymbol{d}_T')'$, which can be computed recursively via

$$\boldsymbol{d}_t = \boldsymbol{\Sigma}_a^{-1}\boldsymbol{a}_{t,0} + \sum_{j=1}^{q} \boldsymbol{\theta}_j' \boldsymbol{d}_{t+j}, \quad t = T, \ldots, 1, \quad \text{with} \quad \boldsymbol{d}_{T+1} = \cdots = \boldsymbol{d}_{T+q} = \boldsymbol{0}.$$

This corresponds to the *back forecasting* of Box, Jenkins, and Reinsel (2008) for univariate ARMA models, and it demonstrates the nature of backward prediction in evaluating the exact likelihood function.

Next, the precision matrix $\boldsymbol{Q} = \boldsymbol{\Xi}^{-1} + \boldsymbol{P}'\boldsymbol{\Omega}^{-1}\boldsymbol{P} = \boldsymbol{\Xi}^{-1} + \boldsymbol{P}'(\boldsymbol{\Theta}')^{-1}(\boldsymbol{I}_T \otimes \boldsymbol{\Sigma}_a^{-1})\boldsymbol{\Theta}\boldsymbol{P}$ requires the calculation of $\boldsymbol{\Theta}^{-1}\boldsymbol{P}$, which can also be computed recursively as follows. Let $\boldsymbol{K} = \boldsymbol{\Theta}^{-1}\boldsymbol{P}$. Then, $\boldsymbol{\Theta}\boldsymbol{K} = \boldsymbol{P}$ so that $\boldsymbol{K}$ can be computed via the same method used to calculate $\boldsymbol{a}_{t,0}$ with $\boldsymbol{\Phi}\boldsymbol{Z}$ replaced by $\boldsymbol{P}$.

From Equation (3.71), we have $\hat{\boldsymbol{A}} = \boldsymbol{\Theta}^{-1}(\boldsymbol{\Phi Z} - \boldsymbol{P}\hat{\boldsymbol{U}}_0)$ when $\boldsymbol{U}_0$ is set equal to $\hat{\boldsymbol{U}}_0$. Applying the identity in Equation (3.80) with $\boldsymbol{U}_0$ replaced by $\hat{\boldsymbol{U}}_0$, we obtain the identity

$$\boldsymbol{W}'\boldsymbol{\Omega}^{-1}\boldsymbol{W} - \hat{\boldsymbol{U}}_0'\boldsymbol{Q}\hat{\boldsymbol{U}}_0 = \hat{\boldsymbol{A}}'(\boldsymbol{I}_T \otimes \boldsymbol{\Sigma}_a^{-1})\hat{\boldsymbol{A}} + \hat{\boldsymbol{U}}_0'\boldsymbol{\Xi}^{-1}\hat{\boldsymbol{U}}_0, \tag{3.81}$$

which can also be obtained by direct substitution of $\boldsymbol{X} = \boldsymbol{\Phi Z} = \boldsymbol{\Theta}\hat{\boldsymbol{A}} + \boldsymbol{P}\hat{\boldsymbol{U}}_0$ into Equation (3.75). Using the result of $\boldsymbol{\Xi}^{-1}$, the definition $\boldsymbol{U}_0 = (\boldsymbol{Z}_*', \boldsymbol{A}_*')'$ and some algebra, we have

$$\boldsymbol{U}_0'\boldsymbol{\Xi}^{-1}\boldsymbol{U}_0 = \boldsymbol{A}_*'(\boldsymbol{I}_q \otimes \boldsymbol{\Sigma}_a^{-1})\boldsymbol{A}_* + (\boldsymbol{Z}_* - \boldsymbol{\Psi}'\boldsymbol{A}_*)'\boldsymbol{\Delta}^{-1}(\boldsymbol{Z}_* - \boldsymbol{\Psi}'\boldsymbol{A}_*), \tag{3.82}$$

where the first term on the right side corresponds to the marginal distribution of $\boldsymbol{A}_*$ whereas the second term corresponds to the conditional density of $\boldsymbol{Z}_*$ given $\boldsymbol{A}_*$. Using Equation (3.82) with $\boldsymbol{U}_0$ replaced by $\hat{\boldsymbol{U}}_0$, we obtain an alternative expression for Equation (3.75) as

$$\begin{aligned}
&\boldsymbol{W}'\boldsymbol{\Omega}^{-1}\boldsymbol{W} - \hat{\boldsymbol{U}}_0'\boldsymbol{Q}\hat{\boldsymbol{U}}_0 \\
&\quad = \hat{\boldsymbol{A}}'(\boldsymbol{I}_T \otimes \boldsymbol{\Sigma}_a^{-1})\hat{\boldsymbol{A}} + \hat{\boldsymbol{A}}_*'(\boldsymbol{I}_q \otimes \boldsymbol{\Sigma}_a^{-1})\hat{\boldsymbol{A}}_* + (\hat{\boldsymbol{Z}}_* - \boldsymbol{\Psi}'\hat{\boldsymbol{A}}_*)'\boldsymbol{\Delta}^{-1}(\hat{\boldsymbol{Z}}_* - \boldsymbol{\Psi}'\hat{\boldsymbol{A}}_*) \\
&\quad = \sum_{t=1-q}^{T} \hat{\boldsymbol{a}}_t'\boldsymbol{\Sigma}_a^{-1}\hat{\boldsymbol{a}}_t + (\hat{\boldsymbol{Z}}_* - \boldsymbol{\Psi}'\hat{\boldsymbol{A}}_*)'\boldsymbol{\Delta}^{-1}(\hat{\boldsymbol{Z}}_* - \boldsymbol{\Psi}'\hat{\boldsymbol{A}}_*),
\end{aligned} \tag{3.83}$$

where $\hat{\boldsymbol{a}}_t$ are elements of $\hat{\boldsymbol{A}}$ and $\hat{\boldsymbol{A}}_*$ and are the residuals of the exact likelihood function obtained by

$$\hat{\boldsymbol{a}}_t = \boldsymbol{z}_t - \sum_{i=1}^{p} \boldsymbol{\phi}_i \boldsymbol{z}_{t-i} + \sum_{j=1}^{q} \boldsymbol{\theta}_j \hat{\boldsymbol{a}}_{t-j}, \quad t = 1, \ldots, T, \tag{3.84}$$

where the re-sample values of $\boldsymbol{z}_t$ and $\boldsymbol{a}_t$ are from $\hat{\boldsymbol{U}}_0$ given in Equation (3.75).

In summary, Equations (3.74), (3.75), and (3.84) provide convenient expressions for calculating the exact likelihood function of a stationary and invertible VARMA(p, q) model. There exist alternative approaches to express the exact likelihood function of a VARMA(p, q) model. For instance, we shall consider an approach in the next section that uses an innovation algorithm.

3.10 INNOVATIONS APPROACH TO EXACT LIKELIHOOD FUNCTION

The exact likelihood function of a stationary and invertible VARMA(p, q) model can also be obtained by the formation of the one-step innovations and their covariance matrices of the data $\{\boldsymbol{z}_t | t = 1, \ldots, T\}$. This approach can be treated as a predictive approach and a similar method has been considered by Ansley (1979) for the univariate ARMA model. See, also, Brockwell and Davies (1987, Chapter 11) and the references therein. The approach is also related to the Kalman filtering; see, for instance, Tsay (2010, Chapter 11).

Again, for simplicity, assume that $E(\boldsymbol{z}_t)=\mathbf{0}$ and let $m=\max\{p,q\}$, the maximum degree of the AR and MA matrix polynomials. Consider the data matrix of a VARMA(p, q) model in Equation (3.71). We can rewrite it as

$$\boldsymbol{W}=\boldsymbol{\Phi}\boldsymbol{Z}=\boldsymbol{\Theta}\boldsymbol{A}+\boldsymbol{P}\boldsymbol{U}_0,$$

where $\boldsymbol{W}=(\boldsymbol{w}_1',\ldots,\boldsymbol{w}_T')'$ with $\boldsymbol{w}_t=\boldsymbol{z}_t-\sum_{i=1}^{t-1}\boldsymbol{\phi}_i\boldsymbol{z}_{t-i}$ for $t=1,\ldots,p$ and $\boldsymbol{w}_t=\boldsymbol{z}_t-\sum_{i=1}^{p}\boldsymbol{\phi}_i\boldsymbol{z}_{t-i}$ for $t>p$. The covariance matrix of $\boldsymbol{W}$ is

$$\boldsymbol{\Gamma}_w=E(\boldsymbol{W}\boldsymbol{W}')=\boldsymbol{\Theta}(\boldsymbol{I}_T\otimes\boldsymbol{\Sigma}_a)\boldsymbol{\Theta}'+\boldsymbol{P}\boldsymbol{\Xi}\boldsymbol{P}'.$$

From the definition, we have $\mathrm{Cov}(\boldsymbol{w}_t,\boldsymbol{w}_{t-\ell})=\mathbf{0}$ for $|\ell|>m$, because $\boldsymbol{w}_t=\boldsymbol{a}_t-\sum_{j=1}^{q}\boldsymbol{\theta}_j\boldsymbol{a}_{t-j}$ is a VMA(q) process for $t>p$. Therefore, the $kT\times kT$ covariance matrix $\boldsymbol{\Gamma}_w$ has nonzero blocks only in a band about the main diagonal blocks of maximum bandwidth m. In fact, the block bandwidth is q after the first m block rows. Applying the Cholesky decomposition on $\boldsymbol{\Gamma}_w$, we obtain a block decomposition of the form

$$\boldsymbol{\Gamma}_w=\boldsymbol{G}\boldsymbol{D}\boldsymbol{G}' \tag{3.85}$$

where $\boldsymbol{G}$ is a lower triangular block-band matrix with bandwidth corresponding to that of $\boldsymbol{\Gamma}_w$ and with diagonal blocks being $\boldsymbol{I}_k$. The matrix $\boldsymbol{D}$ is block diagonal, given by $\boldsymbol{D}=\mathrm{diag}\{\boldsymbol{\Sigma}_{1|0},\boldsymbol{\Sigma}_{2|1},\ldots,\boldsymbol{\Sigma}_{T|T-1}\}$, where $\boldsymbol{\Sigma}_{t|t-1}$ are $k\times k$ covariance matrices. We shall discuss the block Cholesky decomposition and associated recursive calculation later. It suffices here to note that $\boldsymbol{Z}=\boldsymbol{\Phi}^{-1}\boldsymbol{W}$ so that the covariance matrix of $\boldsymbol{Z}$ is

$$\boldsymbol{\Gamma}_z=\boldsymbol{\Phi}^{-1}\boldsymbol{\Gamma}_w(\boldsymbol{\Phi}')^{-1}=\boldsymbol{\Phi}^{-1}\boldsymbol{G}\boldsymbol{D}\boldsymbol{G}'(\boldsymbol{\Phi}')^{-1}.$$

Since $\boldsymbol{G}$ and $\boldsymbol{\Phi}$ are both lower triangular matrices with diagonal blocks being $\boldsymbol{I}_k$, we also have $|\boldsymbol{\Gamma}_z|=|\boldsymbol{D}|=\prod_{t=1}^{T}|\boldsymbol{\Sigma}_{t|t-1}|$.

Next, defining $\boldsymbol{A}^*=\boldsymbol{G}^{-1}\boldsymbol{W}=\boldsymbol{G}^{-1}\boldsymbol{\Phi}\boldsymbol{Z}\equiv(\boldsymbol{a}_{1|0}',\boldsymbol{a}_{2|1}',\ldots,\boldsymbol{a}_{T|T-1}')'$, we have $\mathrm{Cov}(\boldsymbol{A}^*)=\boldsymbol{D}$ and

$$\boldsymbol{Z}'\boldsymbol{\Gamma}_z^{-1}\boldsymbol{Z}=(\boldsymbol{A}^*)'\boldsymbol{D}^{-1}\boldsymbol{A}^*=\sum_{t=1}^{T}\boldsymbol{a}_{t|t-1}'\boldsymbol{\Sigma}_{t|t-1}^{-1}\boldsymbol{a}_{t|t-1}.$$

Note that the transformation from $\boldsymbol{Z}$ to $\boldsymbol{A}^*$ is lower triangular, each vector $\boldsymbol{a}_{t|t-1}$ is equal to $\boldsymbol{z}_t$ minus a linear combination of $\{\boldsymbol{z}_{t-1},\ldots,\boldsymbol{z}_1\}$ and $\boldsymbol{a}_{t|t-1}$ is uncorrelated with $\{\boldsymbol{z}_{t-1}, \ldots, \boldsymbol{z}_1\}$. (See further discussion on the decomposition in the next section.) It follows that $\boldsymbol{a}_{t|t-1}$ is the one-step ahead prediction error of $\boldsymbol{z}_t$ given the model and the data $\{\boldsymbol{z}_1, \ldots, \boldsymbol{z}_{t-1}\}$, that is, $\boldsymbol{a}_{t|t-1} = \boldsymbol{z}_t - E(\boldsymbol{z}_t|F_{t-1})$, where F_{t-1} denotes the information available at time $t-1$ (inclusive). The covariance matrix $\boldsymbol{\Sigma}_{t|t-1} = \text{Cov}(\boldsymbol{a}_{t|t-1})$ is the covariance matrix of the one-step ahead forecast error. Based on the aforementioned discussion, the exact likelihood function for $\boldsymbol{z}_t$ given the data is

$$L(\boldsymbol{\beta}; \boldsymbol{Z}) = \left[\prod_{t=1}^{T} |\boldsymbol{\Sigma}_{t|t-1}|^{-1/2}\right] \exp\left[-\frac{1}{2}\sum_{t=1}^{n} \boldsymbol{a}'_{t|t-1}\boldsymbol{\Sigma}^{-1}_{t|t-1}\boldsymbol{a}_{t|t-1}\right], \tag{3.86}$$

where, for simplicity, the term involving 2π is omitted and $\boldsymbol{\beta}$ is the vector of parameters of the VARMA(p, q) model. This likelihood function can be evaluated recursively. Details are given next.

3.10.1 Block Cholesky Decomposition

The block Cholesky decomposition of Equation (3.85) can be obtained in a similar manner as the traditional Cholesky decomposition with multiple linear regressions replaced by multivariate linear regressions. Specifically, let $\boldsymbol{a}_{1|0} = \boldsymbol{w}_1 = \boldsymbol{z}_1$ and $\boldsymbol{\Sigma}_{1|0} = \text{Cov}(\boldsymbol{a}_{1|0})$. For the second row block, consider the multivariate linear regression $\boldsymbol{w}_2 = \boldsymbol{\theta}^*_{2,1}\boldsymbol{w}_1 + \boldsymbol{a}_{2|1}$. Then, we have $\boldsymbol{\theta}^*_{2,1} = \text{Cov}(\boldsymbol{w}_2, \boldsymbol{w}_1)[\text{Cov}(\boldsymbol{w}_1)]^{-1}$ and $\boldsymbol{a}_{2|1}$ being the one-step ahead innovation of $\boldsymbol{w}_2$ given $\boldsymbol{x}_1$. Here, $\boldsymbol{a}_{2|1}$ is uncorrelated with $\boldsymbol{a}_{1|0}$ and $\boldsymbol{\Sigma}_{2|1} = \text{Cov}(\boldsymbol{a}_{2|1})$. In general, for the tth row block ($t > 1$), we have

$$\boldsymbol{w}_t = \sum_{j=1}^{\min\{m,t-1\}} \boldsymbol{\theta}^*_{t-1,j}\boldsymbol{w}_{t-j} + \boldsymbol{a}_{t|t-1}, \tag{3.87}$$

where it is understood that $\boldsymbol{w}_t$ does not depend on $\boldsymbol{w}_{t-\ell}$ for $\ell > m$ and m is replaced by q for $t > p$. The residual covariance matrix is $\boldsymbol{\Sigma}_{t|t-1} = \text{Cov}(\boldsymbol{a}_{t|t-1})$. From Equation (3.87), the block Cholesky decomposition of Equation (3.85) can be written as a linear transformation

$$\boldsymbol{\Theta}^*\boldsymbol{W} = \boldsymbol{A}^*,$$

where $\boldsymbol{A}^* = (\boldsymbol{a}'_{1|0}, \boldsymbol{a}'_{2|1}, \ldots, \boldsymbol{a}'_{T|T-1})'$ and the row blocks of $\boldsymbol{\Theta}^*$ are given next:

1. The first row block is $[\boldsymbol{I}_k, \boldsymbol{0}_{k,k(T-1)}]$.
2. For $1 < t \leq m$, the tth row block is

$$[-\boldsymbol{\theta}^*_{t-1,t-1}, -\boldsymbol{\theta}^*_{t-1,t-2}, \ldots, -\boldsymbol{\theta}^*_{t-1,1}, \boldsymbol{I}_k, \boldsymbol{0}_{k,k(T-t)}].$$

3. For $t > m$, the tth row block is

$$[\mathbf{0}_{k,k(t-q-1)}, -\boldsymbol{\theta}^*_{t-1,q}, \ldots, -\boldsymbol{\theta}^*_{t-1,1}, \boldsymbol{I}_k, \mathbf{0}_{k,k(T-t)}].$$

The $\boldsymbol{G}$ matrix of Equation (3.85) is simply $[\boldsymbol{\Theta}^*]^{-1}$ because $\boldsymbol{X} = [\boldsymbol{\Theta}^*]^{-1}\boldsymbol{A}^*$. Note that the blocks of matrix $\boldsymbol{G}$ are the coefficient matrices of the multivariate linear regression

$$\boldsymbol{w}_t = \sum_{j=1}^{\min\{m,t-1\}} \boldsymbol{g}_{t-1,j}\boldsymbol{a}_{t-j|t-j-1} + \boldsymbol{a}_{t|t-1}, \tag{3.88}$$

which is equivalent to Equation (3.87).

Equation (3.88) shows clearly the nice properties of using innovation approach to exact likelihood function of a VARMA(p, q) model. First, since the regressors $\boldsymbol{a}_{t-j|t-j-1}$ are mutually uncorrelated, the coefficient matrices of the multivariate linear regression in Equation (3.88) can be obtained easily. More specifically, the coefficient matrix $\boldsymbol{g}_{t-1,j}$ is given by

$$\boldsymbol{g}_{t-1,j} = \text{Cov}(\boldsymbol{w}_t, \boldsymbol{a}_{t-j|t-j-1})[\text{Cov}(\boldsymbol{a}_{t-j|t-j-1})]^{-1}, \quad j = 1, \ldots, \min\{m, t-1\}, \tag{3.89}$$

where it is understood that m is replaced by q for $t > p$. Second, the new one-step ahead innovation is

$$\boldsymbol{a}_{t|t-1} = \boldsymbol{w}_t - \sum_{j=1}^{\min\{m,t-1\}} \boldsymbol{g}_{t-1,j}\boldsymbol{a}_{t-j|t-j-1},$$

and its covariance matrix is

$$\text{Cov}(\boldsymbol{a}_{t|t-1}) = \boldsymbol{\Sigma}_{t|t-1} = \boldsymbol{\Gamma}_0^w - \sum_{j=1}^{\min\{m,t-1\}} \boldsymbol{g}_{t-1,j}\boldsymbol{\Sigma}_{t-j|t-j-1}\boldsymbol{g}'_{t-1,j}, \tag{3.90}$$

where $\boldsymbol{\Gamma}_\ell^w$ denotes the lag ℓ autocovariance matrix of $\boldsymbol{w}_t$.

Equations (3.89) and (3.90) provide a nice recursive procedure to evaluate the exact likelihood function in Equation (3.86). The initial values of the recursion can be obtained from the definition of $\boldsymbol{w}_t$. Specifically, consider $\boldsymbol{w}_t$ for $t = 1, \ldots, m$, where $m = \max\{p, q\}$. We have

$$\begin{bmatrix} \boldsymbol{w}_1 \\ \boldsymbol{w}_2 \\ \vdots \\ \boldsymbol{w}_m \end{bmatrix} = \begin{bmatrix} \boldsymbol{I}_k & \mathbf{0}_k & \cdots & \mathbf{0}_k \\ -\boldsymbol{\phi}_1 & \boldsymbol{I}_k & \cdots & \mathbf{0}_k \\ \vdots & \vdots & \ddots & \vdots \\ -\boldsymbol{\phi}_{m-1} & -\boldsymbol{\phi}_{m-2} & \cdots & \boldsymbol{I}_k \end{bmatrix} \begin{bmatrix} \boldsymbol{z}_1 \\ \boldsymbol{z}_2 \\ \vdots \\ \boldsymbol{z}_m \end{bmatrix},$$

where it is understood that $\boldsymbol{\phi}_i = \mathbf{0}$ for $i > p$. Therefore, $\text{Cov}(\boldsymbol{w}_t, \boldsymbol{w}_i)$, where $1 \leq t, i \leq m$, can be obtained from the autocovariance matrices $\boldsymbol{\Gamma}_i$ of $\boldsymbol{z}_t$ with $i = 0, \ldots, m$. These latter quantities can be obtained by using Equation (3.47) of Section 3.5 for a stationary and invertible VARMA(p, q) model. For $t > p$, $\boldsymbol{w}_t$ follows a VMA(q) model and its autocovariance matrices are easily available as discussed in Section 3.1. The recursion is then as follows:

(1) For $t = 1$, we have $\boldsymbol{x}_1 = \boldsymbol{z}_1$ so that $\boldsymbol{a}_{1|0} = \boldsymbol{z}_1$ and $\boldsymbol{\Sigma}_{1|0} = \boldsymbol{\Gamma}_0$.
(2) For $t = 2$, we have

$$\begin{aligned}
\boldsymbol{g}_{1,1} &= \text{Cov}(\boldsymbol{w}_2, \boldsymbol{a}_{1|0})[\text{Cov}(\boldsymbol{a}_{1|0})]^{-1} = \text{Cov}(\boldsymbol{z}_2 - \boldsymbol{\phi}_1 \boldsymbol{z}_1, \boldsymbol{z}_1)[\text{Cov}(\boldsymbol{z}_1)]^{-1} \\
&= (\boldsymbol{\Gamma}_1 - \boldsymbol{\phi}_1 \boldsymbol{\Gamma}_0)\boldsymbol{\Gamma}_0^{-1} = \boldsymbol{\Gamma}_1 \boldsymbol{\Gamma}_0^{-1} - \boldsymbol{\phi}_1. \\
\boldsymbol{a}_{2|1} &= \boldsymbol{w}_2 - \boldsymbol{g}_{1,1} \boldsymbol{a}_{1|0}. \\
\boldsymbol{\Sigma}_{2|1} &= \text{Cov}(\boldsymbol{w}_2) - \boldsymbol{g}_{1,1} \boldsymbol{\Sigma}_{1|0} \boldsymbol{g}_{1,1}'.
\end{aligned}$$

(3) For $t = 3, \ldots, m$, we have

$$\begin{aligned}
\boldsymbol{g}_{t-1,j} &= \text{Cov}(\boldsymbol{w}_t, \boldsymbol{a}_{t-j|t-j-1})[\text{Cov}(\boldsymbol{a}_{t-j|t-j-1})]^{-1}, \quad j = t-1, \ldots, 1. \\
\boldsymbol{a}_{t|t-1} &= \boldsymbol{w}_t - \sum_{j=1}^{t-1} \boldsymbol{g}_{t-1,j} \boldsymbol{a}_{t-j|t-j-1}. \\
\boldsymbol{\Sigma}_{t|t-1} &= \text{Cov}(\boldsymbol{w}_t) - \sum_{j=1}^{t-1} \boldsymbol{g}_{t-1,j} \boldsymbol{\Sigma}_{t-j|t-j-1} \boldsymbol{g}_{t-1,j}'.
\end{aligned}$$

The quantities $\boldsymbol{g}_{t-1,j}$ can be obtained by a backward recursion:
(a) For $j = t - 1$, we have

$$\boldsymbol{g}_{t-1,t-1} = \text{Cov}(\boldsymbol{w}_t, \boldsymbol{a}_{1|0})[\text{Cov}(\boldsymbol{a}_{1|0})]^{-1} = \text{Cov}(\boldsymbol{w}_t, \boldsymbol{w}_1)\boldsymbol{\Sigma}_{1|0}^{-1},$$

and (b) for $j = t - 2, \ldots, 1$, let $v = t - j$. Then, $v = 2, \ldots, t - 1$ and

$$\begin{aligned}
\boldsymbol{g}_{t-1,t-v} &= \text{Cov}(\boldsymbol{w}_t, \boldsymbol{a}_{v|v-1})[\text{Cov}(\boldsymbol{a}_{v|v-1})]^{-1} \\
&= \text{Cov}\left(\boldsymbol{w}_t, \boldsymbol{w}_v - \sum_{i=1}^{v-1} \boldsymbol{g}_{v-1,i} \boldsymbol{a}_{v-i|v-i-1}\right) \boldsymbol{\Sigma}_{v|v-1}^{-1} \\
&= \left[\text{Cov}(\boldsymbol{w}_t, \boldsymbol{w}_v) - \sum_{i=1}^{v-1} \text{Cov}(\boldsymbol{w}_t, \boldsymbol{a}_{v-i|v-i-1}) \boldsymbol{g}_{v-1,i}'\right] \boldsymbol{\Sigma}_{v|v-1}^{-1} \\
&= \left[\text{Cov}(\boldsymbol{w}_t, \boldsymbol{w}_{t-j}) - \sum_{i=1}^{v-1} \boldsymbol{g}_{t-1,t-v+i} \boldsymbol{\Sigma}_{v-i|v-i-1} \boldsymbol{g}_{v-1,i}'\right] \boldsymbol{\Sigma}_{v|v-1}^{-1}.
\end{aligned}$$

(4) For $t > m$, we have $\boldsymbol{w}_t = \boldsymbol{a}_t - \sum_{i=1}^{q} \boldsymbol{\theta}_i \boldsymbol{a}_{t-i}$, $\text{Cov}(\boldsymbol{w}_t) = \boldsymbol{\Sigma}_a + \sum_{i=1}^{q} \boldsymbol{\theta}_i \boldsymbol{\Sigma}_a \boldsymbol{\theta}_i'$, and $\text{Cov}(\boldsymbol{w}_t, \boldsymbol{w}_j) = \boldsymbol{\Gamma}_{t-j}^{w}$, which is available in Section 3.1 where $j > m$. The recursion can be simplified and becomes

$$
\begin{aligned}
\boldsymbol{g}_{t-1,q} &= \boldsymbol{\Gamma}_q^{w} \boldsymbol{\Sigma}_{t-q|t-q-1}^{-1} \\
\boldsymbol{g}_{t-1,i} &= \left[\boldsymbol{\Gamma}_i^{w} - \sum_{j=i+1}^{q} \boldsymbol{g}_{t-1,j} \boldsymbol{\Sigma}_{t-j|t-j-1} \boldsymbol{g}_{t-i-1,j-i}' \right] \boldsymbol{\Sigma}_{t-i|t-i-1}^{-1}, \quad i = q-1, ., 1. \\
\boldsymbol{a}_{t|t-1} &= \boldsymbol{w}_t - \sum_{i=1}^{q} \boldsymbol{g}_{t-1,i} \boldsymbol{a}_{t-i|t-i-1}. \\
\boldsymbol{\Sigma}_{t|t-1} &= \boldsymbol{\Gamma}_0^{w} - \sum_{i=1}^{q} \boldsymbol{g}_{t-1,i} \boldsymbol{\Sigma}_{t-i|t-i-1} \boldsymbol{g}_{t-1,i}'.
\end{aligned}
$$

The first two recursive equations hold because $\boldsymbol{w}_t$ is a VMA series for $t > m$ so that (1) $\text{Cov}(\boldsymbol{w}_t, \boldsymbol{a}_{t-q|t-q-1} = \text{Cov}(\boldsymbol{w}_t, \boldsymbol{w}_{t-q})$ and (2) for $i = q-1, \ldots, 1$,

$$
\begin{aligned}
\text{Cov}(\boldsymbol{w}_t, \boldsymbol{a}_{t-i|t-i-1}) &= \text{Cov}\left(\boldsymbol{w}_t, \boldsymbol{w}_{t-i} - \sum_{v=1}^{q} \boldsymbol{g}_{t-i-1,v} \boldsymbol{a}_{t-i-v|t-i-v-1} \right) \\
&= \text{Cov}\left(\boldsymbol{w}_t, \boldsymbol{w}_{t-i} - \sum_{v=1}^{q-i} \boldsymbol{g}_{t-i-1,v} \boldsymbol{a}_{t-i-v|t-i-v-1} \right) \\
&= \text{Cov}\left(\boldsymbol{w}_t, \boldsymbol{w}_{t-i} - \sum_{j=i+1}^{q} \boldsymbol{g}_{t-i-1,j-i} \boldsymbol{a}_{t-j|t-j-1} \right) \\
&= \boldsymbol{\Gamma}_i^{w} - \sum_{j=i+1}^{q} \text{Cov}\left(\boldsymbol{w}_t, \boldsymbol{a}_{t-j|t-j-1} \right) \boldsymbol{g}_{t-i-1,j-i}' \\
&= \boldsymbol{\Gamma}_i^{w} - \sum_{j=i+1}^{q} \boldsymbol{g}_{t-1,j} \boldsymbol{\Sigma}_{t-j|t-j-1} \boldsymbol{g}_{t-i-1,j-i}'.
\end{aligned}
$$

3.11 ASYMPTOTIC DISTRIBUTION OF MAXIMUM LIKELIHOOD ESTIMATES

Consider a stationary and invertible VARMA(p, q) time series $\boldsymbol{z}_t$. For simplicity, assume that $E(\boldsymbol{z}_t) = \boldsymbol{0}$ and the innovations $\{\boldsymbol{a}_t\}$ is a white noise series with mean zero, positive-definite covariance matrix $\boldsymbol{\Sigma}_a$, and finite fourth moments. Let $\boldsymbol{\beta} = \text{vec}[\boldsymbol{\phi}_1, \ldots, \boldsymbol{\phi}_p, \boldsymbol{\theta}_1, \ldots, \boldsymbol{\theta}_q]$ be the vector of AR and MA parameters. The asymptotic properties of $\hat{\boldsymbol{\beta}}$ have been investigated in the literature by several authors, for example, Dunsmuir and Hannan (1976), Hannan and Deistler (1988,

Chapter 4), Rissanen and Caines (1979), and the references therein. The main results are summarized next.

Theorem 3.5 Let z_t be a k-dimensional zero-mean, stationary, and invertible VARMA(p, q) process. Assume that the innovation $\boldsymbol{a}_t$ satisfies (a) $E(\boldsymbol{a}_t|F_{t-1}) = \mathbf{0}$, (b) $\text{Cov}(\boldsymbol{a}_t|F_{t-1}) = \boldsymbol{\Sigma}_a$, which is positive-definite, and (c) the fourth moments of $\boldsymbol{a}_t$ are finite, where F_{t-1} is the σ-field generated by $\{\boldsymbol{a}_{t-1}, \boldsymbol{a}_{t-2}, \ldots\}$. Then, as $T \to \infty$

1. $\hat{\boldsymbol{\beta}} \to_{a.s.} \boldsymbol{\beta}$,
2. $T^{1/2}(\hat{\boldsymbol{\beta}} - \boldsymbol{\beta}) \to_d N(\mathbf{0}, \boldsymbol{V}^{-1})$,

where T denotes the sample size, $\boldsymbol{V}$ is the asymptotic information matrix of $\boldsymbol{\beta} = \text{vec}[\boldsymbol{\phi}_1, \ldots, \boldsymbol{\phi}_p, \boldsymbol{\theta}_1, \ldots, \boldsymbol{\theta}_q]$, and the subscripts $a.s.$ and d denote convergence with probability 1 and convergence in distribution, respectively.

Theorem 3.5 can be proven via the standard arguments using Taylor expansion and a martingale central limit theorem. We only provide an outline here. Readers are referred to the prior references for details. Define $\boldsymbol{N}_t = [\boldsymbol{u}_{t-1}, \ldots, \boldsymbol{u}_{t-p}, \boldsymbol{v}_{t-1}, \ldots, \boldsymbol{v}_{t-q}]'$, where

$$\boldsymbol{u}_t = \sum_{i=1}^{q} \boldsymbol{\theta}_i \boldsymbol{u}_{t-i} + (\boldsymbol{z}_t' \otimes \boldsymbol{I}_k), \quad \boldsymbol{v}_t = \sum_{i=1}^{q} \boldsymbol{\theta}_i \boldsymbol{v}_{t-i} - (\boldsymbol{a}_t' \otimes \boldsymbol{I}_k).$$

We see that $\boldsymbol{N}_t$ is a $k^2(p+q) \times k$ random matrix and $\boldsymbol{u}_t$ and $\boldsymbol{v}_t$ are $k \times k^2$ random processes. Both $\boldsymbol{u}_t$ and $\boldsymbol{v}_t$ are derived processes via differentiation, that is,

$$\boldsymbol{u}_{t-i}' = \frac{-\partial \boldsymbol{a}_t'}{\partial \text{vec}(\boldsymbol{\phi}_i)}, \quad \boldsymbol{v}_{t-j}' = \frac{-\partial \boldsymbol{a}_t'}{\partial \text{vec}(\boldsymbol{\theta}_j)}$$

for the VARMA(p, q) process $\boldsymbol{z}_t$. The asymptotic information matrix $\boldsymbol{V}$ is given by

$$\boldsymbol{V} = E(\boldsymbol{N}_t \boldsymbol{\Sigma}_a^{-1} \boldsymbol{N}_t') = \lim_{T \to \infty} T^{-1} E[-\partial^2 \ell(\boldsymbol{\beta}; \boldsymbol{Z}) / \partial \boldsymbol{\beta} \partial \boldsymbol{\beta}'],$$

where $\ell(\boldsymbol{\beta}, \boldsymbol{Z})$ denotes the log-likelihood function of the data. For example, consider the VAR(1) model $\boldsymbol{z}_t = \boldsymbol{\phi}_1 \boldsymbol{z}_{t-1} + \boldsymbol{a}_t$. We have $\boldsymbol{u}_t = \boldsymbol{z}_t' \otimes \boldsymbol{I}_k$ and $\boldsymbol{N}_t = \boldsymbol{u}_{t-1}'$ so that

$$\boldsymbol{V} = E(\boldsymbol{N}_t \boldsymbol{\Sigma}_a^{-1} \boldsymbol{N}_t') = E(\boldsymbol{z}_{t-1} \boldsymbol{z}_{t-1}' \otimes \boldsymbol{\Sigma}_a^{-1}) = \boldsymbol{\Gamma}_0 \otimes \boldsymbol{\Sigma}_a^{-1}.$$

In this equation, we use $\boldsymbol{\Sigma}_a^{-1} = 1 \otimes \boldsymbol{\Sigma}_a^{-1}$. For this particular instance, the asymptotic information matrix is the same as that in Theorem 2.1 for the VAR(1) model, where the estimates are listed as $\text{vec}(\boldsymbol{\phi}_1')$.

Applying Taylor expansion of the vector of partial derivatives of the log-likelihood function with respect to $\boldsymbol{\beta}$, one can show that

$$T^{1/2}(\hat{\boldsymbol{\beta}} - \boldsymbol{\beta}) \approx \left[T^{-1}\sum_{t=1}^{T} \boldsymbol{N}_t \boldsymbol{\Sigma}_a^{-1} \boldsymbol{Z}_t'\right]^{-1} T^{-1/2}\sum_{t=1}^{T} \boldsymbol{N}_t \boldsymbol{\Sigma}_a^{-1} \boldsymbol{a}_t$$
$$\to_d N(\boldsymbol{0}, \boldsymbol{V}^{-1}),$$

where $T^{-1}\sum_{t=1}^{T} \boldsymbol{N}_t \boldsymbol{\Sigma}_a^{-1} \boldsymbol{N}_t' \to_p \boldsymbol{V}$ (convergence in probability) as $T \to \infty$ and

$$T^{-1/2}\sum_{t=1}^{T} \boldsymbol{N}_t \boldsymbol{\Sigma}_a^{-1} \boldsymbol{a}_t \to_d N(\boldsymbol{0}, \boldsymbol{V}), \quad \text{as } T \to \infty.$$

The latter result is obtained by using a martingale central limit theorem because

$$\text{Cov}(\boldsymbol{N}_t \boldsymbol{\Sigma}_a^{-1} \boldsymbol{a}_t) = E(\boldsymbol{N}_t \boldsymbol{\Sigma}_a^{-1} \boldsymbol{a}_t \boldsymbol{a}_t' \boldsymbol{\Sigma}_a^{-1} \boldsymbol{N}_t') = E(\boldsymbol{N}_t \boldsymbol{\Sigma}_a^{-1} \boldsymbol{N}_t') = \boldsymbol{V},$$

based on the property that $E(\boldsymbol{a}_t \boldsymbol{a}_t' | F_{t-1}) = \boldsymbol{\Sigma}_a$.

3.11.1 Linear Parameter Constraints

In practice, the parameters of a VARMA(p, q) model often satisfy certain linear constraints, for example, some of them are set to 0. We can express the linear constraints as

$$\boldsymbol{\beta} = \boldsymbol{R}\boldsymbol{\gamma}, \tag{3.91}$$

where $\boldsymbol{R}$ is a known $k^2(p+q) \times s$ constraint matrix and $\boldsymbol{\gamma}$ is an s-dimensional vector of unrestricted parameters, where $s < k^2(p+q)$. With $s < k^2(p+q)$, the linear constraints in Equation (3.91) are equivalent to the condition $\boldsymbol{S}\boldsymbol{\beta} = \boldsymbol{0}$, where $\boldsymbol{S}$ is a $[k^2(p+q)-s] \times k^2(p+q)$ known matrix determined by $\boldsymbol{R}$. If the constraints are simply for simplification by setting some parameters to 0, then $\boldsymbol{R}$ and $\boldsymbol{S}$ are simply some selection matrices.

Under the constraints, the likelihood function of $\boldsymbol{z}_t$ is a function of the unrestricted parameter $\boldsymbol{\gamma}$, and the normal equations can be obtained via

$$\frac{\partial \ell}{\partial \boldsymbol{\gamma}} = \frac{\partial \boldsymbol{\beta}'}{\partial \boldsymbol{\gamma}} \frac{\partial \ell}{\partial \boldsymbol{\beta}} = \boldsymbol{R}' \frac{\partial \ell}{\boldsymbol{\beta}} = \boldsymbol{R}' \boldsymbol{N} (\boldsymbol{\Theta}')^{-1} (\boldsymbol{I}_T \otimes \boldsymbol{\Sigma}_a^{-1}) \boldsymbol{A}, \tag{3.92}$$

and the approximate Hessian matrix is

$$-\frac{\partial^2 \ell}{\partial \boldsymbol{\gamma} \partial \boldsymbol{\gamma}'} = -\boldsymbol{R}' \left(\frac{\partial^2 \ell}{\partial \boldsymbol{\gamma} \partial \boldsymbol{\gamma}'}\right) \boldsymbol{R} \approx \boldsymbol{R}' \boldsymbol{Z} (\boldsymbol{\Theta}')^{-1} (\boldsymbol{I}_T \otimes \boldsymbol{\Sigma}_a^{-1}) \boldsymbol{\Theta}^{-1} \boldsymbol{N}' \boldsymbol{R}. \tag{3.93}$$

The maximum likelihood estimation thus continues to apply and the asymptotic covariance of $\hat{\gamma}$ is

$$\boldsymbol{V}_\gamma = \lim_{T\to\infty} E(\boldsymbol{R}'\boldsymbol{N}_t\boldsymbol{\Sigma}_a^{-1}\boldsymbol{N}_t'\boldsymbol{R}) = \boldsymbol{R}'\boldsymbol{V}\boldsymbol{R}.$$

More specifically, we have

$$T^{1/2}(\hat{\gamma}-\gamma) \to_d \quad N[\mathbf{0}, (\boldsymbol{R}'\boldsymbol{V}\boldsymbol{R})^{-1}].$$

Let $\tilde{\boldsymbol{a}}_t$ be the residuals under the restricted model, which is often called reduced model under the simplification constraints. The MLE of the residual covariance matrix if $\hat{\boldsymbol{\Sigma}}_\gamma = T^{-1}\sum_{t=1}^{t} \tilde{\boldsymbol{a}}_t\tilde{\boldsymbol{a}}_t'$. We can then perform likelihood ratio test or Wald test to make statistical inference concerning the VARMA model. For instance, we can test the hypothesis $H_0 : \boldsymbol{\beta} = \boldsymbol{R}\gamma$ versus $H_a : \boldsymbol{\beta} \neq \boldsymbol{R}\gamma$. The likelihood ratio test is $-T\log(|\hat{\boldsymbol{\Sigma}}_a|/|\hat{\boldsymbol{\Sigma}}_\gamma|)$. Under H_0, this test statistic is asymptotically distributed as chi-squared with $k^2(p+q)-s$ degrees of freedom, where s is the dimension of γ. See, for instance, Kohn (1979).

3.12 MODEL CHECKING OF FITTED VARMA MODELS

Similar to the VAR case of Chapter 2, model checking of a fitted VARMA(p,q) model focuses on analysis of the residuals

$$\hat{\boldsymbol{a}}_t = \boldsymbol{z}_t - \hat{\boldsymbol{\phi}}_0 - \sum_{i=1}^{p} \hat{\boldsymbol{\phi}}_i\boldsymbol{z}_{t-i} + \sum_{j=1}^{q} \hat{\boldsymbol{\theta}}_j\hat{\boldsymbol{a}}_{t-j}, \tag{3.94}$$

where $\hat{\boldsymbol{\phi}}_i$ and $\hat{\boldsymbol{\theta}}_j$ are the MLE of $\boldsymbol{\phi}_i$ and $\boldsymbol{\theta}_j$. If the fitted VARMA model is adequate, the residual series $\{\hat{\boldsymbol{a}}_t\}$ should behave as a k-dimensional white noises. Thus, we continue to apply the multivariate Ljung–Box statistics (or Portmanteau statistic) of Chapter 2 to check the serial and cross-correlations of the residuals.

The lag ℓ autocovariance matrix of the residual $\hat{\boldsymbol{a}}_t$ is defined as

$$\hat{\boldsymbol{C}}_\ell = \frac{1}{T}\sum_{t=\ell+1}^{T} \hat{\boldsymbol{a}}_t\hat{\boldsymbol{a}}_{t-\ell} \equiv [\hat{C}_{\ell,ij}], \tag{3.95}$$

and the lag ℓ residual cross-correlation matrix $\hat{\boldsymbol{R}}_\ell$ as

$$\hat{\boldsymbol{R}}_\ell = \hat{\boldsymbol{D}}^{-1}\hat{\boldsymbol{C}}_\ell\hat{\boldsymbol{D}}^{-1},$$

where $\hat{\boldsymbol{D}}$ is the diagonal matrix of the residual standard errors, that is, $\hat{\boldsymbol{D}} = \text{diag}\{\hat{C}_{0,11}^{1/2}, \ldots, \hat{C}_{0,kk}^{1/2}\}$. Let $\boldsymbol{R}_\ell$ be the lag ℓ cross-correlation matrix of $\boldsymbol{a}_t$. The null hypothesis of interest is $H_0 : \boldsymbol{R}_1 = \cdots = \boldsymbol{R}_m = \boldsymbol{0}$ and the Ljung–Box statistic is

$$\begin{aligned} Q_k(m) &= T^2 \sum_{\ell=1}^{m} \frac{1}{T-\ell} tr(\hat{\boldsymbol{R}}_\ell' \hat{\boldsymbol{R}}_0^{-1} \hat{\boldsymbol{R}}_\ell \hat{\boldsymbol{R}}_0^{-1}) \\ &= T^2 \sum_{\ell=1}^{m} \frac{1}{T-\ell} tr(\hat{\boldsymbol{C}}_\ell' \hat{\boldsymbol{C}}_0^{-1} \hat{\boldsymbol{C}}_\ell \hat{\boldsymbol{C}}_0^{-1}). \end{aligned} \tag{3.96}$$

Under the assumption that the $\boldsymbol{z}_t$ follows a VARMA(p, q) model and $\boldsymbol{a}_t$ is a white noise series with mean zero, positive-definite covariance, and finite fourth moments, the Ljung–Box statistic of Equation (3.96) is asymptotically distributed as chi-square with $k^2(m-p-q)$ degrees of freedom. See Hosking (1980), Poskitt and Tremayne (1982), Li (2004), and Li and McLeod (1981).

3.13 FORECASTING OF VARMA MODELS

Similar to the VAR models of Chapter 2, we use the minimum mean-squared error criterion to discuss the forecasts of a VARMA(p, q) time series $\boldsymbol{z}_t$. For simplicity, we assume the parameters are known. The effects of using estimated parameters on VARMA predictions have been studied by Yamamoto (1981).

Assume that the forecast origin is h and let F_h denote the information available at h. For one-step ahead prediction, we have, from the model, that

$$\begin{aligned} \boldsymbol{z}_h(1) &= E(\boldsymbol{z}_{h+1}|F_h) \\ &= \boldsymbol{\phi}_0 + \sum_{i=1}^{p} \boldsymbol{\phi}_i \boldsymbol{z}_{h+1-i} - \sum_{j=1}^{q} \boldsymbol{\theta}_j \boldsymbol{a}_{h+1-j}, \end{aligned}$$

and the associated forecast error is

$$\boldsymbol{e}_h(1) = \boldsymbol{z}_{h+1} - \boldsymbol{z}_h(1) = \boldsymbol{a}_{h+1}.$$

The covariance matrix of the one-step ahead forecast error is $\text{Cov}[\boldsymbol{e}_h(1)] = \text{Cov}(\boldsymbol{a}_{h+1}) = \boldsymbol{\Sigma}_a$. For two-step ahead prediction, we have

$$\begin{aligned} \boldsymbol{z}_h(2) &= E(\boldsymbol{z}_{h+2}|F_h) \\ &= \boldsymbol{\phi}_0 + \boldsymbol{\phi}_1 \boldsymbol{z}_h(1) + \sum_{i=2}^{p} \boldsymbol{\phi}_i \boldsymbol{z}_{h+2-i} + \sum_{j=2}^{q} \boldsymbol{\theta}_j \boldsymbol{a}_{h+2-j}, \end{aligned}$$

where we have used $E(\boldsymbol{a}_{h+i}|F_h) = \boldsymbol{0}$ for $i > 0$. This result is obtained by the simple rule that

$$E(\boldsymbol{z}_{h+i}|F_h) = \begin{cases} \boldsymbol{z}_{h+i} & \text{if } i \le 0 \\ \boldsymbol{z}_h(i) & \text{if } i > 0, \end{cases} \qquad E(\boldsymbol{a}_{h+i}|F_h) = \begin{cases} \boldsymbol{a}_{h+i} & \text{if } i \le 0 \\ \mathbf{0} & \text{if } i > 0. \end{cases} \tag{3.97}$$

For the VARMA(p, q) model, the simple rule in Equation (3.97) shows that

$$\boldsymbol{z}_h(\ell) = \boldsymbol{\phi}_0 + \sum_{i=1}^{p} \boldsymbol{\phi}_i \boldsymbol{z}_h(\ell - i), \quad \ell > q. \tag{3.98}$$

Therefore, for $\ell > q$, the points forecasts of a VARMA(p, q) model satisfy the matrix polynomial equation

$$\boldsymbol{\phi}(B)\boldsymbol{z}_h(\ell) = \boldsymbol{\phi}_0 \quad \text{or} \quad \boldsymbol{\phi}(B)[\boldsymbol{z}_h(\ell) - \boldsymbol{\mu}] = \mathbf{0},$$

where $\boldsymbol{\mu} = E(\boldsymbol{z}_t)$. The points forecasts of a VARMA(p, q) model thus can be calculated recursively for $\ell > q$.

Turn to forecast errors. Here, it is most convenient to use the MA representation of $\boldsymbol{z}_t$. See Equation (3.38). The ℓ-step ahead forecast error is

$$\boldsymbol{e}_h(\ell) = \boldsymbol{a}_{h+\ell} + \boldsymbol{\psi}_1 \boldsymbol{a}_{h+\ell-1} + \cdots + \boldsymbol{\psi}_{\ell-1} \boldsymbol{a}_{h+1}. \tag{3.99}$$

Consequently, the covariance matrix of ℓ-step ahead forecast error is

$$\text{Cov}[\boldsymbol{e}_h(\ell)] = \boldsymbol{\Sigma}_a + \boldsymbol{\psi}_1 \boldsymbol{\Sigma}_a \boldsymbol{\psi}_1' + \cdots + \boldsymbol{\psi}_{h-1} \boldsymbol{\Sigma}_a \boldsymbol{\psi}_{h-1}'. \tag{3.100}$$

An important implication of the prior two equations is that a stationary VARMA process is mean-reverting. Equation (3.99) shows that the forecast error approaches the stochastic part of $\boldsymbol{z}_t$ as ℓ increases because $\boldsymbol{\psi}_j \to \mathbf{0}$. That is, $\boldsymbol{z}_h(\ell) \to \boldsymbol{\mu}$ as $\ell \to \infty$. Furthermore, Equation (3.100) shows that $\text{Cov}[\boldsymbol{e}_h(\ell)]$ converges to $\text{Cov}(\boldsymbol{z}_t)$ as $\ell \to \infty$. This is understandable because the dynamic dependence of a stationary and invertible VARMA model decay exponentially to 0 as the time lag increases. Equation (3.100) also provides an efficient way to compute the covariance matrices of forecast errors.

From Equations (3.99) and (3.100), the forecast error variance decomposition discussed in Chapter 2 for VAR models continues to apply for the VARMA forecasting. Furthermore, using the MA representation, one can perform the impulse response analysis of a fitted VARMA model in exactly the same manner as that for a fitted VAR model.

The results of forecasting discussed in this section suggest that the stationary VARMA models are mainly for short-term predictions. Their long-term predictions are essentially the sample mean. This mean-reverting property has many important implications in applications, especially in finance.

3.13.1 Forecasting Updating

Based on the MA representation, the ℓ-step ahead forecast of $\boldsymbol{z}_{h+\ell}$ at the forecast origin h is

$$\boldsymbol{z}_h(\ell) = E(\boldsymbol{z}_{h+\ell}|F_h) = \boldsymbol{\psi}_\ell \boldsymbol{a}_h + \boldsymbol{\psi}_{\ell+1}\boldsymbol{a}_{h-1} + \cdots, \tag{3.101}$$

and the $(\ell - 1)$-step ahead forecast of $\boldsymbol{z}_{h+\ell}$ at the forecast origin $t = h + 1$ is

$$\boldsymbol{z}_{h+1}(\ell - 1) = \boldsymbol{\psi}_{\ell-1}\boldsymbol{a}_{h+1} + \boldsymbol{\psi}_\ell \boldsymbol{a}_h + \boldsymbol{\psi}_{\ell+1}\boldsymbol{a}_{h-1} + \cdots. \tag{3.102}$$

Subtracting Equation (3.101) from Equation (3.102), we obtain an updating equation for VARMA prediction,

$$\boldsymbol{z}_{h+1}(\ell - 1) = \boldsymbol{z}_h(\ell) + \boldsymbol{\psi}_{\ell-1}\boldsymbol{a}_{h+1}. \tag{3.103}$$

Since $\boldsymbol{a}_{h+1}$ is the new information of the time series available at time $h + 1$, Equation (3.103) simply says that the new forecast for $\boldsymbol{z}_{h+\ell}$ at time $h + 1$ is the forecast at time h plus the new information at time $h + 1$ weighted by $\boldsymbol{\psi}_{\ell-1}$. This result is a direct generalization of the updating formula of univariate ARMA models.

3.14 TENTATIVE ORDER IDENTIFICATION

Before introducing structural specification in the next chapter, we consider a method that can help identify the order (p, q) of a VARMA model without estimating any mixed VARMA model. The method focuses on identifying the orders p and q. It does not address the issue of identifiability of a VARMA model.

The method considered is based on the extended cross-correlation matrices of Tiao and Tsay (1983) and is a generalization of the extended autocorrelation function (EACF) of Tsay and Tiao (1984) for the univariate time series. Consider the stationary and invertible VARMA(p, q) model of Equation (3.23). If consistent estimates of the AR matrix polynomial $\boldsymbol{\phi}(B)$ can be obtained, we can transform the VARMA process into a VMA series for which the cross-correlation matrices can be used to specify the order q.

3.14.1 Consistent AR Estimates

For simplicity, we use the simple VARMA(1,q) model to introduce an iterated vector autoregressions that can provide consistent estimates of the AR coefficients in the presence of MA polynomial. However, the method applies to the general VARMA(p, q) model because it is based on the moment equations of the model.

Assume that z_t is a stationary, zero-mean VARMA(1,q) process. That is, the model for z_t is

$$z_t = \phi_1 z_{t-1} + a_t - \theta_1 a_{t-1} - \cdots - \theta_q a_{t-q}. \tag{3.104}$$

Our goal is to obtain a consistent estimate of the AR coefficient matrix ϕ_1. For the VARMA(1,q) process in Equation (3.104), the moment equation is

$$\Gamma_\ell = \phi_1 \Gamma_{\ell-1}, \quad \ell > q,$$

where Γ_ℓ is the lag ℓ autocovariance matrix of z_t.

Consider first the VAR(1) fitting:

$$z_t = \phi_1^{(0)} z_{t-1} + a_t^{(0)}, \tag{3.105}$$

where the superscript (0) is used to denote the original VAR(1) fitting. Postmultiplying the Equation (3.105) by z'_{t-1} and taking expectation, we have

$$\Gamma_1 = \phi_1^{(0)} \Gamma_0, \quad \text{or} \quad \phi_1^{(0)} = \Gamma_1 \Gamma_0^{-1}. \tag{3.106}$$

This is the normal equation of the VAR(1) model for $\phi_1^{(0)}$. From the moment equation of a VARMA(1,q) model, we have that $\phi_1^{(0)} = \phi_1$ if $q = 0$ and $\phi_1^{(0)} \neq \phi_1$ if $q > 0$. The residual of the model is

$$a_t^{(0)} = z_t - \phi_1^{(0)} z_{t-1}.$$

Next, consider the first iterated VAR(1) fitting:

$$\begin{aligned} z_t &= \phi_1^{(1)} z_{t-1} + \gamma_1^{(1)} a_{t-1}^{(0)} + a_t^{(1)}, \\ &= \phi_1^{(1)} z_{t-1} + \gamma_1^{(1)} [z_{t-1} - \phi_1^{(0)} z_{t-2}] + a_t^{(1)}, \end{aligned} \tag{3.107}$$

where $a_{t-1}^{(0)}$ is the lag-1 residual of Equation (3.105). Postmultiplying Equation (3.107) by z'_{t-2} and taking expectation, we have

$$\Gamma_2 = \phi_1^{(1)} \Gamma_1 + \gamma_1^{(1)} [\Gamma_1 - \phi_1^{(0)} \Gamma_0].$$

Using Equation (3.106), we can simplify the aforementioned equation to obtain

$$\Gamma_2 = \phi_1^{(1)} \Gamma_1 \quad \text{or} \quad \phi_1^{(1)} = \Gamma_2 \Gamma_1^{-1}. \tag{3.108}$$

From the moment equation of the VARMA(1,q) model, we have $\phi_1^{(1)} = \phi_1$ if $q \leq 1$, but $\phi_1^{(1)} \neq \phi_1$ if $q > 1$. The first iterated VAR(1) fitting, therefore, provides a consistent estimate of ϕ_1 for a VARMA(1,q) model with $q \leq 1$. In this sense, the first iterated VAR(1) fitting can be regarded as a regression approach to solving the generalized multivariate Yule–Walker equations for VARMA(1,1) models.

Next, consider the second iterated VAR(1) fitting:

$$\begin{aligned} \boldsymbol{z}_t &= \boldsymbol{\phi}_1^{(2)} \boldsymbol{z}_{t-1} + \boldsymbol{\gamma}_2^{(2)} \boldsymbol{a}_{t-2}^{(0)} + \boldsymbol{\gamma}_1^{(2)} \boldsymbol{a}_{t-1}^{(1)} + \boldsymbol{a}_t^{(2)} \\ &= \boldsymbol{\phi}_1^{(2)} \boldsymbol{z}_{t-1} + \boldsymbol{\gamma}_2^{(2)} [\boldsymbol{z}_{t-2} - \boldsymbol{\phi}_1^{(0)} \boldsymbol{z}_{t-3}] \\ &\quad + \boldsymbol{\gamma}_1^{(2)} \{\boldsymbol{z}_{t-1} - \boldsymbol{\phi}_1^{(1)} \boldsymbol{z}_{t-2} - \boldsymbol{\gamma}_1^{(1)} [\boldsymbol{z}_{t-2} - \boldsymbol{\phi}_1^{(0)} \boldsymbol{z}_{t-3}]\} + \boldsymbol{a}_t^{(2)}. \end{aligned} \tag{3.109}$$

Note that both prior residuals are used with lag-2 and lag-1, respectively. Postmultiplying Equation (3.109) by $\boldsymbol{z}'_{t-3}$ and taking expectation, we have

$$\boldsymbol{\Gamma}_3 = \boldsymbol{\phi}_1^{(2)} \boldsymbol{\Gamma}_2 + \boldsymbol{\gamma}_2^{(2)} [\boldsymbol{\Gamma}_1 - \boldsymbol{\phi}_1^{(0)} \boldsymbol{\Gamma}_0] + \boldsymbol{\gamma}_1^{(2)} \{\boldsymbol{\Gamma}_2 - \boldsymbol{\phi}_1^{(1)} \boldsymbol{\Gamma}_1 - \boldsymbol{\gamma}_1^{(1)} [\boldsymbol{\Gamma}_1 - \boldsymbol{\phi}_1^{(0)} \boldsymbol{\Gamma}_0]\}.$$

Using Equations (3.108) and (3.106), we have

$$\boldsymbol{\Gamma}_3 = \boldsymbol{\phi}_1^{(2)} \boldsymbol{\Gamma}_2, \quad \text{or} \quad \boldsymbol{\phi}_1^{(2)} = \boldsymbol{\Gamma}_3 \boldsymbol{\Gamma}_2^{-1}, \tag{3.110}$$

which, again, solves the generalized multivariate Yule–Walker equations for VARMA(1,2) models. In other words, the second iterated VAR(1) fitting provides a consistent estimate of ϕ_1 for VARMA(1,q) model with $q \leq 2$.

Obviously, the iterated VAR(1) fitting can be repeated. In general, the jth iterated VAR(1) fitting is

$$\boldsymbol{z}_t = \boldsymbol{\phi}_1^{(j)} \boldsymbol{z}_{t-1} + \sum_{v=1}^{j} \boldsymbol{\gamma}_v^{(j)} \boldsymbol{a}_{t-v}^{(j-v)} + \boldsymbol{a}_t^{(j)}, \tag{3.111}$$

and we have

$$\boldsymbol{\Gamma}_{j+1} = \boldsymbol{\phi}_1^{(j)} \boldsymbol{\Gamma}_j, \quad \text{or} \quad \boldsymbol{\phi}_1^{(j)} = \boldsymbol{\Gamma}_{j+1} \boldsymbol{\Gamma}_j^{-1}. \tag{3.112}$$

Therefore, the jth iterated VAR(1) fitting provides a consistent estimate of ϕ_1, namely, $\boldsymbol{\phi}_1 = \boldsymbol{\Gamma}_{j+1} \boldsymbol{\Gamma}_j^{-1}$ if $\boldsymbol{z}_t$ follows a VARMA(1,q) model with $q \leq j$.

For a stationary zero-mean VARMA(p, q) model, the jth VAR(p) fitting is given by

$$\boldsymbol{z}_t = \sum_{i=1}^{p} \boldsymbol{\phi}_i^{(j)} \boldsymbol{z}_{t-i} + \sum_{v=1}^{j} \boldsymbol{\gamma}_v^{(j)} \boldsymbol{a}_{t-v}^{(j-v)} + \boldsymbol{a}_t^{(j)}, \tag{3.113}$$

starting with the traditional VAR(p) fitting. It is easy to show, by mathematical induction, that the AR coefficient matrices $\boldsymbol{\phi}_i^{(j)}$ satisfy the generalized multivariate

Yule–Walker equations

$$\begin{bmatrix} \boldsymbol{\Gamma}_j & \boldsymbol{\Gamma}_{j-1} & \cdots & \boldsymbol{\Gamma}_{j+1-p} \\ \boldsymbol{\Gamma}_{j+1} & \boldsymbol{\Gamma}_j & \cdots & \boldsymbol{\Gamma}_{j+2-p} \\ \vdots & \vdots & \ddots & \vdots \\ \boldsymbol{\Gamma}_{j+p-1} & \boldsymbol{\Gamma}_{j+p-2} & \cdots & \boldsymbol{\Gamma}_j \end{bmatrix} \begin{bmatrix} \boldsymbol{\phi}_1^{(j)} \\ \boldsymbol{\phi}_2^{(j)} \\ \vdots \\ \boldsymbol{\phi}_p^{(j)} \end{bmatrix} = \begin{bmatrix} \boldsymbol{\Gamma}_{j+1} \\ \boldsymbol{\Gamma}_{j+2} \\ \vdots \\ \boldsymbol{\Gamma}_{j+p} \end{bmatrix}. \qquad (3.114)$$

Consequently, the LS estimate of $\boldsymbol{\phi}_i^{(j)}$ is a consistent estimate of $\boldsymbol{\phi}_i$ if $q \le j$ and the matrix involved is invertible.

3.14.2 Extended Cross-Correlation Matrices

Using the consistency results of iterated VAR fitting, Tiao and Tsay (1983) define extended cross-correlation matrices that can be used to specify the order (p, q) of a VARMA model. Again, we use VARMA(1,q) models to introduce the idea, which applies equally well to the general VARMA models. Consider the jth iterated VAR(1) coefficient matrix $\boldsymbol{\phi}_1^{(j)}$. We have $\boldsymbol{\phi}_1^{(j)} = \boldsymbol{\phi}_1$ provided that $q \le j$. Define the jth transformed process

$$\boldsymbol{w}_{1,t}^{(j)} = \boldsymbol{z}_t - \boldsymbol{\phi}_1^{(j)} \boldsymbol{z}_{t-1}, \qquad (3.115)$$

where the subscript 1 of $\boldsymbol{w}_{1,t}^{(j)}$ is used to denote that the transformed series is designed for AR order $p = 1$. Clearly, from the consistency of $\boldsymbol{\phi}_1^{(j)}$, $\boldsymbol{w}_{1,t}^{(j)}$ follows the VMA(q) model

$$\boldsymbol{w}_{1,t}^{(j)} = \boldsymbol{a}_t - \boldsymbol{\theta}_1 \boldsymbol{a}_{t-1} - \cdots - \boldsymbol{\theta}_a \boldsymbol{a}_{t-q},$$

provided that $q \le j$. Therefore, the cross-correlation matrices of $\boldsymbol{w}_{1,t}^{(j)}$ satisfy the condition that $\boldsymbol{\rho}_\ell(w_{1t}^{(j)}) = \boldsymbol{0}$ for $\ell \ge q + 1$ and $j \ge q$, where the argument $w_{1t}^{(j)}$ is used to signify the process under study. To summarize systematically the information of the transformed series $\boldsymbol{w}_{1,t}^{(j)}$ for order determination, consider the following table of the cross-correlation matrices of $\boldsymbol{w}_{1,t}^{(j)}$ series for $j = 0, 1, \ldots$:

Series	Cross correlation matrices				
$\boldsymbol{w}_{1,t}^{(0)}$	$\boldsymbol{\rho}_1(w_{1t}^{(0)})$	$\boldsymbol{\rho}_2(w_{1t}^{(0)})$	$\boldsymbol{\rho}_3(w_{1t}^{(0)})$	$\cdots$	$\boldsymbol{\rho}_\ell(w_{1t}^{(0)})$
$\boldsymbol{w}_{1,t}^{(1)}$	$\boldsymbol{\rho}_1(w_{1t}^{(1)})$	$\boldsymbol{\rho}_2(w_{1t}^{(1)})$	$\boldsymbol{\rho}_3(w_{1t}^{(1)})$	$\cdots$	$\boldsymbol{\rho}_\ell(w_{1t}^{(1)})$
$\boldsymbol{w}_{1,t}^{(2)}$	$\boldsymbol{\rho}_1(w_{1t}^{(2)})$	$\boldsymbol{\rho}_2(w_{1t}^{(2)})$	$\boldsymbol{\rho}_3(w_{1t}^{(2)})$	$\cdots$	$\boldsymbol{\rho}_\ell(w_{1t}^{(2)})$
$\boldsymbol{w}_{1,t}^{(3)}$	$\boldsymbol{\rho}_1(w_{1t}^{(3)})$	$\boldsymbol{\rho}_2(w_{1t}^{(3)})$	$\boldsymbol{\rho}_3(w_{1t}^{(3)})$	$\cdots$	$\boldsymbol{\rho}_\ell(w_{1t}^{(3)})$

For $q = 0$, that is, $\boldsymbol{z}_t$ is a VARMA(1,0) process, all cross-correlation matrices of the table are $\mathbf{0}$. For $q=1$, that is, $\boldsymbol{z}_t$ is a VARMA(1,1) process, we have (a) $\boldsymbol{\rho}_v(\boldsymbol{w}_{1,t}^{(0)}) \neq \mathbf{0}$ for some v because $\boldsymbol{\phi}_1^{(0)} \neq \boldsymbol{\phi}_1$, (b) the lag-1 cross-correlation matrix $\boldsymbol{\rho}_1(\boldsymbol{w}_{1,t}^{(j)})$ is not 0 for $j \geq 1$ because $\boldsymbol{w}_{1,t}^{(j)}$ is a VMA(1) series, and (c) $\boldsymbol{\rho}_v(\boldsymbol{w}_{1,t}^{(j)}) = \mathbf{0}$ for $v > 1$ and $j \geq 1$. These three properties imply a nice pattern in the prior table. Specifically, the first column block consists of nonzero correlation matrices, the first row block contains some nonzero correlation matrices, and all remaining correlation matrices are 0.

This type of pattern continues. For $q = 2$, we have (a) $\boldsymbol{\rho}_v(\boldsymbol{w}_{1,t}^{(1)})$ should be nonzero for some $v \geq 2$ because $\boldsymbol{\phi}_1^{(1)}$ is not consistent with $\boldsymbol{\phi}_1$, (b) $\boldsymbol{\rho}_2(\boldsymbol{w}_{1,t}^{(2)}) \neq \mathbf{0}$ because $\boldsymbol{w}_{1,t}^{(2)}$ is a VMA(2) series, and (c) $\boldsymbol{\rho}_v(\boldsymbol{w}_{1,t}^{(j)}) = \mathbf{0}$ for $v > 2$ and $j \geq 2$. In general, for a VARMA(1,q) series $\boldsymbol{z}_t$, we have (a) $\boldsymbol{\rho}_v(\boldsymbol{w}_{1,t}^{(q-1)})$ should be nonzero for some $v > q$ because $\boldsymbol{\phi}_1^{(q-1)} \neq \boldsymbol{\phi}_1$, (b) $\boldsymbol{\rho}_q(\boldsymbol{w}_{1,t}^{(q)}) \neq \mathbf{0}$ because $\boldsymbol{w}_{1,t}^{(q)}$ is a VMA(q) series, and (c) $\boldsymbol{\rho}_v(\boldsymbol{w}_{1,t}^{(j)}) = \mathbf{0}$ for all $v \geq q+1$ and $j \geq q$.

To utilize the aforementioned properties, Tiao and Tsay (1983) focus on the diagonal elements of the prior table and define the lag-j first extended cross-correlation matrix of $\boldsymbol{z}_t$ as the lag-j cross-correlation matrix of the transformed series $\boldsymbol{w}_{1,t}^{(j-1)}$, where $j > 0$. More specifically, the lag-1 first extended cross-correlation matrix of $\boldsymbol{z}_t$ is the lag-1 cross-correlation matrix of $\boldsymbol{w}_{1,t}^{(0)} = \boldsymbol{z}_t - \boldsymbol{\phi}_1^{(0)}\boldsymbol{z}_{t-1}$, the lag-2 first extended cross-correlation matrix of $\boldsymbol{z}_t$ is the lag-2 cross-correlation matrix of $\boldsymbol{w}_{1,t}^{(1)} = \boldsymbol{z}_t - \boldsymbol{\phi}_1^{(1)}\boldsymbol{z}_{t-1}$, and so on. Denoting the lag-j first extended cross-correlation matrix by $\boldsymbol{\rho}_j(1)$, we have $\boldsymbol{\rho}_j(1) = \boldsymbol{\rho}_j(\boldsymbol{w}_{1,t}^{(j-1)})$. From the consistency of $\boldsymbol{\phi}_1^{(j)}$, we have $\boldsymbol{\rho}_j(1) = \mathbf{0}$ for VARMA(1,q) process with $q < j$. This characteristic can be used to specify the order q of a VARMA(1,q) series.

The prior idea can be extended to the general VARMA(p, q) model. Specifically, to specify the order q of a VARMA(p, q) model, we consider the iterated VAR(p) fitting and the associated transformed series

$$\boldsymbol{w}_{p,t}^{(j)} = \boldsymbol{z}_t - \sum_{i=1}^{p} \boldsymbol{\phi}_i^{(j)} \boldsymbol{z}_{t-i}, \quad t = p+j, \ldots, T. \tag{3.116}$$

Define the lag-j pth extended cross-correlation matrix of $\boldsymbol{z}_t$ as

$$\boldsymbol{\rho}_j(p) = \boldsymbol{\rho}_j(\boldsymbol{w}_{p,t}^{(j-1)}). \tag{3.117}$$

Then, we have $\boldsymbol{\rho}_j(p) = \mathbf{0}$ for $j > q$. Tiao and Tsay (1983) arrange the extended cross-correlation matrices $\boldsymbol{\rho}_j(p)$ in a two-way table with row denoting the AR order p and column denoting the MA order q. Based on the property of extended cross-correlation matrices defined in Equation (3.117), one can identify the order (p, q) by checking for a left-upper vertex of zero cross-correlation matrices.

Remark: Strictly speaking, Tiao and Tsay (1983) did not use the transformed series $\boldsymbol{w}_{m,t}^{(0)}$ and defined the lag-j mth extended cross-correlation matrix as the lag-j cross-correlation matrix of $\boldsymbol{w}_{m,t}^{(j)}$. In this sense, they employed the subdiagonal elements of the table of cross-correlation matrices of $\boldsymbol{w}_{1,t}^{(j)}$. As a consequence of the shift in definition, Tiao and Tsay (1983) suggest to check for the left-upper vertex of a triangular pattern in the extended cross-correlation matrix table to identify the order (p, q). As it will be seen later, the method employed in this chapter looks for the left-upper vertex of a rectangle of large p-values to identify the order (p, q). □

3.14.3 A Summary Two-Way Table

In this chapter, we consider an alternative approach to summarize the information of the transformed series $\boldsymbol{w}_{p,t}^{(j)}$ of Equation (3.116). This alternative approach is motivated by two main considerations. First, the two-way table of extended cross-correlation matrices suggested by Tiao and Tsay (1983) becomes harder to comprehend when the dimension k is large, because each extended cross-correlation matrix is a $k \times k$ matrix. We like to use a summary statistic that does not require matrix presentation. Second, as stated before, the extended cross-correlation matrices focus on the diagonal elements of the cross-correlation matrices of $\boldsymbol{w}_{p,t}^{(j)}$ for $j = 0, 1, \ldots$. As such, they only use one cross-correlation matrix from each transformed series $\boldsymbol{w}_{p,t}^{(j)}$. Other cross-correlation matrices of $\boldsymbol{w}_{p,t}^{(j)}$ should also contain valuable information concerning the MA order q. These two considerations led us to employ the multivariate Portmanteau test statistics. Details are described next.

To introduce the proposed summary statistics, we again use the transformed series $\boldsymbol{w}_{1,t}^{(j)}$ shown before. Consider the transformed series $\boldsymbol{w}_{1,t}^{(0)}$. This series is used mainly to check $q = 0$. Therefore, instead of using $\boldsymbol{\rho}_1(w_{1t}^{(0)})$ alone, we consider the hypotheses

$$H_0 : \boldsymbol{\rho}_1(w_{1t}^{(0)}) = \cdots = \boldsymbol{\rho}_\ell(w_{1t}^{(0)}) = \mathbf{0} \quad \text{versus} \quad H_a : \boldsymbol{\rho}_v(w_{1t}^{(0)}) \neq \mathbf{0}, \quad \text{for some } v,$$

where ℓ is a prespecified number of lags. A natural test statistic to use is the multivariate Ljung–Box statistic

$$Q_{1:\ell}^{(1)} = T^2 \sum_{v=1}^{\ell} \frac{1}{T - v} tr[\boldsymbol{\Gamma}_v'(w_{1t}^{(0)})\boldsymbol{\Gamma}_0^{-1}(w_{1t}^{(0)})\boldsymbol{\Gamma}_v(w_{1t}^{(0)})\boldsymbol{\Gamma}_0^{-1}(w_{1t}^{(0)})],$$

where $\boldsymbol{\Gamma}_v(w_{1t}^{(0)})$ denotes the lag-v autocovariance matrix of $\boldsymbol{w}_{1,t}^{(0)}$ and the subscript $1 : \ell$ is used to denote the number of lags employed. Under the null hypothesis that $\boldsymbol{z}_t$ follows a VARMA(1,0) model, $Q_{1:\ell}^{(1)}$ is asymptotically distributed as a chi-square with $k^2\ell$ degrees of freedom. Denote the p-value of $Q_{1:\ell}^{(1)}$ by p_{10}.

Consider next the transformed series $\boldsymbol{w}_{1,t}^{(1)}$. This transformed series is used mainly to check $q = 1$. Therefore, we consider the hypotheses

$$H_0 : \boldsymbol{\rho}_2(w_{1t}^{(1)}) = \cdots = \boldsymbol{\rho}_\ell(w_{1t}^{(1)}) = \mathbf{0} \quad \text{versus}$$
$$H_a : \boldsymbol{\rho}_v(w_{1t}^{(1)}) \neq \mathbf{0}, \quad \text{for some } v > 1,$$

and employ the Ljung–Box statistic

$$Q_{2:\ell}^{(1)} = T^2 \sum_{v=2}^{\ell} \frac{1}{T-v} tr[\boldsymbol{\Gamma}'_v(w_{1t}^{(1)})\boldsymbol{\Gamma}_0^{-1}(w_{1t}^{(1)})\boldsymbol{\Gamma}_v(w_{1t}^{(1)})\boldsymbol{\Gamma}_0^{-1}(w_{1t}^{(1)})].$$

Under the null hypothesis that $\boldsymbol{z}_t$ is a VARMA(1,1) process, $Q_{2:\ell}^{(1)}$ is asymptotically a chi-square with $k^2(\ell-1)$ degrees of freedom. Denote the p-value of $Q_{2:\ell}^{(1)}$ by p_{11}. This procedure can be repeated. For the jth transformed series $\boldsymbol{w}_{1,t}^{(j)}$, our goal is to check $q = j$. The hypotheses of interest are

$$H_0 : \boldsymbol{\rho}_{j+1}(w_{1t}^{(j)}) = \cdots = \boldsymbol{\rho}_\ell(w_{1t}^{(j)}) = \mathbf{0} \quad \text{versus}$$
$$H_a : \boldsymbol{\rho}_v(w_{1t}^{(j)}) \neq \mathbf{0}, \quad \text{for some } v > j,$$

and we employ the Ljung–Box statistic

$$Q_{(j+1):\ell}^{(1)} = T^2 \sum_{v=j+1}^{\ell} \frac{1}{T-v} tr[\boldsymbol{\Gamma}'_v(w_{1t}^{(j)})\boldsymbol{\Gamma}_0^{-1}(w_{1t}^{(j)})\boldsymbol{\Gamma}_v(w_{1t}^{(j)})\boldsymbol{\Gamma}_0^{-1}(w_{1t}^{(j)})].$$

Under the null hypothesis that $\boldsymbol{z}_t$ follows a VARMA(1,j) model, $Q_{(j+1):\ell}^{(1)}$ is asymptotically a chi-square with $k^2(\ell-j)$ degrees of freedom. Denote the p-value of $Q_{(j+1):\ell}^{(1)}$ by p_{1j}. Clearly, by examining the sequence of p-values p_{1j} for $j = 0, 1, \ldots$, for example comparing them with the type I error α, one can identify the order q for a VARMA(1,q) series.

Again, the idea of Ljung–Box statistics and their p-values applies to the transformed series $\boldsymbol{w}_{m,t}^{(j)}$ for $j = 0, 1, \ldots$, where m is a positive integer. The hypotheses of interest are

$$H_0 : \boldsymbol{\rho}_{j+1}(w_{mt}^{(j)}) = \cdots = \boldsymbol{\rho}_\ell(w_{mt}^{(j)}) = \mathbf{0} \quad \text{versus}$$
$$H_a : \boldsymbol{\rho}_v(w_{mt}^{(j)}) \neq \mathbf{0}, \quad \text{for some } v > j,$$

and we employ the Ljung–Box statistic

$$Q_{(j+1):\ell}^{(m)} = T^2 \sum_{v=j+1}^{\ell} \frac{1}{T-v} tr[\boldsymbol{\Gamma}'_v(w_{mt}^{(j)})\boldsymbol{\Gamma}_0^{-1}(w_{mt}^{(j)})\boldsymbol{\Gamma}_v(w_{mt}^{(j)})\boldsymbol{\Gamma}_0^{-1}(w_{mt}^{(j)})]. \quad (3.118)$$

TABLE 3.1 Two-Way Table of Multivariate Q-Statistics and the Associated Table of p-Values, where $Q^{(m)}_{(j+1):\ell}$ is Defined in Equation (3.118)

AR Order	MA Order: q				
p	0	1	2	$\cdots$	$\ell-1$
	(a) Multivariate Ljung–Box Statistics				
0	$Q^{(0)}_{1:\ell}$	$Q^{(0)}_{2:\ell}$	$Q^{(0)}_{3:\ell}$	$\cdots$	$Q^{(0)}_{(\ell-1):\ell}$
1	$Q^{(1)}_{1:\ell}$	$Q^{(1)}_{2:\ell}$	$Q^{(1)}_{3:\ell}$	$\cdots$	$Q^{(1)}_{(\ell-1):\ell}$
$\vdots$	$\vdots$	$\vdots$	$\vdots$		$\vdots$
m	$Q^{(m)}_{1:\ell}$	$Q^{(m)}_{2:\ell}$	$Q^{(m)}_{3:\ell}$	$\cdots$	$Q^{(m)}_{(\ell-1):\ell}$
	(b) p-Values of Multivariate Ljung–Box Statistics				
0	p_{00}	p_{01}	p_{02}	$\cdots$	$p_{0,\ell-1}$
1	p_{10}	p_{11}	p_{12}	$\cdots$	$p_{1,\ell-1}$
$\vdots$	$\vdots$	$\vdots$	$\vdots$		$\vdots$
m	p_{m0}	p_{m1}	p_{m2}	$\cdots$	$p_{m,\ell-1}$

Under the null hypothesis that $\boldsymbol{z}_t$ follows a VARMA(m,j) model, $Q^{(m)}_{(j+1):\ell}$ is asymptotically a chi-square with $k^2(\ell-j)$ degrees of freedom. Denote the p-value of $Q^{(m)}_{(j+1):\ell}$ by p_{mj}. We can use the sequence of p-values, namely, $\{p_{m0},\ldots,p_{mj}\}$, to identify the order q of a VARMA(m,q) model.

Finally, since p and q are unknown, we consider a two-way table of the p-values of the multivariate Ljung–Box statistics $Q^{(v)}_{(j+1):\ell}$ for $j=0,1,\ldots,\ell$ and $v=0,1,\ldots,m$, where ℓ is a maximum MA order entertained and m is the maximum AR order entertained. It is understood that for $v=0$, the Ljung–Box statistics apply to the cross-correlation matrices of time series $\boldsymbol{z}_t$ because $v=0$ corresponds to VARMA(0,q) models. Denote the p-value of $Q^{(0)}_{(j+1):\ell}$ by p_{0j}. We refer to such a table as the two-way p-value table, which is shown in Table 3.1(b). By comparing elements of the two-way p-value table with the type I error α, for example 0.05, one can identify the order (p,q) of a VARMA series.

Table 3.2 shows the asymptotic pattern of the two-way p-value table for a VARMA(2,1) model, where “X" denotes a statistically significant entry and “O" denotes insignificant entry. From the table, the coordinates of the left-upper vertex of “O" correspond to the order (2,1).

Example 3.8 Consider the two-dimensional VARMA(2,1) model used in Example 3.7. We generate 400 observations from the model, compute the extended cross-correlation matrices of the sample, and obtain the two-way p-value table. The resulting table is given in Table 3.3, which, as expected, shows that the order of the time series is indeed (2,1). □

TABLE 3.2 The Asymptotic Behavior of the Two-Way p-Value Table for a k-Dimensional VARMA(2,1) Time Series

AR Order	MA Order: q				
p	0	1	2	⋯	$\ell - 1$
0	X	X	X	⋯	X
1	X	X	X	⋯	X
2	X	O	O	⋯	O
3	X	O	O	⋯	O
4	X	O	O	⋯	O

TABLE 3.3 A Two-Way p-Value Table for Extended Cross-Correlation Matrices of Example 3.8

AR	MA Order: q						
p	0	1	2	3	4	5	6
0	0.00	0.00	0.00	0.00	0.00	0.00	0.00
1	0.00	0.00	0.00	0.00	0.00	0.00	0.58
2	0.00	0.99	0.98	0.84	0.71	0.99	0.80
3	0.00	0.99	0.07	0.54	0.08	0.99	0.93
4	0.00	1.00	0.97	0.98	0.92	0.99	0.93
5	0.04	1.00	0.83	0.93	0.97	0.98	0.98

The sample size is 400.

R Demonstration: Extended cross-correlation matrices.

```
> p1=matrix(c(.816,-1.116,-.623,1.074),2,2)
> p2=matrix(c(-.643,.615,.592,-.133),2,2)
> phi=cbind(p1,p2)
> t1=matrix(c(0,-.801,-1.248,0),2,2)
> Sig=matrix(c(4,2,2,5),2,2)
> m1=VARMAsim(400,arlags=c(1,2),malags=c(1),phi=phi,
             theta=t1,sigma=Sig)
> zt=m1$series
> m2=Eccm(zt,maxp=5,maxq=6)
p-values table of Extended Cross-correlation Matrices:
Column: MA order
Row    : AR order
         0        1        2        3        4        5    6
0 0.00e+00 0.00e+00 0.00e+00 0.00e+00 0.00e+00 0.00e+00 0.00
1 0.00e+00 0.00e+00 0.00e+00 0.00e+00 0.00e+00 7.32e-04 0.58
2 2.78e-15 9.88e-01 9.78e-01 8.44e-01 7.14e-01 9.89e-01 0.80
3 5.31e-08 9.94e-01 7.04e-02 5.39e-01 8.39e-02 9.91e-01 0.93
4 1.95e-04 9.98e-01 9.72e-01 9.80e-01 9.17e-01 9.94e-01 0.93
5 3.66e-02 1.00e+00 8.30e-01 9.29e-01 9.72e-01 9.82e-01 0.98
> names(m2)
[1] "pEccm"  "vEccm"  "ARcoef" <== Details of output.
```

TABLE 3.4 A Two-Way p-Value Table for Extended Cross-Correlation Matrices Applied to the U.S. Hog Data

AR Order	MA Order: q						
p	0	1	2	3	4	5	6
0	0.00	0.00	0.00	0.00	0.00	0.00	0.00
1	0.04	0.67	0.69	1.00	1.00	1.00	0.79
2	0.84	0.86	0.93	1.00	1.00	1.00	0.99
3	0.98	1.00	1.00	1.00	1.00	1.00	0.99
4	1.00	1.00	1.00	1.00	1.00	1.00	1.00
5	1.00	1.00	1.00	1.00	1.00	1.00	1.00

The sample size is 82.

Example 3.9 To demonstrate the performance of extended cross-correlation matrices in practice, we consider the five-dimensional U.S. hog data that have been widely used in the multivariate time series literature. See Quenouille (1957), Box and Tiao (1977), Tiao and Tsay (1983), and the references therein. These are annual data of (a) hog numbers, (b) hog prices in dollars per head, (c) corn supply in bushels, (d) corn prices in dollars per bushels, and (e) farm wage rate. The data were linearly coded by Quenouille and we employ the logged data. The sample size is 82. Table 3.4 shows the two-way p-value table of extended cross-correlation matrices for the hog data. From the table, it is clear that a VAR(2) or a VARMA(1,1) model is recommended for the series. The VARMA(1,1) specification is in agreement with that of Tiao and Tsay (1983) even though we modified the definition of extended cross-correlations and used the multivariate Ljung–Box statistics. Also shown in the R demonstration is the result of VAR order selection. Both BIC and HQ criteria select order 2, but AIC selects order 8, when the maximum order was fixed at 9. □

R Demonstration: U.S. hog data.

```
> da=read.table("ushog.txt",header=T)
> head(da)
   hogsup  hogpri cornsup cornpri   wages
1 6.28786 6.39192 6.80239 6.85013 6.58203
 .....
> m1=Eccm(da,maxp=5,maxq=6)
p-values table of Extended Cross-correlation Matrices:
Column: MA order
Row    : AR order
         0        1        2        3        4        5    6
0 0.00e+00 0.00e+00 0.00e+00 0.00e+00 0.00e+00 4.48e-13 0.00
1 3.81e-02 6.73e-01 6.88e-01 9.97e-01 9.98e-01 9.98e-01 0.79
2 8.43e-01 8.58e-01 9.27e-01 1.00e+00 1.00e+00 9.97e-01 0.99
3 9.77e-01 1.00e+00 1.00e+00 1.00e+00 1.00e+00 1.00e+00 0.99
4 1.00e+00 9.99e-01 1.00e+00 1.00e+00 1.00e+00 1.00e+00 1.00
```

```
5 1.00e+00 9.84e-01 1.00e+00 1.00e+00 1.00e+00 1.00e+00 1.00
> VARorder(da,maxp=9) ## Use VAR models only.
selected order: aic =  9
selected order: bic =  2
selected order: hq =  2
M statistic and its p-value
       Mstat         pv
 [1,] 442.75 0.000e+00
 [2,]  83.41 3.306e-08
 [3,]  34.49 9.784e-02
 [4,]  43.27 1.306e-02
 [5,]  20.53 7.183e-01
 [6,]  28.85 2.704e-01
 [7,]  37.82 4.812e-02
 [8,]  31.78 1.645e-01
 [9,]  19.31 7.821e-01
```

3.15 EMPIRICAL ANALYSIS OF VARMA MODELS

We demonstrate applications of VARMA models via considering two examples in this section. We also compare the results with VAR models. The data used are from the Federal Reserve Bank of St. Louis (FRED Economic Data).

3.15.1 Personal Income and Expenditures

Consider first the monthly personal consumption expenditure (PCE) and disposable personal income (DSPI) of the United States from January 1959 to March 2012 for 639 observations. The data are in billions of dollars and seasonally adjusted. The two original series are nonstationary for obvious reasons so that we focus on the growth rates of PCE and DSPI, that is, the difference of logged data. Let z_{1t} and z_{2t} denote the growth rates, in percentages, of PCE and DSPI, respectively. Figure 3.6 shows the time plots of the two growth rate series. The plots indicate the existence of some outlying observations in the data, especially in the disposable income series. However, we shall not discuss outliers in this analysis.

We begin our analysis by employing the VAR models. To this end, we employ the order specification methods of Chapter 2 to select the AR order. From the attached output, the orders selected by AIC, BIC, and HQ criteria are 8, 3, and 3, respectively. The sequential chi-square statistics suggest a VAR(6) model. For simplicity, we chose the VAR(3) model. The fitted model, after removing an insignificant AR coefficient, is

$$\begin{aligned}\boldsymbol{z}_t &= \begin{bmatrix} 0.403 \\ 0.396 \end{bmatrix} + \begin{bmatrix} -0.154 & 0.131 \\ 0.151 & -0.194 \end{bmatrix} \boldsymbol{z}_{t-1} + \begin{bmatrix} 0.000 & 0.128 \\ 0.198 & -0.124 \end{bmatrix} \boldsymbol{z}_{t-2} \\ &+ \begin{bmatrix} 0.052 & 0.129 \\ 0.349 & -0.100 \end{bmatrix} \boldsymbol{z}_{t-3} + \hat{\boldsymbol{a}}_t, \quad \hat{\boldsymbol{\Sigma}}_a = \begin{bmatrix} 0.295 & 0.109 \\ 0.109 & 0.468 \end{bmatrix}. \end{aligned} \tag{3.119}$$

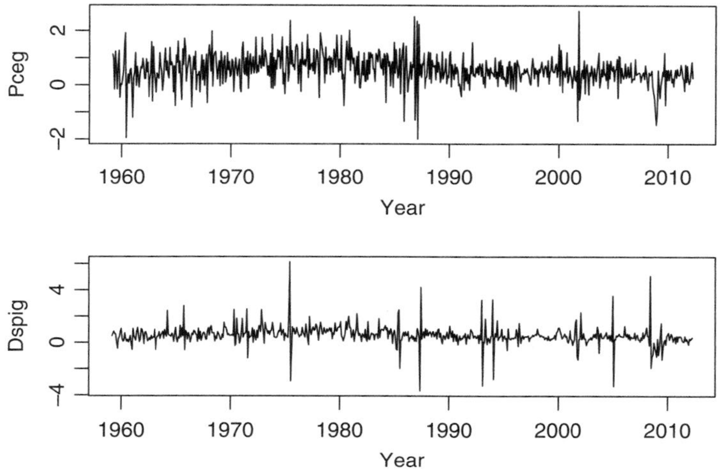

FIGURE 3.6 Time plots of the monthly growth rates, in percentages, of U.S. personal consumption expenditures and disposable personal income from February 1959 to March 2012.

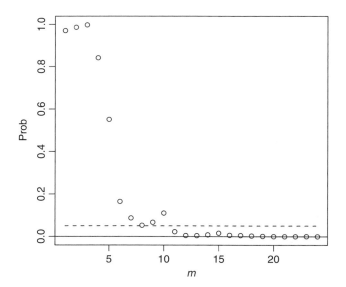

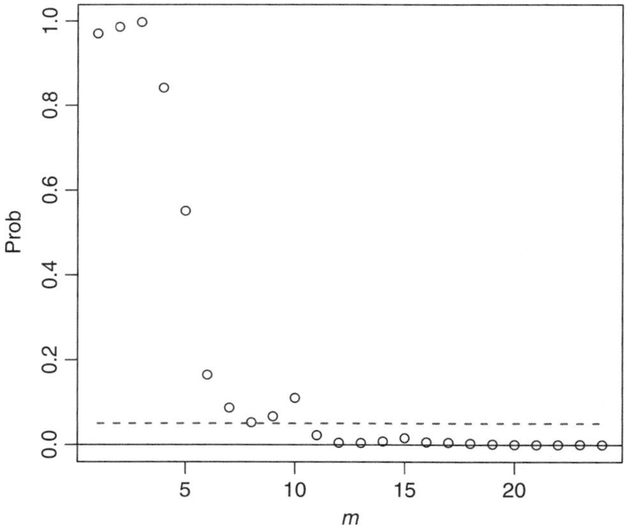

FIGURE 3.7 p-values of the multivariate Ljung–Box statistics for residuals of a VAR(3) model fitted to the monthly growth rates, in percentages, of U.S. personal consumption expenditures and disposable personal income. The data span is from February 1959 to March 2012.

Standard errors of the estimates are given in the R demonstration. Figure 3.7 shows the plot of p-values of the multivariate Ljung–Box statistics for the residuals of the VAR(3) model in Equation (3.119). The model successfully handled the lower-order dynamic correlations but it does not take care of some higher-order cross-correlations. Based on the fitted VAR(3) model, the two growth rate series are

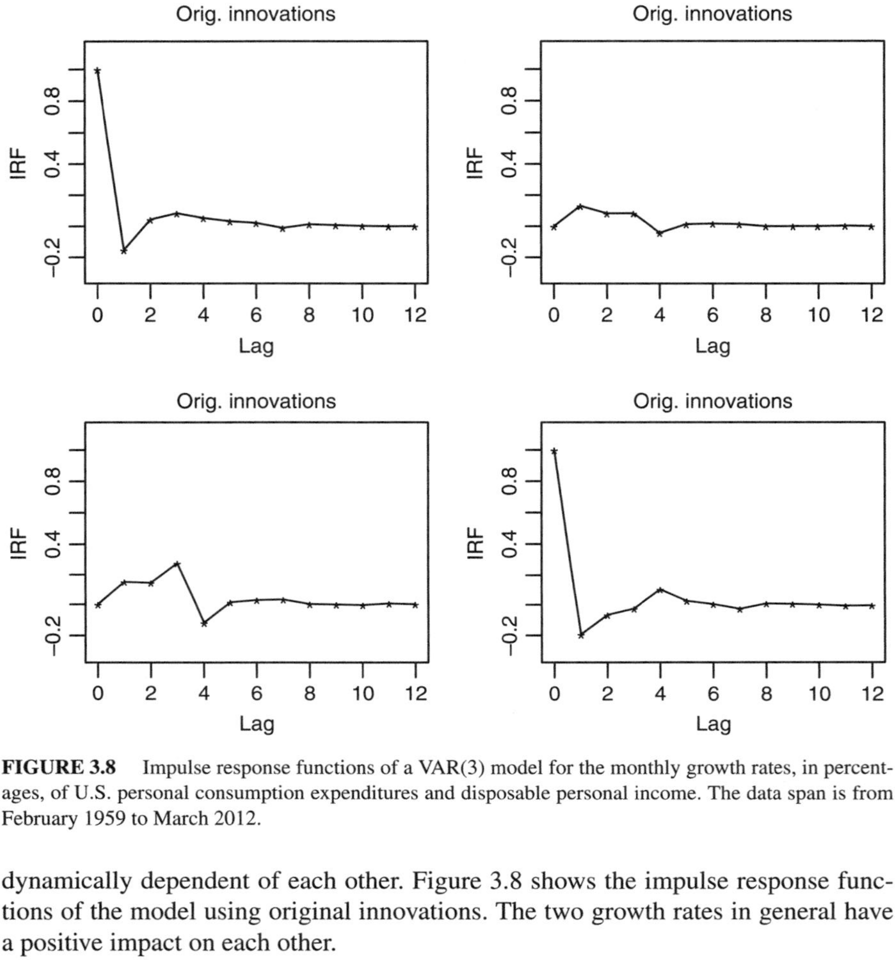

FIGURE 3.8 Impulse response functions of a VAR(3) model for the monthly growth rates, in percentages, of U.S. personal consumption expenditures and disposable personal income. The data span is from February 1959 to March 2012.

dynamically dependent of each other. Figure 3.8 shows the impulse response functions of the model using original innovations. The two growth rates in general have a positive impact on each other.

Turn to VARMA models. We applied the extended cross-correlation matrices to specify the order. From the p-values of Table 3.5, the approach specified a VARMA(3,1) model for the two series when the conventional type I error of 5% is used. Other possible choices include VARMA(2,3), VARMA(1,3), and VARMA(6,0). These alternative choices either employ more parameters or are not as clear-cut as the VARMA(3,1) model.

Using conditional MLE method and removing three insignificant parameters, we obtained the fitted model

$$\boldsymbol{z}_t = \begin{bmatrix} 0.017 \\ 0.000 \end{bmatrix} + \begin{bmatrix} 0.485 & 0.315 \\ 0.549 & 0.266 \end{bmatrix} \boldsymbol{z}_{t-1} + \begin{bmatrix} 0.000 & 0.094 \\ 0.141 & -0.094 \end{bmatrix} \boldsymbol{z}_{t-2}$$
$$+ \begin{bmatrix} 0.000 & 0.077 \\ 0.253 & -0.116 \end{bmatrix} \boldsymbol{z}_{t-3} + \hat{\boldsymbol{a}}_t - \begin{bmatrix} 0.662 & 0.225 \\ 0.423 & 0.536 \end{bmatrix} \hat{\boldsymbol{a}}_{t-1}, \quad (3.120)$$

TABLE 3.5 The p-value Table of Extended Cross-Correlation Matrices for the Monthly Growth Rates of U.S. Personal Consumption Expenditures and Disposable Personal Income

AR	MA Order: q						
p	0	1	2	3	4	5	6
0	0.0000	0.0000	0.0000	0.0000	0.0000	0.0001	0.0120
1	0.0000	0.0005	0.0003	0.0874	0.2523	0.2738	0.7914
2	0.0000	0.0043	0.0054	0.9390	0.4237	0.3402	0.8482
3	0.0000	0.8328	0.9397	0.9965	0.9376	0.9100	0.8193
4	0.0003	0.9643	0.9797	0.9937	0.9701	0.9810	0.9620
5	0.0150	1.0000	1.0000	1.0000	0.9995	0.9997	0.9851
6	0.1514	1.0000	1.0000	1.0000	1.0000	1.0000	0.9985

The data span is from February 1959 to March 2012.

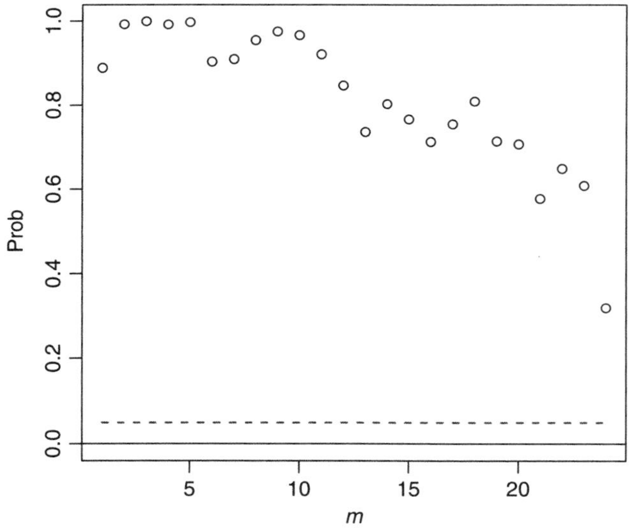

FIGURE 3.9 p-values of the multivariate Ljung–Box statistics for residuals of a VARMA(3,1) model fitted to the monthly growth rates, in percentages, of U.S. personal consumption expenditures and disposable personal income. The data span is from February 1959 to March 2012.

where the covariance matrix of $\hat{\boldsymbol{a}}_t$ is

$$\hat{\boldsymbol{\Sigma}}_a = \begin{bmatrix} 0.281 & 0.092 \\ 0.092 & 0.445 \end{bmatrix}.$$

The standard errors of the estimates are given in the R demonstration. Figure 3.9 shows the p-values of the multivariate Ljung–Box statistics for the residuals of the model in Equation (3.120). From the plot, the model has successfully described the

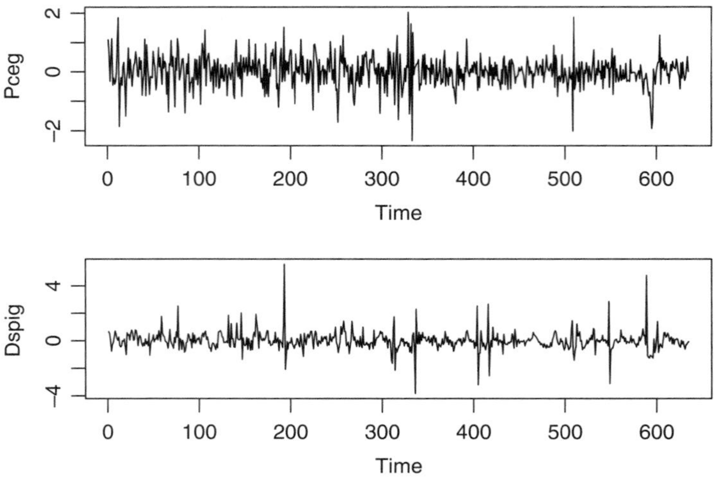

FIGURE 3.10 Residual series of the VARMA(3,1) model of Equation (3.120) for the monthly growth rates, in percentages, of U.S. personal consumption expenditures and disposable personal income. The data span is from February 1959 to March 2012.

dynamic dependence of the data. One cannot reject the null hypothesis that there exist no cross-correlations in the residuals. Figure 3.10 shows the time plots of the residuals of the fitted VARMA(3,1) model. As mentioned before, there exist some outlying observations in the data.

3.15.1.1 Model Comparison

It is interesting to compare the two fitted models for the growth rates of PCE and DSPI. First, the AIC and BIC of the fitted VARMA(3,1) model are lower than those of the VAR(3) model in Equation (3.119). Second, the multivariate Ljung–Box statistics show that the VARMA(3,1) model has no significant residual cross-correlations. The results thus indicate that the VARMA(3,1) model provides a better description of the dynamic dependence of the two growth rate series. Third, Figure 3.11 shows the impulse response functions of the VARMA(3,1) model in Equation (3.120). Compared with the impulse responses in Figure 3.8, we see that some subtle differences exist between the two models, even though their general patterns are similar. For instance, the impulse responses of the PCE to its own shocks have three consecutive increases for the VARMA(3,1) model after the initial drop. For the VAR(3) model, there are only two consecutive increases. Fourth, the diagonal elements of the residual covariance matrices of the two models show that the residuals of the VARMA(3,1) model have smaller variances. In addition, the eigenvalues of the MA matrix are 0.913 and 0.284, respectively, indicating that the fitted VARMA(3,1) model is indeed invertible.

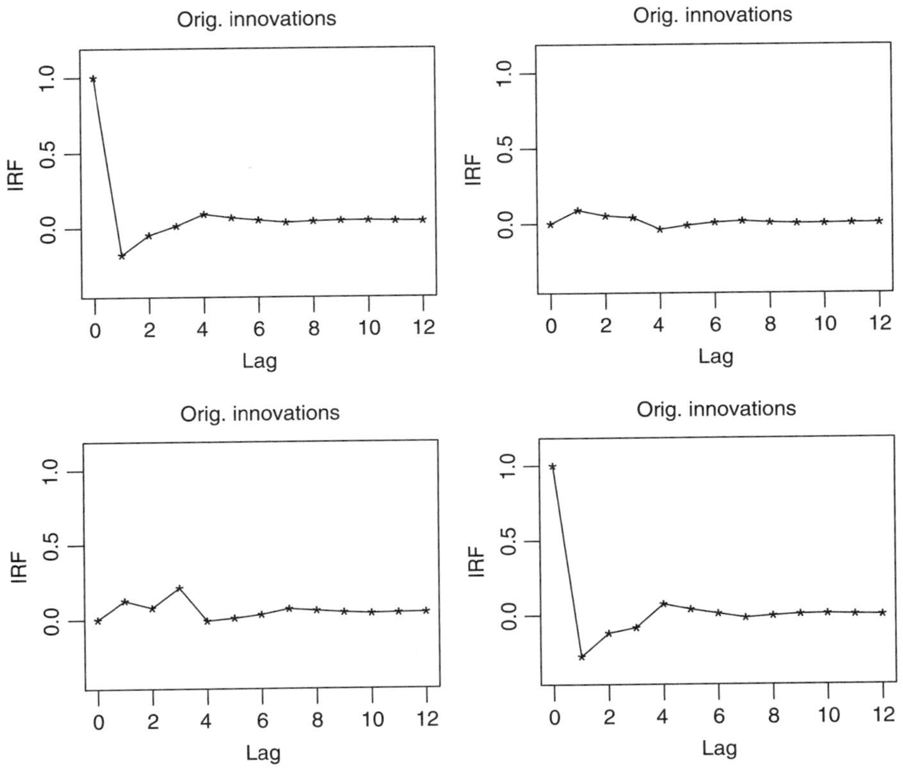

FIGURE 3.11 Impulse response functions of a VARMA(3,1) model for the monthly growth rates, in percentages, of U.S. personal consumption expenditures and disposable personal income. The data span is from February 1959 to March 2012.

R Demonstration: Consumption expenditures and income. Output edited.

```
> z1=diff(log(da1$pce)); z2=diff(log(da2$dspi))
> zt=cbind(z1,z2)*100
> colnames(zt) <- c("pceg","dspig")
> VARorder(zt)
selected order: aic =  8
selected order: bic =  3
selected order: hq =  3
Summary table:
       p     AIC     BIC      HQ    M(p) p-value
 [1,]  0 -1.8494 -1.8494 -1.8494  0.0000  0.0000
 ...
 [4,]  3 -2.0291 -1.9452 -1.9965 68.5648  0.0000
 [5,]  4 -2.0386 -1.9268 -1.9952 13.5685  0.0088
 [6,]  5 -2.0485 -1.9087 -1.9942 13.7797  0.0080
```

```
 [7,]  6 -2.0544 -1.8867 -1.9893 11.2718  0.0237
 [8,]  7 -2.0544 -1.8588 -1.9785  7.6679  0.1045
 [9,]  8 -2.0546 -1.8310 -1.9678  7.7194  0.1024
...
[14,] 13 -2.0423 -1.6789 -1.9013  7.7812  0.0999
> m1=VAR(zt,3)  ## fit a VAR(3) model
AIC =  -2.0325; BIC =  -1.948644; HQ  =  -1.999946
> m1a=refVAR(m1,thres=1)  ## refine the VAR(3) model
Constant term:
Estimates:  0.4026613 0.3960725
Std.Error:  0.04297581 0.05873444
AR coefficient matrix
AR( 1 )-matrix
       [,1]   [,2]
[1,] -0.154  0.131
[2,]  0.151 -0.194
standard error
       [,1]   [,2]
[1,] 0.0404 0.0315
[2,] 0.0515 0.0402
AR( 2 )-matrix
      [,1]   [,2]
[1,] 0.000  0.128
[2,] 0.198 -0.124
standard error
       [,1]   [,2]
[1,] 0.0000 0.0309
[2,] 0.0518 0.0411
AR( 3 )-matrix
       [,1]   [,2]
[1,] 0.0524  0.129
[2,] 0.3486 -0.100
standard error
       [,1]   [,2]
[1,] 0.0398 0.0316
[2,] 0.0508 0.0406

Residuals cov-mtx:
          [,1]      [,2]
[1,] 0.2948047 0.1088635
[2,] 0.1088635 0.4681942

AIC =  -2.035606; BIC =  -1.958738; HQ  =  -2.005766
> MTSdiag(m1a)  ## model checking
Ljung-Box Statistics:
          m       Q(m)   p-value
 [1,]   1.000     0.531     0.97
 [2,]   2.000     1.820     0.99
 [3,]   3.000     2.692     1.00
```

```
 [4,]    4.000     10.446      0.84
 [5,]    5.000     18.559      0.55
....
[23,]   23.000    142.923      0.00
[24,]   24.000    149.481      0.00
> Eccm(zt,maxp=6,maxq=6)
p-values table of Extended Cross-correlation Matrices:
Column: MA order
Row    : AR order
       0      1      2      3      4      5      6
0 0.0000 0.0000 0.0000 0.0000 0.0000 0.0001 0.0120
1 0.0000 0.0005 0.0003 0.0874 0.2523 0.2738 0.7914
2 0.0000 0.0043 0.0054 0.9390 0.4237 0.3402 0.8482
3 0.0000 0.8328 0.9397 0.9965 0.9376 0.9100 0.8193
4 0.0003 0.9643 0.9797 0.9937 0.9701 0.9810 0.9620
5 0.0150 1.0000 1.0000 1.0000 0.9995 0.9997 0.9851
6 0.1514 1.0000 1.0000 1.0000 1.0000 1.0000 0.9985
>
> m2=VARMA(zt,p=3,q=1)  ## fit a VARMA(3,1) model
aic=  -2.097311; bic=  -1.971527
> m2a=refVARMA(m2,thres=0.8) # refine the fit
aic=  -2.100075; bic=  -1.981279
> m2b=refVARMA(m2a,thres=1) # refine further the fit.
Coefficient(s):
        Estimate  Std. Error  t value Pr(>|t|)
         0.01693     0.01158    1.462 0.143744
pceg     0.48453     0.09151    5.295 1.19e-07 ***
dspig    0.31547     0.05524    5.711 1.12e-08 ***
dspig    0.09407     0.03873    2.429 0.015143 *
dspig    0.07706     0.03753    2.053 0.040048 *
pceg     0.54854     0.14828    3.699 0.000216 ***
dspig    0.26599     0.13219    2.012 0.044195 *
pceg     0.14120     0.06197    2.278 0.022703 *
dspig   -0.09406     0.06094   -1.544 0.122667
pceg     0.25305     0.06108    4.143 3.43e-05 ***
dspig   -0.11638     0.05955   -1.954 0.050654 .
        -0.66154     0.07139   -9.267  < 2e-16 ***
        -0.22501     0.04607   -4.883 1.04e-06 ***
        -0.42284     0.13842   -3.055 0.002253 **
        -0.53556     0.13021   -4.113 3.90e-05 ***
---
Estimates in matrix form:
Constant term:
Estimates:  0.01692972 0
AR coefficient matrix
AR( 1 )-matrix
      [,1]  [,2]
[1,] 0.485 0.315
[2,] 0.549 0.266
```

```
AR( 2 )-matrix
      [,1]    [,2]
[1,] 0.000  0.0941
[2,] 0.141 -0.0941
AR( 3 )-matrix
      [,1]    [,2]
[1,] 0.000  0.0771
[2,] 0.253 -0.1164
MA coefficient matrix
MA( 1 )-matrix
      [,1]  [,2]
[1,] 0.662 0.225
[2,] 0.423 0.536

Residuals cov-matrix:
           [,1]       [,2]
[1,] 0.28073730 0.09236968
[2,] 0.09236968 0.44521036
----
aic=  -2.103228; bic=  -1.998409
> MTSdiag(m2b)
> names(m2b)
 [1] "data"  "coef"  "secoef"  "ARorder"  "MAorder"  "cnst"
 [7] "residuals"  "Phi"  "Theta"  "Sigma"  "aic"  "bic"
> phi=m2b$Phi; theta=m2b$Theta; sig=m2b$Sigma
> VARMAirf(Phi=phi,Theta=theta,Sigma=sig,orth=F)
```

3.15.2 Housing Starts and Mortgage Rate

In this section, we consider the monthly housing starts and the 30-year conventional mortgage rate of the United States from April 1971 to March 2012 for 492 observations. The housing starts data denote the total new privately owned housing units started, measured in thousands of units, and are seasonally adjusted. The mortgage rates are in percentage, not seasonally adjusted. The data are obtained from the Federal Reserve Bank of St. Louis. In our analysis, we divided the housing starts by 1000 so that they are measured in millions of units. Individual component series show that there exist strong serial correlations in the data so that we analyze the change series. More specifically, we employ $\boldsymbol{z}_t = (1 - B)\boldsymbol{x}_t$, where $\boldsymbol{x}_t = (x_{1t}, x_{2t})'$ with x_{1t} being the housing starts measured in millions of units and x_{2t} being the 30-year mortgage rate.

Figure 3.12 shows the time plots of the two change series. The mortgage rates were highly volatile in the early 1980. If VAR models were entertained, AIC, HQ and the sequence chi-square test all selected a VAR(4) model, but BIC chose a VAR(2) model. We consider a VAR(4) model. After removing some highly insignificant estimates, we obtained the model

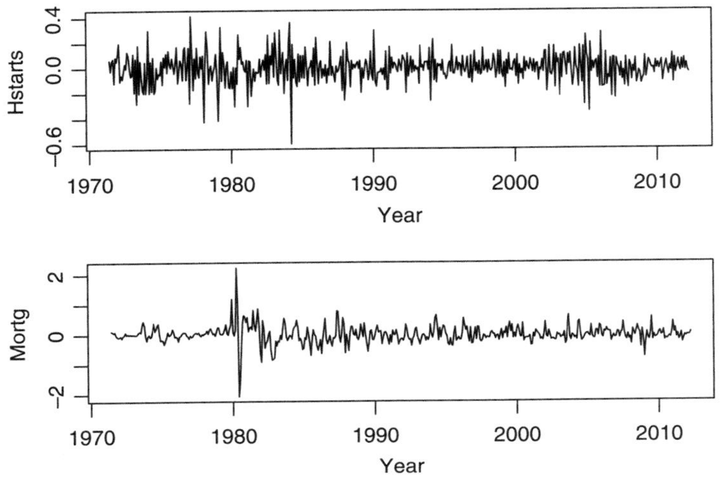

FIGURE 3.12 Time plots of changes in monthly housing starts and 30-year mortgage rate. The data span is from April 1971 to March 2012.

$$\boldsymbol{z}_t = \begin{bmatrix} -0.006 \\ 0.000 \end{bmatrix} + \begin{bmatrix} -0.43 & -0.06 \\ 0.27 & 0.58 \end{bmatrix} \boldsymbol{z}_{t-1} + \begin{bmatrix} -0.16 & -0.05 \\ 0.15 & -0.34 \end{bmatrix} \boldsymbol{z}_{t-2}$$
$$+ \begin{bmatrix} 0.00 & -0.03 \\ 0.00 & 0.08 \end{bmatrix} \boldsymbol{z}_{t-3} + \begin{bmatrix} 0.07 & -0.05 \\ 0.00 & 0.07 \end{bmatrix} \boldsymbol{z}_{t-4} + \boldsymbol{a}_t, \tag{3.121}$$

where the standard errors of the estimates are given in the R demonstration and the residual covariance matrix is

$$\hat{\boldsymbol{\Sigma}}_a = \begin{bmatrix} 0.0108 & -0.0028 \\ -0.0028 & 0.0636 \end{bmatrix}.$$

Model checking indicates that the model provides a reasonably good fit, except for possible seasonality in the data. The p-values of the multivariate Ljung–Box statistics, after adjusting the degrees of freedom due to estimation, are shown in Figure 3.13.

Turn to VARMA models. The summary p-values of the extended cross-correlation matrices are given in Table 3.6. From the table, we identify VARMA(1,2) and VARMA(2,1) as two candidate models for further analysis. After some simplification, we obtained the VARMA(2,1) model

$$\boldsymbol{z}_t = \begin{bmatrix} 0.00 & 0.08 \\ -0.27 & 0.40 \end{bmatrix} \boldsymbol{z}_{t-1} + \begin{bmatrix} 0.00 & -0.09 \\ 0.00 & -0.28 \end{bmatrix} \boldsymbol{z}_{t-2} + \boldsymbol{a}_t$$
$$- \begin{bmatrix} 0.43 & 0.15 \\ -0.55 & -0.17 \end{bmatrix} \boldsymbol{a}_{t-1} \tag{3.122}$$

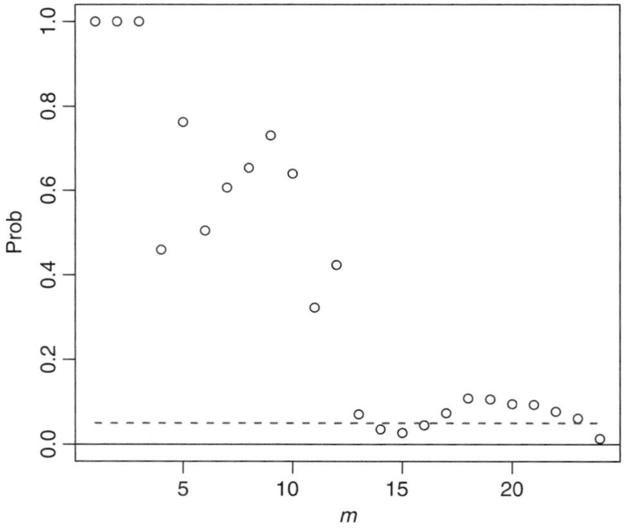

FIGURE 3.13 The p-values of multivariate Ljung–Box statistics for the residuals of the VAR(4) model in Equation (3.121). The degrees of freedom of the chi-square distributions are adjusted for the number of parameters used.

TABLE 3.6 The p-Value Table of the Extended Cross-Correlation Matrices for the Change Series of Housing Starts and 30-Year Mortgage Rate from May 1971 to March 2012

AR	MA Order: q						
p	0	1	2	3	4	5	6
0	0.0000	0.0091	0.1380	0.5371	0.7427	0.7291	0.6922
1	0.0000	0.0715	0.5708	0.8020	0.4297	0.4235	0.5915
2	0.0816	0.2858	0.9936	0.9557	0.9869	0.9099	0.7486
3	0.6841	0.9870	0.9993	0.9977	0.9999	0.9971	0.9894
4	0.9975	0.9999	1.0000	1.0000	0.9999	0.9921	0.9769
5	0.9994	0.9999	1.0000	1.0000	1.0000	0.9983	0.9429
6	1.0000	1.0000	1.0000	1.0000	1.0000	1.0000	0.9995

where the standard errors of the estimates are given in the R demonstration and the residual covariance matrix is

$$\hat{\boldsymbol{\Sigma}}_a = \begin{bmatrix} 0.0111 & -0.0028 \\ -0.0028 & 0.0637 \end{bmatrix}.$$

Again, model checking shows that the model provides a reasonable fit. The p-values of the multivariate Ljung–Box statistics of the residuals, after adjusting the degrees of freedom for the chi-square distributions, are shown in Figure 3.14.

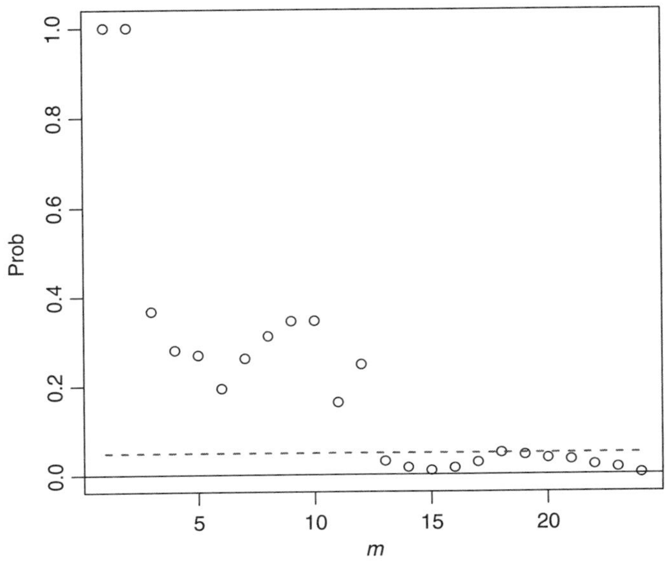

FIGURE 3.14 The p-values of multivariate Ljung–Box statistics for the residuals of the VARMA(2,1) model in Equation (3.122). The degrees of freedom of the chi-square distributions are adjusted for the number of parameters used.

If a VARMA(1,2) model is entertained, we obtained the model

$$\boldsymbol{z}_t = \begin{bmatrix} 0.67 & 0 \\ 1.75 & 0 \end{bmatrix} \boldsymbol{z}_{t-1} + \boldsymbol{a}_t - \begin{bmatrix} 1.12 & 0.07 \\ 1.48 & -0.59 \end{bmatrix} \boldsymbol{a}_{t-1} + \begin{bmatrix} 0.33 & 0.00 \\ 1.01 & -0.15 \end{bmatrix} \boldsymbol{a}_{t-2}, \tag{3.123}$$

where, again, the standard errors are in the R demonstration and the residual covariance matrix is

$$\hat{\boldsymbol{\Sigma}}_a = \begin{bmatrix} 0.0108 & -0.0030 \\ -0.0030 & 0.0647 \end{bmatrix}.$$

Similar to the prior two models, this model also provides a decent fit for the data.

3.15.2.1 Model Comparison

The three models in Equations (3.121)–(3.123) all provide reasonable fit to the data. As a matter of fact, the three models are close to each other. For instance, their residual covariance matrices are rather close. We can also consider the cumulative impulse response functions of the models. Figure 3.15 and Figure 3.16 show the cumulative impulse response functions of the VAR(4) and VARMA(2,1) model,

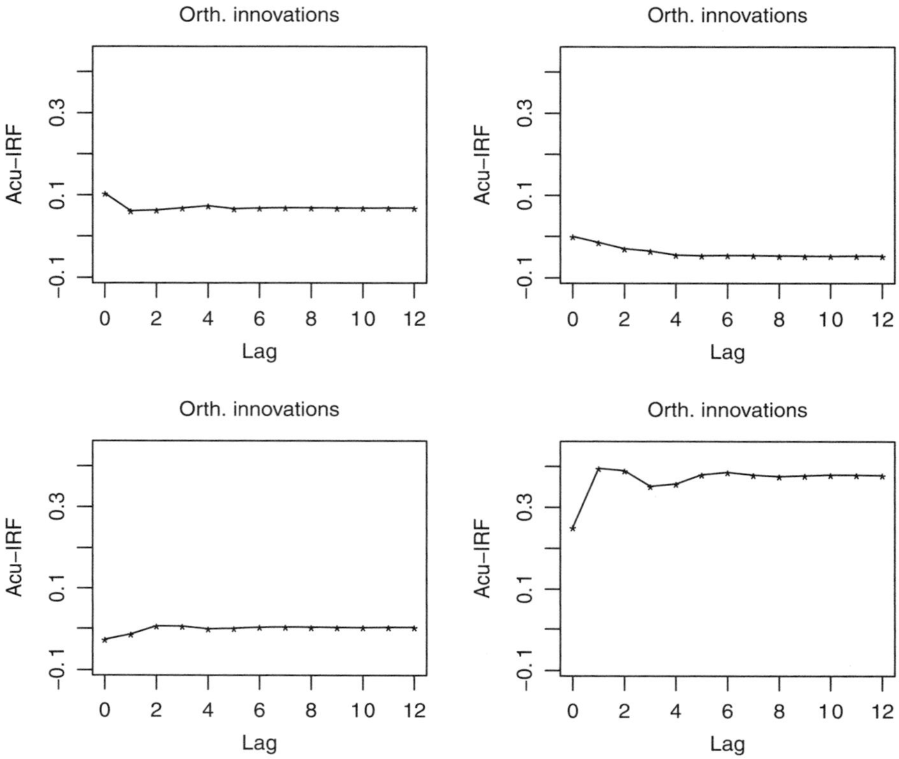

FIGURE 3.15 Cumulative impulse response functions of the VAR(4) model in Equation (3.121) with orthogonalized innovations.

respectively. From the plots, it is clear that the dynamic dependence implied by the models is close to each other. The same is true for the fitted VARMA(1,2) model in Equation (3.123). The impulse responses show that the dynamic dependence of the change series of housing starts is rather weak. The cumulative impact of the changes to its own innovations is small.

The AIC for the fitted VAR(4), VARMA(2,1) and VARMA(1,2) models are -7.237, -7.233 and -7.239, respectively. They are also close to each other, supporting the closeness of the models.

R Demonstration: Housing starts and mortgage rate. Output edited.

```
> da=read.table("m-hsmort7112.txt",header=T)
> zt=da[,3:4]; colnames(zt) <- c("hs","mort")
> dzt=diffM(zt)
> VARorder(dzt)
selected order: aic =  4
```

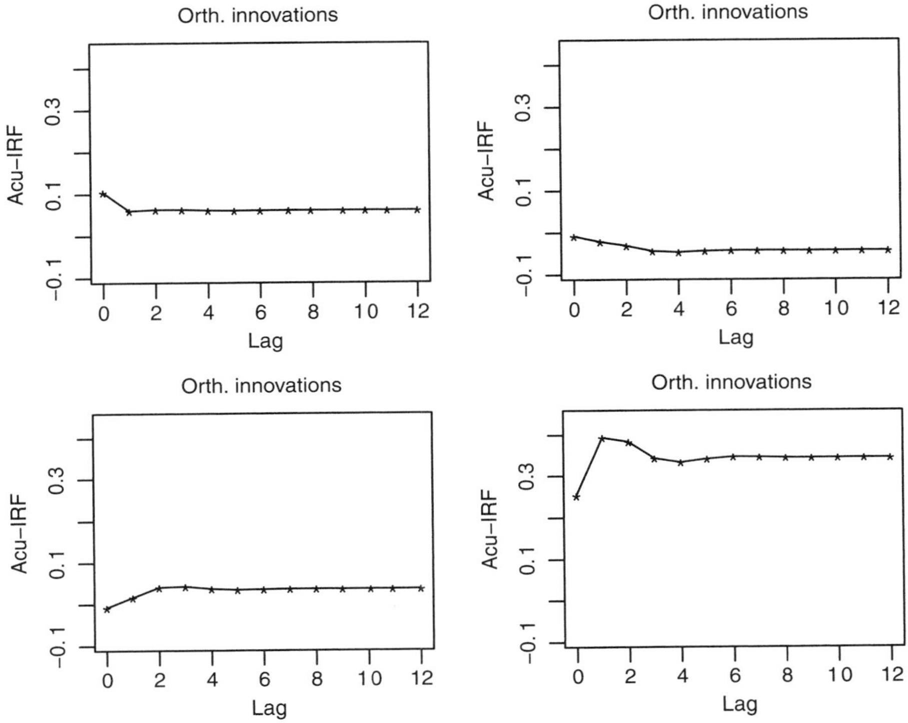

FIGURE 3.16 Cumulative impulse response functions of the VARMA(2,1) model in Equation (3.122) with orthogonalized innovations.

```
selected order: bic =  2
selected order: hq =  4
Summary table:
      p     AIC     BIC      HQ     M(p) p-value
 [1,] 0 -6.7523 -6.7523 -6.7523   0.0000  0.0000
 [2,] 1 -7.0547 -7.0205 -7.0412 151.1868  0.0000
 [3,] 2 -7.1856 -7.1173 -7.1588  69.5808  0.0000
 [4,] 3 -7.2019 -7.0994 -7.1616  15.3277  0.0041
 [5,] 4 -7.2172 -7.0805 -7.1635  14.8115  0.0051
 [6,] 5 -7.2055 -7.0346 -7.1384   2.1248  0.7128
 [7,] 6 -7.1963 -6.9911 -7.1157   3.2756  0.5128
 [8,] 7 -7.1916 -6.9523 -7.0976   5.3665  0.2517
> m1=VAR(dzt,4)                  #### VAR(4) model
> m1a=refVAR(m1,thres=1) ### model refinement.
Constant term:
Estimates:  -0.005936007 0
Std.Error:  0.004772269 0
AR coefficient matrix
```

```
AR( 1 )-matrix
       [,1]    [,2]
[1,] -0.429 -0.0582
[2,]  0.272  0.5774
standard error
      [,1]   [,2]
[1,] 0.045 0.0189
[2,] 0.108 0.0457
AR( 2 )-matrix
       [,1]    [,2]
[1,] -0.164 -0.0511
[2,]  0.146 -0.3400
standard error
       [,1]   [,2]
[1,] 0.0451 0.0219
[2,] 0.1088 0.0529
AR( 3 )-matrix
     [,1]    [,2]
[1,]    0 -0.0293
[2,]    0  0.0815
standard error
     [,1]   [,2]
[1,]    0 0.0216
[2,]    0 0.0521
AR( 4 )-matrix
       [,1]   [,2]
[1,] 0.0742 -0.051
[2,] 0.0000  0.071
standard error
       [,1]   [,2]
[1,] 0.0418 0.0192
[2,] 0.0000 0.0459
  Residuals cov-mtx:
             [,1]         [,2]
[1,]  0.010849747 -0.002798639
[2,] -0.002798639  0.063620722

det(SSE) =  0.0006824364
AIC =  -7.236888; BIC =  -7.125781; HQ  =  -7.193256
> MTSdiag(m1a)
> Eccm(dzt,maxp=6,maxq=6)
p-values table of Extended Cross-correlation Matrices:
Column: MA order
Row   : AR order
       0      1      2      3      4      5      6
0 0.0000 0.0091 0.1380 0.5371 0.7427 0.7291 0.6922
1 0.0000 0.0715 0.5708 0.8020 0.4297 0.4235 0.5915
2 0.0816 0.2858 0.9936 0.9557 0.9869 0.9099 0.7486
3 0.6841 0.9870 0.9993 0.9977 0.9999 0.9971 0.9894
```

```
4 0.9975 0.9999 1.0000 1.0000 0.9999 0.9921 0.9769
5 0.9994 0.9999 1.0000 1.0000 1.0000 0.9983 0.9429
6 1.0000 1.0000 1.0000 1.0000 1.0000 1.0000 0.9995
> m2=VARMA(dzt,p=2,q=1)          ## VARMA(2,1) model.
aic=  -7.219037; bic=  -7.099382
> m2a=refVARMA(m2,thres=0.8)
aic=  -7.227145; bic=  -7.133131
> m2b=refVARMA(m2a,thres=1)
aic=  -7.229817; bic=  -7.14435
> m2c=refVARMA(m2b,thres=1)
Coefficient(s):
       Estimate  Std. Error  t value Pr(>|t|)
 [1,]   0.08394     0.05219    1.608 0.107790
 [2,]  -0.08552     0.02350   -3.639 0.000274 ***
 [3,]  -0.27400     0.22236   -1.232 0.217859
 [4,]   0.39608     0.10794    3.669 0.000243 ***
 [5,]  -0.27786     0.06063   -4.583 4.58e-06 ***
 [6,]  -0.43025     0.04284  -10.042  < 2e-16 ***
 [7,]  -0.15004     0.05695   -2.634 0.008430 **
 [8,]   0.54597     0.24839    2.198 0.027950 *
 [9,]   0.17413     0.11151    1.562 0.118396
---
Estimates in matrix form:
Constant term:
Estimates:  0 0
AR coefficient matrix
AR( 1 )-matrix
       [,1]   [,2]
[1,]  0.000 0.0839
[2,] -0.274 0.3961
AR( 2 )-matrix
     [,1]    [,2]
[1,]    0 -0.0855
[2,]    0 -0.2779
MA coefficient matrix
MA( 1 )-matrix
       [,1]   [,2]
[1,]  0.430  0.150
[2,] -0.546 -0.174

Residuals cov-matrix:
             [,1]         [,2]
[1,]  0.011051912 -0.002784296
[2,] -0.002784296  0.063733777
----
aic=  -7.232599; bic=  -7.155679
> MTSdiag(m2c)
> m3=VARMA(dzt,p=1,q=2)    ## VARMA(1,2) model
aic=  -7.229278; bic=  -7.109624
```

```
> m3a=refVARMA(m3,thres=0.6)
Coefficient(s):
      Estimate  Std. Error  t value Pr(>|t|)
 [1,]   0.66542     0.08646    7.696 1.40e-14 ***
 [2,]   1.74526     1.11287    1.568   0.1168
 [3,]  -1.11687     0.09628  -11.600  < 2e-16 ***
 [4,]  -0.07449     0.01325   -5.620 1.91e-08 ***
 [5,]   0.32496     0.05643    5.758 8.49e-09 ***
 [6,]  -1.47964     1.11408   -1.328   0.1841
 [7,]   0.58974     0.04805   12.273  < 2e-16 ***
 [8,]   1.00768     0.53292    1.891   0.0586 .
 [9,]   0.15153     0.09133    1.659   0.0971 .
---
Estimates in matrix form:
Constant term:
Estimates:  0 0
AR coefficient matrix
AR( 1 )-matrix
      [,1] [,2]
[1,] 0.665    0
[2,] 1.745    0
MA coefficient matrix
MA( 1 )-matrix
     [,1]    [,2]
[1,] 1.12  0.0745
[2,] 1.48 -0.5897
MA( 2 )-matrix
       [,1]   [,2]
[1,] -0.325  0.000
[2,] -1.008 -0.152

Residuals cov-matrix:
             [,1]         [,2]
[1,]  0.010840482 -0.002977037
[2,] -0.002977037  0.064684274
----
aic=  -7.238765
bic=  -7.161844
> MTSdiag(m3a)
> resi=m1a$residuals
> mq(resi,adj=14)  ## Adjust the Chi-square degrees of freedom.
```

3.16 APPENDIX

In this appendix, we consider a VAR(1) representation of the VARMA model. Any VARMA(p, q) model can be expressed as a VAR(1) model with expanded dimension. Consider the mean-adjusted series $\tilde{z}_t$ in Equation (3.37). Define

$$\boldsymbol{Z}_t = \begin{bmatrix} \tilde{\boldsymbol{z}}_t \\ \tilde{\boldsymbol{z}}_{t-1} \\ \vdots \\ \tilde{\boldsymbol{z}}_{t-p+1} \\ \boldsymbol{a}_t \\ \vdots \\ \boldsymbol{a}_{t-q+1} \end{bmatrix}_{k(p+q)\times 1}, \quad \boldsymbol{A}_t = \begin{bmatrix} \boldsymbol{a}_t \\ \boldsymbol{0} \\ \vdots \\ \boldsymbol{0} \\ \boldsymbol{a}_t \\ \vdots \\ \boldsymbol{0} \end{bmatrix}_{k(p+q)\times 1},$$

where $\boldsymbol{0}$ is the k-dimensional vector of zeros, and

$$\boldsymbol{\Phi} = \begin{bmatrix} \boldsymbol{\Phi}_{11} & \boldsymbol{\Phi}_{12} \\ \boldsymbol{\Phi}_{21} & \boldsymbol{\Phi}_{22} \end{bmatrix}_{k(p+q)\times k(p+q)},$$

where $\boldsymbol{\Phi}_{11}$ is the companion matrix of $\boldsymbol{\phi}(B)$, $\boldsymbol{\Phi}_{21}$ is a $kq \times kp$ matrix of zeros, and

$$\boldsymbol{\Phi}_{12} = \begin{bmatrix} -\boldsymbol{\theta}_1 & \cdots & -\boldsymbol{\theta}_{q-1} & -\boldsymbol{\theta}_q \\ \boldsymbol{0}_k & \cdots & \boldsymbol{0}_k & \boldsymbol{0}_k \\ \vdots & & \vdots & \vdots \\ \boldsymbol{0}_k & \cdots & \boldsymbol{0}_k & \boldsymbol{0}_k \end{bmatrix}, \quad \boldsymbol{\Phi}_{22} = \begin{bmatrix} \boldsymbol{0}_{k\times k(q-1)} & \boldsymbol{0}_k \\ \boldsymbol{I}_{k(q-1)} & \boldsymbol{0}_{k(q-1)\times k} \end{bmatrix},$$

where $\boldsymbol{0}_k$ is the $k \times k$ matrix of zeros. The VARMA(p, q) model $\tilde{\boldsymbol{z}}_t$ then becomes

$$\boldsymbol{Z}_t = \boldsymbol{\Phi}\boldsymbol{Z}_{t-1} + \boldsymbol{A}_t, \tag{3.124}$$

which is a $k(p+q)$-dimensional VAR(1) model. Equation (3.124) provides a state–space model representation for the VARMA(p, q) process $\boldsymbol{z}_t$. From the definition of $\boldsymbol{Z}_t$, we have $\boldsymbol{z}_t = [\boldsymbol{I}_k, \boldsymbol{0}]\boldsymbol{Z}_t$, which can be treated as the observation equation of a state–space model.

To demonstrate the application of the VAR(1) model representation in Equation (3.124), we use it to obtain the theoretical autocovariance matrices $\boldsymbol{\Gamma}_\ell$ of a given stationary VARMA(p, q) process $\boldsymbol{z}_t$. Let $\boldsymbol{\Gamma}_Z$ and $\boldsymbol{\Sigma}_A$ be the covariance matrices of $\boldsymbol{Z}_t$ and $\boldsymbol{A}_t$, respectively. Applying the result of VAR(1) model in Chapter 2, we have

$$\boldsymbol{\Gamma}_Z = \boldsymbol{\Phi}\boldsymbol{\Gamma}_Z\boldsymbol{\Phi}' + \boldsymbol{\Sigma}_A. \tag{3.125}$$

Using vectorization, we have

$$\text{vec}(\boldsymbol{\Gamma}_Z) = [\boldsymbol{I}_{k^2(p+q)^2} - \boldsymbol{\Phi} \otimes \boldsymbol{\Phi}]^{-1}\text{vec}(\boldsymbol{\Sigma}_A), \tag{3.126}$$

where the inverse exists because $\boldsymbol{z}_t$, hence $\boldsymbol{Z}_t$, is a stationary process. It remains to connect $\boldsymbol{\Gamma}_Z$ to the cross-covariance matrices $\boldsymbol{\Gamma}_\ell$ of $\boldsymbol{z}_t$.

Partitioning $\boldsymbol{\Gamma}_Z$ in the same way as $\boldsymbol{\Phi}$ and using Lemma 3.2, we have

$$\boldsymbol{\Gamma}_Z = \begin{bmatrix} \boldsymbol{\Gamma}_{Z,11} & \boldsymbol{\Gamma}_{Z,12} \\ \boldsymbol{\Gamma}'_{Z,12} & \boldsymbol{\Gamma}_{Z,22} \end{bmatrix},$$

where

$$\boldsymbol{\Gamma}_{Z,11} = \begin{bmatrix} \boldsymbol{\Gamma}_0 & \boldsymbol{\Gamma}_1 & \cdots & \boldsymbol{\Gamma}_{p-1} \\ \boldsymbol{\Gamma}_{-1} & \boldsymbol{\Gamma}_0 & \cdots & \boldsymbol{\Gamma}_{p-2} \\ \vdots & \vdots & \ddots & \vdots \\ \boldsymbol{\Gamma}_{1-p} & \boldsymbol{\Gamma}_{2-p} & \cdots & \boldsymbol{\Gamma}_0 \end{bmatrix},$$

$$\boldsymbol{\Gamma}_{Z,12} = \begin{bmatrix} \boldsymbol{\Sigma}_a & \boldsymbol{\psi}_1\boldsymbol{\Sigma}_a & \cdots & \boldsymbol{\psi}_{q-1}\boldsymbol{\Sigma}_a \\ \mathbf{0}_k & \boldsymbol{\Sigma}_a & \cdots & \boldsymbol{\psi}_{q-2}\boldsymbol{\Sigma}_a \\ \vdots & & \ddots & \vdots \\ \mathbf{0}_k & \mathbf{0}_k & \cdots & \boldsymbol{\psi}_{q-p}\boldsymbol{\Sigma}_a \end{bmatrix},$$

and $\boldsymbol{\Gamma}_{Z,22} = \boldsymbol{I}_q \otimes \boldsymbol{\Sigma}_a$.

Based on the prior derivation, one can use Equation (3.126) to obtain $\boldsymbol{\Gamma}_0, \boldsymbol{\Gamma}_1, \ldots, \boldsymbol{\Gamma}_{p-1}$ for a stationary VARMA(p, q) process $\boldsymbol{z}_t$. The higher-order auto- and cross-covariance matrices $\boldsymbol{\Gamma}_\ell$ of $\boldsymbol{z}_t$ can then be obtained recursively by the moment equations of $\boldsymbol{z}_t$.

EXERCISES

3.1 Consider the monthly log returns of CRSP decile portfolios 1, 2, and 5 from January 1961 to September 2011.

- **(a)** Specify a VMA model for the three-dimensional log returns.
- **(b)** Estimate the specified VMA model via the conditional maximum likelihood method. Refine the model so that t-ratio of each estimate is greater than 1.645. Write down the fitted model.
- **(c)** Use the fitted model to forecast the log returns of October and November of 2011 at the forecast origin September 2011. Obtain both the point and 95% interval forecasts.

3.2 Suppose that $\boldsymbol{z}_t$ is a k-dimensional weakly stationary, zero-mean time series following the VARMA(2,q) model

$$\boldsymbol{z}_t = \boldsymbol{\phi}_1 \boldsymbol{z}_{t-1} + \boldsymbol{\phi}_2 \boldsymbol{z}_{t-2} + \boldsymbol{a}_t - \sum_{j=1}^{q} \boldsymbol{\theta}_j \boldsymbol{a}_{t-j},$$

where $\{\boldsymbol{a}_t\}$ is a white noise series with positive-definite covariance matrix $\boldsymbol{\Sigma}_a$. Let $\boldsymbol{\Gamma}_j$ be the lag-j autocovariance matrix of $\boldsymbol{z}_t$.

(a) Consider the VAR(2) fitting

$$\boldsymbol{z}_t = \boldsymbol{\phi}_1^{(0)} \boldsymbol{z}_{t-1} + \boldsymbol{\phi}_2^{(0)} \boldsymbol{z}_{t-2} + \boldsymbol{a}_t^{(0)}.$$

Show that the ordinary LS estimates of $\boldsymbol{\phi}_i^{(0)}$ (i = 1, 2) satisfy the system of equations

$$[\boldsymbol{\Gamma}_1, \boldsymbol{\Gamma}_2] = [\boldsymbol{\phi}_1^{(0)}, \boldsymbol{\phi}_2^{(0)}] \begin{bmatrix} \boldsymbol{\Gamma}_0 & \boldsymbol{\Gamma}_1 \\ \boldsymbol{\Gamma}_1' & \boldsymbol{\Gamma}_0 \end{bmatrix}.$$

(b) Let $\boldsymbol{a}_t^{(0)}$ be the residual of the fitted VAR(2) model via the ordinary LS method. Discuss the properties of the autocovariance matrices of $\boldsymbol{a}_t^{(0)}$ when $q = 0$ and when $q = 1$.

(c) Consider the model

$$\boldsymbol{z}_t = \boldsymbol{\phi}_1^{(1)} \boldsymbol{z}_{t-1} + \boldsymbol{\phi}_2^{(1)} \boldsymbol{z}_{t-1} + \boldsymbol{\gamma}_1^{(1)} \boldsymbol{a}_{t-1}^{(0)} + \boldsymbol{a}_t^{(1)}.$$

Show that the ordinary LS estimates of $\boldsymbol{\phi}_i^{(1)}$ satisfy the system of equations

$$[\boldsymbol{\Gamma}_2, \boldsymbol{\Gamma}_3] = [\boldsymbol{\phi}_1^{(1)}, \boldsymbol{\phi}_2^{(1)}] \begin{bmatrix} \boldsymbol{\Gamma}_1 & \boldsymbol{\Gamma}_2 \\ \boldsymbol{\Gamma}_0 & \boldsymbol{\Gamma}_1 \end{bmatrix}.$$

(d) Let $\boldsymbol{a}_t^{(1)}$ be the residual of the LS fit of the model in part (c). Discuss the properties of the autocovariance matrices of $\boldsymbol{a}_t^{(1)}$ when $q = 0$, 1, and 2.

3.3 Consider the quarterly U.S. Federal government debt held by (a) foreign and international investors and (b) by the Federal Reserve Banks, in billions of dollars, from 1970.I to 2012.III. Let $\boldsymbol{z}_t$ be the quarterly growth rate series of the debt, that is, the first difference of the log debt.

(a) Specify a VMA model for the growth series of debt.

(b) Fit the specified VMA model and perform model checking.

(c) Write down the fitted model.

(d) Use the model to produce one-step to three-step ahead forecasts of the growth rates of U.S. Federal debt, using the last data point as the forecast origin.

3.4 Consider two components of U.S. monthly industrial production index from December 1963 to December 2012. The two components are (a) nondurable consumer goods and (b) materials. The data are in columns 3 and 6 of the file `m-ip3comp.txt` and are obtained from the Federal Reserve Bank of St. Louis.

(a) Construct the percentage growth rate series of the two components, that is, first difference of the log data times 100. Denote the series by $\boldsymbol{z}_t$.

(b) Specify a simple VARMA model for $\boldsymbol{z}_t$ using 5% as the type I error.

(c) Fit the specified model and remove the parameter estimates with t-ratios less than 1.96. Write down the fitted model.

(d) Perform model checking of the fitted model. Is the model adequate? Why?

(e) Compute the impulse response functions of the fitted model.

3.5 Consider two components of U.S. monthly industrial production index from December 1963 to December 2012. The two components are (a) business equipments and (b) materials. The data are in columns 5 and 6 of the file `m-ip3comp.txt`.

(a) Construct the percentage growth rate series of the two components and denote the series by $\boldsymbol{z}_t$.

(b) Build a VARMA model for $\boldsymbol{z}_t$, including model refinements with threshold 1.645 and model checking.

(c) Write down the fitted model. Is there any Granger causality in the fitted model? Why?

(d) Use the fitted model to obtain the one-step to four-step ahead predictions of the two growth rate series.

3.6 Consider the bivariate VMA(1) model

$$\boldsymbol{z}_t = \boldsymbol{a}_t - \begin{bmatrix} -0.3 & 0.2 \\ 1.1 & 0.6 \end{bmatrix} \boldsymbol{a}_{t-1},$$

where $\boldsymbol{a}_t$ is a Gaussian white noise series with mean zero and identity covariance matrix.

(a) Generate 300 observations from the VMA(1) model. Fit the VMA(1) model by both the conditional and exact likelihood methods. Write down the fitted models.

(b) To gain insight into VMA modeling, conduct a simulation study using the given VMA(1) model with sample size $T = 300$ and 1000 iterations. For each iteration, fit the VMA(1) model via the conditional maximum likelihood method and compute the Portmanteau statistic $Q_2(5)$ of the residuals. Store the coefficient estimate $\hat{\boldsymbol{\theta}}_1$ and the $Q_2(5)$ statistic over the 1000 iterations. Compute the sample means, standard errors, and the 2.5th and 97.5th percentiles of each parameter estimate and the $Q_2(5)$ statistic. Compare the summary statistics with their asymptotic counterparts.

REFERENCES

Ansley, C. F. (1979). An algorithm for the exact likelihood of a mixed autoregressive moving average process. *Biometrika*, **66**: 59–65.

Box, G. E. P., Jenkins, G. M., and Reinsel, G. (2008). *Time Series Analysis: Forecasting and Control*. 4th Edition. John Wiley & Sons, Inc, Hoboken, NJ.

Box, G. E. P. and Tiao, G. C. (1973). *Bayesian Inference in Statistical Analysis*. Addison-Wesley, Reading, MA.

Box, G. E. P. and Tiao, G. C. (1977). A canonical analysis of multiple time series. *Biometrika*, **64**: 355–365.

Brockwell, P. J. and Davies, R. A. (1987). *Time Series: Theory and Methods*. Springer, New York.

Dunsmuir, W. and Hannan, E. J. (1976). Vector linear time series models. *Advanced Applied Probability*, **8**: 339–364.

Hannan, E. J. and Deistler, M. (1988). *The Statistical Theory of Linear Systems*. John Wiley & Sons, Inc, New York. Republished in 2012 as *Classics in Applied Mathematics*, Society for Industrial and Applied Mathematics.

Hillmer, S. C. and Tiao, G. C. (1979). Likelihood function of stationary multiple autoregressive moving-average models. *Journal of the American Statistical Association*, **74**: 652–660.

Hosking, J. R. M. (1980). The multivariate portmanteau statistic. *Journal of the American Statistical Association*, **75**: 602–607.

Hosking, J. R. M. (1981). Lagrange-multiplier tests of multivariate time series model. *Journal of the Royal Statistical Society, Series B*, **43**: 219–230.

Kohn, R. (1979). Asymptotic estimation and hypothesis results for vector linear time series models. *Econometrica*, **47**: 1005–1030.

Li, W. K. (2004). *Diagnostic Checks in Time Series*. Chapman & Hall/CRC, Boca Raton, FL.

Li, W. K. and McLeod, A. I. (1981). Distribution of the residual autocorrelations in multivariate time series models. *Journal of the Royal Statistical Society, Series B*, **43**: 231–239.

Mittnik, S. (1990). Computation of theoretical autocovariance matrices of multivariate autoregressive moving-average time series. *Journal of the Royal Statistical Society, Series B*, **52**: 151–155.

Nicholls, D. F. and Hall, A. D. (1979). The exact likelihood function of multivariate autoregressive moving average models. *Biometrika*, **66**: 259–264.

Poskitt, D. S. and Tremayne, A. R. (1982). Diagnostic tests for multiple time series models. *Annals of Statistics*, **10**: 114–120.

Quenouille, M. H. (1957). *The Analysis of Multiple Time Series*. Griffin, London.

Reinsel, G. (1993). *Elements of Multivariate Time Series Analysis*. Springer-Verlag, New York.

Rissanen, J. and Caines, P. E. (1979). The strong consistency of maximum likelihood estimators for ARMA processes. *Annals of Statistics*, **7**: 297–236.

Tiao, G. C. and Tsay, R. S. (1983). Multiple time series modeling and extended sample cross-correlations. *Journal of Business & Economic Statistics*, **1**: 43–56.

Tsay, R. S. (1991). Two canonical forms for vector ARMA processes. *Statistica Sinica*, **1**: 247–269.

Tsay, R. S. (2010). *Analysis of Financial Time Series*. 3rd Edition. John Wiley & Sons, Inc, Hoboken, NJ.

Tsay, R. S. and Tiao, G. C. (1984). Consistent estimates of autoregressive parameters and extended sample autocorrelation function for stationary and nonstationary ARMA models. *Journal of the American Statistical Association*, **79**: 84–96.

Tunnicliffe-Wilson, G. (1973). The estimation of parameters in multivariate time series models. *Journal of the Royal Statistical Society, Series B*, **35**: 76–85.

Yamamoto, T. (1981). Prediction of multivariate autoregressive moving-average models. *Biometrika*, **68**: 485–492.

CHAPTER 4

Structural Specification of VARMA Models

Structural specification seeks to find the underlying structure of a multivariate linear time series so that a well-defined VARMA model can be identified. The specified model overcomes the difficulty of identifiability mentioned in Chapter 3 and can reveal the hidden structure of the system. To a certain degree, structural specification is closely related to dimension reduction and variable selection in multivariate analysis. It is an effective way to modeling multivariate linear time series.

For simplicity, we focus on a k-dimensional zero-mean stationary multivariate linear time series

$$\boldsymbol{z}_t = \sum_{i=0}^{\infty} \boldsymbol{\psi}_i \boldsymbol{a}_{t-i}, \tag{4.1}$$

where $\boldsymbol{\psi}_0 = \boldsymbol{I}_k$ and $\{\boldsymbol{a}_t\}$ is an iid sequence of random vectors with mean zero and positive-definite covariance matrix $\boldsymbol{\Sigma}_a$. In theory, the methods discussed continue to apply for unit-root nonstationary time series, but the limiting distributions of the statistics involved will be different. The stationarity assumption implies that $\sum_{i=0}^{\infty} \|\boldsymbol{\psi}_i\|^2 < \infty$, where $\|\boldsymbol{A}\|$ is a matrix norm such as the largest singular value of the matrix $\boldsymbol{A}$. There are two methods available to perform structural specification for multivariate linear time series. The first method uses the idea of *Kronecker index* and the second method applies the concept of *scalar component model* (SCM). See, for instance, Tsay (1991) and the references therein. In real applications, both methods can be implemented using the *canonical correlation analysis*, which is a statistical method useful in multivariate analysis.

For a k-dimensional VARMA time series $\boldsymbol{z}_t$, the Kronecker index approach seeks to specify the maximum order of the AR and MA polynomials for each component. Specifically, the Kronecker index approach specifies an index for each

Multivariate Time Series Analysis: With R and Financial Applications,
First Edition. Ruey S. Tsay.

component z_{it}. These Kronecker indices jointly identify a VARMA model for $\boldsymbol{z}_t$. The order for component z_{it} is in the form (p_i, p_i). Therefore, the Kronecker index approach specifies the maximum order of the AR and MA polynomials for each component z_{it}. The scalar component approach, on the other hand, seeks to specify the general order (p_i, q_i) for each scalar component. In this sense, the SCM approach can be regarded as a refinement over the Kronecker index approach. However, the SCM is a bit harder to understand and employs more statistical tests to achieve its goal. We shall compare the two approaches in this chapter.

Remark: A word of caution is in order. The goal of structural specification is essentially to find the *skeleton* of a multivariate linear time series. As such, the dynamic dependence of individual components in the system is carefully investigated and this leads to the use of rather cumbersome notation with multiple subscripts and/or superscripts. Readers are advised to focus on the concepts rather than the notation on the first read. □

4.1 THE KRONECKER INDEX APPROACH

We start with the Kronecker index approach. At the time index t, define the *past* vector, $\boldsymbol{P}_{t-1}$, and *future* vector, $\boldsymbol{F}_t$, respectively as

$$\boldsymbol{P}_{t-1} = \left(\boldsymbol{z}'_{t-1}, \boldsymbol{z}'_{t-2}, \cdots\right)', \quad \boldsymbol{F}_t = \left(\boldsymbol{z}'_t, \boldsymbol{z}'_{t+1}, \cdots\right)'.$$

In addition, define the *Hankel matrix*, $\boldsymbol{H}_\infty$, as

$$\boldsymbol{H}_\infty = \text{Cov}\left(\boldsymbol{F}_t, \boldsymbol{P}_{t-1}\right) = E\left(\boldsymbol{F}_t \boldsymbol{P}'_{t-1}\right) = \begin{bmatrix} \boldsymbol{\Gamma}_1 & \boldsymbol{\Gamma}_2 & \boldsymbol{\Gamma}_3 & \cdots \\ \boldsymbol{\Gamma}_2 & \boldsymbol{\Gamma}_3 & \boldsymbol{\Gamma}_4 & \cdots \\ \boldsymbol{\Gamma}_3 & \boldsymbol{\Gamma}_4 & \boldsymbol{\Gamma}_5 & \cdots \\ \vdots & \vdots & \vdots & \vdots \end{bmatrix}, \tag{4.2}$$

where the subscript of $\boldsymbol{H}$ indicates that it is an infinite-dimensional matrix and $\boldsymbol{\Gamma}_j$ is the lag-j autocovariance matrix of $\boldsymbol{z}_t$. A special feature of the Hankel matrix is that, looking at blocks of nonoverlapping k rows or k columns, the second block row is a subset of the first block row, the second block column is a subset of the first block column, and so on. This is referred to as the *Toeplitz* form of the matrix. As discussed in Chapter 1, for the linear system in Equation (4.1), we have

$$\boldsymbol{\Gamma}_i = \sum_{j=0}^{\infty} \boldsymbol{\psi}_{i+j} \boldsymbol{\Sigma}_a \boldsymbol{\psi}'_j,$$

so that the Hankel matrix is uniquely defined for the linear system in Equation (4.1). The use of vector ARMA models to model multivariate linear systems can be justified by the following lemma.

Lemma 4.1 For the linear vector process $\boldsymbol{z}_t$ in Equation (4.1), Rank$(\boldsymbol{H}_\infty) = m$ is finite if and only if $\boldsymbol{z}_t$ follows a VARMA model.

Proof. If $\boldsymbol{z}_t$ follows the VARMA(p, q) model

$$\boldsymbol{\phi}(B)\boldsymbol{z}_t = \boldsymbol{\theta}(B)\boldsymbol{a}_t,$$

then we have $\boldsymbol{\phi}(B)\boldsymbol{\Gamma}_j = \boldsymbol{0}$ for $j > q$, where the back-shift operator operates on the subscript j of $\boldsymbol{\Gamma}_j$, that is, $B\boldsymbol{\Gamma}_j = \boldsymbol{\Gamma}_{j-1}$. These are the moment equations of $\boldsymbol{z}_t$ discussed in Chapter 3. Since q is finite, the moment equations of VARMA models and the Toeplitz structure of $\boldsymbol{H}_\infty$ jointly imply that the rank of $\boldsymbol{H}_\infty$ is finite. On the other hand, if Rank$(\boldsymbol{H}_\infty) = m < \infty$, then there exists non-negative integer q such that the $(q+1)$th row block $[\boldsymbol{\Gamma}_{q+1}, \boldsymbol{\Gamma}_{q+2}, \cdots]$ is a linear combination of the first q blocks, namely, $\{[\boldsymbol{\Gamma}_i, \boldsymbol{\Gamma}_{i+1}, \cdots]\}_{i=1}^{q}$. In other words, there exists matrices $\{\boldsymbol{\phi}_i\}_{i=1}^{q}$ such that

$$\boldsymbol{\Gamma}_j = \sum_{i=1}^{q} \boldsymbol{\phi}_i \boldsymbol{\Gamma}_{j-i}, \quad j > q,$$

where the special feature of Hankel matrix is used. This implies that, from the definition of $\boldsymbol{H}_\infty$, $\boldsymbol{z}_{t+\ell} - \sum_{i=1}^{q} \boldsymbol{\phi}_i \boldsymbol{z}_{t+\ell-i}$ is uncorrelated with the past vector $\boldsymbol{P}_{t-1}$ for $\ell \geq q$. Consequently, $\boldsymbol{z}_t$ should follow a VARMA(q, q) model. □

Remark: The order (q, q) discussed in the proof of Lemma 4.1 is the maximum order of a VARMA model. As it will be seen later, we can construct a well-defined lower-order VARMA model under the assumption that Rank$(\boldsymbol{H}_\infty)$ is finite. □

Based on Lemma 4.1, we shall assume, in the sequel, that Rank$(\boldsymbol{H}_\infty)$ is finite so that $\boldsymbol{z}_t$ is a VARMA process. To seek the model structure, it helps to obtain a basis for the row space of the Hankel matrix. To this end, we can make use of the Toeplitz form of the Hankel matrix to examine its row dependence starting with the first row. That is, we can search a basis that expands the space generated by the rows of $\boldsymbol{H}_\infty$ starting with the first row. Denote the $[(i-1)k + j]$th row of $\boldsymbol{H}_\infty$ by $h(i, j)$, where $j = 1, \cdots, k$ and $i = 1, 2, \cdots$. From the definition, $h(i, j) = \mathrm{E}(z_{j,t+i-1}\boldsymbol{P}'_{t-1})$, which measures the linear dependence of $z_{j,t+i-1}$ on the past vector $\boldsymbol{P}_{t-1}$. We say that $h(i, j)$ is a predecessor of $h(u, v)$ if $(i-1)k + j < (u-1)k + v$. Using the Toeplitz form, it is easy to obtain the following result.

Lemma 4.2 For the Hankel matrix in Equation (4.2), if $h(i, j)$ is a linear combination of its predecessors $\{h(i_1, j_1), \cdots, h(i_s, j_s)\}$, then $h(i+1, j)$ is a linear combination of $\{h(i_1+1, j_1), \cdots, h(i_s+1, j_s)\}$.

We can now define the *Kronecker index* associated with the jth component of $\boldsymbol{z}_t$.

Definition 4.1 For the jth component z_{jt}, the *Kronecker index* k_j is defined as the smallest non-negative integer i such that $h(i+1, j)$ is linearly dependent of its predecessors.

For instance, if $z_{1t} = a_{1t}$ is a white noise component, then the first row $h(1,1)$ of $\boldsymbol{H}_\infty$ is a zero row vector, so that $k_1 = 0$. As another example, consider the bivariate VAR(1) model

$$\boldsymbol{z}_t = \boldsymbol{\phi}_1 \boldsymbol{z}_{t-1} + \boldsymbol{a}_t, \quad \text{with} \quad \boldsymbol{\phi}_1 = \begin{bmatrix} 0.2 & 0.3 \\ -0.6 & 1.1 \end{bmatrix}, \tag{4.3}$$

where $\boldsymbol{\Sigma}_a = \boldsymbol{I}_2$. Since each component z_{it} (for $i = 1$ and 2) follows a univariate ARMA(2,1) model with AR polynomial $(1 - 1.3B + 0.4B^2)$, it has nonzero autocorrelations. See properties of VAR models discussed in Chapter 2. The first two rows of $\boldsymbol{H}_\infty$ are nonzero. On the other hand, from the moment equations $\boldsymbol{\Gamma}_j - \boldsymbol{\phi}_1 \boldsymbol{\Gamma}_{j-1} = \boldsymbol{0}$ for $j > 0$, we have $\boldsymbol{\Gamma}_j = \boldsymbol{\phi}_1 \boldsymbol{\Gamma}_{j-1}$ for $j > 1$ so that the second row block of $\boldsymbol{H}_\infty$ can be obtained by premultiplying $\boldsymbol{\phi}_1$ to the first row block. Consequently, Rank$(\boldsymbol{H}_\infty) = 2$, and the first two rows $h(1,1)$ and $h(1,2)$ span the row space of $\boldsymbol{H}_\infty$. Consequently, the third and fourth rows $h(2,1)$ and $h(2,2)$ are linearly dependent of their predecessors $\{h(1,1), h(1,2)\}$, and we have $k_1 = 1$ and $k_2 = 1$.

As a third example, consider the bivariate model

$$\boldsymbol{z}_t = \boldsymbol{a}_t - \boldsymbol{\theta}_1 \boldsymbol{a}_{t-1}, \quad \text{with} \quad \boldsymbol{\theta}_1 = \begin{bmatrix} 0.5 & 0.1 \\ 0.2 & 0.4 \end{bmatrix}, \tag{4.4}$$

where $\boldsymbol{\Sigma}_a = \boldsymbol{I}_2$. The Kronecker indices for z_{it}, in this particular instance, are also $k_1 = 1$ and $k_2 = 1$, because $\boldsymbol{\Gamma}_1$ is a full-rank matrix and $\boldsymbol{\Gamma}_j = \boldsymbol{0}$ for $j > 1$. Consider next the bivariate model

$$\boldsymbol{z}_t - \boldsymbol{\phi}_1 \boldsymbol{z}_{t-1} = \boldsymbol{a}_t - \boldsymbol{\theta}_1 \boldsymbol{a}_{t-1}, \quad \text{with} \quad \boldsymbol{\Gamma}_1 = \begin{bmatrix} 0.36 & 1.76 \\ 0.64 & 4.15 \end{bmatrix},$$

where $\boldsymbol{\phi}_1$ and $\boldsymbol{\theta}_1$ are given in Equations (4.3) and (4.4), respectively, Cov$(\boldsymbol{a}_t) = \boldsymbol{I}_2$, and $\boldsymbol{\Gamma}_1$ is obtained via the method discussed in Chapter 3. For this particular VARMA(1,1) model, $\boldsymbol{\Gamma}_1$ is full rank and $\boldsymbol{\Gamma}_j = \boldsymbol{\phi}_1 \boldsymbol{\Gamma}_{j-1}$ for $j > 1$. Consequently, we also have $k_1 = 1$ and $k_2 = 2$ because the second row block of $\boldsymbol{H}_\infty$ can be obtained by premultiplying the first row block by $\boldsymbol{\phi}_1$. From these simple examples, we see that Kronecker index is concerned with linear lagged dependence of $\boldsymbol{z}_t$. The actual form of linear dependence, that is, AR or MA, does not matter. It pays to figure out the Kronecker indices for some simple VARMA models.

To aid further discussion, it is helpful to consider the Hankel matrix $\boldsymbol{H}_\infty$ as follows:

Block	Component	$\boldsymbol{P}'_{t-1} = [\boldsymbol{z}'_{t-1}, \boldsymbol{z}'_{t-2}, \cdots]$
1	z_{1t}	$h(1,1)$
	z_{2t}	$h(1,2)$
	$\vdots$	$\vdots$
	z_{kt}	$h(1,k)$
2	$z_{1,t+1}$	$h(2,1)$
	$z_{2,t+1}$	$h(2,2)$
	$\vdots$	$\vdots$
	$z_{k,t+1}$	$h(2,k)$
$\vdots$	$\vdots$	$\vdots$
k_j	$z_{1,t+k_j-1}$	$h(k_j,1)$
	$z_{2,t+k_j-1}$	$h(k_j,2)$
	$\vdots$	$\vdots$
	$z_{k,t+k_j-1}$	$h(k_j,k)$
k_j+1	$z_{1,t+k_j}$	$h(k_j+1,1)$
	$\vdots$	$\vdots$
	$z_{j-1,t+k_j}$	$h(k_j+1,j-1)$
	$z_{j,t+k_j}$	$h(k_j+1,j)$

By the definition, the Kronecker index of z_{jt} being k_j says that $h(k_j+1,j)$ is linearly dependent of its predecessors. From the prior display of the Hankel matrix $\boldsymbol{H}_\infty$, this in turn implies the existence of constants $\{\alpha^*_{u,i,j}|u=1,\cdots,k_j; i=1,\cdots,k\}$ if $k_j>0$ and $\{\alpha^*_{k_j+1,i,j}|i=1,\ldots,j-1\}$ if $j>1$ such that

$$h(k_j+1,j) = \sum_{i=1}^{j-1} \alpha^*_{k_j+1,i,j} h(k_j+1,i) + \sum_{u=1}^{k_j}\sum_{i=1}^{k} \alpha^*_{u,i,j} h(u,i), \tag{4.5}$$

where it is understood that a summation is 0 if its upper limit is smaller than its lower limit. In Equation (4.5), the first term on the right side is summing over component i of $\boldsymbol{z}_{t+k_j}$ in the future vector $\boldsymbol{F}_t$ and the second term summing over $\boldsymbol{z}_{t+\ell}$ in $\boldsymbol{F}_t$ with $0 \le \ell < k_j$. In other words, the first term is summing over the (k_j+1)th block row, whereas the second term over the block rows $1,\ldots,k_j$. As it will be seen shortly, the first term on the right side of Equation (4.5) represents the contemporaneous dependence of $z_{j,t}$ on $z_{i,t}$, where $i<j$.

Definition 4.2 The collection $\{k_j\}_{j=1}^k$ is called the set of Kronecker indices of $\boldsymbol{z}_t$. The sum $m=\sum_{j=1}^k k_j$ is called the McMillan degree of $\boldsymbol{z}_t$.

Consider next the main question of how to obtain a well-defined VARMA model from the Kronecker indices of a given Hankel matrix $\boldsymbol{H}_\infty$. To this end, we can simplify Equation (4.5) by considering $\{k_i\}_{i=1}^k$ simultaneously. Rearranging terms in Equation (4.5) according to the second argument of $h(i,j)$, that is, according to the components of $\boldsymbol{z}_t$, we have

$$h(k_j+1,j) = \sum_{i=1}^{j-1}\sum_{u=1}^{k_j+1} \alpha^*_{u,i,j} h(u,i) + \sum_{i=j}^{k}\sum_{u=1}^{k_j} \alpha^*_{u,i,j} h(u,i). \tag{4.6}$$

However, for each component z_{it}, $h(u,i)$ is a linear combination of its own predecessors if $u > k_i$. Therefore, Equation (4.6) can be simplified as

$$h(k_j+1,j) = \sum_{i=1}^{j-1}\sum_{u=1}^{k_j+1\wedge k_i} \alpha_{u,i,j} h(u,i) + \sum_{i=j}^{k}\sum_{u=1}^{k_j\wedge k_i} \alpha_{u,i,j} h(u,i), \quad j=1,\cdots,k, \tag{4.7}$$

where $u \wedge v = \min\{u,v\}$ and $\alpha_{u,i,j}$ are real numbers and are linear functions of $\{\alpha^*_{\ell,r,s}\}$. These k linear combinations provide a way to specify a VARMA model.

Note that the number of $\alpha_{u,i,j}$ coefficients in Equation (4.7) is

$$\delta_j = \sum_{i=1}^{j-1} \min(k_j+1,k_i) + k_j + \sum_{i=j+1}^{k} \min(k_j,k_i). \tag{4.8}$$

Equation (4.7) describes the linear dependence (both contemporaneous and dynamic) of $z_{j,t+k_j}$ in the vector time series. By stationarity, the equation provides the model structure of z_{jt} in the vector time series.

Using Lemma 4.2 and considering Equation (4.7) jointly for $j=1,\cdots,k$, we obtain the following results:

1. The collection of the following rows

   $$\mathcal{B} = \{h(1,1),\cdots,h(k_1,1);h(1,2),\cdots,h(k_2,2);\cdots;h(1,k),\cdots,h(k_k,k)\}$$

   forms a basis for the row space of $\boldsymbol{H}_\infty$ because all other rows are linear combinations of elements in $\mathcal{B}$. In other words, $\mathcal{B}$ is the collection of the rows of $\boldsymbol{H}_\infty$ that appear on the right side of Equation (4.7) for $j=1,\ldots,k$. Note that if $k_j=0$ then no row in the form of $h(\ell,j)$ belongs to $\mathcal{B}$.
2. Rank$(\boldsymbol{H}_\infty) = \sum_{j=1}^k k_j$. That is, the sum of Kronecker indices is the rank of $\boldsymbol{H}_\infty$. Thus, for the linear system $\boldsymbol{z}_t$, the McMillan degree is the rank of its Hankel matrix.
3. From the definition of Kronecker index, the basis $\mathcal{B}$ consists of the first m linearly independent rows of $\boldsymbol{H}_\infty$, starting from the top of the matrix.

Remark: From the definition, the Kronecker index k_j depends on the ordering of the components of $\boldsymbol{z}_t$. However, the rank of the Hankel matrix $\boldsymbol{H}_\infty$ is invariant to the ordering of components of $\boldsymbol{z}_t$. Thus, the set of Kronecker indices $\{k_j\}_{j=1}^k$ is invariant to the ordering of components of $\boldsymbol{z}_t$. □

4.1.1 A Predictive Interpretation

The linear combination in Equation (4.7) has a nice predictive interpretation. Let $\boldsymbol{\alpha}_j$ be an infinitely dimensional, real-valued vector with $\alpha_{u,i}^{(j)}$ as its $[(u-1)k+i]$th element. For each component z_{jt}, we can use Equation (4.7), which is derived from Kronecker indices, to construct a vector $\boldsymbol{\alpha}_j$ as follows:

1. Let $\alpha_{k_j+1,j}^{(j)}=1$. That is, the $(k_j \times k + j)$th element of $\boldsymbol{\alpha}_j$ is one.
2. For each $h(u,i)$ appearing in the right side of Equation (4.7), let $\alpha_{u,i}^{(j)} = -\alpha_{u,i,j}$.
3. For all other (u,i), let $\alpha_{u,i}^{(j)}=0$.

By Equation (4.7), we have

$$\boldsymbol{\alpha}_j' \boldsymbol{H}_\infty = \boldsymbol{0}. \tag{4.9}$$

To see the implication of this result, we recall that $\boldsymbol{H}_\infty = \mathrm{E}(\boldsymbol{F}_t \boldsymbol{P}_{t-1}')$ and that the $(k_j \times k + j)$th element of $\boldsymbol{F}_t$ is $z_{j,t+k_j}$. Define $w_{j,t+k_j} = \boldsymbol{\alpha}_j' \boldsymbol{F}_t$. Then, from Equations (4.9) and (4.2), $w_{j,t+k_j}$ is uncorrelated with the past vector $\boldsymbol{P}_{t-1}$ of $\boldsymbol{z}_t$. Thus, corresponding to each Kronecker index k_j, there is a linear combination of the future vector $\boldsymbol{F}_t$ that is uncorrelated with the past $\boldsymbol{P}_{t-1}$.

Furthermore, using the MA representation in Equation (4.1), $w_{j,t+k_j}$ is a linear function of $\{\boldsymbol{a}_\ell | \ell \le t+k_j\}$, whereas $\boldsymbol{P}_{t-1}$ is a linear function of $\{\boldsymbol{a}_\ell | \ell \le t-1\}$. It is then easily seen that, from the zero correlation between $w_{j,t+k_j}$ and $\boldsymbol{P}_{t-1}$, $w_{j,t+k_j}$ must be a linear function of $\{\boldsymbol{a}_\ell | t \le \ell \le t+k_j\}$. Thus, we have

$$w_{j,t+k_j} = \sum_{i=0}^{k_j} \boldsymbol{u}_i^{(j)} \boldsymbol{a}_{t+k_j-i}, \tag{4.10}$$

where $\boldsymbol{u}_i^{(j)}$s are k-dimensional row vectors such that

$$\boldsymbol{u}_0^{(j)} = [\alpha_{k_j+1,1}^{(j)}, \cdots, \alpha_{k_j+1,j-1}^{(j)}, 1, 0, \cdots, 0]$$

with 1 being in the jth position and it is understood that $\alpha_{k_j+1,i}^{(j)}=0$ if $k_i < k_j + 1$ and $i < j$. In general, $\boldsymbol{u}_i^{(j)}$ are functions of elements of the ψ-weights $\boldsymbol{\psi}_i$s and nonzero elements of $\boldsymbol{\alpha}_j$. Equation (4.10) says that the scalar process $w_{j,t+k_j}$ is at most an MA(k_j) process.

Next, from the definitions of $\boldsymbol{\alpha}_j$ and $w_{j,t+k_j}$, we have

$$w_{j,t+k_j} = z_{j,t+k_j} + \sum_{i=1}^{j-1} \sum_{u=1}^{\min(k_j+1,k_i)} \alpha_{u,i}^{(j)} z_{i,t+u-1} + \sum_{i=j}^{k} \sum_{u=1}^{\min(k_j,k_i)} \alpha_{u,i}^{(j)} z_{i,t+u-1}.$$

Combining with Equation (4.10) and noting that $\alpha_{k_j+1,i}^{(j)} = 0$ if $k_i < k_j + 1$ and $i < j$, we obtain

$$\begin{aligned} z_{j,t+k_j} &+ \sum_{i=1}^{j-1} \sum_{u=1}^{\min(k_j+1,k_i)} \alpha_{u,i}^{(j)} z_{i,t+u-1} + \sum_{i=j}^{k} \sum_{u=1}^{\min(k_j,k_i)} \alpha_{u,i}^{(j)} z_{i,t+u-1} \\ &= a_{j,t+k_j} + \sum_{i<j, k_i<k_j+1} \alpha_{k_j+1,i}^{(j)} a_{i,t+k_j} + \sum_{i=1}^{k_j} \boldsymbol{u}_i^{(j)} \boldsymbol{a}_{t+k_j-i}. \end{aligned} \tag{4.11}$$

Taking the conditional expectation of Equation (4.11) given $\boldsymbol{P}_{t-1}$, we have

$$\begin{aligned} z_{j,t+k_j|t-1} &+ \sum_{i=1}^{j-1} \sum_{u=1}^{\min(k_j+1,k_i)} \alpha_{u,i}^{(j)} z_{i,t+u-1|t-1} \\ &+ \sum_{i=j}^{k} \sum_{u=1}^{\min(k_j,k_i)} \alpha_{u,i}^{(j)} z_{i,t+u-1|t-1} = 0, \end{aligned} \tag{4.12}$$

where $z_{i,t+\ell|t-1} = E(z_{i,t+\ell}|\boldsymbol{P}_{t-1})$ is the conditional expectation of $z_{i,t+\ell}$ given $\boldsymbol{P}_{t-1}$. Let $\boldsymbol{F}_{t|t-1} = E(\boldsymbol{F}_t|\boldsymbol{P}_{t-1})$. The prior equation shows that, for each Kronecker index k_j, there exists a linear relationship among the forecasts in $\boldsymbol{F}_{t|t-1}$. Since k_j is the smallest non-negative integer for Equation (4.12) to hold, one can interpret k_j as the number of forecasts $z_{j,t|t-1}, \cdots, z_{j,t+k_j-1|t-1}$ needed to compute the forecasts $z_{j,t+\ell}$ for any forecast horizon ℓ. Obviously, to compute $z_{j,t+\ell|t-1}$, one also needs forecasts $z_{i,t+u|t-1}$ with $i \neq j$. However, these quantities are taken care of by the other Kronecker indices k_i with $i \neq j$. In view of this, the McMillan degree m is the minimum number of quantities needed to compute all elements in $\boldsymbol{F}_{t|t-1}$. The Kronecker index k_j is the minimum number of those quantities that the component $z_{j,t}$ contributes.

4.1.2 A VARMA Specification

By the stationarity of $\boldsymbol{z}_t$, Equation (4.11) can be rewritten as

$$
\begin{aligned}
z_{j,t} + \sum_{i=1}^{j-1} \sum_{u=1}^{\min(k_j+1,k_i)} \alpha_{u,i}^{(j)} z_{i,t+u-1-k_j} + \sum_{i=j}^{k} \sum_{u=1}^{\min(k_j,k_i)} \alpha_{u,i}^{(j)} z_{i,t+u-1-k_j} \\
= a_{j,t} + \sum_{i<j, k_i<k_j+1} \alpha_{k_j+1,i}^{(j)} a_{i,t} + \sum_{i=1}^{k_j} \boldsymbol{u}_i^{(j)} \boldsymbol{a}_{t-i}.
\end{aligned} \tag{4.13}
$$

Note that in Equation (4.13), the number of coefficients $\alpha_{u,i}^{(j)}$ on the left-hand side is δ_j given in Equation (4.8) and the number of elements of $\boldsymbol{u}_i^{(j)}$s in the right-hand side is $k_j \times k$.

By considering Equation (4.13) jointly for $j = 1, \cdots, k$, we obtain a VARMA model for the $\boldsymbol{z}_t$ process as

$$
\boldsymbol{\Xi}_0 \boldsymbol{z}_t + \sum_{i=1}^{p} \boldsymbol{\Xi}_i \boldsymbol{z}_{t-i} = \boldsymbol{\Xi}_0 \boldsymbol{a}_t + \sum_{i=1}^{p} \boldsymbol{\Omega}_i \boldsymbol{a}_{t-i}, \tag{4.14}
$$

where $p = \max_j\{k_j\}$, $\boldsymbol{\Xi}_0$ is a lower triangular matrix with unit diagonal elements, and its (j,i)th element is $\alpha_{k_j+1,i}^{(j)}$ for $i < j$ so that the (j,i)th element is unknown only if $k_j + 1 \leq k_i$, and the coefficient matrices $\boldsymbol{\Xi}_i$ and $\boldsymbol{\Omega}_i$ are given in Equation (4.13) for $i = 1, \cdots, p$. More specifically, we have the following:

1. For $\boldsymbol{\Omega}_i$ matrix with $i > 0$: (a) the jth row is 0 if $k_j < i \leq p$; (b) all the other rows are unknown and must be estimated.
2. For $\boldsymbol{\Xi}_i$ matrix with $i > 0$: (a) the jth row is 0 if $k_j < i \leq p$; (b) the (j,j)th element is unknown if $i \leq k_j$; (c) the (j,ℓ)th element with $j \neq \ell$ is unknown only if $k_i + i > k_j$.

Equation (4.14) gives rise to a VARMA representation for $\boldsymbol{z}_t$, the model for z_{jt} contains δ_j unknown parameters in the AR polynomials and $k \times k_j$ unknown parameters in the MA polynomials, where δ_j is defined in Equation (4.8). In summary, for the linear stationary process $\boldsymbol{z}_t$ in Equation (4.1) with Kronecker indices $\{k_j | j = 1, \cdots, k\}$ such that $m = \sum_{j=1}^{k} k_j < \infty$, one can specify a VARMA representation to describe the process. Such a representation is given by Equation (4.14) and contains

$$
N = m(1+k) + \sum_{j=1}^{k} \left[\sum_{i=1}^{j-1} \min(k_j+1, k_i) + \sum_{i=j+1}^{k} \min(k_j, k_i) \right] \tag{4.15}
$$

unknown parameters in the AR and MA polynomials, where it is understood that the summation is 0 if its lower limit is greater than its upper limit.

4.1.3 An Illustrative Example

To better understand the results of the preceding sections, we consider a simple example. Suppose that $\boldsymbol{z}_t$ is three-dimensional and the Kronecker indices are $\{k_1 = 3, k_2 = 1, k_3 = 2\}$. Here, the basis $\mathcal{B}$ for the row space of the Hankel matrix $\boldsymbol{H}_\infty$ of $\boldsymbol{z}_t$ is

$$\begin{aligned}\mathcal{B} &= \{h(1,1), h(2,1), h(3,1); h(1,2); h(1,3), h(2,3)\}\\ &= \{h(1,1), h(1,2), h(1,3); h(2,1), h(2,3); h(3,1)\},\end{aligned}$$

where the second representation is based on the row number of the Hankel matrix. Therefore, we have the following results:

1. The first linearly dependent row of $\boldsymbol{H}_\infty$ is $h(2,2)$ that provides a model for z_{2t} as

$$\begin{aligned}z_{2t} &+ \alpha_{2,1}^{(2)} z_{1t} + \alpha_{1,3}^{(2)} z_{3,t-1} + \alpha_{1,2}^{(2)} z_{2,t-1} + \alpha_{1,1}^{(2)} z_{1,t-1} = a_{2,t} + \alpha_{2,1}^{(2)} a_{1,t}\\ &+ \boldsymbol{u}_1^{(2)} \boldsymbol{a}_{t-1},\end{aligned}$$

   where, as defined before, $\boldsymbol{u}_1^{(2)}$ is a k-dimensional row vector. Based on Lemma 4.2, the rows $h(\ell,2)$ with $\ell \geq 2$ are all linearly dependent of their predecessors so that they can be removed from $\boldsymbol{H}_\infty$ for any further consideration.
2. The next linearly dependent row is $h(3,3)$ that gives a model for z_{3t} as

$$\begin{aligned}z_{3t} &+ \alpha_{3,1}^{(3)} z_{1t} + \alpha_{2,3}^{(3)} z_{3,t-1} + \alpha_{2,1}^{(3)} z_{1,t-1} + \alpha_{1,3}^{(3)} z_{3,t-2} + \alpha_{1,2}^{(3)} z_{2,t-2}\\ &+ \alpha_{1,1}^{(3)} z_{1,t-2} = a_{3t} + \alpha_{3,1}^{(3)} a_{1t} + \boldsymbol{u}_1^{(3)} \boldsymbol{a}_{t-1} + \boldsymbol{u}_2^{(3)} \boldsymbol{a}_{t-2}.\end{aligned}$$

   Again, by Lemma 4.2, we can remove all rows $h(\ell,3)$ with $\ell \geq 3$ from $\boldsymbol{H}_\infty$ for any further consideration.
3. The next linearly dependent row is $h(4,1)$ that shows a model for z_{1t} as

$$\begin{aligned}z_{1t} &+ \alpha_{3,1}^{(1)} z_{1,t-1} + \alpha_{2,3}^{(1)} z_{3,t-2} + \alpha_{2,1}^{(1)} z_{1,t-2} + \alpha_{1,3}^{(1)} z_{3,t-3} + \alpha_{1,2}^{(1)} z_{2,t-3}\\ &+ \alpha_{1,1}^{(1)} z_{1,t-3} = a_{1t} + \boldsymbol{u}_1^{(1)} \boldsymbol{a}_{t-1} + \boldsymbol{u}_2^{(1)} \boldsymbol{a}_{t-2} + \boldsymbol{u}_3^{(1)} \boldsymbol{a}_{t-3}.\end{aligned}$$

The preceding three equations can be seen from the sketch of the Hankel matrix shown in Table 4.1.

TABLE 4.1 A Sketch of Model Specification Using Kronecker Indices, where the Indices Are ($k_1 = 3, k_2 = 1, k_3 = 2$) and a Blank Space in the Third Column Indicates a Linearly Independent Row

(a) Find the first linearly dependent row		
	$h(1,1)$	
Block 1	$h(1,2)$	
	$h(1,3)$	
	$h(2,1)$	
Block 2	$h(2,2)$	Linearly dependent ($\Rightarrow$ model for z_{2t})
(b) Find the next linearly dependent row		
	$h(1,1)$	
Block 1	$h(1,2)$	
	$h(1,3)$	
	$h(2,1)$	
Block 2	$h(2,3)$	
	$h(3,1)$	
Block 3	$h(3,3)$	Linearly dependent ($\Rightarrow$ model for z_{3t})
(c) Find the next linearly dependent row		
	$h(1,1)$	
Block 1	$h(1,2)$	
	$h(1,3)$	
	$h(2,1)$	
Block 2	$h(2,3)$	
Block 3	$h(3,1)$	
Block 4	$h(4,1)$	Linearly dependent ($\Rightarrow$ model for z_{1t})

Putting the three equations together, we have a VARMA representation

$$\begin{bmatrix} 1 & 0 & 0 \\ \alpha_{2,1}^{(2)} & 1 & 0 \\ \alpha_{3,1}^{(3)} & 0 & 1 \end{bmatrix} \boldsymbol{z}_t + \begin{bmatrix} \alpha_{3,1}^{(1)} & 0 & 0 \\ \alpha_{1,1}^{(2)} & \alpha_{1,2}^{(2)} & \alpha_{1,3}^{(2)} \\ \alpha_{2,1}^{(3)} & 0 & \alpha_{2,3}^{(3)} \end{bmatrix} \boldsymbol{z}_{t-1} + \begin{bmatrix} \alpha_{2,1}^{(1)} & 0 & \alpha_{2,3}^{(1)} \\ 0 & 0 & 0 \\ \alpha_{1,1}^{(3)} & \alpha_{1,2}^{(3)} & \alpha_{1,3}^{(3)} \end{bmatrix} \boldsymbol{z}_{t-2}$$

$$+ \begin{bmatrix} \alpha_{1,1}^{(1)} & \alpha_{1,2}^{(1)} & \alpha_{1,3}^{(1)} \\ 0 & 0 & 0 \\ 0 & 0 & 0 \end{bmatrix} \boldsymbol{z}_{t-3} = \begin{bmatrix} 1 & 0 & 0 \\ \alpha_{2,1}^{(2)} & 1 & 0 \\ \alpha_{3,1}^{(3)} & 0 & 1 \end{bmatrix} \boldsymbol{a}_t + \begin{bmatrix} u_{1,1}^{(1)} & u_{1,2}^{(1)} & u_{1,3}^{(1)} \\ u_{1,1}^{(2)} & u_{1,2}^{(2)} & u_{1,3}^{(2)} \\ u_{1,1}^{(3)} & u_{1,2}^{(3)} & u_{1,3}^{(3)} \end{bmatrix} \boldsymbol{a}_{t-1}$$

$$+ \begin{bmatrix} u_{2,1}^{(1)} & u_{2,2}^{(1)} & u_{2,3}^{(1)} \\ 0 & 0 & 0 \\ u_{2,1}^{(3)} & u_{2,2}^{(3)} & u_{2,3}^{(3)} \end{bmatrix} \boldsymbol{a}_{t-2} + \begin{bmatrix} u_{3,1}^{(1)} & u_{3,2}^{(1)} & u_{3,3}^{(1)} \\ 0 & 0 & 0 \\ 0 & 0 & 0 \end{bmatrix} \boldsymbol{a}_{t-3},$$

where $u_{i,j}^{(v)}$ is the jth element of $\boldsymbol{u}_i^{(v)}$. In practice, the notation is not important and we can summarize the specified model as

$$\begin{bmatrix} 1 & 0 & 0 \\ X & 1 & 0 \\ X & 0 & 1 \end{bmatrix} \boldsymbol{z}_t + \begin{bmatrix} X & 0 & 0 \\ X & X & X \\ X & 0 & X \end{bmatrix} \boldsymbol{z}_{t-1} + \begin{bmatrix} X & 0 & X \\ 0 & 0 & 0 \\ X & X & X \end{bmatrix} \boldsymbol{z}_{t-2} + \begin{bmatrix} X & X & X \\ 0 & 0 & 0 \\ 0 & 0 & 0 \end{bmatrix} \boldsymbol{z}_{t-3}$$
$$= \begin{bmatrix} 1 & 0 & 0 \\ X & 1 & 0 \\ X & 0 & 1 \end{bmatrix} \boldsymbol{a}_t + \begin{bmatrix} X & X & X \\ X & X & X \\ X & X & X \end{bmatrix} \boldsymbol{a}_{t-1} + \begin{bmatrix} X & X & X \\ 0 & 0 & 0 \\ X & X & X \end{bmatrix} \boldsymbol{a}_{t-2} + \begin{bmatrix} X & X & X \\ 0 & 0 & 0 \\ 0 & 0 & 0 \end{bmatrix} \boldsymbol{a}_{t-3},$$

where X denotes an unknown parameter that requires estimation. The total number of unknown parameters is $N = 6(1+3) + 10 = 34$.

Remark: We have developed an R script that provides the specification of a VARMA model for a given set of Kronecker indices. The command is `Kronspec` in the `MTS` package. For demonstration, consider a three-dimensional series $\boldsymbol{z}_t$ with Kronecker indices $\{2, 1, 1\}$. The results are given next:

```
> kdx=c(2,1,1)
> Kronspec(kdx)
Kronecker indices:  2 1 1
Dimension:  3
Notation:
 0: fixed to 0
 1: fixed to 1
 2: estimation
AR coefficient matrices:
     [,1] [,2] [,3] [,4] [,5] [,6] [,7] [,8] [,9]
[1,]    1    0    0    2    0    0    2    2    2
[2,]    2    1    0    2    2    2    0    0    0
[3,]    2    0    1    2    2    2    0    0    0
MA coefficient matrices:
     [,1] [,2] [,3] [,4] [,5] [,6] [,7] [,8] [,9]
[1,]    1    0    0    2    2    2    2    2    2
[2,]    2    1    0    2    2    2    0    0    0
[3,]    2    0    1    2    2    2    0    0    0
```

where "0" denotes a zero element, "1" denotes the coefficient being 1 used in the coefficient matrix Ξ_0, and "2" denotes parameters that require estimation. In the demonstration, we use the notation of Equation (4.14). The Kronecker indices are $\{2, 1, 1\}$ so that (i) the overall model of the process is a VARMA(2,2) model; (ii) the orders of the AR and MA polynomials for z_{it} are 2, 1, 1, respectively, for $i = 1, 2, 3$; (iii) the second and third components depend simultaneously on the first component; and (iv) the two elements $\Xi_{1,12} = \Xi_{1,13} = 0$ are set to zero because they are redundant given that $\Omega_{1,12}$ and $\Omega_{1,13}$ are in the model, where we have used the notation $\Xi_1 = [\Xi_{1,ij}]$ and $\Omega_1 = [\Omega_{1,ij}]$. □

4.1.4 The Echelon Form

The model representation of Equation (4.14) is a canonical form for the z_t process. It is referred to as a reversed Echelon form and has some nice properties that we discuss next.

4.1.4.1 Degree of Individual Polynomial

Let $A_{jv}(B)$ be the (j, v)th element of the matrix polynomial $\boldsymbol{A}(B)$. Let deg$[A_{jv}(B)]$ be the degree of the polynomial $A_{jv}(B)$. Then, the degree of each polynomial in $\boldsymbol{\Xi}(B) = \boldsymbol{\Xi}_0 + \sum_{i=1}^{p} \boldsymbol{\Xi}_i B^i$ of Equation (4.14) is deg$[\Xi_{jv}(B)] = k_j$ for all $v = 1, \cdots, k$. In other words, the Kronecker index k_j is the degree of all the polynomials in the jth row of $\boldsymbol{\Xi}(B)$. The same result holds for the individual polynomials in $\boldsymbol{\Omega}(B)$ of Equation (4.14). In fact, k_j is the maximum order of $\Xi_{jv}(B)$ and $\Omega_{jv}(B)$. The actual order might be smaller after estimation or further analysis.

4.1.4.2 Number of Unknown Coefficients of the Individual Polynomial

Let n_{jv} be the number of unknown coefficients of $\Xi_{jv}(B)$ in $\boldsymbol{\Xi}(B)$ of Equation (4.14). Then, from the structure of $\boldsymbol{\Xi}(B)$, we have

$$n_{jv} = \begin{cases} \min(k_j, k_v) & \text{if } j \le v \\ \min(k_j + 1, k_v) & \text{if } j > v. \end{cases} \tag{4.16}$$

Similarly, let m_{jv} be the number of unknown coefficients of $\Omega_{jv}(B)$ in $\boldsymbol{\Omega}(B)$. Then, we have

$$m_{jv} = \begin{cases} k_j & \text{if } j \le v \text{ or } (j > v \text{ and } k_j \ge k_v) \\ k_j + 1 & \text{if } j > v \text{ and } k_j < k_v. \end{cases} \tag{4.17}$$

Both n_{jv} and m_{jv} include the lower triangular elements in $\boldsymbol{\Xi}_0$, if they exist.

4.1.4.3 Structure of the Individual Polynomial

Denote by $A_{jv}^{(i)}$ the (j, v)th element of the matrix $\boldsymbol{A}^{(i)}$. Using the degree and the equation form discussed earlier, one can easily specify the exact form of each individual polynomial in $\boldsymbol{\Xi}(B)$ of Equation (4.14). Specifically, we have

$$\Xi_{jj}(B) = 1 + \sum_{i=1}^{k_j} \Xi_{jj}^{(i)} B^i, \quad j = 1, \cdots, k \tag{4.18}$$

$$\Xi_{jv}(B) = \sum_{i=k_j+1-n_{jv}}^{k_j} \Xi_{jv}^{(i)} B^i, \quad j \ne v, \tag{4.19}$$

where n_{jv} is defined in Equation (4.16). For the polynomial in $\mathbf{\Omega}(B)$, the result is

$$\Omega_{jj}(B) = 1 + \sum_{i=1}^{k_j} \Omega_{jj}^{(i)} B^i, \quad j = 1, \cdots, k \tag{4.20}$$

$$\Omega_{jv}(B) = \sum_{i=k_j+1-m_{jv}}^{k_j} \Omega_{jv}^{(i)} B^i, \quad \text{if} \quad j \neq v, \tag{4.21}$$

where m_{jv} is defined in Equation (4.17).

The preceding results show that, for a k-dimensional linear process $\boldsymbol{z}_t$ of Equation (4.1), the Kronecker indices $\{k_j | j = 1, \cdots, k\}$ specify a VARMA representation (4.14) for $\boldsymbol{z}_t$. This VARMA specification is well-defined in the sense that (a) all the unknown parameters in the AR and MA matrix polynomials are identified and (b) each individual polynomial is specifically given. In the literature, such a VARMA representation is called a reversed Echelon form (Hannan and Deistler, 1988) and has the following nice properties.

Theorem 4.1 Suppose that $\boldsymbol{z}_t$ is a k-dimensional stationary time series of Equation (4.1) with Kronecker indices $\{k_j | j = 1, \cdots, k\}$ such that $m = \sum_{j=1}^{k} k_j < \infty$. Then, $\boldsymbol{z}_t$ follows the VARMA model in Equation (4.14) with $\mathbf{\Xi}(B)$ and $\mathbf{\Omega}(B)$ specified by Equations (4.16)–(4.21). Furthermore, $\mathbf{\Xi}(B)$ and $\mathbf{\Omega}(B)$ are left coprime, and deg[$|\mathbf{\Xi}(B)|$] + deg[$|\mathbf{\Omega}(B)|$] $\leq 2m$.

4.1.5 The Example Continued

For the three-dimensional example of Subsection 4.1.3, the number of unknown coefficients in the individual polynomials is given by

$$[n_{jv}] = \begin{bmatrix} 3 & 1 & 2 \\ 2 & 1 & 1 \\ 3 & 1 & 2 \end{bmatrix} \quad \text{and} \quad [m_{jv}] = \begin{bmatrix} 3 & 3 & 3 \\ 2 & 1 & 1 \\ 3 & 2 & 2 \end{bmatrix}.$$

Since $\mathbf{\Xi}_0 = \mathbf{\Omega}_0$, the total number of unknown coefficients in the AR and MA polynomials is different from the sum of all n_{jv} and m_{jv}, as the latter counts the unknown coefficients in $\mathbf{\Xi}_0$ twice.

4.2 THE SCALAR COMPONENT APPROACH

Turn to the second approach for structural specification of a vector time series $\boldsymbol{z}_t$. This approach is referred to as the scalar component model (SCM) approach.

4.2.1 Scalar Component Models

SCM, developed by Tiao and Tsay (1989), generalizes the concept of model equation of a VARMA model to search for simplifying structure of the data. Consider the VARMA(p, q) model

$$\boldsymbol{\phi}(B)\boldsymbol{z}_t = \boldsymbol{\theta}(B)\boldsymbol{a}_t, \tag{4.22}$$

where $\boldsymbol{z}_t$ and $\boldsymbol{a}_t$ are defined in Equation (4.1), and $\boldsymbol{\phi}(B) = \boldsymbol{I} - \sum_{i=1}^{p} \boldsymbol{\phi}_i B^i$ and $\boldsymbol{\theta}(B) = \boldsymbol{I} - \sum_{i=1}^{q} \boldsymbol{\theta}_i B^i$ are matrix polynomials of finite order p and q, respectively. Let $\boldsymbol{A}^{(i)}$ be the ith row of the matrix $\boldsymbol{A}$. Then, the equation for the ith component z_{it} of $\boldsymbol{z}_t$ is

$$\boldsymbol{I}^{(i)}\boldsymbol{z}_t - \sum_{j=1}^{p} \boldsymbol{\phi}_j^{(i)}\boldsymbol{z}_{t-j} = \boldsymbol{I}^{(i)}\boldsymbol{a}_t - \sum_{j=1}^{q} \boldsymbol{\theta}_j^{(i)}\boldsymbol{a}_{t-j}.$$

One way to interpret the prior equation is as follows: Given the row vector $\boldsymbol{v}_0^{(i)} = \boldsymbol{I}^{(i)}$, there exist p k-dimensional row vectors $\{\boldsymbol{v}_j^{(i)}\}_{j=1}^{p}$ with $\boldsymbol{v}_j^{(i)} = -\boldsymbol{\phi}_j^{(i)}$ such that the linear combination

$$w_{it} = \boldsymbol{v}_0^{(i)}\boldsymbol{z}_t + \sum_{j=1}^{p} \boldsymbol{v}_j^{(i)}\boldsymbol{z}_{t-j}$$

is uncorrelated with $\boldsymbol{a}_{t-\ell}$ for $\ell > q$ because we also have

$$w_{it} = \boldsymbol{v}_0^{(i)}\boldsymbol{a}_t - \sum_{j=1}^{q} \boldsymbol{\theta}_j^{(i)}\boldsymbol{a}_{t-j},$$

which is a linear combination of $\{\boldsymbol{a}_{t-j}\}_{j=0}^{q}$. In other words, the process w_{it}, which is a linear combination of $\{\boldsymbol{z}_{t-j}\}_{j=0}^{p}$, has a finite memory in the sense that it is uncorrelated with $\boldsymbol{a}_{t-\ell}$ for $\ell > q$. For a k-dimensional process $\boldsymbol{z}_t$, there are k such linearly independent processes of order (p, q). However, we would like to keep the order (p, q) of each linear combination as low as possible in applications and in model building.

Definition 4.3 Suppose that $\boldsymbol{z}_t$ is a stationary linear vector process of Equation (4.1). A nonzero linear combination of $\boldsymbol{z}_t$, denoted by $y_t = \boldsymbol{v}_0\boldsymbol{z}_t$, is a scalar component of order (r, s) if there exist r k-dimensional row vectors $\{\boldsymbol{v}_i\}_{i=1}^{r}$ such that (a) $\boldsymbol{v}_r \neq \boldsymbol{0}$ if $r > 0$, and (b) the scalar process

$$w_t = y_t + \sum_{i=1}^{r} \boldsymbol{v}_i\boldsymbol{z}_{t-i} \quad \text{satisfies} \quad E(\boldsymbol{a}_{t-\ell}w_t) \begin{cases} = \boldsymbol{0} & \text{if } \ell > s \\ \neq \boldsymbol{0} & \text{if } \ell = s. \end{cases}$$

From the definition, $y_t = \boldsymbol{v}_0 \boldsymbol{z}_t$ is a SCM of $\boldsymbol{z}_t$ if the scalar process w_t is uncorrelated with the past vector $\boldsymbol{P}_{t-\ell}$ for $\ell > s$, but correlated with $\boldsymbol{P}_{t-s}$. The requirements of $\boldsymbol{v}_r \neq \boldsymbol{0}$, if $r > 0$, and $E(\boldsymbol{a}_{t-s} w_t) \neq \boldsymbol{0}$ are used to reduce the order (r, s).

Using Equation (4.1) for $\boldsymbol{z}_{t-i}$ and collecting the coefficient vectors of $\boldsymbol{a}_{t-j}$, we can write the scalar process w_t as

$$w_t = \boldsymbol{v}_0 \boldsymbol{a}_t + \sum_{j=1}^{s} \boldsymbol{u}_j \boldsymbol{a}_{t-i}, \tag{4.23}$$

where $\boldsymbol{u}_j$s are k-dimensional row vectors, $\boldsymbol{u}_s \neq \boldsymbol{0}$ if $s > 0$, and it is understood that the summation is 0 if its upper limit is smaller than its lower limit. Thus, an SCM of order (r, s) implies that there exists a nonzero linear combination of $\boldsymbol{z}_t, \cdots, \boldsymbol{z}_{t-r}$ which is also a linear combination of $\boldsymbol{a}_t, \cdots, \boldsymbol{a}_{t-s}$. With this interpretation, it is seen that a Kronecker index k_j of $\boldsymbol{z}_t$ implies the existence of an SCM(k_j, k_j) of $\boldsymbol{z}_t$.

Note that y_t being an SCM of order (r, s) does not necessarily imply that y_t follows a univariate ARMA(r, s) model. The SCM is a concept within the vector framework and it uses all the components of $\boldsymbol{z}_t$ in describing a model. On the other hand, a univariate model of y_t only depends on its own past y_{t-j} for $j > 0$. Also, from the definition, the order (r, s) of an SCM y_t is not unique. For example, multiplying w_{t-m} with $m > 0$ by a nonzero constant c, then adding it to w_t, we obtain, from Equation (4.23), a new scalar process

$$w_t^* = w_t + c w_{t-m} = \boldsymbol{v}_0 \boldsymbol{a}_t + \sum_{j=1}^{s} \boldsymbol{u}_j \boldsymbol{a}_{t-j} + c \left(\boldsymbol{v}_0 \boldsymbol{a}_{t-m} + \sum_{j=1}^{s} \boldsymbol{u}_j \boldsymbol{a}_{t-m-j} \right),$$

which is uncorrelated with $\boldsymbol{a}_{t-\ell}$ for $\ell > s + m$. This type of redundancies should be eliminated so that we require that the order (r, s) of an SCM satisfies the condition that $r + s$ is as small as possible. Note that, even with the requirement on the sum $r + s$, the order of a given SCM is still not unique. For instance, consider the model

$$\boldsymbol{z}_t - \begin{bmatrix} 0 & 0 \\ 2 & 0 \end{bmatrix} \boldsymbol{z}_{t-1} = \boldsymbol{a}_t,$$

which can be written equivalently as

$$\boldsymbol{z}_t = \boldsymbol{a}_t - \begin{bmatrix} 0 & 0 \\ -2 & 0 \end{bmatrix} \boldsymbol{a}_{t-1}.$$

It is easily seen that $z_{2t} = [0, 1]\boldsymbol{z}_t$ is an SCM of order (1,0) or (0,1). Here, both orders satisfy $r + s = 1$, the lowest possible value. This type of nonuniqueness does not cause problems in model specification because the sum $r + s$, which is fixed, plays an important role in the SCM approach. We shall discuss this issue in the next subsection.

4.2.2 Model Specification Via Scalar Component Models

Suppose that $y_{it} = \boldsymbol{v}_{0,i}\boldsymbol{z}_t$ is an SCM(p_i, q_i) of $\boldsymbol{z}_t$, where $i = 1, \cdots, k$. These k SCMs are linearly independent if the $k \times k$ matrix $\boldsymbol{T}' = [\boldsymbol{v}'_{0,1}, \cdots, \boldsymbol{v}'_{0,k}]$ is nonsingular, that is, $|\boldsymbol{T}| \neq 0$. For a k-dimensional process $\boldsymbol{z}_t$ in Equation (4.1), a set of k linearly independent SCMs determines a VARMA model for $\boldsymbol{z}_t$. From the definition, for each SCM y_{it} there exist p_i k-dimensional row vector $\{\boldsymbol{v}_{j,i}\}_{j=1}^{p_i}$ such that the scalar process $w_{it} = \sum_{\ell=0}^{p_i} \boldsymbol{v}_{\ell,i}\boldsymbol{z}_{t-\ell}$ is uncorrelated with $\boldsymbol{a}_{t-j}$ for $j > q_i$. Let $\boldsymbol{w}_t = (w_{1t}, \cdots, w_{kt})'$, $r = \max\{p_i\}$, and $s = \max\{q_i\}$. We have

$$\boldsymbol{w}_t = \boldsymbol{T}\boldsymbol{z}_t + \sum_{\ell=1}^{r} \boldsymbol{G}_\ell \boldsymbol{z}_{t-\ell}, \tag{4.24}$$

where $\boldsymbol{G}'_\ell = [\boldsymbol{v}_{\ell,1}, \cdots, \boldsymbol{v}_{\ell,k}]$ with $\boldsymbol{v}_{\ell,i} = \boldsymbol{0}$ for $p_i < \ell \leq r$. Furthermore, from Equation (4.23), $\boldsymbol{w}_t$ can also be written as

$$\boldsymbol{w}_t = \boldsymbol{T}\boldsymbol{a}_t + \sum_{\ell=1}^{s} \boldsymbol{U}_\ell \boldsymbol{a}_{t-\ell}, \tag{4.25}$$

where $\boldsymbol{U}_\ell = [\boldsymbol{u}_{\ell,1}, \cdots, \boldsymbol{u}_{\ell,k}]$ is a $k \times k$ matrix whose ith row is 0 if $q_i < \ell \leq s$. Combining Equations (4.24) and (4.25), we have a VARMA(r, s) model for $\boldsymbol{z}_t$. Furthermore, the row structure of the coefficient matrices for the specified model is available. More specifically, we have

$$\boldsymbol{T}\boldsymbol{z}_t + \sum_{\ell=1}^{r} \boldsymbol{G}_\ell \boldsymbol{z}_{t-\ell} = \boldsymbol{T}\boldsymbol{a}_t + \sum_{\ell=1}^{s} \boldsymbol{U}_\ell \boldsymbol{a}_{t-\ell} \tag{4.26}$$

such that

1. The ith row of $\boldsymbol{G}_\ell$ is 0 if $p_i < \ell \leq r$
2. The ith row of $\boldsymbol{U}_\ell$ is 0 if $q_i < \ell \leq s$
3. Some further reduction in the number of parameters is possible under certain circumstances

The last result is due to certain identifiable redundant parameters between AR and MA components in Equation (4.26), which we shall discuss in the next subsection. From Equation (4.26), a VARMA(r, s) model for $\boldsymbol{z}_t$ is obtained.

Note that by inserting $\boldsymbol{T}^{-1}\boldsymbol{T}$ in the front of $\boldsymbol{z}_{t-\ell}$ and $\boldsymbol{a}_{t-\ell}$ in Equation (4.26), one obtains a VARMA(r, s) model for the transformed process $\boldsymbol{y}_t$

$$(\boldsymbol{I} - \boldsymbol{\varphi}_1 B - \cdots - \boldsymbol{\varphi}_r B^r)\boldsymbol{y}_t = (\boldsymbol{I} - \boldsymbol{\Theta}_1 B - \cdots - \boldsymbol{\Theta}_s B^s)\boldsymbol{b}_t, \tag{4.27}$$

where $\boldsymbol{b}_t = \boldsymbol{T}\boldsymbol{a}_t$, $\boldsymbol{\varphi}_i = -\boldsymbol{G}_i\boldsymbol{T}^{-1}$ and $\boldsymbol{\Theta}_j = -\boldsymbol{U}_j\boldsymbol{T}^{-1}$. Because multiplication from right does not change the structure of a zero row, we see that $\boldsymbol{\varphi}_i$ and $\boldsymbol{\Theta}_j$ have the same row structure as those of $\boldsymbol{G}_i$ and $\boldsymbol{U}_j$, respectively, for $i = 1, \cdots, r$ and $j = 1, \cdots, s$. From the model in Equation (4.27), it is clear that the order (p_i, q_i) of an SCM signifies that one needs $p_i + q_i$ unknown rows to describe the structure of y_{it} in the VARMA model of $\boldsymbol{y}_t$. Here, by *unknown* row we mean that its parameters require estimation. This is used in contrast with the other rows that are known to be zero.

4.2.3 Redundant Parameters

In this subsection, we consider the possible redundant parameters in the VARMA representation of Equation (4.27) and discuss a method that can easily identify such parameters when they exist. It is worth mentioning that redundant parameters can occur even without over-specifying the overall order (r, s) of Equation (4.27).

Suppose that the orders (p_1, q_1) and (p_2, q_2) of the first two SCM's y_{1t} and y_{2t} satisfy $p_2 > p_1$ and $q_2 > q_1$. In this case, we can write the model structure for y_{1t} and y_{2t} as

$$y_{it} - \left[\boldsymbol{\varphi}_1^{(i)} B + \cdots + \boldsymbol{\varphi}_{p_i}^{(i)} B^{p_i}\right] \boldsymbol{y}_t = b_{it} - \left[\boldsymbol{\Theta}_1^{(i)} B + \cdots + \boldsymbol{\Theta}_{q_i}^{(i)} B^{q_i}\right] \boldsymbol{b}_t, \tag{4.28}$$

where $i = 1, 2$ and $\boldsymbol{A}^{(i)}$ denotes the ith row of the matrix $\boldsymbol{A}$. Now for $i = 2$ we see from Equation (4.28) that y_{2t} is related to $y_{1,t-1}, \cdots, y_{1,t-p_2}$ and $b_{1,t-1}, \cdots, b_{1,t-q_2}$ via

$$(\varphi_{1,21} B + \cdots + \varphi_{p_2,21} B^{p_2}) y_{1t} - (\Theta_{1,21} B + \cdots + \Theta_{q_2,21} B^{q_2}) b_{1t}, \tag{4.29}$$

where $A_{v,ij}$ denotes the (i, j)th element of the matrix $\boldsymbol{A}_v$. On the other hand, from Equation (4.28) with $i = 1$, we have

$$B^\ell(y_{1t} - b_{1t}) = \left[\boldsymbol{\varphi}_1^{(1)} B + \cdots + \boldsymbol{\varphi}_{p_1}^{(1)} B^{p_1}\right] \boldsymbol{y}_{t-\ell} - \left[\Theta_1^{(1)} B + \cdots + \boldsymbol{\Theta}_{q_1}^{(1)} B^{q_1}\right] \boldsymbol{b}_{t-\ell}. \tag{4.30}$$

Therefore, if all the $\boldsymbol{y}$s and $\boldsymbol{b}$s on the right-hand side of Equation (4.30) are in the component model for y_{2t}, then either the coefficient of $y_{1,t-\ell}$ or that of $b_{1,t-\ell}$ is redundant given that the other is in the model. Consequently, if $p_2 > p_1$ and $q_2 > q_1$, then for each pair of parameters $(\varphi_{\ell,21}, \Theta_{\ell,21})$, $\ell = 1, \cdots, \min\{p_2 - p_1, q_2 - q_1\}$, only one of them is needed.

The preceding method of spotting redundant parameters in a VARMA model of Equation (4.27) is referred to as the *rule of elimination* in Tiao and Tsay (1989). In general, by considering an ARMA model constructed from SCMs and applying the rule of elimination in a pairwise fashion, all redundant parameters of the model structure for y_{it} in Equation (4.27) can be eliminated. By applying the rule of elimination to each pair of SCMs, we obtain

$$\eta_i = \sum_{v=1}^{k} \max[0, \min\{p_i - p_v, q_i - q_v\}]. \tag{4.31}$$

Putting all results together, we see that the total number of unknown parameters in the coefficient matrices of Equation (4.27) is

$$P = k \times \sum_{i=1}^{k} (p_i + q_i) - \sum_{i=1}^{k} \eta_i, \tag{4.32}$$

which can be much smaller than $k^2(r + s)$ of a k-dimensional zero-mean VARMA(r, s) model. This parameter count does not consider the parameters in the transformation matrix T. We shall consider the transformation matrix later.

Example 4.1 Suppose that $\boldsymbol{z}_t$ is a bivariate linear process with two linearly independent scalar components $y_{1t} \sim$ SCM(1,0) and $y_{2t} \sim$ SCM(2,1). In this case, we have $r = \max(1, 2) = 2$ and $s = \max(0, 1) = 1$. The model for $\boldsymbol{y}_t = (y_{1t}, y_{2t})'$ is a VARMA(2,1) model. Since y_{1t} is SCM(1,0), we have

$$y_{1t} = \varphi_{1,11} y_{1,t-1} + \varphi_{1,12} y_{2,t-1} + b_{1t}, \tag{4.33}$$

where, again, $\varphi_{\ell,ij}$ denotes the (i, j)th element of the matrix $\boldsymbol{\varphi}_\ell$. Since y_{2t} is SCM(2,1), we have

$$\begin{aligned} y_{2t} &= \varphi_{1,21} y_{1,t-1} + \varphi_{1,22} y_{2,t-1} + \varphi_{2,21} y_{1,t-2} + \varphi_{2,22} y_{2,t-2} + b_{2t} \\ &\quad - \Theta_{1,21} b_{1,t-1} - \Theta_{1,22} b_{2,t-1}. \end{aligned} \tag{4.34}$$

By time-invariance of the system, Equation (4.33) gives

$$y_{1,t-1} = \varphi_{1,11} y_{1,t-2} + \varphi_{1,12} y_{2,t-2} + b_{1,t-1}. \tag{4.35}$$

Note that all terms of Equation (4.35) appear in the right-hand side of Equation (4.34). This situation occurs because the orders of the two SCMs satisfy the condition that $p_2 > p_1$ and $q_2 > q_1$. Consequently, one can substitute either $y_{1,t-1}$ or $b_{1,t-1}$ of Equation (4.34) by Equation (4.35) to simplify the model. In fact, the parameters $\varphi_{1,21}$ and $\Theta_{1,21}$ are not identifiable in the sense that one of them is redundant given the other. Therefore, one can fix one of these two parameters to zero to simplify the model structure. □

4.2.4 VARMA Model Specification

Results of the preceding subsections enable us to specify an estimable VARMA model for the linear vector process $\boldsymbol{z}_t$ of Equation (4.1) if k linearly independent SCMs are given. Obviously, the order (p_i, q_i) of a given SCM must satisfy the condition that p_i+q_i is as small as possible. In this subsection, we provide a demonstration. Suppose that $k=4$ and $\boldsymbol{z}_t$ has four linearly independent SCMs of orders (0,0), (0,1), (1,0), and (2,1). Since $\max\{p_i\}=2$ and $\max\{q_i\}=1$, $\boldsymbol{z}_t$ is a VARMA(2,1) process. Furthermore, one can easily write down the specified model for the transformed series $\boldsymbol{y}_t=\boldsymbol{T}\boldsymbol{z}_t$, where $\boldsymbol{T}$ is the matrix of SCMs. The model is given by

$$(\boldsymbol{I} - \boldsymbol{\varphi}_1 B - \boldsymbol{\varphi}_2 B^2)\boldsymbol{y}_t = (\boldsymbol{I} - \boldsymbol{\Theta}_1 B)\boldsymbol{b}_t,$$

where the coefficient matrices are

$$\boldsymbol{\varphi}_1 = \begin{bmatrix} 0 & 0 & 0 & 0 \\ 0 & 0 & 0 & 0 \\ X & X & X & X \\ X & X & X & X \end{bmatrix}, \quad \boldsymbol{\varphi}_2 = \begin{bmatrix} 0 & 0 & 0 & 0 \\ 0 & 0 & 0 & 0 \\ 0 & 0 & 0 & 0 \\ X & X & X & X \end{bmatrix},$$

$$\boldsymbol{\Theta}_1 = \begin{bmatrix} 0 & 0 & 0 & 0 \\ X & X & X & X \\ 0 & 0 & 0 & 0 \\ 0 & X & 0 & X \end{bmatrix},$$

where 0 denotes a zero parameter and X denotes an unknown parameter. Note that the (4,1) and (4,3) elements of $\boldsymbol{\Theta}_1$ are set to 0 because, by applying the rule of elimination, they are redundant once the (4,1) and the (4,3) elements of $\boldsymbol{\varphi}_1$ are in the model. Consequently, in this particular instance modeling, the transformed series $\boldsymbol{y}_t$ would involve 18 parameters in the coefficient matrices instead of 48 parameters for an unrestricted VARMA(2,1) model.

4.2.5 The Transformation Matrix

In practice, the transformation matrix $\boldsymbol{T}$ must be estimated. Tiao and Tsay (1989) use a two-stage procedure. First, one estimates $\boldsymbol{T}$ in the process of identifying SCM models; see Section 4.5. Second, using the estimated $\boldsymbol{T}$, one can transform the data to $\boldsymbol{y}_t$ and perform further estimation using the $\boldsymbol{y}_t$ process. See Equation (4.27). The model for $\boldsymbol{z}_t$ can then be obtained from that of $\boldsymbol{y}_t$ via the transformation matrix $\boldsymbol{T}$. Limited experience indicates that this two-stage procedure works reasonably well.

A more direct approach, suggested by Athanasopoulos and Vahid (2008), is to perform a joint estimation of the transformation matrix and the model parameters.

This can be achieved by making use of properties of the SCM models. Specifically, the following two properties of SCMs for $\boldsymbol{z}_t$ are relevant:

1. Multiplying a given SCM(p_1, q_1) component by a nonzero constant does not change the order (p_1, q_1).
2. Consider two SCMs, say $y_{1t} \sim$ SCM(p_1, q_1) and $y_{2t} \sim$ SCM(p_2, q_2), satisfying $p_1 \geq p_2$ and $q_1 \geq q_2$. The linear combination $y_{1t} + cy_{2t}$ is an SCM component of order (p_1, q_1), where c is an arbitrary real number.

Property one follows directly from the definition of SCM, whereas Property two is a special case of Lemma 1 of Tiao and Tsay (1989) with $v = 0$. Property two can be used to eliminate some elements of the transformation matrix $\boldsymbol{T}$.

Consider the transformation matrix $\boldsymbol{T}' = [\boldsymbol{v}'_{0,1}, \ldots, \boldsymbol{v}'_{0,k}]$. For simplicity in notation, we write $\boldsymbol{T} = [T_{ij}]$, where the ith row $\boldsymbol{T}_{i.} = (T_{i1}, \ldots, T_{ik})$ corresponds to an SCM of order (p_i, q_i). Using Property one, we can set an element of the ith row $\boldsymbol{T}_{i.}$ to 1. [This eliminates one parameter from each row from estimation.] In addition, since $\boldsymbol{T}$ is nonsingular, one can choose proper locations so that the element 1 appears in different columns for different rows. Let the location of 1 in the ith row be (i, i_o). Next, consider the ith and jth rows of $\boldsymbol{T}$ with orders (p_i, q_i) and (p_j, q_j). If $p_i \leq p_j$ and $q_i \leq q_j$, then we can let $T_{j,i_o} = 0$ by using Property two. This is so because $\boldsymbol{T}_{j.} - T_{ji} \times \boldsymbol{T}_{i.}$ remains an SCM of order (p_j, q_j).

To demonstrate, suppose that $k = 4$ and $\boldsymbol{z}_t$ has four linearly independent SCMs of orders (0,0), (0,1), (1,0), and (2,1), respectively. Assume further that the transformation matrix $\boldsymbol{T}$, after normalization of each row, is

$$\boldsymbol{T} = \begin{bmatrix} T_{11} & T_{12} & T_{13} & 1 \\ 1 & T_{22} & T_{23} & T_{24} \\ T_{31} & 1 & T_{33} & T_{34} \\ T_{41} & T42 & 1 & T_{44} \end{bmatrix}. \tag{4.36}$$

Applying the two properties of SCMs, we can simplify the transformation matrix T as follows without affecting the SCM models of $\boldsymbol{z}_t$,

$$\boldsymbol{T} = \begin{bmatrix} T_{11} & T_{12} & T_{13} & 1 \\ 1 & T_{22} & T_{23} & 0 \\ T_{21} & 1 & T_{33} & 0 \\ 0 & 0 & 1 & 0 \end{bmatrix}. \tag{4.37}$$

For instance, T_{24} can be set to 0 because $\boldsymbol{v}_{0,2} - T_{24}\boldsymbol{v}_{0,1}$ remains an SCM of order (0,1). Similarly, T_{41} can be set to 0 because $\boldsymbol{v}_{0,4} - T_{41}\boldsymbol{v}_{0,2}$ is still an SCM of order (2,1).

From the aforementioned discussion and demonstration, whenever the orders of any two SCMs are nested, namely, $p_i \leq p_j$ and $q_i \leq q_j$, one can simplify, via Property two, the transformation matrix $\boldsymbol{T}$ by eliminating a nonzero parameter without altering the row structure of the SCM specification. More specifically, suppose that the orders of SCM's $\boldsymbol{y}_t$ of $\boldsymbol{z}_t$ are (p_i, q_i) for $i=1,\cdots,k$. Then, to obtain further simplification in the transformation matrix $\boldsymbol{T}$, one can simply examine the $k(k+1)/2$ pairs of SCMs. For any nested pair, by using Property two, one can identify a zero parameter in the transformation $\boldsymbol{T}$. Mathematically, the total number of zero parameters identified by such a procedure is

$$\tau = \sum_{i=1}^{k-1} \sum_{j=i+1}^{k} \text{Ind}[\min(p_j - p_i, q_j - q_i) \geq 0],$$

where Ind(.) is an indicator operator that assumes the value 1 if its argument is true and the value 0, otherwise.

4.3 STATISTICS FOR ORDER SPECIFICATION

Turn to data analysis. For a given data set $\{\boldsymbol{z}_t | t=1,\cdots,T\}$, we discuss methods to identify the Kronecker indices and the orders of SCMs. It turns out that the technique of canonical correlation analysis can be used to obtain structural specification for both approaches. For details of canonical correlation analysis, readers are referred to the traditional textbooks on multivariate statistical analysis, for example, Chapter 10 of Johnson and Wichern (2007). See also the review in Section 4.8.

In the literature, canonical correlation analysis has been used to identify Kronecker indices by Akaike (1976), Cooper and Wood (1982), and Tsay (1989). A canonical correlation ρ between two random vectors $\boldsymbol{P}$ and $\boldsymbol{F}$ can be obtained from the eigenvalue–eigenvector analysis:

$$\boldsymbol{\Sigma}_{pp}^{-1}\boldsymbol{\Sigma}_{pf}\boldsymbol{\Sigma}_{ff}^{-1}\boldsymbol{\Sigma}_{fp}\boldsymbol{v}_p = \rho^2\boldsymbol{v}_p, \quad \boldsymbol{\Sigma}_{ff}^{-1}\boldsymbol{\Sigma}_{fp}\boldsymbol{\Sigma}_{pp}^{-1}\boldsymbol{\Sigma}_{pf}\boldsymbol{v}_f = \rho^2\boldsymbol{v}_f, \tag{4.38}$$

where $\boldsymbol{\Sigma}_{fp} = \text{Cov}(\boldsymbol{F}, \boldsymbol{P})$ and other matrices are defined similarly, and $\boldsymbol{v}_f$ and $\boldsymbol{v}_p$ are eigenvectors associated with the eigenvalue ρ^2. The variable $X = \boldsymbol{v}_f'\boldsymbol{F}$ and $Y = \boldsymbol{v}_p'\boldsymbol{P}$ are the corresponding canonical variates. The canonical correlation ρ is the absolute value of the cross-correlation between X and Y, that is, $\rho = |\text{corr}(X, Y)|$. In practice, sample covariance matrices of $\boldsymbol{F}$ and $\boldsymbol{P}$ are used to perform canonical correlation analysis.

There exist other statistical methods for finding Kronecker indices and orders of SCM for a linear vector time series. As it will be seen later, the idea of the statistics used is to verify the rank of certain sample matrices of $\boldsymbol{z}_t$. We use canonical correlation analysis for simplicity, especially since the analysis is invariant to nonsingular linear transformation.

4.3.1 Reduced Rank Tests

Canonical correlations can be used to test for linear dependence between variables, especially when the random variables involved are normally distributed. Let p and f be the dimension of $\boldsymbol{P}$ and $\boldsymbol{F}$, respectively. Without loss of generality, we assume that $f \leq p$. Suppose that $\boldsymbol{X} = (\boldsymbol{F}', \boldsymbol{P}')'$ is a $(p+f)$-dimensional normal random vector and we are interested in testing the null hypothesis $H_0 : \boldsymbol{\Sigma}_{fp} = \mathbf{0}$. This is equivalent to testing for the independence between $\boldsymbol{P}$ and $\boldsymbol{F}$, which, in turn, is equivalent to testing $\boldsymbol{\beta} = \mathbf{0}$ in the multivariate linear regression

$$\boldsymbol{F}_i' = \boldsymbol{P}_i'\boldsymbol{\beta}' + \boldsymbol{E}_i', \quad i = 1, \cdots, T.$$

Under the regression framework, a natural test statistic to use is the likelihood ratio statistic, which can be written as

$$\text{LR} = -\left[T - 1 - \frac{1}{2}(p + f + 1)\right] \sum_{i=1}^{f} \ln(1 - \hat{\rho}_i^2),$$

where T is the sample size and $\hat{\rho}_1^2 \geq \hat{\rho}_2^2 \geq \cdots \geq \hat{\rho}_f^2$ are the ordered squared sample canonical correlations between $\boldsymbol{F}$ and $\boldsymbol{P}$. See, for instance, Johnson and Wichern (2007). Under the assumption of an independent random sample, the LR statistic has an asymptotic chi-square distribution with $p \times f$ degrees of freedom.

Suppose that the null hypothesis $H_0 : \boldsymbol{\Sigma}_{fp} = \mathbf{0}$ is rejected. It is then natural to examine the magnitude of the individual canonical correlations. Since the canonical correlations are ordered from the largest to the smallest, we can begin by assuming that only the smallest canonical correlation, in modulus, is zero and the remaining $(f-1)$ canonical correlations are nonzero. In other words, we are interested in testing

$$H_0 : \rho_{f-1}^2 > 0 \quad \text{and} \quad \rho_f^2 = 0 \quad \text{versus} \quad H_a : \rho_f^2 > 0. \tag{4.39}$$

Bartlett (1939) showed that the hypotheses in Equation (4.39) can be tested by the likelihood ratio criterion using the test statistic

$$C^* = -[T - 1 - \tfrac{1}{2}(p + f - 1)] \ln(1 - \hat{\rho}_f^2), \tag{4.40}$$

which is asymptotically a chi-square distribution with $(p-f+1)$ degrees of freedom. We shall use this likelihood ratio statistic in the sequel.

Let r be the rank of the matrix $\boldsymbol{\Sigma}_{fp}$. Then $0 \leq r \leq f$ under the assumption that $f \leq p$. The test in Equation (4.39) is amount to testing $H_0 : r = f - 1$ versus $H_a : r = f$. In general, we might be interested in testing

$$H_0 : \text{Rank}(\boldsymbol{\Sigma}_{fp}) = r \quad \text{versus} \quad H_a : \text{Rank}(\boldsymbol{\Sigma}_{fp}) > r. \tag{4.41}$$

This is equivalent to testing $H_0 : \rho_r^2 > 0$, $\rho_{r+1}^2 = 0$ versus $H_a : \rho_{r+1}^2 > 0$. Alternatively, it is to test the hull hypothesis that the $f - r$ smallest eigenvalues are 0. The test statistic is

$$\text{LR} = -[T - 1 - \frac{1}{2}(p + f - 1)] \sum_{i=r+1}^{f} \ln(1 - \hat{\rho}_i^2),$$

which follows asymptotically a chi-square distribution with $(p - r)(f - r)$ degrees of freedom.

4.4 FINDING KRONECKER INDICES

As discussed before, Kronecker indices are closely related to the rank and the row dependence of the Hankel matrix of $\boldsymbol{z}_t$. In practice, however, we can only entertain a finite-dimensional Hankel matrix. To this end, we approximate the past vector $\boldsymbol{P}_{t-1}$ by a truncated subset. Let $\boldsymbol{P}_{r,t-1} = (\boldsymbol{z}'_{t-1}, \cdots, \boldsymbol{z}'_{t-r})'$ be a subset of $\boldsymbol{P}_{t-1}$, where r is a properly chosen positive integer. In practice, r is either a prespecified integer or the order of a VAR model for $\boldsymbol{z}_t$ selected by an information criterion, for example, AIC or BIC. The latter choice of r can be justified because under the assumption that $\boldsymbol{z}_t$ follows a VARMA model the rank m of the Hankel matrix $\boldsymbol{H}_\infty$ is finite and, for an invertible vector time series, VAR models can provide good approximation.

Using the Toeplitz property of the Hankel matrix, we can search the Kronecker indices by examining the row dependence one by one starting from the first row. Specifically, we construct a subvector $\boldsymbol{F}_t^*$ of $\boldsymbol{F}_t$ by moving elements one by one from $\boldsymbol{F}_t$ into $\boldsymbol{F}_t^*$, starting with $\boldsymbol{F}_t^* = z_{1t}$. To check the row dependence of $\boldsymbol{H}_\infty$, we employ $\boldsymbol{P}_{r,t-1}$ and $\boldsymbol{F}_t^*$ and use the following procedure:

1. Suppose that the last element of $\boldsymbol{F}_t^*$ is $z_{i,t+h}$ with $h \geq 0$. Perform the canonical correlation analysis between $\boldsymbol{P}_{r,t-1}$ and $\boldsymbol{F}_t^*$. Let $\hat{\rho}$ be the smallest sample canonical correlation in modulus between $\boldsymbol{P}_{r,t-1}$ and $\boldsymbol{F}_t^*$. Let $x_{t+h} = \boldsymbol{v}'_f \boldsymbol{F}_t^*$ and $y_{t-1} = \boldsymbol{v}'_p \boldsymbol{P}_{r,t-1}$ be the corresponding canonical variates. The use of subscripts $t + h$ and $t - 1$ will be explained later.
2. Consider the null hypothesis $H_0 : \rho = 0$ versus the alternative hypothesis $H_a : \rho \neq 0$, where ρ is the smallest canonical correlation in modulus between $\boldsymbol{F}_t^*$ and $\boldsymbol{P}_{r,t-1}$. Here, it is understood that by the nature of the procedure, the second smallest canonical correlation between $\boldsymbol{P}_{r,t-1}$ and $\boldsymbol{F}_t^*$ is nonzero. Consequently, we employ a modified test statistic of Equation (4.40), namely,

$$C = -(T - r) \ln \left(1 - \frac{\hat{\rho}^2}{\hat{d}}\right), \tag{4.42}$$

where T is the sample size, r is prespecified positive integer for $\boldsymbol{P}_{r,t-1}$, and $\hat{d} = 1 + 2\sum_{j=1}^{h} \hat{\rho}_{xx}(j)\hat{\rho}_{yy}(j)$. In Equation (4.42), it is understood that $\hat{d} = 1$ if $h = 0$ and $\hat{\rho}_{xx}(j)$ and $\hat{\rho}_{yy}(j)$ are the lag-j sample autocorrelation coefficients of $\{x_t\}$ and $\{y_t\}$ series, respectively.

3. Compare the test statistic C of Equation (4.42) with a chi-square distribution with $kr - f + 1$ degrees of freedom, where kr and f are the dimensions of $\boldsymbol{P}_{r,t-1}$ and $\boldsymbol{F}_t^*$, respectively.
   (a) If the test statistic C is statistically significant, then there is no linearly dependent row found. Go to Step 4.
   (b) If the test statistic C is not significant, then $z_{i,t+h}$ gives rise to a linearly dependent row of the Hankel matrix. In this case, we find the Kronecker index $k_i = h$ for z_{it} and remove all elements $z_{i,t+s}$ with $s \geq h$ from $\boldsymbol{F}_t$.
4. If $\boldsymbol{F}_t$ reduces to an empty set, stop. Otherwise, augment $\boldsymbol{F}_t^*$ by the next available element of $\boldsymbol{F}_t$ and go to Step 1.

The asymptotic limiting distribution

$$C = -(T-r)\ln\left(1 - \frac{\hat{\rho}^2}{\hat{d}}\right) \sim \chi^2_{kr-f+1}$$

is shown in Tsay (1989). It is a modification of Bartlett test statistic in Equation (4.40). The basic idea of the modification is as follows. Since the last element of $\boldsymbol{F}_t^*$ is $z_{i,t+h}$, the canonical variate x_{t+h} is a linear function of $\{\boldsymbol{z}_{t+h}, \cdots, \boldsymbol{z}_t\}$. On the other hand, the other canonical variate y_{t-1} is a linear function of $\{\boldsymbol{z}_{t-1}, \cdots, \boldsymbol{z}_{t-r}\}$. Therefore, the time lag between x_{t+h} and y_{t-1} is $h+1$. Since the canonical correlation coefficient ρ is the cross-correlation between the canonical variates x_{t+h} and y_{t-1}, we can think of ρ as the lag-$(h+1)$ cross-correlation between the $\{x_t\}$ and $\{y_t\}$ series.

Let $\rho_{xy}(j)$ be the lag-j cross-correlation between x_t and y_t. Under the null hypothesis $H_0 : \rho_{xy}(\ell) = 0$, the asymptotic variance of the sample cross-correlation $\hat{\rho}_{xy}(\ell)$ is

$$\text{Var}\left[\hat{\rho}_{xy}(\ell)\right] \approx T^{-1} \sum_{v=-\infty}^{\infty} \{\rho_{xx}(v)\rho_{yy}(v) + \rho_{xy}(\ell+v)\rho_{yx}(\ell-v)\}. \tag{4.43}$$

See the properties of the sample cross-correlations discussed in Chapter 1. For canonical correlation analysis, under $H_0 : \rho = 0$, we have $\text{Cov}(x_{t+h}, \boldsymbol{P}_{t-1}) = \boldsymbol{0}$. Thus, $\text{Cov}(x_{t+h}, \boldsymbol{z}_{t-j}) = \boldsymbol{0}$ for all $j > 0$. Therefore, $\text{Cov}(x_t, x_{t-j}) = 0$ for all $j \geq h+1$ because x_{t-j} is a linear function of $\boldsymbol{P}_{t-1}$. Consequently, $\rho_{xx}(j) = 0$ for $j \geq h+1$ and the x_t series is an MA(h) process. Using this fact and Equation (4.43), the asymptotic variance of the sample canonical correlation coefficient $\hat{\rho}$ is $\text{Var}(\hat{\rho}) = T^{-1}[1 + 2\sum_{v=1}^{h} \rho_{xx}(v)\rho_{yy}(v)] \equiv \hat{d}$. The test statistic in Equation (4.42) uses the normalized squared canonical correlation instead of the ordinary canonical correlation. The normalization is to take care of the serial correlations in the data. If the data are from a random sample, then there are no serial correlations in the canonical variates and $\hat{d}$ reduces to one, which is the original form used in the literature.

Remark: In the prior development, we assume the time series $\boldsymbol{z}_t$ is homogeneous. If $\boldsymbol{z}_t$ has conditional heteroscedasticity, then the asymptotic variance of $\hat{\rho}$ involves the fourth moments of the canonical variates $\{x_t\}$ and $\{y_t\}$. One must modify the test statistic C accordingly to maintain the limiting chi-square distribution; see Min and Tsay (2005) and Tsay and Ling (2008). □

4.4.1 Application

To illustrate, we consider the logarithms of indices of monthly flour prices in three U.S. cities over the period from August 1972 to November 1980. The cities are Buffalo, Minneapolis, and Kansas City. This data set was analyzed in Tiao and Tsay (1989) and have 100 observations. Figure 4.1 shows the time plots of the three series, and they seem to move in unison. Based on the augmented Dickey–Fuller unit-root test with AR order 3, all three series have a unit root. However, Johansen's cointegration test fails to reject the null hypothesis of no cointegration. The test results are given in Table 4.2. Unit-root and cointegration tests are discussed in Chapter 5. We give the test results here simply because the structural specification methods discussed in this chapter are in effect applicable to unit-root nonstationary time series, even though the Hankel matrix is strictly speaking defined only for stationary time series.

Next, turn to specifying the Kronecker indices of the data. Let $\boldsymbol{z}_t$ be the three-dimensional time series under study. If VAR models are entertained, a VAR(2) model is selected by either the sequential chi-square test of Tiao and Box (1981) or the AIC criterion. Thus, one can use $r \geq 2$ to approximate the past vector $\boldsymbol{P}_{t-1}$, and we choose $r = 3$, that is, $\boldsymbol{P}_{3,t-1} = (\boldsymbol{z}'_{t-1}, \boldsymbol{z}'_{t-2}, \boldsymbol{z}'_{t-3})'$. Following the procedure outlined in the section, we obtain the Kronecker indices as $\{k_1 = 1, k_2 = 1, k_3 = 1\}$. Details of the test statistic C of Equation (4.42) and the associated smallest squared canonical correlations are given in Table 4.3. Based on the Kronecker indices, a VARMA(1,1) model is specified for the data.

Remark: Unit-root test can be carried out in R using the command `adfTest` of the package `fUnitRoots`. Cointegration tests, on the other hand, can be performed using the command `ca.jo` of the `urca` package. The Kronecker index approach to specify and estimate a VARMA model can be carried out via the `MTS` package using commands `Kronid`, `Kronfit`, and `refKronfit`, respectively. We demonstrate `Kronid` next. The estimation will be shown later. □

R Demonstration: Specification of Kronecker indices

```
> zt=read.table("flourc.txt")
> Kronid(zt,plag=3)
h =  0
Component =  1
square of the smallest can. corr. =  0.9403009
    test,   df, &  p-value:
[1] 266.342   9.000   0.000
```

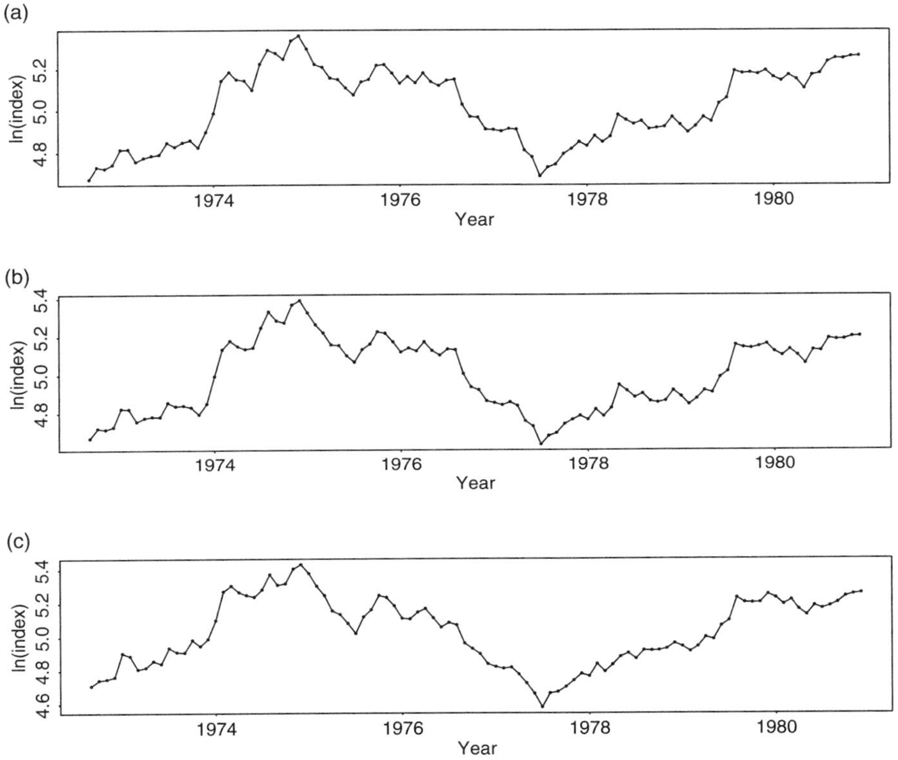

FIGURE 4.1 Time plots of the logarithms of monthly indices of flour prices from August 1972 to November 1980 in three U.S. cities. The cities are (a) Buffalo (b) Minneapolis, and (c) Kansas City.

TABLE 4.2 Unit-Root and Cointegration Tests for the Logarithms of Indices of Monthly Flour Prices in Buffalo, Minneapolis, and Kansas City

(a) Univariate Unit-Root Tests

Series	Test	p-Value	Cons.	AR Coefficients		
z_{1t}	−1.87	0.35	0.28	−0.054	0.159	0.048
z_{2t}	−1.84	0.36	0.26	−0.051	0.258	−0.046
z_{3t}	−1.70	0.43	0.24	−0.047	0.206	−0.005

(b) Johansen Cointegration Tests

Rank	Eigenvalue	Trace	95% CV	Max. Stat	95% CV
0	0.142	22.65	29.68	14.97	20.97
1	0.063	7.68	15.41	6.37	14.07
2	0.013	1.31	2.76	1.31	3.76

TABLE 4.3 Specification of Kronecker Indices for the Three-Dimensional Series of Logarithms of Monthly Flour Price Indices from August 1972 to November 1980

Last Element	Small Eigenvalue	Test	Degrees of Freedom	p-Value	Remark
z_{1t}	0.940	266.3	9	0	
z_{2t}	0.810	156.3	8	0	
z_{3t}	0.761	134.0	7	0	
$z_{1,t+1}$	0.045	4.28	6	0.64	$k_1 = 1$
$z_{2,t+1}$	0.034	3.04	6	0.80	$k_2 = 1$
$z_{3,t+1}$	0.027	2.44	6	0.88	$k_3 = 1$

The past vector is $\boldsymbol{P}_{t-1} = (\boldsymbol{z}'_{t-1}, \boldsymbol{z}'_{t-2}, \boldsymbol{z}'_{t-3})'$ and last element denotes the last element of the future subvector.

```
Component =  2
square of the smallest can. corr. =  0.8104609
    test,   df, &  p-value:
[1] 156.337   8.000   0.000
Component =  3
square of the smallest can. corr. =  0.761341
    test,   df, &  p-value:
[1] 133.959   7.000   0.000
=============
h =  1
Component =  1
Square of the smallest can. corr. =  0.04531181
    test,    df, p-value & d-hat:
[1] 4.281 6.000 0.639 1.007
A Kronecker found
Component =  2
Square of the smallest can. corr. =  0.03435539
    test,    df, p-value & d-hat:
[1] 3.045 6.000 0.803 1.067
A Kronecker found
Component =  3
Square of the smallest can. corr. =  0.02732133
    test,    df, p-value & d-hat:
[1] 2.435 6.000 0.876 1.057
A Kronecker found
============
    Kronecker indexes identified:
[1] 1 1 1
```

4.5 FINDING SCALAR COMPONENT MODELS

We adopt the method of Tiao and Tsay (1989) to identify the SCMs for $\boldsymbol{z}_t$. For the linear process $\boldsymbol{z}_t$ of Equation (4.1), we define an extended $k(m+1)$-dimensional vector process $\boldsymbol{Y}_{m,t}$ as

$$\boldsymbol{Y}_{m,t} = (\boldsymbol{z}_t', \boldsymbol{z}_{t-1}', \cdots, \boldsymbol{z}_{t-m}')', \tag{4.44}$$

where m is a non-negative integer. To search for SCMs of $\boldsymbol{z}_t$, Tiao and Tsay (1989) employ a two-way table of covariance matrices of the $\boldsymbol{Y}_{m,t}$ series. Specifically, given (m, j), where $m \geq 0$ and $j \geq 0$, consider the covariance matrix

$$\begin{aligned}\boldsymbol{\Gamma}(m,j) &= \text{Cov}(\boldsymbol{Y}_{m,t}, \boldsymbol{Y}_{m,t-j-1}) \\ &= \begin{bmatrix} \boldsymbol{\Gamma}_{j+1} & \boldsymbol{\Gamma}_{j+2} & \boldsymbol{\Gamma}_{j+3} & \cdots & \boldsymbol{\Gamma}_{j+1+m} \\ \boldsymbol{\Gamma}_{j} & \boldsymbol{\Gamma}_{j+1} & \boldsymbol{\Gamma}_{j+2} & \cdots & \boldsymbol{\Gamma}_{j+m} \\ \vdots & \vdots & \vdots & \ddots & \vdots \\ \boldsymbol{\Gamma}_{j+1-m} & \boldsymbol{\Gamma}_{j+2-m} & \boldsymbol{\Gamma}_{j+3-m} & \cdots & \boldsymbol{\Gamma}_{j+1} \end{bmatrix},\end{aligned} \tag{4.45}$$

which is a $k(m+1) \times k(m+1)$ matrix of autocovariance matrices of $\boldsymbol{z}_t$. The key to understand SCMs is to study the impact of a SCM on the singularity of the matrices $\boldsymbol{\Gamma}(m, j)$ for $m \geq 0$ and $j \geq 0$. This is equivalent to making use of the moment equations of a stationary VARMA models discussed in Chapter 3.

From the definition, $\boldsymbol{\Gamma}(m, j)$ is a square matrix. This is purely for simplicity. As a matter of fact, Tiao and Tsay (1989) define a general matrix

$$\boldsymbol{\Gamma}(m,h,j) = \text{Cov}(\boldsymbol{Y}_{m,t}, \boldsymbol{Y}_{h,t-j-1}) = E(\boldsymbol{Y}_{m,t}\boldsymbol{Y}_{h,t-j-1}'),$$

where $h \geq m$. All the results discussed later continue to hold if $\boldsymbol{\Gamma}(m, j)$ of Equation (4.45) is replaced by $\boldsymbol{\Gamma}(m, h, j)$ with $h \geq m$.

4.5.1 Implication of Scalar Component Models

Let $\boldsymbol{u}$ be a $k(m+1)$-dimensional row vector of real numbers. We said that $\boldsymbol{u}$ is a *singular left vector* of $\boldsymbol{\Gamma}(m, j)$ if $\boldsymbol{u}\boldsymbol{\Gamma}(m, j) = \boldsymbol{0}$. We use row vector instead of the conventional column vector for simplicity in notation.

Suppose that $y_t = \boldsymbol{v}_0\boldsymbol{z}_t$ is an SCM(r, s) of $\boldsymbol{z}_t$. By definition, there exist r k-dimensional row vector $\boldsymbol{v}_i$ $(i = 1, \cdots, r)$ such that $w_t = \sum_{i=0}^{r} \boldsymbol{v}_i\boldsymbol{z}_{t-i}$ is uncorrelated with $\boldsymbol{a}_{t-j}$ for $j > s$. Consequently, postmultiplying the SCM structure in Definition 4.3 by $\boldsymbol{z}_{t-j}'$ and taking expectation, we have

$$\sum_{i=0}^{r} \boldsymbol{v}_i\boldsymbol{\Gamma}_{j-i} = \boldsymbol{0}, \quad \text{for} \quad j > s. \tag{4.46}$$

Let $\boldsymbol{v} = (\boldsymbol{v}_0, \cdots, \boldsymbol{v}_r)$ be the $k(r+1)$-dimensional row vector consisting of all $\boldsymbol{v}_i$s of the SCM. Then, we have

$$w_t = \boldsymbol{v}\boldsymbol{Y}_{r,t}, \tag{4.47}$$

where $\boldsymbol{Y}_{r,t}$ is defined in Equation (4.44) with m replaced by r. By Equation (4.46), the existence of $y_t \sim$ SCM(r, s) implies that the matrix $\boldsymbol{\Gamma}(r, j)$ is singular for $j \geq s$

and $\boldsymbol{v}$ is the corresponding singular left vector. Furthermore, assume that the order (r, s) satisfies the condition that $r + s$ is as small as possible. Then, the row vector $\boldsymbol{v}$ of y_t is not a singular left vector of $\boldsymbol{\Gamma}(r, s-1)$; otherwise, the SCM order can be reduced to $(r, s-1)$.

To aid further discussion, we define an extended $k(\ell+1)$-dimensional row vector of $\boldsymbol{v}$ as

$$\boldsymbol{v}(\ell, r, g) = (\boldsymbol{0}_{1g}, \boldsymbol{v}, \boldsymbol{0}_{2g}), \tag{4.48}$$

where ℓ and g are integers such that $\ell \geq r$ and $g > 0$, r is associated with the original row vector $\boldsymbol{v}$, and $\boldsymbol{0}_{1g}$ and $\boldsymbol{0}_{2g}$ are respectively $k(g-1)$-dimensional and $k(\ell+1-r-g)$-dimensional row vector of zeros. For instance, $\boldsymbol{v}(r+1, r, 1) = (\boldsymbol{v}, 0, \cdots, 0)$ is a $k(r+2)$-dimensional row vector and $\boldsymbol{v}(r, r, 1) = \boldsymbol{v}$.

Next, consider the matrices $\boldsymbol{\Gamma}(m, s)$ with $m > r$. Using the definition of Equation (4.48), we have

$$w_t = \boldsymbol{v}\boldsymbol{Y}_{r,t} = \boldsymbol{v}(m, r, 1)\boldsymbol{Y}_{m,t}.$$

Consequently, the existence of $y_t \sim$ SCM(r, s) implies that the matrix $\boldsymbol{\Gamma}(m, s)$ is singular for $m > r$. Furthermore, it is easy to see that y_t does not imply any singularity of $\boldsymbol{\Gamma}(r-1, s)$; otherwise, y_t would be an SCM of order $(r-1, s)$.

Finally, consider the matrices $\boldsymbol{\Gamma}(m, j)$ with $m > r$ and $j > s$. First, let us focus on the matrix $\boldsymbol{\Gamma}(r+1, s+1)$ given by

$$\boldsymbol{\Gamma}(r+1, s+1) = E(\boldsymbol{Y}_{r+1,t}\boldsymbol{Y}'_{r+1,t-s-2}).$$

In this particular instance, the scalar component y_t introduces two singular left vectors because

(a) $w_t = \boldsymbol{v}(r+1, r, 1)\boldsymbol{Y}_{r+1,t}$ is uncorrelated with $\boldsymbol{a}_{t-j}$ for $j > s$
(b) $w_{t-1} = \boldsymbol{v}(r+1, r, 2)\boldsymbol{Y}_{r+1,t}$ is uncorrelated with $\boldsymbol{a}_{t-j}$ for $j > s+1$

Thus, a SCM(r, s) y_t gives rise to two singular left vectors of the matrix $\boldsymbol{\Gamma}(r+1, s+1)$. In other words, moving from the $\boldsymbol{\Gamma}(m, s)$ matrix to the $\boldsymbol{\Gamma}(r+1, s+1)$ matrix increases the number of singular left vectors by one. By the same argument, y_t also introduces two singular left vectors for $\boldsymbol{\Gamma}(r+1, j)$ with $j \geq s+1$ and for $\boldsymbol{\Gamma}(m, s+1)$ with $m \geq r+1$. In general, for the matrix $\boldsymbol{\Gamma}(m, j)$ with $m > r$ and $j > s$, the SCM y_t introduces $h = \min\{m-r+1, j-s+1\}$ singular left vectors.

We summarize the results of previous discussions into a theorem.

Theorem 4.2 Suppose that $\boldsymbol{z}_t$ is a stationary linear process of Equation (4.1) and its Hankel matrix is of finite dimension. Suppose also that $y_t = \boldsymbol{v}_0\boldsymbol{z}_t$ follows an SCM(r, s) structure with the associated row vector $\boldsymbol{v}$. Let $\boldsymbol{v}(\ell, r, g)$ be the extended row vector of $\boldsymbol{v}$ defined in Equation (4.48). Then,

TABLE 4.4 Numbers of Singular Left Vectors of the $\Gamma(m,j)$ Matrices Induced by (a) a Scalar Component Model $y_{1t} \sim$ SCM(1,0) and (b) a Scalar Component Model $y_{2t} \sim$ SCM(0,1)

(a)				j				(b)				j			
m	0	1	2	3	4	5	⋯	m	0	1	2	3	4	5	⋯
0	0	0	0	0	0	0	⋯	0	0	1	1	1	1	1	⋯
1	1	1	1	1	1	1	⋯	1	0	1	2	2	2	2	⋯
2	1	2	2	2	2	2	⋯	2	0	1	2	3	3	3	⋯
3	1	2	3	3	3	3	⋯	3	0	1	2	3	4	4	⋯
4	1	2	3	4	4	4	⋯	4	0	1	2	3	4	5	⋯
⋮	⋮	⋮	⋮					⋮	⋮	⋮	⋮				

(a) For $j \geq s$, $\boldsymbol{v}$ is a singular left vector of $\boldsymbol{\Gamma}(r,j)$.
(b) For $m > r$, $\boldsymbol{v}(m,r,1)$ is a singular left vector of $\boldsymbol{\Gamma}(m,s)$.
(c) For $m > r$ and $j > s$, $\boldsymbol{\Gamma}(m,j)$ has $h = \min\{m-r+1, j-s+1\}$ singular left vectors, namely, $\boldsymbol{v}(m,r,g)$ with $g=1,\cdots,h$.
(d) For $m < r$ and $j < s$, the vectors $\boldsymbol{v}(m,r,g)$ are not singular left vectors of $\boldsymbol{\Gamma}(m,j)$.

Example 4.2 Suppose $\boldsymbol{z}_t$ is a linear vector process of Equation (4.1) and y_{1t} is a SCM(1,0) of $\boldsymbol{z}_t$ and y_{2t} is an SCM(0,1) of $\boldsymbol{z}_t$. Also, suppose that y_{it} are not a SCM(0,0). Then, the number of singular left vectors of $\boldsymbol{\Gamma}(m,j)$ induced by y_{it} is given in parts (a) and (b) of Table 4.4, respectively. The diagonal increasing pattern is clearly seen from the table. In addition, the coordinates (m,j) of the upper-left vertex of nonzero entries correspond exactly to the order of the SCM. □

4.5.2 Exchangeable Scalar Component Models

There are cases in which an SCM of $\boldsymbol{z}_t$ has two different orders (p_1,q_1) and (p_2,q_2) such that $p_1+q_1=p_2+q_2$. For example, consider the VAR(1) and VMA(1) model

$$\boldsymbol{z}_t - \begin{bmatrix} 0 & -2 \\ 0 & 0 \end{bmatrix} \boldsymbol{z}_{t-1} = \boldsymbol{a}_t \quad \Leftrightarrow \quad \boldsymbol{z}_t = \boldsymbol{a}_t - \begin{bmatrix} 0 & 2 \\ 0 & 0 \end{bmatrix} \boldsymbol{a}_{t-1}. \tag{4.49}$$

In this particular case, the scalar series $y_t = (1,0)\boldsymbol{z}_t = z_{1t}$ is both SCM(1,0) and SCM(0,1). This type of SCM orders are referred to as *exchangeable orders* in Tiao and Tsay (1989). SCMs with exchangeable orders have some special properties.

Lemma 4.3 Suppose that $\boldsymbol{z}_t$ is a k-dimensional linear process of Equation (4.1) and y_t is both SCM(p_1,q_1) and SCM(p_2,q_2) of $\boldsymbol{z}_t$, where $p_1+q_1=p_2+q_2$.

Then, there exists a SCM(p_3, q_3) x_t such that $p_3 < p_0 = \max\{p_1, p_2\}$ and $q_3 < q_0 = \max\{q_1, q_2\}$.

Proof. From $y_t \sim$ SCM(p_1, q_1), we have the structure

$$w_t = \boldsymbol{v}_0 \boldsymbol{z}_t + \sum_{i=1}^{p_1} \boldsymbol{v}_i \boldsymbol{z}_{t-i} = \boldsymbol{v}_0 \boldsymbol{a}_t + \sum_{j=1}^{q_1} \boldsymbol{u}_j \boldsymbol{a}_{t-j}.$$

But y_t is also SCM(p_2, q_2) so that there exist p_2 k-dimensional row vectors $\boldsymbol{v}_i^*$ ($i = 1, \cdots, p_2$) and q_2 k-dimensional row vectors $\boldsymbol{u}_j^*$ ($j = 1, \cdots, q_2$) such that

$$w_t^* = \boldsymbol{v}_0 \boldsymbol{z}_t + \sum_{i=1}^{p_2} \boldsymbol{v}_i^* \boldsymbol{z}_{t-i} = \boldsymbol{v}_0 \boldsymbol{a}_t + \sum_{j=1}^{q_2} \boldsymbol{u}_j^* \boldsymbol{a}_{t-j}.$$

Consequently, by subtraction, we obtain

$$\sum_{i=1}^{p_0} \boldsymbol{\delta}_i \boldsymbol{z}_{t-i} = \sum_{j=1}^{q_0} \boldsymbol{\varpi}_j \boldsymbol{a}_{t-j},$$

where some $\boldsymbol{\delta}_i$ is nonzero. Suppose that $\boldsymbol{\delta}_d \neq \boldsymbol{0}$ and $\boldsymbol{\delta}_i = \boldsymbol{0}$ for $i < d$. Then, by the linearity of $\boldsymbol{z}_t$, we also have $\boldsymbol{\varpi}_i = \boldsymbol{0}$ for $i < d$ and $\boldsymbol{\delta}_d = \boldsymbol{\varpi}_d$. Let $x_t = \boldsymbol{\delta}_d \boldsymbol{z}_t$. The prior equation shows that x_t is SCM(p_3, q_3) with $p_3 < p_0$ and $q_3 < q_0$. This completes the proof. $\square$

As an illustration, consider the VAR(1) or VMA(1) model in Equation (4.49). There is indeed a SCM(0,0) in the system as shown by the lemma.

Let $\boldsymbol{v}$ be the row vector associated with $y_t \sim$ SCM(p_1, q_1) and $\boldsymbol{v}^*$ be the row vector associated with $y_t \sim$ SCM(p_2, q_2). Obviously, the first k elements of both $\boldsymbol{v}$ and $\boldsymbol{v}^*$ are $\boldsymbol{v}_0$. Let $\boldsymbol{\delta}$ be the row vector associated with $x_t \sim$ SCM(p_3, q_3). Then, from the proof of Lemma 4.3, we have the following result.

Lemma 4.4 Suppose y_t is an SCM of the linear vector process $\boldsymbol{z}_t$ of Equation (4.1) with exchangeable orders (p_1, q_1) and (p_2, q_2) such that $p_1 + q_1 = p_2 + q_2$ and $p_1 > p_2$. Let x_t be the implied SCM(p_3, q_3) and denote the row vector associated with x_t by $\boldsymbol{\delta}$. Furthermore, let $p_0 = \max(p_1, p_2)$.Then,

$$\boldsymbol{v}(p_0, p_1, 1) = \boldsymbol{v}^*(p_0, p_2, 1) + \sum_{j=1}^{h} \eta_j \boldsymbol{u}(p_0, p_3, j+1),$$

where $h = p_1 - p_2$ and $\{\eta_j | j = 1, \cdots, h\}$ are constants such that $\eta_j \neq 0$ for some $j > 0$.

TABLE 4.5 Numbers of Singular Left Vectors of the $\Gamma(m,j)$ Matrices Induced by a Scalar Component Model y_t with Exchangeable Orders (0,1) and (1,0)

	j						
m	0	1	2	3	4	5	...
0	0	1	1	1	1	1	...
1	1	1	2	2	2	2	...
2	1	2	2	3	3	3	...
3	1	2	3	3	4	4	...
4	1	2	3	4	4	5	...
5	1	2	3	4	5	5	...
⋮	⋮	⋮	⋮	⋮	⋮	⋮	

The result of Lemma 4.4 can be extended to the case where y_t has more than two exchangeable orders. Using Lemma 4.4, we can extend Theorem 4.2 to include cases in which some SCMs have exchangeable orders.

Theorem 4.3 Suppose the scalar component y_t has the exchangeable orders stated in Lemma 4.4. Let $h_1 = \min\{m - p_1 + 1, j - q_1 + 1\}$ and $h_2 = \min\{m - p_2 + 1, j - q_2 + 1\}$. Then, the number of singular left vectors of $\boldsymbol{\Gamma}(m, j)$ induced by y_t is $\max\{h_1, h_2\}$.

Example 4.3 Suppose $\boldsymbol{z}_t$ is a linear vector process of Equation (4.1) and y_t is an SCM of $\boldsymbol{z}_t$ with exchangeable orders (0,1) and (1,0). In addition, y_t is not an SCM(0,0) of $\boldsymbol{z}_t$. Then, the number of singular left vectors of $\boldsymbol{\Gamma}(m, j)$ induced by y_{it} is given in Table 4.5. In this case, the diagonal increasing pattern continues to hold, and the coordinates (m, j) of the two upper-left vertexes of nonzero entries correspond exactly to the orders of the SCM. □

Consider Table 4.4 and Table 4.5. Let $N(m, j)$ be the number of singular left vectors of $\boldsymbol{\Gamma}(m, j)$ induced by the SCM y_t. Define the diagonal difference as

$$d(m, j) = \begin{cases} N(m, j) & \text{if } m = 0 \text{ or } j = 0 \\ N(m, j) - N(m - 1, j - 1) & \text{otherwise.} \end{cases}$$

Then, it is easily seen that all three differenced tables consist of "0" and "1" with the upper-left vertex of the "1" locates precisely at the SCM order. For the case of exchangeable orders, there are two upper-left vertexes of "1" in the differenced table. See Table 4.6.

The special feature discussed earlier concerning the number of singular left vectors of $\boldsymbol{\Gamma}(m, j)$ holds for each SCM(r, s) of $\boldsymbol{z}_t$ provided that the sum $r + s$ is as small as possible. This feature is a consequence of Theorems 4.2 and 4.3. In fact, the converse of Theorems 4.2 and 4.3 also hold. For instance, if there exists a $k(r + 1)$-dimensional row vector $\boldsymbol{v}$ whose first k elements are not all zero such that $\boldsymbol{v}$ and

TABLE 4.6 Diagonal Differences of the Numbers of Singular Left Vectors of the $\Gamma(m,j)$ Matrices Induced by a Scalar Component Model y_t with Exchangeable Orders (0,1) and (1,0)

	j						
m	0	1	2	3	4	5	$\cdots$
0	0	1	1	1	1	1	$\cdots$
1	1	1	1	1	1	1	$\cdots$
2	1	1	1	1	1	1	$\cdots$
3	1	1	1	1	1	1	$\cdots$
4	1	1	1	1	1	1	$\cdots$
5	1	1	1	1	1	1	$\cdots$
$\vdots$	$\vdots$	$\vdots$	$\vdots$	$\vdots$	$\vdots$	$\vdots$	$\ddots$

its extended vector $\boldsymbol{v}(\ell, r, g)$ of Equation (4.48) have the properties (a) to (d) of Theorem 4.2, then $y_t = \boldsymbol{v}_0 \boldsymbol{z}_t$ is an SCM of order (r, s), where $\boldsymbol{v}_0$ is the subvector of $\boldsymbol{v}$ consisting of its first k elements. We shall use these results to find SCMs of $\boldsymbol{z}_t$.

4.5.3 Searching for Scalar Components

The objective here is to search for k linearly independent SCMs of the linear vector process $\boldsymbol{z}_t$, say SCM(p_i, q_i) $(i = 1, \cdots, k)$, such that $p_i + q_i$ are as small as possible. To this end, let $\ell = m + j$ and we study the number of singular left vectors of $\boldsymbol{\Gamma}(m, j)$ using the following sequence:

1. Start with $\ell = 0$, that is, $(m, j) = (0, 0)$
2. Increase ℓ by 1 and for a fixed ℓ,
   - **(a)** Start with $j = 0$ and $m = \ell$
   - **(b)** Increase j by 1 until $j = \ell$

For a given order (m, j), we perform the canonical correlation analysis between $\boldsymbol{Y}_{m,t}$ and $\boldsymbol{Y}_{m,t-j-1}$ to identify the number of singular left vectors of $\boldsymbol{\Gamma}(m, j)$, which turns out to be the number of zero canonical correlations between the two extended random vectors. As in the Kronecker index case, the likelihood ratio test statistic can be used. Specifically, let $\hat{\lambda}_i(j)$ be the i smallest squared canonical correlation between $\boldsymbol{Y}_{m,t}$ and $\boldsymbol{Y}_{m,t-j-1}$, where $i = 1, \cdots, k(m+1)$. To test that there are s zero canonical correlations, Tiao and Tsay (1989) use the test statistic

$$C(j, s) = -(T - m - j) \sum_{i=1}^{s} \ln \left[1 - \frac{\hat{\lambda}_i(j)}{d_i(j)} \right], \tag{4.50}$$

where T is the sample size and $d_i(j)$ is defined as

$$d_i(j) = 1 + 2\sum_{u=1}^{j} \hat{\rho}_u(w_{1t})\hat{\rho}_u(w_{2t}),$$

where $\hat{\rho}_u(w_t)$ is the lag-u sample autocorrelation of the scalar time series w_t, and w_{1t} and w_{2t} are the two canonical variates associated with the canonical correlation $\hat{\lambda}_i(j)$. Under the null hypothesis that there are exactly s zero canonical correlations between $\boldsymbol{Y}_{m,t}$ and $\boldsymbol{Y}_{m,t-j-1}$, the test statistic $C(j, s)$ is asymptotically a chi-square random variable with s^2 degrees of freedom provided that the innovations $\boldsymbol{a}_t$ of Equation (4.1) are multivariate Gaussian. Note that if one uses $\boldsymbol{Y}_{m,t}$ and $\boldsymbol{Y}_{h,t-j-1}$, with $h \geq m$, to perform the canonical correlation analysis, then the degrees of freedom of $C(m, j)$ become $s[(h - m)k + s]$.

In the searching process, once a new SCM(p_i, q_i) y_{it} is found, we must use the results of Theorems 4.2 and 4.3 to remove the singular left vectors of $\boldsymbol{\Gamma}(m, j)$ induced by y_{it} in any subsequent analysis. The search process is terminated when k linearly independent SCMs are found. See Tiao and Tsay (1989) for further details.

4.5.4 Application

Again, we use the logarithms of monthly indices of flour prices in Buffalo, Minneapolis, and Kansas City to demonstrate the analysis. First, we use the test statistics $C(m, j)$ of Equation (4.50) to check the number of zero canonical correlations between the extended vectors $\boldsymbol{Y}_{m,t}$ and $\boldsymbol{Y}_{m,t-j-1}$. The results are given in parts (a) and (b) of Table 4.7. From the table, it is seen that a VARMA(1,1) or VAR(2) is specified for the data. The only minor deviation occurs at $(m, j) = (5, 4)$ with 14 zero canonical correlations. In theory, the number of zero canonical correlations should be 15 for a three-dimensional VARMA(1,1) model. Such a minor deviation can easily happen in real applications.

To gain further insight, we consider the eigenvalues and test statistics for some lower-order (m, j) configurations. The results are given in part (c) of Table 4.7. Here, we use $\boldsymbol{Y}_{m,t}$ and $\boldsymbol{Y}_{m,t-j-1}$ to perform the canonical correlation analysis. The degrees of freedom of the $C(m, j)$ statistics of Equation (4.50) are then s^2, where s is the number of zero canonical correlations under the null hypothesis. From the table, we see that there are two SCM(1,0) components and one SCM(1,1) component when an overall VARMA(1,1) model is specified. The eigenvectors associated with (estimated) zero eigenvalues at the (1,0) and (1,1) positions can be used to obtain an estimate of the transformation matrix:

$$\hat{\boldsymbol{T}} = \begin{bmatrix} 0.33 & -0.63 & 0.26 \\ -0.34 & 0.13 & 0.59 \\ 0.14 & 0.20 & 0.00 \end{bmatrix}. \tag{4.51}$$

If an overall VARMA(2,0) is specified, one can also obtain details of the SCM analysis, including an estimate of transformation matrix. Details are given in the R demonstration.

TABLE 4.7 Analysis of the Three-Dimensional Monthly Indices of Flour Price Using the Scalar Component Method

(a) Number of Singular Left Vectors							(b) Diagonal Difference						
	j							j					
m	0	1	2	3	4	5	m	0	1	2	3	4	5
0	0	0	0	0	0	0	0	0	0	0	0	0	0
1	2	3	3	3	3	3	1	2	3	3	3	3	3
2	3	6	6	6	6	6	2	3	3	3	3	3	3
3	3	6	9	9	9	9	3	3	3	3	3	3	3
4	3	6	9	12	12	12	4	3	3	3	3	3	3
5	3	6	9	12	14	15	5	3	3	3	3	2	3

(c) Summary of Eigenvalues and Test Statistics

m	j	Eigenvalue	$C(m,j)$	Degrees of Freedom	p-Value
0	0	0.747	137.55	1	0.000
0	0	0.815	306.06	4	0.000
0	0	0.938	584.25	9	0.000
1	0	0.003	0.29	1	0.588
1	0	0.081	8.67	4	0.070
1	0	0.271	39.92	9	0.000
0	1	0.598	90.22	1	0.00
0	1	0.799	249.03	4	0.00
0	1	0.866	447.87	9	0.00
1	1	0.003	0.34	1	0.558
1	1	0.033	3.59	4	0.464
1	1	0.057	7.49	9	0.586
1	1	0.787	158.92	16	0.000
1	1	0.874	362.13	25	0.000
1	1	0.944	644.97	36	0.000

The 1% significance level is used.

Remark: The SCM approach can be carried out using the commands `SCMid` and `SCMid2` of the `MTS` package, where SCMid is used to identify an overall model, whereas SCMid2 is used to obtain details of the tests. See Tiao and Tsay (1989) for the two-step order specification. One can directly uses `SCMid2` for a given overall order. □

R Demonstration: SCM analysis of flour data.

```
> SCMid(da,crit=0.01)   <== Find overall order
Column: MA order
```

```
Row   : AR order
Number of zero canonical correlations
   0  1  2  3  4  5
0  0  0  0  0  0  0
1  2  3  3  3  3  3
2  3  6  6  6  6  6
3  3  6  9  9  9  9
4  3  6  9 12 12 12
5  3  6  9 12 14 15
Diagonal Differences:
  0 1 2 3 4 5
0 0 0 0 0 0 0
1 2 3 3 3 3 3
2 3 3 3 3 3 3
3 3 3 3 3 3 3
4 3 3 3 3 3 3
5 3 3 3 3 2 3
> m2=SCMid2(da,maxp=1,maxq=1,crit=0.01) <= Details of
  VARMA(1,1) model
For (pi,qi) = ( 0 , 0 )
Tests:
     Eigvalue St.dev    Test deg p-value
[1,]    0.747      1 137.548   1       0
[2,]    0.815      1 306.056   4       0
[3,]    0.938      1 584.245   9       0
Summary:
Number of SCMs detected:  0
Cumulative SCMs found 0
For (pi,qi) = ( 1 , 0 )
Tests:
     Eigvalue St.dev   Test deg p-value
[1,]    0.003      1  0.294   1   0.588
[2,]    0.081      1  8.672   4   0.070
[3,]    0.271      1 39.922   9   0.000
Summary:
Number of SCMs detected:  2
Found  2  new SCMs
Updated transformation matrix:
      [,1]   [,2]
V1  0.332 -0.338
V2 -0.628  0.125
V3  0.260  0.588
Cumulative SCMs found 2
For (pi,qi) = ( 0 , 1 )
Tests:
     Eigvalue St.dev    Test deg p-value
[1,]    0.598      1  90.222   1       0
[2,]    0.799      1 249.032   4       0
[3,]    0.866      1 447.866   9       0
```

```
Summary:
Number of SCMs detected:  0
Cumulative SCMs found 2
For (pi,qi) = ( 1 , 1 )
Tests:
     Eigvalue St.dev    Test deg p-value
[1,]    0.003  0.998   0.343   1   0.558
[2,]    0.033  1.022   3.589   4   0.464
[3,]    0.057  1.448   7.491   9   0.586
[4,]    0.787  1.000 158.919  16   0.000
[5,]    0.874  1.000 362.131  25   0.000
[6,]    0.944  1.000 644.965  36   0.000
Summary:
Number of SCMs detected:  3
Found  1  new SCMs
Transpose of Transformation-matrix:
      [,1]   [,2]  [,3]
V1  0.332 -0.338 0.136
V2 -0.628  0.125 0.200
V3  0.260  0.588 0.004
Cumulative SCMs found 3
> names(m2)
[1] "Tmatrix"  "SCMorder"
> print(round(m2$Tmatrix,3))
         V1     V2    V3
[1,]  0.332 -0.628 0.260
[2,] -0.338  0.125 0.588
[3,]  0.136  0.200 0.004

> SCMid2(da,maxp=2,maxq=0,crit=0.01) # Details of VARMA(2,0)
  model
For (pi,qi) = ( 0 , 0 )
Tests:
     Eigvalue St.dev    Test deg p-value
[1,]    0.747      1 137.548   1       0
[2,]    0.815      1 306.056   4       0
[3,]    0.938      1 584.245   9       0
Summary:
Number of SCMs detected:  0
Cumulative SCMs found 0
For (pi,qi) = ( 1 , 0 )
Tests:
     Eigvalue St.dev   Test deg p-value
[1,]    0.003      1  0.294   1   0.588
[2,]    0.081      1  8.672   4   0.070
[3,]    0.271      1 39.922   9   0.000
Summary:
Number of SCMs detected:  2
Found  2  new SCMs
```

```
Updated transformation matrix:
      [,1]   [,2]
V1  0.332 -0.338
V2 -0.628  0.125
V3  0.260  0.588
Cumulative SCMs found 2
For (pi,qi) = ( 2 , 0 )
Tests:
     Eigvalue St.dev  Test deg p-value
[1,]    0.001      1 0.070   1   0.791
[2,]    0.007      1 0.753   4   0.945
[3,]    0.034      1 4.180   9   0.899
Summary:
Number of SCMs detected:  3
Found  1  new SCMs
Transpose of Transformation-matrix:
      [,1]   [,2]   [,3]
V1  0.332 -0.338 -0.359
V2 -0.628  0.125  0.024
V3  0.260  0.588 -0.039
Cumulative SCMs found 3
```

Finally, for recent developments of using SCM approach, including simplification of the transformation matrix $\boldsymbol{T}$, see Athanasopoulos and Vahid (2008) and Athanasopoulos, Poskitt, and Vahid (2012).

4.6 ESTIMATION

A specified VARMA model via either the Kronecker index or the SCM approach can be estimated by the maximum likelihood method. If some of the estimated parameters are not statistically significant, then one can further refine the model by removing insignificant parameters. However, there exists no unique way to remove insignificant parameters of a fitted VARMA model. In principle, one can adopt an iterated procedure by removing insignificant parameter one at a time. That is, one may remove the least significant parameter and re-estimate the model. Such an iterative approach can easily become impractical when the number of insignificant parameters is large. In practice, one may remove several parameters jointly. For instance, one may remove all parameters with their t-ratios less than a given value. In this case, one should pay attention to the change in the residual variances. Deleting simultaneously multiple parameters may result in a substantial increase in the residual variance if those parameters are highly correlated. Experience often plays a role in achieving model simplification.

For a model specified by the SCM approach, two approaches can be used in estimation. The first approach is to use the transformation matrix obtained from the model specification stage. In this case, the transformed series is used and one can directly apply the VARMA estimation discussed in Chapter 3. The second approach is to perform a joint estimation of the transformation matrix and other parameters. In this case, care must be exercised to ensure that the transformation matrix is properly specified.

A fitted VARMA model must be checked carefully to ensure that it is adequate for the given data set. This is done via model checking. Similar to VAR modeling of Chapter 2, one can apply many model checking statistics to validate the adequacy of a fitted VARMA model. For instance, the multivariate Ljung–Box statistics can be used to check the serial and cross-sectional dependence in the residuals.

A special feature of structural specification discussed in this chapter is that it identifies all estimable parameters in a VARMA model. The methods discussed thus overcome the difficulty of identifiability. On the other hand, both the Kronecker index and the SCM approaches may specify a VARMA model for $\boldsymbol{z}_t$ with leading coefficient matrix that is not an identity matrix. See the coefficient matrix $\boldsymbol{\Xi}_0$ in Equation (4.14) and the transformation matrix $\boldsymbol{T}$ in Equation (4.26). One can easily modify the likelihood function to accommodate such a nonidentity leading coefficient matrix. We have performed the modification in our estimation.

4.6.1 Illustration of the Kronecker Index Approach

To demonstrate structural specification of a multivariate time series, we consider, again, the logarithms of monthly flour price indices of Buffalo, Minneapolis, and Kansas City. The Kronecker indices of the series are $\{1, 1, 1\}$ so that the Kronecker index approach specifies a VARMA(1,1) model for the data with $\boldsymbol{\Xi}_0 = \boldsymbol{I}_3$. Furthermore, using Equation (4.15) and adding three constants, the specified VARMA model contains 18 estimable parameter.

Using the conditional maximum likelihood method, we obtain the model

$$\boldsymbol{z}_t = \boldsymbol{\phi}_0 + \boldsymbol{\phi}_1 \boldsymbol{z}_{t-1} + \boldsymbol{a}_t - \boldsymbol{\theta}_1 \boldsymbol{a}_{t-1}, \tag{4.52}$$

where $\hat{\boldsymbol{\phi}}_0 = (0.22, 0.22, 0.31)'$,

$$\hat{\boldsymbol{\phi}}_1 = \begin{bmatrix} 1.21 & -0.64 & 0.39 \\ 0.26 & 0.26 & 0.44 \\ 0.21 & -0.63 & 1.36 \end{bmatrix}, \quad \hat{\boldsymbol{\theta}}_1 = \begin{bmatrix} 1.38 & -1.80 & 0.42 \\ 1.08 & -1.56 & 0.48 \\ 0.72 & -1.34 & 0.62 \end{bmatrix},$$

and the residual covariance matrix is

$$\hat{\boldsymbol{\Sigma}}_a = \begin{bmatrix} 0.18 & 0.19 & 0.18 \\ 0.19 & 0.21 & 0.19 \\ 0.18 & 0.20 & 0.24 \end{bmatrix} \times 10^{-2}.$$

Standard errors and t-ratios of the estimates are given in the following R output. Since t-ratios are all greater than 1.55 in modulus, we do not seek to remove any parameter from the model. Figure 4.2 shows the time plots of residuals, and

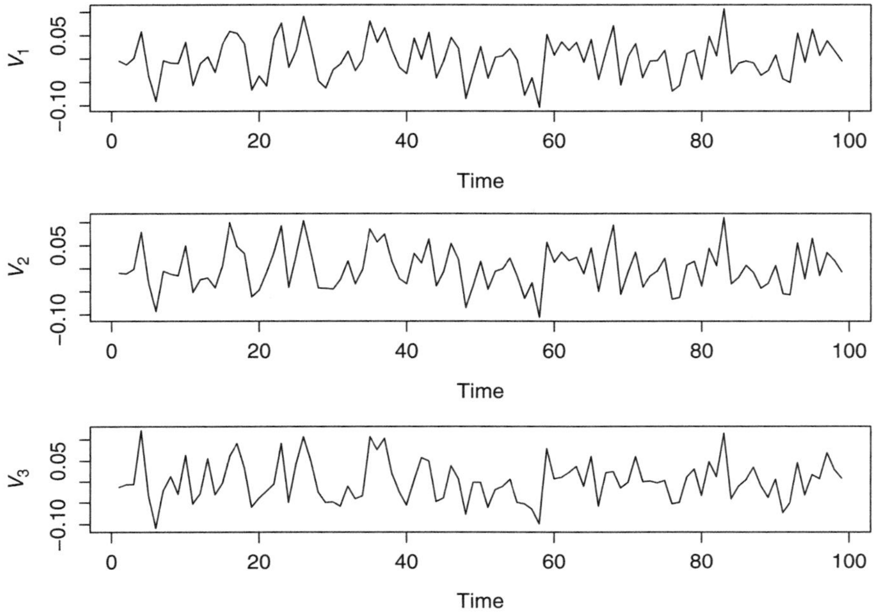

FIGURE 4.2 Time plots of the residuals of the VARMA(1,1) model in Equation (4.52) for the monthly flour price indices of Buffalo, Minneapolis, and Kansas City from August 1972 to November 1980.

Figure 4.3 gives the p-values of the Ljung–Box statistics of the residuals of model (4.52). From the plots, the fitted VARMA(1,1) model appears to be adequate.

R Demonstration: Flour indices.

```
> m2=Kronfit(da,kdx)
Number of parameters:  21
Coefficient(s):
       Estimate  Std. Error  t value Pr(>|t|)
 [1,]   0.22168     0.13084    1.694 0.090204 .
 [2,]   1.20600     0.03674   32.829  < 2e-16 ***
 [3,]  -0.64388     0.13672   -4.709 2.48e-06 ***
 [4,]   0.39133     0.09638    4.060 4.90e-05 ***
 [5,]  -1.37901     0.38257   -3.605 0.000313 ***
 [6,]   1.80145     0.42586    4.230 2.34e-05 ***
 [7,]  -0.41488     0.21581   -1.922 0.054553 .
 [8,]   0.22433     0.14376    1.560 0.118663
 [9,]   0.26157     0.05257    4.975 6.51e-07 ***
[10,]   0.25614     0.15167    1.689 0.091257 .
[11,]   0.43458     0.10479    4.147 3.36e-05 ***
[12,]  -1.07723     0.41779   -2.578 0.009926 **
[13,]   1.56230     0.45350    3.445 0.000571 ***
```

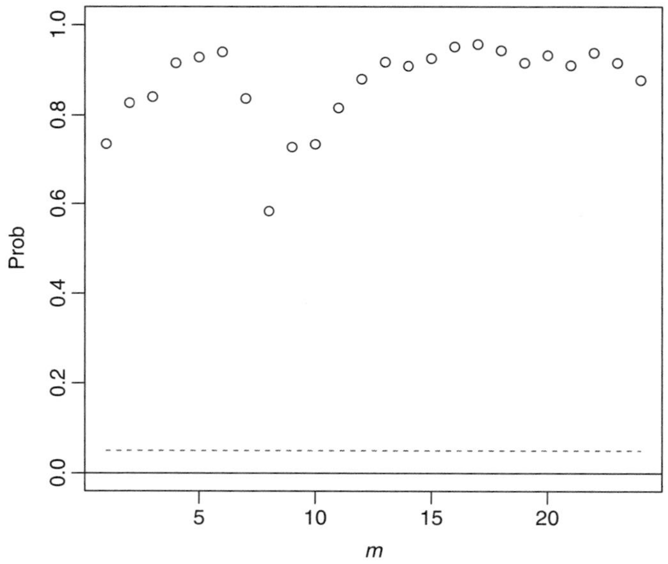

FIGURE 4.3 Plot of p-values of Ljung–Box statistics applied to the residuals of model (4.52).

```
[14,]  -0.47832      0.22941    -2.085 0.037072 *
[15,]   0.31130      0.13413     2.321 0.020298 *
[16,]   0.20979      0.09051     2.318 0.020457 *
[17,]  -0.63253      0.15830    -3.996 6.45e-05 ***
[18,]   1.35839      0.09692    14.016  < 2e-16 ***
[19,]  -0.71978      0.46380    -1.552 0.120681
[20,]   1.34290      0.49119     2.734 0.006258 **
[21,]  -0.62388      0.26936    -2.316 0.020550 *
---
Estimates in matrix form:
Constant term:
Estimates:  0.222 0.224 0.311
AR and MA lag-0 coefficient matrix
     [,1] [,2] [,3]
[1,]    1    0    0
[2,]    0    1    0
[3,]    0    0    1
AR coefficient matrix
AR( 1 )-matrix
      [,1]   [,2]  [,3]
[1,] 1.206 -0.644 0.391
[2,] 0.262  0.256 0.435
[3,] 0.210 -0.633 1.358
MA coefficient matrix
MA( 1 )-matrix
```

```
        [,1]    [,2]   [,3]
[1,] 1.379 -1.801 0.415
[2,] 1.077 -1.562 0.478
[3,] 0.720 -1.343 0.624
  Residuals cov-matrix:
             [,1]        [,2]        [,3]
[1,] 0.001782430 0.001850533 0.001782211
[2,] 0.001850533 0.002053910 0.001960200
[3,] 0.001782211 0.001960200 0.002374997
----
aic=  -22.43436
bic=  -21.88728
> MTSdiag(m2)
```

4.6.2 Illustration of the SCM Approach

Turn to the SCM approach. As shown in Section 4.5.4, the SCM approach specified two possible models for the flour series. They are SCM {(1,0), (1,0), (1,1)} and SCM{(1,0), (1,0), (2,0)}, respectively. Therefore, the approach identified a VARMA(1,1) or a VAR(2) model for the data. We use the VARMA(1,1) model in this section. The model is in the form

$$\boldsymbol{T}\boldsymbol{z}_t = \boldsymbol{g}_0 + \boldsymbol{g}_1\boldsymbol{z}_{t-1} + \boldsymbol{T}\boldsymbol{a}_t - \boldsymbol{\omega}_1\boldsymbol{a}_{t-1}, \tag{4.53}$$

where $\boldsymbol{T}$ is the transformation matrix, $\boldsymbol{g}_0$ is a constant vector, $\boldsymbol{g}_1$ and $\boldsymbol{\omega}_1$ are $k \times k$ matrices and the first two rows of $\boldsymbol{\omega}_1$ are 0. Compared with Equation (4.26), we have $\boldsymbol{g}_1 = -\boldsymbol{G}_1$ and $\boldsymbol{\omega}_1 = -\boldsymbol{U}_1$.

With SCM orders {(1,0), (1,0), (1,1)}, there exists no redundant parameters so that $\boldsymbol{g}_1$ and $\boldsymbol{\omega}_1$ jointly contain 12 estimable parameters. Also, from the initial estimate of the transformation matrix $\hat{\boldsymbol{T}}$ in Equation (4.51), we see that the maximum element, in absolute value, of the first row is -0.63 at the (1,2)th position and the maximum element of the second row is 0.59 at the (2,3)th position. Based on properties of SCM, we can scale the first two SCMs so that the (1,2)th and (2,3)th elements of the transformation matrix are 1. Next, using, again, properties of SCM, we can set (3,2)th and (3,3)th elements of the transformation matrix to 0 because SCM(1,0) is embedded in the SCM(1,1). Furthermore, we can also set the (2,2)th element to 0 because a linear combination of two SCM(1,0) models remains a SCM(1,0) model. [An alternative is to set the (1,3)th element to 0 while estimating the (2,2)th element.] Consequently, the transformation matrix $\boldsymbol{T}$ has three free parameters for estimation and is in the form

$$\hat{\boldsymbol{T}} = \begin{bmatrix} X & 1 & X \\ X & 0 & 1 \\ 1 & 0 & 0 \end{bmatrix},$$

where "X" denotes unknown parameters that require estimation. In summary, with the constant term, the VARMA(1,1) model specified by the SCM approach also requires estimation of 18 parameters.

For the VARMA(1,1) model in Equation (4.53) with 18 estimable parameters, the estimated transformation matrix is

$$\hat{\boldsymbol{T}} = \begin{bmatrix} -0.698 & 1.000 & -0.209 \\ -0.851 & - & 1.000 \\ 1.000 & - & - \end{bmatrix}.$$

The constant term $\hat{\boldsymbol{g}}_0$ is $(0.007, 0.095, 0.209)'$ and other parameters are

$$\hat{\boldsymbol{g}}_1 = \begin{bmatrix} -0.63 & 0.86 & -0.13 \\ -0.74 & -0.11 & 0.98 \\ 1.21 & -0.61 & 0.35 \end{bmatrix}, \quad \hat{\boldsymbol{\omega}}_1 = \begin{bmatrix} - & - & - \\ - & - & - \\ 1.61 & -1.76 & 0.15 \end{bmatrix},$$

and the residual covariance matrix is

$$\hat{\boldsymbol{\Sigma}}_a = \begin{bmatrix} 0.18 & 0.19 & 0.18 \\ 0.19 & 0.21 & 0.20 \\ 0.18 & 0.20 & 0.25 \end{bmatrix} \times 10^{-2}.$$

Figure 4.4 shows the residual plots of the fitted SCM(1,1) model in Equation (4.53), whereas Figure 4.5 provides the p-values of the Ljung–Box statistics for these residual series. The fitted model appears to be adequate because it has successfully removed the dynamic dependence in the three flour indices.

It is interesting to compare the two VARMA(1,1) models specified by the Kronecker index and the SCM approaches for the flour price series. First, the residual covariance matrices of the two fitted VARMA(1,1) models are very close, indicating that the two models provide similar fit. Second, we can rewrite the VARMA(1,1) model of Equation (4.53) in the conventional VARMA format by premultiplying the model by $\hat{\boldsymbol{T}}^{-1}$. The resulting constant term is $(0.21, 0.21, 0.27)$ and the coefficient matrices are

$$\hat{\boldsymbol{\phi}}_1 = \begin{bmatrix} 1.21 & -0.61 & 0.35 \\ 0.27 & 0.30 & 0.38 \\ 0.29 & -0.63 & 1.28 \end{bmatrix}, \quad \hat{\boldsymbol{\theta}}_1 = \begin{bmatrix} 1.61 & -1.76 & 0.15 \\ 1.41 & -1.54 & 0.13 \\ 1.37 & -1.50 & 0.12 \end{bmatrix}.$$

Compared with the estimates in Equation (4.52), we see that the constant term and the AR(1) coefficient matrix of the two models are very close. The two MA(1) coefficient matrices show the same general pattern, but differ in some elements, especially in column 3. The differences, however, are likely to be immaterial as the asymptotic

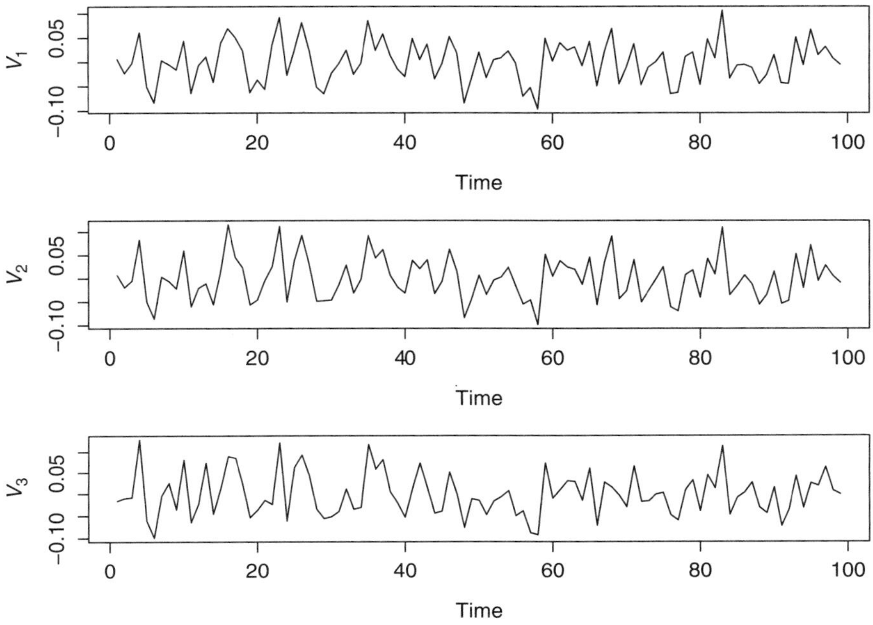

FIGURE 4.4 Time plots of the residuals of the VARMA(1,1) model in Equation (4.53) for the monthly flour price indices of Buffalo, Minneapolis, and Kansas City from August 1972 to November 1980.

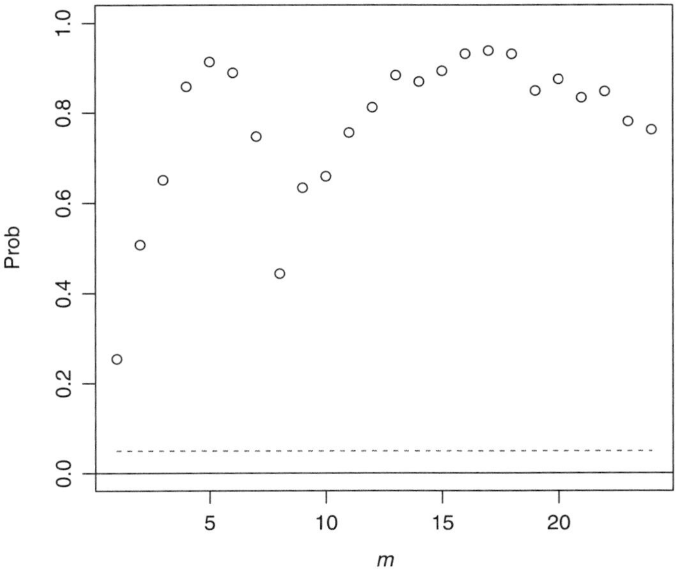

FIGURE 4.5 Plot of p-values of Ljung–Box statistics applied to the residuals of model (4.53).

standard errors of the estimates are relatively large. One can compute the ψ-weight matrices of the two fitted models to verify further that the two models are indeed close to each other. Finally, the AIC criterion prefers slightly the VARMA(1,1) model specified by the SCM approach.

Remark: The estimation of VARMA models specified by the SCM approach is carried out by the commands `SCMfit` and `refSCMfit` of the `MTS` package. The `SCMfit` command requires the orders of identified SCM models and the positions of fixed parameter "1" in the transformation matrix $\boldsymbol{T}$. The SCM orders are in the form of a $k \times 2$ matrix of (p_i, q_i), whereas the positions of "1" are in a locating vector. For instance, the locating vector (2,3,1) denotes the (1,2), (2,3), and (3,1)th elements of $\boldsymbol{T}$ are fixed to 1. See the attached R output for illustration. For given SCMs and the locating vector of the transformation matrix, the structure of the corresponding VARMA model is specified by the command `SCMmod` of the `MTS` package. □

R Demonstration: SCM approach. Output edited.

```
> scms=matrix(c(1,1,1,0,0,1),3,2) % SCM orders
> Tdx=c(2,3,1) % positions of "1" in the transformation matrix
> m1=SCMfit(da,scms,Tdx)
Maximum VARMA order: ( 1 , 1 )
Locations of estimable parameters: Transformation Matrix
     [,1] [,2] [,3]
[1,]    2    1    2 % 2 denotes estimable parameter
[2,]    2    0    1
[3,]    1    0    0
AR parameters
     [,1] [,2] [,3]
[1,]    2    2    2
[2,]    2    2    2
[3,]    2    2    2
MA parameters
     [,1] [,2] [,3]
[1,]    0    0    0
[2,]    0    0    0
[3,]    2    2    2
Estimates in matrix form:
Constant term:
Estimates:  0.007 0.095 0.209
AR and MA lag-0 coefficient matrix
       [,1] [,2]   [,3]
[1,] -0.698    1 -0.209
[2,] -0.851    0  1.000
[3,]  1.000    0  0.000
AR coefficient matrix
AR( 1 )-matrix
```

```
        [,1]   [,2]   [,3]
[1,] -0.634  0.859 -0.134
[2,] -0.742 -0.111  0.982
[3,]  1.213 -0.608  0.351
MA coefficient matrix
MA( 1 )-matrix
      [,1]   [,2]  [,3]
[1,] 0.000  0.000 0.000
[2,] 0.000  0.000 0.000
[3,] 1.609 -1.763 0.146
  Residuals cov-matrix:
            [,1]        [,2]        [,3]
[1,] 0.001793900 0.001880137 0.001828587
[2,] 0.001880137 0.002108857 0.002042209
[3,] 0.001828587 0.002042209 0.002498876
----
aic=  -22.46955; bic=  -22.07878
> MTSdiag(m1)
> zt=da*100
> m2=SCMfit(zt,scms,Tdx)
> m3=refSCMfit(m2,thres=0.3)
> m4=refSCMfit(m3,thres=1.645)
```

Remark: For the flour index example, the estimation of VARMA(1,1) model via the SCM approach can be further simplified by removing some insignificant parameters. To see this, we multiply the indices by 100 for numerical stability. Interested readers can carry out the refinement commands shown in the aforementioned R demonstration. □

4.7 AN EXAMPLE

In this section, we apply both the Kronecker index approach and the SCM approach to the quarterly GDP growth rates of United Kingdom, Canada, and United States from the second quarter of 1980 to the second quarter of 2011 for 125 observations. This three-dimensional time series has been analyzed before in Chapters 2 and 3. Our goals here are (a) to demonstrate the efficacy of structural specification in multivariate time series analysis and (b) to compare the two approaches of structural specification discussed in this chapter. Figure 4.6 shows the time plots of the quarterly GDP growth rates of United Kingdom, Canada, and United States.

4.7.1 The SCM Approach

Let $\boldsymbol{z}_t$ be the three-dimensional series of quarterly GDP growth rates of United Kingdom, Canada, and United States. We start our analysis with the SCM approach. Table 4.8 summarizes the number of zero canonical correlations, its diagonal differences, and some test statistics for $\boldsymbol{z}_t$. In this particular instance, the table of diagonal

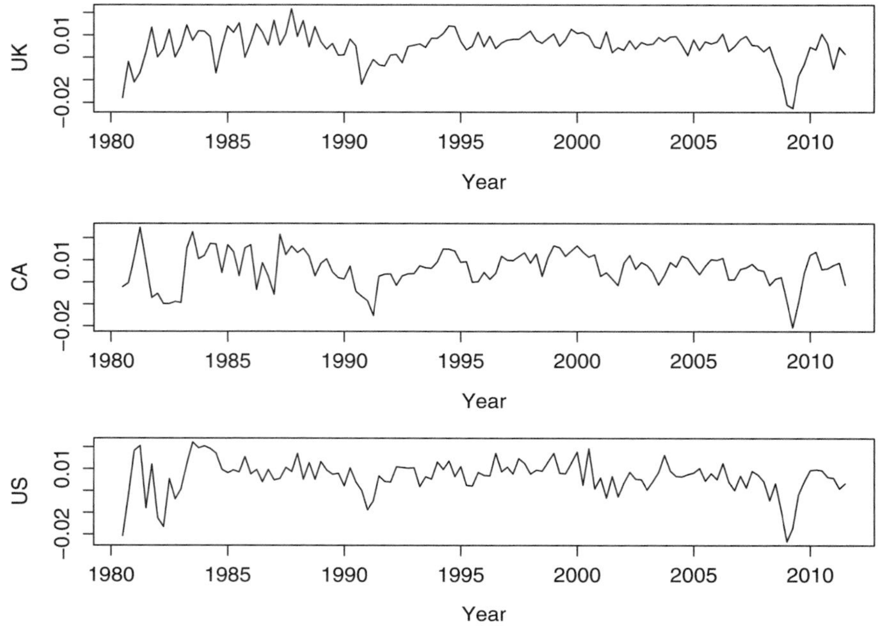

FIGURE 4.6 Time plots of the quarterly GDP growth rates of United Kingdom, Canada, and United States from the second quarter 1980 to the second quarter 2011.

differences suggests an overall VARMA(2,1) model for $\boldsymbol{z}_t$; see part (b) of the table. On the other hand, the table of the number of singular left vectors shows that (i) there is an SCM(0,0) component, (ii) there exists a new SCM(1,0) component, and (iii) there also exists a new SCM(0,1) component. See part (a) of the table. These results suggest a VARMA(1,1) model.

Next, we consider the second-stage analysis of the SCM approach with maximum AR order 2 and maximum MA order 1. The summary of eigenvalues and test statistics is also shown in Table 4.8. See part (c) of the table. From the table, we see that the SCM approach specifies SCM{(0,0), (1,0), (0,1)} for the GDP series. See the following R demonstration for further detail. A tentative estimate of the transformation matrix is given as

$$\hat{\boldsymbol{T}} = \begin{bmatrix} -0.585 & -0.347 & 1.000 \\ -0.673 & 0.104 & -0.357 \\ -0.572 & 0.704 & -0.090 \end{bmatrix}. \quad (4.54)$$

The three SCMs of orders {(0,0), (1,0), (0,1)} specify a VARMA(1,1) model for the GDP growth rates. The model can be written as

$$\boldsymbol{T}\boldsymbol{z}_t = \boldsymbol{g}_0 + \boldsymbol{g}_1\boldsymbol{z}_{t-1} + \boldsymbol{T}\boldsymbol{a}_t - \boldsymbol{\omega}_1\boldsymbol{a}_{t-1}, \quad (4.55)$$

where $\boldsymbol{g}_1$ and $\boldsymbol{\omega}_1$ are in the form

TABLE 4.8 Analysis of the Quarterly GDP Growth Rates of United Kingdom, Canada, and United States from 1980.II to 2011.II Using the Scalar Component Method

(a) Number of Singular Left Vectors

m	j = 0	1	2	3	4	5
0	1	2	2	2	3	3
1	2	3	4	4	6	6
2	2	5	6	7	8	8
3	2	5	8	9	10	11
4	3	6	8	11	12	13
5	3	6	9	11	13	15

(b) Diagonal Difference

m	j = 0	1	2	3	4	5
0	1	2	2	2	3	3
1	2	2	2	2	3	3
2	2	3	3	3	3	2
3	2	3	3	3	3	3
4	3	3	3	3	3	3
5	3	3	3	3	2	3

(c) Summary of Eigenvalues and Test Statistics

m	j	Eigenvalue	$C(m,j)$	Degrees of Freedom	p-Value
0	0	0.000	0.03	1	0.874
0	0	0.098	12.86	4	0.012
0	0	0.568	117.71	9	0.000
1	0	0.000	0.02	1	0.903
1	0	0.021	2.70	4	0.609
1	0	0.138	21.05	9	0.012
0	1	0.006	0.76	1	0.384
0	1	0.029	4.37	4	0.358
0	1	0.293	47.28	9	0.000

The 5% significance level is used.

$$\boldsymbol{g}_1 = \begin{bmatrix} 0 & 0 & 0 \\ X & X & X \\ 0 & 0 & 0 \end{bmatrix}, \quad \boldsymbol{\omega}_1 = \begin{bmatrix} 0 & 0 & 0 \\ 0 & 0 & 0 \\ X & X & X \end{bmatrix},$$

with "X" denoting estimable parameters. In addition, based on the tentative estimate of the transformation matrix in Equation (4.54) and making use of the properties of SCM, we can simplify the transformation matrix as

$$\boldsymbol{T} = \begin{bmatrix} X & X & 1 \\ 1 & X & 0 \\ X & 1 & 0 \end{bmatrix},$$

where, again, "X" denotes estimable parameters, the coefficient "1" is used due to normalization, and "0" are obtained via the SCM properties. Consequently, the SCM approach specifies a VARMA(1,1) model for the GDP growth rates and the model contains 13 estimable parameters. Here, the number of parameters includes three constants and four elements in the transformation matrix $\boldsymbol{T}$.

Table 4.9(a) contains the initial estimates of the model in Equation (4.55). Several estimates are not significant at the usual 5% level and we refine the model via an

TABLE 4.9 Estimation of a VARMA(1,1) Model Specified by the SCM Approach for the GDP Growth Rates of United Kingdom, Canada, and United States

	Part (a)			Part (b)		
Parameter	Estimates			Estimates		
g_0'	0.002	0.002	0.002	0	0.002	0
	−0.43	−0.44	1.00	−0.66	−0.43	1.00
T	1.00	0.05	0	1	0	0
	−0.55	1	0	−1.09	1.00	0
	0	0	0	0	0	0
g_1	0.45	0.24	0.02	0.34	0.27	0.00
	0	0	0	0	0	0
	0	0	0	0	0	0
ω_1	0	0	0	0	0	0
	−0.02	−0.31	−0.33	0.00	0.00	−0.41
	2.88	0.36	0.74	2.95	0.26	0.70
$\Sigma \times 10^{-5}$	0.36	3.17	1.58	0.26	3.11	1.50
	0.74	1.58	3.83	0.70	1.50	3.83
AIC		−31.122			−31.188	

The sample period is from 1980.II to 2011.II.

iterated procedure to remove insignificant estimates. The procedure consists of three iterations with thresholds for the t-ratios being 0.8, 1.5, and 1.5, respectively. For instance, our first iteration removes all estimates with t-ratios less than 0.8 in modulus. Table 4.9(b) gives the final parameter estimates of the specified VARMA(1,1) model. The standard errors and t-ratios of the estimates are given in the following R demonstration. From the table, we see that the final VARMA(1,1) model contains seven significant coefficient estimates.

Figure 4.7 shows the p-values of the Ljung–Box statistics for the residuals of the final VARMA(1,1) model in Table 4.9(b). From the plot, the fitted model is adequate as it successfully captures the linear dynamic dependence in the data. Even including the transformation matrix, this final VARMA(1,1) model is parsimonious because it only employs seven estimated coefficients. As shown in the following R demonstration, the SCM approach starts with 13 parameters, instead of 21 for a three-dimensional VARMA(1,1) model. This simple example demonstrates that the SCM approach can achieve a high degree of parsimony in modeling multivariate time series.

Of particular interest of the fitted model is that the linear combination of GDP growth rates $y_t = \text{US} - 0.66\text{UK} - 0.43\text{CA}$ is a white noise series, that is, SCM(0,0). Such information is not revealed by the analyses in Chapter 2 or Chapter 3.

R Demonstration: SCM analysis of GDP growth rates. Output edited.

```
> da=read.table("q-gdp-ukcaus.txt",header=T)
> gdp=log(da[,3:5])
```

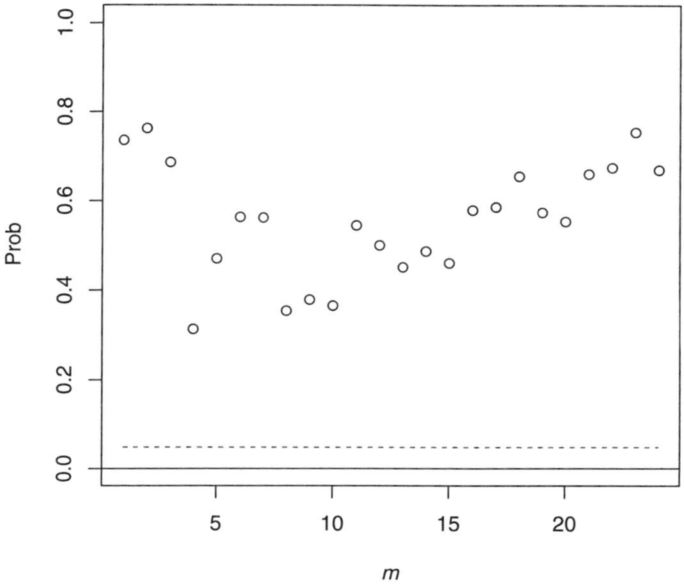

FIGURE 4.7 Plot of p-values of Ljung–Box statistics applied to the residuals of the model in part (b) of Table 4.9.

```
> zt=diffM(gdp)
> colnames(zt) <- c("uk","ca","us")
> SCMid(zt)
Column: MA order
Row    : AR order
Number of zero canonical correlations
   0  1  2  3  4  5
0  1  2  2  2  3  3
1  2  3  4  4  6  6
2  2  5  6  7  8  8
3  2  5  8  9 10 11
4  3  6  8 11 12 13
5  3  6  9 11 13 15
Diagonal Differences:
  0 1 2 3 4 5
0 1 2 2 2 3 3
1 2 2 2 2 3 3
2 2 3 3 3 3 2
3 2 3 3 3 3 3
4 3 3 3 3 3 3
5 3 3 3 3 2 3
> SCMid2(zt,maxp=2,maxq=1)
For (pi,qi) = ( 0 , 0 )
```

```
Tests:
     Eigvalue St.dev    Test deg p-value
[1,]    0.000      1   0.025   1   0.874
[2,]    0.098      1  12.863   4   0.012
[3,]    0.568      1 117.706   9   0.000
Number of SCMs detected:  1
Newly detected SCMs:
       [,1]
[1,] -0.585
[2,] -0.347
[3,]  1.000
Cumulative SCMs found 1
For (pi,qi) = ( 1 , 0 )
Tests:
     Eigvalue St.dev   Test deg p-value
[1,]    0.000      1  0.015   1    0.903
[2,]    0.021      1  2.701   4    0.609
[3,]    0.138      1 21.049   9    0.012
Number of SCMs detected:  2
Found  1  new SCMs
Transformation-matrix:
       [,1]   [,2]   [,3]
[1,] -0.585 -0.347  1.000
[2,] -0.673  0.104 -0.357
Cumulative SCMs found 2
For (pi,qi) = ( 0 , 1 )
Tests:
     Eigvalue St.dev   Test deg p-value
[1,]    0.006  1.008  0.758   1    0.384
[2,]    0.029  1.016  4.373   4    0.358
[3,]    0.293  1.000 47.283   9    0.000
Number of SCMs detected:  2
Found  1  new SCMs
Transpose of Transformation-matrix:  % Preliminary estimate of
                                     % t(T).
     [,1]   [,2]   [,3]
uk -0.585 -0.673 -0.088
ca -0.347  0.104 -0.033
us  1.000 -0.357 -0.063
Cumulative SCMs found 3
> print(round(m1$Tmatrix,3))
         uk     ca     us
[1,] -0.585 -0.347  1.000
[2,] -0.673  0.104 -0.357
[3,] -0.088 -0.033 -0.063
> scms=matrix(c(0,1,0,0,0,1),3,2) % SCM orders
> Tdx=c(3,1,2)  % Positions of "1" in the transformation
  matrix
> m1=SCMfit(zt,scms,Tdx) % estimation
```

```
Number of parameters: 13
Coefficient(s):
        Estimate  Std. Error  t value Pr(>|t|)
 [1,]  0.0016050   0.0008439    1.902  0.05718 .
 [2,] -0.4298622   0.2137662   -2.011  0.04434 *
 [3,] -0.4421370   0.1946831   -2.271  0.02314 *
 [4,]  0.0017418   0.0009978    1.746  0.08088 .
 [5,]  0.0517923   0.3149038    0.164  0.86936
 [6,]  0.4532122   0.0917305    4.941 7.78e-07 ***
 [7,]  0.2427519   0.1119083    2.169  0.03007 *
 [8,]  0.0147471   0.1151393    0.128  0.89809
 [9,]  0.0032545   0.0016821    1.935  0.05302 .
[10,] -0.5525652   0.2760356   -2.002  0.04531 *
[11,]  0.0162190   0.1565082    0.104  0.91746
[12,]  0.3128620   0.1849435    1.692  0.09071 .
[13,]  0.3311366   0.1047870    3.160  0.00158 **
---
aic=  -31.12173; bic=  -30.91809
> m2=refSCMfit(m1,thres=0.8) % Simplification.
> m3=refSCMfit(m2,thres=1.5)
> m4=refSCMfit(m3,thres=1.5)
Maximum VARMA order: ( 1 , 1 )
Number of parameters:  7
Coefficient(s):
       Estimate  Std. Error  t value Pr(>|t|)
[1,] -0.6574063   0.2071234   -3.174   0.0015 **
[2,] -0.4306281   0.1675213   -2.571   0.0102 *
[3,]  0.0021545   0.0005339    4.035 5.45e-05 ***
[4,]  0.3434554   0.0585798    5.863 4.54e-09 ***
[5,]  0.2645765   0.0507830    5.210 1.89e-07 ***
[6,] -1.0863097   0.1141818   -9.514  < 2e-16 ***
[7,]  0.4130114   0.0755228    5.469 4.53e-08 ***
---
Estimates in matrix form:
Constant term:
Estimates:  0 0.002 0
AR and MA lag-0 coefficient matrix
       [,1]   [,2] [,3]
[1,] -0.657 -0.431    1
[2,]  1.000  0.000    0
[3,] -1.086  1.000    0
AR coefficient matrix
AR( 1 )-matrix
      [,1]  [,2] [,3]
[1,] 0.000 0.000    0
[2,] 0.343 0.265    0
[3,] 0.000 0.000    0
MA coefficient matrix
MA( 1 )-matrix
```

TABLE 4.10 Specification of Kronecker Indices for the Quarterly GDP Growth Rates of United Kingdom, Canada, and United States from 1980.II to 2011.II

Last Element	Small Canonical Correction	Test	Degrees of Freedom	p-Value	Remark
z_{1t}	0.478	76.72	12	0	
z_{2t}	0.250	33.79	11	0	
z_{3t}	0.113	14.03	10	0.172	$k_3 = 0$
$z_{1,t+1}$	0.073	10.11	9	0.341	$k_1 = 1$
$z_{2,t+1}$	0.081	9.76	9	0.370	$k_2 = 1$

The past vector used is $\boldsymbol{P}_t = (\boldsymbol{z}'_{t-1}, \ldots, \boldsymbol{z}'_{t-4})'$ and last element denotes the last element of the future vector.

```
     [,1] [,2]   [,3]
[1,]    0    0  0.000
[2,]    0    0  0.000
[3,]    0    0 -0.413
Residuals cov-matrix:
             [,1]         [,2]         [,3]
[1,] 2.947774e-05 2.557159e-06 6.978774e-06
[2,] 2.557159e-06 3.109328e-05 1.504319e-05
[3,] 6.978774e-06 1.504319e-05 3.826788e-05
----
aic=  -31.18843; bic=  -31.12055
```

4.7.2 The Kronecker Index Approach

In this section, we apply the Kronecker index approach to modeling the GDP growth rates of United Kingdom, Canada, and United States. Using the past vector $\boldsymbol{P}_t = (\boldsymbol{z}'_{t-1}, \ldots, \boldsymbol{z}'_{t-4})'$ and the test statistics of Equation (4.42), we obtain Kronecker indices $\{1,1,0\}$ for the three GDP growth rates. See the summary in Table 4.10. Therefore, the Kronecker index approach also specifies a VARMA(1,1) model, which can be written as

$$\boldsymbol{\Xi}_0 \boldsymbol{z}_t = \boldsymbol{\phi}_0 + \boldsymbol{\phi}_1 \boldsymbol{z}_{t-1} + \boldsymbol{\Xi}_0 \boldsymbol{a}_t - \boldsymbol{\theta}_1 \boldsymbol{a}_{t-1}, \tag{4.56}$$

where $\boldsymbol{\phi}_0$ denotes the constant vector and $\boldsymbol{\Xi}_0$ is in the form

$$\boldsymbol{\Xi}_0 = \begin{bmatrix} 1 & 0 & 0 \\ 0 & 1 & 0 \\ X & X & 1 \end{bmatrix},$$

which consists of two estimable parameters. These two parameters give rise to a white noise linear combination because $k_3 = 0$. The $\boldsymbol{\phi}_1$ and $\boldsymbol{\theta}_1$ matrices of Equation (4.56) are in the form

$$\boldsymbol{\phi}_1 = \begin{bmatrix} X & X & 0 \\ X & X & 0 \\ 0 & 0 & 0 \end{bmatrix}, \quad \boldsymbol{\theta}_1 = \begin{bmatrix} X & X & X \\ X & X & X \\ 0 & 0 & 0 \end{bmatrix},$$

where, again, "X" denotes estimable parameters. Note that the last column of $\boldsymbol{\phi}_1$ is set to 0 because of the redundant parameters. Specifically, with the Kronecker index $k_3 = 0$, the dependence of z_{1t} and z_{2t} on the lag-1 value of z_{3t} can be either in $z_{3,t-1}$ or $a_{3,t-1}$, but not both. (This is possible because $\boldsymbol{\Xi}_0$ is used in the model.) The Echelon form automatically sets the coefficients of $z_{3,t-1}$ to 0. In summary, the Kronecker index approach specifies a VARMA(1,1) model in Equation (4.56) for the three-dimensional GDP growth rates with 15 estimable parameters (including the three constants).

Table 4.11(a) gives the initial estimates of the VARMA(1,1) model in Equation (4.56). Similar to the SCM approach, some of the estimates are not statistically significant at the usual 5% level. We adopt a similar procedure as that of the SCM approach to refine the model. Table 4.11(b) shows the final estimates of the VARMA(1,1) model. The model is also parsimonious with 9 estimated coefficients. Figure 4.8 shows the p-values of Ljung–Box statistics applied to the residuals of the final fitted model in Table 4.11(b). From the plot, this VARMA(1,1) model also successfully describes the dynamic dependence among the three GDP growth rates. The model is adequate. Figure 4.9 shows the time plot of three residual series. The variabilities of the residuals might be time-varying. However, we shall not discuss multivariate volatility models in this chapter.

R Demonstration: GDP growth rates via Kronecker index approach. Output edited.

```
> Kronid(zt,plag=4)   # specify the Kronecker indices
h =  0
Component =  1
square of the smallest can. corr. =  0.4780214
    test,   df, &  p-value:
[1] 76.715 12.000  0.000
Component =  2
square of the smallest can. corr. =  0.2498917
    test,   df, &  p-value:
[1] 33.786 11.000  0.000
Component =  3
square of the smallest can. corr. =  0.1130244
    test,   df, &  p-value:
[1] 14.033 10.000  0.172
A Kronecker index found
```

TABLE 4.11 Estimation of a VARMA(1,1) Model Specified by the Kronecker Index Approach for the GDP Growth Rates of United Kingdom, Canada, and United States

	Part (a)			Part (b)		
Parameter	Estimates			Estimates		
ϕ_0'	0.001	0.003	0.001	0.002	0.003	0.002
	1	0	0	1	0	0
Ξ_0	0	1	0	0	1	0
	−0.28	−0.61	1	0	−0.77	1
	0.44	0.31	0	0.41	0.23	0
ϕ_1	0.06	0.46	0	0	0.46	0
	0	0	0	0	0	0
	0.04	0.16	0.02	0	0	0
θ_1	−0.27	0.07	−0.38	−0.36	0	−0.32
	0	0	0	0	0	0
	2.83	0.32	0.74	2.90	0.37	0.80
$\Sigma \times 10^{-5}$	0.32	3.06	1.46	0.37	3.10	1.47
	0.74	1.46	3.75	0.80	1.47	3.79
AIC		−31.078			−31.130	

The sample period is from 1980.II to 2011.II.

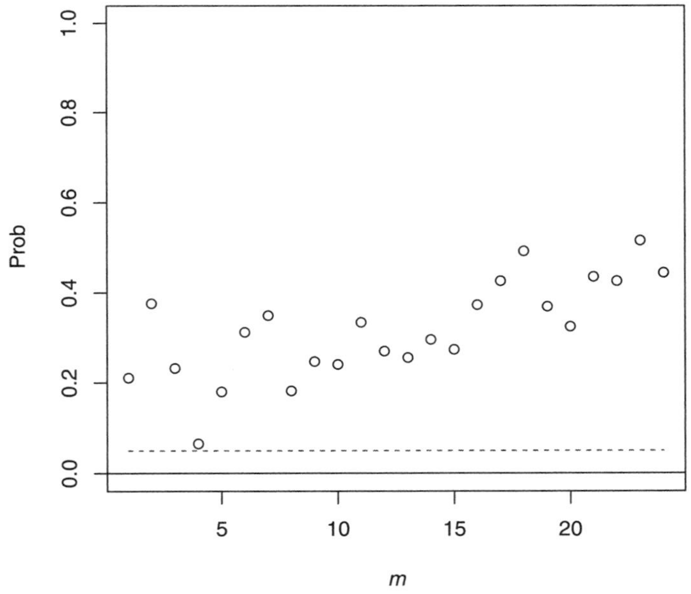

FIGURE 4.8 Plot of p-values of Ljung–Box statistics applied to the residuals of the model in part (b) of Table 4.11.

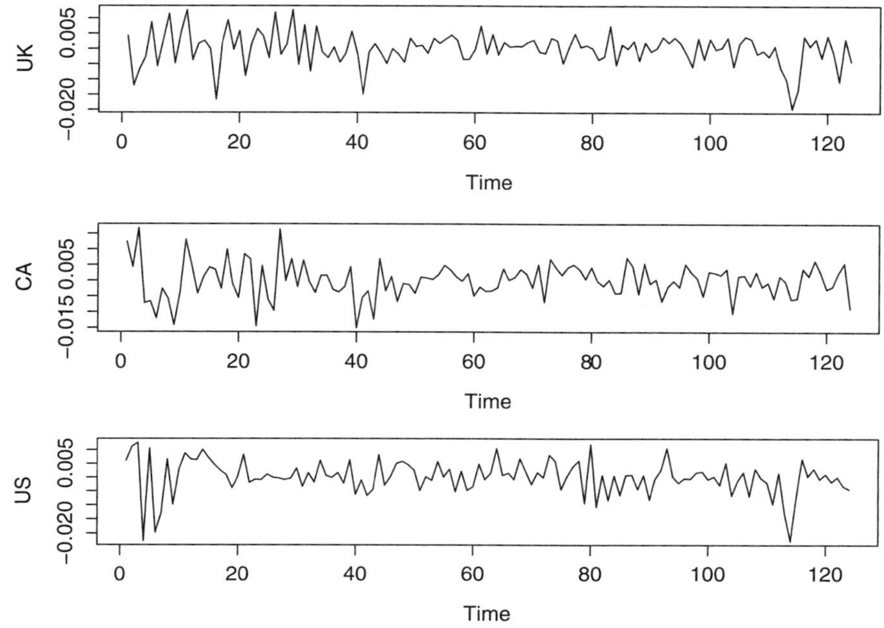

FIGURE 4.9 Time plots of the residuals of the model in part (b) of Table 4.11.

```
=============
h =  1
Component =  1
Square of the smallest can. corr. =  0.07259501
    test,     df, p-value & d-hat:
[1] 10.114  9.000  0.341  0.873
A Kronecker found
Component =  2
Square of the smallest can. corr. =  0.08127047
    test,     df, p-value & d-hat:
[1] 9.759 9.000 0.370 1.011
A Kronecker found
============
Kronecker indexes identified:
[1] 1 1 0

> kdx=c(1,1,0)    % Kronecker indices
> m1=Kronfit(zt,kdx)  % Estimation
     [,1] [,2] [,3] [,4] [,5] [,6]  % Xi_0 and phi_1
[1,]    1    0    0    2    2    0 % 2 denotes estimable
                                      parameter
[2,]    0    1    0    2    2    0
[3,]    2    2    1    0    0    0
```

```
     [,1] [,2] [,3] [,4] [,5] [,6] % Xi_0 and theta_1
[1,]    1    0    0    2    2    2
[2,]    0    1    0    2    2    2
[3,]    2    2    1    0    0    0
Number of parameters:  15
Coefficient(s):
        Estimate  Std. Error  t value Pr(>|t|)
 [1,]  0.0011779   0.0007821    1.506 0.132059
 [2,]  0.4359168   0.2009202    2.170 0.030037 *
 [3,]  0.3115727   0.1457037    2.138 0.032484 *
 [4,] -0.0349292   0.2364209   -0.148 0.882547
 [5,] -0.1574241   0.1607823   -0.979 0.327524
 [6,] -0.0202607   0.0860744   -0.235 0.813909
 [7,]  0.0031470   0.0010095    3.117 0.001825 **
 [8,]  0.0550135   0.2587264    0.213 0.831614
 [9,]  0.4642734   0.1883191    2.465 0.013688 *
[10,]  0.2659062   0.2924283    0.909 0.363190
[11,] -0.0685227   0.2593264   -0.264 0.791600
[12,]  0.3804101   0.0947375    4.015 5.93e-05 ***
[13,]  0.0014249   0.0008304    1.716 0.086169 .
[14,] -0.2773537   0.2279021   -1.217 0.223609
[15,] -0.6050206   0.1707929   -3.542 0.000396 ***
----
aic=  -31.07752; bic=  -30.73812
> m2=refKronfit(m1,thres=0.8) % Simplification
> m3=refKronfit(m2,thres=1.5)
Number of parameters:  9
Coefficient(s):
        Estimate  Std. Error  t value Pr(>|t|)
 [1,]  0.0018025   0.0006393    2.819 0.004813 **
 [2,]  0.4141641   0.0687452    6.025 1.70e-09 ***
 [3,]  0.2290893   0.0632500    3.622 0.000292 ***
 [4,]  0.0034813   0.0007941    4.384 1.17e-05 ***
 [5,]  0.4552192   0.0725278    6.276 3.46e-10 ***
 [6,]  0.3591829   0.0873272    4.113 3.90e-05 ***
 [7,]  0.3228183   0.0756492    4.267 1.98e-05 ***
 [8,]  0.0018911   0.0007950    2.379 0.017373 *
 [9,] -0.7707567   0.0967124   -7.970 1.55e-15 ***
---
Estimates in matrix form:
Constant term:
Estimates:  0.002 0.003 0.002
AR and MA lag-0 coefficient matrix
     [,1]   [,2] [,3]
[1,]    1  0.000    0
[2,]    0  1.000    0
[3,]    0 -0.771    1
AR coefficient matrix
AR( 1 )-matrix
```

```
       [,1]  [,2] [,3]
[1,] 0.414 0.229    0
[2,] 0.000 0.455    0
[3,] 0.000 0.000    0
MA coefficient matrix
MA( 1 )-matrix
       [,1] [,2]   [,3]
[1,]  0.000    0  0.000
[2,] -0.359    0 -0.323
[3,]  0.000    0  0.000
Residuals cov-matrix:
             [,1]         [,2]         [,3]
[1,] 2.902578e-05 3.650548e-06 7.956844e-06
[2,] 3.650548e-06 3.098022e-05 1.469477e-05
[3,] 7.956844e-06 1.469477e-05 3.786774e-05
----
aic=  -31.1301; bic=  -30.92647
> MTSdiag(m3)
```

4.7.3 Discussion and Comparison

In this section, we have demonstrated the use of structural specification in multivariate time series analysis. Both the Kronecker index and SCM approaches can reveal the dynamic structure of the data and achieve parsimony in parameterization. The results show that structural specification can simplify the modeling process in multivariate time series analysis. For the series of quarterly GDP growth rates of United Kingdom, Canada, and United States, both approaches specify an adequate and parsimonious model. Both approaches show that linear combinations of the growth rates can behave as a white noise series. On the surface, the white noise combination suggested by the Kronecker index approach is

$$y_{1t} = \text{US}_t - 0.771\text{CA}_t,$$

whereas that suggested by the SCM approach is

$$y_{2t} = \text{US}_t - 0.657\text{UK}_t - 0.431\text{CA}_t.$$

Figure 4.10 shows the sample autocorrelation functions of y_{1t} and y_{2t}. From the plots, all ACFs of the two series are small and within the two standard-error limits. The plots thus confirm that y_{1t} and y_{2t} are indeed white noise series. Figure 4.11 shows the scatter plot of y_{1t} versus y_{2t}. As expected, the two white noise series are highly correlated. Therefore, the two approaches reveal essentially the same structure of the data.

To facilitate further comparison, we can rewrite the VARMA(1,1) models built via the SCM and Kronecker index approaches in the conventional VARMA form. For the SCM approach, the final model in Table 4.9(b) becomes

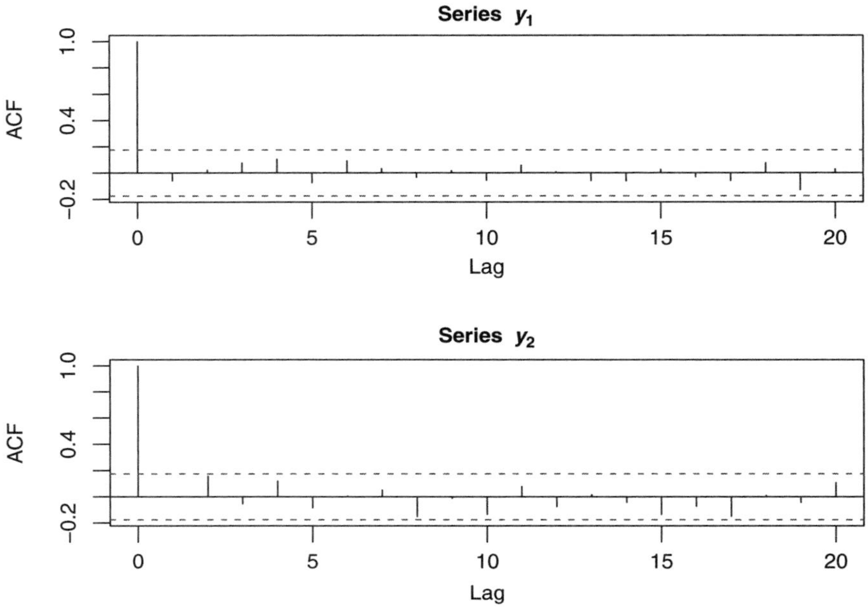

FIGURE 4.10 Sample autocorrelation functions of two white noise linear combinations of the quarterly GDP growth rates of United Kingdom, Canada, and United States: upper panel: $y_{1t} = \mathrm{US}_t - 0.771\mathrm{CA}_t$; lower panel: $y_{2t} = \mathrm{US}_t - 0.657\mathrm{UK}_t - 0.431\mathrm{CA}_t$.

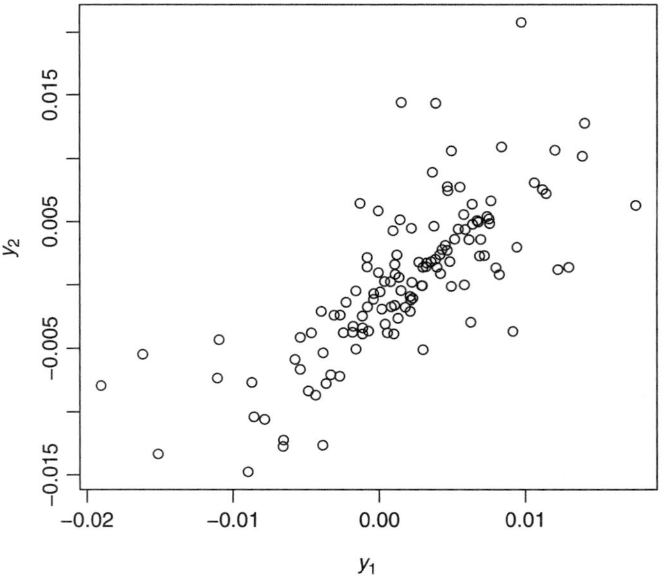

FIGURE 4.11 Scatter plot of two white noise linear combinations of the quarterly GDP growth rates of United Kingdom, Canada, and United States.

$$\boldsymbol{z}_t - \begin{bmatrix} 0.34 & 0.26 & 0 \\ 0.37 & 0.29 & 0 \\ 0.39 & 0.30 & 0 \end{bmatrix} \boldsymbol{z}_{t-1} = \begin{bmatrix} 0.0021 \\ 0.0023 \\ 0.0024 \end{bmatrix} + \boldsymbol{a}_t - \begin{bmatrix} 0 & 0 & 0 \\ 0 & 0 & -0.41 \\ 0 & 0 & -0.18 \end{bmatrix} \boldsymbol{a}_{t-1}. \quad (4.57)$$

On the other hand, the final model of the Kronecker index approach in Table 4.11(b) is

$$\boldsymbol{z}_t - \begin{bmatrix} 0.41 & 0.23 & 0 \\ 0 & 0.46 & 0 \\ 0 & 0.35 & 0 \end{bmatrix} \boldsymbol{z}_{t-1} = \begin{bmatrix} 0.0018 \\ 0.0035 \\ 0.0046 \end{bmatrix} + \boldsymbol{a}_t - \begin{bmatrix} 0 & 0 & 0 \\ -0.36 & 0 & -0.32 \\ -0.28 & 0 & -0.25 \end{bmatrix} \boldsymbol{a}_{t-1}. \quad (4.58)$$

Comparing Equations (4.57) and (4.58), it is clear that the two models are close. It seems that the model in Equation (4.58) puts the lag-1 dependence of components z_{2t} and z_{3t} on z_{1t} at the MA part, whereas the model in Equation (4.57) puts the dependence in the AR part. The magnitudes of these lag-1 dependence are close. As one would expect, the impulse response functions of the two VARMA(1,1) models are also close to each other.

Next, the sample means of the three GDP growth rates are 0.0052, 0.0062, and 0.0065, respectively, for United Kingdom, Canada, and United States. The sample means implied by the model in Equation (4.57) are 0.0058, 0.0063, and 0.0066, respectively. As expected, these values are close to the sample means. The implied means of the model in Equation (4.58) are 0.0056, 0.0064, and 0.0068, respectively. They are also close to the sample means.

Finally, the AIC criterion slightly prefers the VARMA(1,1) model built via the SCM approach. This is likely due to the fact that the model uses fewer coefficients. Theoretically speaking, the SCM approach can be regarded as a refinement over the Kronecker index approach. The Kronecker index approach specifies the maximum order between the AR and MA polynomials for each component series z_{it}. On the other hand, the SCM approach separates the AR and MA orders for each scalar component. For the GDP growth rates, the order is either 0 or 1 so that the difference between the two approaches is small. The difference between the two approaches could become larger when the orders (p_i, q_i) of the scalar components are higher.

4.8 APPENDIX: CANONICAL CORRELATION ANALYSIS

In this appendix, we briefly introduce canonical correlation analysis between two random vectors with a joint multivariate normal distribution. The basic theory of the analysis was developed by Hotelling (1935) and (1936).

Assume that $\boldsymbol{X}$ and $\boldsymbol{Y}$ are p-dimensional and q-dimensional random vectors such that the joint variate $\boldsymbol{Z} = (\boldsymbol{X}', \boldsymbol{Y}')'$ has a joint distribution with mean zero and positive-definite covariance matrix $\boldsymbol{\Sigma}$. Without loss of generality, assume that $p \leq q$. Partition $\boldsymbol{\Sigma}$ as

$$\boldsymbol{\Sigma} = \begin{bmatrix} \boldsymbol{\Sigma}_{xx} & \boldsymbol{\Sigma}_{xy} \\ \boldsymbol{\Sigma}_{yx} & \boldsymbol{\Sigma}_{yy} \end{bmatrix}.$$

Consider an arbitrary linear combination $U = \boldsymbol{\alpha}'\boldsymbol{X}$ of the components of $\boldsymbol{X}$ and an arbitrary linear combination $V = \boldsymbol{\gamma}'\boldsymbol{Y}$ of the components of $\boldsymbol{Y}$. Canonical correlation analysis seeks to find U and V that have maximum correlation.

Since scaling does not change correlation, one typically normalizes the arbitrary vectors such that U and V have unit variance, that is,

$$1 = E(U^2) = \boldsymbol{\alpha}'\boldsymbol{\Sigma}_{xx}\boldsymbol{\alpha}, \tag{4.59}$$

$$1 = E(V^2) = \boldsymbol{\gamma}'\boldsymbol{\Sigma}_{yy}\boldsymbol{\gamma}. \tag{4.60}$$

Since $E(U) = E(V) = 0$, the correlation between U and V is

$$E(UV) = \boldsymbol{\alpha}'\boldsymbol{\Sigma}_{xy}\boldsymbol{\gamma}. \tag{4.61}$$

The problem then becomes finding $\boldsymbol{\alpha}$ and $\boldsymbol{\gamma}$ to maximize Equation (4.61) subject to the constraints in Equations (4.59) and (4.60). Let

$$\psi = \boldsymbol{\alpha}'\boldsymbol{\Sigma}_{xy}\boldsymbol{\gamma} - \tfrac{1}{2}\lambda(\boldsymbol{\alpha}'\boldsymbol{\Sigma}_{xx}\boldsymbol{\alpha} - 1) - \tfrac{1}{2}\omega(\boldsymbol{\gamma}'\boldsymbol{\Sigma}_{yy}\boldsymbol{\gamma} - 1), \tag{4.62}$$

where λ and ω are Lagrange multipliers. Differentiating ψ with respect to elements of $\boldsymbol{\alpha}$ and $\boldsymbol{\gamma}$ and setting the vectors of derivative to 0, we obtain

$$\frac{\partial \psi}{\partial \boldsymbol{\alpha}} = \boldsymbol{\Sigma}_{xy}\boldsymbol{\gamma} - \lambda\boldsymbol{\Sigma}_{xx}\boldsymbol{\alpha} = \boldsymbol{0} \tag{4.63}$$

$$\frac{\partial \psi}{\partial \boldsymbol{\gamma}} = \boldsymbol{\Sigma}'_{xy}\boldsymbol{\alpha} - \omega\boldsymbol{\Sigma}_{yy}\boldsymbol{\gamma} = \boldsymbol{0}. \tag{4.64}$$

Premultiplying Equation (4.63) by $\boldsymbol{\alpha}'$ and Equation (4.64) by $\boldsymbol{\gamma}'$ gives

$$\boldsymbol{\alpha}'\boldsymbol{\Sigma}_{xy}\boldsymbol{\gamma} - \lambda\boldsymbol{\alpha}'\boldsymbol{\Sigma}_{xx}\boldsymbol{\alpha} = 0 \tag{4.65}$$

$$\boldsymbol{\gamma}'\boldsymbol{\Sigma}'_{xy}\boldsymbol{\alpha} - \omega\boldsymbol{\gamma}'\boldsymbol{\Sigma}_{yy}\boldsymbol{\gamma} = 0. \tag{4.66}$$

Using Equations (4.59) and (4.60), we have $\lambda = \omega = \boldsymbol{\alpha}'\boldsymbol{\Sigma}_{xy}\boldsymbol{\gamma}$. Consequently, Equations (4.63) and (4.64) can be written as

$$-\lambda\boldsymbol{\Sigma}_{xx}\boldsymbol{\alpha} + \boldsymbol{\Sigma}_{xy}\boldsymbol{\gamma} = \boldsymbol{0} \tag{4.67}$$

$$\boldsymbol{\Sigma}_{yx}\boldsymbol{\alpha} - \lambda\boldsymbol{\Sigma}_{yy}\boldsymbol{\gamma} = \boldsymbol{0}, \tag{4.68}$$

where we use $\mathbf{\Sigma}'_{xy} = \mathbf{\Sigma}_{yx}$. In matrix form, the prior equation is

$$\begin{bmatrix} -\lambda\mathbf{\Sigma}_{xx} & \mathbf{\Sigma}_{xy} \\ \mathbf{\Sigma}_{yx} & -\lambda\mathbf{\Sigma}_{yy} \end{bmatrix} \begin{bmatrix} \boldsymbol{\alpha} \\ \boldsymbol{\gamma} \end{bmatrix} = \mathbf{0}. \tag{4.69}$$

Since $\mathbf{\Sigma}$ is nonsingular, the necessary condition under which the prior equation has a nontrivial solution is that the matrix on the left must be singular; that is,

$$\begin{vmatrix} -\lambda\mathbf{\Sigma}_{xx} & \mathbf{\Sigma}_{xy} \\ \mathbf{\Sigma}_{yx} & -\lambda\mathbf{\Sigma}_{yy} \end{vmatrix} = 0. \tag{4.70}$$

Since $\mathbf{\Sigma}$ is $(p+q)$ by $(p+q)$, the determinant is a polynomial of degree $p+q$. Denote the solutions of the determinant by $\lambda_1 \geq \lambda_2 \geq \cdots \geq \lambda_{p+q}$.

From Equation (4.65), we see that $\lambda = \boldsymbol{\alpha}'\mathbf{\Sigma}_{xy}\boldsymbol{\gamma}$ is the correlation between $U = \boldsymbol{\alpha}'\boldsymbol{X}$ and $V = \boldsymbol{\gamma}'\boldsymbol{Y}$ when $\boldsymbol{\alpha}$ and $\boldsymbol{\gamma}$ satisfy Equation (4.69) for some λ. Since we want the maximum correlation, we choose $\lambda = \lambda_1$. Denote the solution to (4.69) for $\lambda = \lambda_1$ by $\boldsymbol{\alpha}_1$ and $\boldsymbol{\gamma}_1$, and let $U_1 = \boldsymbol{\alpha}'_1\boldsymbol{X}$ and $V_1 = \boldsymbol{\gamma}'_1\boldsymbol{Y}$. Then U_1 and V_1 are normalized linear combinations of $\boldsymbol{X}$ and $\boldsymbol{Y}$, respectively, with maximum correlation. This completes the discussion of the first canonical correlation and canonical variates U_1 and V_1.

We can continue to introduce second canonical correlation and the associate canonical variates. The idea is to find a linear combination of $\boldsymbol{X}$, say $U_2 = \boldsymbol{\alpha}'\boldsymbol{X}$, that is orthogonal to U_1 and a linear combination of $\boldsymbol{Y}$, say $V_2 = \boldsymbol{\gamma}'\boldsymbol{Y}$, that is orthogonal to V_1 such that the correlation between U_2 and V_2 is maximum. This new pair of linear combinations must satisfy the normalization and orthogonality constraints. They can be obtained by a similar argument as that of the first pair of canonical variates. For details, readers are referred to Anderson (2003, Chapter 12).

In general, we can derive a single matrix equation for $\boldsymbol{\alpha}$ and $\boldsymbol{\gamma}$ as follows. Multiplying Equation (4.67) by λ and premultiplying Equation (4.68) by $\mathbf{\Sigma}_{yy}^{-1}$, we have

$$\lambda\mathbf{\Sigma}_{xy}\boldsymbol{\gamma} = \lambda^2\mathbf{\Sigma}_{xx}\boldsymbol{\alpha}, \tag{4.71}$$

$$\mathbf{\Sigma}_{yy}^{-1}\mathbf{\Sigma}_{yx}\boldsymbol{\alpha} = \lambda\boldsymbol{\gamma}. \tag{4.72}$$

Substitution from Equation (4.72) into Equation (4.71) gives

$$\mathbf{\Sigma}_{xy}\mathbf{\Sigma}_{yy}^{-1}\mathbf{\Sigma}_{yx}\boldsymbol{\alpha} = \lambda^2\mathbf{\Sigma}_{xx}\boldsymbol{\alpha}$$

or

$$(\mathbf{\Sigma}_{xy}\mathbf{\Sigma}_{yy}^{-1}\mathbf{\Sigma}_{yx} - \lambda^2\mathbf{\Sigma}_{xx})\boldsymbol{\alpha} = \mathbf{0}. \tag{4.73}$$

The quantities $\lambda_1^2, \ldots, \lambda_p^2$ satisfy

$$|\boldsymbol{\Sigma}_{xy}\boldsymbol{\Sigma}_{yy}^{-1}\boldsymbol{\Sigma}_{yx} - v\boldsymbol{\Sigma}_{xx}| = 0, \tag{4.74}$$

and the vectors $\boldsymbol{\alpha}_1, \ldots, \boldsymbol{\alpha}_p$ satisfy Equation (4.73) for $\lambda^2 = \lambda_1^2, \ldots, \lambda_p^2$, respectively. Similar argument shows that $\boldsymbol{\gamma}_1, \ldots, \boldsymbol{\gamma}_q$ occur when $\lambda^2 = \lambda_1^2, \ldots, \lambda_q^2$ for the equation

$$\left(\boldsymbol{\Sigma}_{yx}\boldsymbol{\Sigma}_{xx}^{-1}\boldsymbol{\Sigma}_{xy} - \lambda^2\boldsymbol{\Sigma}_{yy}\right)\boldsymbol{\gamma} = \mathbf{0}.$$

Note that Equation (4.73) is equivalent to

$$\boldsymbol{\Sigma}_{xx}^{-1}\boldsymbol{\Sigma}_{xy}\boldsymbol{\Sigma}_{yy}^{-1}\boldsymbol{\Sigma}_{yx}\boldsymbol{\alpha} = \lambda^2\boldsymbol{\alpha}.$$

EXERCISES

4.1 Write down the structure of a three-dimensional VARMA model if the Kronecker indices of the vector time series are $\{1, 2, 1\}$. How many parameters does the model employ if it includes the constant vector? You may exclude the covariance matrix $\boldsymbol{\Sigma}_a$ of the innovations in the calculation.

4.2 Write down the structure of a three-dimensional VARMA model if the Kronecker indices of the vector time series are $\{1, 0, 1\}$.

4.3 Write down the structure of a three-dimensional VARMA model if the vector time series has the following three components: SCM(0,0), SCM(0,1), and SCM(2,1). Are there any redundant parameters? Why?

4.4 Consider the three SCMs of Question 3. If the maximum elements, in absolute value, of the three rows of the transformation matrix $\boldsymbol{T}$ are (1,1), (2,3), and (3,2), respectively. Write down the transformation matrix $\boldsymbol{T}$ that consists of only estimable parameters. How many parameters does the resulting VARMA model contain? Again, you may include the constant vector, but not the $\boldsymbol{\Sigma}_a$ covariance matrix.

4.5 Consider the realized volatilities of the Alcoa stock from January 2, 2003, to May 7, 2004, for 340 observations. The realized volatilities are the sum of squares of intraday m-minute log returns. In this particular instance, we consider three series of realized volatilities obtained, respectively, from 5-minute, 10-minute, and 20-minute intraday log returns. The data are in the file `aa-3rv.txt`. Focus on the log series of the realized volatilities.

  (a) Identify the Kronecker indices of the three-dimensional log series of realized volatilities.

  (b) Use the Kronecker index approach to build a VARMA model for the three-dimensional log realized volatilities. Perform model checking and write down the fitted model.

(c) Identify the SCMs for the three-dimensional log series of realized volatilities.

(d) Use the SCM approach to build a VARMA model for the three-dimensional log realized volatilities. Perform model checking and write down the fitted model.

(e) Compare and contrast the two VARMA models.

4.6 Consider the monthly unemployment rates of Illinois, Michigan, and Ohio from January 1976 to November 2009. The data are seasonally adjusted and can be obtained from the Federal Reserve Bank of St. Louis (FRED). See the file `m-3state-un.txt`. Identity the Kronecker indices of the three-dimensional unemployment rates. No estimation is needed as the data were analyzed in Chapter 2.

4.7 Consider, again, the monthly unemployment rates of the states of Illinois, Michigan, and Ohio in the prior question. Specify three SCMs for the data. No estimation is needed.

REFERENCES

Akaike, H. (1976). Canonical correlation analysis of time series and the use of an information criterion. In R. K. Methra and D. G. Lainiotis (eds.). *Systems Identification: Advances and Case Studies*, pp. 27–96. Academic Press, New York.

Anderson, T. W. (2003). *An Introduction to Multivariate Statistical Analysis*. 3rd Edition. John Wiley & Sons, Inc, Hoboken, NJ.

Athanasopoulos, G. and Vahid, F. (2008). A complete VARMA modeling methodology based on scalar components. *Journal of Time Series Analysis*, **29**: 533–554.

Athanasopoulos, G., Poskitt, D. S., and Vahid, F. (2012). Two canonical VARMA forms: scalar component models vis-a-vis the Echelon form. *Econometric Reviews*, **31**: 60–83.

Bartlett, M. S. (1939). A note on tests of significance in multivariate analysis, *Proceedings of the Cambridge Philosophical Society*, **35**: 180–185.

Cooper, D. M. and Wood, E. F. (1982). Identifying multivariate time series models, *Journal of Time Series Analysis*, **3**: 153–164.

Hannan, E. J. and Deistler, M. (1988). *The Statistical Theory of Linear Systems*. John Wiley & Sons, Inc., New York.

Hotelling, H. (1935). The most predictable criterion. *Journal of Educational Psychology*, **26**: 139–142.

Hotelling, H. (1936). Relations between two sets of variates. *Biometrika*, **28**: 321–377.

Johnson, R. A. and Wichern, D. W. (2007). *Applied Multivariate Statistical Analysis*. 6th Edition. Prentice Hall, Upper Saddle River, NJ.

Min, W. and Tsay, R. S. (2005). On canonical analysis of multivariate time series. *Statistica Sinica*, **15**: 303–323.

Tiao, G. C. and Box, G. E. P. (1981). Modeling multiple time series with applications. *Journal of the American Statistical Association*, **76**: 802–816.

Tiao, G. C. and Tsay, R. S. (1989). Model specification in multivariate time series (with discussion). *Journal of the Royal Statistical Society, Series B*, **51**: 157–213.

Tsay, R. S. (1989). Identifying multivariate time series models. *Journal of Time Series Analysis*, **10**: 357–371.

Tsay, R. S. (1991). Two canonical forms for vector ARMA processes. *Statistica Sinica*, **1**: 247–269.

Tsay, R. S. and Ling, S. (2008). Canonical correlation analysis for the vector AR(1) model with ARCH innovations. *Journal of Statistical Planning and Inference*, **138**: 2826–2836.

CHAPTER 5

Unit-Root Nonstationary Processes

In the previous chapters, we focused our study on weakly stationary processes whose first two moments are time invariant. There are, however, many empirical time series that exhibit time-varying properties. For instance, stock prices and macroeconomic variables tend to exhibit certain stochastic trends so that their first moment is often time-varying. Figure 5.1 shows the time plot of U.S. quarterly real gross domestic product (GDP), in logarithm, from the first quarter of 1948 to the second quarter of 2012. The plot shows a clear upward trend. In this chapter, we study some special multivariate time series that exhibit certain stochastic trends. We refer to such series as unit-root nonstationary processes.

The simplest unit-root nonstationary time series is the univariate random walk, which can be written as

$$z_t = z_{t-1} + a_t, \tag{5.1}$$

where $\{a_t\}$ is an iid sequence of random variates with mean zero and finite variance σ_a^2. The model in Equation (5.1) can be treated as an AR(1) model with $\phi_1 = 1$, that is,

$$z_t = \phi_1 z_{t-1} + a_t, \quad \phi_1 = 1. \tag{5.2}$$

The characteristic equation of this special model is $1 - x = 0$, which has a solution $x = 1$. Consequently, the random walk process z_t is called a unit-root process. The solution $x = 1$ is on the unit circle so that z_t is not a stationary series.

In practice, one might be interested in verifying that an observed time series z_t follows a random walk model. To this end, one may consider the hypothesis testing: $H_0 : \phi_1 = 1$ versus $H_a : \phi_1 < 1$, where ϕ_1 is the coefficient of Equation (5.2) or in the following equation

$$z_t = \phi_0 + \phi_1 z_{t-1} + a_t. \tag{5.3}$$

Multivariate Time Series Analysis: With R and Financial Applications,
First Edition. Ruey S. Tsay.

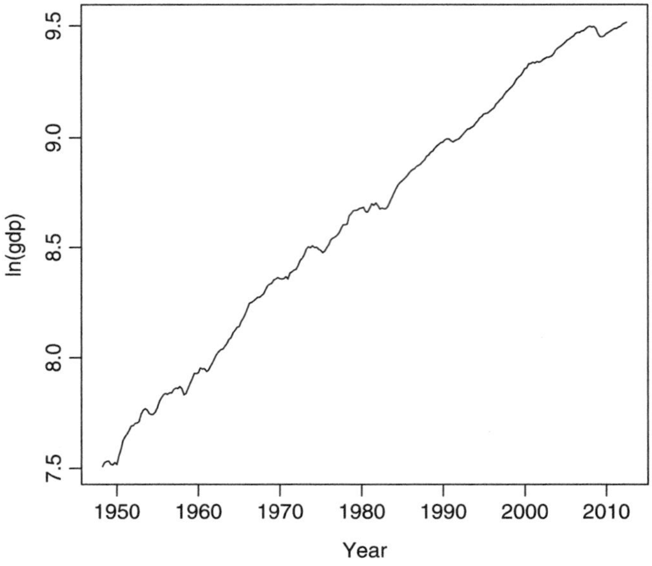

FIGURE 5.1 Time plot of U.S. quarterly real GDP, in logarithm, from 1948.I to 2012.II. The original GDP were in billions of chained 2005 dollars and measured to 1 decimal.

This testing problem is called unit-root testing. It had attracted much attention in the 1980s, and many test statistics are available. See Tsay (2010, Chapter 2), Hamilton (1994, Chapter 15) and the references therein. We shall consider the generalizations of the random walk model and unit-root testing to the multivariate case in this chapter. Our goal is to provide a better understanding of such unit-root nonstationary processes and to consider their applications.

Our study of unit-root nonstationary processes starts with a brief review of some fundamental properties of the random walk model in Equation (5.1). We then consider multivariate generalization of those fundamental properties and discuss multivariate models that are unit-root nonstationary. In particular, we study cointegration and error-correction model representation. Similar to the previous chapters, real examples are used to demonstrate the basic results.

5.1 UNIVARIATE UNIT-ROOT PROCESSES

We begin with a brief review of estimation and testing of unit roots in a univariate time series. There are two useful references available. The first reference is Phillips (1987), which focuses on a single unit root with weakly dependent innovations. This approach to unit root inference has been widely used in the econometric literature. The second reference is Chan and Wei (1988) considering all characteristic roots on the unit circle, such as complex unit roots, and unit roots with higher multiplicities, such as double unit roots. This latter paper assumes that the innovations a_t are martingale differences satisfying $E(|a_t|^{2+\delta}) < \infty$ for some $\delta > 0$. Tsay and Tiao (1990) extend some asymptotic results of Chan and Wei (1988) to the multivariate time series, again, allowing for all characteristic roots on the unit circle and with higher multiplicities.

5.1.1 Motivation

To motivate the unit-root theory, we consider the simplest case

$$z_t = z_{t-1} + \epsilon_t, \tag{5.4}$$

where $z_0 = 0$ and $\{\epsilon_t\}$ is a sequence of iid standard Gaussian random variables, that is, $\epsilon_t \sim N(0, 1)$. Suppose that the realization $\{z_1, \ldots, z_T\}$ is available and we wish to verify that the coefficient of z_{t-1} in Equation (5.4) is indeed 1. To this end, we consider the null hypothesis H_0: $\pi = 1$ versus the alternative hypothesis $H_a : \pi < 1$, where π is the coefficient of the AR(1) model

$$z_t = \pi z_{t-1} + e_t, \quad t = 1, \ldots, T,$$

where e_t denotes the error terms. The ordinary least-squares estimate (LSE) of π is

$$\hat{\pi} = \frac{\sum_{t=1}^{T} z_t z_{t-1}}{\sum_{t=1}^{T} z_{t-1}^2}.$$

The deviation of $\hat{\pi}$ from 1 can be expressed as

$$\hat{\pi} - 1 = \frac{\sum_{t=1}^{T} z_{t-1}\epsilon_t}{\sum_{t=1}^{T} z_{t-1}^2}. \tag{5.5}$$

To understand the asymptotic properties of the statistic $\hat{\pi} - 1$ in Equation (5.5), we need to study the limiting properties of z_t and both the numerator and denominator of Equation (5.5).

For the simple model in Equation (5.4), we have

$$z_t = \sum_{i=1}^{t} \epsilon_i = \epsilon_1 + \epsilon_2 + \cdots + \epsilon_t. \tag{5.6}$$

Therefore, $z_t \sim N(0, t)$. Furthermore,

$$z_t^2 = (z_{t-1} + \epsilon_t)^2 = z_{t-1}^2 + 2z_{t-1}\epsilon_t + \epsilon_t^2.$$

Therefore,

$$z_{t-1}\epsilon_t = \frac{z_t^2 - z_{t-1}^2 - \epsilon_t^2}{2}. \tag{5.7}$$

Summing over Equation (5.7) for $t = 1, \ldots, T$ and using $z_0 = 0$, we obtain

$$\sum_{t=1}^{T} z_{t-1}\epsilon_t = \frac{z_T^2}{2} - \frac{1}{2}\sum_{t=1}^{T}\epsilon_t^2.$$

Since $z_T \sim N(0, T)$, we divide the prior equation by T and obtain

$$\frac{1}{T}\sum_{t=1}^{T} z_{t-1}\epsilon_t = \frac{1}{2}\left[\left(\frac{z_T}{\sqrt{T}}\right)^2 - \frac{1}{T}\sum_{t=1}^{T}\epsilon_t^2\right]. \tag{5.8}$$

By Law of Large Numbers, the second term of Equation (5.8) converges in probability to 1, which is the variance of ϵ_t. On the other hand, $z_T/\sqrt{T} \sim N(0, 1)$. Therefore, the limiting distribution of the numerator of Equation (5.5) is

$$\frac{1}{T}\sum_{t=1}^{T} z_{t-1}\epsilon_t \Rightarrow \frac{1}{2}(\chi_1^2 - 1), \tag{5.9}$$

where χ_1^2 denotes a chi-square distribution with 1 degree of freedom.

It remains to study the denominator of Equation (5.5). Here, the fact that z_t is a partial sum of iid random variables plays an important role. See Equation (5.6). For any real number $r \in [0, 1]$, define the partial sum

$$\tilde{X}_T(r) = \frac{1}{\sqrt{T}} z_{[Tr]} = \frac{1}{\sqrt{T}}\sum_{t=1}^{[Tr]}\epsilon_t, \tag{5.10}$$

where $[Tr]$ denotes the integer part of $T \times r$, that is, the largest integer less than or equal to Tr. In particular, we have $\tilde{X}_T(1) = (1/\sqrt{T})z_T$, which is $N(0, 1)$. It turns out the limiting distribution of $\tilde{X}_T(r)$ is available in the literature. More specifically, for the z_t process in Equation (5.4),

$$\tilde{X}_T(r) \Rightarrow W(r), \quad \text{as} \quad T \to \infty,$$

where $W(r)$ is a standard Brownian motion or Wiener process and $\Rightarrow$ denotes weak convergence or convergence in distribution. A stochastic process $W(t)$ is a standard Brownian motion or Wiener process if (a) $W(0) = 0$, (b) the mapping $t : [0, 1] \to W(t)$ is continuous almost surely, and (c) $W(t)$ has independent increments with $W(t + h) - W(t)$ being independent of $W(t)$ and distributed as $N(0, h)$, where $h > 0$. See, for instance, Billingsley (2012, Section 37). The prior result is called the Donsker's Theorem; see Donsker (1951). The denominator of Equation (5.5) then becomes an integral of the Standard Brownian motion $W(r)$. Details are given in the next section as we generalize the innovation ϵ_t of z_t in Equation (5.4) to a linear stationary process.

Using the properties of standard Brownian motion, we have $W(1) \sim N(0,1)$ so that the result of Equation (5.9) can be written as

$$\frac{1}{T}\sum_{t=1}^{T} z_{t-1}\epsilon_t \Rightarrow \frac{1}{2}[W(1)^2 - 1].$$

This is the form commonly used in the literature and will be used in the rest of the chapter.

5.1.2 Unit Root with Stationary Innovations

Consider the discrete-time process $\{z_t\}$ generated by

$$z_t = \pi z_{t-1} + y_t, \quad t = 1, 2, \ldots, \tag{5.11}$$

where $\pi = 1$, z_0 is a fixed real number, and y_t is a stationary time series to be defined shortly. It will become clear later that the starting value z_0 has no effect on the limiting distributions discussed in this chapter.

Define the partial sum of $\{y_t\}$ as

$$S_t = \sum_{i=1}^{t} y_i. \tag{5.12}$$

For simplicity, we define $S_0 = 0$. Then, $z_t = S_t + z_0$ for $\pi = 1$. A fundamental result to unit-root theory is the limiting behavior of S_t. To this end, one must properly standardize the partial sum S_T as $T \to \infty$. It is common in the literature to employ the average variance of S_T given by

$$\sigma^2 = \lim_{T\to\infty} E(T^{-1}S_T^2), \tag{5.13}$$

which is assumed to exist and positive. Define

$$X_T(r) = \frac{1}{\sqrt{T}\sigma} S_{[Tr]}, \quad 0 \le r \le 1, \tag{5.14}$$

where, as before, $[Tr]$ denotes the integer part of Tr. Under certain conditions, $X_T(r)$ is shown to converge weakly to the well-known standard Brownian motion or the Wiener process. This is commonly referred to as the *functional central limit theorem.*

Assumption A. Assume that $\{y_t\}$ is a stationary time series such that (a) $E(y_t) = 0$ for all t, (b) $\sup_t E(|y_t|^\beta) < \infty$ for some $\beta > 2$, (c) the average variance σ^2 of Equation (5.13) exists and is positive, and (d) y_t is strong mixing with mixing coefficients α_m that satisfy

$$\sum_{m=1}^{\infty} \alpha_m^{1-2/\beta} < \infty.$$

5.1.2.1 Strong Mixing

Strong mixing is a measure of serial dependence of a time series $\{y_t\}$. Let $F_{-\infty}^q$ and F_r^∞ be the σ-field generated by $\{y_q, y_{q-1}, \ldots\}$ and $\{y_r, y_{r+1}, \ldots\}$, respectively. That is, $F_{-\infty}^q = F\{y_q, y_{q-1}, \ldots\}$ and $F_r^\infty = F\{y_r, y_{r+1}, \ldots\}$. We say that y_t satisfies a strong mixing condition if there exists a positive function $\alpha(.)$ satisfying $\alpha_n \to 0$ as $n \to \infty$ so that

$$|P(A \cap B) - P(A)P(B)| < \alpha_{r-q}, \quad A \in F_{-\infty}^q, \quad B \in F_r^\infty.$$

If y_t is strong mixing, then the serial dependence between y_t and y_{t-h} approaches 0 as h increases. The strong mixing condition is in general hard to verify. It suffices to say that the stationary and invertible ARMA(p, q) process considered in this book satisfies the strong mixing condition, where p and q are finite nonnegative integers. This is easy to understand as the autocorrelations of such an ARMA process decay exponentially.

The following two theorems are widely used in unit-root study. See Herrndorf (1984) and Billingsley (1999) for more details.

5.1.2.2 Functional Central Limit Theorem (FCLT)

If $\{y_t\}$ satisfies Assumption A, then $X_T(r) \Rightarrow W(r)$, where $W(r)$ is a standard Brownian motion for $r \in [0, 1]$ and $\Rightarrow$ denotes weak convergence, that is, convergence in distribution.

5.1.2.3 Continuous Mapping Theorem

If $X_T(r) \Rightarrow W(r)$ and $h(.)$ is a continuous functional on D[0,1], the space of all real valued functions on [0,1] that are right continuous at each point on [0,1] and have finite left limits, then $h(X_T(r)) \Rightarrow h(W(r))$ as $T \to \infty$.

Remark: Instead of the strong mixing condition, one can work directly with the linear process $y_t = \sum_{i=0}^{\infty} \psi_i a_{t-i}$, where $\psi_0 = 1$ and the coefficients satisfy $\sum_{v=1}^{\infty} v|\psi_v| < \infty$. The innovation a_t is further assumed to have finite fourth moment. The FCLT continues to apply under these new conditions. Again stationary ARMA processes meet the ψ-weight condition as the ψ_v decays exponentially. □

The ordinary LSE of π in Equation (5.11) is

$$\hat{\pi} = \frac{\sum_{t=1}^{T} z_{t-1} z_t}{\sum_{t=1}^{T} z_{t-1}^2},$$

and its variance is estimated by

$$\text{Var}(\hat{\pi}) = \frac{s^2}{\sum_{t=1}^{T} z_{t-1}^2},$$

where s^2 is the residual variance given by

$$s^2 = \frac{1}{T-1} \sum_{t=1}^{T} (z_t - \hat{\pi} z_{t-1})^2. \tag{5.15}$$

The usual t-ratio for testing the null hypothesis $H_0 : \pi = 1$ versus $H_a : \pi < 1$ is given by

$$t_\pi = \left(\sum_{t=1}^{T} z_{t-1}^2 \right)^{1/2} \frac{\hat{\pi} - 1}{s} = \frac{\sum_{t=1}^{T} z_{t-1} y_t}{s\sqrt{\sum_{t=1}^{T} z_{t-1}^2}}. \tag{5.16}$$

We have the following basic results of unit-root process z_t.

Theorem 5.1 Suppose that $\{y_t\}$ satisfies Assumption A and $\sup_t E|y_t|^{\beta+\eta} < \infty$, where $\beta > 2$ and $\eta > 0$, then as $T \to \infty$, we have

(a) $T^{-2} \sum_{t=1}^{T} z_{t-1}^2 \Rightarrow \sigma^2 \int_0^1 W(r)^2 dr$

(b) $T^{-1} \sum_{t=1}^{T} z_{t-1} y_t \Rightarrow \sigma^2/2 \left[W(1)^2 - \sigma_y^2/\sigma^2 \right]$

(c) $T(\hat{\pi} - 1) \Rightarrow (\frac{1}{2})(W(1)^2 - (\sigma_y^2/\sigma^2))/\int_0^1 W(r)^2 dr$

(d) $\hat{\pi} \to_p 1$, where $\to_p$ denotes convergence in probability

(e) $t_\pi \Rightarrow (\sigma/(2\sigma_y))[W(1)^2 - (\sigma_y^2/\sigma^2)]/\left[\int_0^1 W(r)^2 dr\right]^{1/2}$,

where σ^2 and σ_y^2 are defined as

$$\sigma^2 = \lim_{T\to\infty} E(T^{-1} S_T^2), \quad \sigma_y^2 = \lim_{T\to\infty} T^{-1} \sum_{t=1}^{T} E(y_t^2).$$

Proof. For part (a), we have

$$\begin{aligned}
\frac{1}{T^2}\sum_{t=1}^{T} z_{t-1}^2 &= T^{-2}\sum_{t=1}^{T}(S_{t-1}+z_0)^2 \\
&= T^{-2}\sum_{t=1}^{T}(S_{t-1}^2+2z_0S_{t-1}+z_0^2) \\
&= \sigma^2\sum_{t=1}^{T}\left(\frac{1}{\sigma\sqrt{T}}S_{t-1}\right)^2\frac{1}{T}+2z_0\sigma T^{-1/2}\sum_{t=1}^{T}\left(\frac{1}{\sigma\sqrt{T}}S_{t-1}\right)\frac{1}{T}+\frac{z_0^2}{T} \\
&= \sigma^2\sum_{t=1}^{T}\int_{(t-1)/T}^{t/T}\left(\frac{1}{\sigma\sqrt{T}}S_{[Tr]}\right)^2 dr \\
&\quad + \frac{2z_0\sigma}{T^{1/2}}\sum_{t=1}^{T}\int_{(t-1)/T}^{t/T}\frac{1}{\sigma\sqrt{T}}S_{[Tr]}dr+\frac{z_0^2}{T} \\
&= \sigma^2\int_0^1 X_T^2(r)dr+2z_0\sigma T^{-1/2}\int_0^1 X_T(r)dr+\frac{z_0^2}{T} \\
&\Rightarrow \sigma^2\int_0^1 W(r)^2dr, \quad T\to\infty.
\end{aligned}$$

For part (b), we have

$$\begin{aligned}
T^{-1}\sum_{t=1}^{T}z_{t-1}y_t &= T^{-1}\sum_{t=1}^{T}(S_{t-1}+z_0)y_t \\
&= T^{-1}\sum_{t=1}^{T}S_{t-1}y_t+z_0\bar{y} \\
&= T^{-1}\sum_{t=1}^{T}\frac{1}{2}(S_t^2-S_{t-1}^2-y_t^2)+z_0\bar{y} \\
&= (2T)^{-1}S_T^2-(2T)^{-1}\sum_{t=1}^{T}y_t^2+z_0\bar{y} \\
&= \frac{\sigma^2}{2}X_T(1)^2-\frac{1}{2}T^{-1}\sum_{t=1}^{T}y_t^2+z_0\bar{y} \\
&\Rightarrow \frac{\sigma^2}{2}\left[W(1)^2-\frac{\sigma_y^2}{\sigma^2}\right],
\end{aligned}$$

because $\bar{y}\to 0$ and $T^{-1}\sum_{t=1}^{T}y_t^2\to\sigma_y^2$ almost surely as $T\to\infty$.

Part (c) follows parts (a) and (b) and the Continuous Mapping Theorem. Part (d) follows part (c). □

When $y_t = \epsilon_t \sim N(0,1)$, part (b) of Theorem 5.1, as expected, reduces to Equation (5.9). From part (d) of Theorem 5.1, the LSE $\hat{\pi}$ converges to 1 at the rate T^{-1}, not the usual rate $T^{-1/2}$. This is referred to as the super consistency in the unit-root literature. Using the fast convergence rate of $\hat{\pi}$, parts (a) and (b), and Equation (5.15), we have

$$\begin{aligned} s^2 &= \frac{1}{T-1}\sum_{t=1}^{T}(z_t - \hat{\pi} z_{t-1})^2 \\ &= \frac{1}{T-1}\sum_{t=1}^{T}[(z_t - z_{t-1}) + (1-\hat{\pi})z_{t-1}]^2 \\ &= \frac{1}{T-1}\sum_{t=1}^{T} y_t^2 + \frac{2(1-\hat{\pi})}{T-1}\sum_{t=1}^{T} z_{t-1}y_t + \frac{(1-\hat{\pi})^2}{T-1}\sum_{t-1}^{T} z_{t-1}^2 \\ &\to_p \sigma_y^2, \end{aligned}$$

because the last two terms vanish as $T \to \infty$. Thus, the t-ratio can be written as

$$\begin{aligned} t_\pi &= \frac{\sum_{t=1}^{T} z_{t-1}y_t}{\sigma_y(\sum_{t=1}^{T} z_{t-1}^2)^{0.5}}, \\ &= \frac{\sigma^{-2}T^{-1}\sum_{t=1}^{T} z_{t-1}a_t}{\sigma_y[(\sigma^{-2}T^{-1})^2\sum_{t=1}^{T} z_{t-1}^2]^{1/2}}. \end{aligned}$$

Using parts (a) and (b) of Theorem 5.1 and the Continuous Mapping Theorem, we have

$$t_\pi \to_d \frac{\sigma/2\sigma_y[W(1)^2 - \sigma_y^2/\sigma^2]}{[\int_0^1 W(r)^2 dr]^{1/2}}.$$

Using similar methods as those of Theorem 5.1, one can further establish the following results, which are useful when fitting models with a constant or time trend.

Theorem 5.2 Under the same assumption as Theorem 5.1, we have

(a) $T^{-3/2}\sum_{t=1}^{T} t y_t \Rightarrow \sigma\left[W(1) - \int_0^1 W(r)dr\right]$

(b) $T^{-3/2}\sum_{t=1}^{T} z_{t-1} \Rightarrow \sigma \int_0^1 W(r)dr$

(b) $T^{-5/2}\sum_{t=1}^{T} t z_{t-1} \Rightarrow \sigma \int_0^1 rW(r)dr$

(c) $T^{-3}\sum_{t=1}^{T} t z_{t-1}^2 \Rightarrow \sigma^2 \int_0^1 rW^2(r)dr$,

In what follows, we consider unit-root properties for some special time series. Since y_t is stationary with zero mean, we have

$$E(S_T^2) = T\gamma_0 + 2\sum_{i=1}^{T-1}(T-i)\gamma_i, \tag{5.17}$$

where γ_i is the lag-i autocovariance of y_t. Moreover, for a stationary linear process y_t with MA representation $y_t = \sum_{i=0}^{\infty} \psi_i a_{t-i}$, where $\psi_0 = 1$, we have $\gamma_i = \sigma_a^2 \sum_{j=0}^{\infty} \psi_j \psi_{j+i}$ for $i = 0, 1, \ldots$. In this particular case, one can show that

$$\begin{aligned}\sigma^2 &= \lim_{T\to\infty} E(T^{-1}S_T^2) = \gamma_0 + 2\sum_{i=1}^{\infty}\gamma_i \\ &= \sigma_a^2\left(\sum_{i=0}^{\infty}\psi_i^2 + 2\sum_{i=1}^{\infty}\sum_{j=0}^{\infty}\psi_j\psi_{i+j}\right) = [\sigma_a\psi(1)]^2,\end{aligned} \tag{5.18}$$

where $\psi(1) = \sum_{i=0}^{\infty} \psi_i$.

5.1.3 AR(1) Case

Consider the simple AR(1) model $z_t = \pi z_{t-1} + a_t$, that is, $y_t = a_t$ being an iid sequence of random variables with mean zero and variance $\sigma_a^2 > 0$. In this case, it is easy to see that, from Equation (5.17),

$$\sigma^2 = \lim_{T\to\infty} E(T^{-1}S_T^2) = \sigma_a^2, \quad \sigma_y^2 = \sigma_a^2.$$

Consequently, for the random walk model in Equation (5.1), we have

1. $T^{-2}\sum_{t=1}^{T} z_{t-1}^2 \Rightarrow \sigma_a^2 \int_0^1 W^2(r)dr$
2. $T^{-1}\sum_{t=1}^{T} z_{t-1}(z_t - z_{t-1}) \Rightarrow \sigma_a^2/2[W^2(1) - 1]$
3. $T(\hat{\pi} - 1) \Rightarrow 0.5[W^2(1) - 1]/\int_0^1 W^2(r)dr$
4. $t_\pi \Rightarrow 0.5[W^2(1) - 1]/[\int_0^1 W^2(r)dr]^{1/2}$.

The critical values of t_π has been tabulated by several authors. See, for instance, Fuller (1976, Table 8.5.2).

5.1.4 AR(p) Case

We start with the AR(2) case in which $(1-B)(1-\phi B)z_t = a_t$, where $|\phi| < 1$. The model can be written as

$$z_t = z_{t-1} + y_t, \quad y_t = \phi y_{t-1} + a_t.$$

For the stationary AR(1) process y_t, $\sigma_y^2 = \sigma_a^2/(1-\phi^2)$ and $\gamma_i = \phi^i\gamma_0$. Thus, by Equation (5.17), $\sigma^2 = \sigma_a^2/(1-\phi)^2$. Consequently, the limiting distributions discussed depend on the AR(1) coefficient ϕ. For instance, the t-ratio of $\hat{\pi}$ becomes

$$t_\pi \Rightarrow \frac{\frac{\sqrt{1+\phi}}{2\sqrt{1-\phi}}\left[W(1)^2 - \frac{1-\phi}{1+\phi}\right]}{[\int_0^1 W(r)^2 dr]^{1/2}}.$$

Consequently, it is difficult to use t_π in unit-root testing because the asymptotic critical values of t_π depend on the nuisance parameter ϕ. This type of dependence continues to hold for the general AR(p) process y_t. To overcome this difficulty, Said and Dickey (1984) consider the *augmented Dickey–Fuller* test statistic.

For an AR(p) process, $\phi(B)z_t = a_t$ with $p > 1$, we focus on the case that $\phi(B) = \phi^*(B)(1-B)$, where $\phi^*(B)$ is a stationary AR polynomial. This z_t process contains a unit root. Let $\phi^*(B) = 1 - \sum_{i=1}^{p-1} \phi_i^* B^i$. The model becomes $\phi(B)z_t = \phi^*(B)$ $(1-B)z_t = (1-B)z_t - \sum_{i=1}^{p-1} \phi_i^*(1-B)z_{t-i} = a_t$. Testing for a unit root in $\phi(B)$ is equivalent to testing $\pi = 1$ in the model

$$z_t = \pi z_{t-1} + \sum_{j=1}^{p-1} \phi_j^*(z_{t-j} - z_{t-j-1}) + a_t.$$

Or equivalently, the same as testing for $\pi - 1 = 0$ in the model

$$\Delta z_t = (\pi - 1)z_{t-1} + \sum_{j=1}^{p-1} \phi_j^* \Delta z_{t-j} + a_t,$$

where $\Delta z_t = z_t - z_{t-1}$. The prior model representation is the univariate version of error-correction form. It is easy to verify that (a) $\pi - 1 = -\phi(1) = \sum_{i=1}^{p} \phi_i - 1$ and $\phi_j^* = -\sum_{i=j+1}^{p} \phi_i$. In practice, the linear model

$$\Delta z_t = \beta z_{t-1} + \sum_{j=1}^{p-1} \phi_j^* \Delta z_{t-j} + a_t, \tag{5.19}$$

where $\beta = \pi - 1$, is used. LSE of β can then be used in unit-root testing. Specifically, testing $H_o : \pi = 1$ versus $H_a : \pi < 1$ is equivalent to testing $H_o : \beta = 0$ versus $H_a : \beta < 0$. It can be shown that the t-ratio of $\hat{\beta}$ (against 0) has the same limiting distribution as t_π in the random-walk case. In other words, for an AR(p) model with $p > 1$, by including the lagged variables of Δz_t in the linear regression of Equation (5.19), one can remove the nuisance parameters in unit-root testing. This is the well-known augmented Dickey–Fuller unit-root test. Furthermore, the limiting distribution of the LSE $\hat{\phi}_i^*$ in Equation (5.19) is the same as that of fitting an AR($p - 1$) model to Δz_t. In other words, limiting properties of the estimates for the stationary part remain unchanged when we treat the unit root as known *a priori*.

Remark: In our discussion, we assume that there is no constant term in the model. One can include a constant term in the model and obtain the associated limiting distribution for unit-root testing. The limiting distribution will be different, but the idea remains the same. □

5.1.5 MA(1) Case

Next, assume that $z_t = z_{t-1} + y_t$, where $y_t = a_t - \theta a_{t-1}$ with $|\theta| < 1$. In this case, we have $\gamma_0 = (1+\theta^2)\sigma_a^2$, $\gamma_1 = -\theta\sigma_a^2$, and $\gamma_i = 0$ for $i > 1$. Consequently, $\sigma_y^2 = (1+\theta^2)\sigma_a^2$ and, by Equation (5.17), $\sigma^2 = (1-\theta)^2\sigma_a^2$. The limiting distributions of unit-root statistics become

1. $T^{-2}\sum_{t=1}^T z_{t-1}^2 \Rightarrow (1-\theta)^2\sigma_a^2 \int_0^1 W^2(r)dr$,
2. $T^{-1}\sum_{t=1}^T z_{t-1}(z_t - z_{t-1}) \Rightarrow (1-\theta)^2\sigma_a^2/2[W^2(1) - 1 + \theta^2/(1-\theta)^2]$,
3. $T(\hat{\pi} - 1) \Rightarrow 1 - \theta/2\sqrt{1+\theta^2}[W^2(1) - 1 + \theta^2/(1-\theta)^2]/\int_0^1 W^2(r)dr$
4. $t_\pi \to_d 1 - \theta/2\sqrt{1+\theta^2}[W^2(1) - 1 + \theta^2/(1-\theta)^2]/[\int_0^1 W^2(r)dr]^{1/2}$.

From the results, it is clear that when θ is close to 1 the asymptotic behavior of t_π is rather different from that of the case when y_t is a white noise series. This is not surprising because when θ approaches 1, the z_t process is close to being a white noise series. This might explain the severe size distortions of Phillips–Perron unit-root test statistics seen in Table 1 of Phillips and Perron (1988).

5.1.6 Unit-Root Tests

Suppose that the univariate time series z_t follows the AR(p) model $\phi(B)z_t = a_t$, where $\phi(B) = (1-B)\phi^*(B)$ such that $\phi^*(1) \neq 0$. That is, z_t has a single unit root. In this subsection, we summarize the framework of unit-root tests commonly employed in the literature. See, for example, Dickey and fuller (1979). Three models are often employed in the test. They are

1. No constant:

$$\Delta z_t = \beta z_{t-1} + \sum_{i=1}^{p-1} \phi_i^* \Delta z_{t-i} + a_t. \tag{5.20}$$

2. With constant:

$$\Delta z_t = \alpha + \beta z_{t-1} + \sum_{i=1}^{p-1} \phi_i^* \Delta z_{t-i} + a_t. \tag{5.21}$$

3. With constant and time trend:

$$\Delta z_t = \omega_0 + \omega_1 t + \beta z_{t-1} + \sum_{i=1}^{p-1} \phi_i^* \Delta z_{t-i} + a_t. \tag{5.22}$$

The null hypothesis of interest is $H_o : \beta = 0$ and the alternative hypothesis is $H_a : \beta < 0$. The test statistic is the t-ratio of the LSE of β.

The limiting distribution of the t-ratio of the LSE of β in Equation (5.20) is the same as that of t_π discussed before; see Said and Dickey (1984). On the other hand, the limiting distributions of the t-ratio of the LSE of β in Equations (5.21) and (5.22) are the same as the t-ratio of testing $\pi = 1$ in the models

$$z_t = \alpha + \pi z_{t-1} + y_t, \tag{5.23}$$

and

$$z_t = \omega_0 + \omega_1 t + \pi z_{t-1} + y_t, \tag{5.24}$$

respectively. These limiting distributions can be derived using the results of Theorem 5.2. They are functions of the standard Brownian motion under the null hypothesis of a single unit root. Interested readers are referred to Hamilton (1994, Chapter 17) for detail. Critical values of these t-ratios have been obtained via simulation in the literature. See, for example, Fuller (1976, Table 8.5.2).

Phillips and Perron (1988) employ Equations (5.11), (5.23), and (5.24) to perform unit-root tests. They use modified t-ratio of the LSE of π. The modification is to take care of the linear dependence and/or conditional heteroscedasticity in the y_t process. Briefly speaking, they proposed nonparametric estimates for σ_y^2 and σ^2 so that their impact on the test statistics can be mitigated. For example, σ^2 can be estimated using the result in Equation (5.17).

Remark: The augmented Dickey–Fuller test is available in the package `fUnitRoots`. Many other unit-root tests, including Phillips and Perron (1988) tests, are available in the package `urca` of R. □

5.1.7 Example

Consider the series of U.S. quarterly gross domestic product (GDP) from the first quarter of 1948 to the second quarter of 2012. The original GDP data were obtained from the Federal Reserve Bank at St Louis. The data were in billions of chained 2005 dollars measured to 1 decimal and seasonally adjusted. We employ the logarithm of GDP in our analysis. For augmented Dickey–Fuller test, the three models discussed in Equations (5.20)–(5.22) are denoted by the subcommand `type` being **nc**, **c**, **ct**, respectively. For Phillips and Perron test, we only consider the case with a constant. All tests fail to reject the null hypothesis of a unit root in the U.S. log GDP series.

R Demonstration: Unit-root testing. Output edited.

```
> library(fUnitRoots)
> da=read.table("q-ungdp-4812.txt",header=T)
> dim(da)
[1] 258   4
```

```
> head(da)
  year mon    unemp      gdp
1 1948   1 3.733333 7.507580
> library(fUnitRoots)
> m1=ar(diff(gdp),method="mle") % Find error-correction lags
> m1$order
[1] 3
> adfTest(gdp,lags=3)
Title: Augmented Dickey-Fuller Test
Test Results:
  PARAMETER:
    Lag Order: 3
  STATISTIC:
    Dickey-Fuller: 5.5707
  P VALUE:
    0.99
> adfTest(gdp,lags=3,type="c")
Title: Augmented Dickey-Fuller Test
Test Results:
  PARAMETER:
    Lag Order: 3
  STATISTIC:
    Dickey-Fuller: -2.0176
  P VALUE:
    0.3056
> adfTest(gdp,lags=3,type="ct")
Title: Augmented Dickey-Fuller Test
Test Results:
  PARAMETER:
    Lag Order: 3
  STATISTIC:
    Dickey-Fuller: -1.625
  P VALUE:
    0.7338
> library(urca)
> urppTest(gdp) % Phillips and Perron test
Title: Phillips-Perron Unit Root Test
Test Results:
    Test regression with intercept
  Coefficients:
              Estimate Std. Error t value Pr(>|t|)
  (Intercept) 0.031192   0.008801   3.544 0.000468 ***
  y.l1        0.997289   0.001018 979.834  < 2e-16 ***
  ---
  Value of test-statistic, type: Z-alpha  is: -0.7218
             aux. Z statistics
  Z-tau-mu             2.7627
```

5.2 MULTIVARIATE UNIT-ROOT PROCESSES

The unit-root results of Section 5.1 can be generalized to the multivariate case. We start with the definition of a k-dimensional standard Brownian motion. Let $\boldsymbol{W}(r) = (W_1(r), \ldots, W_k(r))'$ be a k-dimensional process, where $W_i(r)$ is a function from $[0, 1] \to R$, with R being the real number.

Definition 5.1 The k-dimensional standard Brownian motion $\boldsymbol{W}(r)$ is a continuous-time process from $[0, 1]$ to R^k satisfying the following conditions:

1. $\boldsymbol{W}(0) = \boldsymbol{0}$.
2. For any sequence $\{r_i | i = 1, \ldots, n\}$ such that $0 \leq r_1 < r_2 < \cdots < r_n \leq 1$, the increments $\boldsymbol{W}(r_2) - \boldsymbol{W}(r_1)$, $\boldsymbol{W}(r_3) - \boldsymbol{W}(r_2), \cdots, \boldsymbol{W}(r_n) - \boldsymbol{W}(r_{n-1})$ are independent multivariate Gaussian random vectors with $[\boldsymbol{W}(t) - \boldsymbol{W}(s)] \sim N[\boldsymbol{0}, (t-s)\boldsymbol{I}_k]$, where $t > s$.
3. For any given realization, $\boldsymbol{W}(r)$ is continuous in r with probability 1. □

Let $\{\boldsymbol{\epsilon}_t\}$ be a sequence of iid k-dimensional random vectors with mean zero and $\text{Var}(\boldsymbol{\epsilon}_t) = \boldsymbol{I}_k$. Define the partial sum

$$\tilde{\boldsymbol{X}}_T(r) = \frac{1}{\sqrt{T}} \sum_{t=1}^{[Tr]} \boldsymbol{\epsilon}_t.$$

The generalization of the Donsker's Theorem of Section 5.1 then becomes

$$\tilde{\boldsymbol{X}}_T(r) \Rightarrow \boldsymbol{W}(r).$$

In particular, $\tilde{\boldsymbol{X}}_T(1) \Rightarrow \boldsymbol{W}(1)$. The generalization of the limiting distribution

$$T^{-1} \sum_{t=1}^{T} z_{t-1}\epsilon_t \Rightarrow \frac{1}{2}[W(1)^2 - 1],$$

to the multivariate case is a bit more complicated. It turns out to be

$$\frac{1}{T} \sum_{t=1}^{T} (\boldsymbol{z}_{t-1}\boldsymbol{\epsilon}_t' + \boldsymbol{\epsilon}_t \boldsymbol{z}_{t-1}') \Rightarrow \boldsymbol{W}(1)[\boldsymbol{W}(1)]' - \boldsymbol{I}_k. \tag{5.25}$$

Phillips (1988) derived an alternative expression using differential of Brownian motion. The result is

$$\frac{1}{T} \sum_{t=1}^{T} \boldsymbol{z}_{t-1}\boldsymbol{\epsilon}_t' \Rightarrow \int_0^1 \boldsymbol{W}(r)[d\boldsymbol{W}(r)]'. \tag{5.26}$$

The definition of the differential $d\boldsymbol{W}(r)$ and the derivation of Equation (5.26) are relatively involved. Interested readers are referred to Phillips (1988) for detail. Comparing Equations (5.25) and (5.26), we have

$$\int_0^1 W_i(r)dW_j(r) + \int_0^1 W_j(r)dW_i(r) = W_i(1)W_j(1), \quad i \neq j,$$

where $W_i(r)$ is the ith component of $\boldsymbol{W}(r)$, and

$$\int_0^1 W_i(r)dW_i(r) = \frac{1}{2}[W_i^2(1) - 1].$$

Next, let $\{\boldsymbol{a}_t\}$ be a sequence of iid k-dimensional random vectors with mean zero and positive-definite covariance matrix $\boldsymbol{\Sigma}_a$. We can write $\boldsymbol{\Sigma}_a = \boldsymbol{P}_a\boldsymbol{P}_a'$, where $\boldsymbol{P}_a$ is a $k \times k$ matrix. For instance, $\boldsymbol{P}_a$ can be the Cholesky factor or the positive-definite square-root matrix of $\boldsymbol{\Sigma}_a$. Define the partial sum of $\boldsymbol{a}_t$ as

$$\boldsymbol{X}_T^a(r) = \frac{1}{\sqrt{T}}\sum_{t=1}^{[Tr]} \boldsymbol{a}_t.$$

We can think of $\boldsymbol{a}_t = \boldsymbol{P}_a\boldsymbol{\epsilon}_t$ so that

$$\boldsymbol{X}_T^a(r) = \boldsymbol{P}_a\frac{1}{\sqrt{T}}\sum_{t=1}^{[Tr]} \boldsymbol{\epsilon}_t = \boldsymbol{P}_a\tilde{\boldsymbol{X}}_T(r).$$

Consequently, we have

$$\boldsymbol{X}_T^a(r) \Rightarrow \boldsymbol{P}_a\boldsymbol{W}(r). \tag{5.27}$$

Since $\boldsymbol{W}(r) \sim N(\boldsymbol{0}, r\boldsymbol{I}_k)$, we have $\boldsymbol{P}_a\boldsymbol{W}(r) \sim N(\boldsymbol{0}, r\boldsymbol{P}_a\boldsymbol{P}_a') = N(\boldsymbol{0}, r\boldsymbol{\Sigma}_a)$. In this sense, we say that $\boldsymbol{P}_a\boldsymbol{W}(r)$ is a k-dimensional Brownian motion with covariance matrix $r\boldsymbol{\Sigma}_a$.

Finally, we consider the linear stationary innovations $\boldsymbol{y}_t$ and study the generalization of Theorems 5.1 and 5.2 to the multivariate case. To this end, we adopt the approach of Phillips and Solo (1992). Consider the k-dimensional vector process

$$\boldsymbol{z}_t = \boldsymbol{z}_{t-1} + \boldsymbol{y}_t, \tag{5.28}$$

where $\boldsymbol{z}_t = (z_{1t}, \ldots, z_{kt})'$ and $\boldsymbol{y}_t = (y_{1t}, \ldots, y_{kt})'$ such that $\boldsymbol{y}_t$ is a linear process

$$\boldsymbol{y}_t = \sum_{i=0}^{\infty} \boldsymbol{\psi}_i\boldsymbol{a}_{t-i}, \tag{5.29}$$

where $\{\boldsymbol{a}_t\}$ is a sequence of iid random vectors with mean $\mathbf{0}$ and positive-definite covariance matrix $\text{Var}(\boldsymbol{a}_t) = \boldsymbol{\Sigma}_a$, $\boldsymbol{\psi}_0 = \boldsymbol{I}_k$, and $\boldsymbol{\psi}_v = [\psi_{v,ij}]$ is a $k \times k$ real-valued matrix satisfying

$$\sum_{v=1}^{\infty} v|\psi_{v,ij}| < \infty, \tag{5.30}$$

for $i, j = 1, \cdots, k$. We shall further assume that $\boldsymbol{a}_t$ has finite fourth moments.

Define the partial sum of $\{\boldsymbol{y}_t\}$ process as

$$\boldsymbol{S}_t = \sum_{i=1}^{t} \boldsymbol{y}_i. \tag{5.31}$$

Again, for simplicity we assume $\boldsymbol{S}_0 = \mathbf{0}$.

Lemma 5.1 For the linear process $\boldsymbol{y}_t$ in Equation (5.29) with the ψ-weights satisfying the summable condition in Equation (5.30), the partial sum can be written as

$$\boldsymbol{S}_t = \boldsymbol{\psi}(1) \sum_{v=1}^{t} \boldsymbol{a}_v + \boldsymbol{\eta}_t - \boldsymbol{\eta}_0,$$

where $\boldsymbol{\psi}(1) = \sum_{v=0}^{\infty} \boldsymbol{\psi}_v$ and $\boldsymbol{\eta}_t = \sum_{s=0}^{\infty} \boldsymbol{\alpha}_s \boldsymbol{a}_{t-s}$ for $\boldsymbol{\alpha}_s = \sum_{v=0}^{s} \boldsymbol{\psi}_v - \boldsymbol{\psi}(1)$.

Proof of Lemma 5.1 is given in the Appendix. It suffices to say that the coefficient matrices $\{\boldsymbol{\alpha}_s | s = 0, \ldots, \infty\}$ are absolutely summable. The result of Lemma 5.1 is a multivariate generalization of the Beveridge–Nelson decomposition.

Using the partial sum in Equation (5.31), we define

$$\boldsymbol{X}_T(r) = \frac{1}{\sqrt{T}} \boldsymbol{S}_{[Tr]}, \quad 0 \leq r \leq 1. \tag{5.32}$$

By Lemma 5.1, we have

$$\boldsymbol{X}_T(r) = \frac{1}{\sqrt{T}} \left(\boldsymbol{\psi}(1) \sum_{i=1}^{[Tr]} \boldsymbol{a}_i + \boldsymbol{\eta}_{[Tr]} - \boldsymbol{\eta}_0 \right).$$

Under the assumption of finite fourth moment and using Chebyshev's inequality, one can show that

$$\sup_{r \in [0,1]} \frac{1}{\sqrt{T}} |\eta_{i,[Tr]} - \eta_{i,0}| \to_p 0,$$

for $i = 1, \ldots, k$, where $\eta_{i,t}$ is the ith component of $\boldsymbol{\eta}_t$. See, for instance, Hamilton (1994, Chapter 17). Next, by the multivariate Donsker's Theorem in Equation (5.27), we have

$$\boldsymbol{X}_T(r) \to_p \boldsymbol{\psi}(1)\boldsymbol{P}_a\tilde{\boldsymbol{X}}_T(r) \Rightarrow \boldsymbol{\psi}(1)\boldsymbol{P}_a\boldsymbol{W}(r), \tag{5.33}$$

where $\boldsymbol{\psi}(1)\boldsymbol{P}_1\boldsymbol{W}(r)$ is distributed as $N(\boldsymbol{0}, r[\boldsymbol{\psi}(1)\boldsymbol{\Sigma}_a\boldsymbol{\psi}(1)'])$. This generalizes the functional central limit theorem to multivariate time series with stationary linear innovation process $\boldsymbol{y}_t$. Generalizations of other limiting properties can be obtained similarly with some requiring more care because of matrix multiplication. We summarize the results into Theorem 5.3.

Theorem 5.3 Let $\boldsymbol{y}_t$ be a stationary linear process in Equation (5.29) with the ψ-weight coefficients satisfying Equation (5.30) and $\{\boldsymbol{a}_t\}$ being an iid sequence of random vectors with mean zero, positive-definite covariance matrix $\boldsymbol{\Sigma}_a = \boldsymbol{P}_a\boldsymbol{P}_a'$, and finite fourth moments. Let $\boldsymbol{S}_t = \boldsymbol{z}_t$ be the partial sum of $\boldsymbol{y}_t$ defined in Equation (5.31) and T the sample size. Define $\boldsymbol{\Lambda} = \boldsymbol{\psi}(1)\boldsymbol{P}_a$, where $\boldsymbol{\psi}(1) = \sum_{v=0}^{\infty}\boldsymbol{\psi}_v$. Then

(a) $T^{-1/2}\boldsymbol{z}_T \Rightarrow \boldsymbol{\Lambda}\boldsymbol{W}(1)$

(b) $T^{-1}\sum_{t=1}^{T}(\boldsymbol{z}_{t-s}\boldsymbol{y}_t' + \boldsymbol{y}_{t-s}\boldsymbol{z}_{t-1}') \Rightarrow \begin{cases} \boldsymbol{\Xi} - \boldsymbol{\Gamma}_0 & \text{for } s = 0 \\ \boldsymbol{\Xi} + \sum_{v=-s+1}^{s-1}\boldsymbol{\Gamma}_v & \text{for } s = 1, 2, \ldots, \end{cases}$
where $\boldsymbol{\Xi} = \boldsymbol{\Lambda}\boldsymbol{W}(1)[\boldsymbol{W}(1)]'\boldsymbol{\Lambda}'$ and $\boldsymbol{\Gamma}_j$ is the lag-j autocovariance matrix of $\boldsymbol{y}_t$.

(c) $T^{-1}\sum_{t=1}^{T}\boldsymbol{z}_{t-1}\boldsymbol{y}_t' \Rightarrow \boldsymbol{\Lambda}\left(\int_0^1\boldsymbol{W}(r)[d\boldsymbol{W}(r)]'\right)\boldsymbol{\Lambda}' + \sum_{v=1}^{\infty}\boldsymbol{\Gamma}_v'$

(d) $T^{-3/2}\sum_{t=1}^{T}\boldsymbol{z}_{t-1} \Rightarrow \boldsymbol{\Lambda}\int_0^1\boldsymbol{W}(r)dr$

(e) $T^{-3/2}\sum_{t=1}^{T}t\boldsymbol{y}_{t-1} \Rightarrow \boldsymbol{\Lambda}\left(\boldsymbol{W}(1) - \int_0^1\boldsymbol{W}(r)dr\right)$

(f) $T^{-2}\sum_{t=1}^{T}\boldsymbol{z}_{t-1}\boldsymbol{z}_{t-1}' \Rightarrow \boldsymbol{\Lambda}\left(\int_0^1\boldsymbol{W}(r)[\boldsymbol{W}(r)]'dr\right)\boldsymbol{\Lambda}'$

(g) $T^{-5/2}\sum_{t=1}^{T}t\boldsymbol{z}_{t-1} \Rightarrow \boldsymbol{\Lambda}\int_0^1 r\boldsymbol{W}(r)dr$

(h) $T^{-3}\sum_{t=1}^{T}t\boldsymbol{z}_{t-1}\boldsymbol{z}_{t-1}' \Rightarrow \boldsymbol{\Lambda}\left(\int_0^1 r\boldsymbol{W}(r)[\boldsymbol{W}(r)]'dr\right)\boldsymbol{\Lambda}'$.

5.2.1 An Alternative Model Representation

To encompass the possibility of unit roots in a k-dimensional time series $\boldsymbol{z}_t$, we write the VARMA(p, q) model as

$$\boldsymbol{\Phi}(B)\boldsymbol{z}_t = \boldsymbol{c}(t) + \boldsymbol{\Theta}(B)\boldsymbol{a}_t, \tag{5.34}$$

where $\boldsymbol{\Phi}(B) = \boldsymbol{I}_k - \boldsymbol{\Phi}_1B - \cdots - \boldsymbol{\Phi}_pB^p$, $\boldsymbol{c}(t) = \boldsymbol{c}_0 + \boldsymbol{c}_1t$, $\boldsymbol{\Theta}(B) = \boldsymbol{I}_k - \boldsymbol{\Theta}_1B - \cdots - \boldsymbol{\Theta}_qB^q$, and $\{\boldsymbol{a}_t\}$ is a sequence of iid random vectors with mean zero and positive-definite covariance matrix $\boldsymbol{\Sigma}_a$. The AR and MA order p and q are nonnegative integers and $\boldsymbol{c}_0$ and $\boldsymbol{c}_1$ are k-dimensional constant vectors. Furthermore, we assume that the VARMA model in Equation (5.34) is identifiable; see conditions discussed in Chapters 3 and 4. However, we allow 1 to be a solution of the determinant equation $|\boldsymbol{\Phi}(B)| = 0$ of the AR matrix polynomial of the VARMA model. If $|\boldsymbol{\Phi}(1)| = 0$, then $\boldsymbol{z}_t$ has, at least, a unit root so that $\boldsymbol{z}_t$ is unit-root nonstationary. In real applications, the VARMA process $\boldsymbol{z}_t$ typically does not involve any time trend so that $\boldsymbol{c}_1 = \boldsymbol{0}$. We use the general form here to facilitate cointegration tests to be discussed later in the chapter.

Let $\Delta \boldsymbol{z}_t = \boldsymbol{z}_t - \boldsymbol{z}_{t-1} = (1 - B)\boldsymbol{z}_t$ be the first differenced series of $\boldsymbol{z}_t$. This implies that $\Delta \boldsymbol{z}_t$ consists of the increments of the components of $\boldsymbol{z}_t$. We shall rewrite the AR matrix polynomial so that we can explicitly study the unit-root structure of $\boldsymbol{z}_t$. The new model representation can be done in several ways. We consider an approach commonly used in the literature. To begin, consider the special case of $p = 3$. Here, the AR matrix polynomial can be rewritten as

$$\begin{aligned}
\boldsymbol{\Phi}(B) &= \boldsymbol{I}_k - \boldsymbol{\Phi}_1 B - \boldsymbol{\Phi}_2 B^2 - \boldsymbol{\Phi}_3 B^3 \\
&= \boldsymbol{I}_k - \boldsymbol{\Phi}_1 B - \boldsymbol{\Phi}_2 B^2 - \boldsymbol{\Phi}_3 B^2 + \boldsymbol{\Phi}_3 B^2 - \boldsymbol{\Phi}_3 B^3 \\
&= \boldsymbol{I}_k - \boldsymbol{\Phi}_1 B - (\boldsymbol{\Phi}_2 + \boldsymbol{\Phi}_3) B^2 + \boldsymbol{\Phi}_3 B^2 (1 - B) \\
&= \boldsymbol{I}_k - \boldsymbol{\Phi}_1 B - (\boldsymbol{\Phi}_2 + \boldsymbol{\Phi}_3) B + (\boldsymbol{\Phi}_2 + \boldsymbol{\Phi}_3) B - (\boldsymbol{\Phi}_2 + \boldsymbol{\Phi}_3) B^2 \\
&\quad + \boldsymbol{\Phi}_3 B^2 (1 - B) \\
&= \boldsymbol{I}_k - (\boldsymbol{\Phi}_1 + \boldsymbol{\Phi}_2 + \boldsymbol{\Phi}_3) B + (\boldsymbol{\Phi}_2 + \boldsymbol{\Phi}_3) B (1 - B) \\
&\quad + \boldsymbol{\Phi}_3 B^2 (1 - B).
\end{aligned} \tag{5.35}$$

The second equality of the prior equation holds because we simply subtract and add $\boldsymbol{\Phi}_3 B^2$ to $\boldsymbol{\Phi}(B)$. We do so in order to obtain $\boldsymbol{\Phi}_3 B^2(1 - B)$ shown in the right side of step 3. The fourth equality holds because we subtract and add $(\boldsymbol{\Phi}_2 + \boldsymbol{\Phi}_3)B$ to the equation. Define

$$\begin{aligned}
\boldsymbol{\Pi}^* &= \boldsymbol{\Phi}_1 + \boldsymbol{\Phi}_2 + \boldsymbol{\Phi}_3 \\
\boldsymbol{\Phi}_2^* &= -\boldsymbol{\Phi}_3 \\
\boldsymbol{\Phi}_1^* &= -(\boldsymbol{\Phi}_2 + \boldsymbol{\Phi}_3).
\end{aligned}$$

Equation (5.35) then becomes

$$\boldsymbol{\Phi}(B) = \boldsymbol{I}_k - \boldsymbol{\Pi}^* B - [\boldsymbol{\Phi}_1^* B + \boldsymbol{\Phi}_2^* B^2](1 - B). \tag{5.36}$$

Using Equation (5.36), a general VARMA(3, q) model can be expressed as

$$\boldsymbol{z}_t = \boldsymbol{\Pi}^* \boldsymbol{z}_{t-1} + \boldsymbol{\Phi}_1^* \Delta \boldsymbol{z}_{t-1} + \boldsymbol{\Phi}_2^* \Delta \boldsymbol{z}_{t-2} + \boldsymbol{c}(t) + \boldsymbol{\Theta}(B) \boldsymbol{a}_t. \tag{5.37}$$

Next, subtracting $\boldsymbol{z}_{t-1}$ from both sides of Equation (5.37), we obtain

$$\Delta \boldsymbol{z}_t = \boldsymbol{\Pi} \boldsymbol{z}_{t-1} + \boldsymbol{\Phi}_1^* \Delta \boldsymbol{z}_{t-1} + \boldsymbol{\Phi}_2^* \Delta \boldsymbol{z}_{t-2} + \boldsymbol{c}(t) + \boldsymbol{\Theta}(B) \boldsymbol{a}_t, \tag{5.38}$$

where $\boldsymbol{\Pi} = \boldsymbol{\Pi}^* - \boldsymbol{I}_k = -\boldsymbol{\Phi}(1)$.

The techniques used to derive Equations (5.35)–(5.38) continue to apply to the general VARMA(p, q) model in Equation (5.34). In general, for VARMA(p, q) models, define

$$\begin{aligned} \boldsymbol{\Pi} &= -\boldsymbol{\Phi}(1) = \boldsymbol{\Phi}_1 + \cdots + \boldsymbol{\Phi}_p - \boldsymbol{I}_k \\ \boldsymbol{\Phi}_j^* &= -(\boldsymbol{\Phi}_{j+1} + \cdots + \boldsymbol{\Phi}_p), \quad j = 1, \ldots, p-1. \end{aligned} \tag{5.39}$$

Then, we have

$$\boldsymbol{\Phi}(B) = \boldsymbol{I}_k - \boldsymbol{\Pi}^* B - (\boldsymbol{\Phi}_1^* B + \cdots + \boldsymbol{\Phi}_{p-1}^* B^{p-1})(1 - B), \tag{5.40}$$

where $\boldsymbol{\Pi}^* = \boldsymbol{\Phi}_1 + \cdots + \boldsymbol{\Phi}_p = \boldsymbol{\Pi} + \boldsymbol{I}_k$. Equation (5.40) can also be shown by equating the coefficient matrices of B^i on both sides of the equation, leading to the identities defined in Equation (5.39). The VARMA model can, therefore, be written as

$$\boldsymbol{z}_t = \boldsymbol{\Pi}^* \boldsymbol{z}_{t-1} + \sum_{j=1}^{p-1} \boldsymbol{\Phi}_j^* \Delta \boldsymbol{z}_{t-j} + \boldsymbol{c}(t) + \boldsymbol{\Theta}(B) \boldsymbol{a}_t, \tag{5.41}$$

where $\boldsymbol{\Pi}^* = \boldsymbol{\Phi}_1 + \cdots + \boldsymbol{\Phi}_p = \boldsymbol{\Pi} + \boldsymbol{I}_k$, or equivalently, we have

$$\Delta \boldsymbol{z}_t = \boldsymbol{\Pi} \boldsymbol{z}_{t-1} + \sum_{j=1}^{p-1} \boldsymbol{\Phi}_j^* \Delta \boldsymbol{z}_{t-j} + \boldsymbol{c}(t) + \boldsymbol{\Theta}(B) \boldsymbol{a}_t. \tag{5.42}$$

Let $\boldsymbol{\Phi}^*(B) = \boldsymbol{I}_k - \boldsymbol{\Phi}_1^* B - \cdots - \boldsymbol{\Phi}_{p-1}^* B^{p-1}$. We assume that all solutions of the determinant equation $|\boldsymbol{\Phi}^*(B)| = 0$ are outside the unit circle. That is, $|\boldsymbol{\Phi}^*(x)| \neq 0$ if $|x| \leq 1$.

If $\boldsymbol{z}_t$ contains a unit root, then $|\boldsymbol{\Phi}(1)| = \boldsymbol{0}$. By the definition of $\boldsymbol{\Pi}$ in Equation (5.39), we have $|\boldsymbol{\Pi}| = 0$. On the other hand, $|\boldsymbol{\Pi}| = 0$ does not necessarily imply that $|\boldsymbol{\Phi}(1)| = 0$. Of course, $\boldsymbol{\Pi} = \boldsymbol{0}$ implies that $\boldsymbol{\Phi}_1 + \cdots + \boldsymbol{\Phi}_p = \boldsymbol{I}_k$ and $|\boldsymbol{\Phi}(1)| = 0$.

From Equation (5.40), we have

$$\begin{aligned} \boldsymbol{\Phi}(B) &= \boldsymbol{I}_k - \boldsymbol{I}_k B + \boldsymbol{I}_k B - \boldsymbol{\Pi}^* B - (\boldsymbol{\Phi}_1^* B + \cdots + \boldsymbol{\Phi}_{p-1}^* B^{p-1})(1 - B) \\ &= \boldsymbol{I}_k(1 - B) - (\boldsymbol{\Pi}^* - \boldsymbol{I}_k) B - (\boldsymbol{\Phi}_1^* B + \cdots + \boldsymbol{\Phi}_{p-1}^* B^{p-1})(1 - B) \\ &= \boldsymbol{\Phi}^*(B)(1 - B) - \boldsymbol{\Pi} B. \end{aligned} \tag{5.43}$$

This expression also leads to Equation (5.42) and will be used later.

5.2.2 Unit-Root VAR Processes

To gain further insight into unit-root properties and to have a closed-form expression for parameter estimates, we focus on the VAR(p) processes in this section. Also, we assume that there is no time trend so that $\boldsymbol{c}(t) = \boldsymbol{c}_0$. In other words, we assume $\boldsymbol{z}_t$ follows the VAR(p) model

$$\boldsymbol{z}_t = \boldsymbol{c}_0 + \boldsymbol{\Phi}_1 \boldsymbol{z}_{t-1} + \cdots + \boldsymbol{\Phi}_p \boldsymbol{z}_{t-p} + \boldsymbol{a}_t. \tag{5.44}$$

In this particular case, Equations (5.41) and (5.42) become

$$\boldsymbol{z}_t = \boldsymbol{c}_0 + \boldsymbol{\Pi}^* \boldsymbol{z}_{t-1} + \sum_{j=1}^{p-1} \boldsymbol{\Phi}_j^* \Delta \boldsymbol{z}_{t-j} + \boldsymbol{a}_t, \tag{5.45}$$

and

$$\Delta \boldsymbol{z}_t = \boldsymbol{c}_0 + \boldsymbol{\Pi} \boldsymbol{z}_{t-1} + \sum_{j=1}^{p-1} \boldsymbol{\Phi}_j^* \Delta \boldsymbol{z}_{t-j} + \boldsymbol{a}_t, \tag{5.46}$$

respectively, where $\boldsymbol{\Pi}^*$, $\boldsymbol{\Phi}_j^*$, and $\boldsymbol{\Pi}$ are given in Equation (5.39).

The null hypothesis of interest here is $H_0 : \boldsymbol{\Pi} = \boldsymbol{0}$, or equivalently $\boldsymbol{\Pi}^* = \boldsymbol{\Phi}_1 + \cdots + \boldsymbol{\Phi}_p = \boldsymbol{I}_k$. In other words, we are interested in testing the hypothesis that each component z_{it} of $\boldsymbol{z}_t$ has a unit root. We shall employ Equation (5.45) to perform LS estimation.

As discussed in Chapter 2, LSEs of parameters in Equation (5.45) can be estimated component-by-component. Thus, we consider the component model

$$z_{it} = c_{0i} + \boldsymbol{\Pi}_i^* \boldsymbol{z}_{t-1} + \boldsymbol{\Phi}_{1,i}^* \Delta \boldsymbol{z}_{t-1} + \cdots + \boldsymbol{\Phi}_{p-1,i}^* \Delta \boldsymbol{z}_{t-p+1} + a_{it}, \tag{5.47}$$

where c_{0i} is the ith element of $\boldsymbol{c}_0$, $\boldsymbol{\Pi}_i^*$ is ith row of $\boldsymbol{\Pi}^*$, and $\boldsymbol{\Phi}_{j,i}^*$ denotes the ith row of $\boldsymbol{\Phi}_j^*$. Here, row vector is used for a coefficient matrix. Note that under the null hypothesis, $\boldsymbol{\Pi}^* = \boldsymbol{I}_k$ so that $\boldsymbol{\Pi}_i^*$ is simply the ith unit vector in R^k. Following Chapter 2, we express Equation (5.47) as

$$z_{it} = \boldsymbol{x}_t' \boldsymbol{\beta}_i + a_{it},$$

where $\boldsymbol{x}_t' = (1, \boldsymbol{z}_{t-1}', \Delta \boldsymbol{z}_{t-1}', \ldots, \Delta \boldsymbol{z}_{t-p+1}')$ and $\boldsymbol{\beta}_i' = (c_{0i}, \boldsymbol{\Pi}_i^*, \boldsymbol{\Phi}_{1,i}^*, \ldots, \boldsymbol{\Phi}_{p-1,i}^*)$. Let $\hat{\boldsymbol{\beta}}_i$ be the ordinary LSE estimate of $\boldsymbol{\beta}_i$. Then, we have

$$\hat{\boldsymbol{\beta}}_i - \boldsymbol{\beta}_i = \left(\sum_{t=1}^{T} \boldsymbol{x}_t \boldsymbol{x}_t' \right)^{-1} \left(\sum_{t=1}^{T} \boldsymbol{x}_t a_{it} \right). \tag{5.48}$$

Case I. ***z_t has no drift.***
To obtain the asymptotic properties of $\hat{\boldsymbol{\beta}}_i$, we make use of Theorem 5.3. To this end, we further assume $\boldsymbol{c}_0 = \mathbf{0}$, that is, there is no drift in the VAR process $\boldsymbol{z}_t$. Using the notation of Theorem 5.3, we have $\boldsymbol{y}_t = \Delta \boldsymbol{z}_t$, which is a stationary VAR($p-1$) process with $\boldsymbol{\psi}(B) = [\boldsymbol{\Phi}^*(B)]^{-1}$. Also, the matrix $\boldsymbol{\Lambda}$ becomes $\boldsymbol{\Lambda} = [\boldsymbol{\Phi}^*(1)]^{-1}\boldsymbol{P}_a$, where $\boldsymbol{P}_a$ satisfies $\boldsymbol{P}_a\boldsymbol{P}_a' = \boldsymbol{\Sigma}_a = \text{Cov}(\boldsymbol{a}_t)$. Under the unit-root null hypothesis, elements of $\sum_{t=1}^T \boldsymbol{x}_t\boldsymbol{x}_t'$ are of different stochastic orders so that they must be normalized accordingly. Defining

$$\boldsymbol{N}_T = \text{diag}\{T^{1/2}, T\boldsymbol{I}_k, T^{1/2}\boldsymbol{I}_{k(p-1)}\},$$

we have

$$\boldsymbol{N}_T[\hat{\boldsymbol{\beta}}_i - \boldsymbol{\beta}_i] = \left(\boldsymbol{N}_T^{-1}\sum_{t=1}^T \boldsymbol{x}_t\boldsymbol{x}_t'\boldsymbol{N}_T^{-1}\right)^{-1}\left(\boldsymbol{N}_T^{-1}\sum_{t=1}^T \boldsymbol{x}_t a_{it}\right).$$

Let $\boldsymbol{\Gamma}(p-1)$ be the covariance matrix of the $k(p-1)$-dimensional process $(\boldsymbol{y}_{t-1}', \ldots, \boldsymbol{y}_{t-p+1}')'$, which exists because $\boldsymbol{y}_t = \Delta\boldsymbol{z}_t$ is a stationary VAR($p-1$) process. Applying parts (a), (b), (c), (d), and (f) of Theorem 5.3 and properties of stationary VAR process, we obtain the following results.

Corollary 5.1 Assume that the VAR(p) process $\boldsymbol{z}_t$ follows the model in Equation (5.45) with $\boldsymbol{\Pi}^* = \boldsymbol{I}_k$ and $\boldsymbol{c}_0 = \mathbf{0}$. The noise series $\boldsymbol{a}_t$ is a sequence of iid random vectors with mean zero, positive-definite covariance matrix $\boldsymbol{\Sigma}_a$, and finite fourth moments. Then,

(a) $\boldsymbol{N}_T^{-1}(\sum_{t=1}^T \boldsymbol{x}_t\boldsymbol{x}_t')\boldsymbol{N}_T^{-1} \Rightarrow \begin{bmatrix} \boldsymbol{Q} & \mathbf{0} \\ \mathbf{0} & \boldsymbol{\Gamma}(p-1)\end{bmatrix}$,

(b) $\boldsymbol{N}_T^{-1}\sum_{t=1}^T \boldsymbol{x}_t a_{it} \Rightarrow \begin{bmatrix} \boldsymbol{g}_1 \\ \boldsymbol{g}_2 \end{bmatrix}$,

(c) $\boldsymbol{N}_T[\hat{\boldsymbol{\beta}}_i - \boldsymbol{\beta}_i] \Rightarrow \begin{bmatrix} \boldsymbol{Q}^{-1}\boldsymbol{g}_1 \\ \{\boldsymbol{\Gamma}(p-1)\}^{-1}\boldsymbol{g}_2 \end{bmatrix}$,

where

$$\boldsymbol{Q} = \begin{bmatrix} 1 & \left[\int_0^1 \boldsymbol{W}(r)dr\right]'\boldsymbol{\Lambda}' \\ \boldsymbol{\Lambda}\int_0^1 \boldsymbol{W}(r)dr & \boldsymbol{\Lambda}\left\{\int_0^1 \boldsymbol{W}(r)[\boldsymbol{W}(r)]'dr\right\}\boldsymbol{\Lambda}' \end{bmatrix},$$

$$\boldsymbol{g}_1 = \begin{bmatrix} \boldsymbol{\Pi}_i^*\boldsymbol{P}_a\boldsymbol{W}(1) \\ \boldsymbol{\Lambda}\left\{\int_0^1 \boldsymbol{W}(r)[d\boldsymbol{W}(r)]'\right\}\boldsymbol{P}_a'\boldsymbol{\Pi}_i^* \end{bmatrix},$$

$$\boldsymbol{g}_2 \sim N[\mathbf{0}, \sigma_{ii}\boldsymbol{\Gamma}(p-1)], \quad \sigma_{ii} = E(a_{it}^2).$$

Corollary 5.1 has several important implications for making inference using Equation (5.47). The asymptotic distribution of $\hat{\boldsymbol{\Pi}}_i^*$ is non-Gaussian so that the test statistic, for example, the F-ratio, of the null hypothesis $H_0 : \boldsymbol{\Pi}_i^* = \boldsymbol{e}_i$ with $\boldsymbol{e}_i$ being the ith unit vector of R^k, has a nonstandard limiting distribution. Critical values of such test statistics need to be obtained by simulation. Next, under the null hypothesis, $\boldsymbol{z}_{t-1}$ is unit-root nonstationary and $\Delta \boldsymbol{z}_t$ is stationary. The diagonal block structure of Corollary 5.1(a) says that $\boldsymbol{z}_{t-1}$ and $\Delta \boldsymbol{z}_{t-j}$ $(j = 1, \ldots, p-1)$ are asymptotically uncorrelated. See the definition of $\boldsymbol{x}_t$. This is understandable because a stationary time series cannot depend on a unit-root nonstationary series. Corollary 5.1(c) implies that the ordinary LSEs of $\boldsymbol{\Phi}_j^*$ of Equation (5.47) have the same asymptotic distribution as the model

$$\Delta \boldsymbol{z}_t = c_{0i} + \boldsymbol{\Phi}_{1,i}^* \Delta \boldsymbol{z}_{t-1} + \cdots + \boldsymbol{\Phi}_{p-1,i}^* \Delta \boldsymbol{z}_{t-p+1} + a_{it}. \tag{5.49}$$

Therefore, test statistic of any linear hypothesis that involves only $\boldsymbol{\Phi}_j^*$ has the usual asymptotic chi-square distribution as that of Equation (5.49). In other words, one does not need to concern about Brownian motion if the goal is simply to making inference on $\boldsymbol{\Phi}_j^*$. In this case, either Equation (5.47) or Equation (5.49) can be used in estimation. Of course, one may prefer Equation (5.49) in finite sample due to numerical consideration.

Corollary 5.1 also shows that the convergence rate of $\hat{\boldsymbol{\Pi}}_i^*$ is T, whereas those of other estimate is $T^{1/2}$. Consequently, test statistic of a single linear hypothesis that involves $\boldsymbol{\Pi}_i^*$ and $\boldsymbol{\Phi}_j^*$ would be dominated asymptotically by the behavior of $\hat{\Phi}_{j,i}^*$. Indeed, the limiting distribution of such a test statistic is the same as that of treating $\boldsymbol{\Pi}_i^* = \boldsymbol{e}_i$. For example, consider the ordinary LSEs of the VAR(p) model in Equation (5.44). Using the transformation in Equation (5.39), we have

$$\begin{aligned}
\hat{\boldsymbol{\Phi}}_p &= -\hat{\boldsymbol{\Phi}}_{p-1}^*, \\
\hat{\boldsymbol{\Phi}}_j &= \hat{\boldsymbol{\Phi}}_j^* - \hat{\boldsymbol{\Phi}}_{j-1}^*, \quad j = p-1, \ldots, 2 \\
\hat{\boldsymbol{\Phi}}_1 &= \hat{\boldsymbol{\Pi}}^* + \hat{\boldsymbol{\Phi}}_1^*.
\end{aligned} \tag{5.50}$$

It is then clear that the convergence rate of $\hat{\boldsymbol{\Phi}}_j$ for $j = 1, \ldots, p$ is $T^{1/2}$. Consequently, ordinary LSEs of Equation (5.44) have a limiting Gaussian distribution and the usual test statistics concerning coefficients $\boldsymbol{\Phi}_j$ continue to apply. This justifies the conventional inference used in the statistical packages. In the literature, this property has been used. For instance, Tsay (1984) shows that the information criteria such as AIC continue to apply to univariate autoregressive models with characteristic roots on the unit circle. Finally, it should be noted that any inference involving $\boldsymbol{\Pi}^*$ explicitly would have the nonstandard limiting distribution.

Case II. ***Some components of z_t contain drift.***

Consider a univariate random walk with a nonzero constant,

$$z_t = c + z_{t-1} + a_t,$$

where $\{a_t\}$ represents a sequence of iid random variables with mean zero and variance $\sigma_a^2 > 0$. The model can be rewritten as

$$z_t = z_0 + ct + w_t, \quad \text{with} \quad w_t = \sum_{i=1}^{t} a_i,$$

where z_0 is the initial value and w_t is a random walk. See, for instance, Tsay (2010, Chapter 2). Using $T^{-(v+1)} \sum_{t=1}^{T} t^v \to 1/(v+1)$ for non-negative integer v and properties of pure random walk series w_t, it is seen that asymptotic properties of z_t is dominated by the time trend ct. This feature continues to hold for the multivariate case. However, we need to isolate the drift, if any, in the multivariate series $\boldsymbol{z}_t$ to provide a better understanding of the asymptotic behavior of $\boldsymbol{z}_t$. In the literature, unit-root VAR models with nonzero drift have been investigated by Chan (1989), West (1988), Sims, Stock and Watson (1990), among others.

A unit-root VAR(p) model with drift can be written as

$$\boldsymbol{z}_t = \boldsymbol{c}_0 + \boldsymbol{\Phi}_1 \boldsymbol{z}_{t-1} + \cdots + \boldsymbol{\Phi}_p \boldsymbol{z}_{t-p} + \boldsymbol{a}_t, \tag{5.51}$$

where $\boldsymbol{c}_0 \neq \boldsymbol{0}$, $|\boldsymbol{\Phi}(1)| = 0$, and $\{\boldsymbol{a}_t\}$ is a sequence of iid random vectors with mean zero, positive-definite covariance matrix $\boldsymbol{\Sigma}_a$, and finite fourth moments. In addition, $\Delta \boldsymbol{z}_t = (1 - B)\boldsymbol{z}_t$ is stationary so that there exists no double unit roots or other characteristic roots on the unit circle. This VAR(p) model can be rewritten in the form of Equation (5.46) with $\boldsymbol{\Pi} = \sum_{i=1}^{p} \boldsymbol{\Phi}_i - \boldsymbol{I}_k$. Under the null hypothesis and model assumptions considered in this section, $\boldsymbol{\Pi} = \boldsymbol{0}$ and all solutions of the polynomial equation $|\boldsymbol{\Phi}^*(B)| = 0$ are outside the unit cirlce. Let $\boldsymbol{\mu} = (\mu_1, \ldots, \mu_k)' = E(\Delta \boldsymbol{z}_t)$. By properties of VAR models in Chapter 2, we have

$$\boldsymbol{\mu} = (\boldsymbol{I} - \boldsymbol{\Phi}_1^* - \cdots - \boldsymbol{\Phi}_{p-1}^*)^{-1} \boldsymbol{c}_0.$$

Consequently, we obtain

$$\begin{aligned} \Delta \boldsymbol{z}_t &= \boldsymbol{\mu} + \boldsymbol{y}_t, \\ \boldsymbol{y}_t &= \boldsymbol{\psi}(B) \boldsymbol{a}_t, \quad \text{with} \quad \boldsymbol{\psi}(B) = [\boldsymbol{\Phi}^*(B)]^{-1}. \end{aligned} \tag{5.52}$$

Thus, $\boldsymbol{y}_t$ is a stationary VAR($p-1$) process following the model

$$(\boldsymbol{I}_k - \boldsymbol{\Phi}_1^* B - \cdots - \boldsymbol{\Phi}_{p-1}^* B^{p-1}) \boldsymbol{y}_t = \boldsymbol{a}_t,$$

where all solutions of $|\boldsymbol{\Phi}^*(B)| = 0$ are outside the unit circle.

Equation (5.52) implies that

$$\boldsymbol{z}_t = \boldsymbol{z}_0 + \boldsymbol{\mu} t + \boldsymbol{y}_1 + \boldsymbol{y}_2 + \cdots + \boldsymbol{y}_t, \tag{5.53}$$

where $\boldsymbol{z}_0$ denotes the initial value of $\boldsymbol{z}_t$. Since $\boldsymbol{c}_0 \neq 0$, $\boldsymbol{\mu} \neq \mathbf{0}$ and without loss of generality, we assume that $\mu_k \neq 0$. To see the effects of time trend, we consider the transformations

$$
\begin{aligned}
z_{k,t}^* &= z_{k,t} \\
z_{i,t}^* &= z_{i,t} - (\mu_1/\mu_k) z_{k,t}, \quad i = 1, \ldots, k-1.
\end{aligned}
$$

From Equation (5.53), we have, for $i = 1, \ldots, k-1$,

$$
\begin{aligned}
z_{i,t}^* &= (z_{i,0} + \mu_i t + y_{i,1} + \cdots + y_{i,t}) - \frac{\mu_i}{\mu_k}(z_{k,0} + \mu_k t + y_{k,1} + \cdots + y_{k,t}) \\
&\equiv z_{i,0}^* + s_{i,t}^*,
\end{aligned}
$$

where $s_{i,t}^* = y_{i,1}^* + \cdots + y_{i,t}^*$ with $y_{i,j}^* = y_{i,j} - (\mu_i/\mu_k) y_{k,j}$. There is no time trend in $z_{i,t}^*$.

Define the $(k-1)$-dimensional processes $\boldsymbol{z}_t^* = (z_{1,t}^*, \ldots, z_{k-1,t}^*)'$ and $\boldsymbol{y}_t^* = (y_{1,t}^*, \ldots, y_{k-1,t}^*)'$. It is then easily seen that

$$
\boldsymbol{y}_t^* = \boldsymbol{\psi}^*(B)\boldsymbol{a}_t,
$$

where $\boldsymbol{\psi}^*(B)$ is a $(k-1) \times k$ matrix polynomial given by

$$
\boldsymbol{\psi}^*(B) = \boldsymbol{D}\boldsymbol{\psi}(B),
$$

where $\boldsymbol{D} = [\boldsymbol{I}_{k-1}, -\boldsymbol{\delta}]$ with $\boldsymbol{\delta} = (\mu_1, \ldots, \mu_{k-1})'/\mu_k$. Since $\{v\boldsymbol{\psi}_v\}_{v=0}^{\infty}$ is absolutely summable, $\{v\boldsymbol{\psi}_v^*\}_{v=0}^{\infty}$ is also absolutely summable. With the newly defined variables, Equation (5.45) can be expressed as

$$
\boldsymbol{z}_t = \boldsymbol{c} + \boldsymbol{\omega}\boldsymbol{z}_{t-1}^* + \boldsymbol{\gamma} z_{k,t-1} + \boldsymbol{\Phi}_1^* \boldsymbol{y}_{t-1} + \cdots + \boldsymbol{\Phi}_{p-1}^* \boldsymbol{y}_{t-p+1} + \boldsymbol{a}_t, \tag{5.54}
$$

where $\boldsymbol{\omega}$ is a $k \times (k-1)$ matrix of coefficients and $\boldsymbol{\gamma}$ is a k-dimensional vector of coefficients. This equation decomposes $\boldsymbol{z}_t$ into the constant term $\boldsymbol{c}$, the $(k-1)$-dimensional driftless integrated regressors $\boldsymbol{z}_{t-1}^*$, the zero-mean stationary regressors $\boldsymbol{y}_{t-j}$, and a term dominated by a time trend $y_{n,t-1}$. Clearly the ordinary LSEs of the coefficients matrices in Equation (5.44) or in Equation (5.45) can be obtained from those of Equation (5.54).

The ith equation of Equation (5.54) is

$$
z_{it} = c_i + \boldsymbol{\omega}_i \boldsymbol{z}_{t-1}^* + \gamma_i z_{k,t-1} + \boldsymbol{\Phi}_{1,i}^* \boldsymbol{y}_{t-1} + \cdots + \boldsymbol{\Phi}_{1,p-1}^* \boldsymbol{y}_{t-p+1} + a_{it}, \tag{5.55}
$$

where, as before, $\boldsymbol{A}_i$ denotes the ith row vector of the matrix $\boldsymbol{A}$. Define $\boldsymbol{x}_t' = (1, \boldsymbol{z}_{t-1}^{*'}, z_{k,t-1}, \boldsymbol{y}_{t-1}', \ldots, \boldsymbol{y}_{t-p+1}')$, $\boldsymbol{\beta}' = (c_i, \boldsymbol{\omega}_i, \gamma_i, \boldsymbol{\Phi}_{1,i}^*, \ldots, \boldsymbol{\Phi}_{p-1,i}^*)$, the $(kp+1) \times (kp+1)$ normalization matrix

$$\boldsymbol{N}_T = \text{diag}\{T^{1/2}, T\boldsymbol{I}_{k-1}, T^{3/2}, T^{1/2}\boldsymbol{I}_{k(p-1)}\},$$

and $\boldsymbol{\Lambda}^* = \boldsymbol{\psi}^*(1)\boldsymbol{P}_a$, where $\boldsymbol{\Sigma}_a = \boldsymbol{P}_a\boldsymbol{P}_a'$. We can apply Theorem 5.3 and the property $T^{-(v+1)}\sum_{t=1}^T t = 1/(v+1)$ for non-negative integer v to obtain the following result.

Corollary 5.2 For the unit-root VAR(p) process in Equation (5.44) with nonzero drift $\boldsymbol{c}_0$, assume that $c_k \neq 0$ and the conditions of Theorem 5.3 hold. Then,

(a) $\boldsymbol{N}_T^{-1}(\sum_{t=1}^T \boldsymbol{x}_t\boldsymbol{x}_t')\boldsymbol{N}_T^{-1} \Rightarrow \text{diag}\{\boldsymbol{Q}, \boldsymbol{\Gamma}(p-1)\}$, where $\boldsymbol{\Gamma}(p-1)$ is the covariance matrix of $(\boldsymbol{y}_t', \ldots, \boldsymbol{y}_{t-p+2}')'$ and

$$\boldsymbol{Q} = \begin{bmatrix} 1 & \left[\int_0^1 \boldsymbol{W}(r)dr\right]' \boldsymbol{\Lambda}^{*'} & u_k/2 \\ \boldsymbol{\Lambda}^* \int_0^1 \boldsymbol{W}(r)dr & \boldsymbol{\Lambda}^* \left\{\int_0^1 \boldsymbol{W}(r)[\boldsymbol{W}(r)]'dt\right\} \boldsymbol{\Lambda}^{*'} & u_k \boldsymbol{\Lambda}^* \int r\boldsymbol{W}(r)dt \\ u_k/2 & u_k \left[\int_0^1 r\boldsymbol{W}(r)dr\right]' \boldsymbol{\Lambda}^{*'} & u_k^2/3 \end{bmatrix}.$$

(b) $\boldsymbol{N}_T^{-1}\sum_{t=1}^T \boldsymbol{x}_t a_{it} \Rightarrow (g_1, \boldsymbol{g}_2', g_3, \boldsymbol{g}_4')'$, where $\boldsymbol{g}_4 \sim N[\boldsymbol{0}, \sigma_{ii}\boldsymbol{\Gamma}(p-1)]$ and g_1 and g_3 are also Gaussian. However, $\boldsymbol{g}_2$ consists of functions of the multivariate standard Brownian motion.

(c) $\boldsymbol{N}_T(\hat{\boldsymbol{\beta}} - \boldsymbol{\beta}) \Rightarrow \begin{bmatrix} \boldsymbol{Q}^{-1}\boldsymbol{g} \\ \boldsymbol{K}^{-1}\boldsymbol{g}_4 \end{bmatrix}$, where $\boldsymbol{K} = \boldsymbol{\Gamma}(p-1)$ and $\boldsymbol{g} = (g_1, \boldsymbol{g}_2', g_3)'$.

The implications of Corollary 5.2 are similar to those of Corollary 5.1. The main difference between the two corollaries is the asymptotic normality of the LSE of $\boldsymbol{\gamma}$ in Equation (5.54).

5.3 SPURIOUS REGRESSIONS

Care must be exercised if both the dependent variable and regressors of a linear regression are unit-root nonstationary. Consider the linear regression

$$z_t = \boldsymbol{x}_t'\boldsymbol{\beta} + e_t,$$

where z_t and some elements of $\boldsymbol{x}_t$ are unit-root nonstationary. If there does not exist some true value of $\boldsymbol{\beta}$ for which the residual $e_t = z_t - \boldsymbol{x}_t'\boldsymbol{\beta}$ is a stationary series, then the ordinary LSE of $\boldsymbol{\beta}$ is likely to produce spurious results. This phenomenon was first discovered in a simulation study by Granger and Newbold (1974) and has been studied by Phillips (1986).

To understand the spurious regression, it suffices to employ two independent random walks,

$$\begin{bmatrix} z_{1t} \\ z_{2t} \end{bmatrix} = \begin{bmatrix} 1 & 0 \\ 0 & 1 \end{bmatrix} \begin{bmatrix} z_{1,t-1} \\ z_{2,t-1} \end{bmatrix} + \begin{bmatrix} a_{1t} \\ a_{2t} \end{bmatrix}, \tag{5.56}$$

where $\{\boldsymbol{a}_t\}$ is a sequence of iid two-dimensional random vectors with mean zero and positive-definite covariance matrix $\boldsymbol{\Sigma}_a = \text{diag}\{\sigma_1^2, \sigma_2^2\}$. Consider the linear regression

$$z_{1t} = \alpha + \beta z_{2t} + e_t \tag{5.57}$$

for which the true parameter values are $\alpha = \beta = 0$. However, based on unit-root properties discussed in Theorems 5.1 and 5.3, the LSE of the regression (5.57) satisfy

$$\begin{bmatrix} T^{-1/2}\hat{\alpha} \\ \hat{\beta} \end{bmatrix} \Rightarrow \begin{bmatrix} \sigma_a g_1 \\ \dfrac{\sigma_1}{\sigma_2} g_2 \end{bmatrix}, \tag{5.58}$$

where

$$\begin{bmatrix} g_1 \\ g_2 \end{bmatrix} = \begin{bmatrix} 1 & \int_0^1 W_2(r)dr \\ \int_0^1 W_2(r)dr & \int_0^1 W_2^2(2)dr \end{bmatrix}^{-1} \begin{bmatrix} \int_0^1 W_1(r)dr \\ \int_0^1 W_1(r)W_2(2)dr \end{bmatrix},$$

where $W_1(r)$ and $W_2(r)$ are two independent scalar standard Brownian motions.

From Equation (5.58), it is seen that the LSEs are not consistent. In fact, $\hat{\beta}$ is a random variable and $\hat{\alpha}$ diverges. The latter needs to be divided by $T^{1/2}$ to become a random variable with well-specified distribution. Consequently, the usual t-ratio of testing the hypothesis $H_0 : \beta = 0$ is likely to be rejected even though the true value of β is 0. Another feature of the spurious regression is that the residual $\hat{e}_t = z_{1t} - \hat{\alpha} - \hat{\beta} z_{2t}$ would behave like an integrated process because $\Delta \hat{e}_t$ is a function of $\Delta \boldsymbol{z}_t$. Consequently, the conventional residual mean squares $s^2 = \sum_{t=1}^T \hat{e}_t^2/(T-2)$ also diverges.

To avoid spurious regression, one should always check the stationarity of the residuals when both the dependent variable and the regressors of a linear regression are likely to be unit-root nonstationary. For the simple linear regression in Equation (5.57), one can employ the linear regression model with time series errors to make inference concerning β. See, for instance, Tsay (2010, Chapter 2). Alternatively, one can model all the time series jointly using multivariate models discussed in this book. Differencing all components simultaneously may lead to over-differencing, but the methods of cointegration discussed later can be used.

5.4 MULTIVARIATE EXPONENTIAL SMOOTHING

A vector model for which all components have a single unit root is the multivariate exponential smoothing model. Similar to the univariate case, this model is a special VARMA(1,1) model in the form

$$\boldsymbol{z}_t = \boldsymbol{z}_{t-1} + \boldsymbol{a}_t - \boldsymbol{\theta}\boldsymbol{a}_{t-1}, \tag{5.59}$$

where all eigenvalues of $k \times k$ matrix $\boldsymbol{\theta}$ are less than 1 in absolute value. In most cases, the eigenvalues of $\boldsymbol{\theta}$ are positive. Using $(\boldsymbol{I}_k - \boldsymbol{\theta}B)^{-1} = \boldsymbol{I}_k + \boldsymbol{\theta}B + \boldsymbol{\theta}^2 B^2 + \cdots$, we can rewrite Equation (5.59) as

$$(\boldsymbol{I}_k + \boldsymbol{\theta}B + \boldsymbol{\theta}^2 B^2 + \cdots)(\boldsymbol{I}_k - \boldsymbol{I}_k B)\boldsymbol{z}_t = \boldsymbol{a}_t.$$

Consequently, we have

$$\boldsymbol{z}_t = (\boldsymbol{I}_k - \boldsymbol{\theta})\boldsymbol{z}_{t-1} + (\boldsymbol{I}_k - \boldsymbol{\theta})\boldsymbol{\theta}\boldsymbol{z}_{t-2} + (\boldsymbol{I}_k - \boldsymbol{\theta})\boldsymbol{\theta}^2\boldsymbol{z}_{t-3} + \cdots + \boldsymbol{a}_t.$$

Suppose the forecast origin is T. From the prior equation, it is seen that the one-step ahead prediction of $\boldsymbol{z}_{T+1}$ is

$$\begin{aligned} \boldsymbol{z}_T(1) &= (\boldsymbol{I}_k - \boldsymbol{\theta})\boldsymbol{z}_T + (\boldsymbol{I}_k - \boldsymbol{\theta})\boldsymbol{\theta}\boldsymbol{z}_{T-1} + (\boldsymbol{I}_k - \boldsymbol{\theta})\boldsymbol{\theta}^2\boldsymbol{z}_{T-2} + \cdots \\ &= (\boldsymbol{I}_k - \boldsymbol{\theta})[\boldsymbol{z}_T + \boldsymbol{\theta}\boldsymbol{z}_{T-1} + \boldsymbol{\theta}^2\boldsymbol{z}_{T-2} + \cdots]. \end{aligned} \tag{5.60}$$

The point forecasts are simply a weighted average of the past observations with weights decaying exponentially. The leading factor $(\boldsymbol{I}_k - \boldsymbol{\theta})$ exists to ensure that the sum of weights is $\boldsymbol{I}_k$. Compared with the univariate exponential smoothing, we see that the weights of multivariate model decay exponentially in matrix form as such the weights decay exponentially in eigenvalues (hence, determinants), but not in individual elements.

To demonstrate, suppose that $k = 2$ and

$$\boldsymbol{\theta} = \begin{bmatrix} 0.3 & 0.3 \\ -0.5 & 1.1 \end{bmatrix}.$$

In this particular case, we have $|\boldsymbol{\theta}| = 0.48$, $|\boldsymbol{\theta}^2| = (0.48)^2 = 0.2304$, etc. However,

$$\boldsymbol{\theta}^2 = \begin{bmatrix} -0.06 & 0.42 \\ -0.70 & 1.06 \end{bmatrix}, \quad \boldsymbol{\theta}^3 = \begin{bmatrix} -0.228 & 0.444 \\ -0.740 & 0.956 \end{bmatrix}.$$

Clearly, the (i, j)th elements θ_{ij}^{ℓ} do not decay exponentially as ℓ increases.

Example 5.1 Figure 5.2 shows the time plots of the daily closing values of the S&P 500 index and the CBOE VIX index from January 5, 2004, to August 23, 2012, for 2177 observations. The contemporaneous negative relationship between the two series is clearly seen. Predictions of these two important financial indices are of interest in many financial applications. Given the random walk characteristic of the series, multivariate exponential smoothing can provide a decent approximation in modeling the data. Therefore, we apply the bivariate exponential smoothing model to the series. This is equivalent to fitting a VMA(1) model to the first differenced data. The

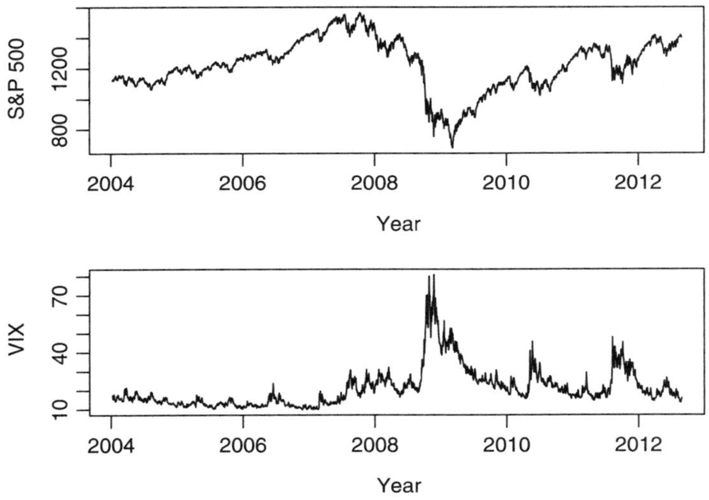

FIGURE 5.2 Time plots of the daily closing values of the S&P 500 index and the VIX index from January 5, 2004, to August 23, 2012.

one-sample t-test fails to reject the null hypothesis of mean zero for both differenced series, indicating that there exists no drift in the two index series. The fitted VMA(1) model is

$$\boldsymbol{\Delta z}_t = \boldsymbol{a}_t - \begin{bmatrix} 0.066 & -0.717 \\ 0.000 & 0.195 \end{bmatrix} \boldsymbol{a}_{t-1}, \qquad (5.61)$$

where θ_{21} is fixed to 0 as it is statistically insignificant at the traditional 5% level and the residual covariance matrix is

$$\hat{\Sigma}_a = \begin{bmatrix} 217.10 & -23.44 \\ -23.44 & 3.67 \end{bmatrix}.$$

Model checking shows that some minor serial and cross-correlations remain in the residuals of the model in Equation (5.61). Figure 5.3 plots the residual serial and cross-correlations of the multivariate exponential smoothing model. Some marginally significant correlations are seen from the plots. Consequently, the bivariate exponential smoothing model is only an approximation for use in forecasting. The contemporaneous negative correlation is also clearly seen in the cross-correlation plots.

Based on the fitted integrated moving-average (IMA) model in Equation (5.61), the first three weight matrices for forecasting are

$$\begin{bmatrix} 0.934 & 0.717 \\ 0.000 & 0.805 \end{bmatrix}, \quad \begin{bmatrix} 0.062 & -0.530 \\ 0.000 & 0.157 \end{bmatrix}, \quad \begin{bmatrix} 0.004 & -0.148 \\ 0.000 & 0.031 \end{bmatrix},$$

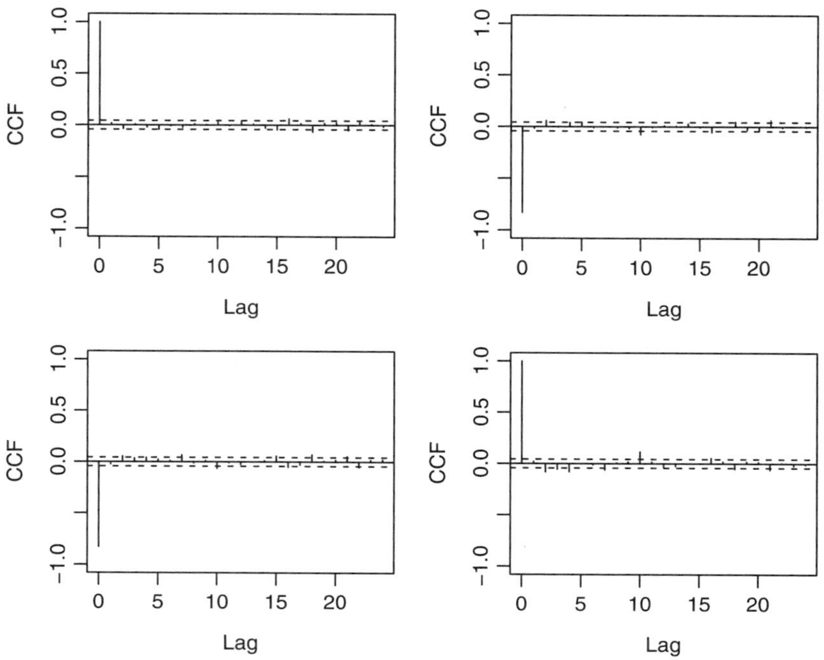

FIGURE 5.3 Residual serial and cross-correlations of the bivariate exponential model in Equation (5.61) for the daily S&P 500 index and the VIX index from January 5, 2004, to August 23, 2012.

respectively. See Equation (5.60). These weights decay quickly by the factor $\hat{\boldsymbol{\theta}}$. From the weights, the predictions of the S&P 500 index depend on lagged values of both indices, yet those of the VIX index depend only on its own past values. The eigenvalues of $\hat{\boldsymbol{\theta}}$ are 0.195 and 0.066 so that the weights decay quickly. □

Finally, the marginal model for z_{it} of the multivariate exponential smoothing model in Equation (5.59) is a univariate exponential smoothing model because any linear combination of $\boldsymbol{a}_t$ and $\boldsymbol{a}_{t-1}$ follows a univariate MA(1) model. Similar to the univariate case and demonstrated by Example 5.1, putting multivariate exponential smoothing models in the VARMA framework has some advantages. First, one can fit a VMA(1) model to the first differenced series to obtain multivariate exponential smoothing model. In this way, the decaying rates of the weights for prediction are data dependent, not fixed *a priori*. Second, model checking enables us to check the validity of using multivariate exponential smoothing models.

5.5 COINTEGRATION

Unit roots play an important role in time series analysis. The k-dimensional process $\boldsymbol{z}_t$ discussed so far is either stationary or contains k unit roots. There are situations under which $\boldsymbol{z}_{it}$ are unit-root nonstationary, but $\boldsymbol{z}_t$ does not contain k unit roots. We shall study these cases in this section.

In the econometric literature, a time series z_t is said to be an integrated process of order 1, that is, an $I(1)$ process, if $(1-B)z_t$ is stationary and invertible. In general, a univariate time series z_t is an $I(d)$ process if $(1-B)^d z_t$ is stationary and invertible, where $d > 0$. The order d is referred to as the order of integration or the multiplicity of a unit root. A stationary and invertible time series is said to be an $I(0)$ process.

Consider a multivariate process $\boldsymbol{z}_t$. If z_{it} are $I(1)$ processes, but a nontrivial linear combination $\boldsymbol{\beta}'\boldsymbol{z}_t$ is an $I(0)$ series, then $\boldsymbol{z}_t$ is said to be cointegrated of order 1. In general, if z_{it} are $I(d)$ nonstationary and $\boldsymbol{\beta}'\boldsymbol{z}_t$ is $I(h)$ with $h < d$, then $\boldsymbol{z}_t$ is cointegrated. In real applications, the case of $d = 1$ and $h = 0$ is of major interest. Thus, cointegration often means that a linear combination of individually unit-root nonstationary time series becomes a stationary and invertible series. The linear combination vector $\boldsymbol{\beta}$ is called a cointegrating vector.

Suppose that $\boldsymbol{z}_t$ is unit-root nonstationary such that the marginal models for z_{it} have a unit root. If $\boldsymbol{\beta}$ is a $k \times m$ matrix of full rank m, where $m < k$, such that $\boldsymbol{w}_t = \boldsymbol{\beta}'\boldsymbol{z}_t$ is $I(0)$, then $\boldsymbol{z}_t$ is a cointegrated series with m cointegrating vectors, which are the columns of $\boldsymbol{\beta}$. This means that there are $k-m$ unit roots in $\boldsymbol{z}_t$. For the given full-rank $k \times m$ matrix $\boldsymbol{\beta}$ with $m < k$, let $\boldsymbol{\beta}_\perp$ be a $k \times (k-m)$ full-rank matrix such that $\boldsymbol{\beta}'\boldsymbol{\beta}_\perp = \boldsymbol{0}$. Then, $\boldsymbol{n}_t = \boldsymbol{\beta}_\perp'\boldsymbol{z}_t$ is unit-root nonstationary. The components n_{it} $[i = 1, \cdots, (k-m)]$ are referred to as the *common trends* of $\boldsymbol{z}_t$. We shall discuss methods for finding cointegrating vectors and common trends later.

Cointegration implies a long-term stable relationship between variables in forecasting. Since $\boldsymbol{w}_t = \boldsymbol{\beta}'\boldsymbol{z}_t$ is stationary, it is mean-reverting so that the ℓ-step ahead forecast of $\boldsymbol{w}_{T+\ell}$ at the forecast origin T satisfies

$$\hat{\boldsymbol{w}}_T(\ell) \to_p E(\boldsymbol{w}_t) \equiv \boldsymbol{\mu}_w, \quad \ell \to \infty.$$

This implies that $\boldsymbol{\beta}'\hat{\boldsymbol{z}}_T(\ell) \to \boldsymbol{\mu}_w$ as ℓ increases. Thus, point forecasts of $\boldsymbol{z}_t$ satisfy a long-term stable constraint.

5.5.1 An Example of Cointegration

To understand cointegration, we consider a simple example. Suppose that the bivariate process $\boldsymbol{z}_t$ follows the model

$$\begin{bmatrix} z_{1t} \\ z_{2t} \end{bmatrix} - \begin{bmatrix} 0.5 & -1.0 \\ -0.25 & 0.5 \end{bmatrix} \begin{bmatrix} z_{1,t-1} \\ z_{2,t-1} \end{bmatrix} = \begin{bmatrix} a_{1t} \\ a_{2t} \end{bmatrix} - \begin{bmatrix} 0.2 & -0.4 \\ -0.1 & 0.2 \end{bmatrix} \begin{bmatrix} a_{1,t-1} \\ a_{2,t-1} \end{bmatrix},$$

where the covariance matrix $\boldsymbol{\Sigma}_a$ of the shock $\boldsymbol{a}_t$ is positive-definite. For simplicity, assume that $\boldsymbol{\Sigma}_a = \boldsymbol{I}$. The prior VARMA(1,1) model, from Tsay (2010, Chapter 8), is not weakly stationary because the two eigenvalues of the AR coefficient matrix are 0 and 1. One can easily verify that $\boldsymbol{\Phi}(1) \neq \boldsymbol{0}$, but

$$|\boldsymbol{\Phi}(1)| = \left|\boldsymbol{I}_2 - \begin{bmatrix} 0.5 & -1.0 \\ -0.25 & 0.5 \end{bmatrix}\right| = \begin{vmatrix} 0.5 & 1.0 \\ 0.25 & 0.5 \end{vmatrix} = 0.$$

Rewrite the model as

$$\begin{bmatrix} 1-0.5B & B \\ 0.25B & 1-0.5B \end{bmatrix} \begin{bmatrix} z_{1t} \\ z_{2t} \end{bmatrix} = \begin{bmatrix} 1-0.2B & 0.4B \\ 0.1B & 1-0.2B \end{bmatrix} \begin{bmatrix} a_{1t} \\ a_{2t} \end{bmatrix}. \tag{5.62}$$

Premultiplying the aforementioned equation by the adjoint matrix

$$\begin{bmatrix} 1-0.5B & -B \\ -0.25B & 1-0.5B \end{bmatrix},$$

we obtain

$$\begin{bmatrix} 1-B & 0 \\ 0 & 1-B \end{bmatrix} \begin{bmatrix} z_{1t} \\ z_{2t} \end{bmatrix} = \begin{bmatrix} 1-0.7B & -0.6B \\ -0.15B & 1-0.7B \end{bmatrix} \begin{bmatrix} a_{1t} \\ a_{2t} \end{bmatrix}. \tag{5.63}$$

Therefore, each component z_{it} of the model is unit-root nonstationary and follows an ARIMA(0,1,1) model.

However, we can consider a linear transformation by defining

$$\begin{bmatrix} y_{1t} \\ y_{2t} \end{bmatrix} = \begin{bmatrix} 1.0 & -2.0 \\ 0.5 & 1.0 \end{bmatrix} \begin{bmatrix} z_{1t} \\ z_{2t} \end{bmatrix} \equiv \boldsymbol{L}\boldsymbol{z}_t,$$

$$\begin{bmatrix} b_{1t} \\ b_{2t} \end{bmatrix} = \begin{bmatrix} 1.0 & -2.0 \\ 0.5 & 1.0 \end{bmatrix} \begin{bmatrix} a_{1t} \\ a_{2t} \end{bmatrix} \equiv \boldsymbol{L}\boldsymbol{a}_t.$$

The VARMA model of the transformed series $\boldsymbol{y}_t$ can be obtained as follows:

$$\begin{aligned} \boldsymbol{L}\boldsymbol{z}_t &= \boldsymbol{L}\boldsymbol{\Phi}\boldsymbol{z}_{t-1} + \boldsymbol{L}\boldsymbol{a}_t - \boldsymbol{L}\boldsymbol{\Theta}\boldsymbol{a}_{t-1} \\ &= \boldsymbol{L}\boldsymbol{\Phi}\boldsymbol{L}^{-1}\boldsymbol{L}\boldsymbol{z}_{t-1} + \boldsymbol{L}\boldsymbol{a}_t - \boldsymbol{L}\boldsymbol{\Theta}\boldsymbol{L}^{-1}\boldsymbol{L}\boldsymbol{a}_{t-1} \\ &= \boldsymbol{L}\boldsymbol{\Phi}\boldsymbol{L}^{-1}(\boldsymbol{L}\boldsymbol{z}_{t-1}) + \boldsymbol{b}_t - \boldsymbol{L}\boldsymbol{\Theta}\boldsymbol{L}^{-1}\boldsymbol{b}_{t-1}. \end{aligned}$$

Thus, the model for $\boldsymbol{y}_t$ is

$$\begin{bmatrix} y_{1t} \\ y_{2t} \end{bmatrix} - \begin{bmatrix} 1.0 & 0 \\ 0 & 0 \end{bmatrix} \begin{bmatrix} y_{1,t-1} \\ y_{2,t-1} \end{bmatrix} = \begin{bmatrix} b_{1t} \\ b_{2t} \end{bmatrix} - \begin{bmatrix} 0.4 & 0 \\ 0 & 0 \end{bmatrix} \begin{bmatrix} b_{1,t-1} \\ b_{2,t-1} \end{bmatrix}. \tag{5.64}$$

From the prior model, we see that (a) y_{1t} and y_{2t} are not dynamically related, except for the concurrent correlation between b_{1t} and b_{2t}, (b) y_{1t} follows a univariate ARIMA(0,1,1) model, and (c) y_{2t} is a stationary series. In fact, y_{2t} is a white noise series. Consequently, there is only one unit root in $\boldsymbol{z}_t$ even though both z_{it}

are unit-root nonstationary. In other words, the unit roots in z_{it} are from the same source y_{1t}, which is referred to as the *common trend* of z_{it}. The linear combination $y_{2t} = (0.5, 1)\boldsymbol{z}_t$ is stationary so that $(0.5, 1)'$ is a cointegrating vector for $\boldsymbol{z}_t$. If the cointegration relationship is imposed, the forecasts $\boldsymbol{z}_T(\ell)$ must satisfy the constraint $(0.5, 1)\boldsymbol{z}_T(\ell) = 0$.

5.5.2 Some Justifications of Cointegration

The fact that a linear combination of several unit-root nonstationary time series can become a stationary series was observed in the literature. Box and Tiao (1977) show this feature clearly with a five-dimensional time series using canonical correlation analysis. The term *cointegration*, however, was first used in Granger (1983). The concept of cointegration has attracted much attention in the econometric literature for several reasons. For example, the theory of purchasing power parity (PPP) lends itself naturally to cointegration. Basically, PPP theory states that, apart from transportation costs, goods should be sold for the same effective price in two countries. As another example, suppose that z_{1t} is a unit-root time series and z_{2t} is a rational forecast of future values of z_{1t}, then z_{1t} and z_{2t} should be cointegrated. See, for instance, Campbell and Shiller (1988) concerning dividend-price ratio in finance.

5.6 AN ERROR-CORRECTION FORM

The model in Equation (5.63) is in the form

$$\Delta \boldsymbol{z}_t = (\boldsymbol{I}_2 - \boldsymbol{\theta}_1 B)\boldsymbol{a}_t, \quad \text{with} \quad \boldsymbol{\theta}_1 = \begin{bmatrix} 0.7 & 0.6 \\ 0.15 & 0.7 \end{bmatrix}. \tag{5.65}$$

As expected, $\Delta \boldsymbol{z}_t$ is stationary because it follows a VMA(1) model. However, the eigenvalues of $\boldsymbol{\theta}_1$ are 1 and 0.4, indicating that the VMA(1) model is not invertible. As a matter of fact, it is easily seen that $\boldsymbol{\theta}(1) \neq 0$, but $|\boldsymbol{\theta}(1)| = 0$. The noninvertibility arises because we over-differenced the series. For this particular instance, $\boldsymbol{z}_t$ has a single unit root, but $(1 - B)\boldsymbol{z}_t$ imposes two unit roots. The unit root in $\boldsymbol{\theta}_1$ is induced by the over-difference. This simple example demonstrates that over-differencing results in a noninvertible model.

Noninvertibility has some implications in practice. First, a noninvertible VARMA model cannot be approximated by a finite-order VAR model. To demonstrate, consider again the over-differenced model in Equation (5.65). The AR representation of $\Delta \boldsymbol{z}_t$ is

$$(\boldsymbol{I}_2 + \boldsymbol{\theta}_1 B + \boldsymbol{\theta}_1^2 B^2 + \cdots)\Delta \boldsymbol{z}_t = \boldsymbol{a}_t,$$

so that the coefficient matrix of $\Delta \boldsymbol{z}_{t-j}$ is $\boldsymbol{\theta}_1^j$. Since 1 is an eigenvalue of $\boldsymbol{\theta}_1$, it is also an eigenvalue of $\boldsymbol{\theta}_1^j$ for all $j \geq 1$. Therefore, $\boldsymbol{\theta}_1^j$ does not converge to $\boldsymbol{0}$ as j

increases. As such, one cannot truncate the AR representation. Second, as discussed in Chapter 3, the exact maximum likelihood method is recommended to estimate a noninvertible VARMA model because the impact of the initial values such as $\boldsymbol{a}_0$ on the likelihood function does not decay as time index t increases.

To avoid the noninvertibility, Engle and Granger (1987) proposed an error-correction form of multivariate time series that keeps the MA structure of the model. To illustrate, consider the model in Equation (5.62). Moving the AR(1) part of the model to the right-hand side of the equation and subtracting $\boldsymbol{z}_{t-1}$ from the model, we have

$$\begin{aligned}
\Delta \boldsymbol{z}_t &= \left(\begin{bmatrix} 0.5 & -1 \\ -0.25 & 0.5 \end{bmatrix} - \begin{bmatrix} 1 & 0 \\ 0 & 1 \end{bmatrix}\right) \boldsymbol{z}_{t-1} - \boldsymbol{a}_t - \boldsymbol{\theta}_1 \boldsymbol{a}_{t-1} \\
&= \begin{bmatrix} -0.5 & -1 \\ -0.25 & -0.5 \end{bmatrix} \boldsymbol{z}_{t-1} + \boldsymbol{a}_t - \boldsymbol{\theta}_1 \boldsymbol{a}_{t-1} \\
&= \begin{bmatrix} -1 \\ -0.5 \end{bmatrix} [0.5, 1] \boldsymbol{z}_{t-1} + \boldsymbol{a}_t - \boldsymbol{\theta}_1 \boldsymbol{a}_{t-1} \\
&= \begin{bmatrix} -1.0 \\ -0.5 \end{bmatrix} y_{2,t-1} + \boldsymbol{a}_t - \boldsymbol{\theta}_1 \boldsymbol{a}_t.
\end{aligned}$$

This is an *error-correction* form for the model, which has an invertible MA structure, but uses $\boldsymbol{z}_{t-1}$ in the right-hand side of the model. The $\boldsymbol{z}_{t-1}$ term is referred to as the error-correction term and its coefficient matrix is of rank 1 representing the number of cointegrating vectors of the system. More importantly, the cointegrating vector $(0.5, 1)'$ is shown so that the stationary series $\Delta \boldsymbol{z}_t$ depends on the lagged value of the stationary series $y_{2t} = (0.5, 1)\boldsymbol{z}_t$. The coefficient vector $(-1.0, -0.5)'$ highlights the dependence of $\Delta \boldsymbol{z}_t$ on the cointegrating series $y_{2,t-1}$.

The property discussed earlier can be easily extended to the general unit-root nonstationary VARMA(p, q) process in Equation (5.34). As a matter of fact, we have derived the error-correction form in Equation (5.42). For completeness and for ease in reference, we summarize the error-correction form as follows. Consider the general VARMA(p, q) model in Equation (5.34). Assume that $|\boldsymbol{\Phi}(x)| \neq 0$ for $|x| \leq 1$ except $|\boldsymbol{\Phi}(1)| = 0$, so that the k-dimensional process $\boldsymbol{z}_t$ is unit-root nonstationary. In addition, assume that $|\boldsymbol{\Theta}(x)| \neq 0$ for $|x| \leq 1$ so that $\boldsymbol{z}_t$ is invertible and $\boldsymbol{c}(t) = \boldsymbol{c}_0 + \boldsymbol{c}_1 t$. Define

$$\begin{aligned}
\boldsymbol{\Phi}_j^* &= -(\boldsymbol{\Phi}_{j+1} + \cdots + \boldsymbol{\Phi}_p), \quad j = 1, \ldots, p-1, \\
\boldsymbol{\Pi} &= -\boldsymbol{\Phi}(1) = \boldsymbol{\Phi}_1 + \cdots + \boldsymbol{\Phi}_p - \boldsymbol{I}_k.
\end{aligned} \tag{5.66}$$

Then, the VARMA(p, q) model can be rewritten as

$$\Delta \boldsymbol{z}_t = \boldsymbol{\Pi} \boldsymbol{z}_{t-1} + \sum_{i=1}^{p-1} \boldsymbol{\Phi}_i^* \Delta \boldsymbol{z}_{t-i} + \boldsymbol{c}(t) + \boldsymbol{\Theta}(B) \boldsymbol{a}_t. \tag{5.67}$$

This model representation is referred to as an error-correction form of a VARMA model. Equation (5.67) is often called an error-correction model (ECM) for the unit-root nonstationary series $\boldsymbol{z}_t$. Note that the MA part $\boldsymbol{\Theta}(B)$ and the deterministic component $\boldsymbol{c}(t)$ of the model remain unchanged. Consequently, the ECM representation is invertible.

Remark: There are many ways to write an error-correction form. For instance, instead of using $\boldsymbol{z}_{t-1}$, one can subtract $\boldsymbol{z}_{t-p}$ from a VARMA(p, q) model and obtain another error-correction form as

$$\Delta \boldsymbol{z}_t = \boldsymbol{\Pi} \boldsymbol{z}_{t-p} + \sum_{i=1}^{p-1} \boldsymbol{\phi}_i^* \Delta \boldsymbol{z}_{t-i} + \boldsymbol{c}(t) + \boldsymbol{\Theta}(B)\boldsymbol{a}_t, \tag{5.68}$$

where $\boldsymbol{\Pi} = -\boldsymbol{\Phi}(1)$ and the $\boldsymbol{\phi}_i^*$ are given by

$$\boldsymbol{\phi}_j^* = \sum_{i=1}^{j} \boldsymbol{\Phi}_i - \boldsymbol{I}_k, \quad j = 1, \ldots, p-1.$$

The result can be easily verified by equating the coefficient matrices of B^i in Equation (5.68). This alternative ECM is also invertible. As a matter of fact, one can use $\boldsymbol{z}_{t-j}$ for a finite j on the right-hand side to obtain an ECM form for a VARMA(p, q) model. In the literature, only the forms in Equations (5.67) and (5.68) are used. □

Since $\boldsymbol{\Phi}(1)$ is a singular matrix for a cointegrated system, $\boldsymbol{\Pi} = -\boldsymbol{\Phi}(1)$ is not full rank. Assume that Rank($\boldsymbol{\Pi}$) $= m > 0$. Then, there exist $k \times m$ matrices $\boldsymbol{\alpha}$ and $\boldsymbol{\beta}$ of rank m such that $\boldsymbol{\Pi} = \boldsymbol{\alpha}\boldsymbol{\beta}'$. This decomposition, however, is not unique. In fact, for any $m \times m$ orthogonal matrix $\boldsymbol{P}$ such that $\boldsymbol{P}\boldsymbol{P}' = \boldsymbol{I}$, it is easy to see that

$$\boldsymbol{\alpha}\boldsymbol{\beta}' = \boldsymbol{\alpha}\boldsymbol{P}\boldsymbol{P}'\boldsymbol{\beta}' = (\boldsymbol{\alpha}\boldsymbol{P})(\boldsymbol{\beta}\boldsymbol{P})'. \tag{5.69}$$

Thus, $\boldsymbol{\alpha}\boldsymbol{P}$ and $\boldsymbol{\beta}\boldsymbol{P}$ are also of rank m and may serve as another decomposition of $\boldsymbol{\Pi}$. Consequently, some additional normalization is needed to uniquely identify $\boldsymbol{\alpha}$ and $\boldsymbol{\beta}$. We shall return to this point later.

Since the stationary series $\Delta \boldsymbol{z}_t$ cannot depend directly on the unit-root nonstationary series $\boldsymbol{z}_{t-1}$, the factorization $\boldsymbol{\Pi} = \boldsymbol{\alpha}\boldsymbol{\beta}'$ must give rise to stationary linear combinations of $\boldsymbol{z}_t$. Indeed, defining the m-dimensional process $\boldsymbol{w}_t = \boldsymbol{\beta}'\boldsymbol{z}_t$, we have

$$\Delta \boldsymbol{z}_t = \boldsymbol{\alpha} \boldsymbol{w}_{t-1} + \sum_{i=1}^{p-1} \boldsymbol{\Phi}_i^* \Delta \boldsymbol{z}_{t-i} + \boldsymbol{c}(t) + \boldsymbol{\Theta}(B)\boldsymbol{a}_t, \tag{5.70}$$

where $\boldsymbol{\alpha}$ is a full-rank $k \times m$ matrix. The nontrivial dependence of $\Delta \boldsymbol{z}_t$ on $\boldsymbol{w}_{t-1}$ suggests that the latter must also be stationary. Therefore, the columns of the matrix

$\boldsymbol{\beta}$ are cointegrating vectors. Thus, we have m cointegrating vectors for $\boldsymbol{z}_t$ and the observed series $\boldsymbol{z}_t$ has $k - m$ unit roots. In other words, there are $k - m$ common trends in $\boldsymbol{z}_t$.

5.7 IMPLICATIONS OF COINTEGRATING VECTORS

In this section, we consider the implications of cointegrations in a k-dimensional time series. For simplicity, we assume that $\boldsymbol{z}_t$ follows the VARMA model in Equation (5.34) with $\boldsymbol{c}(t) = \boldsymbol{c}_0 + \boldsymbol{c}_1 t$ and, as before, $\boldsymbol{a}_t$ being a sequence of iid random vectors with mean zero, positive-definite covariance matrix $\boldsymbol{\Sigma}_a$, and finite fourth moments. The model is assumed to be invertible and cointegrated with cointegrating matrix $\boldsymbol{\beta}$, which is a $k \times m$ matrix, where $0 < m < k$.

5.7.1 Implications of the Deterministic Term

The ECM in Equation (5.70) for the model considered becomes

$$\Delta \boldsymbol{z}_t = \boldsymbol{\alpha} \boldsymbol{w}_{t-1} + \sum_{i=1}^{p-1} \boldsymbol{\Phi}_i^* \Delta \boldsymbol{z}_{t-i} + \boldsymbol{c}_0 + \boldsymbol{c}_1 t + \boldsymbol{\Theta}(B) \boldsymbol{a}_t. \tag{5.71}$$

Since both $\Delta \boldsymbol{z}_t$ and $\boldsymbol{w}_t$ are stationary, we define

$$E(\Delta \boldsymbol{z}_t) = \boldsymbol{\mu} \quad \text{and} \quad E(\boldsymbol{w}_t) = \boldsymbol{\mu}_w.$$

Taking expectation of Equation (5.71), we obtain

$$\boldsymbol{\mu} = \boldsymbol{\alpha} \boldsymbol{\mu}_w + \sum_{i=1}^{p-1} \boldsymbol{\Phi}_i^* \boldsymbol{\mu} + \boldsymbol{c}_0 + \boldsymbol{c}_1 t.$$

Consequently,

$$\boldsymbol{\Phi}^*(1) \boldsymbol{\mu} = \boldsymbol{\alpha} \boldsymbol{\mu}_w + \boldsymbol{c}_0 + \boldsymbol{c}_1 t, \tag{5.72}$$

where $\boldsymbol{\Phi}^*(1) = \boldsymbol{I}_k - \boldsymbol{\Phi}_1^* - \cdots - \boldsymbol{\Phi}_{p-1}^*$, which is nonsingular because $|\boldsymbol{\Phi}^*(1)| \neq 0$. From Equation (5.72), it is clear that, under the stationarity assumption of $\Delta \boldsymbol{z}_t$ and $\boldsymbol{w}_t$, $\boldsymbol{c}_1 = \boldsymbol{0}$ and the constant term $\boldsymbol{c}_0$ must satisfy the constraint

$$\boldsymbol{c}_0 = \boldsymbol{\Phi}^*(1) \boldsymbol{\mu} - \boldsymbol{\alpha} \boldsymbol{\mu}_w.$$

If $\boldsymbol{\mu}_w$ is orthogonal to $\boldsymbol{\alpha}$, then $\boldsymbol{\alpha} \boldsymbol{\mu}_w = \boldsymbol{0}$ and $\boldsymbol{c}_0 = \boldsymbol{\Phi}^*(1) \boldsymbol{\mu}$. In this particular case, $\boldsymbol{c}_1 = \boldsymbol{0}$ and $\boldsymbol{c}_0$ acts exactly as the constant term of a stationary VARMA model with AR matrix polynomial $\boldsymbol{\Phi}^*(B)$. Thus, $\boldsymbol{\mu}$ becomes the drift of $\boldsymbol{z}_t$.

On the other hand, if $\boldsymbol{\alpha}\boldsymbol{\mu}_w \neq \mathbf{0}$, then

$$\boldsymbol{\mu} = [\boldsymbol{\Phi}^*(1)]^{-1}(\boldsymbol{\alpha}\boldsymbol{\mu}_w + \boldsymbol{c}_0).$$

In this case, unless $\boldsymbol{\alpha}\boldsymbol{\mu}_w = -\boldsymbol{c}_0$, the series $\boldsymbol{z}_t$ will have a drift parameter $\boldsymbol{\mu}$.

In a rare event, if one allows the cointegrated process $\boldsymbol{w}_t$ to be trend-stationary, then Equation (5.72) implies that

$$\boldsymbol{\mu} = [\boldsymbol{\Phi}^*(1)]^{-1}(\boldsymbol{\alpha}\boldsymbol{\mu}_w + \boldsymbol{c}_0 + \boldsymbol{c}_1 t).$$

In this case, $\boldsymbol{\mu}_w$ must satisfy the condition that $\boldsymbol{\alpha}\boldsymbol{\mu}_w + \boldsymbol{c}_0 + \boldsymbol{c}_1 t$ is time invariant. If this latter condition holds, then $\boldsymbol{\mu}$ may or may not be 0. Consequently, $\boldsymbol{z}_t$ may or may not have a drift term. If $\boldsymbol{\mu} = \mathbf{0}$, then $\boldsymbol{\alpha}\boldsymbol{\mu}_w + \boldsymbol{c}_0 + \boldsymbol{c}_1 t = \mathbf{0}$. Since the $k \times m$ matrix $\boldsymbol{\alpha}$ is of full rank m, $\boldsymbol{\alpha}'\boldsymbol{\alpha}$ is a nonsingular $m \times m$ matrix. Consequently, under the condition $\boldsymbol{\mu} = \mathbf{0}$, we have, by Equation (5.72), that

$$\boldsymbol{\mu}_w = -(\boldsymbol{\alpha}'\boldsymbol{\alpha})^{-1}\boldsymbol{\alpha}'(\boldsymbol{c}_0 + \boldsymbol{c}_1 t). \tag{5.73}$$

This result confirms that $\boldsymbol{w}_t$ is trend-stationary provided that $\boldsymbol{c}_1 \neq \mathbf{0}$.

5.7.2 Implications for Moving-Average Representation

Under the cointegration framework, $\Delta\boldsymbol{z}_t$ is stationary. By Wold decomposition, we have

$$\Delta\boldsymbol{z}_t = \boldsymbol{\mu} + \boldsymbol{y}_t, \quad \text{with} \quad \boldsymbol{y}_t = \sum_{i=0}^{\infty} \boldsymbol{\psi}_i \boldsymbol{a}_{t-i}, \tag{5.74}$$

where $\boldsymbol{\mu} = E(\Delta\boldsymbol{z}_t)$, $\{\boldsymbol{a}_t\}$ is a sequence of uncorrelated innovations with mean zero and positive-definite covariance matrix $\boldsymbol{\Sigma}_a$. We further assume that $\boldsymbol{a}_t$ has finite fourth moments and $\{v\boldsymbol{\psi}_v | v = 0, \ldots, \infty\}$ is absolutely summable, where $\boldsymbol{\psi}_0 = \boldsymbol{I}_k$. Let $\boldsymbol{\psi}(1) = \sum_{i=0}^{\infty} \boldsymbol{\psi}_i$. Here, the mean of $\boldsymbol{y}_t$ is 0. Denoting the initial value of $\boldsymbol{z}_t$ by $\boldsymbol{z}_0$, we have

$$\boldsymbol{z}_t = \boldsymbol{z}_0 + \boldsymbol{\mu} t + \sum_{i=1}^{t} \boldsymbol{y}_t.$$

By Lemma 5.1, we have

$$\boldsymbol{z}_t = \boldsymbol{z}_0 + \boldsymbol{\mu} t + \boldsymbol{\psi}(1) \sum_{v=1}^{t} \boldsymbol{a}_v + \boldsymbol{\eta}_t - \boldsymbol{\eta}_0, \tag{5.75}$$

where $\boldsymbol{\eta}_t = \sum_{s=0}^{\infty} \boldsymbol{\alpha}_s \boldsymbol{a}_{t-s}$ is a stationary process and $\boldsymbol{\alpha}_s = \sum_{v=0}^{s} \boldsymbol{\psi}_v - \boldsymbol{\psi}(1)$.

Assume further that $\boldsymbol{z}_t$ is cointegrated with cointegrating matrix $\boldsymbol{\beta}$, which is a $k \times m$ full-rank matrix with $0 < m < k$. Premultiplying Equation (5.75) by $\boldsymbol{\beta}'$, we obtain

$$\boldsymbol{w}_t = \boldsymbol{\beta}'\boldsymbol{z}_t = \boldsymbol{\beta}'(\boldsymbol{z}_0 - \boldsymbol{\eta}_0) + \boldsymbol{\beta}'\boldsymbol{\mu}t + \boldsymbol{\beta}'\boldsymbol{\psi}(1)\sum_{v=1}^{t} \boldsymbol{a}_v + \boldsymbol{\beta}'\boldsymbol{\eta}_t. \tag{5.76}$$

Since $\boldsymbol{w}_t$ is stationary, but $\sum_{v=0}^{t} \boldsymbol{a}_v$ is an $I(1)$ process, we obtain

$$\boldsymbol{\beta}'\boldsymbol{\psi}(1) = \boldsymbol{0} \quad \text{and} \quad \boldsymbol{\beta}'\boldsymbol{\mu} = \boldsymbol{0}. \tag{5.77}$$

From $\boldsymbol{\beta}'\boldsymbol{\psi}(1) = \boldsymbol{0}$ and $\boldsymbol{\beta} \neq \boldsymbol{0}$, $\boldsymbol{\psi}(1)$ must be singular. That is, $|\boldsymbol{\psi}(1)| = 0$. This implies that $\boldsymbol{y}_t = \Delta\boldsymbol{z}_t - \boldsymbol{\mu}$ is noninvertible. This provides yet another proof of over-differencing when $m < k$.

5.8 PARAMETERIZATION OF COINTEGRATING VECTORS

As mentioned before, the cointegrating vectors in $\boldsymbol{\beta}$ are not uniquely defined. Several parameterizations have been used in the literature to render the model identifiable. A good parameterization provides helpful interpretations and simplifies estimation of error-correction models. An approach considered by Phillips (1991) is to rewrite the cointegrating matrix as

$$\boldsymbol{\beta} = \begin{bmatrix} \boldsymbol{I}_m \\ \boldsymbol{\beta}_1 \end{bmatrix}, \tag{5.78}$$

where $\boldsymbol{\beta}_1$ is an arbitrary $(k-m) \times m$ matrix. This parameterization can be achieved by reordering the components of $\boldsymbol{z}_t$ if necessary and is similar to those commonly used in statistical factor models. The basic idea behind Equation (5.78) is that the cointegrating matrix $\boldsymbol{\beta}$ is of full rank m and satisfies Equation (5.69).

Partition $\boldsymbol{z}_t = (\boldsymbol{z}_{1t}', \boldsymbol{z}_{2t}')'$, where the dimensions of $\boldsymbol{z}_{1t}$ and $\boldsymbol{z}_{2t}$ are m and $(k-m)$, respectively. Then, Equation (5.78) implies

$$\boldsymbol{w}_t^* = [\boldsymbol{I}_m, \boldsymbol{\beta}_1']\boldsymbol{z}_t = \boldsymbol{z}_{1t} + \boldsymbol{\beta}_1'\boldsymbol{z}_{2t},$$

is stationary. Here we use the super script "*" to denote that the components of $\boldsymbol{z}_t$ might have been rearranged. Using Equation (5.74) and similar partitions for $\boldsymbol{\mu}$ and $\boldsymbol{y}_t$, we also have

$$\boldsymbol{z}_{2t} = \boldsymbol{z}_{2,t-1} + \boldsymbol{\mu}_2 + \boldsymbol{y}_{2t}.$$

Under the assumption that $\boldsymbol{z}_t$ has m cointegrating vectors, it is easy to see that $\boldsymbol{z}_{2t}$ consists of $k - m$ $I(1)$ processes and there is no cointegration relationship between components of $\boldsymbol{z}_{2t}$. In this sense, $\boldsymbol{z}_{2t}$ represents the $k - m$ common trends of $\boldsymbol{z}_t$.

In some cases, one may fix the upper $m \times m$ submatrix of $\boldsymbol{\beta}$ as given and estimate the remaining $(k - m) \times m$ submatrix provided that the upper submatrix is nonsingular. One can choose this parameterization based on the preliminary estimate of $\boldsymbol{\beta}$ from cointegration tests discussed in the next section.

5.9 COINTEGRATION TESTS

From the error-correction representation in Equation (5.67) and the discussions of the prior section, the matrix $\boldsymbol{\Pi}$ plays an important role in cointegration study. If $\text{Rank}(\boldsymbol{\Pi}) = 0$, then $\boldsymbol{\Pi} = \boldsymbol{0}$ and there is no cointegrating vector, implying that the system is not cointegrated. Testing for cointegration thus focuses on checking the rank of $\boldsymbol{\Pi}$.

5.9.1 The Case of VAR Models

Because of its simplicity, much work in the literature concerning cointegration tests focuses on VAR models. Johansen's method is perhaps the best-known approach to cointegration tests for VAR models. See Johansen (1991) and Johansen and Katarina (1990).

Consider the Gaussian VAR(p) model with a trend component,

$$\boldsymbol{z}_t = \boldsymbol{c}(t) + \sum_{i=1}^{p} \boldsymbol{\phi}_i \boldsymbol{z}_{t-i} + \boldsymbol{a}_t, \tag{5.79}$$

where $\boldsymbol{c}(t) = \boldsymbol{c}_0 + \boldsymbol{c}_1 t$, $\boldsymbol{c}_i$ are constant vectors, and $\{\boldsymbol{a}_t\}$ is a sequence of iid Gaussian random vectors with mean zero and positive-definite covariance matrix $\text{Cov}(\boldsymbol{a}_t) = \boldsymbol{\Sigma}_a$. For this VAR($p$) model, the ECM of Equation (5.67) becomes

$$\Delta \boldsymbol{z}_t = \boldsymbol{c}_0 + \boldsymbol{c}_1 t + \boldsymbol{\Pi} \boldsymbol{z}_{t-1} + \sum_{i=1}^{p-1} \boldsymbol{\Phi}_i^* \Delta \boldsymbol{z}_{t-i} + \boldsymbol{a}_t. \tag{5.80}$$

Let m be the rank of $\boldsymbol{\Pi}$. There are two cases of interest.

1. $\text{Rank}(\boldsymbol{\Pi}) = 0$: This implies that $\boldsymbol{\Pi} = \boldsymbol{0}$. Thus, there is no cointegrating vector. In this case, $\boldsymbol{z}_t$ has k unit roots and we can work directly on the differenced series $\Delta \boldsymbol{z}_t$, which is a VAR($p - 1$) process.
2. $\text{Rank}(\boldsymbol{\Pi}) = m > 0$: In this case, $\boldsymbol{z}_t$ has m cointegrating vectors and $k - m$ unit roots. As discussed before, there are $k \times m$ full-rank matrices $\boldsymbol{\alpha}$ and $\boldsymbol{\beta}$ such that

$$\boldsymbol{\Pi} = \boldsymbol{\alpha}\boldsymbol{\beta}'.$$

The vector series $\boldsymbol{w}_t = \boldsymbol{\beta}'\boldsymbol{z}_t$ is an $I(0)$ process, which is referred to as the cointegrating series, and $\boldsymbol{\alpha}$ denotes the impact of the cointegrating series on $\Delta\boldsymbol{z}_t$. Let $\boldsymbol{\beta}_\perp$ be a $k \times (k-m)$ full-rank matrix such that $\boldsymbol{\beta}'_\perp\boldsymbol{\beta} = \boldsymbol{0}$. Then, $\boldsymbol{d}_t = \boldsymbol{\beta}'_\perp\boldsymbol{z}_t$ has $k-m$ unit roots and can be considered as the $k-m$ common trends of $\boldsymbol{z}_t$.

5.9.2 Specification of Deterministic Terms

As discussed in Section 5.7.1, the deterministic function $\boldsymbol{c}(t) = \boldsymbol{c}_0 + \boldsymbol{c}_1 t$ has important implication in a cointegrated system. See also Johansen (1995). Thus, the form of $\boldsymbol{c}(t)$ needs to be addressed in cointegration tests.

1. $\boldsymbol{c}(t) = \boldsymbol{0}$: In this case, there is no constant term in the ECM model of Equation (5.80). Thus, the components of $\boldsymbol{z}_t$ are $I(1)$ processes without drift and $\boldsymbol{w}_t = \boldsymbol{\beta}'\boldsymbol{z}_t$ has mean zero.
2. $\boldsymbol{c}_0 = \boldsymbol{\alpha}\boldsymbol{d}_0$ and $\boldsymbol{c}_1 = \boldsymbol{0}$: This is a case of restricted constant. The ECM model becomes

$$\Delta\boldsymbol{z}_t = \boldsymbol{\alpha}(\boldsymbol{\beta}'\boldsymbol{z}_{t-1} + \boldsymbol{d}_0) + \sum_{i=1}^{p-1} \boldsymbol{\Phi}_i^* \Delta\boldsymbol{z}_{t-i} + \boldsymbol{a}_t.$$

   Here, the components of $\boldsymbol{z}_t$ are $I(1)$ series without drift and the cointegrating series $\boldsymbol{w}_t$ has a nonzero mean $\boldsymbol{d}_0$.
3. $\boldsymbol{c}_1 = \boldsymbol{0}$ and $\boldsymbol{c}_0$ is unrestricted: This is a case of unrestricted constant. The ECM becomes

$$\Delta\boldsymbol{z}_t = \boldsymbol{c}_0 + \boldsymbol{\alpha}\boldsymbol{\beta}'\boldsymbol{z}_{t-1} + \sum_{i=1}^{p-1} \boldsymbol{\Phi}_i^* \Delta\boldsymbol{z}_{t-i} + \boldsymbol{a}_t.$$

   The $\boldsymbol{z}_t$ series are $I(1)$ with drift $\boldsymbol{c}_0$ and $\boldsymbol{w}_t$ may have a nonzero mean.
4. $\boldsymbol{c}(t) = \boldsymbol{c}_0 + \boldsymbol{\alpha}\boldsymbol{d}_1 t$: This is a case of restricted trend. The ECM becomes

$$\Delta\boldsymbol{z}_t = \boldsymbol{c}_0 + \boldsymbol{\alpha}(\boldsymbol{\beta}'\boldsymbol{z}_{t-1} + \boldsymbol{d}_1 t) + \sum_{i=1}^{p-1} \boldsymbol{\Phi}_i^* \Delta\boldsymbol{z}_{t-i} + \boldsymbol{a}_t.$$

   The $\boldsymbol{z}_t$ series are $I(1)$ with drift $\boldsymbol{c}_0$ and the cointegrating series $\boldsymbol{w}_t$ has a linear trend term $\boldsymbol{d}_1 t$.
5. $\boldsymbol{c}(t) = \boldsymbol{c}_0 + \boldsymbol{c}_1 t$: In this case, the component series of $\boldsymbol{z}_t$ are $I(1)$ with a quadratic trend and $\boldsymbol{w}_t$ has a linear trend.

The limiting distributions, hence the critical values, of cointegrating tests depend on the specification of $c(t)$. Simulations are used to obtain various critical values for cointegration tests.

5.9.3 Review of Likelihood Ratio Tests

We start with a brief review of likelihood ratio test under multivariate normal distribution. Consider a random vector $\boldsymbol{z}$ that follows a multivariate normal distribution with mean zero and positive-definite covariance matrix $\boldsymbol{\Sigma}_z$. Suppose that $\boldsymbol{z}' = (\boldsymbol{x}', \boldsymbol{y}')$, where the dimensions of $\boldsymbol{x}$ and $\boldsymbol{y}$ are p and q, respectively. Without loss of generality, assume that $q \leq p$. We partition $\boldsymbol{\Sigma}_z$ accordingly as

$$\boldsymbol{\Sigma}_z = \begin{bmatrix} \boldsymbol{\Sigma}_{xx} & \boldsymbol{\Sigma}_{xy} \\ \boldsymbol{\Sigma}_{yx} & \boldsymbol{\Sigma}_{yy} \end{bmatrix}.$$

Suppose also that the null hypothesis of interest is that $\boldsymbol{x}$ and $\boldsymbol{y}$ are uncorrelated. That is, we are interested in testing $H_o : \boldsymbol{\Sigma}_{xy} = \boldsymbol{0}$ versus $H_a : \boldsymbol{\Sigma}_{xy} \neq \boldsymbol{0}$. This is equivalent to testing the coefficient matrix $\boldsymbol{\Pi}$ being 0 in the multivariate linear regression

$$\boldsymbol{x}_i = \boldsymbol{\Pi} \boldsymbol{y}_i + \boldsymbol{e}_i,$$

where $\boldsymbol{e}_i$ denotes the error term. The likelihood ratio test of $\boldsymbol{\Pi} = \boldsymbol{0}$ has certain optimal properties and is the test of choice in practice.

Assume that the available random sample is $\{\boldsymbol{z}_i\}_{i=1}^T$. Under the null hypothesis $\boldsymbol{\Sigma}_{xy} = 0$ so that the maximum likelihood estimate of $\boldsymbol{\Sigma}_z$ is

$$\hat{\boldsymbol{\Sigma}}_o = \begin{bmatrix} \hat{\boldsymbol{\Sigma}}_{xx} & \boldsymbol{0} \\ \boldsymbol{0} & \hat{\boldsymbol{\Sigma}}_{yy} \end{bmatrix},$$

where $\hat{\boldsymbol{\Sigma}}_{xx} = 1/T \sum_{i=1}^T \boldsymbol{x}_i \boldsymbol{x}_i'$ and $\hat{\boldsymbol{\Sigma}}_{yy} = 1/T \sum_{i=1}^T \boldsymbol{y}_i \boldsymbol{y}_i'$. If $E(\boldsymbol{z}_t) \neq \boldsymbol{0}$, the mean-correction is needed in the aforementioned covariance matrix estimators. The maximized likelihood function under the null hypothesis is

$$\ell_o \propto |\hat{\boldsymbol{\Sigma}}_o|^{-T/2} = (|\hat{\boldsymbol{\Sigma}}_{xx}||\hat{\boldsymbol{\Sigma}}_{yy}|)^{-T/2},$$

see, for example, Johnson and Wichern (2007, pp. 172). On the other hand, under the alternative, the is no constraint on the covariance matrix so that the maximum likelihood estimate of $\boldsymbol{\Sigma}_z$ is

$$\hat{\boldsymbol{\Sigma}}_a = \frac{1}{T} \sum_{i=1}^{T} \begin{bmatrix} \boldsymbol{x}_i \\ \boldsymbol{y}_i \end{bmatrix} [\boldsymbol{x}_i', \boldsymbol{y}_i'] \equiv \begin{bmatrix} \hat{\boldsymbol{\Sigma}}_{xx} & \hat{\boldsymbol{\Sigma}}_{xy} \\ \hat{\boldsymbol{\Sigma}}_{yx} & \hat{\boldsymbol{\Sigma}}_{yy} \end{bmatrix}.$$

The maximized likelihood function under the alternative hypothesis is

$$\ell_a \propto |\hat{\mathbf{\Sigma}}_a|^{-T/2} = (|\hat{\mathbf{\Sigma}}_{xx}||(\hat{\mathbf{\Sigma}}_{yy} - \hat{\mathbf{\Sigma}}_{yx}\hat{\mathbf{\Sigma}}_{xx}^{-1}\hat{\mathbf{\Sigma}}_{xy})|)^{-T/2}.$$

The likelihood ratio test statistic is therefore

$$L = \frac{\ell_o}{\ell_a} = \left(\frac{|\hat{\mathbf{\Sigma}}_a|}{|\hat{\mathbf{\Sigma}}_o|}\right)^{T/2} = \left(|\boldsymbol{I} - \hat{\mathbf{\Sigma}}_{yy}^{-1}\hat{\mathbf{\Sigma}}_{yx}\hat{\mathbf{\Sigma}}_{xx}^{-1}\hat{\mathbf{\Sigma}}_{xy}|\right)^{T/2}.$$

One rejects the null hypothesis if L is small. Next, let $\{\lambda_i\}_{i=1}^q$ be the eigenvalues of the matrix $\hat{\mathbf{\Sigma}}_{yy}^{-1}\hat{\mathbf{\Sigma}}_{yx}\hat{\mathbf{\Sigma}}_{xx}^{-1}\hat{\mathbf{\Sigma}}_{xy}$. Then, $\{1 - \lambda_i\}_{i=1}^q$ are the eigenvalues of $\boldsymbol{I} - \hat{\mathbf{\Sigma}}_{yy}^{-1}\hat{\mathbf{\Sigma}}_{yx}\hat{\mathbf{\Sigma}}_{xx}^{-1}\hat{\mathbf{\Sigma}}_{xy}$. Consequently, the negative log likelihood ratio statistic is

$$\begin{aligned}
\text{LR} &= -\frac{T}{2}\ln\left(|\boldsymbol{I} - \hat{\mathbf{\Sigma}}_{yy}^{-1}\hat{\mathbf{\Sigma}}_{yx}\hat{\mathbf{\Sigma}}_{xx}^{-1}\hat{\mathbf{\Sigma}}_{xy}|\right) \\
&= -\frac{T}{2}\ln\left(\prod_{i=1}^q(1-\lambda_i)\right) = -\frac{T}{2}\sum_{i=1}^q \ln(1-\lambda_i).
\end{aligned} \tag{5.81}$$

One rejects the null hypothesis if the test statistic LR is large.

Note that $\{\lambda_i\}$ are the squared sample canonical correlation coefficients between $\boldsymbol{x}$ and $\boldsymbol{y}$. Thus, the likelihood ratio test for $\mathbf{\Pi} = \mathbf{0}$ is based on the canonical correlation analysis between $\boldsymbol{x}$ and $\boldsymbol{y}$.

5.9.4 Cointegration Tests of VAR Models

Return to cointegration test, which essentially is to test the rank of the matrix $\mathbf{\Pi}$ in Equation (5.80). Since $\mathbf{\Pi}$ is related to the covariance matrix between $\boldsymbol{z}_{t-1}$ and $\Delta\boldsymbol{z}_t$, we can apply the likelihood ratio test to check its rank. To simplify the procedure, it pays to concentrate out the effects of $\boldsymbol{c}(t)$ and $\Delta\boldsymbol{z}_{t-i}$ from Equation (5.80) before estimating $\mathbf{\Pi}$. To this end, consider the next two linear regressions

$$\Delta\boldsymbol{z}_t = \boldsymbol{c}(t) + \sum_{i=1}^{p-1}\boldsymbol{\varpi}_i\Delta\boldsymbol{z}_{t-i} + \boldsymbol{u}_t, \tag{5.82}$$

$$\boldsymbol{z}_{t-1} = \boldsymbol{c}(t) + \sum_{i=1}^{p-1}\boldsymbol{\varpi}_i^*\Delta\boldsymbol{z}_{t-i} + \boldsymbol{v}_t, \tag{5.83}$$

where it is understood that the form of $\boldsymbol{c}(t)$ is prespecified, and $\boldsymbol{u}_t$ and $\boldsymbol{v}_t$ denote the error terms. These two regressions can be estimated by the least-squares method. Let $\hat{\boldsymbol{u}}_t$ and $\hat{\boldsymbol{v}}_t$ are the residuals of Equations (5.82) and (5.83), respectively. Then, we have the regression

$$\hat{\boldsymbol{u}}_t = \boldsymbol{\Pi}\hat{\boldsymbol{v}}_t + \boldsymbol{e}_t, \tag{5.84}$$

where $\boldsymbol{e}_t$ denotes the error term. The LSEs of $\boldsymbol{\Pi}$ are identical between Equations (5.80) and (5.84). Let

$$H(0) \subset H(1) \subset \cdots \subset H(k)$$

be the nested models such that under $H(m)$ there are m cointegrating vectors in $\boldsymbol{z}_t$. That is, under $H(m)$, rank($\boldsymbol{\Pi}$) $= m$. In particular, we have $\boldsymbol{\Pi} = \boldsymbol{0}$ under $H(0)$.

Applying the results of Subsection 5.9.3, we define

$$\hat{\boldsymbol{\Sigma}}_{00} = \frac{1}{T}\sum_{t=1}^{T}\hat{\boldsymbol{u}}_t\hat{\boldsymbol{u}}_t', \quad \hat{\boldsymbol{\Sigma}}_{11} = \frac{1}{T}\sum_{t=1}^{T}\hat{\boldsymbol{v}}_t\hat{\boldsymbol{v}}_t', \quad \hat{\boldsymbol{\Sigma}}_{01} = \frac{1}{T}\sum_{t=1}^{T}\hat{\boldsymbol{u}}_t\hat{\boldsymbol{v}}_t'.$$

Let $\lambda_1 \geq \lambda_2 \geq \cdots \geq \lambda_k \geq 0$ be the ordered eigenvalues of the sample matrix $\hat{\boldsymbol{\Sigma}}_{11}^{-1}\hat{\boldsymbol{\Sigma}}_{10}\hat{\boldsymbol{\Sigma}}_{00}^{-1}\hat{\boldsymbol{\Sigma}}_{01}$, and $\boldsymbol{g}_i$ be the eigenvector associated with eigenvalue λ_i. That is,

$$\hat{\boldsymbol{\Sigma}}_{11}^{-1}\hat{\boldsymbol{\Sigma}}_{10}\hat{\boldsymbol{\Sigma}}_{00}^{-1}\hat{\boldsymbol{\Sigma}}_{01}\boldsymbol{g}_i = \lambda_i\boldsymbol{g}_i. \tag{5.85}$$

Equivalently, we have

$$\hat{\boldsymbol{\Sigma}}_{10}\hat{\boldsymbol{\Sigma}}_{00}^{-1}\hat{\boldsymbol{\Sigma}}_{01}\boldsymbol{g}_i = \lambda_i\hat{\boldsymbol{\Sigma}}_{11}\boldsymbol{g}_i, \quad i = 1, \cdots, k.$$

Furthermore, the eigenvectors can be normalized such that

$$\boldsymbol{G}'\hat{\boldsymbol{\Sigma}}_{11}^{-1}\boldsymbol{G} = \boldsymbol{I},$$

where $\boldsymbol{G} = [\boldsymbol{g}_1, \cdots, \boldsymbol{g}_k]$ is the matrix of eigenvectors.

Remark: The eigenvalues are non-negative and the eigenvectors are mutually orthogonal with respect to $\hat{\boldsymbol{\Sigma}}_{11}$ because

$$\left[\hat{\boldsymbol{\Sigma}}_{11}^{-1/2}\hat{\boldsymbol{\Sigma}}_{10}\hat{\boldsymbol{\Sigma}}_{00}^{-1}\hat{\boldsymbol{\Sigma}}_{01}\hat{\boldsymbol{\Sigma}}_{11}^{-1/2}\right]\left(\hat{\boldsymbol{\Sigma}}_{11}^{1/2}\boldsymbol{g}_i\right) = \lambda_i\left(\hat{\boldsymbol{\Sigma}}_{11}^{1/2}\boldsymbol{g}_i\right), \quad i = 1, \cdots, k.$$

In the prior equation, the matrix inside the square brackets is symmetric. □

Consider the nested hypotheses:

$$H_0 : m = m_0 \quad \text{versus} \quad H_a : m > m_0,$$

where $m = \text{Rank}(\mathbf{\Pi})$ and m_0 is a given integer between 0 and $k-1$ with k being the dimension of $\boldsymbol{z}_t$. Johansen's trace statistic is defined as

$$L_{tr}(m_0) = -(T-kp) \sum_{i=m_0+1}^{k} \ln(1-\lambda_i), \tag{5.86}$$

where λ_i are the eigenvalues defined in Equation (5.85). If $\text{rank}(\mathbf{\Pi}) = m_0$, then the m_0 smallest eigenvalues should be 0, that is, $\lambda_{m_0+1} = \cdots = \lambda_k = 0$, and the test statistic should be small. On the other hand, if $\text{rank}(\mathbf{\Pi}) > m_0$, then some of the eigenvalues in $\{\lambda_i\}_{i=m_0+1}^{k}$ are nonzero and the test statistic should be large. Because of the existence of unit roots, the limiting distribution of $L_{tr}(\cdot)$ statistic is not chi-square. It is a function of standard Brownian motions. Specifically, the limiting distribution of the test statistic $L_{tr}(m_0)$ has been derived and is given by $L_{tr}(m_0) \Rightarrow$

$$\text{tr}\left\{\left[\int_0^1 \boldsymbol{W}_v(r)d\boldsymbol{W}_v(r)'\right]' \left[\int_0^1 \boldsymbol{W}_v(r)\boldsymbol{W}_v(r)'dr\right]^{-1} \left[\int_0^1 \boldsymbol{W}_v(r)d\boldsymbol{W}_v(r)'\right]\right\}, \tag{5.87}$$

where $v = m_0$ and $\boldsymbol{W}_v(r)$ is a v-dimensional standard Brownian motion. This distribution depends only on m_0 and not on any parameters or the order of the VAR model. Thus, the critical values can be tabulated via simulation. See, for instance, Johansen (1988) and Reinsel and Ahn (1992).

The trace statistic $L_{tr}(\cdot)$ can be applied sequentially, starting with $m_0 = 0$, to identify the rank of $\mathbf{\Pi}$, that is, the number of cointegrating vectors. Johansen also proposes a likelihood ratio test based on the maximum eigenvalue to identify the number of cointegrating vectors. Consider the hypotheses

$$H_o : m = m_0 \quad \text{versus} \quad H_a : m = m_0 + 1,$$

where m and m_0 are defined as before in Equation (5.86). The test statistic is

$$L_{\max}(m_0) = -(T-kp)\ln(1-\lambda_{m_0+1}). \tag{5.88}$$

Similarly to the case of $L_{tr}(m_0)$ statistic, the limiting distribution of $L_{\max}(m_0)$ is not chi-square. It is also a function of the standard Brownian motion, and the critical values of the test statistic have been obtained via simulation.

Since the eigenvectors $\boldsymbol{g}_i$ give rise to linear combinations of $\boldsymbol{z}_t$, they provide an estimate of the cointegrating vectors. In other words, for a given $m_0 = \text{Rank}(\mathbf{\Pi})$, an estimate of $\boldsymbol{\beta}$ is $\hat{\boldsymbol{\beta}} = [\boldsymbol{g}_1, \cdots, \boldsymbol{g}_{m_0}]$. See Equation (5.85).

Remark: In our discussion, we use the ECM in Equation (5.67). The same conclusion continues to hold if one uses the ECM in Equation (5.68). In applications, we use the package `urca` with command `ca.jo` to perform cointegration analysis. This command implements Johansen's cointegration test. The command allows users to choose the ECM in (5.67) or (5.67). The ECM in Equation (5.67) is referred to in `urca` by the subcommand `spec = c("transitory")`, whereas that in Equation (5.68) by the subcommand `spec = c("longrun")`. The choice between test statistics $L_{tr}(m_0)$ in Equation (5.86) and $L_{\max}(m_0)$ in Equation (5.88) is by the subcommand `type = c("trace")` and `type = c("eigen")`, respectively. Also, the order p is denoted by K in the command `ca.jo`. Finally, the specification of the deterministic term $\boldsymbol{c}(t)$ is by the subcommand `ecdet`. The choices include `"none"`, `"const"`, and `"trend"`. □

5.9.5 An Illustration

Consider the monthly yields of Moody's seasoned corporate Aaa and Baa bonds from July 1954 to February 2005 for 609 observations. The data were obtained from Federal Reserve Bank of St. Louis. Figure 5.4 shows the time plots of the bond yields. As expected, the two series move in a parallel manner. The augmented Dickey–Fuller unit-root test confirms that the two bond yields are unit-root nonstationary. See the following R demonstration. Assuming that the bivariate series of bond yields follows a VAR model, we applied the information criteria to specify the order. The order

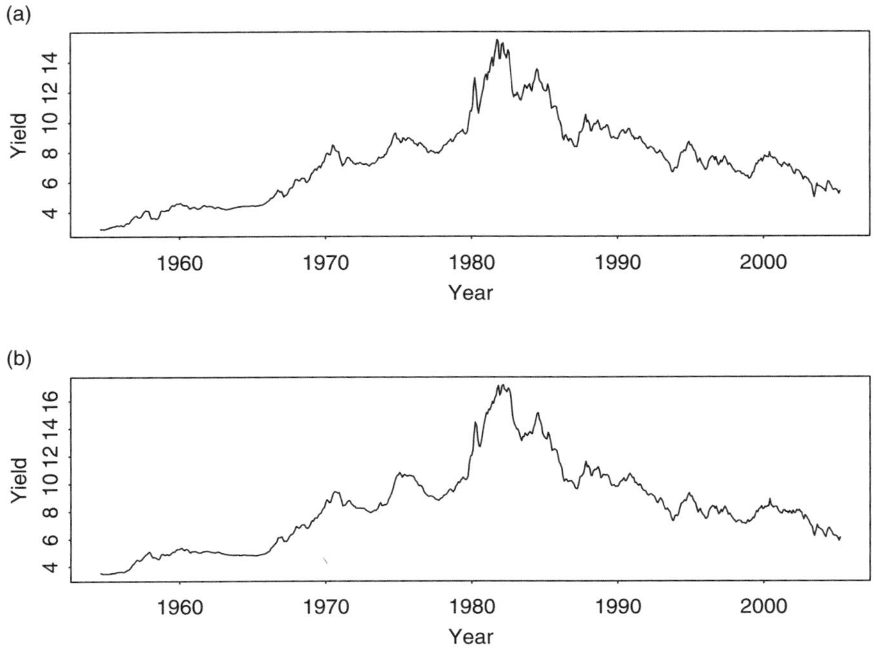

FIGURE 5.4 Time plots of monthly Moody's seasoned (a) Aaa and (b) Baa bond yields from July 1954 to February 2005.

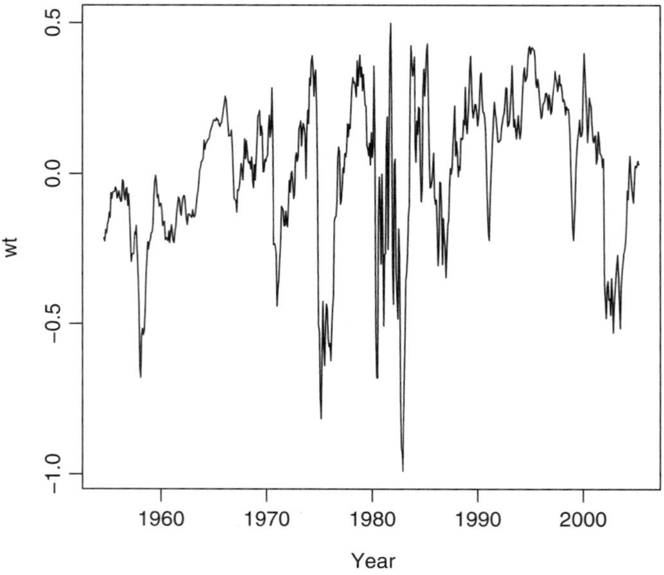

FIGURE 5.5 Time plot of a cointegrated series between monthly Moody's seasoned Aaa and Baa bond yields from July 1954 to February 2005.

$p = 3$ is selected by both BIC and HQ. Therefore, we employ a VAR(3) model in the cointegration test. Since the mean vector of the differenced series $\Delta \boldsymbol{z}_t$ is not significantly different from 0, we do not consider the constant in the test.

For this particular instance, the two eigenvalues are 0.055 and 0.005, respectively. If the test statistics $L_{\max}(m_0)$ of Equation (5.88) is used, we obtain that $L_{\max}(1) = 2.84$ and $L_{\max}(0) = 34.19$. Compared with critical values, we reject $r = 0$, but cannot reject $r = 1$. Therefore, the $\boldsymbol{\Pi}$ matrix of ECM is of rank 1 and there is a cointegrating vector. If the trace statistics $L_{tr}(m_0)$ of Equation (5.86), we have $L_{tr}(1) = 2.84$ and $L_{tr}(0) = 37.03$. Again, the null hypothesis of a cointegration is not rejected. The cointegration series is $w_t = (1, -0.886)\boldsymbol{z}_t$. The ADF test of w_t confirms that the series does not have any unit root. The test statistic is -4.61 with p-value 0.01. Figure 5.5 shows the time plot of the cointegrated series w_t. As expected, the series shows the characteristics of a stationary time series.

R Demonstration: Cointegration test. Output edited.

```
> require(fUnitRoots)
> require(urca)
> require(MTS)
> da=read.table("m-bnd.txt")
> head(da)
    V1 V2 V3   V4   V5
1 1954  7  1 2.89 3.50
```

```
2 1954  8  1 2.87 3.49
> tail(da)
     V1 V2 V3   V4   V5
608 2005  2  1 5.20 5.82
609 2005  3  1 5.40 6.06
> bnd=da[,4:5]
> colnames(bnd) <- c("Aaa","Baa")
> m1=VARorder(bnd)
selected order: aic =  11
selected order: bic =  3
selected order: hq =  3
> pacf(bnd[,1]); pacf(bnd[,2])
> adfTest(bnd[,1],lags=3,type="c")
Title: Augmented Dickey-Fuller Test
Test Results:
  PARAMETER:
    Lag Order: 3
  STATISTIC:
    Dickey-Fuller: -1.7007
  P VALUE:
    0.425
> adfTest(bnd[,2],lags=2,type="c")
Title: Augmented Dickey-Fuller Test
Test Results:
  PARAMETER:
    Lag Order: 2
  STATISTIC:
    Dickey-Fuller: -1.6221
  P VALUE:
    0.4544
> m2=ca.jo(bnd,K=2,ecdet=c("none")) % use maximum eigenvalue
> summary(m2) % use z(t-k) in ECM
######################
# Johansen-Procedure #
######################
Test type: maximal eigenvalue (lambda max), with linear trend
Eigenvalues (lambda): 0.054773196 0.004665298

Values of teststatistic and critical values of test:
          test 10pct  5pct  1pct
r <= 1 |  2.84  6.50  8.18 11.65
r = 0  | 34.19 12.91 14.90 19.19
Eigenvectors, normalised to first column:(Co-integration
                                           relations)
           Aaa.l2    Baa.l2
Aaa.l2  1.0000000  1.000000
Baa.l2 -0.8856789 -2.723912

Weights W: (This is the loading matrix)
```

```
           Aaa.l2       Baa.l2
Aaa.d -0.04696894 0.002477064
Baa.d  0.04046524 0.002139536

> m3=ca.jo(bnd,K=2,ecdet=c("none"),spec=c("transitory"))
> summary(m3)  % use the z(t-1) in ECM.
######################
# Johansen-Procedure #
######################
Test type: maximal eigenvalue (lambda max), with linear trend

Eigenvalues (lambda): 0.054773196 0.004665298

Values of teststatistic and critical values of test:
          test 10pct  5pct  1pct
r <= 1 |  2.84  6.50  8.18 11.65
r = 0  | 34.19 12.91 14.90 19.19

Eigenvectors, normalised to first column:(Co-integration
                                          relations)
           Aaa.l1     Baa.l1
Aaa.l1  1.0000000  1.000000
Baa.l1 -0.8856789 -2.723912

> m4=ca.jo(bnd,K=2,ecdet=c("none"),type=c("trace"),
  spec=c("transitory"))
> summary(m4)  % use trace test statistic
######################
# Johansen-Procedure #
######################
Test type: trace statistic , with linear trend
Eigenvalues (lambda):0.054773196 0.004665298

Values of teststatistic and critical values of test:
          test 10pct  5pct  1pct
r <= 1 |  2.84  6.50  8.18 11.65
r = 0  | 37.03 15.66 17.95 23.52

> wt=bnd[,1]-0.886*bnd[,2] % co-integrating series
> adfTest(wt,lags=3,type='c')  % Reject unit root in w(t)
Title: Augmented Dickey-Fuller Test
Test Results:
  PARAMETER:
    Lag Order: 3
  STATISTIC:
    Dickey-Fuller: -4.6054
  P VALUE:
    0.01
```

Remark: The `urca` package also includes the residual-based cointegration tests of Phillips and Ouliaris (1990). The command is `ca.po`. For the bond yield data, the residual-based tests also detect a cointegration relationship. □

5.9.6 Cointegration Tests of VARMA Models

The cointegration tests discussed in the prior section can be generalized to the VARMA models using the ECM representation in Equation (5.67). The estimation involved, however, becomes more intensive. In this section, we consider an approximation approach that continues to employ LS estimation. The approach modifies Equations (5.82) and (5.83) as

$$\Delta \boldsymbol{z}_t = \boldsymbol{c}(t) + \sum_{i=1}^{p-1} \boldsymbol{\Phi}_i^* \Delta \boldsymbol{z}_{t-i} + \sum_{j=1}^{q} \boldsymbol{\Theta}_j \hat{\boldsymbol{a}}_{t-j} + \boldsymbol{u}_t, \tag{5.89}$$

$$\boldsymbol{z}_{t-1} = \boldsymbol{c}(t) + \sum_{i=1}^{p-1} \boldsymbol{\Phi}_i^* \Delta \boldsymbol{z}_{t-i} + \sum_{j=1}^{q} \boldsymbol{\Theta}_j \hat{\boldsymbol{a}}_{t-j} + \boldsymbol{v}_t, \tag{5.90}$$

where $\hat{\boldsymbol{a}}_t$ is an estimate of $\boldsymbol{a}_t$. In practice, $\hat{\boldsymbol{a}}_t$ can be obtained by fitting a long VAR model to $\boldsymbol{z}_t$. Since an invertible VARMA model can be approximated by a VAR model, the residual $\hat{\boldsymbol{a}}_t$ should be an appropriate approximation of $\boldsymbol{a}_t$ provided that the VAR order is properly chosen.

Let $\hat{\boldsymbol{u}}_t$ and $\hat{\boldsymbol{v}}_t$ be the residuals of Equations (5.89) and (5.90), respectively. We can then continue to employ Equation (5.84) and compute the test statistics $L_{tr}(m_0)$ and $L_{\max}(m_0)$ of Equations (5.86) and (5.88) to perform cointegration tests. This approach is only an approximation and it should work well when the sample size T is sufficiently large.

5.10 ESTIMATION OF ERROR-CORRECTION MODELS

We discuss maximum likelihood estimation of ECM in this section, assuming that the number of cointegrating vectors is known. We further assume that the innovations $\{\boldsymbol{a}_t\}$ are Gaussian. This is equivalent to using the quasi maximum likelihood estimates (QMLE). The orders p and q of the model can be specified by using the methods discussed in Chapter 4. See, for instance, Tiao and Tsay (1989).

5.10.1 VAR Models

We start with the VAR(p) models. Two cases are entertained in this section. In the first case, the cointegrating matrix $\boldsymbol{\beta}$ is known so that the cointegrating process $\boldsymbol{w}_t = \boldsymbol{\beta}' \boldsymbol{z}_t$ is available. In this particular case, the model reduces to

$$\Delta \boldsymbol{z}_t = \boldsymbol{\alpha} \boldsymbol{w}_{t-1} + \boldsymbol{c}(t) + \sum_{i=1}^{p-1} \boldsymbol{\Phi}_i^* \Delta \boldsymbol{z}_{t-i} + \boldsymbol{a}_t, \tag{5.91}$$

which can be estimated by the ordinary least-squares method. The estimates have the usual asymptotic normal distribution and the conventional approach can be used to make statistical inference.

In the second case, the $\boldsymbol{\beta}$ is unknown. Here, the model becomes

$$\Delta \boldsymbol{z}_t = \boldsymbol{\alpha}\boldsymbol{\beta}' \boldsymbol{z}_{t-1} + \boldsymbol{c}(t) + \sum_{i=1}^{p-1} \boldsymbol{\Phi}_i^* \Delta \boldsymbol{z}_{t-i} + \boldsymbol{a}_t, \tag{5.92}$$

which involves products of parameters and requires nonlinear estimation. We shall use the parameterization of Section 5.8 with $\boldsymbol{\beta}' = [\boldsymbol{I}_m, \boldsymbol{\beta}_1']$, where $\boldsymbol{\beta}_1$ is a $(k-m) \times m$ matrix; see Equation (5.78). We use QMLE to estimate the model. In practice, the initial estimate of β from cointegration tests and the results of the previous case can be used to start the nonlinear estimation. With the presence of $\boldsymbol{z}_{t-1}$, the limiting distribution of the estimate of $\boldsymbol{\Pi}$ involves functions of Brownian motion.

To demonstrate, we consider, again, the monthly Moody's Aaa and Baa bond yields used in cointegration tests. With $p = 3$ and $m = 1$, the cointegration tests gave the initial cointegrating vector as $(1, -0.886)'$. If we treat $w_t = (1, -0.886)\boldsymbol{z}_t$ as given, then the ECM can be estimated by the least-squares method. The fitted model is

$$\begin{aligned} \Delta \boldsymbol{z}_t &= \begin{bmatrix} -0.001 \\ 0.064 \end{bmatrix} w_t + \begin{bmatrix} 0.452 & -0.001 \\ 0.293 & 0.204 \end{bmatrix} \Delta \boldsymbol{z}_{t-1} \\ &\quad + \begin{bmatrix} -0.300 & 0.054 \\ -0.151 & 0.028 \end{bmatrix} \Delta \boldsymbol{z}_{t-2} + \boldsymbol{a}_t, \ \hat{\boldsymbol{\Sigma}}_a = \frac{1}{100} \begin{bmatrix} 4.01 & 3.17 \\ 3.17 & 3.11 \end{bmatrix}. \end{aligned} \tag{5.93}$$

The standard errors of the estimates are given in the following R demonstration. Several estimates are not statistically significant at the conventional 5% level. Removing insignificant parameters, we refine the model as

$$\begin{aligned} \Delta \boldsymbol{z}_t &= \begin{bmatrix} 0.000 \\ 0.063 \end{bmatrix} w_t + \begin{bmatrix} 0.448 & 0.000 \\ 0.286 & 0.212 \end{bmatrix} \Delta \boldsymbol{z}_{t-1} \\ &\quad + \begin{bmatrix} -0.256 & 0.00 \\ -0.129 & 0.00 \end{bmatrix} \Delta \boldsymbol{z}_{t-2} + \boldsymbol{a}_t, \ \hat{\boldsymbol{\Sigma}}_a = \frac{1}{100} \begin{bmatrix} 4.01 & 3.17 \\ 3.17 & 3.11 \end{bmatrix} \end{aligned} \tag{5.94}$$

Model checking indicates that some minor serial and cross-correlations exist at lag 6, but the model can serve as a reasonable approximation to the dynamic dependence of the system. It is interesting to see that the first element of $\boldsymbol{\alpha}$, α_{11}, is not statistically significant, implying that the first differenced series of Aaa bond yields is unit-root stationary. Thus, the monthly Aaa bond yield represents the common trend in the bond yield system.

Finally, we estimate the cointegrating vector $(1, \beta_1)'$, where β_1 is unknown, simultaneously with other parameters. The result of this nonlinear estimation is

$$\Delta \boldsymbol{z}_t = \begin{bmatrix} -0.001 \\ 0.064 \end{bmatrix} [1, -0.887]\boldsymbol{z}_{t-1} + \begin{bmatrix} 0.452 & -0.001 \\ 0.293 & 0.204 \end{bmatrix} \Delta \boldsymbol{z}_{t-1}$$

$$+ \begin{bmatrix} -0.300 & 0.054 \\ -0.151 & 0.028 \end{bmatrix} \Delta \boldsymbol{z}_{t-2} + \boldsymbol{a}_t, \; \hat{\boldsymbol{\Sigma}}_a = \frac{1}{100} \begin{bmatrix} 4.00 & 3.16 \\ 3.16 & 3.10 \end{bmatrix}. \qquad (5.95)$$

As expected, this model is very close to Equation (5.93) because the estimated cointegrating vector is essentially the same as that obtained in the cointegration test. This is not surprising as the sample size is relatively large and the dimension is low.

Remark: Estimation of ECM for VAR processes can be carried out by the commands `ECMvar1` and `ECMvar` of the `MTS` package. The first command treats the cointegrated series $\boldsymbol{w}_t$ as given whereas the second command performs joint estimation of the cointegrating vectors and other parameters. □

R Demonstration: Estimation of ECM-VAR models. Output edited.

```
> wt=bnd[,1]-0.886*bnd[,2]
> m1=ECMvar1(bnd,3,wt)   ## Co-integrated series is given.
alpha:
            Aaa    Baa
[1,] -0.000976 0.0636
standard error
       [,1]   [,2]
[1,] 0.0347 0.0306
AR coefficient matrix
AR( 1 )-matrix
      Aaa       Baa
Aaa 0.452 -0.00144
Baa 0.293  0.20386
standard error
       [,1]   [,2]
[1,] 0.0879 0.1008
[2,] 0.0774 0.0887
AR( 2 )-matrix
       Aaa    Baa
Aaa -0.300 0.0536
Baa -0.151 0.0275
standard error
       [,1]   [,2]
[1,] 0.0860 0.0940
[2,] 0.0757 0.0827
-----
Residuals cov-mtx:
           Aaa        Baa
Aaa 0.04008513 0.03167097
Baa 0.03167097 0.03105743
```

```
-----
AIC =  -8.294184; BIC =  -8.221741
> m2=refECMvar1(m1) ## Refine the model
alpha:
     [,1]   [,2]
[1,]    0 0.0627
standard error
     [,1]   [,2]
[1,]    1 0.0304
AR coefficient matrix
AR( 1 )-matrix
      [,1]  [,2]
[1,] 0.448 0.000
[2,] 0.286 0.212
standard error
       [,1]   [,2]
[1,] 0.0393 1.0000
[2,] 0.0746 0.0855
AR( 2 )-matrix
       [,1] [,2]
[1,] -0.256    0
[2,] -0.129    0
standard error
       [,1] [,2]
[1,] 0.0393    1
[2,] 0.0382    1
-----
Residuals cov-mtx:
           [,1]       [,2]
[1,] 0.04010853 0.03168199
[2,] 0.03168199 0.03106309
-----
AIC =  -8.306263; BIC =  -8.262797
> m3=ECMvar(bnd,3,beta,include.const=F) % Joint estimation
Order p:  3  Co-integrating rank:  1
Number of parameters:  11
alpha:
          [,1]
[1,] -0.000786
[2,]  0.063779
standard error
       [,1]
[1,] 0.0354
[2,] 0.0311
beta:
       [,1]
[1,]  1.000
[2,] -0.887
standard error
```

```
         [,1]
[1,] 1.0000
[2,] 0.0055
AR coefficient matrix
AR( 1 )-matrix
        [,1]      [,2]
[1,] 0.452 -0.00141
[2,] 0.293  0.20392
standard error
         [,1]   [,2]
[1,] 0.0896 0.103
[2,] 0.0786 0.090
AR( 2 )-matrix
         [,1]    [,2]
[1,] -0.300 0.0536
[2,] -0.151 0.0276
standard error
         [,1]    [,2]
[1,] 0.0860 0.0939
[2,] 0.0757 0.0827
-----
Residuals cov-mtx:
          Aaa          Baa
Aaa 0.0399535 0.03156630
Baa 0.0315663 0.03095427
-----
AIC =   -8.297496;   BIC =   -8.217807
```

5.10.2 Reduced Regression Method

An alternative approach to estimate ECM for a VAR process is to use reduced-rank regression method. For a given $m = \text{Rank}(\boldsymbol{\Pi})$, the ECM becomes a reduced rank multivariate regression and can be estimated accordingly. As mentioned in Section 5.9.4, an estimate of $\boldsymbol{\beta}$ is $\hat{\boldsymbol{\beta}} = [\boldsymbol{g}_1, \cdots, \boldsymbol{g}_m]$. From this estimate, one can form the normalized estimator $\hat{\boldsymbol{\beta}}_c$ by imposing the appropriate normalization and identifying constraints. For instance, use the parameterization of Section 5.8.

Next, given $\hat{\boldsymbol{\beta}}_c$, we can obtain estimates of other parameters by the multivariate linear regression

$$\Delta \boldsymbol{z}_t = \boldsymbol{\alpha}_c(\hat{\boldsymbol{\beta}}_c' \boldsymbol{z}_{t-1}) + \sum_{i=1}^{p-1} \boldsymbol{\phi}_i^* \Delta \boldsymbol{z}_{t-i} + \boldsymbol{c}(t) + \boldsymbol{a}_t,$$

where the subscript "c" of $\boldsymbol{\alpha}$ is used to signify the dependence on $\hat{\boldsymbol{\beta}}_c$. The maximized likelihood function based on the m cointegrating vectors is

$$\ell^{-2/T} \propto |\hat{\boldsymbol{\Sigma}}_{00}| \prod_{i=1}^{m} (1 - \lambda_i).$$

The estimates of the orthogonal complements of $\boldsymbol{\alpha}_c$ and $\boldsymbol{\beta}_c$ are given by

$$\hat{\boldsymbol{\alpha}}_{c,\perp} = \hat{\boldsymbol{\Sigma}}_{00}^{-1}\hat{\boldsymbol{\Sigma}}_{11}[\boldsymbol{g}_{m+1}, \cdots, \boldsymbol{g}_k] \tag{5.96}$$

$$\hat{\boldsymbol{\beta}}_{c,\perp} = \hat{\boldsymbol{\Sigma}}_{11}[\boldsymbol{g}_{m+1}, \cdots, \boldsymbol{g}_k]. \tag{5.97}$$

Finally, premultiplying the ECM by the orthogonal complement $\boldsymbol{\alpha}'_{\perp}$, we obtain

$$\boldsymbol{\alpha}'_{\perp}(\Delta \boldsymbol{z}_t) = \boldsymbol{\alpha}'_{\perp}\boldsymbol{\alpha}\boldsymbol{\beta}'\boldsymbol{z}_{t-1} + \sum_{i=1}^{p} \boldsymbol{\alpha}'_{\perp}\boldsymbol{\phi}_i^* \Delta \boldsymbol{z}_{t-i} + \boldsymbol{\alpha}'_{\perp}[\boldsymbol{c}(t) + \boldsymbol{a}_t].$$

Since $\boldsymbol{\alpha}'_{\perp}\boldsymbol{\alpha} = \boldsymbol{0}$, we see that

$$\boldsymbol{\alpha}'_{\perp}(\Delta \boldsymbol{z}_t) = \sum_{i=1}^{p} \boldsymbol{\alpha}'_{\perp}\boldsymbol{\phi}_i^*(\Delta \boldsymbol{z}_{t-i}) + \boldsymbol{\alpha}'_{\perp}[\boldsymbol{c}(t) + \boldsymbol{a}_t],$$

which does not contain any error-correction term. Consequently, $\boldsymbol{y}_t = \boldsymbol{\alpha}'_{\perp}\boldsymbol{z}_t$ has $k - m$ unit roots and represents the common trends of $\boldsymbol{z}_t$. From Equations (5.96) and (5.97), using $\boldsymbol{y}_t$ as the command trend is consistent with that discussed in Section 5.9.

5.10.3 VARMA Models

The QMLE of the ECM for VAR processes discussed in the prior subsection can be extended to the ECM in Equation (5.67) for VARMA processes. Again, we use the parameterization $\boldsymbol{\beta}' = [\boldsymbol{I}_m, \boldsymbol{\beta}'_1]$ of Section 5.8. The general model then becomes

$$\Delta \boldsymbol{z}_t = \boldsymbol{\alpha}\boldsymbol{\beta}'\boldsymbol{z}_{t-1} + \sum_{i=1}^{p-1} \boldsymbol{\Phi}_i^* \Delta \boldsymbol{z}_{t-i} + \boldsymbol{c}(t) - \sum_{j=1}^{q} \boldsymbol{\Theta}_j \boldsymbol{a}_{t-j} + \boldsymbol{a}_t. \tag{5.98}$$

In estimation, the innovations $\boldsymbol{a}_t$ can be evaluated by similar methods as those of Chapter 3 for VARMA models with some minor modifications. The initial values of the MA parameters can be obtained by a two-step procedure. In the first step, a higher-order VAR model is estimated to obtain $\hat{\boldsymbol{a}}_t$. At the second stage, $\hat{\boldsymbol{a}}_t$ is treated as given and the model in Equation (5.98) is estimated by the least-squares method with $\boldsymbol{\beta}'\boldsymbol{z}_{t-1}$ replaced by $\boldsymbol{w}_t$ from the cointegration test.

Remark: The ECM derived in Section 5.2.1 uses the traditional VARMA model representation in which the AR matrix polynomial starts with $\boldsymbol{I}_k$. The same derivation continues to hold if one replaces $\boldsymbol{I}_k$ by any well-defined $k \times k$ matrix $\boldsymbol{\Phi}_0$. More specifically, Equation (5.40) becomes

$$\begin{aligned}\boldsymbol{\Phi}(B) &= \boldsymbol{\Phi}_0 - \boldsymbol{\Phi}_1 B - \cdots - \boldsymbol{\Phi}_p B^p \\ &= \boldsymbol{\Phi}_0 - \boldsymbol{\Pi}^* - (\boldsymbol{\Phi}_1^* B + \cdots + \boldsymbol{\Phi}_{p-1}^* B^{p-1})(1 - B),\end{aligned} \tag{5.99}$$

where $\mathbf{\Pi}^*$ and $\mathbf{\Phi}_i^*$ remain unchanged, but Equation (5.42) becomes

$$\mathbf{\Phi}_0 \Delta \boldsymbol{z}_t = \mathbf{\Pi} \boldsymbol{z}_{t-1} + \sum_{j=1}^{p-1} \mathbf{\Phi}_j^* \Delta \boldsymbol{z}_{t-j} + \boldsymbol{c}(t) + \mathbf{\Theta}(B) \boldsymbol{a}_t, \tag{5.100}$$

where $\mathbf{\Pi} = \mathbf{\Phi}_1 + \cdots + \mathbf{\Phi}_p - \mathbf{\Phi}_0$, which, again, is given by $\mathbf{\Pi} = -\mathbf{\Phi}(1)$. □

5.11 APPLICATIONS

In this section, we consider an example of cointegration analysis, using two U.S. quarterly macroeconomic variables and two interest rates. The data employed are

- z_{1t}: Logarithm of the U.S. real gross national product (GNP) measured in billions of chained 2005 dollars and seasonally adjusted
- z_{2t}: Rate of U.S. 3-month treasury bills obtained by simple average of the monthly rates within a quarter
- z_{3t}: Logarithm of the U.S. M1 money stock obtained from the average of monthly data, which are measured in billions of dollars and seasonally adjusted
- z_{4t}: Rate of U.S. 10-year constant maturity interest rate obtained by the same method as that of z_{2t}.

The sample period is from the first quarter of 1959 to the second quarter of 2012 for 214 observations. The data were downloaded from the Federal Reserve Bank of St. Louis. Figure 5.6 shows the time plots of the four series. As expected, the real GNP and M1 money stock show a clear upward trend but the two interest rate series do not. The interest rate series, on the other hand, move in unison with the shorten rates appearing to be more volatile.

Univariate unit-root tests, for example, the augmented Dickey–Fuller test, confirm that all four series are unit-root nonstationary. To apply cointegration tests, we employ VAR models. Based on the Akaike information criterion, a VAR(6) model is selected. Both BIC and HQ criteria select a VAR(2) model, whereas the sequential chi-square tests also support a VAR(6) model. In what follows, we use a VAR(5) model in cointegration tests and a VAR(6) specification in estimation. The discrepancy in model specifications occurs because our analysis indicates that using $K = 5$ in Johansen's test provides a better description of the cointegrated processes. Since both real GNP and M1 money stock exhibit a upward drift, the constant vector of the differenced data is not 0. With $K = 5$ and $c(t) = \boldsymbol{c}_0$, the cointegration tests show that the eigenvalues are 0.1922, 0.1357, 0.0400, and 0.0201, respectively. The test statistics and some critical values are given in Table 5.1. From the table, it is seen that we cannot reject the null hypothesis of $m = 2$. The two cointegrating vectors are $\boldsymbol{g}_1 = (1, -0.282, -0.792, 0.313)'$ and $\boldsymbol{g}_2 = (1, -0.780, -0.673, 0.773)'$, respectively. These eigenvectors are normalized so that the first element of each vector is 1.

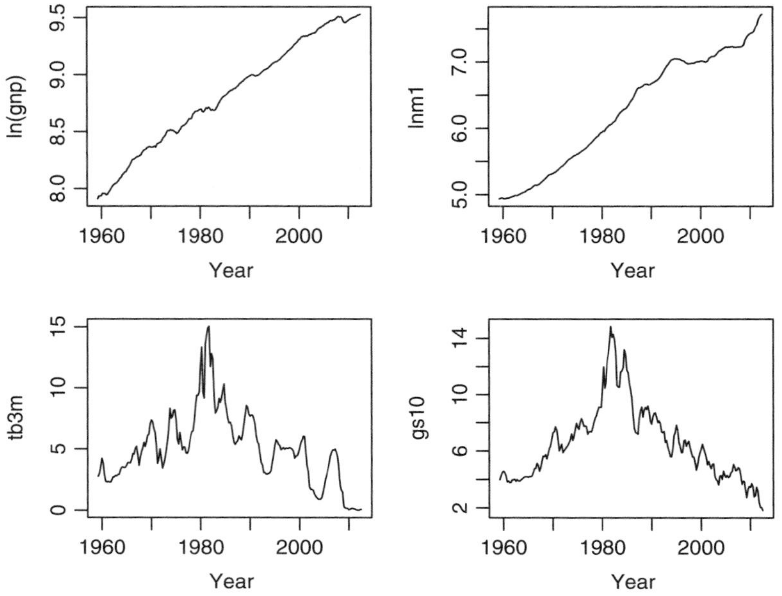

FIGURE 5.6 Time plots of four U.S. quarterly macroeconomic and interest rate variables. The four series are logged gross national product, 3-month treasury bills, logged M1 money stock, and 10-year constant maturity interest rate, respectively. The sample period is from 1959 to the second quarter of 2012.

TABLE 5.1 Cointegration Tests for a Four-Dimensional U.S. Quarterly Time Series Consisting of Two Macroeconomic Variables and Two Interest Rates Based on a VAR(5) Model

H_0 :	$L_{\max}$	5%	1%	L_{tr}	5%	1%
$m \leq 3$	4.25	9.24	12.97	4.25	9.24	12.97
$m \leq 2$	8.53	15.67	20.20	12.79	19.96	24.60
$m \leq 1$	30.47	22.00	26.81	43.26	34.91	41.07
$m = 0$	44.61	28.14	33.24	87.87	53.12	60.16

The sample period is from the first quarter of 1959 to the second quarter of 2012 with 214 observations.

Let $\hat{\beta} = [\boldsymbol{g}_1, \boldsymbol{g}_2]$ and $\boldsymbol{w}_t = (w_{1t}, w_{2t})' = \hat{\boldsymbol{\beta}}' \boldsymbol{z}_t$. Figure 5.7 shows the time plots and sample autocorrelation functions of cointegrated w_{it} series. The plots show that w_{it} series indeed exhibit behavior of a stationary time series. The augmented Dickey–Fuller test further confirms that these two integrating series are unit-root stationary. The test statistics are -3.87 and -4.37, respectively, for w_{1t} and w_{2t}. These tests are based on a univariate AR(6) model and have p-value close to 0.01, which implies that the unit-root null hypothesis is rejected at the conventional 5% level.

Treating the cointegrated process $\boldsymbol{w}_t$ as given, we estimate the ECM-VAR(6) model for the four-dimensional quarterly series. The model fits the data well. Figure 5.8 plots the p-values of Ljung–Box test statistics for the residuals of the

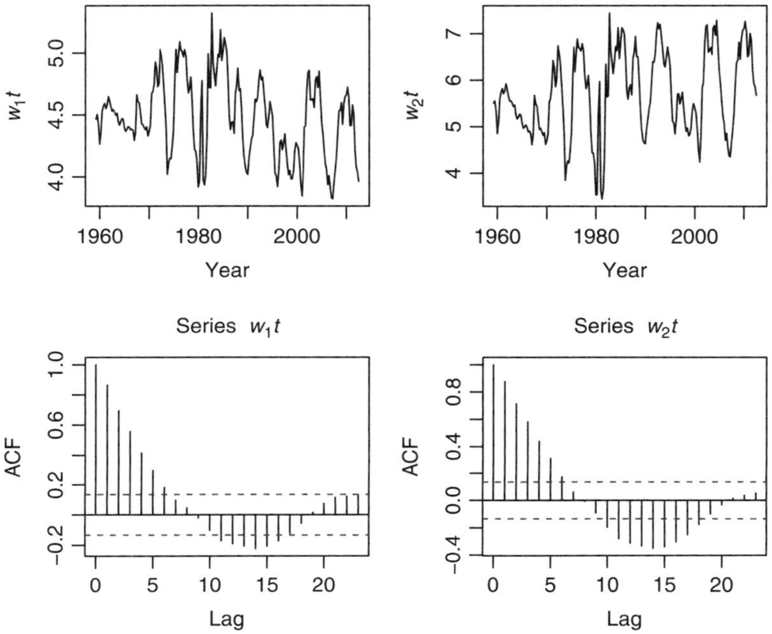

FIGURE 5.7 Time plots and sample autocorrelations of the cointegrated series for the U.S. quarterly time series of macroeconomic variables and interest rates.

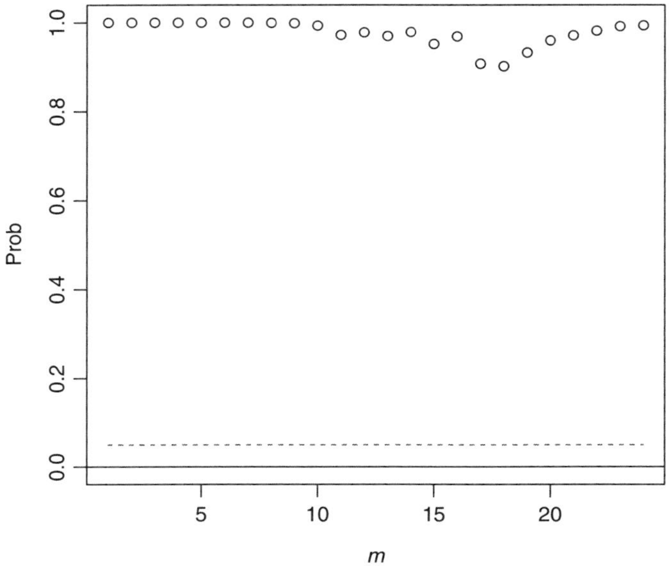

FIGURE 5.8 Plot of p-values of the Ljung–Box statistics for the residuals of an ECM-VAR model for the U.S. quarterly time series of macroeconomic variables and interest rates.

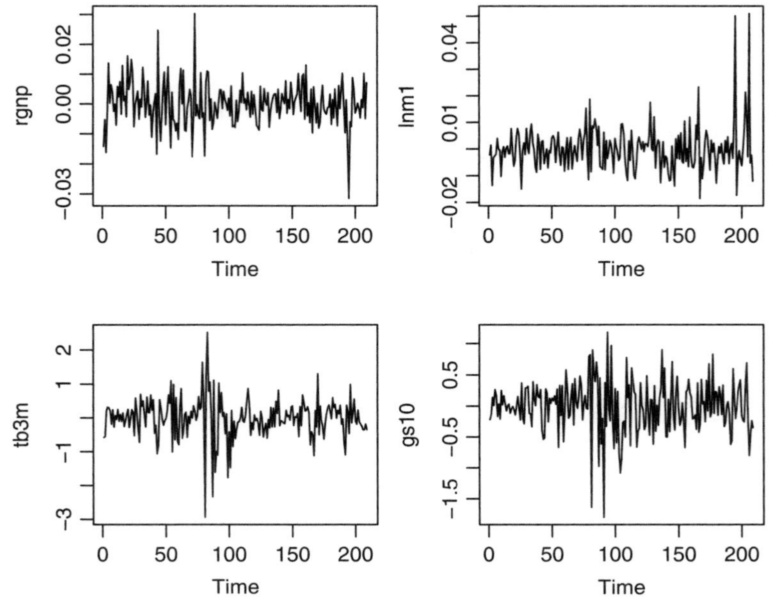

FIGURE 5.9 Time plots of the residual series of an ECM-VAR model for the U.S. quarterly time series of macroeconomic variables and interest rates.

fitted ECM model, whereas Figure 5.9 shows the time plots of the corresponding residuals. The latter plots suggest that the volatility of the residuals is likely to be time-varying. In other words, the residuals have conditional heteroscedasticity, which can be handled by multivariate volatility models of Chapter 7.

The fitted ECM-VAR(6) model contains many parameters some of which are statistical insignificant, making it hard to provide any meaningful model interpretation. The model must be simplified. For simplicity, we simply remove those parameters with t-ratio less than 0.8 in modulus. The choice of 0.8 is arbitrary. We use it because we do not wish to simultaneously delete too many parameters. The resulting model is still complicated, but we have

$$\hat{\boldsymbol{\alpha}}' = \begin{bmatrix} 0.0084 & 0 & 0.0053 & 0 \\ 0 & 0.172 & 0 & -0.150 \end{bmatrix},$$

where all estimates are statistically significant at the conventional 5% level. This particular structure shows that (a) Δz_{1t} and Δz_{3t} depend on $w_{1,t-1}$ but not on $w_{2,t-1}$, and (b) Δz_{2t} and Δz_{4t}, on the other hand, depend on $w_{2,t-1}$ but not on $w_{1,t-1}$. Since z_{1t} and z_{3t} are macroeconomic variables whereas z_{2t} and z_{4t} are interest rate series, we see that macroeconomic variables and interest rates require different adjustments in the error-correction representation. It is easy to see that

$$\hat{\alpha}'_{\perp} = \begin{bmatrix} 0.0053 & 0 & -0.0084 & 0 \\ 0 & 0.150 & 0 & 0.172 \end{bmatrix}.$$

Since $\hat{\alpha}'_{\perp} z_t$ gives rise to common trends, we see that the two common trends consist separately of macroeconomic variables and interest rates. This is not surprising in view of the characteristics shown by the time plots in Figure 5.6.

R Demonstration: Cointegration analysis. Output edited.

```
> require(MTS)
> da=read.table("q-4macro.txt",header=T)
> head(da)
  year mon day   rgnp     tb3m     gs10     m1sk
1 1959   1   1 2725.1 2.773333 3.990000 139.3333
> tail(da)
    year mon day    rgnp       tb3m     gs10     m1sk
214 2012   4   1 13768.7 0.08666667 1.823333 2252.500
> zt=cbind(log(da$rgnp),da$tb3m,log(da$m1sk),da$gs10)
> colnames(zt) <- c("rgnp","tb3m","lnm1","gs10")
> m1=VARorderI(zt) % Order selection using different sample
sizes
selected order: aic =  6
selected order: bic =  2
selected order: hq =  2
M statistic and its p-value
        Mstat        pv
 [1,] 4512.65 0.000e+00
 [2,]  150.07 0.000e+00
 [3,]   47.98 4.776e-05
 [4,]   32.28 9.206e-03
 [5,]   19.24 2.564e-01
 [6,]   55.17 3.335e-06
 [7,]   22.17 1.379e-01
 [8,]   27.65 3.479e-02
 [9,]   22.97 1.147e-01
[10,]   27.71 3.423e-02
[11,]   15.20 5.097e-01
[12,]   18.21 3.118e-01
[13,]   29.04 2.368e-02
> require(fUnitRoots); require(urca)
> m2=ca.jo(zt,K=5,ecdet=c("const"),spec=c("transitory"))
> summary(m2)
######################
# Johansen-Procedure #
######################
Test type: maximal eigenvalue statistic (lambda max),
without linear trend and constant in cointegration.
Eigenvalues (lambda):
```

```
[1] 1.922074e-01 1.356749e-01 4.000707e-02 2.014771e-02
   -7.820871e-18

Values of teststatistic and critical values of test:
          test 10pct  5pct  1pct
r <= 3 |  4.25  7.52  9.24 12.97
r <= 2 |  8.53 13.75 15.67 20.20
r <= 1 | 30.47 19.77 22.00 26.81
r = 0  | 44.61 25.56 28.14 33.24

Eigenvectors, normalised to first column:(the cointegration
relations)
            rgnp.l1    tb3m.l1     lnm1.l1       gs10.l1   constant
rgnp.l1   1.0000000  1.0000000  1.00000000  1.000000000  1.0000000
tb3m.l1  -0.2819887 -0.7797076 -0.01416840 -0.045476665  0.0396296
lnm1.l1  -0.7919986 -0.6729377 -0.53816014 -0.761616376 -0.1692782
gs10.l1   0.3133461  0.7725252  0.03191589  0.005341667 -0.1914556
constant -3.7262634 -5.9350635 -5.60843447 -3.891464213 -6.4644984

> m3=ca.jo(zt,K=5,ecdet=c("const"),spec=c("transitory"),
  type=c("trace"))
> summary(m3)
######################
# Johansen-Procedure #
######################
Test type: trace statistic,
   without linear trend and constant in cointegration
Values of teststatistic and critical values of test:
          test 10pct  5pct  1pct
r <= 3 |  4.25  7.52  9.24 12.97
r <= 2 | 12.79 17.85 19.96 24.60
r <= 1 | 43.26 32.00 34.91 41.07
r = 0  | 87.87 49.65 53.12 60.16

> w1t=zt[,1]-0.282*zt[,2]-0.792*zt[,3]+0.313*zt[,4]
> w2t=zt[,1]-0.780*zt[,2]-0.673*zt[,3]+0.773*zt[,4]
> adfTest(w1t,lags=6,type="c")
Title: Augmented Dickey-Fuller Test
Test Results:  STATISTIC:
    Dickey-Fuller: -3.8739
  P VALUE:    0.01
> adfTest(w2t,lags=6,type="c")
Title: Augmented Dickey-Fuller Test
Test Results:  STATISTIC:
    Dickey-Fuller: -4.3688
  P VALUE:    0.01
> wt=cbind(w1t,w2t)
> m3=ECMvar1(zt,6,wt,include.const=T) % Estimation
alpha:
        w1t       w2t
rgnp 0.00974 -0.000622
```

```
tb3m 0.11632  0.139180
lnm1 0.00809 -0.000992
gs10 0.05296 -0.161531
standard error
        [,1]    [,2]
[1,] 0.00348 0.00113
[2,] 0.28304 0.09213
[3,] 0.00397 0.00129
[4,] 0.19919 0.06484
constant term:
   rgnp    tb3m    lnm1    gs10
-0.0325 -1.5371 -0.0263  0.5819
standard error
[1] 0.0120 0.9763 0.0137 0.6870
AR coefficient matrix
AR( 1 )-matrix
       rgnp      tb3m    lnm1      gs10
rgnp  0.193  0.00422 -0.0914 -0.00233
tb3m 14.982  0.45321  0.4497 -0.08414
lnm1 -0.069 -0.00208  0.4453 -0.00471
gs10  7.415  0.06947  0.9398  0.13990
standard error
       [,1]    [,2]   [,3]    [,4]
[1,] 0.0711 0.00120 0.0600 0.00153
[2,] 5.7750 0.09744 4.8721 0.12405
[3,] 0.0809 0.00137 0.0683 0.00174
[4,] 4.0641 0.06857 3.4286 0.08730
 .....
AR( 5 )-matrix
        rgnp      tb3m    lnm1     gs10
rgnp -0.0693 -7.74e-06 -0.0105 -0.00221
tb3m  6.9776  3.49e-01  6.1031 -0.30925
lnm1  0.0901  7.11e-05  0.0987  0.00281
gs10 -4.0396  1.29e-01  7.3253 -0.30182
standard error
       [,1]    [,2]   [,3]    [,4]
[1,] 0.0672 0.00108 0.0665 0.00155
[2,] 5.4602 0.08787 5.4045 0.12577
[3,] 0.0765 0.00123 0.0757 0.00176
[4,] 3.8426 0.06184 3.8033 0.08851
-----
Residuals cov-mtx:
              rgnp          tb3m          lnm1          gs10
rgnp  5.349817e-05  0.0010139262 -7.142742e-06  0.0006583782
tb3m  1.013926e-03  0.3529833289 -1.493661e-04  0.1646324845
lnm1 -7.142742e-06 -0.0001493661  6.933185e-05 -0.0001368415
gs10  6.583782e-04  0.1646324845 -1.368415e-04  0.1748132824

AIC =  -21.99356; BIC =  -20.54651
> MTSdiag(m3)  % Model checking
[1] "Covariance matrix:"
```

```
           rgnp      tb3m       lnm1       gs10
rgnp  5.38e-05  0.00102 -7.18e-06  0.000662
tb3m  1.02e-03  0.35468 -1.50e-04  0.165424
lnm1 -7.18e-06 -0.00015  6.97e-05 -0.000137
gs10  6.62e-04  0.16542 -1.37e-04  0.175654
CCM at lag:  0
        [,1]     [,2]     [,3]     [,4]
[1,]  1.000  0.2333 -0.1173  0.2153
[2,]  0.233  1.0000 -0.0302  0.6628
[3,] -0.117 -0.0302  1.0000 -0.0393
[4,]  0.215  0.6628 -0.0393  1.0000
> m4=refECMvar1(m3,thres=0.8)  # Refinement
alpha:
        [,1]   [,2]
[1,] 0.00836  0.000
[2,] 0.00000  0.172
[3,] 0.00528  0.000
[4,] 0.00000 -0.150
standard error
        [,1]   [,2]
[1,] 0.00233 1.0000
[2,] 1.00000 0.0702
[3,] 0.00224 1.0000
[4,] 1.00000 0.0483
constant term:
[1] -0.0301 -1.2372 -0.0196  0.7577
standard error
[1] 0.00976 0.36863 0.00954 0.25755
AR coefficient matrix
AR( 1 )-matrix
       [,1]      [,2]     [,3]      [,4]
[1,]  0.196  0.00415 -0.0893 -0.00219
[2,] 14.986  0.39745  0.0000  0.00000
[3,] -0.062 -0.00227  0.4339 -0.00445
[4,]  7.860  0.06760  0.0000  0.13433
standard error
       [,1]     [,2]    [,3]     [,4]
[1,] 0.0696 0.00112 0.0598 0.00145
[2,] 5.5469 0.07110 1.0000 1.00000
[3,] 0.0757 0.00117 0.0631 0.00159
[4,] 3.8226 0.06224 1.0000 0.08164
....
```

5.12 DISCUSSION

Cointegration is an interesting concept and has attracted much interest in various scientific fields. It has many applications. For instance, it has been widely used as a theoretical justification for pairs trading in finance. See, for instance, Tsay (2010, Chapter 8). However, cointegration also has its shares of weakness. First, it does not address the rate of achieving long-term equilibrium. This is related to the speed of mean reverting. For instance, if the cointegrating series $\boldsymbol{w}_t = \boldsymbol{\beta}'\boldsymbol{z}_t$ has a characteristic root that is close to unit circle, then the cointegration relationship may take a long time to achieve. Second, cointegration tests are scale invariant. For instance, multiplying the component series of $\boldsymbol{z}_t$ by any $k \times k$ nonsingular matrix does not change the result of a cointegration test. However, scaling can be very important in practice. Consider the following two simple examples.

Example 5.2 Suppose $\{\boldsymbol{a}_t\}$ is a sequence of iid bivariate normal random vectors with mean zero and covariance matrix $\text{Cov}(\boldsymbol{a}_t) = \boldsymbol{I}_2$. Let $\boldsymbol{y}_0 = \boldsymbol{0}$ and define $\boldsymbol{y}_t = \boldsymbol{y}_{t-1} + \boldsymbol{a}_t$. That is, the components of $\boldsymbol{y}_t$ are univariate random walk processes. Let $z_{1t} = 10{,}000y_{1t} + (1/10{,}000)y_{2t}$ and $z_{2t} = 10{,}000y_{1t} - (1/10{,}000)y_{2t}$. Clearly, the $\boldsymbol{z}_t$ series has two unit roots, that is, z_{1t} and z_{2t} are not cointegrated. However, because of the scaling effects, for any reasonable sample size commonly encountered in practice z_{1t} and z_{2t} should move closely together. In fact, for moderate T, the ℓ-step ahead point forecasts $z_{1,T}(\ell)$ and $z_{2,T}(\ell)$ should be very close to each other for all ℓ. Consequently, a rejection of cointegration does not necessarily imply that the two forecast series are far apart. □

Example 5.3 Suppose that $\{\boldsymbol{a}_t\}$ are the same as those of Example 5.2. However, construct the $\boldsymbol{y}_t$ series via $y_{1t} = 0.9y_{1,t-1} + a_{1t}$ and $y_{2t} = y_{2,t-1} + a_{2t}$. Define $z_{1t} = (1/10{,}000)y_{2t} - 10{,}000y_{1t}$ and $z_{2t} = (1/10{,}000)y_{2t} + 10{,}000y_{1t}$. In this case, $\boldsymbol{z}_t$ has a single unit root and its two components series are cointegrated with cointegrating $(1, -1)'$. However, the ℓ-step ahead forecasts $z_{1,T}(\ell)$ and $z_{2,T}(\ell)$ can be very different for moderate T and ℓ. This example demonstrates that a cointegrated system may take a long period of time to show the cointegrating relationship in forecasting. □

5.13 APPENDIX

Proof of Lemma 5.1. The partial sum is

$$
\begin{aligned}
\boldsymbol{S}_t &= \sum_{s=1}^{t} \boldsymbol{y}_s = \sum_{s=1}^{t}\sum_{j=0}^{\infty} \boldsymbol{\psi}_j \boldsymbol{a}_{s-j} \\
&= [\boldsymbol{\psi}_0\boldsymbol{a}_t + \boldsymbol{\psi}_1\boldsymbol{a}_{t-1} + \boldsymbol{\psi}_2\boldsymbol{a}_{t-2} + \cdots + \boldsymbol{\psi}_t\boldsymbol{a}_0 + \boldsymbol{\psi}_{t+1}\boldsymbol{a}_{-1} + \cdots] \\
&\quad + [\boldsymbol{\psi}_0\boldsymbol{a}_{t-1} + \boldsymbol{\psi}_1\boldsymbol{a}_{t-2} + \boldsymbol{\psi}_2\boldsymbol{a}_{t-3} + \cdots + \boldsymbol{\psi}_{t-1}\boldsymbol{a}_0 + \boldsymbol{\psi}_t\boldsymbol{a}_{-1} + \cdots] \\
&\quad + \cdots + [\boldsymbol{\psi}_0\boldsymbol{a}_1 + \boldsymbol{\psi}_1\boldsymbol{a}_0 + \boldsymbol{\psi}_2\boldsymbol{a}_{-1} + \cdots] \\
&= \boldsymbol{\psi}_0\boldsymbol{a}_t + (\boldsymbol{\psi}_0 + \boldsymbol{\psi}_1)\boldsymbol{a}_{t-1} + (\boldsymbol{\psi}_0 + \boldsymbol{\psi}_1 + \boldsymbol{\psi}_2)\boldsymbol{a}_{t-2} + \cdots \\
&\quad + (\boldsymbol{\psi}_0 + \cdots + \boldsymbol{\psi}_{t-1})\boldsymbol{a}_1 + (\boldsymbol{\psi}_1 + \cdots + \boldsymbol{\psi}_t)\boldsymbol{a}_0 + (\boldsymbol{\psi}_2 + \cdots + \boldsymbol{\psi}_{t+1})\boldsymbol{a}_{-1} + \cdots
\end{aligned}
$$

$$
\begin{aligned}
&= \boldsymbol{\psi}(1)\boldsymbol{a}_t - \left(\sum_{v=1}^{\infty}\boldsymbol{\psi}_v\right)\boldsymbol{a}_t + \boldsymbol{\psi}(1)\boldsymbol{a}_{t-1} - \left(\sum_{v=2}^{\infty}\boldsymbol{\psi}_v\right)\boldsymbol{a}_{t-1} + \cdots \\
&\quad + \boldsymbol{\psi}(1)\boldsymbol{a}_1 - \left(\sum_{v=t}^{\infty}\boldsymbol{\psi}_v\right)\boldsymbol{a}_1 \\
&\quad + \left(\sum_{v=1}^{\infty}\boldsymbol{\psi}_v\right)\boldsymbol{a}_0 - \left(\sum_{v=t+1}^{\infty}\boldsymbol{\psi}_v\right)\boldsymbol{a}_0 \\
&\quad + \left(\sum_{v=2}^{\infty}\boldsymbol{\psi}_v\right)\boldsymbol{a}_{-1} - \left(\sum_{v=t+2}^{\infty}\boldsymbol{\psi}_v\right)\boldsymbol{a}_{-1} + \cdots .
\end{aligned}
$$

Therefore,

$$
\boldsymbol{S}_t = \boldsymbol{\psi}(1)\sum_{s=1}^{t}\boldsymbol{a}_t + \boldsymbol{\eta}_t - \boldsymbol{\eta}_0,
$$

where

$$
\begin{aligned}
\boldsymbol{\eta}_t &= -\left(\sum_{v=1}^{\infty}\boldsymbol{\psi}_v\right)\boldsymbol{a}_t - \left(\sum_{v=2}^{\infty}\boldsymbol{\psi}_v\right)\boldsymbol{a}_{t-1} - \left(\sum_{v=3}^{\infty}\boldsymbol{\psi}_v\right)\boldsymbol{a}_{t-2} - \cdots \\
\boldsymbol{\eta}_0 &= -\left(\sum_{v=1}^{\infty}\boldsymbol{\psi}_v\right)\boldsymbol{a}_0 - \left(\sum_{v=2}^{\infty}\boldsymbol{\psi}_v\right)\boldsymbol{a}_{-1} - \left(\sum_{v=3}^{\infty}\boldsymbol{\psi}_v\right)\boldsymbol{a}_{-2} - \cdots
\end{aligned}
$$

From the prior derivation, we have $\boldsymbol{\eta}_t = \sum_{j=0}^{\infty}\boldsymbol{\alpha}_j\boldsymbol{a}_{t-j}$, where $\boldsymbol{\alpha}_j = -\sum_{v=j+1}^{\infty}\boldsymbol{\psi}_v$. Writing $\boldsymbol{\alpha}_v = [\alpha_{v,ij}]$ for $i, j = 1, \ldots, k$, we see that

$$
\begin{aligned}
\sum_{v=1}^{\infty}|\alpha_{v,ij}| &= |\psi_{1,ij} + \psi_{2,ij} + \cdots| + |\psi_{2,ij} + \psi_{3,ij} + \cdots| + \cdots \\
&\le [|\psi_{1,ij}| + |\psi_{2,ij}| + \ldots] + [|\psi_{3,ij}| + |\psi_{4,ij}| + \cdots] + \cdots \\
&= |\psi_{1,ij}| + 2|\psi_{2,ij}| + 3|\psi_{3,ij}| + \cdots \\
&= \sum_{v=1}^{\infty} v|\psi_{v,ij}| < \infty.
\end{aligned}
$$

Therefore, $\{\boldsymbol{\alpha}_v\}_{v=1}^{\infty}$ is absolutely summable. □

EXERCISES

5.1 Derive the error-correction form in Equation (5.68) for a VAR(p) model.

5.2 Consider the quarterly real GDP of United Kingdom, Canada, and the United States from the first quarter of 1980 to the second quarter of 2011. The data are available from the Federal Reserve Bank of St. Louis (FRED). Let z_t be the log GDP series.

(a) Information criteria suggest a VAR(3) model for z_t. Based on a VAR(3) model, is the series z_t cointegrated at the 5% level? Why?

(b) How many cointegrating vectors are there? Write down the cointegrating vector, if any.

(c) Compute the cointegrated series w_t if any. Perform univariate unit-root test to confirm the stationarity of the components of w_t. Draw your conclusion at the 5% significance level.

(d) Build an ECM-VAR model for z_t. Perform model checking and write down the fitted model.

5.3 Consider the daily closing prices of the stocks of Billiton Ltd. of Australia and Vale S.A. of Brazil with tick symbols BHP and VALE, respectively. The data are obtained from Yahoo Finance and the sample period is from July 1, 2002 to March 31, 2006. We use the log-adjusted prices in this exercise. The adjustment is made for stock splits and for dividends. See Tsay (2010, Chapter 8), where the data were used to demonstrate pairs trading. The data file is `d-bhpvale-0206.txt`. Let z_t be the log-adjusted closing prices.

(a) Are the two log prices series cointegrated? Why?

(b) What is the cointegrating vector, if any?

(c) Build an ECM-VAR model for z_t.

5.4 Consider the U.S. monthly personal consumption expenditures (PCE) and disposable personal income (DSPI) from January 1959 to March 2012. The data are from FRED of the Federal Reserve Bank of St. Louis, in billions of dollars, and seasonally adjusted. Let z_t be the log series of PCE and DSPI. Data are also in the files `m-pce.txt` and `m-dspi.txt`.

(a) Are the components of z_t cointegrated? Why?

(b) What is the cointegrating vector, if any?

(c) Perform univariate analysis of the cointegrated series to confirm its stationarity.

(d) Build an ECM-VAR model for z_t, including model checking. Write down the fitted model.

5.5 Consider the U.S. populations of men and women from January 1948 to December 2012. The data are available from FRED of the Federal Reserve Bank of St. Louis and are in thousands. See also the files `m-popmen.txt` and `m-popwomen.txt`. Let z_t be the log series of the U.S. monthly populations of men and women.

(a) Are the two population series cointegrated? Why?

(b) Build VARMA model for z_t, including model checking. Write down the fitted model.

(c) Use the fitted model to produce one-step to three-step ahead predictions of the z_t series at the forecast origin December 2012.

5.6 Consider the annual real gross domestic products per capita of four OECD countries. The countries are (a) United States, (b) Germany, (c) United Kingdom, and (d) France. The real GDP are in 2011 U.S. dollars, and the sample period is from 1960 to 2011 for 52 observations. The data are available from FRED of the Federal Reserve Bank of St. Louis and in the file `a-rgdp-per-4.txt`. Let z_t be the log series of the data.

(a) Use $K = 2$ to perform cointegration tests. Are the log gdp series cointegrated? Why?

(b) What is the cointegrating vector, if any?

(c) Perform unit-root test to confirm the stationarity of the cointegrated process.

(d) Build a ECM-VAR model for z_t, including model checking. Write down the fitted model.

5.7 Consider, again, the annual real gross domestic products per capita for United States. and United Kingdom of Problem 6. Use multivariate exponential smoothing to produce one-step ahead forecasts of the two real GDP series at the forecast origin year 2011.

REFERENCES

Ahn, S. K. and Reinsel, G. C. (1990). Estimation for partially nonstationary multivariate autoregressive models. *Journal of the American Statistical Association*, **85**: 849–856.

Billingsley, P. (1999). *Convergence of Probability Measures*. 2nd Edition. John Wiley & Sons, Inc, New York.

Billingsley, P. (2012). *Probability and Measures*. Anniversary Edition. John Wiley & Sons, Inc, Hoboken, N.J.

Box, G. E. P. and Tiao, G. C. (1977). A canonical analysis of multiple time series. *Biometrika*, **64**: 355–366.

Campbell, J. Y. and Shiller, R. J. (1988). The dividend-price ratio and expectations of future dividends and discount factors. *Review of Financial Studies*, **1**: 195–228.

Chan, N. H. (1989). Asymptotic inference for unstable autoregressive time series with drifts. *Journal of Statistical Planning and Inference*, **23**: 301–312.

Chan, N. H. and Wei, C. Z. (1988). Limiting distributions of least squares estimates of unstable autoregressive processes. *Annals of Statistics*, **16**: 367–401.

Dickey, D. A. and Fuller, W. A. (1979). Distribution of the estimators for autoregressive time series with a unit root. *Journal of the American Statistical Association*, **74**: 427–431.

Donsker, M. (1951). An invariance principle for certain probability limit theorems. *Memoirs American Mathematical Society* No. 6.

Engle, R. F. and Granger, C. W. J. (1987). Cointegration and error correction: Representation, estimation and testing. *Econometrica*, **55**, 251–276.

Fuller, W. A. (1976). *Introduction to Statistical Time Series*. John Wiley & Sons, Inc, New York.

Granger, C. W. J. (1983). Cointegrated variables and error-correcting models. Unpublished discussion paper 83–13. Department of Economics, University of California San Diego.

Granger, C. W. J. and Newbold, P. (1974). Spurious regressions in econometrics. *Journal of Econometrics*, **2**: 111–120.

Hamilton, J. D. (1994). *Time Series Analysis*. Princeton University Press, Princeton, NJ.

Herrndorf, N. (1984). A functional central limit theorem for weakly dependent sequences of random variables. *Annuals of Probability*, **12**: 141–153.

Johansen, S. (1988). Statistical analysis of cointegration vectors. *Journal of Economic Dynamics and Control*, **12**: 251–254.

Johansen, S. (1991). Estimation and hypothesis testing of cointegration vectors in Gaussian vector autoregressive models. *Econometrica*, **59**: 1551–1580.

Johansen, S. (1995). *Likelihood Based Inference in Cointegrated Vector Error Correction Models*. Oxford University Press, Oxford, UK.

Johansen, S. and Juselius, K. (1990). Maximum likelihood estimation and inference on cointegration-with applications to the demand for money. *Oxford Bulletin of Economics and Statistics*, **52**: 169–210.

Johnson, R. A. and Wichern, D. W. (2007) *Applied Multivariate Statistical Analysis*. 6th Edition. Prentice Hall, Upper Saddle River, NJ.

Phillips, P. C. B. (1986). Understanding spurious regressions in econometrics. *Journal of Econometrics*, **33**: 311–340.

Phillips, P. C. B. (1987). Time series regression with a unit root. *Econometrica*, **55**: 277–301.

Phillips, P. C. B. (1988). Weak convergence of sample covariance matrices to stochastic integrals via martingale approximations. *Econometric Theory*, **4**: 528–533.

Phillips, P. C. B. (1991). Optical inference in cointegrated systems. *Econometrica*, **59**: 283–306.

Phillips, P. C. B. and Ouliaris, S. (1990). Asymptotic properties of residual based tests for co-integration. *Econometrica*, **58**: 165–193.

Phillips, P. C. B. and Perron, P. (1988). Testing for a unit root in time series regression. *Biometrika*, **75**: 335–346.

Phillips, P. C. B. and Solo, V. (1992). Asymptotics for linear processes. *Annals of Statistics*, **20**: 971–1001.

Reinsel, G. C. and Ahn, S. K. (1992). Vector autoregressive models with unit roots and reduced tank structure: estimation, likelihood ratio test, and forecasting. *Journal of Time Series Analysis*, **13**: 133–145.

Said, S. E. and Dickey, D. A. (1984). Testing for unit roots in autoregressive-moving average of unknown order. *Biometrika*, **71**: 599–607.

Sims, C. A., Stock, J. H., and Watson, M. W. (1990). Inference in linear time series models with some unit roots. *Econometrica*, **58**: 113–144.

Tiao, G. C. and Tsay, R. S. (1989). Model specification in time series (with discussion). *Journal of the Royal Statistical Society, Series B*, **51**: 157–213.

Tsay, R. S. (1984). Order selection in nonstationary autoregressive models. *Annals of Statistics*, **12**: 1425–1433.

Tsay, R. S. (2010). *Analysis of Financial Time Series*. 3rd Edition. John Wiley & Sons, Inc, Hoboken, NJ.

Tsay, R. S. and Tiao, G. C. (1990). Asymptotic properties of multivariate nonstationary processes with applications to autoregressions. *Annals of Statistics*, **18**: 220–250.

West, K. D. (1988). Asymptotic normality, when regressors have a unit root. *Econometrica*, **56**: 1397–1417.

CHAPTER 6

Factor Models and Selected Topics

In this chapter, we consider some selected topics in multivariate time series analysis that are of practical importance and theoretical interest. The topics discussed include seasonal models, principal component analysis (PCA), use of exogenous variables, factor models, missing values, and classification and clustering analysis. For some topics, the treatment is brief as they are either well known or still under intensive investigation in the literature. For instance, the treatment of seasonal models is brief because it involves direct extensions of the univariate models. Special attention is paid to factor models. We provide a comprehensive description of the factor models available in the literature and discuss some of their properties and limitations. We also consider asymptotic principal component analysis. For clustering analysis, we focus on the model-based approach in which a univariate AR model is used for all series in a given cluster. More elaborate models for classification can be entertained at the expense of increased computation and harder prior elicitation. Again, real examples are used to demonstrate the analyses.

6.1 SEASONAL MODELS

Seasonality occurs in many economic, financial, and environmental time series. The quarterly earnings per share of a company tend to exhibit an annual cyclic pattern. The unemployment rates of a country often show a summer effect as many students seek summer intern jobs. Of course, daily temperature of a given location in the United States, critical to demands of electricity and natural gas, is seasonal.

Many economic data published in the United States are seasonally adjusted. However, it is well known that the sample spectral density function of a seasonally adjusted series via most seasonal adjustment methods, for example, the model-based X-12 procedure, has a trough at the seasonal frequency, implying that some

Multivariate Time Series Analysis: With R and Financial Applications,
First Edition. Ruey S. Tsay.

seasonality remains in the data. See, for instance, the U.S. monthly unemployment rate analyzed in Tsay (2013, Chapter 3). In some applications, it is necessary to consider seasonal models even for seasonally adjusted data.

A direct generalization of the useful univariate *airline* model for the vector seasonal time series $\boldsymbol{z}_t$ is

$$(1-B)(1-B^s)\boldsymbol{z}_t = (\boldsymbol{I}_k - \boldsymbol{\theta}B)(\boldsymbol{I}_k - \boldsymbol{\Theta}B^s)\boldsymbol{a}_t, \tag{6.1}$$

where all eigenvalues of $\boldsymbol{\theta}$ and $\boldsymbol{\Theta}$ are less than 1 in modulus, $s > 1$ denotes the number of seasons within a year, and $\{\boldsymbol{a}_t\}$ is a sequence of iid random vectors with mean zero and positive-definite covariance matrix $\boldsymbol{\Sigma}_a$. For instance, we typically have $s = 4$ for quarterly data and $s = 12$ for monthly data. For the model in Equation (6.1), $(1 - B^s)$ is referred to as the seasonal difference, and $(1 - B)$ the regular difference. We refer to the model as the seasonal airline model. Rewriting Equation (6.1) as

$$\boldsymbol{z}_t = \frac{\boldsymbol{I}_k - \boldsymbol{\theta}B}{1-B} \times \frac{\boldsymbol{I}_k - \boldsymbol{\Theta}B^s}{1-B^s}\boldsymbol{a}_t,$$

we see that the model continues to represent double exponential smoothing with one smoothing for the regular dependence and another one for the seasonal dependence.

Let $\boldsymbol{w}_t = (1-B)(1-B^s)\boldsymbol{z}_t$ and $\boldsymbol{\Gamma}_{w,\ell}$ be the lag-ℓ autocovariance matrix of $\boldsymbol{w}_t$. It is straightforward to see that $\boldsymbol{w}_t = (\boldsymbol{I}_k - \boldsymbol{\theta}B - \boldsymbol{\Theta}B^s + \boldsymbol{\theta}\boldsymbol{\Theta}B^{s+1})\boldsymbol{a}_t$ and

1. $\boldsymbol{\Gamma}_{w,0} = (\boldsymbol{\Sigma}_a + \boldsymbol{\Theta}\boldsymbol{\Sigma}_a\boldsymbol{\Theta}') + \boldsymbol{\theta}(\boldsymbol{\Sigma}_a + \boldsymbol{\Theta}\boldsymbol{\Sigma}_a\boldsymbol{\Theta}')\boldsymbol{\theta}'$
2. $\boldsymbol{\Gamma}_{w,1} = -\boldsymbol{\theta}(\boldsymbol{\Sigma}_a + \boldsymbol{\Theta}\boldsymbol{\Sigma}_a\boldsymbol{\Theta}')$
3. $\boldsymbol{\Gamma}_{w,s-1} = \boldsymbol{\Theta}\boldsymbol{\Sigma}_a\boldsymbol{\theta}'$
4. $\boldsymbol{\Gamma}_{w,s} = -(\boldsymbol{\Theta}\boldsymbol{\Sigma}_a + \boldsymbol{\theta}\boldsymbol{\Theta}\boldsymbol{\Sigma}_a\boldsymbol{\theta}')$
5. $\boldsymbol{\Gamma}_{w,s+1} = \boldsymbol{\theta}\boldsymbol{\Theta}\boldsymbol{\Sigma}_a$
6. $\boldsymbol{\Gamma}_{w,\ell} = \boldsymbol{0}$, otherwise.

Since matrix multiplication is generally not commutable, further simplification in the formulas for $\boldsymbol{\Gamma}_{w,\ell}$ is not always possible. Let $\boldsymbol{\rho}_{w,\ell}$ be the lag-ℓ cross-correlation matrix of $\boldsymbol{w}_t$. The aforementioned results show that the $\boldsymbol{w}_t$ series has nonzero cross-correlation matrices at lags 1, $s-1$, s, and $s+1$. Thus, $\boldsymbol{w}_t$ has the same nonzero lag dependence as its univariate counterpart, but it does not share the same symmetry in correlations, because $\boldsymbol{\rho}_{w,s-1} \neq \boldsymbol{\rho}_{w,s+1}$ in general.

In some applications, the seasonal airline model can be generalized to

$$(1-B)(1-B^s)\boldsymbol{z}_t = \boldsymbol{\theta}(B)\boldsymbol{\Theta}(B)\boldsymbol{a}_t, \tag{6.2}$$

where both $\boldsymbol{\theta}(B)$ and $\boldsymbol{\Theta}(B)$ are matrix polynomials of order q and Q, respectively, with $q < s$. Limited experience indicates that $Q = 1$ is often sufficient in applications.

Example 6.1 Consider the monthly housing data of the United States from January 1963 to July 2012. The two series employed are

1. z_{1t}: Logarithm of new homes sold in thousands of units (new residential sales)
2. z_{2t}: Logarithm of the total new privately owned housing units started in thousands of units (new residential construction)

The original data are downloaded from the Federal Reserve Bank of St. Louis and are not seasonally adjusted. We took log transformation of the data in our analysis to keep the variability stable. Figure 6.1 shows the time plots of the two log time series. A strong seasonal pattern is clearly seen from the plots. The impact of 2007 subprime financial crisis and the subsequent prolong decline in the housing market are also visible.

Figure 6.2 shows the cross-correlation matrices of $\boldsymbol{w}_t = (1 - B)(1 - B^{12})\boldsymbol{z}_t$. From the plots, the main nonzero cross-correlation matrices are at lags 1, 11, 12, and 13 with some additional minor cross dependence at lags 2 and 3. These plots give some justifications for employing a multiplicative seasonal model. To begin, we consider the univariate seasonal models and obtain

$$w_{1t} = (1 - 0.21B)(1 - 0.90B^{12})a_{1t} \tag{6.3}$$

$$w_{2t} = (1 - 0.33B + 0.04B^2 + 0.11B^3)(1 - 0.86B^{12})a_{2t}, \tag{6.4}$$

where w_{it} is the ith component of $\boldsymbol{w}_t$, and the residual variances are 0.0069 and 0.0080, respectively, for a_{1t} and a_{2t}. Both estimates in Equation (6.3) are statistically

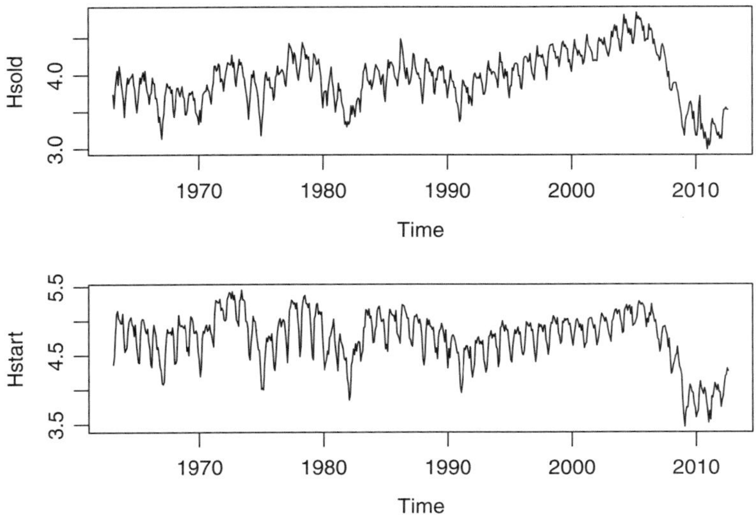

FIGURE 6.1 Time plots of monthly U.S. housing data from January 1963 to July 2012: (a) new homes sold and (b) new privately owned housing units started. Both series are in logarithm.

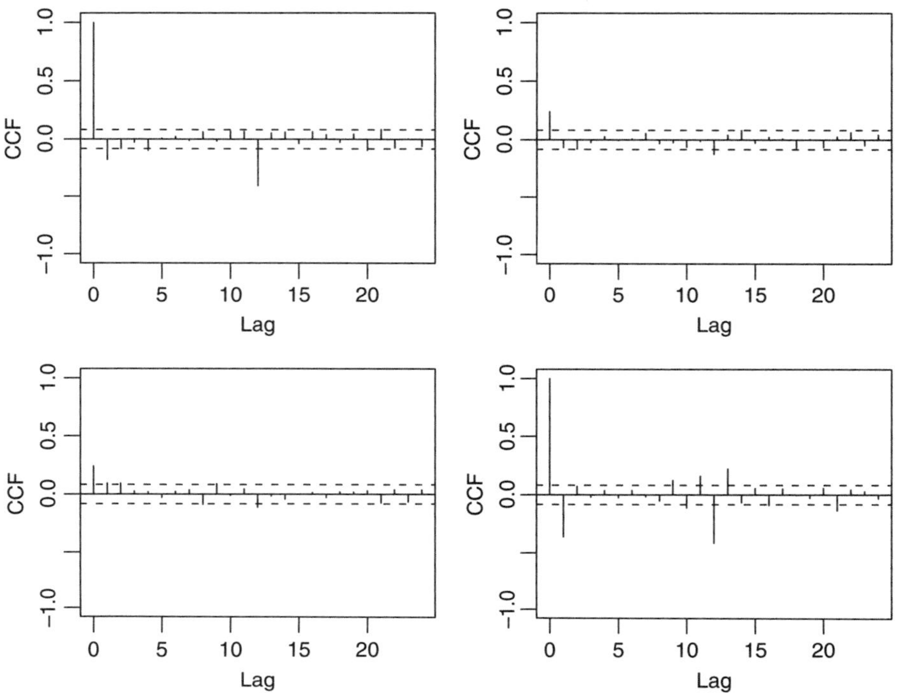

FIGURE 6.2 Sample cross-correlation coefficients of $\boldsymbol{w}_t = (1 - B)(1 - B^{12})\boldsymbol{z}_t$ for the U.S. logged housing data.

significant at the usual 5% level. The standard errors of the regular MA coefficients in Equation (6.4) are 0.042, 0.042, and 0.041, respectively, implying that the lag-2 coefficient is not statistically significant at the conventional 5% level.

Turn to multivariate seasonal analysis. Based on the cross-correlation matrices in Figure 6.2 and the results of univariate models, we employ the model

$$(1 - B)(1 - B^{12})\boldsymbol{z}_t = (\boldsymbol{I}_k - \boldsymbol{\theta}_1 B - \boldsymbol{\theta}_2 B^2 - \boldsymbol{\theta}_3 B^3)(\boldsymbol{I}_k - \boldsymbol{\Theta} B^{12})\boldsymbol{a}_t. \quad (6.5)$$

The parameter estimates and those of a refined version of the model in Equation (6.5) are given in Table 6.1. The refinement is to remove some insignificant parameters. Both AIC and BIC prefer the refined model over the unconstrained one. Standard errors of the coefficient estimates are given in the following R demonstration. From the table, we see that large coefficients appear in $\boldsymbol{\theta}_1$ and $\boldsymbol{\Theta}$, indicating that the fitted models are close to that of a multivariate airline model. Figure 6.3 shows the time plots of the residuals of the reduced model in Table 6.1, whereas Figure 6.4 plots the cross-correlation matrices of the residuals. The fitted model appears to be adequate as there are no major significant cross-correlations in the residuals.

Comparing with the univariate seasonal models, we make the following observations.

TABLE 6.1 Estimation Results of Seasonal Models for Monthly Housing Data

Parameter	Full Model		Reduced Model		Switched	
$\boldsymbol{\theta}_1$	0.246	−0.168	0.253	−0.168	0.246	−0.168
	−0.290	0.416	−0.292	0.416	−0.297	0.416
$\boldsymbol{\theta}_2$	0.028	0.034	0	0	0.022	0.040
	−0.168	0.048	−0.168	0	−0.168	0.048
$\boldsymbol{\theta}_3$	0.063	−0.067	0.076	−0.068	0.057	−0.062
	−0.085	−0.074	−0.085	−0.073	−0.085	−0.068
$\boldsymbol{\Theta}$	0.850	0.033	0.850	0.033	0.850	0.037
	0.044	0.824	0.044	0.824	0.029	0.848
$10^2 \times \boldsymbol{\Sigma}_a$	0.694	0.263	0.696	0.266	0.694	0.261
	0.263	0.712	0.266	0.714	0.261	0.712
AIC and BIC	−10.012	−9.892	−10.017	−9.920	−10.009	−9.889

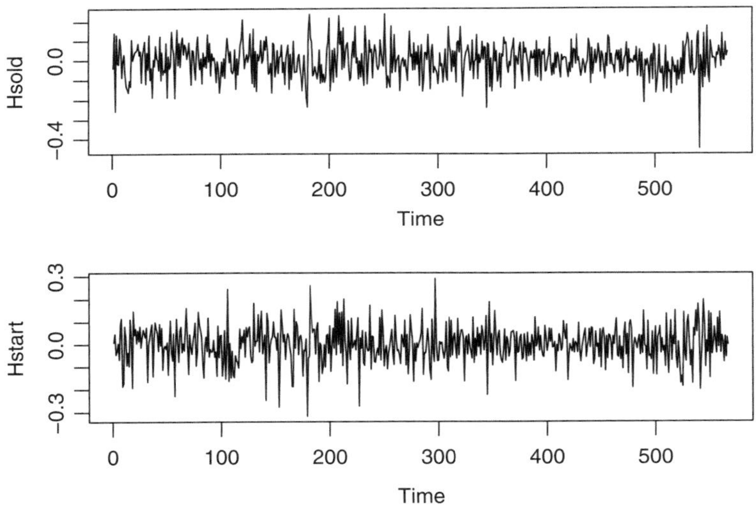

FIGURE 6.3 Residuals of the reduced model in Table 6.1 for the monthly housing data in logarithms.

(a) The model for z_{1t} (houses sold) in the VARMA model is close to the univariate model in Equation (6.3). For instance, the residual variances of the two models are close and elements of $\hat{\boldsymbol{\theta}}_3$ are small.

(b) The VARMA model seems to provide a better fit for z_{2t} (housing starts) because the VARMA model has a smaller residual variance.

(c) The VARMA model shows that there exists a feedback relationship between houses sold and housing starts. The cross dependence is strong at lag 1, but only marginal at the seasonal lag. Finally, predictions and forecast error variance decomposition of the fitted seasonal model can be carried out in the same way as that of the nonseasonal VARMA models. Details are omitted. □

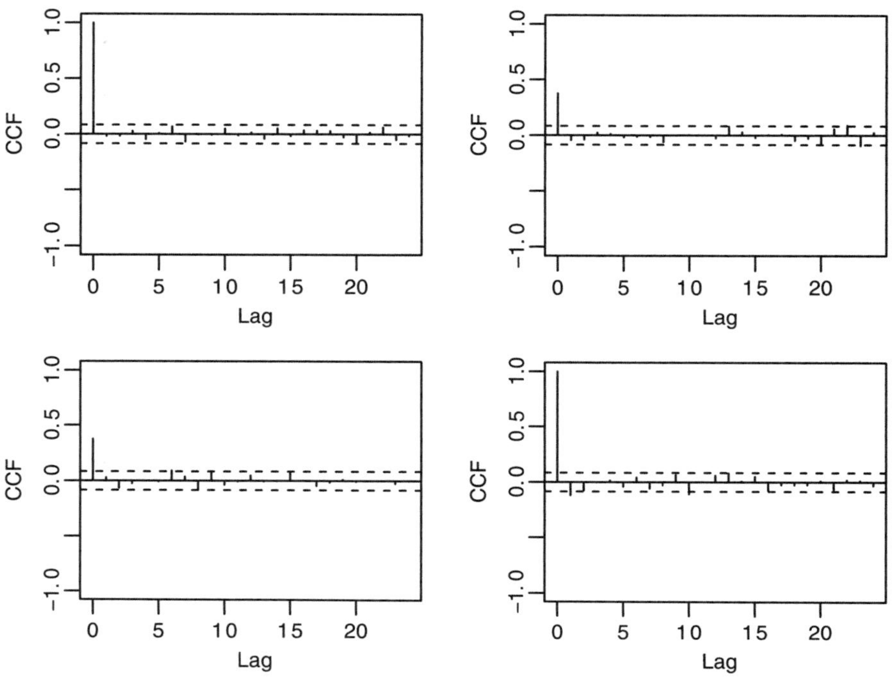

FIGURE 6.4 Cross-correlations of the residuals of the reduced model in Table 6.1 for the U.S. monthly housing data.

Remark: Since matrix multiplication is not commutable, an alternative to the seasonal model in Equation (6.2) is

$$(1-B)(1-B^s)\boldsymbol{z}_t = \boldsymbol{\Theta}(B)\boldsymbol{\theta}(B)\boldsymbol{a}_t. \tag{6.6}$$

Autocovariance matrices of this model are different from those of the model in Equation (6.2). For instance, one can easily derive the autocovariance matrices for the differenced series $\boldsymbol{w}_t$ of Equation (6.6). The number of nonzero lags remain the same as before, but the actual matrices differ. For example, in this particular case, $\boldsymbol{\Gamma}_{w,1}$ becomes $-(\boldsymbol{\theta}\boldsymbol{\Sigma}_a + \boldsymbol{\Theta}\boldsymbol{\theta}\boldsymbol{\Sigma}_a\boldsymbol{\Theta}')$. In practice, it is not easy to distinguish between the two seasonal models from the sample cross-correlation matrices, but one can fit two different seasonal VARMA models and compare the fits to choose a model. For the monthly U.S. housing data, the parameter estimates of the model

$$(1-B)(1-B^{12})\boldsymbol{z}_t = (\boldsymbol{I}_k - \boldsymbol{\Theta}B^{12})(\boldsymbol{I}_k - \boldsymbol{\theta}_1 B - \boldsymbol{\theta}_2 B^2 - \boldsymbol{\theta}_3 B^3)\boldsymbol{a}_t$$

are given in the last two columns of Table 6.1 under the heading `switched`. Based on AIC and BIC, the model in Equation (6.5) is preferred. However, for this particular instance, the difference between the two fitted models is small. □

Remark: The commands used to estimate seasonal VARMA models in the `MTS` package are `sVARMA` and `refsVARMA`. These commands use the conditional maximum likelihood method. The model is specified by the regular order (p, d, q) and seasonal order (P, D, Q) with s denoting the seasonality. The model in Equation (6.6) is specified by the subcommand `switch = TRUE`. See the following R demonstration. □

R Demonstration: Estimation of seasonal VARMA models. Output edited.

```
> m3=sVARMA(zt,order=c(0,1,3),sorder=c(0,1,1),s=12)
Number of parameters:  16
initial estimates:  0.1985901 -0.08082791 0.05693362 ....
Coefficient(s):
       Estimate  Std. Error  t value Pr(>|t|)
 [1,]   0.24587     0.04621    5.321 1.03e-07 ***
 [2,]  -0.16847     0.04483   -3.758 0.000171 ***
 [3,]   0.02834     0.05220    0.543 0.587194
  ....
[15,]   0.04428     0.02444    1.812 0.070012 .
[16,]   0.82424     0.02860   28.819  < 2e-16 ***
---
Regular MA coefficient matrix
MA( 1 )-matrix
       [,1]   [,2]
[1,]  0.246 -0.168
[2,] -0.290  0.416
MA( 2 )-matrix
        [,1]   [,2]
[1,]  0.0283 0.0338
[2,] -0.1678 0.0479
MA( 3 )-matrix
        [,1]    [,2]
[1,]  0.0628 -0.0671
[2,] -0.0850 -0.0736
Seasonal MA coefficient matrix
MA( 12 )-matrix
          [,1]        [,2]
[1,] 0.8498980 0.03322685
[2,] 0.0442752 0.82423827

Residuals cov-matrix:
                    resi m1$residuals
resi         0.006941949  0.002633186
m1$residuals 0.002633186  0.007116746
----
aic=  -10.01172 ; bic=  -9.89168
> m4=refsVARMA(m3,thres=1.2)
Number of parameters:  13
```

```
initial estimates:  0.1985901 -0.08082791 0.01039118  ...
Coefficient(s):
       Estimate  Std. Error  t value Pr(>|t|)
 [1,]   0.25255     0.04521    5.586 2.32e-08 ***
 [2,]  -0.16847     0.04357   -3.866  0.00011 ***
 [3,]   0.07582     0.03910    1.939  0.05247 .
 [4,]  -0.06798     0.03953   -1.720  0.08547 .
 [5,]  -0.29222     0.04659   -6.272 3.56e-10 ***
 [6,]   0.41585     0.04668    8.909  < 2e-16 ***
 [7,]  -0.16783     0.04498   -3.732  0.00019 ***
 [8,]  -0.08497     0.04840   -1.756  0.07912 .
 [9,]  -0.07345     0.04619   -1.590  0.11182
[10,]   0.84990     0.01959   43.378  < 2e-16 ***
[11,]   0.03296     0.02395    1.376  0.16878
[12,]   0.04387     0.02441    1.797  0.07231 .
[13,]   0.82472     0.02846   28.980  < 2e-16 ***
---
Regular MA coefficient matrix
MA( 1 )-matrix
       [,1]   [,2]
[1,]  0.253 -0.168
[2,] -0.292  0.416
MA( 2 )-matrix
       [,1] [,2]
[1,]  0.000    0
[2,] -0.168    0
MA( 3 )-matrix
        [,1]    [,2]
[1,]  0.0758 -0.0680
[2,] -0.0850 -0.0734
Seasonal MA coefficient matrix
MA( 12 )-matrix
           [,1]       [,2]
[1,] 0.84989804 0.03296309
[2,] 0.04387419 0.82471809
  Residuals cov-matrix:
                     resi m1$residuals
resi          0.006964991  0.002655917
m1$residuals 0.002655917  0.007139871
----
aic=  -10.01722; bic=  -9.919685
> m5=sVARMA(zt,order=c(0,1,3),sorder=c(0,1,1),s=12,switch=T)
Regular MA coefficient matrix
MA( 1 )-matrix
       [,1]   [,2]
[1,]  0.246 -0.168
[2,] -0.297  0.416
MA( 2 )-matrix
        [,1]   [,2]
```

```
[1,]  0.0217 0.0397
[2,] -0.1678 0.0479
MA( 3 )-matrix
        [,1]    [,2]
[1,]  0.0565 -0.0615
[2,] -0.0850 -0.0675
Seasonal MA coefficient matrix
MA( 12 )-matrix
           [,1]       [,2]
[1,] 0.84989804 0.03714659
[2,] 0.02909768 0.84752096
  Residuals cov-matrix:
                    resi m1$residuals
resi         0.006938460  0.002605709
m1$residuals 0.002605709  0.007118109
----
aic=  -10.0087; bic=  -9.888657
```

6.2 PRINCIPAL COMPONENT ANALYSIS

PCA is a useful tool in multivariate statistical analysis. It was developed originally for independent Gaussian observations, but has been found to be useful in time series analysis. PCA performs orthogonal rotations of the observed coordinates to seek simplification in model interpretation and/or to achieve dimension reduction in modeling. In this section, we apply PCA to multivariate time series to seek stable linear relationships embedded in the data. The relationship may involve lagged variables. Theory of PCA can be found in most textbooks of multivariate statistical analysis or statistical learning, for example, Johnson and Wichern (2007) and Hastie, Tibshirani, and Friedman (2009).

To seek stable linear relations in a multivariate time series, we apply PCA to the observed series $\boldsymbol{z}_t$ and the residuals $\hat{\boldsymbol{a}}_t$ of a VAR model. The former analysis is to find stable contemporaneous relations whereas the latter for stable lagged relations. We demonstrate the application with a real example. Consider the 4-dimensional monthly time series $\boldsymbol{z}_t = (z_{1t}, \ldots, z_{4t})'$ of U.S. manufacturers data on durable goods, where

1. z_{1t}: New orders (NO)
2. z_{2t}: Total inventory (TI)
3. z_{3t}: Unfilled orders (UO)
4. z_{4t}: Values in shipments (VS)

All measurements are in billions of U.S. dollars and the data are seasonally adjusted. The original data are downloaded from the Federal Reserve Bank of St. Louis and are in millions of dollars, and the sample period is from February 1992 to July 2012

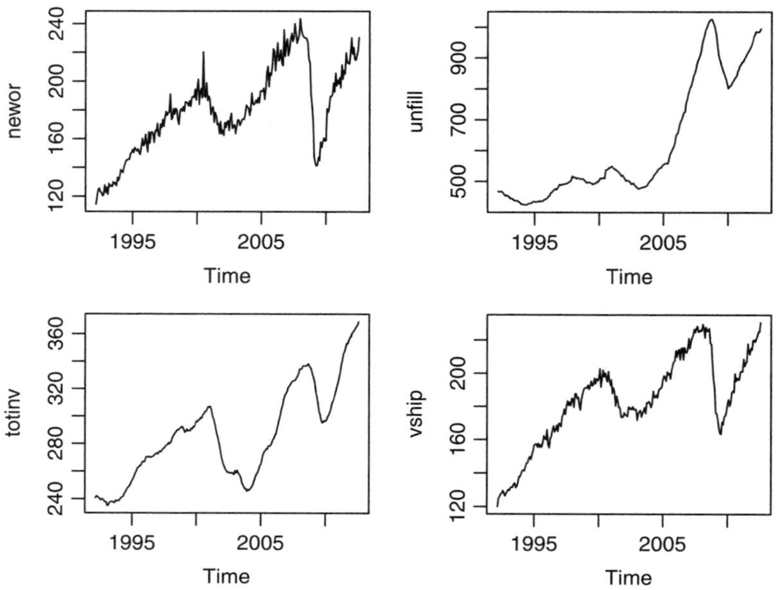

FIGURE 6.5 Monthly series of U.S. manufacturers data on durable goods from February 1992 to July 2012: (a) new orders, (b) total inventory, (c) unfilled orders, and (d) values in shipments. Data are in billions of dollars and seasonally adjusted.

for 246 observations. Figure 6.5 shows the time plots of the four time series. From the plots, the new orders and values in shipments show a similar trending pattern, whereas total inventory and unfilled orders move closely. As expected, all four time series show an upward trend and, hence, are unit-root nonstationary.

If one specifies a VAR model for the $\boldsymbol{z}_t$ series, a VAR(4) model is selected by all order selection methods discussed in Chapter 2. For ease in discussion, we write a VAR(p) model as

$$\boldsymbol{z}_t = \boldsymbol{\phi}_{p,0} + \boldsymbol{\Phi}_{p,1}\boldsymbol{z}_{t-1} + \cdots + \boldsymbol{\Phi}_{p,p}\boldsymbol{z}_{t-p} + \boldsymbol{a}_{p,t}, \tag{6.7}$$

where $\boldsymbol{\phi}_{p,0}$ denotes the constant vector and the subscript p is used to signify the fitted order. Let $\hat{\boldsymbol{a}}_{p,t}$ be the residuals of a VAR(p) model of Equation (6.7). Table 6.2 provides the standard deviation (square root of eigenvalue) and the proportion of variance explained by each principal component for $\boldsymbol{z}_t$ and $\hat{\boldsymbol{a}}_{p,t}$ for p = 1, 2, 3, and 4. Of particular interest from the table is that the results of PCA appear to be stable for the residual series for all four fitted VAR models. The last principal component explains about 0.1% of the total variability.

Table 6.3 provides the loading matrices of the PCA for $\boldsymbol{z}_t$ and the residuals $\hat{\boldsymbol{a}}_{p,t}$ for $p = 1$, 2, and 3. The loading matrix for the residual series $\hat{\boldsymbol{a}}_{4,t}$ of the VAR(4) model is similar to those of the three residual series and, hence, is omitted. Again, the stability of the loading matrices for the residuals of VAR models is remarkable,

TABLE 6.2 Summary of Principal Component Analysis Applied to the Monthly U.S. Manufacturers Data of Durable Goods from 1992.2 to 2012.7

Series	Variable	Principal Components			
z_t	SD	197.00	30.700	12.566	3.9317
	Proportion	0.9721	0.0236	0.0040	0.0004
$\hat{a}_{1,t}$	SD	8.8492	3.6874	1.5720	0.3573
	Proportion	0.8286	0.1439	0.0261	0.0014
$\hat{a}_{2,t}$	SD	8.3227	3.5233	1.1910	0.2826
	Proportion	0.8327	0.1492	0.0171	0.0010
$\hat{a}_{3,t}$	SD	8.0984	3.4506	1.0977	0.2739
	Proportion	0.8326	0.1512	0.01530	0.0010
$\hat{a}_{4,t}$	SD	7.8693	3.2794	1.0510	0.2480
	Proportion	0.8386	0.1456	0.0140	0.0008

$\hat{a}_{p,t}$ denotes the residuals of a VAR(p) model.

TABLE 6.3 Loadings of Principal Component Analysis Applied to the Monthly U.S. Manufacturers Data of Durable Goods from 1992.2 to 2012.7, where TS Stands for Time Series

TS	Loading Matrix				TS	Loading Matrix			
z_t	0.102	0.712	0.342	0.604	$\hat{a}_{1,t}$	0.794	0.161	0.066	0.583
	0.152	0.315	−0.928	0.129		0.058	−0.109	−0.990	0.063
	0.978	−0.182	0.098	−0.006		0.547	−0.592	0.060	−0.588
	0.096	0.600	0.110	−0.786		0.260	0.782	−0.106	−0.557
$\hat{a}_{2,t}$	0.796	0.150	0.017	0.587	$\hat{a}_{3,t}$	0.797	0.143	0.009	0.586
	0.026	−0.070	−0.997	0.012		0.017	−0.063	−0.998	0.007
	0.543	−0.600	0.049	−0.585		0.537	−0.608	0.044	−0.583
	0.267	0.783	−0.055	−0.560		0.274	0.778	−0.048	−0.563

especially for the last principal component. From the table, to a close approximation, the eigenvector associated with the fourth principal component can be written as $h_4 \approx (1, 0, -1, -1)'$.

Next, consider the fitted VAR(1) model, which is

$$z_t = \begin{bmatrix} 0.008 \\ -0.129 \\ -8.348 \\ 2.804 \end{bmatrix} + \begin{bmatrix} 0.686 & -0.027 & -0.001 & 0.357 \\ 0.116 & 0.995 & -0.000 & -0.102 \\ 0.562 & -0.023 & 0.995 & -0.441 \\ 0.108 & 0.023 & -0.003 & 0.852 \end{bmatrix} z_{t-1} + \hat{a}_{1,t}, \qquad (6.8)$$

where all elements of the constant term ϕ_0 are statistically insignificant at the conventional 5% level. Premultiplying Equation (6.8) by h_4', we have

$$h_4' z_t \approx 5.55 + (0.015, -0.027, -0.994, -0.054) z_{t-1} + h_4' \hat{a}_{1,t}.$$

With the small eigenvalue and the fact that means of the residuals $\hat{\boldsymbol{a}}_{1,t}$ are 0, $\boldsymbol{h}_4'\hat{\boldsymbol{a}}_{1,t}$ is essentially being 0. Consequently, with $\boldsymbol{h}_4 \approx (1, 0, -1, -1)'$, the prior equation implies

$$\mathrm{NO}_t - \mathrm{UO}_t - \mathrm{VS}_t + \mathrm{UO}_{t-1} \approx c_4,$$

where c_4 denotes a constant. In other words, PCA of the residuals of the VAR(1) model reveals a stable relation

$$\mathrm{NO}_t - \mathrm{VS}_t - (\mathrm{UO}_t - \mathrm{UO}_{t-1}) \approx c_4 \tag{6.9}$$

Therefore, for the monthly manufacturer data on durable goods, the difference between new orders and values in shipments is roughly equal to some constant plus the change in unfilled orders.

Next, consider the VAR(2) model,

$$\boldsymbol{z}_t = \hat{\boldsymbol{\phi}}_{2,0} + \hat{\boldsymbol{\Phi}}_{2,1}\boldsymbol{z}_{t-1} + \hat{\boldsymbol{\Phi}}_{2,2}\boldsymbol{z}_{t-1} + \hat{\boldsymbol{a}}_{2,t}. \tag{6.10}$$

where the coefficient estimates are given in Table 6.4. Again, PCA of the residuals $\hat{\boldsymbol{a}}_{2,t}$ shows that the smallest eigenvalue is essentially 0 with eigenvector approximately $\boldsymbol{h}_4 = (1, 0, -1, -1)'$. Premultiplying Equation (6.10) by $\boldsymbol{h}_4'$, we get

$$\boldsymbol{h}_4'\boldsymbol{z}_t \approx 2.21 + (0.59, -0.08, -1.57, -0.61)\boldsymbol{z}_{t-1} + (0.01, 0.07, 0.57, -0.01)\boldsymbol{z}_{t-2}.$$

Consequently, ignoring terms with coefficients close to 0, we have

$$z_{1t} - z_{3t} - z_{4t} - 0.59z_{1,t-1} + 1.57z_{3,t-1} + 0.61z_{4,t-1} - 0.57z_{3,t-2} \approx c_1,$$

TABLE 6.4 Summary of the VAR(2) Model for the Four-Dimensional Time Series of Monthly Manufacturers Data on Durable Goods

Parameter	Estimates			
$\hat{\boldsymbol{\phi}}_{2,0}'$	−0.221	3.248	−6.267	3.839
$\hat{\boldsymbol{\Phi}}_{2,1}$	1.033	1.012	−0.638	−0.108
	−0.445	1.549	0.441	0.537
	0.307	0.645	1.005	−0.120
	0.141	0.452	−0.072	0.619
$\hat{\boldsymbol{\Phi}}_{2,2}$	0.243	−1.028	0.634	−0.115
	0.064	−0.568	−0.438	−0.166
	0.247	−0.663	−0.010	−0.336
	−0.016	−0.440	0.070	0.227

where c_1 denotes a constant. Rearranging terms, the prior equation implies

$$z_{1t} - z_{3t} - z_{4t} + z_{3,t-1} \\ - (0.59z_{1,t-1} - 0.57z_{3,t-1} - 0.61z_{4,t-1} + 0.57z_{3,t-2}) \approx c_1.$$

This approximation further simplifies as

$$(z_{1t} - z_{3t} - z_{4t} + z_{3,t-1}) - 0.59(z_{1,t-1} - z_{3,t-1} - z_{4,t-1} + z_{3,t-2}) \approx c,$$

where c is a constant. The linear combination in the parentheses of the prior equation is

$$\text{NO}_t - \text{UO}_t - \text{VS}_t + \text{UO}_{t-1}$$

and its lagged value. This is exactly the linear combination in Equation (6.9). Consequently, PCA of the residuals of the VAR(2) model reveals the same stable relation between the four variables of durable goods. As a matter of fact, the same stable relation is repeated when one applies PCA to the residuals of VAR(3) and VAR(4) models of $\boldsymbol{z}_t$.

R Demonstration: Commands used in PCA.

```
> da=read.table("m-amdur.txt",header=T)
> dur= da[,3:6]/1000
> v0 =princomp(dur)
> summary(v0)
> M0 = matrix(v0$loadings[,1:4],4,4)
> VARorder(dur)   # Find VAR order
> m1=VAR(dur,1)   # Fit  VAR(1)
> v1=princomp(m1$residuals)
> summary(v1)
> M1=matrix(v1$loadings[,1:4],4,4)
> h4=matrix(c(1,0,-1,-1),4,1)
> t(h4)%*%m1$Phi
> t(h4)%*%m1$Ph0
> m2=VAR(dur,2)
> v2=princomp(m2$residuals)
> summary(v2)
> M2=matrix(v2$loadings[,1:4],4,4)
> print(round(M0,3))
> print(round(M1,3))
```

6.3 USE OF EXOGENOUS VARIABLES

In many forecasting exercises, exogenous variables or independent variables are available. These variables, if properly handled, can improve the accuracy in prediction. For instance, the U.S. Conference Board leading economic index (LEI) is an indicator of U.S. economy and can be used to help predict the GDP growth rate. The U.S. purchasing managers index (PMI) of the Institute for Supply Management (ISM) is an indicator of U.S. industrial production. In this section, we discuss two approaches to handle exogenous variables in multivariate time series analysis. For simplicity, we use VAR models in the discussion. The idea continues to apply for VARMA models.

6.3.1 VARX Models

The first approach to include exogenous variables in multivariate time series modeling is to use the vector autoregressive model with exogenous variables. In the literature, this type of models is referred to as the VARX models with X signifying exogenous variables. The term exogenous variable is used loosely here as it may contain independent (or input) variables. Let $\boldsymbol{z}_t$ be a k-dimensional time series and $\boldsymbol{x}_t$ an m-dimensional series of exogenous variables or leading indicators. The general form of a VARX model is

$$\boldsymbol{z}_t = \boldsymbol{\phi}_0 + \sum_{i=1}^{p} \boldsymbol{\phi}_i \boldsymbol{z}_{t-i} + \sum_{j=0}^{s} \boldsymbol{\beta}_j \boldsymbol{x}_{t-j} + \boldsymbol{a}_t, \tag{6.11}$$

where $\boldsymbol{a}_t$ is a sequence of iid random vectors with mean zero and positive-definite covariance matrix $\boldsymbol{\Sigma}_a$, p and s are non-negative integers, $\boldsymbol{\phi}_i$ are the usual VAR coefficient matrices, and $\boldsymbol{\beta}_j$ are $k \times m$ coefficient matrices. This model allows for $\boldsymbol{x}_t$ to affect $\boldsymbol{z}_t$ instantaneously if $\boldsymbol{\beta}_0 \neq \boldsymbol{0}$. The orders p and s of the VARX model in Equation (6.11) can be determined in several ways. For instance, one can use the information criteria or the ideal of partial F-test in multivariate multiple linear regression similar to those discussed in Chapter 2. We use information criteria in our demonstration. Conditional on $\{\boldsymbol{x}_{t-j}|j = 0, \ldots, s\}$, the stationarity condition of $\boldsymbol{z}_t$ is the same as that of the corresponding VAR(p) model. The VARX model can be estimated by the least-squares method and the resulting estimates are asymptotically normally distributed. For demonstration, we consider an example.

Example 6.2 Consider the monthly U.S. regular conventional gas price z_{1t} and No. 2 heating oil price z_{2t} of New York Harbor. Both series are measured in dollars per gallon. These prices depend on the crude oil and natural gas prices. Let x_{1t} be the spot oil price of West Texas Intermediate, dollars per barrel, and x_{2t} the natural gas price of Henry Hub, LA, measured in dollars per million BTU. Thus, we have $\boldsymbol{z}_t = (z_{1t}, z_{2t})'$, $\boldsymbol{x}_t = (x_{1t}, x_{2t})'$, and $k = m = 2$. The sample period is from November 1993 to August 2012. We downloaded the data from the Federal Reserve Bank of St Louis. The original data of $\boldsymbol{z}_t$ are from the Energy Information Administration of U.S. Department of Energy and that of $\boldsymbol{x}_t$ are from the Wall Street Journal of Dow Jones & Company.

Figure 6.6 shows the time plots of the four monthly time series. The left panel shows the regular gas price and the heating oil price, whereas the right panel contains

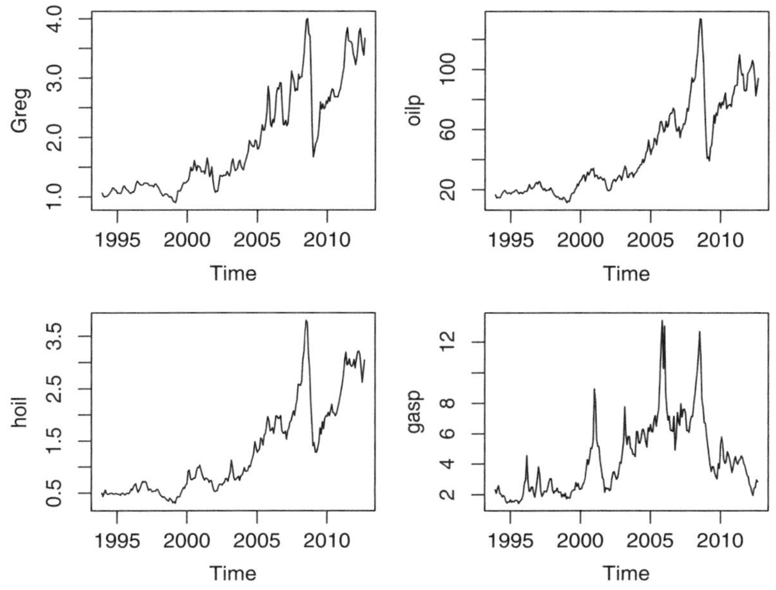

FIGURE 6.6 Time plots of monthly prices of energy series: The upper-left plot is the regular conventional gas price, the lower-left plot is the heating oil price, the upper-right plot is the spot oil price, and the lower-right plot is the natural gas price. The sample period is from November 1993 to August 2012.

the two independent variables. The gasoline and heating oil prices behave in a similar manner as that of the spot oil prices. The relationship between the natural gas prices and the $\boldsymbol{z}_t$ series appears to be more variable. If VAR models are entertained for the $\boldsymbol{z}_t$ series, the BIC and HQ criteria select a VAR(2) and VAR(3), respectively. The AIC, on the other hand, selects a VAR(11) model. The purpose of our demonstration is to employ a VARX model. To this end, we used information criteria to select the orders p and s. Here, the maximum AR order is set to 11 and the maximum lag of exogenous variables is 3. Both BIC and HQ criteria select $(p, s) = (2, 1)$ for the series. If one extends the maximum lags of the input variables, then both BIC and HQ select $(p, s) = (2, 3)$, with (p, s) = (2,1) as a close second. We have estimated both the VARX(2,1) and VARX(2,3) models and found that the results are similar. Therefore, we focus on the simpler VARX(2,1) model.

The selected model is then

$$\boldsymbol{z}_t = \boldsymbol{\phi}_0 + \boldsymbol{\phi}_1 \boldsymbol{z}_{t-1} + \boldsymbol{\phi}_2 \boldsymbol{z}_{t-2} + \boldsymbol{\beta}_0 \boldsymbol{x}_t + \boldsymbol{\beta}_1 \boldsymbol{x}_{t-1} + \boldsymbol{a}_t. \qquad (6.12)$$

Table 6.5 summarizes the estimation results of the VARX(2,1) model in Equation (6.12). The full model of the table gives unconstrained estimates, whereas the reduced model shows the estimates of a simplified model. The simplification is obtained by removing simultaneously all estimates with t-ratio less than 1.0. Figure 6.7 shows the cross-correlation matrices of the residual series of the reduced

TABLE 6.5 Estimation Results of a VARX(2,1) Model for the Monthly Gas and Heating Oil Prices with Spot Oil and Natural Gas Prices as Input Variables

	Full Model				Reduced Model			
Parameter	Estimates		Standard Errors		Estimates		Standard Errors	
$\boldsymbol{\phi}_0'$	0.182	−0.017	0.030	0.019	0.183	0	0.028	0
$\boldsymbol{\phi}_1$	1.041	0.169	0.059	0.080	1.044	0.162	0.054	0.056
	0.005	0.844	0.037	0.050	0	0.873	0	0.031
$\boldsymbol{\phi}_2$	−0.322	−0.008	0.056	0.068	−0.327	0	0.040	0
	0.014	0.012	0.035	0.043	0	0	0	0
$\boldsymbol{\beta}_0$	0.018	0.010	0.001	0.007	0.018	0.010	0.001	0.007
	0.024	0.026	0.001	0.004	0.023	0.026	0.001	0.004
$\boldsymbol{\beta}_1$	−0.015	−0.012	0.002	0.007	−0.015	−0.012	0.002	0.007
	−0.019	−0.029	0.001	0.004	−0.019	−0.029	0.001	0.004
$10^2\boldsymbol{\Sigma}_a$	0.674	0.105			0.674	0.105		
	0.105	0.268			0.105	0.270		

The sample period is from November 1993 to August 2012.

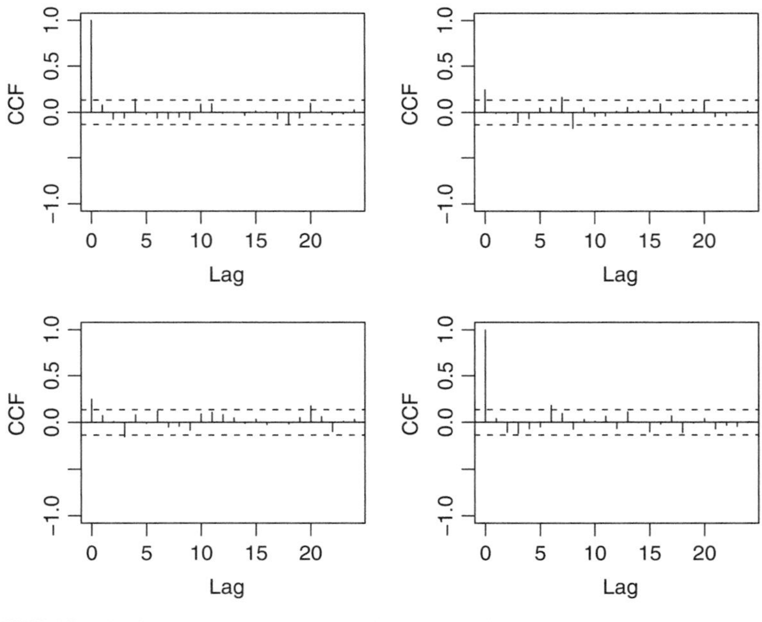

FIGURE 6.7 Residual cross-correlation matrices of the reduced VARX(2,1) model in Table 6.5.

model in Table 6.5. From the cross-correlation matrices, the linear dynamic dependence of the original data is largely removed. However, there are some minor residual correlations at lags 4, 6, and 7. These residual cross- and serial correlations can also be seen from the multivariate Ljung–Box statistics of the residuals. See the plot of

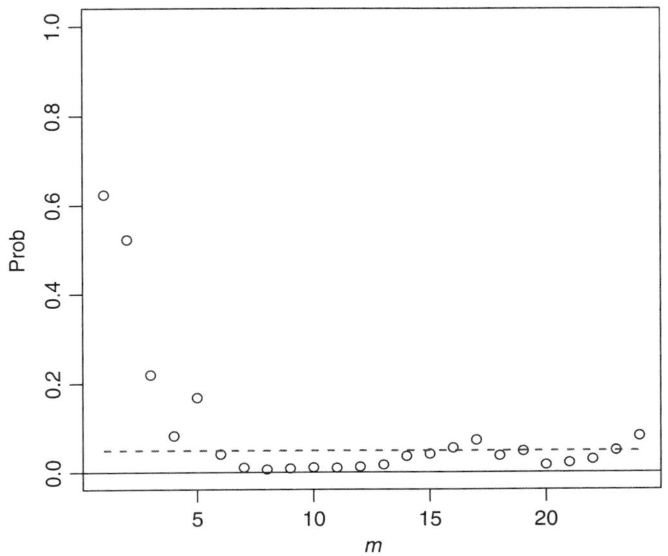

FIGURE 6.8 p-values of multivariate Ljung–Box statistics of the residual series of the reduced VARX(2,1) model in Table 6.5.

p-values of the Ljung–Box statistics in Figure 6.8. The model thus only provides a decent approximation to the underlying structure of the data.

From Table 6.5, we see that the independent variables $\boldsymbol{x}_t$ and $\boldsymbol{x}_{t-1}$ have significant impacts on the monthly regular gas price and heating oil price. More specifically, the spot oil price and its lagged value, x_{1t} and $x_{1,t-1}$, have substantial effects on the regular gas price. But the natural gas price and its lagged value, x_{2t} and $x_{2,t-1}$, only have marginal effects on the regular gas price. On the other hand, both $\boldsymbol{x}_t$ and $\boldsymbol{x}_{t-1}$ affect significantly the heating oil price. The importance of the independent variables $\boldsymbol{x}_t$ and $\boldsymbol{x}_{t-1}$ can also be seen by comparing the VARX(2,1) model in Equation (6.12) with a pure VAR(2) model. The residual covariance matrix of a VAR(2) model for $\boldsymbol{z}_t$ is

$$\hat{\boldsymbol{\Sigma}}_{\text{var}} = 10^2 \begin{bmatrix} 1.327 & 0.987 \\ 0.987 & 1.463 \end{bmatrix}.$$

Clearly, the residual variances of the VAR(2) model are much larger than those of the VARX(2,1) model. □

Remark: Analysis of VARX models is carried out by the commands `VARXorder`, `VARX`, and `refVARX` of the `MTS` package. `VARXorder` provides information criteria for data, `VARX` performs least-squares estimation of a specified VARX(p, s) model, and `refVARX` refines a fitted VARX model by removing insignificant estimates. Prediction can be obtained via the command `VARXpred`, which requires input of the exogenous variables. □

R Demonstration: VARX modeling. Output edited.

```
> da=read.table("m-gasoil.txt",header=T)
> head(da)
  year mon  Greg  hoil   oilp gasp
1 1993  11 1.066 0.502 16.699 2.32
 ....
6 1994   4 1.027 0.479 16.380 2.04
> zt=da[,3:4]; xt=da[,5:6]
> VARorder(zt)
selected order: aic =  11
selected order: bic =  2
selected order: hq =  3
> VARXorder(zt,exgo,maxp=11,maxm=2) ## order selection
selected order(p,s): aic =  11 1
selected order(p,s): bic =  2 1
selected order(p,s): hq =  2 1
> m1=VARX(zt,2,xt,1) ## Full model estimation
constant term:
est:  0.1817 -0.017
 se:  0.0295 0.0186
AR( 1 ) matrix
      Greg  hoil
Greg 1.041 0.169
hoil 0.005 0.844
standard errors
      [,1] [,2]
[1,] 0.059 0.08
[2,] 0.037 0.05
AR( 2 ) matrix
       Greg   hoil
Greg -0.322 -0.008
hoil  0.014  0.012
standard errors
      [,1]  [,2]
[1,] 0.056 0.068
[2,] 0.035 0.043
Coefficients of exogenous
lag- 0  coefficient matrix
      oilp  gasp
Greg 0.018 0.010
hoil 0.024 0.026
standard errors
      [,1]  [,2]
[1,] 0.001 0.007
```

```
[2,] 0.001 0.004
lag- 1  coefficient matrix
       oilp   gasp
Greg -0.015 -0.012
hoil -0.019 -0.029
standard errors
      [,1]  [,2]
[1,] 0.002 0.007
[2,] 0.001 0.004
Information criteria:
AIC:  -10.83396; BIC:  -10.55981
> cov(m2$residuals)
             Greg        hoil
Greg 0.006736072 0.001049648
hoil 0.001049648 0.002678639
> m1a=refVARX(m1,thres=1) ## Model refinement
constant term:
est:  0.1828 0
 se:  0.028 1
AR( 1 ) matrix
      [,1]  [,2]
[1,] 1.044 0.162
[2,] 0.000 0.873
standard errors
      [,1]  [,2]
[1,] 0.054 0.056
[2,] 1.000 0.031
AR( 2 ) matrix
       [,1] [,2]
[1,] -0.327    0
[2,]  0.000    0
standard errors
     [,1] [,2]
[1,] 0.04    1
[2,] 1.00    1
Coefficients of exogenous
lag- 0  coefficient matrix
      [,1]  [,2]
[1,] 0.018 0.010
[2,] 0.023 0.026
standard errors
      [,1]  [,2]
[1,] 0.001 0.007
[2,] 0.001 0.004
lag- 1  coefficient matrix
       [,1]   [,2]
[1,] -0.015 -0.012
[2,] -0.019 -0.029
standard errors
```

```
        [,1]  [,2]
[1,] 0.002 0.007
[2,] 0.001 0.004
Information criteria:
AIC:   -10.87015
BIC:   -10.67216
> MTSdiag(m1a)   ## Model checking
[1] "Covariance matrix:"
        Greg    hoil
Greg 0.00674 0.00105
hoil 0.00105 0.00270
```

6.3.2 Regression Model

The second approach to include independent variables in multivariate time series analysis is to use multivariate linear regression models with time series errors. The model can be written as

$$\boldsymbol{\phi}(B)(\boldsymbol{z}_t - \boldsymbol{\beta}\boldsymbol{w}_t) = \boldsymbol{a}_t, \tag{6.13}$$

where $\boldsymbol{a}_t$ is defined in Equation (6.11), $\boldsymbol{\phi}(B) = \boldsymbol{I}_k - \sum_{i=1}^{p} \boldsymbol{\phi}_i B^i$ is a matrix polynomial with non-negative degree p, $\boldsymbol{w}_t$ is a v-dimensional vector of independent variables, and $\boldsymbol{\beta}$ is a $k \times v$ coefficient matrix. In most applications, we have $w_{1t} = 1$ for all t so that a constant term is included in the model. Here, $\boldsymbol{w}_t$ may contain lagged values of the observed independent variable $\boldsymbol{x}_t$. The model in Equation (6.13) is nonlinear in parameters and differs from the VARX model of Equation (6.11) in several ways, even if $\boldsymbol{w}_t = (1, \boldsymbol{x}_t')'$. To see this, assume that $\boldsymbol{w}_t = (1, \boldsymbol{x}_t')'$ so that $v = m+1$. Partition the coefficient matrix $\boldsymbol{\beta}$ as $\boldsymbol{\beta} = [\boldsymbol{\beta}_1, \boldsymbol{\beta}_2]$ with $\boldsymbol{\beta}_1$ being a column vector. Then, we can rewrite the model in Equation (6.13) as

$$\begin{aligned} \boldsymbol{z}_t &= \sum_{i=1}^{p} \boldsymbol{\phi}_i \boldsymbol{z}_{t-i} + \boldsymbol{\phi}(1)\boldsymbol{\beta}_1 + \boldsymbol{\phi}(B)\boldsymbol{\beta}_2 \boldsymbol{x}_t + \boldsymbol{a}_t \\ &\equiv \boldsymbol{\phi}_0 + \sum_{i=1}^{p} \boldsymbol{\phi}_i \boldsymbol{z}_{t-i} + \sum_{j=0}^{p} \boldsymbol{\gamma}_j \boldsymbol{x}_{t-j} + \boldsymbol{a}_t, \end{aligned} \tag{6.14}$$

where $\boldsymbol{\phi}_0 = \boldsymbol{\phi}(1)\boldsymbol{\beta}_1$, $\boldsymbol{\gamma}_0 = \boldsymbol{\beta}_2$, and $\boldsymbol{\gamma}_j = -\boldsymbol{\phi}_j \boldsymbol{\beta}_2$ for $j = 1, \ldots, p$. From Equation (6.14), $\boldsymbol{z}_t$ also depends on the lagged values of $\boldsymbol{x}_t$ provided $p > 0$. Thus, the regression model with time series errors of Equation (6.14) can be regarded as a special case of the VARX model in Equation (6.11). On the other hand, due to its parameter constraints, the model in Equation (6.13) requires nonlinear estimation. The likelihood method is commonly used.

In applications, a two-step procedure is often used to specify a regression model with time series errors. In the first step, the multivariate multiple linear regression

$$\boldsymbol{z}_t = \boldsymbol{c} + \boldsymbol{\beta}\boldsymbol{x}_t + \boldsymbol{e}_t$$

is fitted by the ordinary least-squares method to obtain the residuals $\hat{\boldsymbol{e}}_t$. In the second step, information criteria are used to select a VAR model for $\hat{\boldsymbol{e}}_t$.

Example 6.2 ***(continued)*** To demonstrate, we again consider the monthly prices of regular gas and heating oil with spot price of crude oil and price of natural gas as independent variables. We start with fitting the multivariate linear regression model,

$$\boldsymbol{z}_t = \begin{bmatrix} 0.627 \\ -0.027 \end{bmatrix} + \begin{bmatrix} 0.029 & -0.007 \\ 0.030 & -0.008 \end{bmatrix} \boldsymbol{x}_t + \hat{\boldsymbol{e}}_t,$$

and use the residual series $\hat{\boldsymbol{e}}_t$ to specify the VAR order. All three information criteria employed identify a VAR(2) model. Thus, we use the model

$$(\boldsymbol{I}_2 - \boldsymbol{\phi}_1 B - \boldsymbol{\phi}_2 B^2)(\boldsymbol{z}_t - \boldsymbol{c} - \boldsymbol{\beta}\boldsymbol{x}_t) = \boldsymbol{a}_t, \quad (6.15)$$

where $\boldsymbol{c}$ is a two-dimensional constant vector. Table 6.6 summarizes the estimation results of the model in Equation (6.15). The full model denotes the maximum likelihood estimates with bivariate Gaussian distribution and the reduced model is obtained by setting simultaneously all estimates with t-ratio less than 1.2 to 0. The choice of the threshold 1.2 is based on the AIC as several threshold values were used. Figure 6.9 shows the residual cross-correlation matrices of the reduced model in Table 6.6, whereas Figure 6.10 plots the p-values of the multivariate Ljung–Box statistics for the residuals of the reduced model. Both figures indicate that the reduced model in Table 6.6 is adequate in describing the linear dynamic dependence of the two price series. □

It is interesting to compare the reduced VARX(2,1) model of Table 6.5 with that of Table 6.6. First, the VARX model employs 13 coefficients, whereas the regression model only uses 9 coefficients. The regression model is more parsimonious.

TABLE 6.6 Estimation Results of the Regression Model with Time Series Errors in Equation (6.15) for the Monthly Prices of Regular Gas and Heating Oil with Spot Price of Crude Oil and Price of Natural Gas as Independent Variable

	Full Model				Reduced Model			
Parameter	Estimate		Standard Errors		Estimate		Standard Errors	
$\boldsymbol{c}'$	0.650	−0.025	0.049	0.043	0.679	0	0.036	0
$\boldsymbol{\beta}$	0.028	−0.005	0.001	0.008	0.027	0	0.001	0
	0.029	−0.004	0.001	0.006	0.028	0	0.001	0
$\boldsymbol{\phi}_1$	1.012	−0.071	0.069	0.113	1.009	0	0.068	0
	0.070	0.949	0.044	0.071	0.045	0.973	0.030	0.066
$\boldsymbol{\phi}_2$	−0.326	0.186	0.068	0.110	−0.305	0.166	0.064	0.070
	−0.046	−0.138	0.044	0.072	0	−0.128	0	0.073
$10^2\boldsymbol{\Sigma}_a$	0.92	0.19			0.897	0.171		
	0.19	0.38			0.171	0.355		

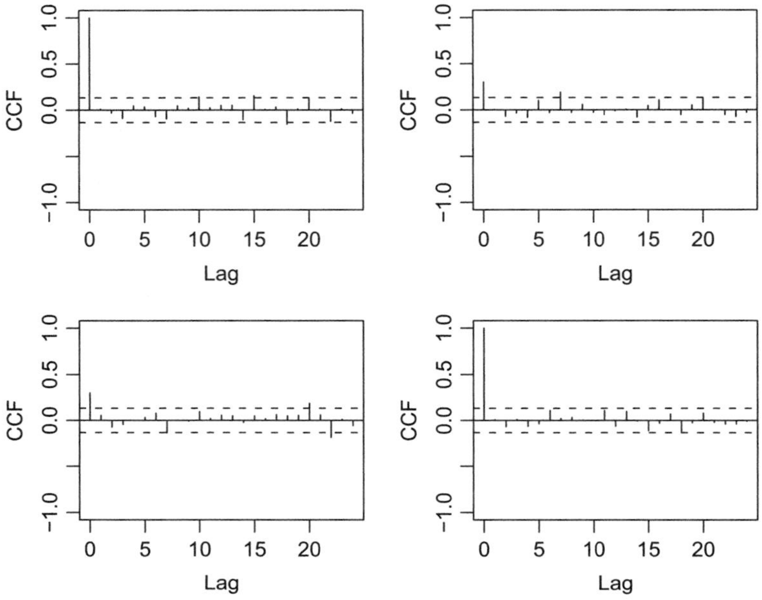

FIGURE 6.9 Residual cross-correlation matrices of the reduced regression model in Table 6.6.

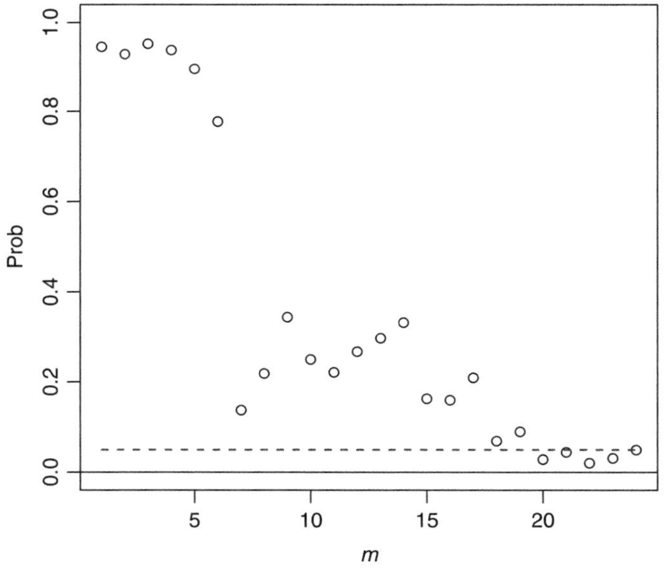

FIGURE 6.10 p-values of multivariate Ljung–Box statistics of the residual series of the reduced regression model in Table 6.6.

However, the information criteria seem to prefer the VARX(2,1) model; see the R demonstrations. Second, the two reduced models are not nested because the regression model implicitly employs $\boldsymbol{x}_{t-2}$. In the form of Equation (6.14), the reduced model of Table 6.6 approximately becomes

$$\begin{aligned}\boldsymbol{z}_t &= \begin{bmatrix} 0.201 \\ -0.030 \end{bmatrix} + \begin{bmatrix} 1.009 & 0 \\ 0.045 & 0.973 \end{bmatrix} \boldsymbol{z}_{t-1} + \begin{bmatrix} -0.305 & 0.166 \\ 0 & -0.128 \end{bmatrix} \boldsymbol{z}_{t-2} \\ &= \begin{bmatrix} 0.027 & 0 \\ 0.028 & 0 \end{bmatrix} \boldsymbol{x}_t - \begin{bmatrix} 0.027 & 0 \\ 0.028 & 0 \end{bmatrix} \boldsymbol{x}_{t-1} + \begin{bmatrix} 0.003 & 0 \\ 0.004 & 0 \end{bmatrix} \boldsymbol{x}_{t-2} + \boldsymbol{a}_t. \end{aligned} \tag{6.16}$$

From the prior model representation, the added impact of $\boldsymbol{x}_{t-2}$ on $\boldsymbol{z}_t$ appears to be small. Consequently, the regression model is approximately a submodel of the VARX(2,1) model. It is then not surprising to see that the residual variances of the VARX(2,1) model are smaller than those of the regression model. Third, the model in Equation (6.16) seems to suggest that, conditional on the spot price of crude oil and lagged values of $\boldsymbol{z}_t$, the prices of regular gas and heating oil do not depend on the price of natural gas. The VARX(2,1) model does not reveal such a structure. However, Equation (6.16) indicates that there exists a feedback relation between prices of regular gas and heating oil. The reduced VARX model, on the other hand, shows that, conditional of $\boldsymbol{x}_t$ and $\boldsymbol{x}_{t-1}$, the price of natural gas does not depend on the past values of the regular gas price.

Remark: Multivariate regression models with time series errors can be estimated by the commands `REGts` and `refREGts` of the `MTS` package. Multivariate multiple linear regression can be estimated by the multivariate linear model command `Mlm`. □

R Demonstration: Regression model with time series errors. Output edited.

```
> da=read.table("m-gasoil.txt",header=T)
> zt=da[,3:4]; xt=da[,5:6]
> m1=Mlm(zt,xt)
[1] "LSE of parameters"
[1] "    est     s.d.    t-ratio    prob"
          [,1]        [,2]   [,3]     [,4]
[1,]   0.62686 0.022768 27.53 0.0000
[2,]   0.02864 0.000424 67.59 0.0000
[3,] -0.00675 0.005222 -1.29 0.1977
[4,] -0.02659 0.016308 -1.63 0.1044
[5,]   0.02961 0.000303 97.56 0.0000
[6,] -0.00846 0.003740 -2.26 0.0246
> VARorder(m1$residuals)   # Order selection
selected order: aic =   2
selected order: bic =   2
selected order: hq =   2
> m3=REGts(zt,2,xt) # Estimation
```

```
Number of parameters:  14
=======
Coefficient matrix for constant + exogenous variable
Estimates:
       [,1]  [,2]   [,3]
[1,]  0.650 0.028 -0.005
[2,] -0.025 0.029 -0.004
Standard errors:
      [,1]  [,2]  [,3]
[1,] 0.049 0.001 0.008
[2,] 0.043 0.001 0.006
VAR coefficient matrices:
AR( 1 ) coefficient:
      [,1]   [,2]
[1,] 1.012 -0.071
[2,] 0.070  0.949
standard errors:
      [,1]  [,2]
[1,] 0.069 0.113
[2,] 0.044 0.071
AR( 2 ) coefficient:
       [,1]   [,2]
[1,] -0.326  0.186
[2,] -0.046 -0.138
standard errors:
      [,1]  [,2]
[1,] 0.068 0.110
[2,] 0.044 0.072
Residual Covariance matrix:
       Greg   hoil
Greg 0.0092 0.0019
hoil 0.0019 0.0038
============
Information criteria:
AIC:  -10.26927; BIC:  -10.05605
> m3c=refREGts(m3,thres=1.2)  # Model refinement
Number of parameters:  9
=======
Coefficient matrix for constant + exogenous variable
Estimates:
      [,1]  [,2] [,3]
[1,] 0.679 0.027    0
[2,] 0.000 0.028    0
Standard errors:
      [,1]  [,2] [,3]
[1,] 0.036 0.001    1
[2,] 1.000 0.001    1
VAR coefficient matrices:
AR( 1 ) coefficient:
```

```
       [,1]   [,2]
[1,] 1.009 0.000
[2,] 0.045 0.973
standard errors:
       [,1]   [,2]
[1,] 0.068 1.000
[2,] 0.030 0.066
AR( 2 ) coefficient:
        [,1]    [,2]
[1,] -0.305  0.166
[2,]  0.000 -0.128
standard errors:
       [,1]   [,2]
[1,] 0.064 0.070
[2,] 1.000 0.073
Residual Covariance matrix:
       Greg   hoil
Greg 0.0089 0.0017
hoil 0.0017 0.0035
============
Information criteria:
AIC:   -10.37776; BIC:   -10.24069
> MTSdiag(m3c)   # Model checking
```

6.4 MISSING VALUES

Missing values occur often in applications of multivariate time series analysis. For instance, the IMF data of M1 money supply of United States contain multiple missing values; see FRED of the Federal Reserve Bank of St. Louis. For a k-dimensional series $\boldsymbol{z}_t$, a particular observation, say $\boldsymbol{z}_h$, may be completely or partially missing. In this section, we discuss a simple approach to fill in $\boldsymbol{z}_h$, assuming that the model for $\boldsymbol{z}_t$ is given. When the model for $\boldsymbol{z}_t$ is unknown, one can use an iterative procedure to handle the missing values and model building. The method discussed here can be treated as part of such an iterative procedure. Also, we assume that the missing values occur at random.

A convenient way to estimate a missing value of $\boldsymbol{z}_t$ is to use the VAR model representation of $\boldsymbol{z}_t$, namely,

$$\boldsymbol{\pi}(B)\boldsymbol{z}_t = \boldsymbol{c} + \boldsymbol{a}_t, \tag{6.17}$$

where $\boldsymbol{\pi}(B) = \boldsymbol{I}_k - \sum_{i=1}^{\infty} \boldsymbol{\pi}_i B^i$, $\boldsymbol{c}$ is a k-dimensional constant, and $\boldsymbol{a}_t$ is a sequence of iid random vectors with mean zero and positive-definite covariance matrix $\boldsymbol{\Sigma}_a$. We assume that $\boldsymbol{z}_t$ is invertible. For an invertible VARMA(p, q) model, $\boldsymbol{\phi}(B)\boldsymbol{z}_t = \boldsymbol{\phi}_0 + \boldsymbol{\theta}(B)\boldsymbol{a}_t$, we have $\boldsymbol{\pi}(B) = [\boldsymbol{\theta}(B)]^{-1}\boldsymbol{\phi}(B)$ and $\boldsymbol{c} = [\boldsymbol{\theta}(1)]^{-1}\boldsymbol{\phi}_0$. See, for instance, Chapter 3. The model representation in Equation (6.17) is rather general. It includes unit-root nonstationary series. It can also be extended to include the VARX models discussed in Section 6.3.

Suppose the available sample consists of $\{\boldsymbol{z}_1, \ldots, \boldsymbol{z}_T\}$, where $\boldsymbol{z}_1$ and $\boldsymbol{z}_T$ are not missing for obvious reasons and T is the sample size. We divide the discussion into two cases depending on whether or not $\boldsymbol{z}_h$ is completely missing.

6.4.1 Completely Missing

For the model in Equation (6.17), the following observations are related explicitly to $\boldsymbol{z}_h$:

$$\boldsymbol{z}_h = \boldsymbol{c} + \sum_{i=1}^{h-1} \boldsymbol{\pi}_i \boldsymbol{z}_{h-i} + \boldsymbol{a}_h \tag{6.18}$$

$$\boldsymbol{z}_{h+1} = \boldsymbol{c} + \boldsymbol{\pi}_1 \boldsymbol{z}_h + \sum_{i=2}^{h} \boldsymbol{\pi}_i \boldsymbol{z}_{h+1-i} + \boldsymbol{a}_{h+1} \tag{6.19}$$

$$\boldsymbol{z}_{h+2} = \boldsymbol{c} + \boldsymbol{\pi}_1 \boldsymbol{z}_{h+1} + \boldsymbol{\pi}_2 \boldsymbol{z}_h + \sum_{i=3}^{h+1} \boldsymbol{\pi}_i \boldsymbol{z}_{h+2-i} + \boldsymbol{a}_{h+2}$$

$$\vdots = \vdots$$

$$\boldsymbol{z}_{h+j} = \boldsymbol{c} + \sum_{i=1}^{j-1} \boldsymbol{\pi}_i \boldsymbol{z}_{h+j-i} + \boldsymbol{\pi}_j \boldsymbol{z}_h + \sum_{i=j+1}^{h-j-1} \boldsymbol{\pi}_i \boldsymbol{z}_{h+j-i} + \boldsymbol{a}_{h+j} \tag{6.20}$$

$$\vdots = \vdots$$

$$\boldsymbol{z}_T = \boldsymbol{c} + \sum_{i=1}^{T-h-1} \boldsymbol{\pi}_i \boldsymbol{z}_{T-i} + \boldsymbol{\pi}_{T-h} \boldsymbol{z}_h + \sum_{i=T-h+1}^{T-1} \boldsymbol{\pi}_i \boldsymbol{z}_{T-i} + \boldsymbol{a}_T. \tag{6.21}$$

This system of equations appears to be complicated, but it simplifies for a VAR(p) model for which $\boldsymbol{\pi}_j = \boldsymbol{0}$ for $j > p$ and the system only involves the first $p+1$ equations.

The prior system of equations gives rise to a multiple linear regression with $\boldsymbol{z}_h$ as the coefficient vector. Specifically, we can combine terms that do not involve $\boldsymbol{z}_h$ in each equation to define

$$\boldsymbol{y}_h = \boldsymbol{c} + \sum_{i=1}^{h-1} \boldsymbol{\pi}_i \boldsymbol{z}_{h-i}$$

$$\boldsymbol{y}_{h+1} = \boldsymbol{z}_{h+1} - \boldsymbol{c} - \sum_{i=2}^{h} \boldsymbol{\pi}_i \boldsymbol{z}_{h+1-i}$$

$$\vdots = \vdots$$

$$
\begin{aligned}
\boldsymbol{y}_{h+j} &= \boldsymbol{z}_{h+j} - \boldsymbol{c} - \sum_{i=1}^{j-1} \boldsymbol{\pi}_i \boldsymbol{z}_{h+j-i} - \sum_{i=j+1}^{h-j-1} \boldsymbol{\pi}_i \boldsymbol{z}_{h+j-i} \qquad (6.22)\\
\vdots &= \vdots \\
\boldsymbol{y}_T &= \boldsymbol{z}_T - \boldsymbol{c} - \sum_{i=1}^{T-h-1} \boldsymbol{\pi}_i \boldsymbol{z}_{T-i} - \sum_{i=T-h+1}^{T-1} \boldsymbol{\pi}_i \boldsymbol{z}_{T-i}.
\end{aligned}
$$

The multivariate multiple linear regression then becomes

$$
\boldsymbol{y}_h = \boldsymbol{I}_k \boldsymbol{z}_h - \boldsymbol{a}_h \qquad (6.23)
$$

$$
\boldsymbol{y}_{h+j} = \boldsymbol{\pi}_j \boldsymbol{z}_h + \boldsymbol{a}_{h+j}, \quad j = 1, \ldots, T-h. \qquad (6.24)
$$

Since $-\boldsymbol{a}_h$ and $\boldsymbol{a}_h$ have the sample distribution for the time series models considered in this book, the aforementioned equations form a (nonhomogeneous) multiple linear regression. Let $\boldsymbol{\Sigma}_a^{1/2}$ be a positive-definite square root matrix of $\boldsymbol{\Sigma}_a$. We define $\boldsymbol{a}_v^* = \boldsymbol{\Sigma}_a^{-1/2}\boldsymbol{a}_v$ and $\boldsymbol{y}_v^* = \boldsymbol{\Sigma}_a^{-1/2}\boldsymbol{y}_v$ for $v = h, \ldots, T$ and $\boldsymbol{\pi}_i^* = \boldsymbol{\Sigma}_a^{-1/2}\boldsymbol{\pi}_i$ for $i = 0, 1, \ldots$ where it is understood that $\boldsymbol{\pi}_0 = \boldsymbol{I}_k$. Then, $\text{Cov}(\boldsymbol{a}_v^*) = \boldsymbol{I}_k$ and we have a homogeneous multiple linear regression

$$
\begin{bmatrix} \boldsymbol{y}_h^* \\ \boldsymbol{y}_{h+1}^* \\ \vdots \\ \boldsymbol{y}_T^* \end{bmatrix} = \begin{bmatrix} \boldsymbol{\pi}_0^* \\ \boldsymbol{\pi}_1^* \\ \vdots \\ \boldsymbol{\pi}_{T-h}^* \end{bmatrix} \boldsymbol{z}_h + \begin{bmatrix} -\boldsymbol{a}_h^* \\ \boldsymbol{a}_{h+1}^* \\ \vdots \\ \boldsymbol{a}_T^* \end{bmatrix}. \qquad (6.25)
$$

Consequently, a consistent estimate of $\boldsymbol{z}_h$ is

$$
\hat{\boldsymbol{z}}_h = \left[\sum_{i=0}^{T-h} \boldsymbol{\pi}_i^* (\boldsymbol{\pi}_i^*)' \right]^{-1} \sum_{i=0}^{T-h} \boldsymbol{\pi}_i^* (\boldsymbol{y}_{i+h}^*)'. \qquad (6.26)
$$

In practice, the computation of $\boldsymbol{y}_v$ for $v = h+1, \ldots, T$ can be simplified by letting the missing value $\boldsymbol{z}_h = \boldsymbol{0}$. Then, we have

$$
\boldsymbol{y}_v = \boldsymbol{z}_v - \boldsymbol{c} - \sum_{i=1}^{v-1} \boldsymbol{\pi}_i \boldsymbol{z}_{v-i}, \quad v = h+1, \ldots, T.
$$

Example 6.3 Consider the log series of the quarterly GDP of United Kingdom, Canada, and United States from the first quarter of 1980 to the second quarter of 2011 for 126 observations. The three series are shown in Figure 6.11 and, as expected, contain unit roots. There are no missing values in the data. However, for illustrative purpose, we artificially treat some observations as missing and compare the observed values with estimates for the missing ones. Here, we employ a VAR(3) model and

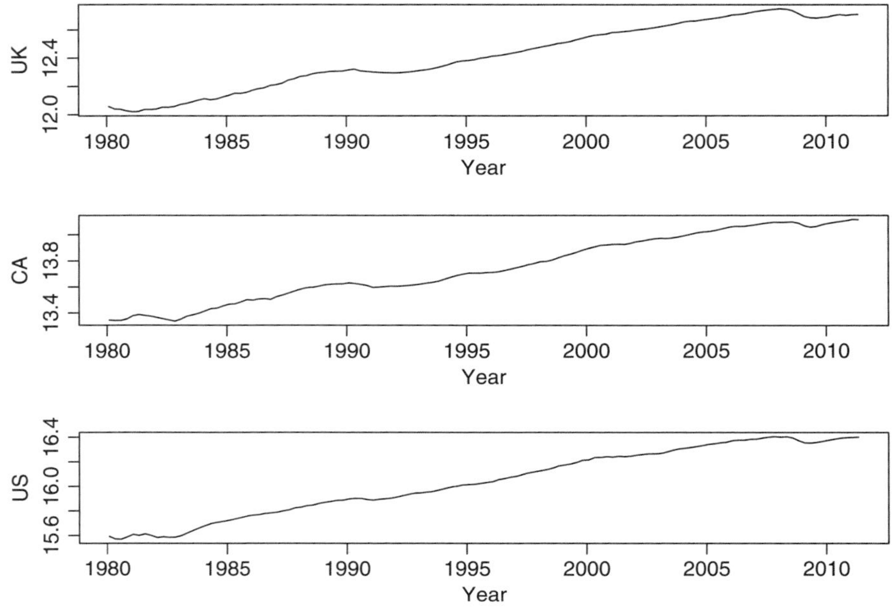

FIGURE 6.11 Time plots of the log GDP series of United Kingdom, Canada, and United States from 1980.I to 2011.II. The original data are in billions of local currencies.

TABLE 6.7 Parameter Estimates of a VAR(3) Model for the Log Series of Quarterly GDP of United Kingdom, Canada, and United States

$\hat{\Phi}_0$		$\hat{\Phi}_1$			$\hat{\Phi}_2$	
−0.052	1.321	0.108	0.062	−0.302	−0.013	−0.026
−0.016	0.367	1.317	0.452	−0.547	−0.514	−0.465
0.144	0.532	0.252	1.117	−0.779	−0.375	−0.107
		$\hat{\Phi}_3$			$10^5 \times \hat{\Sigma}_a$	
	−0.104	−0.087	0.027	2.637	0.287	0.773
	0.223	0.139	0.031	0.287	2.811	1.283
	0.380	0.016	−0.031	0.773	1.283	3.147

The sample period is from 1980.I to 2011.II.

treat the estimates obtained by the full sample as the true model. The parameters of the VAR(3) model are given in Table 6.7. The AR order 3 is selected by AIC and HQ criteria.

If we treat $\boldsymbol{z}_{50}$ as missing, the estimates(true values) are 12.2977(12.2950), 13.6091(13.6090), and 15.9246(15.9250), respectively. If we treat $\boldsymbol{z}_{100}$ as missing, the estimates(true values) become 12.6700(12.6721), 14.0186(14.0205), and 16.3322(16.3322), respectively. The results show that the estimates of missing values are reasonable. Strictly speaking, we use information of the missing values in the estimation because the model is estimated using the full sample. As stated before, one needs to iterate between estimation of missing values and model building to handle missing values in real applications. □

Remark: Missing values are estimated by the command `Vmiss` in the `MTS` package. The input variables are the series, π-weights, $\boldsymbol{\Sigma}_a$ and the time index of the missing value. The command for estimating partially missing values is `Vpmiss`, which requires, in addition, a k-dimensional indicator for missing components with 0 for missing and 1 for observed component. See the R demonstration in the next subsection. □

Discussion: An alternative approach to estimate a time series model and missing values simultaneously is to apply the regression model with time series errors of Section 6.3.2. This is particularly so for a stationary vector time series $\boldsymbol{z}_t$. Here, the missing values are completely missing and the time index of each missing data point is denoted by an indicator variable. For instance, if $\boldsymbol{z}_{50}$ is completely missing, we can create an explanatory variable $x_t(50)$ as

$$\boldsymbol{x}_t(50) = \begin{cases} 1 & \text{if } t = 50, \\ 0 & \text{otherwise.} \end{cases}$$

To perform estimation, one can fill in temporarily the missing observations via simple interpolation. For instance, $\boldsymbol{z}_{50}^* = 0.5(\boldsymbol{z}_{49} + \boldsymbol{z}_{51})$. The model then becomes

$$\boldsymbol{\phi}(B)[\boldsymbol{z}_t - \boldsymbol{c} - \boldsymbol{\beta} x_t(50)] = \boldsymbol{\theta}(B)\boldsymbol{a}_t.$$

Estimate of the missing value $\boldsymbol{z}_{50}$ is then obtained by $\hat{\boldsymbol{z}}_{50} = \boldsymbol{z}_{50}^* - \hat{\boldsymbol{\beta}}$. □

6.4.2 Partially Missing

Consider next the case that only some components of $\boldsymbol{z}_h$ are missing. Without loss of generality, we assume that the first m components of $\boldsymbol{z}_h$ are missing and partition $\boldsymbol{z}_h = (\boldsymbol{z}_{1,h}', \boldsymbol{z}_{2,h}')'$, where $\boldsymbol{z}_{1,h} = (z_{1t}, \ldots, z_{mt})'$. In this case, to estimate $\boldsymbol{z}_{1,h}$, most steps discussed in the prior section continue to hold. For instance, the statistics $\boldsymbol{y}_h$ and $\boldsymbol{y}_v$ for $v = h+1, \ldots, T$ in Equations (6.23) and (6.24) remain unchanged. Furthermore, $\boldsymbol{y}_v$ can also by calculated by treating $\boldsymbol{z}_h = \boldsymbol{0}$. Similarly, the transformed quantities $\boldsymbol{y}_v^*$ and $\boldsymbol{\pi}_i^*$ are also unchanged. On the other hand, $\boldsymbol{z}_{2,h}$ may contain information about the missing components in $\boldsymbol{z}_{1,h}$ and we need to make use of such information in estimating the missing values. This can be achieved by using the homogeneous multiple linear regression in Equation (6.25). Specifically, we partition the system of equations, in matrix form, as

$$\boldsymbol{Y}^* = [\boldsymbol{\Omega}_1, \boldsymbol{\Omega}_2] \begin{bmatrix} \boldsymbol{z}_{1,h} \\ \boldsymbol{z}_{2,h} \end{bmatrix} + \boldsymbol{A}^*,$$

where $\boldsymbol{Y}^*$ denotes the $k(T-h+1)$-dimensional vector of the dependent variable in Equation (6.25), $\boldsymbol{\Omega}_1$ is a $k(T-h+1) \times m$ matrix consisting of the first m columns of the $\boldsymbol{\pi}^*$ matrix in Equation (6.25), and $\boldsymbol{\Omega}_2$ denotes the remaining $k(T-h+1) \times (k-m)$ matrix. The $\boldsymbol{A}^*$ vector satisfies Cov($\boldsymbol{A}^*$) = $\boldsymbol{I}_{k(T-h+1)}$. Since $\boldsymbol{z}_{2,h}$ is observed, we can rewrite Equation (6.25) as

$$[\boldsymbol{Y}^* - \boldsymbol{\Omega}_2 \boldsymbol{z}_{2,h}] = \boldsymbol{\Omega}_1 \boldsymbol{z}_{1,y} + \boldsymbol{A}^*. \tag{6.27}$$

Consequently, the least-squares estimate of the missing component $\boldsymbol{z}_{1,h}$ is

$$\hat{\boldsymbol{z}}_{1,h} = (\boldsymbol{\Omega}_1'\boldsymbol{\Omega}_1)^{-1}[\boldsymbol{\Omega}_1'(\boldsymbol{Y}^* - \boldsymbol{\Omega}_2 \boldsymbol{z}_{2,h})]. \tag{6.28}$$

Example 6.3 ***(continued)*** To demonstrate the estimation of partially missing observations in a vector time series, we continue to use the quarterly GDP data of United Kingdom, Canada, and United States. Again, we employ the VAR(3) model in Table 6.7. Suppose that the last two components of $\boldsymbol{z}_{50}$ are missing, but the first component of which is observed. In this case, the estimates(true values) of the missing values are 13.6365(13.6090) and 15.8815(15.9250), respectively. As another illustration, suppose that only the second component of $\boldsymbol{z}_{50}$ is missing. In this particular instance, the estimate and the true value are 13.6098 and 13.6090, respectively. The estimates of missing values appear to be reasonable. □

R Demonstration: Missing values. Output edited.

```
> da=read.table("q-gdp-ukcaus.txt",header=T)
> gdp=log(da[,3:5])
> VARorder(gdp)
selected order: aic =  3
selected order: bic =  2
selected order: hq =  3
> m1=VAR(gdp,3)  # Model estimation
Constant term:
Estimates:  -0.05211161 -0.0163411 0.1438155
Std.Error:  0.04891077 0.05049953 0.0534254
AR coefficient matrix
AR( 1 )-matrix
      [,1]  [,2]   [,3]
[1,] 1.321 0.108 0.0617
[2,] 0.367 1.317 0.4519
[3,] 0.532 0.252 1.1169
standard error
       [,1]   [,2]   [,3]
[1,] 0.0952 0.0973 0.0938
[2,] 0.0983 0.1005 0.0969
[3,] 0.1040 0.1063 0.1025
AR( 2 )-matrix
```

```
        [,1]    [,2]    [,3]
[1,] -0.302 -0.0133 -0.0264
[2,] -0.547 -0.5135 -0.4648
[3,] -0.779 -0.3750 -0.1067
standard error
      [,1]  [,2]  [,3]
[1,] 0.152 0.141 0.126
[2,] 0.157 0.146 0.130
[3,] 0.166 0.154 0.138
AR( 3 )-matrix
        [,1]    [,2]    [,3]
[1,] -0.104 -0.0865  0.0273
[2,]  0.223  0.1385  0.0310
[3,]  0.380  0.0157 -0.0310
standard error
        [,1]   [,2]   [,3]
[1,] 0.0933 0.0900 0.0946
[2,] 0.0963 0.0930 0.0977
[3,] 0.1019 0.0984 0.1034
  Residuals cov-mtx:
              [,1]         [,2]         [,3]
[1,] 2.637286e-05 2.867568e-06 7.732008e-06
[2,] 2.867568e-06 2.811401e-05 1.283314e-05
[3,] 7.732008e-06 1.283314e-05 3.146616e-05

> piwgt=m1$Phi
> sig=m1$Sigma
> cnst=m1$Ph0
> m2=Vmiss(gdp,piwgt,sig,50,cnst)
Estimate of missing value at time index 50
         [,1]
[1,] 12.29765
[2,] 13.60911
[3,] 15.92464
> gdp[50,]  # Observed values
         uk       ca       us
50 12.29495 13.60897 15.92503
> m2=Vmiss(gdp,piwgt,sig,100,cnst)
Estimate of missing value at time index 100
         [,1]
[1,] 12.67002
[2,] 14.01860
[3,] 16.33219
> gdp[100,]  # Observed values
          uk       ca       us
100 12.67209 14.02048 16.33217
> m3=Vpmiss(gdp,piwgt,sig,50,mdx=c(1,0,0),cnst) # Partially
  missing
Estimate of missing value at time index 50
```

```
          [,1]
[1,] 13.63654
[2,] 15.88154
> m3=Vpmiss(gdp,piwgt,sig,50,mdx=c(1,0,1),cnst)
Estimate of missing value at time index 50
          [,1]
[1,] 13.60977
```

6.5 FACTOR MODELS

With advances in computing facilities and data collection, we nowadays often face the challenge of analyzing high-dimensional time series. Most of the methods and models discussed so far might become inefficient when the dimension k of $\boldsymbol{z}_t$ is large. Some alternative methods are needed. On the other hand, empirical experience shows that many time series exhibit similar characteristics such as common trends or common periodic behavior. This observation leads naturally to consider models that can describe the common features in multiple time series. Factor models in which the component series are driven by a small number of factors become popular and have attracted substantial interest in recent years. Since common factors are latent, their identification requires some assumptions. This leads to the development of various factor models. In this section, we briefly discuss some factor models.

6.5.1 Orthogonal Factor Models

The simplest factor model is the traditional orthogonal factor model in which the observed time series $\boldsymbol{z}_t$ is driven by a small number of common factors. The model can be written as

$$\boldsymbol{z}_t = \boldsymbol{L}\boldsymbol{f}_t + \boldsymbol{\epsilon}_t, \tag{6.29}$$

where $\boldsymbol{L}$ is a $k \times m$ loading matrix, $\boldsymbol{f}_t = (f_{1t}, \ldots, f_{mt})'$ with $m \ll k$ is an m-dimensional vector of common factors, and $\boldsymbol{\epsilon}_t$ is a sequence of k-dimensional random vectors with mean zero and Cov($\boldsymbol{\epsilon}_t$) = $\boldsymbol{\Sigma}_\epsilon$ = diag$\{\sigma_1^2, \ldots, \sigma_k^2\}$ with "diag" denoting diagnoal matrix. In Equation (6.29), $\boldsymbol{f}_t$ and $\boldsymbol{\epsilon}_t$ are assumed to be independent and, for identifiability, we require that $E(\boldsymbol{f}_t) = \boldsymbol{0}$ and Cov($\boldsymbol{f}_t$) = $\boldsymbol{I}_m$. The loading matrix $\boldsymbol{L}$ is assumed to be of full column rank; otherwise, the number of common factors can be reduced. In Equation (6.29), we assume, for simplicity, $E(\boldsymbol{z}_t) = \boldsymbol{0}$.

Under the aforementioned assumptions and stationarity, the autocovariance matrices of $\boldsymbol{z}_t$ satisfy

$$\begin{aligned}
\boldsymbol{\Gamma}_z(0) &= \boldsymbol{L}\boldsymbol{\Gamma}_f(0)\boldsymbol{L}' + \boldsymbol{\Sigma}_\epsilon = \boldsymbol{L}\boldsymbol{L}' + \boldsymbol{\Sigma}_\epsilon \\
\boldsymbol{\Gamma}_z(\ell) &= \boldsymbol{L}\boldsymbol{\Gamma}_f(\ell)\boldsymbol{L}', \quad \ell > 0,
\end{aligned} \tag{6.30}$$

where $\mathbf{\Gamma}_y(\ell)$ denotes the lag ℓ autocovariance matrix of a vector time series $\boldsymbol{y}_t$. From Equation (6.30), the autocovariance matrices of an orthogonal factor model $\boldsymbol{z}_t$ are all singular for $\ell > 0$. This leads to the idea of defining the number of common factor m as the maximum of the ranks of $\mathbf{\Gamma}_z(\ell)$ for $\ell > 0$. Statistical methods for exploiting the property in Equation (6.30) were proposed by Peña and Box (1987) and Geweke (1977). A weakness of this approach is that Equation (6.30) holds only for stationary processes $\boldsymbol{z}_t$. Peña and Poncela (2006) extended the model to include unit-root nonstationary processes.

Estimation of the orthogonal factor models in Equation (6.29) can be carried out in two ways. The first approach is to apply PCA to $\boldsymbol{z}_t$ and select the first m PCs corresponding to the largest m eigenvalues as the common factors. From the spectral decomposition, we have

$$\mathbf{\Gamma}_z(0) = \lambda_1^2 \boldsymbol{e}_1\boldsymbol{e}_1' + \lambda_2^2 \boldsymbol{e}_2\boldsymbol{e}_2' + \cdots + \lambda_k^2 \boldsymbol{e}_k\boldsymbol{e}_k',$$

where $\lambda_1^2 \geq \lambda_2^2 \geq \cdots \geq \lambda_k^2$ are the eigenvalues of the covariance matrix $\mathbf{\Gamma}_z(0)$ and $\boldsymbol{e}_i$ is an eigenvector associated with the eigenvalue λ_i^2 and satisfies $\|\boldsymbol{e}_i\|^2 = 1$. Suppose that the first m eigenvalues are large and the other eigenvalues are small, then we can obtain the approximation

$$\mathbf{\Gamma}_z(0) \approx [\lambda_1\boldsymbol{e}_1, \lambda_2\boldsymbol{e}_2, \ldots, \lambda_m\boldsymbol{e}_m] \begin{bmatrix} \lambda_1\boldsymbol{e}_1' \\ \lambda_2\boldsymbol{e}_2' \\ \vdots \\ \lambda_m\boldsymbol{e}_m' \end{bmatrix} + \hat{\mathbf{\Sigma}}_\epsilon \equiv \hat{\boldsymbol{L}}\hat{\boldsymbol{L}}' + \hat{\mathbf{\Sigma}}_\epsilon, \qquad (6.31)$$

where $\hat{\mathbf{\Sigma}}_e = \text{diag}\{u_1^2, \ldots, u_k^2\} = \text{diag}[\mathbf{\Gamma}_z(0)] - \text{diag}[\hat{\boldsymbol{L}}\hat{\boldsymbol{L}}']$. Traditionally, the choice of m is determined subjectively by looking at the scree plot of eigenvalues of $\mathbf{\Gamma}_z(0)$. To avoid the scaling effect, one may choose to standardize the components of $\boldsymbol{z}_t$ before estimating the factor model. This is equivalent to replacing $\mathbf{\Gamma}_z(0)$ by the correlation matrix of $\boldsymbol{z}_t$. For further details of using PCA in factor analysis, see Johnson and Wichern (2007), Peña and Box (1987), and the diffusion index discussed later. The second approach is to use maximum likelihood estimation under the normality assumption and some regularity conditions for model identification. This approach also requires that m is known. Commonly used regularity conditions for maximum likelihood estimation are as follows: (a) $\mathbf{\Sigma}_\epsilon$ is invertible and (b) $\boldsymbol{L}'\mathbf{\Sigma}_\epsilon^{-1}\boldsymbol{L}$ is a diagonal matrix.

Example 6.4 Consider the monthly log returns of stocks for ten U.S. companies from January 2001 to December 2011 for 132 observations. The companies and their return summary statistics are given in Table 6.8. These ten companies can roughly be classified into three industrial sectors, namely, semiconductor, pharmaceutical, and investment banks. The mean returns of the ten stocks are all close to 0, but the log returns have some negative skewness and high excess kurtosis.

TABLE 6.8 Ten U.S. Companies and Some Summary Statistics of Their Monthly Log Returns from January 2001 to December 2011

Sector	Name	Tick	Mean	Variance	Skewness	Kurtosis
Semi-	Texas Instru.	TXN	−0.003	0.012	−0.523	0.948
conductor	Micron Tech.	MU	−0.013	0.027	−0.793	2.007
	Intel Corp.	INTC	−0.000	0.012	−0.584	1.440
	Taiwan Semi.	TSM	0.006	0.013	−0.201	1.091
Pharma-	Pfizer	PFE	−0.003	0.004	−0.284	−0.110
ceutical	Merck & Co.	MRK	−0.003	0.006	−0.471	1.017
	Eli Lilly	LLY	−0.003	0.005	−0.224	1.911
Investment	JPMorgan Chase	JPM	0.000	0.009	−0.466	1.258
Bank	Morgan Stanley	MS	−0.010	0.013	−1.072	4.027
	Goldman Sachs	GS	−0.001	0.008	−0.500	0.455

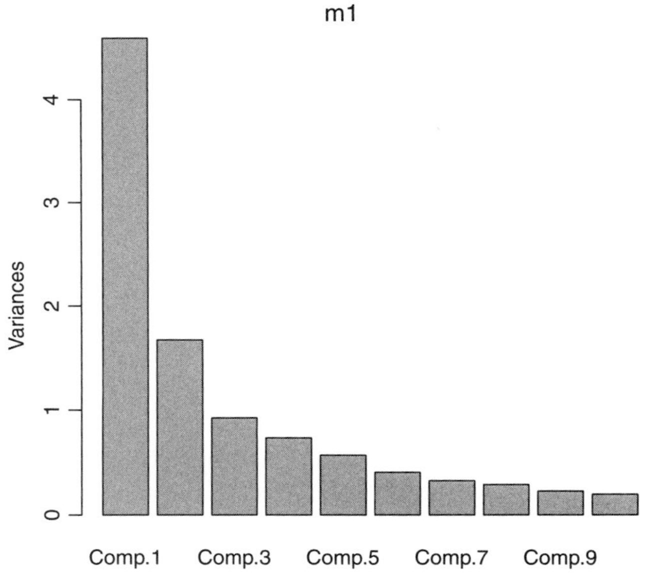

FIGURE 6.12 Scree plot for the monthly log returns of ten U.S. companies given in Table 6.8.

Let $\boldsymbol{z}_t$ be the standardized log returns, where the ith component is defined as $z_{it} = r_{it}/\hat{\sigma}_i$ with r_{it} being the ith return series and $\hat{\sigma}_i$ being the sample standard error of r_{it}. We perform the PCA of $\boldsymbol{z}_t$. Figure 6.12 shows the scree plot of the analysis. From the plot, eigenvalues of the sample correlation matrix of $\boldsymbol{z}_t$ decay exponentially and the first few eigenvalues appear to be substantially larger than the others. Thus, one may select $m = 3$ or 4. We use $m = 3$ for simplicity. The PCA shows that the first three components explain about 72.4% of the total variability of the log returns. See the following R output for details. Estimates of the loading matrix $\boldsymbol{L}$ and the covariance matrix of $\boldsymbol{\epsilon}_t$ are given in Table 6.9.

TABLE 6.9 Estimation Results of Orthogonal Factor Models for the Monthly Log Returns of Ten U.S. Companies Given in Table 6.8, where L_i is the ith Column of L and $\Sigma_{\epsilon,i}$ is the (i, i)th Element of Σ_ϵ

	PCA Approach				Maximum Likelihood Method			
Tick	L_1	L_2	L_3	$\Sigma_{\epsilon,i}$	L_1	L_2	L_3	$\Sigma_{\epsilon,i}$
TXN	0.788	0.198	0.320	0.237	0.714	−0.132	0.417	0.299
MU	0.671	0.361	0.289	0.336	0.579	−0.263	0.315	0.496
INTC	0.789	0.183	0.333	0.232	0.718	−0.130	0.442	0.272
TSM	0.802	0.270	0.159	0.258	0.725	−0.186	0.330	0.331
PFE	0.491	−0.643	−0.027	0.345	0.351	0.580	0.197	0.501
MRK	0.402	−0.689	0.226	0.312	0.282	0.582	0.220	0.533
LLY	0.448	−0.698	0.061	0.309	0.359	0.638	0.184	0.430
JPM	0.724	0.020	−0.345	0.357	0.670	0.067	0.010	0.547
MS	0.755	0.053	−0.425	0.246	0.789	0.052	−0.154	0.352
GS	0.745	0.122	−0.498	0.182	0.879	−0.008	−0.363	0.096

From Equation (6.31), the ith column of the loading matrix $\boldsymbol{L}$ is proportional to the ith normalized eigenvector that, in turn, produces the ith principal component of the data. The estimated loading matrix $\boldsymbol{L}$ in Table 6.9 suggests that (a) the first factor is a weighted average of the log returns with similar weights for stocks in the same industrial sector, (b) the second factor is essentially a weighted difference of the log returns between the semiconductor sector and the pharmaceutical industry, and (c) the third factor represents a weighted difference of log returns between the semiconductor sector and the investment banks. On the other hand, the variances of the noise components are between 18 to 35% of each standardized return series, indicating that marked variability remains in each log return series. Figure 6.13 shows the time plots of the three common factors, that is, $\hat{\boldsymbol{f}}_t$.

We also applied the maximum likelihood estimation of orthogonal factor models to the standardized log returns, assuming $m = 3$. The results are also given in Table 6.9. No factor rotation is used so that one can easily compare the results with those of the principal component approach. Indeed, the estimation results are similar, at least qualitatively. The two estimated loading matrices are close. The main difference appears to be in the variances of the innovations. The maximum likelihood method shows that the variances of individual innovations for MU, PFE, MRK, and JPM are around 50%. This indicates the choice of three common factors might be too low. As a matter of fact, a choice of $m = 5$ appears to provide a better fit in this particular instance. □

R Demonstration: Example of an orthogonal factor model.

```
> da=read.table("m-tenstocks.txt",header=T)
> rtn=log(da[,2:11]+1)   # log returns
> std=diag(1/sqrt(diag(cov(rtn))))
> rtns=as.matrix(rtn)%*%std    # Standardize individual series
```

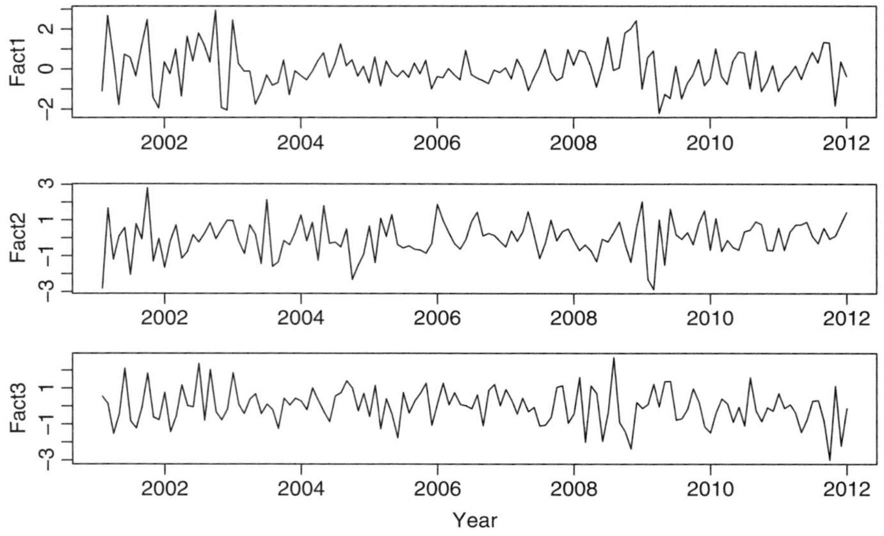

FIGURE 6.13 Estimated common factors for the monthly log returns of ten U.S. companies given in Table 6.8.

```
> m1=princomp(rtns)
> names(m1)
[1] "sdev"    "loadings" "center"  "scale"   "n.obs"    "scores"  "call"
> sdev=m1$sdev    # square root of eigenvalues
> M=m1$loadings
> summary(m1)
Importance of components:
                        Comp.1    Comp.2    Comp.3    Comp.4    Comp.5
Standard deviation     2.14257 1.292276 0.9617832 0.8581358 0.7543204
Proportion of Variance 0.46257 0.168273 0.0932088 0.0742018 0.0573343
Cumulative Proportion  0.46257 0.630839 0.7240474 0.7982492 0.8555835
                        Comp.6     Comp.7    Comp.8    Comp.9   Comp.10
Standard deviation     0.6351810 0.5690478 0.5351137 0.4738664 0.441654
Proportion of Variance 0.0406535 0.0326287 0.0288533 0.0226263 0.019655
Cumulative Proportion  0.8962370 0.9288657 0.9577190 0.9803453 1.000000
> screeplot(m1)
> SD=diag(sdev[1:3])  # Compute the loading matrix
> L=M[,1:3]%*%SD
> print(round(L,3))
        [,1]   [,2]   [,3]
 [1,] -0.788 -0.198 -0.320
 [2,] -0.671 -0.361 -0.289
 [3,] -0.789 -0.183 -0.333
 [4,] -0.802 -0.270 -0.159
 [5,] -0.491  0.643  0.027
 [6,] -0.402  0.689 -0.226
 [7,] -0.448  0.698 -0.061
 [8,] -0.724 -0.020  0.345
```

```
 [9,] -0.755 -0.053  0.425
[10,] -0.745 -0.122  0.498
> LLt=L%*%t(L)
> diag(LLt)
 [1] 0.762621 0.664152 0.767795 0.741557 0.655427 0.687690 0.691248
 [8] 0.643403 0.753815 0.817913
> SigE=1-diag(LLt)
> SigE
 [1] 0.237379 0.335848 0.232205 0.258444 0.344573 0.312310 0.308752
 [8] 0.356597 0.246185 0.182087

> m2=factanal(rtns,3,scores="regression",rotation="none") #MLE
> m2
factanal(x =rtns, factors=3, scores ="regression", rotation ="none")
Uniquenesses:
 [1] 0.299 0.496 0.272 0.331 0.501 0.533 0.430 0.547 0.352 0.096
Loadings:
      Factor1 Factor2 Factor3
 [1,]  0.714  -0.132   0.417
 [2,]  0.579  -0.263   0.315
....
> names(m2)
 [1] "converged" "loadings"  "uniquenesses" "correlation" "criteria"
 [6] "factors"   "dof"       "method"       "scores"      "STATISTIC"
[11] "PVAL"      "n.obs"     "call"
> L2=matrix(m2$loadings,10,3)
> print(round(L2,3))
       [,1]   [,2]   [,3]
 [1,] 0.714 -0.132  0.417
 [2,] 0.579 -0.263  0.315
 [3,] 0.718 -0.130  0.442
 [4,] 0.725 -0.186  0.330
 [5,] 0.351  0.580  0.197
 [6,] 0.282  0.582  0.220
 [7,] 0.359  0.638  0.184
 [8,] 0.670  0.067  0.010
 [9,] 0.789  0.052 -0.154
[10,] 0.879 -0.008 -0.363
```

Discussion: The PCA approach to estimating orthogonal factor models in Equation (6.31) provides only an approximate solution in general. However, the approach can be justified asymptotically if the covariance matrix of the noises is proportional to the $k \times k$ identity matrix, that is, $\boldsymbol{\Sigma}_\epsilon = \sigma^2 \boldsymbol{I}_k$. To see this, we consider the simplest case of $k = 2$, $m = 1$, and $E(\boldsymbol{z}_t) = \mathbf{0}$. In this case, the model is

$$\begin{bmatrix} z_{1t} \\ z_{2t} \end{bmatrix} = \begin{bmatrix} L_1 \\ L_2 \end{bmatrix} f_t + \begin{bmatrix} \epsilon_{1t} \\ \epsilon_{2t} \end{bmatrix}, \tag{6.32}$$

where $\text{Cov}(\boldsymbol{\epsilon}_t) = \sigma^2 \boldsymbol{I}_2$, L_1 and L_2 are real numbers and f_t is the common factor. Suppose the sample size is T. Then, we have

$$\frac{1}{T}\sum_{t=1}^{T} \boldsymbol{z}_t \boldsymbol{z}_t' = \begin{bmatrix} L_1 \\ L_2 \end{bmatrix} \frac{1}{T}\left(\sum_{t=1}^{T} f_t^2\right) [L_1, L_2] + \frac{1}{T}\sum_{t=1}^{T} \boldsymbol{\epsilon}_t \boldsymbol{\epsilon}_t'. \tag{6.33}$$

Under the assumptions of orthogonal factor models, Equation (6.33) is asymptotically equivalent to

$$\boldsymbol{\Gamma}_z(0) = \begin{bmatrix} L_1^2 & L_1 L_2 \\ L_1 L_2 & L_2^2 \end{bmatrix} + \sigma^2 \boldsymbol{I}_2.$$

It is then easy to calculate that the eigenvalues of $\boldsymbol{\Gamma}_z(0)$ are σ^2 and $\sigma^2 + L_1^2 + L_2^2$. Since $L_1^2 + L_2^2 > 0$, the largest eigenvalue of $\boldsymbol{\Gamma}_z(0)$ is $\sigma^2 + L_1^2 + L_2^2$. It is also easy to calculate that the eigenvector associated with the largest eigenvalue is exactly proportional to $\boldsymbol{L}$, the factor loading matrix. □

6.5.2 Approximate Factor Models

The requirement that the covariance matrix of the innovations $\boldsymbol{\epsilon}_t$ being diagonal in Equation (6.29) may be hard to achieve when the dimension k of $\boldsymbol{z}_t$ is large, especially in financial applications in which innovations orthogonal to some known risk factors remain correlated. For this reason, Chamberlain (1983) and Chamberlain and Rothschild (1983) proposed the approximate factor models that allow the innovations $\boldsymbol{\epsilon}_t$ to have a general covariance matrix $\boldsymbol{\Sigma}_\epsilon$. Approximate factor models are commonly used in economics and finance.

Consider an approximate factor model

$$\boldsymbol{z}_t = \boldsymbol{L}\boldsymbol{f}_t + \boldsymbol{\epsilon}_t, \tag{6.34}$$

where $\boldsymbol{\epsilon}_t$ satisfies (a) $E(\boldsymbol{\epsilon}_t) = \boldsymbol{0}$, (b) $\text{Cov}(\boldsymbol{\epsilon}_t) = \boldsymbol{\Sigma}_\epsilon$, and (c) $\text{Cov}(\boldsymbol{\epsilon}_t, \boldsymbol{\epsilon}_s) = \boldsymbol{0}$ for $t \neq s$. A weakness of this model is that, for a finite k, the model is not uniquely identified. For instance, assume that $\boldsymbol{f}_t$ is not serially correlated. Let $\boldsymbol{C}$ be any $k \times k$ orthonormal matrix but $\boldsymbol{I}_k$, and define

$$\boldsymbol{L}^* = \boldsymbol{C}\boldsymbol{L}, \quad \boldsymbol{\epsilon}_t^* = (\boldsymbol{I}_k - \boldsymbol{C})\boldsymbol{L}\boldsymbol{f}_t + \boldsymbol{\epsilon}_t.$$

It is easy to see that we can rewrite the model as

$$\boldsymbol{z}_t = \boldsymbol{L}^*\boldsymbol{f}_t + \boldsymbol{\epsilon}_t^*,$$

which is also an approximate factor model. Forni et al. (2000) show that the approximate factor model can be identified if the m largest eigenvalues of the covariance matrix of $\boldsymbol{z}_t$ diverge to infinity when $k \to \infty$ and the eigenvalues of the noise covariance matrix remain bounded. Therefore, the approximate factor models in Equation (6.34) are only asymptotically identified.

Another line of research emerges recently in the statistical literature for the approximate factor models in Equation (6.34). See Pan and Yao (2008), Lam, Yao, and Bathia (2011), and Lam and Yao (2012). The model considered assumes the same form as Equation (6.34), but requires that (a) $\boldsymbol{L}$ is a $k \times m$ orthonormal matrix so that $\boldsymbol{L}'\boldsymbol{L} = \boldsymbol{I}_m$, (b) $\boldsymbol{\epsilon}_t$ is a white noise series with $E(\boldsymbol{\epsilon}_t) = \boldsymbol{0}$ and $\text{Cov}(\boldsymbol{\epsilon}_t) = \boldsymbol{\Sigma}_\epsilon$ being a general covariance matrix, (c) $\text{Cov}(\boldsymbol{\epsilon}_t, \boldsymbol{f}_s) = \boldsymbol{0}$ for $s \le t$, and (d) no linear combinations of $\boldsymbol{f}_t$ are white noise. Some additional assumptions are needed for consistent estimation of $\boldsymbol{L}$ and $\boldsymbol{f}_t$; see Lam, Yao and Bathia (2011) for details. Lam and Yao (2012) investigates the estimation of the number of common factors m when k is large. This approach effectively treats any nonwhite noise linear combination as a common factor.

By assumptions, there exists a $k \times (k-m)$ orthonormal matrix $\boldsymbol{U}$ such that $[\boldsymbol{L}, \boldsymbol{U}]$ is a $k \times k$ orthonormal matrix and $\boldsymbol{U}'\boldsymbol{L} = \boldsymbol{0}$. Premultiplying Equation (6.34) by $\boldsymbol{U}'$, we have

$$\boldsymbol{U}'\boldsymbol{z}_t = \boldsymbol{U}'\boldsymbol{\epsilon}_t, \tag{6.35}$$

which is a $(k - m)$-dimensional white noise. Based on this white noise property, $\boldsymbol{U}'\boldsymbol{z}_t$ is uncorrelated with $\{\boldsymbol{z}_{t-1}, \boldsymbol{z}_{t-2}, \ldots\}$. In this sense, $\boldsymbol{U}'\boldsymbol{z}_t$ consists of $k - m$ scalar component models of order (0,0), that is, SCM(0,0) of Chapter 4. This argument also applies to other factor models. In Tiao and Tsay (1989), search for SCM(0,0) is carried out by canonical correlation analysis between $\boldsymbol{z}_t$ and $\boldsymbol{Z}_{h,t-1} \equiv (\boldsymbol{z}'_{t-1}, \ldots, \boldsymbol{z}'_{t-h})'$ for some selected positive integer h. Thus, it considers the rank of $\text{Cov}(\boldsymbol{z}_t, \boldsymbol{Z}_{h,t-1}) = [\boldsymbol{\Gamma}_z(1), \ldots, \boldsymbol{\Gamma}_z(h)]$. In Lam and Yao (2012), search for white noise linear combinations is carried out by the eigenanalysis of the matrix $\boldsymbol{G} = \sum_{i=1}^{h} \boldsymbol{\Gamma}_z(i)\boldsymbol{\Gamma}_z(i)'$, which by construction is non-negative definite. Thus, the two methods are highly related. On the other hand, Lam and Yao (2012) also consider the case of $k \to \infty$, whereas Tiao and Tsay (1989) assume k is fixed.

The requirement that no linear combination of $\boldsymbol{f}_t$ is a white noise implies that the orthogonal factor model is not a special case of the approximate factor model of Pan and Yao (2008). This may limit the applicability of the latter model in economics and finance.

6.5.3 Diffusion Index Models

In a sequence of papers, Stock and Watson (2002a, b) consider a diffusion index model for prediction. The model is highly related to factor models and can be written as

$$\boldsymbol{z}_t = \boldsymbol{L}\boldsymbol{f}_t + \boldsymbol{\epsilon}_t, \tag{6.36}$$

$$y_{t+h} = \boldsymbol{\beta}'\boldsymbol{f}_t + e_{t+h}, \tag{6.37}$$

where $\boldsymbol{z}_t = (z_{1t}, \ldots, z_{kt})'$ is a k-dimensional stationary time series, which is observed and $E(\boldsymbol{z}_t) = \boldsymbol{0}$, $\boldsymbol{f}_t = (f_{1t}, \ldots, f_{mt})'$ is an m-dimensional vector of common factors with $E(\boldsymbol{f}_t) = \mathbf{0}$ and Cov($\boldsymbol{f}_t$) = $\boldsymbol{I}_m$, $\boldsymbol{L}$ is a $k \times m$ loading matrix, and $\boldsymbol{\epsilon}_t$ is a sequence of iid random vectors with mean zero and covariance matrix $\boldsymbol{\Sigma}_\epsilon$. Thus, Equation (6.36) is an approximate factor model. The $\{y_t\}$ is a scalar time series of interest, the positive integer h denotes the forecast horizon, $\boldsymbol{\beta} = (\beta_1, \ldots, \beta_m)'$ denotes the coefficient vector, and e_t is a sequence of uncorrelated random variables with mean zero and variance σ_e^2. Equation (6.37) represents the linear equation for h-step ahead prediction of y_{t+h} based on the common factor $\boldsymbol{f}_t$. Components of $\boldsymbol{f}_t$ are the diffusion indices. In applications, Equation (6.37) can be extended to include some preselected predictors. For instance,

$$y_{t+h} = \boldsymbol{\beta}'\boldsymbol{f}_t + \boldsymbol{\gamma}'\boldsymbol{w}_t + e_{t+h}, \tag{6.38}$$

where $\boldsymbol{w}_t$ consists of some predetermined predictors such as y_t and y_{t-1}. Obviously, a constant term can be added to Equation (6.37) if $E(y_t) \neq 0$.

Since $\boldsymbol{f}_t$ is unobserved, inference concerning the diffusion index model in Equations (6.36) and (6.37) is carried out in two steps. In the first step, one extracts the common factor $\boldsymbol{f}_t$ from the observed high-dimensional series $\boldsymbol{z}_t$. Stock and Watson (2002a,b) apply the PCA to extract $\boldsymbol{f}_t$. Denote the sample principal component by $\hat{\boldsymbol{f}}_t$. The second step is to estimate $\boldsymbol{\beta}$ in Equation (6.37) by the ordinary least-squares method with $\boldsymbol{f}_t$ replaced by $\hat{\boldsymbol{f}}_t$.

An important contribution of Stock and Watson (2002a) is that the authors show under rather general conditions that (a) $\hat{\boldsymbol{f}}_t$ converges in probability to $\boldsymbol{f}_t$ (up to possible sign changes) and (b) the forecasts derived from Equation (6.37) with $\hat{\boldsymbol{f}}_t$ converge in probability to the mean-square efficient estimates that could be obtained if the common factor $\boldsymbol{f}_t$ was observable as $k \to \infty$ and $T \to \infty$ jointly with a joint growth rates of k and T, where T denotes the sample size. In short, these authors successfully extend the traditional PCA with finite k to the case in which both k and T approach infinity at some joint growth rates.

The selection of the number of common factors m from $\boldsymbol{z}_t$ has been investigated by several authors. Bai and Ng (2002) extended information criteria such as AIC to account for the effects of increasing k and T. Onatski (2009) proposed a test statistic using random matrix theory. Bai (2003) studied limiting properties of the approximate factor models in Equation (6.36). A criticism of the analysis of diffusion index models discussed so far is that the selection of the common factors $\hat{\boldsymbol{f}}_t$ in Equation (6.36) does not make use of the objective function of predicting y_t. To overcome this weakness, several alternative methods have been proposed in the recent literature, including using partial least squares and some of its variants. On the other hand, Stock and Watson (2002b) demonstrated that the diffusion index performs well in practice. Finally, Heaton and Solo (2003, 2006) provide some useful insights into the diffusion index models in Equations (6.36) and (6.37).

Example 6.5 To illustrate the prediction of diffusion index models, we consider the monthly unemployment rates of the U.S. 50 states from January 1976 to August

2010 for 416 observations. Preliminary analysis indicates the existence of some large outliers and we made some simple adjustments accordingly. The changes are

1. Arizona: Due to a level shift, we subtract 3.3913 from each of the first 92 observations.
2. Louisiana: To adjust for the effect of Hurricane Katrina, we subtracted (6,6,6,1) from the unemployment rates for t from 357 to 360, respectively.
3. Mississippi: To partially remove the effect of Hurricane Katrina, we subtracted (3,2.5,3,2,0.5) from the observed rates for t from 357 to 361, respectively.

Let $\boldsymbol{z}_t$ be the first differenced monthly unemployment rates of the 50 states. Figure 6.14 shows the time plots of the 50 change series in $\boldsymbol{z}_t$. The series are highly related and still contain some aberrant observations.

In our demonstration, we predict y_{t+1} for the first five components of $\boldsymbol{z}_t$ using $\boldsymbol{x}_t = (\boldsymbol{z}_t', \boldsymbol{z}_{t-1}', \boldsymbol{z}_{t-2}', \boldsymbol{z}_{t-3}')'$. Thus, we use 200 regressors in the exercise. The five states used as dependent variables are Alabama (AL), Alaska (AK), Arizona (AZ), Arkansas (AR), and California (CA). The forecast origin is 350. In other words, we use the first 350 observations of $\boldsymbol{z}_t$ to construct $\boldsymbol{x}_t$ and perform the PCA on $\boldsymbol{x}_t$ to derive the diffusion indices. The components of $\boldsymbol{x}_t$ are standardized individually before the PCA. Denote the diffusion indices by f_{it}. We use the prediction equation

$$y_{t+1} = \beta_0 + \sum_{i=1}^{m} \beta_i f_{it} + e_t, \quad t = 5, \ldots, 350 \tag{6.39}$$

to obtain the estimates $\hat{\beta}_i$. The fitted equation is then used to perform one-step ahead predictions for $t = 351, \ldots, 414$. Note that the diffusion indices f_{it} in the forecasting

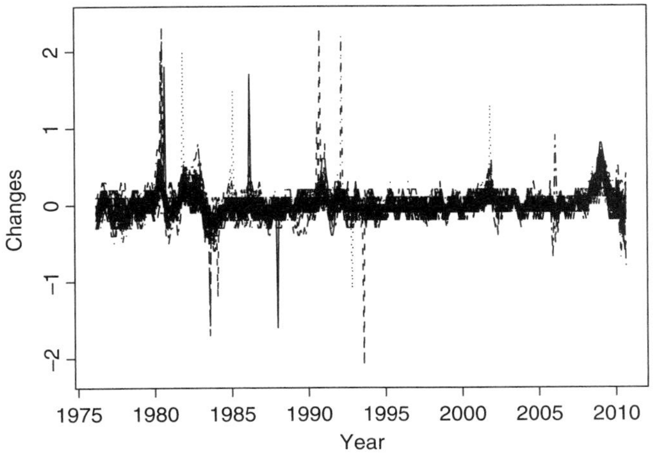

FIGURE 6.14 Time plots of the first differenced series of the monthly unemployment rates of the U.S. 50 states. The time period is from February 1976 to August 2010. The series were adjusted for outliers.

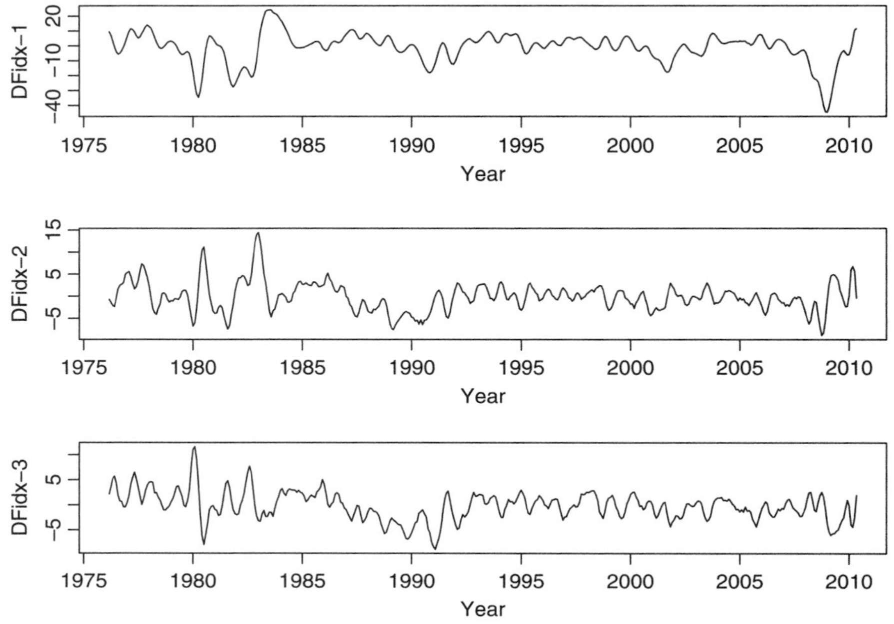

FIGURE 6.15 Time plots of the first three diffusion indices of the differenced series of monthly state unemployment rates. The indices are based on principal component analysis of the first 350 observations with lags 1 to 4.

TABLE 6.10 Mean Squares of Forecast Errors for One-Step Ahead Out-of-Sample Prediction Using Diffusion Index, where m Denotes the Number of Indices Used

	Number of Diffusion Indices m					
State	10	20	30	50	100	200
AL	1.511	1.275	1.136	1.045	1.144	2.751
AK	0.619	0.480	0.530	0.641	0.590	2.120
AZ	1.223	1.274	1.195	1.109	1.738	3.122
AR	0.559	0.466	0.495	0.473	0.546	1.278
CA	0.758	0.694	0.626	0.636	0.663	0.849

The forecast origin is March, 2004. The numbers shown are MSE$\times 10^2$.

period were obtained by the same loadings used in the estimation period. Figure 6.15 shows the time plots of the first three diffusion indices. Based on the PCA, the first diffusion index f_{1t} explains about 50% of the variability in the standardized data. Table 6.10 summarizes the out-of-sample performance of the diffusion index approach for various choices of m. From the table, we made the following observations. First, using all 200 regressors did poorly in prediction. As a matter of fact, it is the worst case for all five states considered. This result confirms that over-fitting fares poorly in out-of-sample prediction. Second, no single choice of m outperforms the others. A choice of m between 20 to 50 seems to work well. In applications, one can use out-of-sample prediction like the one shown in Table 6.10 to select m. The choice depends on y_t, the variable of interest. □

Remark: The out-of-sample prediction using the diffusion index approach of Stock and Watson (2002a) is obtained by the command `SWfore` of the `MTS` package. Details are in the following R demonstration. □

R Demonstration: Forecasting via diffusion index.

```
> da=read.table("m-unempstatesAdj.txt",header=T)
> dim(da)
[1] 416  50
> drate=diffM(da)  # First difference
> dim(drate)
[1] 415  50
> y=drate[5:415,1]  # First y_t series (Alabama)
> length(y)
[1] 411
> x=cbind(drate[4:414,],drate[3:413,],drate[2:412,],
          drate[1:411,]) #z_t series
> dim(x)
[1] 411 200
> m1=SWfore(y,x,350,10)
MSE of out-of-sample forecasts:  0.01510996
> m1=SWfore(y,x,350,20)
MSE of out-of-sample forecasts:  0.01274754
> m1=SWfore(y,x,350,30)
MSE of out-of-sample forecasts:  0.01136177
> m1=SWfore(y,x,350,50)
MSE of out-of-sample forecasts:  0.01044645
> m1=SWfore(drate[5:415,1],x,350,200)
MSE of out-of-sample forecasts:  0.02750807.
```

6.5.4 Dynamic Factor Models

In our discussion of factor models so far, we do not consider explicitly the dynamic dependence of the common factors $\boldsymbol{f}_t$. Forni et al. (2000, 2004, 2005) proposed the dynamic factor model

$$\boldsymbol{z}_t = \boldsymbol{L}(B)\boldsymbol{u}_t + \boldsymbol{\epsilon}_t, \tag{6.40}$$

where $\boldsymbol{\epsilon}_t$ is a white noise series with mean zero and covariance matrix $\boldsymbol{\Sigma}_\epsilon$, $\boldsymbol{u}_t$ is an m-dimensional process of orthonormal white noise, and $\boldsymbol{L}(B) = \boldsymbol{L}_0 + \boldsymbol{L}_1 B + \cdots + \boldsymbol{L}_r B^r$ is a $k \times m$ matrix polynomial of order r (might be infinity), and B

is the back-shift operator such that $B\boldsymbol{u}_t = \boldsymbol{u}_{t-1}$. By orthonormal white noise we mean that $\boldsymbol{u}_t$ satisfies $E(\boldsymbol{u}_t) = \boldsymbol{0}$, $\text{Cov}(\boldsymbol{u}_t) = \boldsymbol{I}_m$, and $\boldsymbol{u}_t$ is serially uncorrelated. Let $\boldsymbol{L}(B) = [L_{ij}(B)]$. Coefficients of each polynomial $L_{ij}(B)$ are assumed to be square summable. Let $\boldsymbol{c}_t = \boldsymbol{L}(B)\boldsymbol{u}_t = (c_{1t}, \ldots, c_{kt})'$, the model becomes

$$\boldsymbol{z}_t = \boldsymbol{c}_t + \boldsymbol{\epsilon}_t. \tag{6.41}$$

Forni et al. (2000) refer to c_{it} and ϵ_{it} as the common component and idiosyncratic component of z_{it}, respectively. The $\boldsymbol{u}_t$ process is referred to as the common shocks.

Assuming that $\boldsymbol{z}_t$ of Equation (6.40) is stationary, Forni et al. (2000) employed the spectral density matrix of $\boldsymbol{z}_t$ to propose an estimation method for the common components $\boldsymbol{c}_t$ when both k and the sample size T go to infinity in some given path. Under some regularity conditions, the authors show that the estimates converge to the true $\boldsymbol{c}_t$ in mean square. Details are given in Forni et al. (2000).

If we assume that the common factor $\boldsymbol{f}_t$ of the approximate factor model in Equation (6.34) follows a stationary VARMA(p, q) model

$$\boldsymbol{\phi}(B)\boldsymbol{f}_t = \boldsymbol{\theta}(B)\boldsymbol{u}_t,$$

where $\boldsymbol{\phi}(B)$ and $\boldsymbol{\theta}(B)$ are $m \times m$ AR and MA matrix polynomial, respectively. Then, the model becomes

$$\boldsymbol{z}_t = \boldsymbol{L}[\boldsymbol{\phi}(B)]^{-1}\boldsymbol{\theta}(B)\boldsymbol{u}_t + \boldsymbol{\epsilon}_t.$$

Therefore, we have $\boldsymbol{L}(B) = \boldsymbol{L}[\boldsymbol{\phi}(B)]^{-1}\boldsymbol{\theta}(B)$. This connection between the two models provides an alternative approach to estimate the dynamic factor models. Specifically, if the common shock $\boldsymbol{u}_t$ is of interest, it can be estimated by first building an approximate factor model for $\boldsymbol{z}_t$, then modeling the estimated latent process $\hat{\boldsymbol{f}}_t$ via VARMA models. The latter can be done by the methods discussed in the previous chapters.

6.5.5 Constrained Factor Models

Empirical applications of the factor models in Equations (6.29) and (6.34) often show that the estimated loading matrix exhibits certain characteristic patterns. For instance, consider the monthly log returns of ten U.S. stocks in Table 6.8. The fitted loading matrices via either PCA or MLE show that for each column of the loading matrices the loading weights are similar for companies in the same industrial sector. See the estimated loadings in Table 6.9. Motivated by the observation and in preference of parsimonious loading matrices for ease in interpretation, Tsai and Tsay (2010) propose a constrained factor model that can explicitly describe the observed patterns. For simplicity, assume that $E(\boldsymbol{z}_t) = \boldsymbol{0}$. A constrained factor model can be written as

$$\boldsymbol{z}_t = \boldsymbol{H}\boldsymbol{\omega}\boldsymbol{f}_t + \boldsymbol{\epsilon}_t, \tag{6.42}$$

where $\boldsymbol{f}_t$ and $\boldsymbol{\epsilon}_t$ are as defined before, $\boldsymbol{H}$ is a known $k \times r$ matrix, and $\boldsymbol{\omega}$ is a $r \times m$ matrix of unknown parameters. The matrix $\boldsymbol{H}$ is a constraint matrix with each column giving rise to a specific constraint. For instance, consider the ten stocks of Example 6.4. These ten stocks belong to three industrial sectors so that we can define $\boldsymbol{H} = [\boldsymbol{h}_1, \boldsymbol{h}_2, \boldsymbol{h}_3]$ with $\boldsymbol{h}_i$ denoting the ith sector. To demonstrate, let

$$\begin{aligned} \boldsymbol{h}_1 &= (1,1,1,1,0,0,0,0,0,0)' \\ \boldsymbol{h}_2 &= (0,0,0,0,1,1,1,0,0,0)' \\ \boldsymbol{h}_3 &= (0,0,0,0,0,0,0,1,1,1)'. \end{aligned} \tag{6.43}$$

Then, $\boldsymbol{h}_1$, $\boldsymbol{h}_2$, and $\boldsymbol{h}_3$ represent the sector of semiconductor, pharmaceutical, and investment bank, respectively. In practice, $\boldsymbol{H}$ is flexible and can be specified using prior information or economic theory.

The number of parameters in the loading matrix of the constrained factor model in Equation (6.42) is $r \times m$, whereas that of the unconstrained factor model is $k \times m$. When r is much smaller than k, the constrained factor model can be much more parsimonious. Under the same assumptions as the traditional factor models, we have

$$\begin{aligned} \boldsymbol{\Gamma}_z(0) &= \boldsymbol{H}\boldsymbol{\omega}\boldsymbol{\omega}'\boldsymbol{H}' + \boldsymbol{\Sigma}_\epsilon, \\ \boldsymbol{\Gamma}_z(\ell) &= \boldsymbol{H}\boldsymbol{\omega}\boldsymbol{\Gamma}_f(\ell)\boldsymbol{\omega}'\boldsymbol{H}', \end{aligned}$$

for the constrained factor model. These properties can be used to estimate the constrained factor models. In general, the constrained model can be also estimated by either the least-squares method or maximum likelihood method. The maximum likelihood method assumes normality and requires the additional constraint $\hat{\boldsymbol{\omega}}'\boldsymbol{H}'\hat{\boldsymbol{\Sigma}}_\epsilon\boldsymbol{H}\hat{\boldsymbol{\omega}}$ is a diagonal matrix. There is no closed-form solution, and an iterated procedure is used. Details are given in Tsai and Tsay (2010). Here, we briefly discuss the least-squares method.

Write the data of constrained factor model in Equation (6.42) as

$$\boldsymbol{Z} = \boldsymbol{F}\boldsymbol{\omega}'\boldsymbol{H}' + \boldsymbol{\epsilon}, \tag{6.44}$$

where $\boldsymbol{Z} = [\boldsymbol{Z}_1, \ldots, \boldsymbol{Z}_T]$ is a $T \times k$ data matrix, $\boldsymbol{F} = [\boldsymbol{f}_1, \ldots, \boldsymbol{f}_T]'$ is the $T \times m$ matrix of common factors, $\boldsymbol{\epsilon} = [\boldsymbol{\epsilon}_1, \ldots, \boldsymbol{\epsilon}_T]'$, and T is the sample size. Let tr($\boldsymbol{A}$) be the trace of matrix $\boldsymbol{A}$. The least-squares approach to estimate $\boldsymbol{F}$ and $\boldsymbol{\omega}$ is to minimize the objective function

$$\ell(\boldsymbol{F}, \boldsymbol{\omega}) = \text{tr}[(\boldsymbol{Z} - \boldsymbol{F}\boldsymbol{\omega}'\boldsymbol{H}')(\boldsymbol{Z} - \boldsymbol{F}\boldsymbol{\omega}'\boldsymbol{H}')'], \tag{6.45}$$

subject to the constraint $\boldsymbol{F}\boldsymbol{F} = T\boldsymbol{I}_m$. Using least-squares theory, we have $\hat{\boldsymbol{\omega}} = \boldsymbol{T}^{-1}(\boldsymbol{H}'\boldsymbol{H})^{-1}\boldsymbol{H}'\boldsymbol{Z}'\boldsymbol{F}$. Plugging $\hat{\boldsymbol{\omega}}$ into Equation (6.45) and using tr$(\boldsymbol{A}\boldsymbol{B}) = tr(\boldsymbol{B}\boldsymbol{A})$, we obtain the concentrated function

$$\ell(\boldsymbol{F}) = \text{tr}(\boldsymbol{Z}\boldsymbol{Z}') - \boldsymbol{T}^{-1}\text{tr}[\boldsymbol{F}'\boldsymbol{Z}\boldsymbol{H}(\boldsymbol{H}'\boldsymbol{H})^{-1}\boldsymbol{H}'\boldsymbol{Z}'\boldsymbol{F}].$$

This objective function is minimized when the second term is maximized. Applying Theorem 6 of Magnus and Neudecker (1999, p. 205) or proposition A.4 of Lütkepohl (2005, p. 672), we have $\hat{\boldsymbol{F}} = [\boldsymbol{g}_1, \ldots, \boldsymbol{g}_m]$, where $\boldsymbol{g}_i$ is an eigenvector of the ith largest eigenvalue λ_i of $\boldsymbol{Z}\boldsymbol{H}(\boldsymbol{H}'\boldsymbol{H})^{-1}\boldsymbol{H}'\boldsymbol{Z}'$. In practice, the eigenvectors are normalized so that $\hat{\boldsymbol{F}}'\hat{\boldsymbol{F}} = T\boldsymbol{I}_m$. The corresponding estimate of $\boldsymbol{\omega}$ becomes $\hat{\boldsymbol{\omega}} = T^{-1}(\boldsymbol{H}'\boldsymbol{H})^{-1}\boldsymbol{H}'\boldsymbol{Z}'\hat{\boldsymbol{F}}$. Finally, the covariance matrix of the noises is estimated by $\hat{\boldsymbol{\Sigma}}_\epsilon = T^{-1}\boldsymbol{Z}'\boldsymbol{Z} - \boldsymbol{H}\hat{\boldsymbol{\omega}}\hat{\boldsymbol{\omega}}'\boldsymbol{H}'$. Properties of the least-squares estimates are studied in Tsai and Tsay (2010) and the references therein. Tsai and Tsay (2010) also consider test statistics for checking constraints in a factor model and a partially constrained factor model.

Example 6.6 Consider again the monthly log returns of ten U.S. companies given in Table 6.8. We apply the constrained factor model to the returns using the constriant matrix $\boldsymbol{H}$ of Equation (6.43) using the least-squares method. The result is summarized in Table 6.11. For ease in comparison, Table 6.11 also reports the result of the principal component approach to the orthogonal factor model as shown in Table 6.9. From the table, we make the following observations. First, in this particular application, we have $r = m = 3$ and $k = 10$. The constrained factor model uses 9 parameters in the matrix $\boldsymbol{\omega}$. Its loading matrix is obtained by $\hat{\boldsymbol{L}} = \boldsymbol{H}\hat{\boldsymbol{\omega}}$. Therefore, the loading matrix applies the same weights to each stock in a given sector. This seems reasonable as the returns are standardized to have unit variance. On the other hand, the orthogonal factor model has 30 parameters in the loading matrix. The weights are close, but not identical for each stock in the same sector. Second, with

TABLE 6.11 Estimation Results of Constrained and Orthogonal Factor Models for the Monthly Log Returns of Ten U.S. Companies Given in Table 6.8, where L_i is the ith Column of L and $\Sigma_{\epsilon,i}$ Is the (i, i)th Element of Σ_ϵ

Stock	Constrained Model: $\hat{\boldsymbol{L}} = \boldsymbol{H}\hat{\boldsymbol{\omega}}$				Orthogonal Model: PCA			
Tick	L_1	L_2	L_3	$\Sigma_{\epsilon,i}$	L_1	L_2	L_3	$\Sigma_{\epsilon,i}$
TXN	0.761	0.256	0.269	0.283	0.788	0.198	0.320	0.237
MU	0.761	0.256	0.269	0.283	0.671	0.361	0.289	0.336
INTC	0.761	0.256	0.269	0.283	0.789	0.183	0.333	0.232
TSM	0.761	0.256	0.269	0.283	0.802	0.270	0.159	0.258
PFE	0.444	−0.675	0.101	0.337	0.491	−0.643	−0.027	0.345
MRK	0.444	−0.675	0.101	0.337	0.402	−0.689	0.226	0.312
LLY	0.444	−0.675	0.101	0.337	0.448	−0.698	0.061	0.309
JPM	0.738	0.055	−0.431	0.267	0.724	0.020	−0.345	0.357
MS	0.738	0.055	−0.431	0.267	0.755	0.053	−0.425	0.246
GS	0.738	0.055	−0.431	0.267	0.745	0.122	−0.498	0.182
e.v.	4.576	1.650	0.883		4.626	1.683	0.932	
	Variability explained: 70.6%				Variability explained: 72.4%			

the constraints, the estimated 3-factor model explains about 70.6% of the total variability. This is very close to the unconstrained 3-factor model, which explains 72.4% of the total variability. This confirms that the constraints are acceptable. Third, with the constraints, the first common factor f_{1t} represents the market factor, which is a weighted average of the three sectors employed. The second factor f_{2t} is essentially a weighted difference between the semiconductor and pharmaceutical sector. The third common factor f_{3t} represents the difference between investment banks and the other two sectors. □

Discussion: At the first glance, the constraint matrix $\boldsymbol{H}$ appears to be awkward. However, in applications, $\boldsymbol{H}$ is flexible and through which one can incorporate into the analysis substantive prior knowledge of the problem at hand. For instance, it is well known that the term structure of interest rates can be approximately described by trend, slope, and curvature. This knowledge can be used to specify $\boldsymbol{H}$. For simplicity, suppose $\boldsymbol{z}_t = (z_{1t}, z_{2t}, z_{3t})'$ with components representing interest rates with short, intermediate, and long maturities, respectively. Then, one can use $\boldsymbol{h}_1 = (1,1,1)'$, $\boldsymbol{h}_2 = (-1,0,1)'$, and $\boldsymbol{h}_3 = (1,-2,1)'$ to represent trend, slope, and curvature, respectively. The choices of $\boldsymbol{H}$ become more flexible when k is large. □

Remark: The least-squares estimation of constrained factor models can be carried out via the command `hfactor` in the `MTS` package. □

R Demonstration: Constrained factor models. Output edited.

```
> da=read.table("m-tenstocks.txt",header=T)
> rtn=log(da[,2:11]+1)  # compute log returns
> h1=c(1,1,1,1,rep(0,6))  # specify the constraints
> h2=c(0,0,0,0,1,1,1,0,0,0)
> h3=c(rep(0,7),1,1,1)
> H=cbind(h1,h2,h3)
> m1=hfactor(rtn,H,3)
[1] "Data are individually standardized"
[1] "First m eigenvalues of the correlation matrix:"
[1] 4.6256602 1.6827255 0.9320882
[1] "Variability explained: "
[1] 0.7240474
[1] "Loadings:"
        [,1]     [,2]     [,3]
 [1,] -0.368 -0.1532 -0.3331
 [2,] -0.313 -0.2792 -0.3000
 [3,] -0.368 -0.1419 -0.3465
 [4,] -0.374 -0.2090 -0.1649
 [5,] -0.229  0.4978  0.0278
 [6,] -0.188  0.5333 -0.2354
 [7,] -0.209  0.5401 -0.0632
```

```
 [8,] -0.338 -0.0153  0.3586
 [9,] -0.352 -0.0411  0.4421
[10,] -0.348 -0.0946  0.5173
[1] "eigenvalues of constrained part:"
[1] 4.576 1.650 0.883
[1] "Omega-Hat"
      [,1]     [,2]   [,3]
[1,] 0.761  0.2556  0.269
[2,] 0.444 -0.6752  0.101
[3,] 0.738  0.0547 -0.431
[1] "Variation explained by the constrained factors:"
[1] 0.7055665
[1] "H*Omega: constrained loadings"
       [,1]     [,2]   [,3]
 [1,] 0.761  0.2556  0.269
 [2,] 0.761  0.2556  0.269
 [3,] 0.761  0.2556  0.269
 [4,] 0.761  0.2556  0.269
 [5,] 0.444 -0.6752  0.101
 [6,] 0.444 -0.6752  0.101
 [7,] 0.444 -0.6752  0.101
 [8,] 0.738  0.0547 -0.431
 [9,] 0.738  0.0547 -0.431
[10,] 0.738  0.0547 -0.431
[1] "Diagonal elements of Sigma_epsilon:"
  TXN    MU  INTC   TSM   PFE   MRK   LLY   JPM    MS    GS
0.283 0.283 0.283 0.283 0.337 0.337 0.337 0.267 0.267 0.267
[1] "eigenvalues of Sigma_epsilon:"
 [1]  0.7632  0.6350  0.4539  0.3968  0.3741  0.2474  0.2051
 [8]  0.0262 -0.0670 -0.0902
```

6.5.6 Asymptotic Principal Component Analysis

In the analysis of panel data or repeated measurements, we often face the situation in which the number of components k is larger than the number of time periods T. This is referred to as a large k, small T problem. It has attracted much attention in the recent statistical and econometric literature. Some theory for statistical inference has been established based on the idea of k approaches infinity or both k and T increase to infinity at some proper rate. For approximate factor models, Connor and Korajczyk (1986, 1988) developed the approach of *asymptotic principal component analysis*. The name asymptotic PCA is somewhat confusing, but the idea is relatively simple. Simply put, asymptotic principal component analysis provides consistent estimation of a factor model under certain assumptions when $k \to \infty$. In this section, we introduce the asymptotic principal component analysis.

Consider, again, the approximate factor model

$$\boldsymbol{z}_t = \boldsymbol{L}\boldsymbol{f}_t + \boldsymbol{\epsilon}_t, \quad t = 1, \ldots, T, \tag{6.46}$$

where $\boldsymbol{z}_t = (z_{1t}, \ldots, z_{kt})'$ is a k-dimensional stationary process with $E(\boldsymbol{z}_t) = \boldsymbol{0}$, $\boldsymbol{f}_t = (f_{1t}, \ldots, f_{mt})'$ is an m-dimensional vector of common factors, $\boldsymbol{L}$ is the $k \times m$ loading matrix, and $\boldsymbol{\epsilon}_t$ are iid random vector with mean zero and covariance matrix $\boldsymbol{\Sigma}_\epsilon$. The common factor $\boldsymbol{f}_t$ and the noise term $\boldsymbol{\epsilon}_t$ are asymptotically uncorrelated, which will be made more precise later. Since $\boldsymbol{f}_t$ is latent, its scale is not uniquely determined. Asymptotic principal component analysis imposes the same scaling approach as Lam and Yao (2012) by assuming that

$$\frac{1}{k}\boldsymbol{L}'\boldsymbol{L} \to \boldsymbol{I}_m, \quad k \to \infty. \tag{6.47}$$

In addition, asymptotic PCA also assumes that, for any fixed t,

$$\frac{1}{k}\sum_{i=1}^{k} \epsilon_{it}^2 \to \sigma^2, \quad k \to \infty. \tag{6.48}$$

With $k > T$, we arrange the data matrix as

$$[\boldsymbol{z}_1, \ldots, \boldsymbol{z}_T] = \boldsymbol{L}[\boldsymbol{f}_1, \ldots, \boldsymbol{f}_T] + [\boldsymbol{\epsilon}_1, \ldots, \boldsymbol{\epsilon}_T].$$

For ease in reference, we write the prior equation as

$$\boldsymbol{D}_z = \boldsymbol{L}\boldsymbol{D}_f + \boldsymbol{D}_\epsilon, \tag{6.49}$$

where $\boldsymbol{D}_z$ is a $k \times T$ data matrix, $\boldsymbol{D}_f$ is an $m \times T$ factor matrix, and $\boldsymbol{D}_\epsilon$ is a $k \times T$ noise matrix. Premultiplying Equation (6.49) by its $\boldsymbol{D}_z'$ and dividing the result by k, we have

$$\frac{1}{k}\boldsymbol{D}_z'\boldsymbol{D}_z = \boldsymbol{D}_f'(\frac{1}{k}\boldsymbol{L}'\boldsymbol{L})\boldsymbol{D}_f + \frac{1}{k}\boldsymbol{D}_\epsilon'\boldsymbol{D}_\epsilon + \frac{1}{k}(\boldsymbol{G} + \boldsymbol{G}'), \tag{6.50}$$

where $\boldsymbol{G} = \boldsymbol{D}_\epsilon'\boldsymbol{L}\boldsymbol{D}_f$. The asymptotic principal component analysis further assumes that

$$\frac{1}{k}\boldsymbol{G} \to \boldsymbol{0}, \quad k \to \infty. \tag{6.51}$$

Equation (6.51) states that the common factor $\boldsymbol{f}_t$ and the noise term $\boldsymbol{\epsilon}_t$ are asymptotically uncorrelated.

Under the assumptions (6.47), (6.48), and (6.51), Equation (6.50) becomes

$$\boldsymbol{\Gamma}_0 = \boldsymbol{D}_f'\boldsymbol{D}_f + \sigma^2\boldsymbol{I}_T, \quad k \to \infty, \tag{6.52}$$

where $\mathbf{\Gamma}_0$ is the $T \times T$ cross-sectional covariance matrix of $\boldsymbol{z}_t$. Since $k > T$, the sample cross-sectional covariance matrix

$$\hat{\mathbf{\Gamma}}_0 = \frac{1}{k}\boldsymbol{D}_z'\boldsymbol{D}_z$$

is positive-definite. The asymptotic principal component analysis is to perform the traditional PCA based on the matrix $\hat{\mathbf{\Gamma}}_0$. From Equation (6.52), eigenvalues of $\mathbf{\Gamma}_0$ are the solutions to the determinant equation

$$0 = |\mathbf{\Gamma}_0 - \lambda \boldsymbol{I}_T| = |\boldsymbol{D}_f'\boldsymbol{D}_f - (\lambda - \sigma^2)\boldsymbol{I}_T|.$$

Thus, eigenvalues of $\boldsymbol{D}_f'\boldsymbol{D}_f$ can be obtained from those of $\mathbf{\Gamma}_0$. Consequently, the eigenvectors corresponding to the m largest eigenvalues of $\mathbf{\Gamma}_0$ can be used to form consistent estimates of the common factor $\boldsymbol{f}_t$. This provides theoretical justification for use of asymptotic principal component analysis in the analysis of factor models.

A simple example is helpful to demonstrate the asymptotic PCA. Consider, again, the case of a single common factor and two time periods. That is, we have

$$[\boldsymbol{z}_1, \boldsymbol{z}_2] = \boldsymbol{L}[f_1, f_2] + [\boldsymbol{\epsilon}_1, \boldsymbol{\epsilon}_2],$$

where $\boldsymbol{L} = (L_1, \ldots, L_k)'$ is a $k \times 1$ matrix of factor loadings. With two observations, we cannot estimate the covariance matrix Cov($\boldsymbol{z}_t$) efficiently if $k > 2$. However, we make use of the fact that k is increasing. That is, the size of the panel is expanding. In this particular instance, Equation (6.50) becomes

$$\frac{1}{k}\begin{bmatrix} \sum_{i=1}^k z_{i1}^2 & \sum_{i=1}^k z_{i1}z_{i2} \\ \sum_{i=1}^k z_{i1}z_{i2} & \sum_{i=1}^k z_{i2}^2 \end{bmatrix} = \begin{bmatrix} f_1 \\ f_2 \end{bmatrix}\left(\frac{1}{k}\sum_{i=1}^k L_i^2\right)[f_1, f_2]$$
$$+ \frac{1}{k}\begin{bmatrix} \sum_{i=1}^k \epsilon_{i1}^2 & \sum_{i=1}^k \epsilon_{i1}\epsilon_{i2} \\ \sum_{i=1}^k \epsilon_{i1}\epsilon_{i2} & \sum_{i=1}^k \epsilon_{i2}^2 \end{bmatrix}.$$

Since $\boldsymbol{\epsilon}_1$ is uncorrelated with $\boldsymbol{\epsilon}_2$ and under the assumptions of Equations (6.47) and (6.48), the prior equation shows, as $k \to \infty$,

$$\mathbf{\Gamma}_0 = \begin{bmatrix} f_1^2 & f_1 f_2 \\ f_1 f_2 & f_2^2 \end{bmatrix} + \sigma^2 \boldsymbol{I}_2.$$

It is then easy to verify that the two eigenvalues of $\mathbf{\Gamma}_0$ are σ^2 and $\sigma^2 + f_1^2 + f_2^2$. In addition, the eigenvector associated with the larger eigenvalue is proportional to $(f_1, f_2)'$. Consequently, one can recover the common factor f_t asymptotically.

Discussion: From the assumption of Equation (6.47), we have

$$\lim_{k\to\infty} \frac{1}{k}\sum_{i=1}^{k} L_{ij}^2 = 1, \quad j = 1, \ldots, m.$$

This implies that, for each $j \in \{1, \ldots, m\}$, L_{ij} should be nonzero infinitely often as k increases. In other words, there exists no positive integer h such that $L_{ij} = 0$ for all $i > h$. In practice, the assumption simply ensures that as the size k of the panel increases, the new individuals of the panel should continue to provide information about the each common factor f_{jt}. This should be obvious if one wishes to obtain a consistent estimate of the common factor.

In our discussion of factor models, we assume $E(\boldsymbol{z}_t) = \mathbf{0}$. For the traditional factor models, this assumption is not critical because the mean of $\boldsymbol{z}_t$ can be estimated consistently by the sample mean when the sample size T increases. On the other hand, for the asymptotic principal component analysis, the assumption becomes essential because T can be quite small. In financial application, $\boldsymbol{z}_t$ may consist of excess returns of multiple assets. In this case, theory may justify the zero-mean assumption. □

Example 6.7 Consider the monthly log returns, without dividends, for the 100 stocks comprising the S&P 100 index in 2011. In this particular instance, we have $T = 12$ and $k = 100$. We apply asymptotic principal component analysis to seek common factors that drove the market in 2011. Ticker symbols for the companies involved are given in Table 6.12. Figure 6.16 shows the scree plot of the asymptotic principal component analysis. From the plot, we chose $m = 3$ common factors because the first three components explain about 81% of the cross-sectional variability. The 12 observations of the common factors are given in the following R demonstration. Figure 6.17 shows the time plots of the loading matrix of the asymptotic principal component analysis, where each panel corresponds to a column of the loading matrix. The time plot of the first loading column shows a big spike at

TABLE 6.12 Ticker Symbols of Components of S&P 100 Index in 2011: In the Order of CRSP Permno

ORCL	MSFT	HON	EMC	DELL	KO	DD	XOM	GD	GE	IBM	PEP
MO	COP	AMGN	SLB	CVX	AAPL	TXN	CVS	UTX	PG	SO	CAT
CL	BMY	WAG	BA	ABT	DOW	LMT	EXC	PFE	EMR	JNJ	MMM
MRK	HNZ	HAL	AEP	RTN	F	DIS	HPQ	BAX	OXY	WMB	WFC
APA	MCD	JPM	UNP	TGT	BK	LLY	WMT	NKE	AXP	INTC	BAC
MDT	FDX	LOW	NSC	VZ	T	USB	HD	MS	APC	C	BHI
CSCO	QCOM	GILD	TWX	SBUX	ALL	SPG	COF	FCX	BRK	NOV	AMZN
EBAY	GS	COST	DVN	UPS	MET	MON	KFT	ACN	CMCSA	GOOG	NWSA
MA	PM	V	UNH								

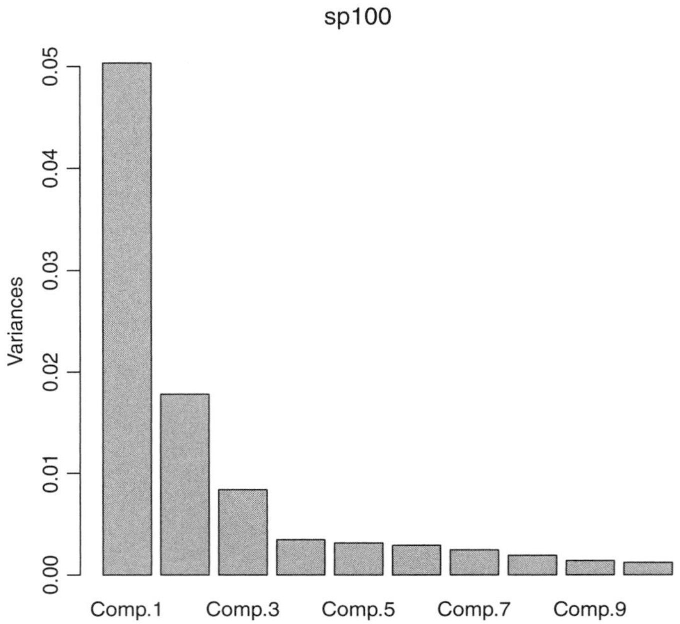

FIGURE 6.16 Scree plot of asymptotic principal component analysis for the monthly log returns of components of S&P 100 index in 2011.

index 71, which corresponds to Citigroup whereas the third loading column has a large value for the Freeport-McMoRan Copper & Gold. It should be noted that the common factors and the loading matrix are not uniquely determined. They can be transformed by any 3×3 orthonormal matrix. □

Remark: Asymptotic principal component analysis is carried out by the command `apca` of the `MTS` package. □

R Demonstration: Asymptotic PCA. Output edited.

```
> rtn=read.table("m-sp100y2011.txt",header=T)
> sp100=apca(rtn,3)
Importance of components:
                          Comp.1    Comp.2     Comp.3     Comp.4     Comp.5
Standard deviation     0.2243715 0.1333795 0.09154043 0.05873897 0.05594621
Proportion of Variance 0.5298019 0.1872217 0.08818687 0.03631037 0.03293967
Cumulative Proportion  0.5298019 0.7170236 0.80521047 0.84152084 0.87446052
                           Comp.6    Comp.7     Comp.8     Comp.9    Comp.10
Standard deviation     0.05387705 0.0495897 0.04381794 0.03738959 0.03489291
Proportion of Variance 0.03054819 0.0258798 0.02020607 0.01471226 0.01281305
Cumulative Proportion  0.90500871 0.9308885 0.95109459 0.96580685 0.97861990
                          Comp.11    Comp.12
```

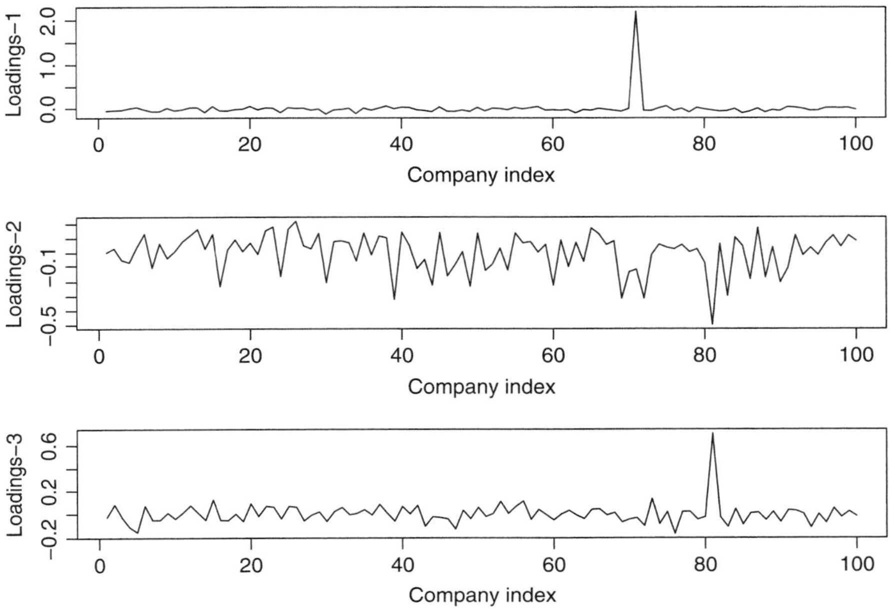

FIGURE 6.17 Time plots of the columns of estimated loading matrix for the monthly log returns of components of S&P 100 index in 2011.

```
Standard deviation      0.03228930 0.03144792
Proportion of Variance 0.01097224 0.01040787
Cumulative Proportion  0.98959213 1.00000000
> screeplot(sp100)
> factors=sp100$factors
> print(round(factors,4))
         [,1]    [,2]    [,3]
 [1,] -0.0136 -0.1309 -0.2613
 [2,] -0.0258  0.2105 -0.9075
 [3,] -0.0206  0.0045  0.1139
 [4,]  0.0079  0.1165 -0.0343
 [5,]  0.9963  0.0505 -0.0107
 [6,]  0.0118  0.0204  0.0208
 [7,] -0.0228 -0.0139  0.0384
 [8,] -0.0337  0.4671  0.2725
 [9,] -0.0208  0.6372 -0.0512
[10,]  0.0167 -0.4830 -0.1069
[11,] -0.0586  0.1457  0.0457
[12,] -0.0117  0.2076  0.0390
> loadings=sp100$loadings
> which.max(loadings[,1])
 C
> which.max(loadings[,3])
FCX
```

6.6 CLASSIFICATION AND CLUSTERING ANALYSIS

Classification and clustering analysis are commonly used in statistical applications concerning many subjects or units. Classification attempts to assign new subjects into some existing groups based on observed characteristics. Clustering analysis, on the other hand, divides subjects into different clusters so that the within-cluster variations are small compared with the variation between clusters. In this section, we discuss these two methods in the content of multivariate time series analysis and demonstrate their applications with some examples.

6.6.1 Clustering Analysis

Consider a panel of k time series each with T observations. Denote the data by $\{z_{it}|i = 1,\ldots,k; t = 1,\ldots,T\}$. This is a balanced panel because all time series have T observations. Suppose that there are H clusters denoted by $\{C_1,\ldots,C_H\}$. These clusters are mutually exclusive so that $C_i \cap C_j$ is an empty set for $i \neq j$. The goal of clustering analysis is to assign each series into one of the clusters based on some similarity measure. There are many methods available in the statistical literature for clustering analysis. See, for instance, Johnson and Wichern (2007). These methods employ certain distance measure and are classified as hierarchical or nonhierarchical clustering methods. The most well-known hierarchical clustering method is perhaps the *Agglomerative hierarchical methods*. For nonhierarchical methods, *k-means* is commonly used. Readers are referred to classical textbooks on multivariate analysis for descriptions of clustering methods; for example, Johnson and Wichern (2007). In this section, we consider the model-based approach to classify multiple time series. Recent developments, especially those with Markov chain Monte Carlo (MCMC) techniques, show that model-based clustering analysis works well in applications. See, for instance, Frühwirth-Schnatter and Kaufmann (2008) and Frühwirth-Schnatter (2006).

For simplicity, we assume that the time series z_{it} are autoregressive processes. This simplifying assumption can be relaxed with some increase in computational intensity. For instance, one can entertain ARMA models and use the Metropolis–Hasting algorithm to handle the MA parameters that are nonlinear in estimation. The model-based clustering method considered assumes that each series in a given cluster follows the same AR(p) model. Specifically, we can write the model as

$$z_{it} = \phi_0^{(h)} + \sum_{v=1}^{p} \phi_v^{(h)} z_{i,t-v} + a_t^{(h)}, \quad i \in C_h; \quad h = 1,\ldots,H, \tag{6.53}$$

$$\boldsymbol{S}_T = (s_{1T}, s_{2T}, \ldots, s_{kT})', \tag{6.54}$$

where T denotes the sample size, $i \in C_h$ denotes that the ith series is in the hth cluster, and $\boldsymbol{S}_T$ is a selection (or allocation) vector with $s_{iT} \in \{1, 2, \ldots, H\}$. Here, $s_{iT} = h$ denotes that z_{it} belongs to the hth cluster, that is, $i \in C_h$. The innovations $\{a_t^{(h)}\}$ in Equation (6.53) are an iid sequence of random variables with mean zero and variance σ_h^2, and the series $\{a_t^{(h)}\}$ and $\{a_t^{(v)}\}$ are independent for $h \neq v$. In this setting, the parameter of the hth cluster is $\boldsymbol{\theta}_h = (\phi_0^{(h)}, \phi_1^{(h)}, \ldots, \phi_p^{(h)}, \sigma_h^2)'$. All series in cluster C_h follow the same AR(p) model. The selection vector $\boldsymbol{S}_T$ in Equation (6.54) is governed by a prior probability $\boldsymbol{p} = (p_1, \ldots, p_H)'$ satisfying $0 < p_h < 1$ and $\sum_{h=1}^{H} p_h = 1$, and information available at time T (inclusive). In the absence of any preference, one can use the uniform prior with $p_h = 1/H$ for all h for all series. We use the subscript T to emphasize that the selection vector is time-varying as it depends on the sample size.

For the clustering model in Equations (6.53) and (6.54), the parameters are $\boldsymbol{\Theta} = (\boldsymbol{\theta}_1, \ldots, \boldsymbol{\theta}_H, \boldsymbol{S}_T)$. If desired, the number of clusters H can also be treated as a parameter and estimated jointly with $\boldsymbol{\Theta}$. In this section, we simply assume H is given *a priori* and try several values of H in an application to select a clustering result. Frühwirth-Schnatter and Kaufmann (2008) use marginal log likelihood to select H. We also assume that the AR order p is given and the same for all clusters. Again, this assumption can be easily relaxed.

6.6.2 Bayesian Estimation

The likelihood function of the clustering model in Equations (6.53) and (6.54) is high dimensional because it involves the selection space for $\boldsymbol{S}_T$. A simpler approach to estimate the model is to use MCMC methods. This is particularly so for the AR model in Equation (6.53) for which all conditional posterior distributions are well known so that random draws of the MCMC iterations can be carried out efficiently.

Let $\boldsymbol{Z}$ denote the collection of the data. A natural decomposition of the estimation problem is as follows:

1. For each cluster C_h, estimate the AR parameters $\boldsymbol{\phi}^{(h)} = (\phi_0^{(h)}, \phi_1^{(h)}, \ldots, \phi_p^{(h)})'$ conditional on the data and other parameters.
2. For each cluster C_h, estimate the residual variance σ_h^2 conditional on all data and other parameters.
3. The conditional posterior distribution of $\boldsymbol{S}_T$ given the data and the other parameters.

In what follows, we discuss the conditional posterior distributions used in implementing MCMC estimation of the clustering models in Equations (6.53) and (6.54).

6.6.2.1 *Posterior Distribution of $\boldsymbol{\phi}^{(h)}$*

For each cluster C_h, let n_h be the number of time series in C_h. We pool these n_h time series together to obtain the conditional posterior distribution of $\boldsymbol{\phi}^{(h)}$. Specifically, for $i \in C_h$, consider the AR(p) data frame

$$\begin{bmatrix} z_{i,p+1} \\ z_{i,p+2} \\ \vdots \\ z_{i,T} \end{bmatrix} = \begin{bmatrix} 1 & z_{i,p} & z_{i,p-1} & \cdots & z_{i,1} \\ 1 & z_{i,p+1} & z_{i,p} & \cdots & z_{i,2} \\ \vdots & \vdots & \vdots & & \vdots \\ 1 & z_{i,T-1} & z_{i,T-2} & \cdots & z_{i,T-p} \end{bmatrix} \begin{bmatrix} \phi_0^{(h)} \\ \phi_1^{(h)} \\ \vdots \\ \phi_p^{(h)} \end{bmatrix} + \begin{bmatrix} a_{i,p+1}^{(h)} \\ a_{i,p+2}^{(h)} \\ \vdots \\ a_{i,T}^{(h)} \end{bmatrix}.$$

Denote this equation as

$$\boldsymbol{Z}_{h,i} = \boldsymbol{X}_{h,i}\boldsymbol{\phi}^{(h)} + \boldsymbol{a}_{h,i}. \tag{6.55}$$

Stacking the data frame for all series in C_h together, we have

$$\begin{bmatrix} \boldsymbol{Z}_{h,1} \\ \vdots \\ \boldsymbol{Z}_{h,n_h} \end{bmatrix} = \begin{bmatrix} \boldsymbol{X}_{h,1} \\ \vdots \\ \boldsymbol{X}_{h,n_h} \end{bmatrix} \boldsymbol{\phi}^{(h)} + \begin{bmatrix} \boldsymbol{a}_{h,1} \\ \vdots \\ \boldsymbol{a}_{h,n_h} \end{bmatrix}.$$

Denote the prior equation as

$$\boldsymbol{Z}_h = \boldsymbol{X}_h\boldsymbol{\phi}^{(h)} + \boldsymbol{a}_h. \tag{6.56}$$

The ordinary least-squares estimate of $\boldsymbol{\phi}^{(h)}$ given the data is

$$\hat{\boldsymbol{\phi}}^{(h)} = (\boldsymbol{X}'_h\boldsymbol{X}_h)^{-1}(\boldsymbol{X}'_h\boldsymbol{Z}_h),$$

with covariance matrix

$$\hat{\boldsymbol{\Sigma}}_h = \sigma_h^2(\boldsymbol{X}'_h\boldsymbol{X}_h)^{-1}.$$

Suppose that the prior distribution of $\boldsymbol{\phi}^{(h)}$ is multivariate normal with mean $\boldsymbol{\phi}_{h,o}$ and covariance matrix $\boldsymbol{\Sigma}_{h,o}$, where the subscript "o" is used to denote prior distribution. Then, under the normality assumption, the poster distribution of $\boldsymbol{\phi}^{(h)}$ given the data and other parameters is also multivariate normal with mean $\boldsymbol{\phi}_*^{(h)}$ and covariance $\boldsymbol{\Sigma}_{h,*}$, where the subscript "*" is used to denote posterior distribution. The posterior mean and covariance matrix are given by

$$\boldsymbol{\Sigma}_{h,*}^{-1} = \boldsymbol{\Sigma}_{h,o}^{-1} + \hat{\boldsymbol{\Sigma}}_h^{-1}, \quad \boldsymbol{\phi}_*^{(h)} = \boldsymbol{\Sigma}_{h,*}[\boldsymbol{\Sigma}_{h,o}^{-1}\boldsymbol{\phi}_{h,o} + \hat{\boldsymbol{\Sigma}}_h^{-1}\hat{\boldsymbol{\phi}}^{(h)}]. \tag{6.57}$$

Equation (6.57) follows directly results of conjugate priors in Bayesian inference; see, for instance, Tsay (2010, Chapter 12). In practice, noninformative prior is often used by specifying $\boldsymbol{\phi}_{h,o} = \boldsymbol{0}$ and $\boldsymbol{\Sigma}_{h,o}$ being a diagonal matrix with large diagonal elements.

6.6.2.2 *Posterior Distribution of* σ_h^2

Given the data and other parameters, σ_h^2 is the variance of $a_t^{(h)}$ in Equation (6.55). Specifically, from Equation (6.56), we have

$$\boldsymbol{a}_h = \boldsymbol{Z}_h - \boldsymbol{X}_h\boldsymbol{\phi}^{(h)},$$

so that the sample estimate of σ_h^2 is

$$\hat{\sigma}_h^2 = \boldsymbol{a}_h'\boldsymbol{a}_h/[n_h(T-p)],$$

where $n_h(T-p)$ is the number of rows of the matrices in Equation (6.56) or the number of effective sample size of cluster C_h. Under the normality assumption, $n_h(T-p)\hat{\sigma}_h^2/\sigma^2$ is distributed as $\chi^2_{n_h(T-p)}$. Equivalently, $n_h(T-p)\hat{\sigma}_h^2/\sigma_h^2$ is distributed as Gamma$[n_h(T-p)/2, 2]$ or $\sigma_h^2/(\boldsymbol{a}_h'\boldsymbol{a}_h)$ is distributed as inverse Gamma$[n_h(T-p)/2, 1/2]$. Depending on the choice of the three equivalent distributions for $\boldsymbol{a}_h'\boldsymbol{a}_h/\sigma_h^2$, one can specify a conjugate prior distribution for σ_h^2. In this section, we use $\sigma_h^2/(\boldsymbol{a}_h'\boldsymbol{a}_h)$ being an inverse Gamma$[n_h(T-p)/2, 1/2]$. The conjugate prior is that σ_h^2 is distributed as an inverse Gamma$[v_o/2, u_o/2]$, where v_o and u_o are positive real numbers. Consequently, the posterior distribution of σ_h^2 is inverse Gamma$[\alpha, \theta]$, where

$$\alpha = [v_o + n_h(T-p)]/2, \quad \theta = \frac{u_o}{2} + \frac{1}{2}(\boldsymbol{a}_h'\boldsymbol{a}_h). \tag{6.58}$$

This result can be found in most Bayesian textbooks. In applications, $u_o = 1$ and v_o a small positive integer are commonly used.

Remark: A random variable Y follows an inverse Gamma$[\alpha, \beta]$ distribution with shape parameter α and scale parameter β if its pdf is

$$f(y|\alpha, \beta) = \frac{\beta^\alpha}{\Gamma(\alpha)} y^{-\alpha-1} \exp(-\beta/y), \quad y \geq 0,$$

where $\Gamma(.)$ denotes the Gamma function. It is easy to see that the random variable $X = 1/Y$ has a pdf

$$\begin{aligned} f(x|\alpha, \beta) &= \frac{\beta^\alpha}{\Gamma(\alpha)} x^{\alpha-1} \exp(-\beta x) \\ &= \frac{1}{\theta^\alpha \Gamma(\alpha)} x^{\alpha-1} \exp(-x/\theta), \end{aligned}$$

where $\theta = 1/\beta$. This latter expression is the pdf of Gamma(α, θ). Therefore, $X = 1/Y$ is distributed as Gamma$(\alpha, 1/\beta)$ if Y is inverse Gamma(α, β). □

6.6.2.3 Posterior Distribution of S_T

Given the data and other parameters, s_{iT} for the ith time series can be determined by posterior probabilities as follows. The likelihood that z_{it} belongs to the hth cluster C_h can be measured by the likelihood function of z_{it} evaluated at the parameters of C_h. Specifically, let $\boldsymbol{a}_{h,i} = (a_{i,p+1}, a_{i,p+2}, \ldots, a_{i,T})'$ be the residuals of z_{it} evaluated by using the parameters of C_h, that is,

$$\boldsymbol{a}_{h,i} = \boldsymbol{Z}_{h,i} - \boldsymbol{X}_{h,i}\boldsymbol{\phi}^{(h)},$$

from Equation (6.55). The corresponding log likelihood function is then

$$\ell(h,i) = \sum_{t=p+1}^{T} \log[f(a_{i,t}|\sigma_h^2)], \tag{6.59}$$

where $f(x|\sigma_h^2)$ denotes the pdf of a normal distribution with mean zero and variance σ_h^2. Let $\boldsymbol{P}_0 = (p_{01}, \ldots, p_{0H})'$ denote the prior probabilities for H clusters, where $\sum_{i=1}^{H} p_{0i} = 1$ and $p_{0i} > 0$. Then, the posterior probability p_{*i} that z_{it} belongs to the hth cluster C_h is given by

$$\log(p_{*i}) \propto \log(p_{0i}) + \ell(h,i), \quad h = 1, \ldots, H, \tag{6.60}$$

where $\propto$ denotes proportional. Consequently, we can draw s_{iT} from $\{1, 2, \ldots, H\}$ using posterior probabilities of Equation (6.60). In applications, the prior probabilities can be either uniform or a random draw from a Dirichlet distribution.

6.6.3 An MCMC Procedure

Using the conditional posterior distributions discussed in the prior section, one can perform model-based clustering analysis using the following procedure.

1. Initialization: One can either draw the initial estimates from the prior distributions or apply the traditional *k-means* method with H clusters to the k time series, each with T observations. The mean series of the hth cluster, called center of the cluster, can be used to obtain initial estimates of an AR(p) model, namely, $\boldsymbol{\phi}_0^{(h)}$ and the residual variance $\hat{\sigma}_{h,0}^2$.
2. For iteration $m = 1, \ldots, M$, where M is a prespecified positive integer:
   (a) Update the classification $\boldsymbol{S}_T^{(m)}$ using the conditional posterior probabilities in Equation (6.60) with parameter estimates $\boldsymbol{\phi}_{m-1}^{(h)}$ and $\sigma_{h,m-1}^2$.
   (b) Update AR coefficient estimates $\boldsymbol{\phi}_m^{(h)}$ by drawing from the conditional posterior distribution in Equation (6.57) with parameters $\boldsymbol{S}_T^{(m)}$ and $\sigma_{h,m-1}^2$.
   (c) Update the residual variances $\sigma_{h,m}^2$ by drawing from the conditional posterior distributions in Equation (6.58) with parameters $\boldsymbol{S}_T^{(m)}$ and $\boldsymbol{\phi}_m^{(h)}$.
3. Remove burns-in sample: Discard the first N iterations of the random draws and use the remaining draws of $M - N$ iterations to make inference.

Under some general regularity conditions, the remaining draws can be treated as a random sample from the joint posterior distribution of the unknown parameters. This is a typically example of MCMC methods commonly used in the statistical literature. Of course, one needs to check the convergence of the MCMC iterations. Readers are referred to Gamerman and Lopes (2006) and the references therein for details. In our demonstration given later, we use several choices of M and N to confirm that the results are stable.

6.6.3.1 *Label Switching*

In the discussion of model-based clustering analysis, we assume implicitly that the clusters are identified. However, in actual implementation, the labeling of the clusters requires some attention because the clusters may not have a natural ordering. In the extreme case, there are $H!$ possible orderings of the H clusters. To overcome the difficulty, one applies the *k-means* method to sort out the labeling problem. Specifically, for each MCMC iteration, we have H sets of AR parameters, namely, $\boldsymbol{\theta}_h^{(m)} = (\phi_{0m}^{(h)}, \phi_{1m}^{(h)}, \ldots, \phi_{pm}^{(h)}, \sigma_{h,m}^2)'$. Consider the AR parameters for iterations $m = N+1, \ldots, M$. We apply the *k-means* method to these AR parameters to obtain H clusters. The center of a resulting cluster provides the posterior means of the AR parameters for that cluster. We can also use the clustering result to identity members of a given cluster.

Example 6.8 To demonstrate the model-based clustering analysis, we consider the monthly unemployment rates of the 50 states in the United States. The data were analyzed in Example 6.5. As before, the unemployment rates for Arizona were adjusted due to a level shift for the first 92 observations and the rates for Louisiana and Mississippi are also adjusted because of the effects of Hurricane Katrina in 2005. Here, we employ an AR(4) model to classify the 50 differenced time series into 4 clusters. Let z_{it} denote the first-differenced series of the monthly unemployment rate of the ith state. These differenced series are shown in Figure 6.14. The sample period is from February 1976 to August 2011 for 415 observations. In this particular application, we have $k = 50$, $T = 415$, $p = 4$, and $H = 4$.

Using noninformative priors,

$$\boldsymbol{\phi}^{(h)} \sim N(\mathbf{0}, 10^4 \boldsymbol{I}_5), \; v_o = 4, \; u_o = 1, \; \boldsymbol{P}_0 = (1,1,1,1)'/4, \tag{6.61}$$

we apply the proposed MCMC method to perform model-based clustering analysis. Let the number of burn-in and total MCMC iterations be N and M, respectively. We use $(N, M) = (2000, 3000)$ and (3000,5000) in this example. The classification results are identical and the associated parameter estimates are very close for these two choices of N and M. The posterior means of the parameters and member states of the clusters are given in Table 6.13. These results shown are based on $(N, M) = (3000, 5000)$. From the table, it is seen that geographical proximity is not a main factor in the classification. As expected, the posterior means of the AR parameters show marked differences between the four clusters. Finally, for

TABLE 6.13 Results of Model-Based Clustering Analysis for the First-Differenced Series of Monthly Unemployment Rates of the 50 States in United States. Sample Period from February 1976 to August 2011

Cluster	Posterior Means of Model Parameters					
h	$\phi_0^{(h)}$	$\phi_1^{(h)}$	$\phi_2^{(h)}$	$\phi_3^{(h)}$	$\phi_4^{(h)}$	$\hat{\sigma}_h^2$
1	0.0013	0.699	0.398	−0.128	−0.189	0.082
2	0.0008	0.498	0.474	−0.001	−0.174	0.069
3	0.0020	0.314	0.298	0.125	−0.026	0.114
4	0.0023	0.391	0.268	0.038	−0.058	0.151

Cluster	Member States of the Cluster
1(18 states)	AL, AK, CO, HI, IL, KY, LA, ME, MI, MO, NV, NY, OR, SC, TN, VT, WV, WY
2(18 states)	AR, CA, CT, ID, KS, MA, MN, MT, NE, NH, NJ, NM, ND, RI, SD, UT, VA, WA
3(6 states)	AZ, FL, IA, MD, PA, TX
4(8 states)	DE, GA, IN, MS, NC, OH, OK, WI

Priors used are given in Equation (6.61). The number of iterations used is (N, M) = (3000,5000).

comparison purpose, we also consider the case of $H = 3$. The resulting marginal log likelihood is lower, supporting the choice of four clusters. See the following R demonstration.

The results of clustering analysis can be used in several ways. For instance, one can use the fitted AR(4) model to produce forecasts of unemployment rates for all states in the same cluster. This provides a parsimonious approach to forecasting. As shown in Wang et al. (2013), such a model-based clustering procedure outperforms many competing models in an out-of-sample forecasting comparison. □

Remark: Model-based clustering is carried out by the program `MBcluster` written by Mr. Yongning Wang. The author wishes to thank Mr. Wang for his assistance and for making the program available. The basic input includes data, p, H, and `mcmc`, which consists of burn-in and the remaining number of iterations of the MCMC procedure. □

R Demonstration: Model-based clustering. Output edited.

```
> da=read.table("m-unempstatesAdj.txt",header=T)
> dim(da)
[1] 416  50
> zt=apply(da,2,diff)
> mcmc=list(burnin=3000,rep=2000)
```

```
> m4=MBcluster(zt,4,4,mcmc=mcmc)
Use default priors

Estimation for Cluster Parameters:
Number of Clusters: K= 4
Number of Lags in AR model: p= 4
               phi 0   phi 1   phi 2       phi 3     phi 4    sigma
Cluster 1 0.00127731 0.69889 0.39798 -0.12770645 -0.188757 0.082096
Cluster 2 0.00080087 0.49824 0.47396 -0.00050977 -0.173598 0.069493
Cluster 3 0.00197971 0.31440 0.29790  0.12533160 -0.025974 0.114214
Cluster 4 0.00226912 0.39076 0.26922  0.03767486 -0.057572 0.150859

Classification Probabilities:
   Cluster 1 Cluster 2 Cluster 3 Cluster 4
AL    0.9995    0.0005    0.0000    0.0000
AK    0.9010    0.0990    0.0000    0.0000
AZ    0.0000    0.0000    0.9945    0.0055
AR    0.0000    1.0000    0.0000    0.0000
.....
WI    0.0000    0.0000    0.0010    0.9990
WY    0.7775    0.2225    0.0000    0.0000
Classification:
Cluster  1 :
Number of members:  18
AL AK CO HI IL KY LA ME MI MO NV NY OR SC TN VT WV WY
Cluster  2 :
Number of members:  18
AR CA CT ID KS MA MN MT NE NH NJ NM ND RI SD UT VA WA
Cluster  3 :
Number of members:  6
AZ FL IA MD PA TX
Cluster  4 :
Number of members:  8
DE GA IN MS NC OH OK WI

Marginal LogLikelihood: 20812.45

> mcmc=list(burnin=3000,rep=2000)  ## 3 clusters
> m1=MBcluster(zt,4,3,mcmc=mcmc)
Use default priors

Estimation for Cluster Parameters:
Number of Clusters: K= 3
Number of Lags in AR model: p= 4
               phi 0   phi 1   phi 2       phi 3    phi 4    sigma
Cluster 1 0.00130630 0.69890 0.39728 -0.12662853 -0.18929 0.082118
Cluster 2 0.00081792 0.49835 0.47399 -0.00016417 -0.17423 0.069498
Cluster 3 0.00218949 0.37009 0.27833  0.06288030 -0.04956 0.135998

Marginal LogLikelihood: 20697.27
```

EXERCISES

6.1 Housing markets in the United States have been under pressure since the 2007 subprime financial crisis. In this problem, we consider the housing starts for the West and South regions of U.S. Census. The data are from FRED of the Federal Reserve Bank of St. Louis, in thousands of units and not seasonally adjusted. The sample period is from January 1959 to December 2012 for 648 observations. See the file `m-houst-nsa.txt`, which contains data for all four regions. Let $\boldsymbol{z}_t$ be the log series of the housing starts for West and South regions.

- **(a)** Build a seasonal VARMA model for $\boldsymbol{z}_t$. Perform model checking and discuss the implications of the model.
- **(b)** Use the fitted model to obtain 1-step to 3-step forecasts of $\boldsymbol{z}_t$ at the forecast origin December 2012.

6.2 Consider the monthly unemployment rate and the industrial production index of United States from January 1967 to December 2012 for 552 observations. The data are seasonally adjusted and obtained from FRED of the Federal Reserve Bank of St. Louis. We focus on modeling $\boldsymbol{z}_t$, which consists of the change series of unemployment rate and industrial production index. We also consider two input variables. They are the PMI composite index of the Institute of Supply Management and total capacity utilization (TCU). These series are also seasonally adjusted. Let $\boldsymbol{x}_t$ be the change series of PMI index and TCU. The data are in the file `m-unippmitcu-6712.txt`.

- **(a)** Build a VAR model for $\boldsymbol{z}_t$, including model checking.
- **(b)** Build a VARX model for $\boldsymbol{z}_t$ using the input variable $\boldsymbol{x}_t$. In model selection, you may use maximum $p = 11$ and maximum $m = 6$. Check the fitted model.
- **(c)** Build a regression model with time series errors for $\boldsymbol{z}_t$ using $\boldsymbol{w}_t = (\boldsymbol{x}_t', \boldsymbol{x}_{t-1}')'$ as the input variable. Check the fitted model.
- **(d)** Compare the three fitted models for $\boldsymbol{z}_t$.

6.3 Consider the monthly data from the Institute of Supply Management from 1988 to 2012 for exactly 300 observations. The variables used are (a) production index of manufacturing, (b) inventories index, (c) new orders index, and (d) supplier deliveries index. The data are available from FRED of the Federal Reserve Bank of St. Louis and are also in the file `m-napm-8812.txt`.

- **(a)** From the data, we have $\boldsymbol{z}_{151} = (53.7, 48.5, 52.0, 53.9)'$. Treat this data point as missing. Build a model for $\boldsymbol{z}_t$ and estimate the missing values.
- **(b)** Suppose that we have $\boldsymbol{z}_{151} = (53.7, 48.5, X, X)'$, where X denotes missing values. Build a model for $\boldsymbol{z}_t$ and estimate the missing values.
- **(c)** Compare the estimates of $z_{151,3}$ and $z_{151,4}$ of parts (a) and (b). Comment on these estimates.

6.4 Consider the monthly simple excess returns of 10 U.S. stocks from January 1990 to December 2003 for 168 observations. The 3-month Treasury bill rate on the secondary market is used to compute the excess returns. The data are in the file `m-excess-10c-9003.txt`. The 10 tick symbols of the stocks are

(a) pharmaceutical: ABT, LLY, MRK, and PFE; (b) automobile: F and GM; (c) oil companies: BP, CVX, RD, and XOM. Let $\boldsymbol{z}_t$ be the 10-dimensional excess return series.

(a) Perform PCA on $\boldsymbol{z}_t$. Obtain the scree plot and select the number of command factors.

(b) Fit an orthogonal factor model to $\boldsymbol{z}_t$ using the PCA approach.

(c) Fit an orthogonal factor model to $\boldsymbol{z}_t$ using the maximum likelihood method.

(d) Fit a constrained factor model, based on the industrial sectors, to $\boldsymbol{z}_t$.

(e) Compare the three fitted factor models.

6.5 Consider monthly simple returns of 40 stocks from NASDAQ and NYSE for years 2002 and 2003. The data are in the file `m-apca40stocks.txt`. The file has 40 columns. The first row of the file contains the CRSP Permno of the stocks.

(a) Perform an asymptotic principal component analysis on the 40 monthly simple stock returns with 4 common factors.

(b) Based on the scree plot, is it reasonable to use 4 common factors? If not, select an alternative number of common factors.

(c) Show the time plots of the associated loading matrix.

6.6 Consider the annual real GDP of 14 countries from 1960 to 2011 for 52 observations. The data are obtained from FRED of the Federal Reserve Bank of St. Louis, in 2005 millions of U.S. dollars, and in the file `a-rgdp-14.txt`. The first row of the file contains names of the 14 countries. Let $\boldsymbol{z}_t$ be the annual real GDP growth rates, in percentages, of the 14 countries. The $\boldsymbol{z}_t$ series has 51 observations. Divide the data into two parts with the second part consisting of the last 6 observations for forecasting purpose. The first part has 45 observations and is used in modeling. In this problem, we focus on 1-step ahead forecasting for United States and South Korea. Do the followings separately for United States and South Korea.

(a) Find a univariate time series model to produce 1-step ahead forecasts and compute the mean square error of forecasts in the forecasting subsample.

(b) Apply diffusion index method with $\boldsymbol{x}_t = (\boldsymbol{z}'_{t-1}, \boldsymbol{z}'_{t-2})'$ and m = 1, 3, 5 to produce forecasts and compute the mean square error of forecasts.

(c) Let v_t be the series of simple average of the 14 GDP growth rates. Here, v_t may represent the global GDP growth rates. Use a bivariate time series model (country and v_t) to produce 1-step ahead forecasts. Also, compute the mean square error of forecasts in the forecasting subsample.

6.7 Again, consider the 14 annual real GDP growth rates, in percentages, of the prior problem.

(a) Perform a model-based clustering analysis with AR order $p = 2$ and 2 clusters.

(b) Standardize the 14 annual real GDP growth rates by computing $x_{it} = z_{it}/\hat{\sigma}_i$, where $\hat{\sigma}_i$ is the sample standard deviation of z_{it}. Repeat the clustering analysis of part (a), but using $\boldsymbol{x}_t$. This type of clustering analysis can be regarded as focusing on the dynamic dependence rather than the variability in classification.

Hint: Use the command `xt=scale(zt,center=F,scale=T)`.

(c) Compare the results of the two clustering analysis.

REFERENCES

Bai, J. (2003). Inferential theory for factor models of large dimensions. *Econometrica* **71**: 135–171.

Bai, J. and Ng, S. (2002). Determining the number of factors in approximate factor models. *Econometrica*, **70**: 191–221.

Chamberlain, G. (1983). Funds, factors, and diversification in arbitrage pricing models. *Econometrica*, **51**: 1305–1323.

Chamberlain, G. and M. Rothschild (1983). Arbitrage, factor structure, and mean-variance analysis on large asset markets. *Econometrica*, **51**: 1281–1304.

Connor, G. and R. A. Korajczyk (1986). Performance measurement with the arbitrage princing theory: a new framework for analysis. *Journal of Financial Economics*, **15**: 373–394.

Connor, G. and R. A. Korajczyk (1988). Risk and return in an equilibrium APT: application of a new test methodology. *Journal of Financial Economics*, **21**: 255–289.

Forni, M., M. Hallin, M. Lippi, and L. Reichlin (2000). The generalized dynamic-factor model: identification and estimation. *The Review of Economics and Statistics*, **82**: 540–554.

Forni, M., M. Hallin, M. Lippi, and L. Reichlin (2004). The generalized dynamic-factor model: consistency and rates. *Journal of Econometrics*, **119**: 231–255.

Forni, M., M. Hallin, M. Lippi, and L. Reichlin (2005). The generalized dynamic factor model: one-sided estimation and forecasting. *Journal of the American Statistical Association*, **100**: 830–840.

Frühwirth-Schnatter, S. (2006). *Finite Mixture and Markov Switching Models*. Springer, New York.

Frühwirth-Schnatter, S. and Kaufmann, S. (2008). Model-based clustering of multiple time series. *Journal of Business & Economic Statistics*, **26**: 78–89.

Gamerman, D. and Lopes, H. F. (2006). *Markov Chain Monte Carlo: Stochastic Simulation for Bayesian Inference*. 2nd Edition. Chapman & Hall/CRC, Boca Raton, FL.

Geweke, J. (1977). The dynamic factor analysis of economic time series models. In D. J. Aigner and A. S. Goldberger (eds.). *Latent Variables in Socio-Economic Models*, pp. 365–383. North-Holland, Amsterdam.

Hastie, T., Tibshirani, R., and Friedman, J. (2009). *The Elements of Statistical Learning: Data Mining, Inference, and Prediction*. 2nd Edition. Springer, New York.

Heaton, C. and Solo, V. (2003). Asymptotic principal components estimation of large factor models. Working paper, Department of Economics, Macquarie University, Australia.

Heaton, C. and Solo, V. (2006). Estimation of approximate factor models: Is it important to have a large number of variables? Working paper, Department of Economics, Macquarie University, Australia.

Johnson, R. A. and Wichern, D. W. (2007). *Applied Multivariate Statistical Analysis*. 6th Edition. Pearson Prentice Hall, Upper Saddle River, NJ.

Lam, C. and Yao, Q. (2012). Factor modeling for high-dimensional time series: inference for the number of factors. *Annals of Statistics* **40**: 694–726.

Lam, C., Yao, Q., and Bathia, N. (2011). Estimation of latent factors for high-dimensional time series. *Biometrika* **98**: 1025–1040.

Lütkepohl, H. (2005). *New Introduction to Multiple Time Series Analysis*. Springer, New York.

Magnus, J. R. and Neudecker, H. (1999). *Matrix Differential Calculus With Applications in Statistics and Econometrics*. Revised Edition. John Wiley & Sons, Inc, New York.

Onatski, A. (2009). Testing hypothesis about the number of factors in large factor models. *Econometrica*, **77**: 1447–1479.

Pan, J. and Yao, Q. (2008), Modeling multiple time series via common factors. *Biometrika*, **95**: 365–379.

Peña, D. and Box, G. E. P. (1987). Identifying a simplifying structure in time series. *Journal of the American Statistical Association*, **82**: 836–843.

Peña, D. and Poncela, P. (2006). Nonstationary dynamic factor analysis. *Journal Statistical Planning and Inference*, **136**: 1237–1257.

Stock, J. H. and Watson, M. W. (2002a). Forecasting using principal components from a large number of predictors. *Journal of the American Statistical Association*, **97**: 1167–1179.

Stock, J. H. and Watson, M. W. (2002b). Macroeconomic forecasting using diffusion indexes. *Journal of Business and Economic Statistics*, **20**: 147–162.

Tiao, G. C. and Tsay, R. S. (1989). Model specification in multivariate time series (with discussion). *Journal of the Royal Statistical Society, Series B*, **51**: 157–213.

Tsai, H. and Tsay, R. S. (2010). Constrained factor models. *Journal of the American Statistical Association*, **105**: 1593–1605.

Tsay, R. S. (2010). *Analysis of Financial Time Series*. 3rd Edition. John Wiley & Sons, Inc, Hoboken, NJ.

Tsay, R. S. (2013). *An Introduction to Analysis of Financial Data with R*. John Wiley & Sons, Hoboken, NJ.

Wang, Y., Tsay, R. S., Ledolter, J., and Shrestha, K. M. (2013). Forecasting high-dimensional time series: a robust clustering approach. *Journal of Forecasting* (to appear).

CHAPTER 7

Multivariate Volatility Models

In previous chapters, we assume the innovations $\boldsymbol{a}_t$ of a multivariate time series $\boldsymbol{z}_t$ are serially uncorrelated and have zero mean and positive-definite covariance matrix. We also assume that the covariance matrix of $\boldsymbol{a}_t$ is time-invariant. Let F_{t-1} denote the σ-field generated by the past data $\{\boldsymbol{z}_{t-i}|i = 1, 2, \ldots\}$. These assumptions imply that $E(\boldsymbol{a}_t|F_{t-1}) = \mathbf{0}$ and $E(\boldsymbol{a}_t\boldsymbol{a}_t'|F_{t-1}) = \boldsymbol{\Sigma}_a > 0$, which is a constant matrix. On the other hand, most financial time series have conditional heteroscedasticity. Let $\boldsymbol{\Sigma}_t = \text{Cov}(\boldsymbol{a}_t|F_{t-1})$ be the conditional covariance matrix of $\boldsymbol{z}_t$ given F_{t-1}. Conditional heteroscedasticity means that $\boldsymbol{\Sigma}_t$ is time-dependent. The dynamic dependence of $\boldsymbol{\Sigma}_t$ is the subject of multivariate volatility modeling and the focus point of this chapter. For ease in reference, we shall refer to $\boldsymbol{\Sigma}_t$ as the volatility matrix of $\boldsymbol{z}_t$.

The volatility matrix has many financial applications. For instance, it is widely used in asset allocation and risk management. Modeling $\boldsymbol{\Sigma}_t$, however, faces two major difficulties. The first difficulty is the curse of dimensionality. For a k-dimensional time series $\boldsymbol{z}_t$, the volatility matrix $\boldsymbol{\Sigma}_t$ consists of k conditional variances and $k(k-1)/2$ conditional covariances. In other words, $\boldsymbol{\Sigma}_t$ consists of $k(k+1)/2$ different time-varying elements. For $k = 30$, $\boldsymbol{\Sigma}_t$ contains 465 different elements. The dimension of $\boldsymbol{\Sigma}_t$ thus increases quadratically with k. The second difficulty is maintaining the positive-definite constraint. The volatility matrix $\boldsymbol{\Sigma}_t$ must be positive-definite for all t. Special attention is needed to maintain this constraint when k is large.

Many multivariate volatility models have been proposed in the literature, including multivariate stochastic volatility and multivariate generalizations of GARCH models. See review articles Asai, McAleer, and Yu (2006) and Bauwens, Laurent, and Rombouts (2006), and the references therein. See also the handbook of volatility by Bauwens, Hafner, and Laurent (2012). The goal of this chapter is to introduce some multivariate volatility models that are applicable in practice, yet relatively

Multivariate Time Series Analysis: With R and Financial Applications,
First Edition. Ruey S. Tsay.

easy to understand in theory. The chapter, therefore, does not cover all multivariate volatility models available in the literature.

Similarly to the univariate case, we decompose a multivariate time series $\boldsymbol{z}_t$ as

$$\boldsymbol{z}_t = \boldsymbol{\mu}_t + \boldsymbol{a}_t, \tag{7.1}$$

where $\boldsymbol{\mu}_t = E(\boldsymbol{z}_t|F_{t-1})$ is the conditional expectation of $\boldsymbol{z}_t$ given F_{t-1} or the *predictable component* of $\boldsymbol{z}_t$. For a linear process, $\boldsymbol{\mu}_t$ follows one of the multivariate models discussed in the previous chapters; nonlinear models can also be used if necessary. The innovation $\boldsymbol{a}_t$ is unpredictable because it is serially uncorrelated. We write the shock $\boldsymbol{a}_t$ as

$$\boldsymbol{a}_t = \boldsymbol{\Sigma}_t^{1/2}\boldsymbol{\epsilon}_t, \tag{7.2}$$

where $\{\boldsymbol{\epsilon}_t\}$ is a sequence of independent and identically distributed random vectors such that $E(\boldsymbol{\epsilon}_t) = \boldsymbol{0}$ and $\text{Cov}(\boldsymbol{\epsilon}_t) = \boldsymbol{I}_k$ and $\boldsymbol{\Sigma}_t^{1/2}$ denotes the positive-definite square-root matrix of $\boldsymbol{\Sigma}_t$. Specifically, let $\boldsymbol{\Sigma}_t = \boldsymbol{P}_t\boldsymbol{\Lambda}_t\boldsymbol{P}_t'$ represent the spectral decomposition of $\boldsymbol{\Sigma}_t$, where $\boldsymbol{\Lambda}_t$ is the diagonal matrix of the eigenvalues of $\boldsymbol{\Sigma}_t$ and $\boldsymbol{P}_t$ denotes the orthonormal matrix of eigenvectors. Then, $\boldsymbol{\Sigma}_t^{1/2} = \boldsymbol{P}_t\boldsymbol{\Lambda}_t^{1/2}\boldsymbol{P}_t'$. There are other ways to parameterize the volatility matrix. For instance, one can use the Cholesky decomposition of $\boldsymbol{\Sigma}_t$. The parameterization, however, does not affect modeling of the volatility matrix $\boldsymbol{\Sigma}_t$. If the innovation $\boldsymbol{\epsilon}_t = (\epsilon_{1t}, \ldots, \epsilon_{kt})'$ is not Gaussian, we further assume that $E(\epsilon_{it}^4)$ is finite for all i. A commonly used non-Gaussian distribution for $\boldsymbol{\epsilon}_t$ is the standardized multivariate Student-t distribution with v degrees of freedom and probability density function (pdf)

$$f(\boldsymbol{\epsilon}_t|v) = \frac{\Gamma[(v+k)/2]}{[\pi(v-2)]^{k/2}\Gamma(v/2)}[1 + (v-2)^{-1}\boldsymbol{\epsilon}_t'\boldsymbol{\epsilon}_t]^{-(v+k)/2}, \tag{7.3}$$

where $\Gamma(\cdot)$ denotes the Gamma function. In this case, the marginal distribution of ϵ_{it} is the univariate standardized Student-t distribution and the pdf of the shock $\boldsymbol{a}_t$ is

$$f(\boldsymbol{a}_t|v, \boldsymbol{\Sigma}_t) = \frac{\Gamma[(v+k)/2]}{[\pi(v-2)]^{k/2}\Gamma(v/2)|\boldsymbol{\Sigma}_t|^{1/2}}[1 + (v-2)^{-1}\boldsymbol{a}_t'\boldsymbol{\Sigma}_t^{-1}\boldsymbol{a}_t]^{-(v+k)/2}.$$

Volatility modeling typically consists of two sets of equations. The first set of equations governs the time evolution of the conditional mean $\boldsymbol{\mu}_t$, whereas the second set describes the dynamic dependence of the volatility matrix $\boldsymbol{\Sigma}_t$. These two sets of equations are, therefore, referred to as the mean and volatility equations. If linear models are entertained, then one can apply the models discussed in the previous chapters to handle $\boldsymbol{\mu}_t$. For most asset return series, the model for $\boldsymbol{\mu}_t$ is relatively simple because $\boldsymbol{\mu}_t$ can either be a constant vector or follow a simple VAR model. The focus of this chapter is therefore on the model for $\boldsymbol{\Sigma}_t$.

The chapter is organized as follows. We investigate in Section 7.1 the problem of testing the presence of conditional heteroscedasticity in a vector time series. Two types of test are employed. We also demonstrate the importance of using robust version of the test statistics when $\boldsymbol{a}_t$ has heavy tails. In Section 7.2, we study the properties of quasi maximum likelihood estimates of a multivariate volatility model. The result is useful as the actual distributions of asset returns are hard to verify in practice. We then discuss diagnostic checking of a fitted multivariate volatility model in Section 7.3. We employ in Section 7.4 the simple exponentially weighted moving average (EWMA) method to calculate time-varying volatility, including estimation of the smoothing parameter. We study in Section 7.5 the simple Baba–Engle–Kraft–Kroner (BEKK) model of Engle and Kroner (1995) and discuss its pros and cons. In Section 7.6, we employ Cholesky decomposition to study the dynamic dependence of $\boldsymbol{\Sigma}_t$. Both Bayesian and non-Bayesian methods are introduced. We study the dynamic conditional correlation (DCC) models in Section 7.7. In Section 7.8, we discuss additional multivariate volatility models based on orthogonal transformation, Section 7.9 introduces copula-based multivariate volatility models. Finally, we introduce principal volatility component analysis in Section 7.10. Again, real examples are used throughout in demonstration.

7.1 TESTING CONDITIONAL HETEROSCEDASTICITY

Consider a k-dimensional time series $\boldsymbol{z}_t$. We start the volatility modeling by discussing two simple tests for checking the presence of conditional heteroscedasticity. To begin, we assume, for simplicity, that $\boldsymbol{\mu}_t$ is known so that the noise process $\boldsymbol{a}_t$ is available. Since volatility is concerned with the second-order moment of $\boldsymbol{a}_t$, the tests considered employ either the $\boldsymbol{a}_t^2$ process or a quadratic function of $\boldsymbol{a}_t$.

7.1.1 Portmanteau Test

If $\boldsymbol{a}_t$ has no conditional heteroscedasticity, then its conditional covariance matrix $\boldsymbol{\Sigma}_t$ is time-invariant. This implies that $\boldsymbol{\Sigma}_t$, hence $\boldsymbol{a}_t^2$, does not depend on the $\boldsymbol{a}_{t-i}^2$ for $i > 0$. Therefore, one can test the hypothesis $H_0 : \boldsymbol{\rho}_1 = \boldsymbol{\rho}_2 = \cdots = \boldsymbol{\rho}_m = \boldsymbol{0}$ versus the alternative H_a: $\boldsymbol{\rho}_i \neq \boldsymbol{0}$ for some i $(1 \leq i \leq m)$, where $\boldsymbol{\rho}_i$ is the lag-i cross-correlation matrix of $\boldsymbol{a}_t^2$. An obvious test statistic to use in this situation is the well-known Ljung–Box statistics

$$Q_k^*(m) = T^2 \sum_{i=1}^{m} \frac{1}{T-i} \boldsymbol{b}_i'(\hat{\boldsymbol{\rho}}_0^{-1} \otimes \hat{\boldsymbol{\rho}}_0^{-1})\boldsymbol{b}_i, \tag{7.4}$$

where T denotes the sample size, k is the dimension of $\boldsymbol{a}_t$, and $\boldsymbol{b}_i = \text{vec}(\hat{\rho}_i{}')$ with $\hat{\boldsymbol{\rho}}_j$ being the lag-j sample cross-correlation matrix of $\boldsymbol{a}_t^2$. Under the null hypothesis that $\boldsymbol{a}_t$ has no conditional heteroscedasticity, $Q_k^*(m)$ is asymptotically distributed as $\chi^2_{k^2 m}$. It can be shown that $Q_k^*(m)$ is asymptotically equivalent to the multivariate generalization of the Lagrange multiplier (LM) test of Engle (1982) for conditional heteroscedasticity. See Li (2004).

Alternatively, one can employ the standardized series

$$e_t = \boldsymbol{a}_t' \boldsymbol{\Sigma}^{-1} \boldsymbol{a}_t - k, \tag{7.5}$$

where $\boldsymbol{\Sigma}$ denotes the unconditional covariance matrix of $\boldsymbol{a}_t$, and consider the hypothesis $H_0 : \rho_1 = \cdots = \rho_m = 0$ versus H_a: $\rho_i \neq 0$ for some i $(1 \leq i \leq m)$, where ρ_i is the lag-i autocorrelation of e_t. The test statistic in this case is the traditional Ljung–Box statistic for the univariate series e_t. That is, $Q^*(m) = T(T+2) \sum_{i=1}^{m} \hat{\rho}_i^2/(T-i)$, where $\hat{\rho}_i$ is the lag-i sample ACF of e_t. In practice, $\boldsymbol{\Sigma}$ is estimated by the sample covariance matrix of $\boldsymbol{a}_t$. Under the null hypothesis that $\boldsymbol{a}_t$ has no conditional heteroscedasticity, $Q^*(m)$ is asymptotically distributed as χ^2_m.

It is easy to see that the two portmanteau tests are asymptotically equivalent when $\boldsymbol{a}_t$ is Gaussian. The multivariate test $Q_k^*(m)$ of Equation (7.4) may fare poorly in finite samples when $\boldsymbol{a}_t$ has heavy tails. Some robust modifications could be helpful. We consider a simple robust modification of Q_k^* in this section. One approach to reduce the effect of heavy tails in statistics is trimming. For volatility testing, we adopt a simple procedure by trimming away data in the upper 5% tail. Specifically, let $q_{0.95}$ be the empirical 95th quantile of the standardized scalar residuals e_t of Equation (7.5). We remove from $\boldsymbol{z}_t$ those observations whose corresponding e_t exceeds $q_{0.95}$ and use the remaining $0.95T$ data points to compute the test statistic of Equation (7.4). Denote the resulting test statistic by $Q_k^r(m)$ with the superscript r signifying robust test. Performance of various test statistics in finite sample is investigated via simulation later. It turns out that the 5% trimming works reasonably well.

7.1.2 Rank-Based Test

Asset returns tend to have heavy tails. Some extreme returns might have marked effects on the performance of the portmanteau statistic Q_k^* of Equation (7.4). To overcome this potential weakness, Dufour and Roy (1985, 1986) consider the rank series of the standardized series e_t in Equation (7.5). Let R_t be the rank of e_t. The lag-ℓ rank autocorrelation of e_t can be defined as

$$\tilde{\rho}_\ell = \frac{\sum_{t=\ell+1}^{T} (R_t - \bar{R})(R_{t-\ell} - \bar{R})}{\sum_{t=1}^{T} (R_t - \bar{R})^2}, \quad \ell = 1, 2, \ldots, \tag{7.6}$$

where

$$\bar{R} = \sum_{t=1}^{T} R_t/T = (T+1)/2,$$

$$\sum_{t=1}^{T} (R_t - \bar{R})^2 = T(T^2 - 1)/12.$$

Dufour and Roy (1985, 1986) showed that the distribution of the rank autocorrelations is the same whenever $\{e_t\}$ are continuous exchangeable random variables. The reason is that all rank permutations in this situation are equally probable. It can also be shown that

$$E(\tilde{\rho}_\ell) = -(T-\ell)/[T(T-1)]$$
$$\text{Var}(\tilde{\rho}_\ell) = \frac{5T^4 - (5\ell+9)T^3 + 9(\ell-2)T^2 + 2\ell(5\ell+8)T + 16\ell^2}{5(T-1)^2T^2(T+1)}.$$

See Moran (1948) and Dufour and Roy (1986). Furthermore, Dufour and Roy (1986) showed that the statistic

$$Q_R(m) = \sum_{i=1}^{m} \frac{[\tilde{\rho}_i - E(\tilde{\rho}_i)]^2}{\text{Var}(\tilde{\rho}_i)} \tag{7.7}$$

is distributed as χ^2_m asymptotically if e_t has no serial dependence.

7.1.3 Simulation

We conduct some simulations to study the finite sample performance of the test statistics

1. $Q_k^*(m)$ of Equation (7.4)
2. $Q_k^r(m)$: robust version of $Q_k^*(m)$ with 5% upper tail trimming
3. $Q^*(m)$ based on the transformed scalar residuals e_t of Equation (7.5)
4. $Q_R(m)$: rank test based on the rank of e_t.

Two multivariate distributions are used in the simulation. The first distribution is the bivariate standard normal, that is, $\boldsymbol{\epsilon}_t \sim N(\mathbf{0}, \boldsymbol{I}_2)$. The second distribution is bivariate Student-t distribution with $\boldsymbol{\Sigma} = \boldsymbol{I}_2$ and 5 degrees of freedom. This second distribution has heavy tails. The sample sizes used are 500, 1000, and 2000. The choices of m are 5 and 10. Table 7.1 provides some summary statistics of the test statistics $Q_k^*(m)$ and $Q_k^r(m)$. For each model and configuration of m and T, the results are based on 10,000 iterations.

From Table 7.1, we make the following observations. First, as expected, the performance of $Q_k^*(m)$ works reasonably well when the innovations are multivariate normal. Second, the $Q_k^*(m)$ statistic of Equation (7.4) fares poorly for the heavy-tailed innovations. In particular, the variability of $Q_k^*(m)$ is too large when the innovations are Student-t with 5 degrees of freedom. Finally, the robust version $Q_k^r(m)$ works very well under both normal and Student-t innovations. Thus, 5% trimming seems a good choice for the multivariate portmanteau test in detecting conditional heteroscedasticity.

TABLE 7.1 Finite Sample Performance of the $Q_k^*(m)$ Statistic of Equation (7.4) and Its Robust Version $Q_k^r(m)$ with 5% Upper Tail Trimming

	$m = 5$					$m = 10$				
T	ave	var	q_{90}	q_{95}	q_{99}	ave	var	q_{90}	q_{95}	q_{99}
True	20	40	28.41	31.41	37.57	40	80	51.81	55.76	63.69
(a) $Q_k^*(m)$: Bivariate Standard Normal Distribution										
500	19.70	42.08	28.17	31.63	38.86	39.73	86.68	52.04	56.38	65.46
1000	19.85	41.81	28.31	31.46	38.13	39.83	84.35	51.92	56.16	64.73
2000	19.95	39.99	28.39	31.38	37.16	39.89	83.16	51.76	56.09	64.50
(b) $Q_k^*(m)$: Bivariate Student-t with $\boldsymbol{\Sigma} = \boldsymbol{I}_2$ and 5 Degrees of Freedom										
500	18.30	210.8	30.96	39.66	70.04	36.77	406.6	56.17	68.34	114.7
1000	18.50	225.2	30.62	39.93	72.37	37.48	590.1	57.12	71.27	121.1
2000	18.89	301.4	31.21	40.48	83.83	37.67	702.0	57.35	71.40	121.0
(c) $Q_k^r(m)$: Bivariate Standard Normal Distribution										
500	19.90	38.78	28.12	31.05	37.58	40.00	78.40	51.88	55.61	63.54
1000	19.91	39.54	28.24	31.31	37.66	39.89	80.24	51.58	55.48	63.63
2000	19.98	40.09	28.56	31.55	37.45	39.98	80.67	51.84	55.45	63.45
(d) $Q_k^r(m)$: Bivariate Student-t with $\boldsymbol{\Sigma} = \boldsymbol{I}_2$ and 5 Degrees of Freedom										
500	19.82	40.88	28.24	31.50	38.01	39.62	81.47	51.49	55.53	64.75
1000	19.90	40.44	28.37	31.39	37.87	39.82	80.85	51.62	55.83	64.44
2000	19.92	40.71	28.51	31.34	37.99	39.96	80.24	51.70	55.56	63.63

The innovations are bivariate standard normal and bivariate standard Student-t with 5 degrees of freedom. The results are based on 10,000 iterations and q_v denotes the vth percentile.

Turn to the scalar transformation of multivariate innovations. Table 7.2 provides some summary statistics of the performance of the test statistic $Q^*(m)$ and the rank-based statistic $Q_R(m)$ for the transformed residual e_t of Equation (7.5). Again, we employ the same bivariate innovations as those in Table 7.1. The results are also based on 10,000 replications. From the table, we make the following observations. First, the scalar transformed residuals seem to provide more stable test statistics in comparison with the $Q_k^*(m)$ statistics of Table 7.1. Second, the $Q^*(m)$ statistics continue to show some deviation from its limiting properties in finite samples even when the innovations are normally distributed. Third, the rank-based test statistic $Q_R(m)$ has nice finite-sample properties. It follows reasonably close to its limiting χ_m^2 distribution for both innovation distributions used.

Finally, to provide a better description of the simulation results, Figure 7.1 shows the density plots of the four test statistics and their asymptotic reference distributions. The simulation here is for $k = 5$, $m = 5$, $T = 2000$, and the innovation distribution is a five-dimensional Student-t distribution with $\boldsymbol{\Sigma} = \boldsymbol{I}_5$ and 5 degrees of freedom. Again, 10,000 iterations were used. The reference limiting distribution for the two multivariate portmanteau statistics is χ_{125}^2, whereas that for the two scalar portmanteau tests is χ_5^2. From the plots, it is clear that the robust $Q_k^r(m)$ statistic follows its

TABLE 7.2 Finite Sample Performance of the $Q^*(m)$ Statistic and the Rank-Based Statistic $Q_R(m)$ for the Scalar Transformed Residuals of Equation (7.5)

	$m=5$					$m=10$				
T	ave	var	q_{90}	q_{95}	q_{99}	ave	var	q_{90}	q_{95}	q_{99}
True	5	10	9.24	11.07	15.09	10	20	15.99	18.31	23.21
(a) $Q^*(m)$: Bivariate Standard Normal Distribution										
500	4.96	10.14	9.13	10.97	15.69	9.91	19.74	15.81	18.27	23.29
1000	4.96	9.84	9.12	11.03	15.02	10.00	20.87	16.05	18.50	23.40
2000	5.01	9.98	9.22	11.13	15.09	9.92	20.03	16.01	18.25	22.73
(b) $Q^*(m)$: Bivariate Student-t with $\boldsymbol{\Sigma} = \boldsymbol{I}_2$ and 5 Degrees of Freedom										
500	4.31	19.40	8.50	11.59	22.58	8.64	39.67	15.68	19.52	32.89
1000	4.40	21.44	8.52	11.83	22.49	8.82	51.98	15.54	20.30	35.07
2000	4.54	37.14	8.63	11.77	25.13	9.00	65.17	15.82	20.47	38.46
(c) $Q_R(m)$: Bivariate Standard Normal Distribution										
500	4.98	10.39	9.15	11.11	15.26	10.06	20.63	16.10	18.53	23.95
1000	4.98	10.28	9.30	11.14	15.46	10.02	20.34	15.93	18.36	23.67
2000	4.99	10.37	9.29	11.30	15.11	9.97	20.33	15.95	18.29	23.53
(d) $Q_R(m)$: Bivariate Student-t with $\boldsymbol{\Sigma} = \boldsymbol{I}_2$ and 5 Degrees of Freedom										
500	4.95	10.03	9.26	10.98	15.02	10.04	20.83	16.02	18.52	24.15
1000	5.05	10.37	9.36	11.30	15.23	9.88	19.38	15.81	18.17	22.78
2000	4.96	9.51	9.14	10.84	14.64	9.99	20.24	16.05	18.36	23.36

The innovations are bivariate standard normal and bivariate standard Student-t with 5 degrees of freedom. The results are based on 10,000 iterations and q_v denotes the vth percentile.

limiting distribution closely, but the $Q_k(m)$ does not. Furthermore, the rank-based scalar portmanteau test clearly outperforms the conventional portmanteau statistic.

In summary, our limited simulation results show that (a) robust version of the multivariate portmanteau test statistics should be used in detecting conditional heteroscedasticity and (b) the rank-based portmanteau test statistics work well for normal innovations as well as for heavy-tailed Student-t innovations. Finally, the traditional multivariate portmanteau statistics encounter marked size distortion in detecting conditional heteroscedasticity when the innovations have heavy tails.

7.1.4 Application

To demonstrate, we apply the test statistics discussed in this section to two time series. The first time series is a five-dimensional iid series generated from $N(\mathbf{0}, \boldsymbol{I}_5)$ with 400 observations and the second series consists of the monthly log returns of IBM stock and the S&P composite index. Figure 7.2 shows the time plots of the two log return series. For the stock returns, we use $\hat{\boldsymbol{\mu}}_t$ being the sample mean. As expected, the test statistics fail to reject the null hypothesis of no conditional heteroscedasticity for the simulated series, but confirm the presence of conditional heteroscedasticity in the monthly log return series.

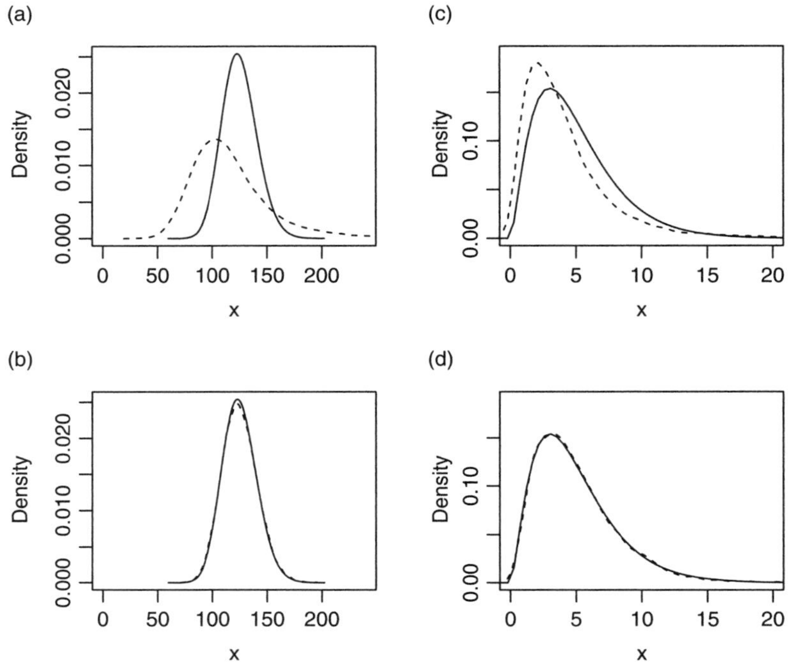

FIGURE 7.1 Density plots for four test statistics for detecting the presence of conditional heteroscedasticity: (a) multivariate $Q_k^*(m)$ and its reference, (b) robust multivariate $Q_k^r(m)$ and its reference, (c) the scalar $Q^*(m)$ and its reference, and (d) rank-based $Q_R(m)$ and its reference. The results are based on simulation with $k = 5$, $m = 5$, $T = 2000$, and 10,000 iterations.

Remark: The volatility test for multivariate time series is carried out by the command `MarchTest` of the `MTS` package. The default option uses 10 lags. □

R Demonstration: Multivariate volatility test.

```
> zt=matrix(rnorm(2000),400,5)
> MarchTest(zt)
Q(m) of squared scalar series(LM test):
Test statistic:  13.2897  p-value:  0.2079223
Rank-based Test:
Test statistic:  6.753778  p-value:  0.7484673
Q_k(m) of squared series:
Test statistic:  280.1069  p-value:  0.09251779
Robust Q_k(m) statistic (5% trimming):
Test statistics:  261.0443  p-value:  0.3027401
> da=read.table("m-ibmsp-6111.txt",header=T)
> rtn=log(da[,2:3]+1)
> at=scale(rtn,center=T,scale=F)  ## remove sample means
```

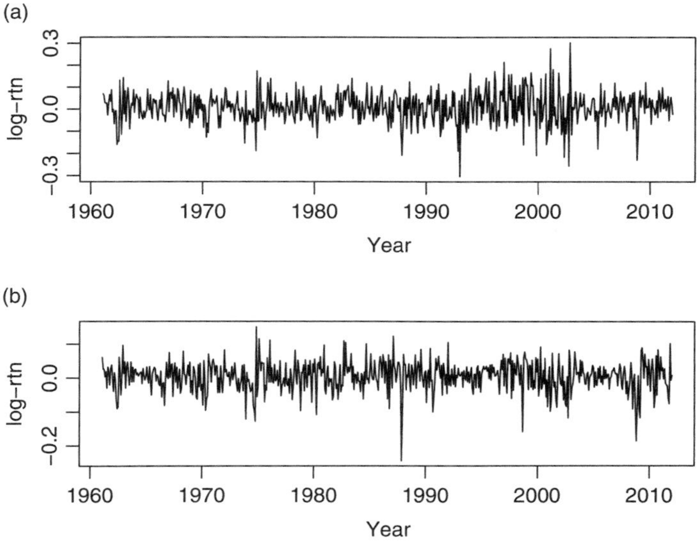

FIGURE 7.2 Time plots of monthly log returns of (a) IBM stock and (b) S&P composite index from January 1961 to December 2011.

```
> MarchTest(at)
Q(m) of squared scalar series(LM test):
Test statistic:  38.06663  p-value:  3.695138e-05
Rank-based Test:
Test statistic:  108.3798  p-value:  0
Q_k(m) of squared series:
Test statistic:  109.4194  p-value:  2.276873e-08
Robust Q_k(m) statistic (5% trimming):
Test statistics:  118.7134  p-value:  9.894441e-10
```

7.2 ESTIMATION OF MULTIVARIATE VOLATILITY MODELS

In this section, we discuss the estimation of a multivariate volatility model. Let $\boldsymbol{\theta}$ denote the parameter vector of a multivariate volatility model. Both $\boldsymbol{\mu}_t$ of Equation (7.1) and $\boldsymbol{\Sigma}_t$ of Equation (7.2) are functions of $\boldsymbol{\theta}$ and $\boldsymbol{F}_{t-1}$. Suppose that $\boldsymbol{\theta}$ is a $p \times 1$ vector and let $\boldsymbol{\Theta}$ be the parameter space in R^p. Assume that there exists an interior point $\boldsymbol{\theta}_o \in \boldsymbol{\Theta}$ such that the entertained multivariate volatility model holds. In other words, $\boldsymbol{\theta}_o$ is a true parameter vector, and we have

$$E(\boldsymbol{z}_t|F_{t-1}) = \boldsymbol{\mu}_t(F_{t-1}, \boldsymbol{\theta}_o) \tag{7.8}$$

$$\text{Var}(\boldsymbol{z}_t|F_{t-1}) = \boldsymbol{\Sigma}_t(F_{t-1}, \boldsymbol{\theta}_o) \tag{7.9}$$

for $t = 1, 2, \ldots$. Under Equation (7.8), the true error vector is $\boldsymbol{a}_t^o = \boldsymbol{z}_t - \boldsymbol{\mu}_t(F_{t-1}, \boldsymbol{\theta}_o)$ and we have $E(\boldsymbol{a}_t^o|F_{t-1}) = \boldsymbol{0}$. In addition, if Equation (7.9) also holds, then $E[\boldsymbol{a}_t^o(\boldsymbol{a}_t^o)'|F_{t-1}] = \boldsymbol{\Sigma}_t(F_{t-1}, \boldsymbol{\theta}_o) \equiv \boldsymbol{\Sigma}_t^o$.

If the innovation $\boldsymbol{\epsilon}_t$ of Equation (7.2) follows the standard multivariate normal distribution, that is, $\boldsymbol{\epsilon}_t \sim N(\boldsymbol{0}, \boldsymbol{I}_k)$, then the (conditional) log likelihood function, except for a constant term, is

$$\ell(\boldsymbol{\theta}) = \sum_{t=1}^{T} \ell_t(\boldsymbol{\theta}), \quad \ell_t(\boldsymbol{\theta}) = -\frac{1}{2}[\log(|\boldsymbol{\Sigma}_t|) + \boldsymbol{a}_t'\boldsymbol{\Sigma}_t^{-1}\boldsymbol{a}_t], \tag{7.10}$$

where T is the sample size and $\boldsymbol{a}_t = \boldsymbol{z}_t - \boldsymbol{\mu}_t$. Other multivariate distributions, such as multivariate Student-t distribution or skewed multivariate Student-t distribution, can also be used. However, the Gaussian likelihood function in Equation (7.10) is often used in practice to obtain quasi maximum likelihood estimate (QMLE) of $\boldsymbol{\theta}$. The use of normal likelihood function can be justified partly because the true distribution of the innovations $\boldsymbol{\epsilon}_t$ is unknown and partly because QMLE has some nice limiting properties. See, for instance, Bollerslev and Wooldridge (1992).

Using the properties of matrix differentials in Section A.6 of Appendix A, the score function of $\ell_t(\boldsymbol{\theta})$ of Equation (7.10) is

$$\boldsymbol{S}(\boldsymbol{\theta}) = \frac{\partial \ell_t(\boldsymbol{\theta})}{\partial \boldsymbol{\theta}'} = \boldsymbol{a}_t'\boldsymbol{\Sigma}_t^{-1}\frac{\partial \boldsymbol{\mu}_t}{\partial \boldsymbol{\theta}'} + \frac{1}{2}[\text{vec}(\boldsymbol{a}_t\boldsymbol{a}_t' - \boldsymbol{\Sigma}_t)]'(\boldsymbol{\Sigma}_t^{-1} \otimes \boldsymbol{\Sigma}_t^{-1})\frac{\partial \text{vec}(\boldsymbol{\Sigma}_t)}{\partial \boldsymbol{\theta}'}, \tag{7.11}$$

where $\partial \boldsymbol{\mu}_t/(\partial \boldsymbol{\theta}')$ is a $k \times p$ matrix with p being the number of parameters in $\boldsymbol{\theta}$ and $\partial \text{vec}(\boldsymbol{\Sigma}_t)/(\partial \boldsymbol{\theta}')$ is a $k^2 \times p$ matrix. For simplicity in notation, we omit the arguments $\boldsymbol{\theta}$ and F_{t-1} from $\boldsymbol{a}_t$, $\boldsymbol{\mu}_t$ and $\boldsymbol{\Sigma}_t$. Under the correct specifications of the first two moments of $\boldsymbol{a}_t$ given F_{t-1}, we have

$$E[\boldsymbol{S}(\boldsymbol{\theta}_o)|F_{t-1}] = \boldsymbol{0}.$$

Thus, the score evaluated at the true parameter vector $\boldsymbol{\theta}_o$ is a vector martingale difference with respect to F_{t-1}.

For asymptotic properties of QMLE, we also need the Hessian matrix $\boldsymbol{H}_t(\boldsymbol{\theta})$ of $\ell_t(\boldsymbol{\theta})$. More specifically, we need $E[\boldsymbol{H}_t(\boldsymbol{\theta}_o)|F_{t-1}]$. Taking the partial derivatives of $\boldsymbol{S}(\boldsymbol{\theta})$ with respect to $\boldsymbol{\theta}$ appears to be complicated. However, since all we need is the expectation of the Hessian matrix evaluated at $\boldsymbol{\theta}_o$, we can make use of $E(\boldsymbol{a}_t^o) = \boldsymbol{0}$. See Equations (7.8) and (7.9). From Equation (7.11), we have

$$\begin{aligned}\boldsymbol{A}_t(\boldsymbol{\theta}_o) &= -E\left[\frac{\partial \boldsymbol{S}(\boldsymbol{\theta}_o)}{\partial \boldsymbol{\theta}}|F_{t-1}\right] = \left[\frac{\partial \boldsymbol{\mu}_t^o}{\partial \boldsymbol{\theta}'}\right]' \boldsymbol{\Sigma}_t^o \left[\frac{\partial \boldsymbol{\mu}_t^o}{\partial \boldsymbol{\theta}'}\right] \\ &\quad + \frac{1}{2}\left[\frac{\partial \text{vec}(\boldsymbol{\Sigma}_t^o)}{\partial \boldsymbol{\theta}'}\right]' [(\boldsymbol{\Sigma}_t^o)^{-1} \otimes (\boldsymbol{\Sigma}_t^o)^{-1}] \left[\frac{\partial \text{vec}(\boldsymbol{\Sigma}_t^o)}{\partial \boldsymbol{\theta}'}\right].\end{aligned} \tag{7.12}$$

If $\boldsymbol{a}_t$ is normally distributed, $\boldsymbol{A}_t(\boldsymbol{\theta}_o)$ is the conditional information matrix. Under some regularity conditions, Bollerslev and Wooldridge (1992) derived the asymptotic properties of the QMLE $\hat{\boldsymbol{\theta}}$ of a multivariate volatility model. Here, QMLE is obtained by maximizing the Gaussian log likelihood function $\ell(\boldsymbol{\theta})$ in Equation (7.10).

Theorem 7.1 Under the regularity conditions of Theorem 2.1 of Bollerslev and Wooldridge (1992), and assume that $\boldsymbol{\theta}_o$ is an interior point of the parameter space $\boldsymbol{\Theta} \subset R^p$ and Equations (7.8) and (7.9) hold for $t = 1, 2, \ldots$, then

$$\sqrt{T}(\hat{\boldsymbol{\theta}} - \boldsymbol{\theta}_o) \Rightarrow_d N(\boldsymbol{0}, \boldsymbol{A}_o^{-1}\boldsymbol{B}_o\boldsymbol{A}_o^{-1}),$$

where $\Rightarrow_d$ denotes convergence in distribution and

$$\boldsymbol{A}_o = \frac{1}{T}\sum_{t=1}^{T} E[\boldsymbol{A}_t(\boldsymbol{\theta}_o)], \quad \boldsymbol{B}_o = \frac{1}{T}\sum_{t=1}^{T} E[\boldsymbol{S}_t'(\boldsymbol{\theta}_o)\boldsymbol{S}_t(\boldsymbol{\theta}_o)].$$

Furthermore, $\hat{\boldsymbol{A}} \to_p \boldsymbol{A}_o$ and $\hat{\boldsymbol{B}} \to_p \boldsymbol{B}_o$, where $\to_p$ denotes convergence in probability and

$$\hat{\boldsymbol{A}} = \frac{1}{T}\sum_{t=1}^{T} \boldsymbol{A}_t(\hat{\boldsymbol{\theta}}), \quad \text{and} \quad \hat{\boldsymbol{B}} = \frac{1}{T}\sum_{t=1}^{T} \boldsymbol{S}_t'(\hat{\boldsymbol{\theta}})\boldsymbol{S}_t(\hat{\boldsymbol{\theta}}).$$

For some specific models, the partial derivatives of $\ell(\boldsymbol{\theta})$ can be obtained explicitly. However, for general multivariate volatility models, numerical methods are often used in estimation. We shall make use of Theorem 7.1 in this chapter.

7.3 DIAGNOSTIC CHECKS OF VOLATILITY MODELS

To verify the adequacy of a fitted multivariate volatility model, one performs some diagnostic checks on the residuals $\hat{\boldsymbol{a}}_t = \boldsymbol{z}_t - \hat{\boldsymbol{\mu}}_t$, where $\hat{\boldsymbol{\mu}}_t$ is the fitted conditional mean of $\boldsymbol{z}_t$. Typically, one uses $\hat{\boldsymbol{a}}_t$ to examine the mean equation and some quadratic functions of $\hat{\boldsymbol{a}}_t$ to validate the volatility equation. Diagnostic checks of a fitted VARMA model discussed in the previous chapters continue to apply to $\hat{\boldsymbol{a}}_t$. Several diagnostic statistics have been proposed in the literature for checking the volatility equation. We discuss two such test statistics in this section.

7.3.1 Ling and Li Statistics

Assume that the innovation $\boldsymbol{\epsilon}_t$ of Equation (7.2) also satisfies (a) $E(\epsilon_{it}^3) = 0$ and $E(\epsilon_{it}^4) = c_1 < \infty$ for $i = 1, \ldots, k$, and (b) $\{\epsilon_{it}\}$ and $\{\epsilon_{jt}\}$ are mutually uncorrelated up to the fourth order for $i \neq j$. Ling and Li (1997) employed $\hat{\boldsymbol{a}}_t$ to propose a model checking statistic for volatility models. Let

$$\hat{e}_t = \hat{\boldsymbol{a}}_t'\hat{\boldsymbol{\Sigma}}_t^{-1}\hat{\boldsymbol{a}}_t \tag{7.13}$$

be a transformed quadratic residual series. If the fitted model is correctly specified, then, by the ergodic theorem,

$$\frac{1}{T}\sum_{t=1}^{T}\hat{e}_t = \frac{1}{T}\sum_{t=1}^{T}\hat{\boldsymbol{a}}_t'\hat{\boldsymbol{\Sigma}}_t^{-1}\hat{\boldsymbol{a}}_t \rightarrow_{a.s.} E(\boldsymbol{a}_t'\boldsymbol{\Sigma}_t^{-1}\boldsymbol{a}_t) = E(\boldsymbol{\epsilon}_t'\boldsymbol{\epsilon}_t) = k,$$

where $\rightarrow_{a.s.}$ denotes almost sure convergence or convergence with probability 1. The lag-l sample autocorrelation of $\hat{e}_t$, therefore, can be defined as

$$\hat{\rho}_l = \frac{\sum_{t=l+1}^{T}(\hat{e}_t - k)(\hat{e}_{t-l} - k)}{\sum_{t=1}^{T}(\hat{e}_t - k)^2}. \tag{7.14}$$

It can also be shown that, under the correct model specification,

$$\frac{1}{T}\sum_{t=1}^{T}(\hat{e}_t - k)^2 \rightarrow_{a.s.} E(\boldsymbol{a}_t'\boldsymbol{\Sigma}_t^{-1}\boldsymbol{a}_t - k)^2,$$

$$E(\boldsymbol{a}_t'\boldsymbol{\Sigma}_t^{-1}\boldsymbol{a}_t - k)^2 = E(\boldsymbol{\epsilon}_t'\boldsymbol{\epsilon}_t)^2 - k^2 = [E(\epsilon_{it}^4) - 1]k \equiv ck,$$

where $c = E(\epsilon_{it}^4) - 1$. Ling and Li (1997) propose using $(\hat{\rho}_1, \ldots, \hat{\rho}_m)$ of the transformed residuals $\hat{e}_t$ to check the fitted volatility model. To this end, they derive the asymptotic joint distribution of $(\hat{\rho}_1, \ldots, \hat{\rho}_m)$. Since the denominator of $\hat{\rho}_l$ in Equation (7.14) converges to a constant, it suffices to consider the numerator in studying the limiting properties of $\hat{\rho}_l$. Let

$$\hat{C}_l = \frac{1}{T}\sum_{t=l+1}^{T}(\hat{e}_t - k)(\hat{e}_{t-l} - k),$$

be the lag-l sample autocovariance of the transformed residual $\hat{e}_t$ and C_l be its theoretical counterpart with $\hat{e}_t$ replaced by $e_t = \boldsymbol{a}_t'\boldsymbol{\Sigma}_t^{-1}\boldsymbol{a}_t$. To investigate the properties of $\hat{C}_l$ as a function of the estimate $\hat{\boldsymbol{\theta}}$, we use the Taylor series expansion

$$\hat{C}_l \approx C_l + \frac{\partial C_l}{\partial \boldsymbol{\theta}'}(\hat{\boldsymbol{\theta}} - \boldsymbol{\theta}).$$

It remains to consider the partial derivatives of C_l with respect to $\boldsymbol{\theta}$. Using the matrix differentials reviewed in Appendix A, we have

$$\begin{aligned} d(e_t) &= d(\boldsymbol{a}_t')\boldsymbol{\Sigma}_t^{-1}\boldsymbol{a}_t + \boldsymbol{a}_t' d(\boldsymbol{\Sigma}_t^{-1})\boldsymbol{a}_t + \boldsymbol{a}_t'\boldsymbol{\Sigma}_t^{-1} d(\boldsymbol{a}_t) \\ &= -[d(\boldsymbol{\mu}_t)\boldsymbol{\Sigma}_t^{-1}\boldsymbol{a}_t + \boldsymbol{a}_t'\boldsymbol{\Sigma}_t^{-1} d(\boldsymbol{\Sigma}_t)\boldsymbol{\Sigma}_t^{-1}\boldsymbol{a}_t + \boldsymbol{a}_t'\boldsymbol{\Sigma}_t^{-1} d(\boldsymbol{\mu}_t)]. \end{aligned}$$

Using $E(\boldsymbol{a}_t^o|F_{t-1}) = 0$ and $\text{Var}(\boldsymbol{a}_t^o|F_{t-1}) = \boldsymbol{\Sigma}_t$ when the model is correctly specified, we have

$$\begin{aligned} E[d(e_t)|F_{t-1}] &= -E[\boldsymbol{a}_t'\boldsymbol{\Sigma}_t^{-1} d(\boldsymbol{\Sigma}_t)\boldsymbol{\Sigma}_t^{-1}\boldsymbol{a}_t|F_{t-1}] \\ &= -E[tr\{d(\boldsymbol{\Sigma}_t)\boldsymbol{\Sigma}_t^{-1}\boldsymbol{a}_t\boldsymbol{a}_t'\boldsymbol{\Sigma}_t^{-1}\}|F_{t-1}] \\ &= -E[tr\{d(\boldsymbol{\Sigma}_t)\boldsymbol{\Sigma}_t^{-1}\}] = -E[d\{\text{vec}(\boldsymbol{\Sigma}_t)\}'\text{vec}(\boldsymbol{\Sigma}_t^{-1})], \end{aligned}$$

where we have used the identity $tr(\boldsymbol{ABD}) = [\text{vec}(\boldsymbol{A}')]'(\boldsymbol{I}_k \otimes \boldsymbol{B})\text{vec}(\boldsymbol{D})$ with $\boldsymbol{B} = \boldsymbol{I}_k$ and some properties of matrix differential. Consequently, by iterative expectation and the ergodic theorem, we obtain

$$\frac{\partial C_l}{\partial \boldsymbol{\theta}'} \to_{a.s.} -\boldsymbol{X}_l', \quad \text{with} \quad \boldsymbol{X}_l = E\left[\frac{\partial \text{vec}(\boldsymbol{\Sigma}_t)}{\partial \boldsymbol{\theta}'}\right]' \text{vec}(\boldsymbol{\Sigma}_t^{-1} e_{t-l}). \tag{7.15}$$

Next, let $\boldsymbol{C}_m = (C_1, \ldots, C_m)'$, $\hat{\boldsymbol{C}}_m = (\hat{C}_1, \ldots, \hat{C}_m)'$, and $\boldsymbol{X}' = [\boldsymbol{X}_1, \ldots, \boldsymbol{X}_m]$. We have

$$\hat{\boldsymbol{C}}_m \approx \boldsymbol{C}_m - \boldsymbol{X}(\hat{\boldsymbol{\theta}} - \boldsymbol{\theta}).$$

Furthermore, define $\hat{\boldsymbol{\rho}}_m = (\hat{\rho}_1, \ldots, \hat{\rho}_m)'$ and $\boldsymbol{\rho}_m = (\rho_1, \ldots, \rho_m)'$, where $\hat{\rho}_l$ is defined in Equation (7.14) and ρ_l is the theoretical counterpart of $\hat{\rho}_l$. Ling and Li (1997) use Theorem 7.1 and prior derivations to obtain the following theorem.

Theorem 7.2 Under the regularity conditions of Theorem 7.1 and the additional conditions of the third and fourth moments of $\boldsymbol{\epsilon}_t$, the transformed quadratic residuals $\hat{e}_t$ provide

$$\begin{aligned} \sqrt{T}\hat{\boldsymbol{C}}_m &\Rightarrow_d \ N[\boldsymbol{0}, (ck)^2\boldsymbol{\Omega}] \\ \sqrt{T}\hat{\boldsymbol{\rho}}_m &\Rightarrow_d \ N(\boldsymbol{0}, \boldsymbol{\Omega}), \end{aligned}$$

where $\boldsymbol{\Omega} = \boldsymbol{I}_m - \boldsymbol{X}(c\boldsymbol{A}_0^{-1} - \boldsymbol{A}_o^{-1}\boldsymbol{B}_o\boldsymbol{A}_o^{-1})\boldsymbol{X}'/(ck)^2$ and $(ck)^2$ can be estimated by $\hat{C}_0^2$.

From Theorem 7.2, the asymptotic variance of the sample ACF $\hat{\rho}_l$ of the transformed residual $\hat{e}_t$ is in general less than $1/T$. However, if $\boldsymbol{\Sigma}_t$ is time-invariant, then $\boldsymbol{X} = \boldsymbol{0}$ and the asymptotic variance of $\hat{\rho}_l$ becomes $1/T$. Also, if $\boldsymbol{\epsilon}_t$ is normally distributed, then $\boldsymbol{\Omega} = \boldsymbol{I}_m - \boldsymbol{X}\boldsymbol{A}_o^{-1}\boldsymbol{X}'/(2k)^2$. Finally, from Theorem 7.2, the test statistic

$$Q_{ll}(m) = T\hat{\boldsymbol{\rho}}_m'\boldsymbol{\Omega}^{-1}\hat{\boldsymbol{\rho}}_m \tag{7.16}$$

is asymptotically distributed as χ^2_m if the volatility model is correctly specified. We shall refer to the test statistic $Q_{ll}(m)$ as the Ling–Li volatility test.

7.3.2 Tse Statistics

Tse (2002) proposes some residual-based statistics for checking a fitted multivariate volatility model. Unlike the test statistic of Ling and Li (1997) who employed the transformed quadratic residual $\hat{e}_t$, which is a scalar series. Tse (2002) focuses on the squares of elements of the (marginally) standardized residuals. Let $\hat{\boldsymbol{\Sigma}}_t$ be the fitted volatility matrix at time t and $\hat{\boldsymbol{a}}_t = \boldsymbol{z}_t - \hat{\boldsymbol{\mu}}_t$ be the residual. Define the ith standardized residual as

$$\hat{\eta}_{it} = \frac{\hat{a}_{it}}{\sqrt{\hat{\sigma}_{ii,t}}}, \quad i = 1, \ldots, k, \tag{7.17}$$

where $\hat{\sigma}_{ii,t}$ is the (i, i)th element of $\hat{\boldsymbol{\Sigma}}_t$. If the model is correctly specified, then $\{\hat{\eta}_{it}\}$ is asymptotically an iid sequence with mean zero and variance 1. Consequently, if the model is correctly specified, then

1. The squared series $\{\hat{\eta}^2_{it}\}$ has no serial correlations for $i = 1, \ldots, k$.
2. The cross-product series $\hat{\eta}_{it}\hat{\eta}_{jt}$ has no serial correlations for $i \neq j$ and $1 \leq i, j \leq k$.

One can use the idea of Lagrange multiplier test to verify the serial correlations in $\hat{\eta}^2_{it}$ and $\hat{\eta}_{it}\hat{\eta}_{jt}$. More specifically, Tse (2002) considers the linear regressions

$$\hat{\eta}^2_{it} - 1 = \hat{\boldsymbol{d}}_{it}'\boldsymbol{\delta}_i + \xi_{it}, \quad i = 1, \ldots, k \tag{7.18}$$

$$\hat{\eta}_{it}\hat{\eta}_{jt} - \hat{\rho}_{ij,t} = \hat{\boldsymbol{d}}_{ij,t}'\boldsymbol{\delta}_{ij} + \xi_{ij,t}, \quad 1 \leq i < j \leq k, \tag{7.19}$$

where $\hat{\boldsymbol{d}}_{it} = (\hat{\eta}^2_{i,t-1}, \ldots, \hat{\eta}^2_{i,t-m})'$, $\hat{\rho}_{ij,t} = \hat{\sigma}_{ij,t}/\sqrt{\hat{\sigma}_{ii,t}\hat{\sigma}_{jj,t}}$ is the fitted correlation coefficient at time t, $\hat{\sigma}_{ij,t}$ is the (i, j)th element of $\hat{\boldsymbol{\Sigma}}_t$, and $\hat{\boldsymbol{d}}_{ij,t} = (\hat{\eta}_{i,t-1}\hat{\eta}_{j,t-1}, \ldots, \hat{\eta}_{i,j-m}\hat{\eta}_{j,t-m})'$. In Equations (7.18) and (7.19), ξ_{it} and $\xi_{ij,t}$ denote the error terms. If the model is correctly specified, then $\boldsymbol{\delta}_i = \boldsymbol{0}$ and $\boldsymbol{\delta}_{ij} = \boldsymbol{0}$ for all i and j.

Let $\hat{\boldsymbol{\delta}}_i$ and $\hat{\boldsymbol{\delta}}_{ij}$ be the ordinary least-squares estimates of $\boldsymbol{\delta}_i$ and $\boldsymbol{\delta}_{ij}$ of Equations (7.18) and (7.19), respectively, and let η_{it} and $\boldsymbol{d}_{it}$ denote the theoretical counterpart of $\hat{\eta}_{it}$ and $\hat{\boldsymbol{d}}_{it}$, respectively. Also, let $\boldsymbol{G} = \boldsymbol{A}_o^{-1}\boldsymbol{B}_o\boldsymbol{A}_o^{-1}$ of Theorem 7.1. Under some regularity conditions, Tse (2002) showed the following two theorems.

Theorem 7.3 If the model is correctly specified and the regularity conditions of Theorem 7.1 hold, then $\sqrt{T}\hat{\boldsymbol{\delta}}_i \Rightarrow N(\mathbf{0}, \boldsymbol{L}_i^{-1}\boldsymbol{\Omega}\boldsymbol{L}_i^{-1})$, where

$$\boldsymbol{L}_i = \text{plim}\left(\frac{1}{T}\sum_{t=1}^{T}\boldsymbol{d}_{it}\boldsymbol{d}_{it}'\right), \quad \boldsymbol{\Omega}_i = c_i\boldsymbol{L}_i - \boldsymbol{Q}_i\boldsymbol{G}\boldsymbol{Q}_i'$$

with plim denoting convergence in probability and

$$\boldsymbol{Q}_i = \text{plim}\left(\frac{1}{T}\sum_{t=1}^{T}\boldsymbol{d}_{it}\frac{\partial\eta_{it}^2}{\partial\boldsymbol{\theta}'}\right), \quad c_i = E[(\eta_{it}^2 - 1)^2].$$

In finite samples, the quantities c_i, $\boldsymbol{L}_i$, and $\boldsymbol{Q}_i$ can be estimated consistently by $\hat{c}_i = T^{-1}\sum_{t=1}^{T}(\hat{\eta}_{it}^2 - 1)^2$, $\hat{\boldsymbol{L}}_i = T^{-1}\sum_{t=1}^{T}\hat{\boldsymbol{d}}_{it}\hat{\boldsymbol{d}}_{it}'$, and $\hat{\boldsymbol{Q}}_i = T^{-1}\sum_{t=1}^{T}\hat{\boldsymbol{d}}_{it}(\partial\hat{\eta}_{it}^2/\partial\boldsymbol{\theta}')$. Based on Theorem 7.3, we define

$$Q_t(i,m) = T\hat{\boldsymbol{\delta}}_i\hat{\boldsymbol{L}}_i\hat{\boldsymbol{\Omega}}^{-1}\hat{\boldsymbol{L}}_i\hat{\boldsymbol{\delta}}_i, \tag{7.20}$$

which is asymptotically distributed as χ_m^2. We can use $Q_t(i,m)$ to check the serial correlations in $\hat{\eta}_{it}^2$ for $i = 1, \ldots, k$.

Theorem 7.4 If the model is correctly specified and the regularity conditions of Theorem 7.1 hold, then $\sqrt{T}\hat{\boldsymbol{\delta}}_{ij} \Rightarrow_d N(\mathbf{0}, \boldsymbol{L}_{ij}^{-1}\boldsymbol{\Omega}_{ij}\boldsymbol{L}_{ij}^{-1})$, where

$$\boldsymbol{L}_{ij} = \text{plim}\left(\frac{1}{T}\sum_{t=1}^{T}\boldsymbol{d}_{ij,t}\boldsymbol{d}_{ij,t}'\right), \quad \boldsymbol{\Omega}_{ij} = c_{ij}\boldsymbol{L}_{ij} - \boldsymbol{Q}_{ij}\boldsymbol{G}\boldsymbol{Q}_{ij}'$$

with

$$\boldsymbol{Q}_{ij} = \text{plim}\left(\frac{1}{T}\sum_{t=1}^{T}\boldsymbol{d}_{ij,t}\frac{\partial(\eta_{it}\eta_{jt} - \rho_{ij,t})}{\partial\boldsymbol{\theta}'}\right), \quad c_{ij} = E[(\eta_{it}\eta_{jt} - \rho_{ij,t})^2].$$

In finite samples, c_{ij}, $\boldsymbol{L}_{ij}$, and $\boldsymbol{Q}_{ij}$ can be estimated by $\hat{c}_{ij} = T^{-1}\sum_{t=1}^{T}(\hat{\eta}_{ij}\hat{\eta}_{jt} - \hat{\rho}_{ij,t})^2$, $\hat{\boldsymbol{L}}_{ij} = T^{-1}\sum_{t=1}^{T}\hat{\boldsymbol{d}}_{ij,t}\hat{\boldsymbol{d}}_{ij,t}'$, and $\hat{\boldsymbol{Q}}_{ij} = T^{-1}\sum_{t=1}^{T}\hat{\boldsymbol{d}}_{ij,t}(\partial(\hat{\eta}_{it}\hat{\eta}_{jt} - \hat{\rho}_{ij,t})/\partial\boldsymbol{\theta}')$, respectively. Define

$$Q_t(ij,m) = T\hat{\delta}_{ij}'\hat{\boldsymbol{L}}_{ij}\hat{\boldsymbol{\Omega}}_{ij}^{-1}\hat{\boldsymbol{L}}_{ij}\hat{\boldsymbol{\delta}}_{ij}, \quad 1 \le i < j \le k, \tag{7.21}$$

which is asymptotically distributed as χ_m^2. We can use $Q_t(ij,m)$ to test the serial correlations in $\hat{\eta}_{it}\hat{\eta}_{jt}$ series.

The test statistics $Q_{ll}(m)$ of Ling and Li (1997) and $Q_t(i,m)$ and $Q_t(ij,m)$ of Tse (2002) all depend on the partial derivatives of the fitted volatility $\hat{\boldsymbol{\Sigma}}_t$ with respect to the parameter vector $\boldsymbol{\theta}$. For some special multivariate volatility models, the derivatives either have a closed-form formula or can be evaluated recursively. For complicated volatility models, numeric methods are needed. If the distribution of the innovation $\boldsymbol{\epsilon}_t$ of Equation (7.2) is known, one might be able to compute the covariance matrices involved in the test statistics via some parametric bootstrap procedures. An obvious complication of using the test statistics of Tse (2002) is the choice of type I error when k is large. Some simulation results in Ling and Li (1997) and in Tse (2002) show that the proposed test statistics work reasonably well in finite samples, but the simulations only involve small k.

7.4 EXPONENTIALLY WEIGHTED MOVING AVERAGE (EWMA)

We now turn to some specific multivariate volatility models providing positive-definite volatility matrices. A simple approach to estimate time-varying volatilities is the EWMA method. Again, let $\hat{\boldsymbol{a}}_t$ be the residuals of the mean equation. The EWMA model for volatility is

$$\hat{\boldsymbol{\Sigma}}_t = \lambda\hat{\boldsymbol{\Sigma}}_{t-1} + (1-\lambda)\hat{\boldsymbol{a}}_{t-1}\hat{\boldsymbol{a}}'_{t-1}, \tag{7.22}$$

where $0 < \lambda < 1$ denotes the decaying rate or the persistence parameter. If one starts the recursion with a positive-definite matrix $\hat{\boldsymbol{\Sigma}}_0$, then the volatility matrix $\hat{\boldsymbol{\Sigma}}_t$ is positive-definite for all t. An obvious choice of $\hat{\boldsymbol{\Sigma}}_0$ is the sample covariance matrix of $\hat{\boldsymbol{a}}_t$ provided that $T > k$. This is the initial estimate used in this chapter. In practice, the parameter λ can be fixed *a priori* or estimated by the QMLE. Limited experience indicates that $\hat{\lambda} \approx 0.96$ for many asset return series. This model is extremely parsimonious and the resulting volatility matrices are easy to update, but it tends to be rejected by diagnostic checks in an application. This is not surprising as it is hard to imagine that a single decaying parameter can adequately govern the time decay of all conditional variances and covariances.

Example 7.1 Consider the monthly log returns of CRSP decile 1, 2, and 5 portfolios from January 1961 to September 2011. Preliminary analysis indicates there exist some minor serial correlations in the return series so that we use a VAR(1) model for the mean equation. In other words, denoting the monthly log returns by $\boldsymbol{z}_t$, we employ $\boldsymbol{\mu}_t = \boldsymbol{\phi}_0 + \boldsymbol{\phi}_1\boldsymbol{z}_{t-1}$. The fitted mean equation is

$$\hat{\boldsymbol{\mu}}_t = \begin{bmatrix} 0.0064 \\ 0.0070 \\ 0.0073 \end{bmatrix} + \begin{bmatrix} -0.194 & 0.224 & 0.008 \\ -0.232 & 0.366 & -0.042 \\ -0.313 & 0.452 & 0.003 \end{bmatrix} \boldsymbol{z}_{t-1}.$$

The test statistics of Section 7.1 confirm the existence of conditional heteroscedasticity in the return series. All p-values of the four test statistics used are close to 0. We then fit an EWMA model to the residual series $\hat{\boldsymbol{a}}_t = \boldsymbol{z}_t - \hat{\boldsymbol{\mu}}_t$. The QMLE

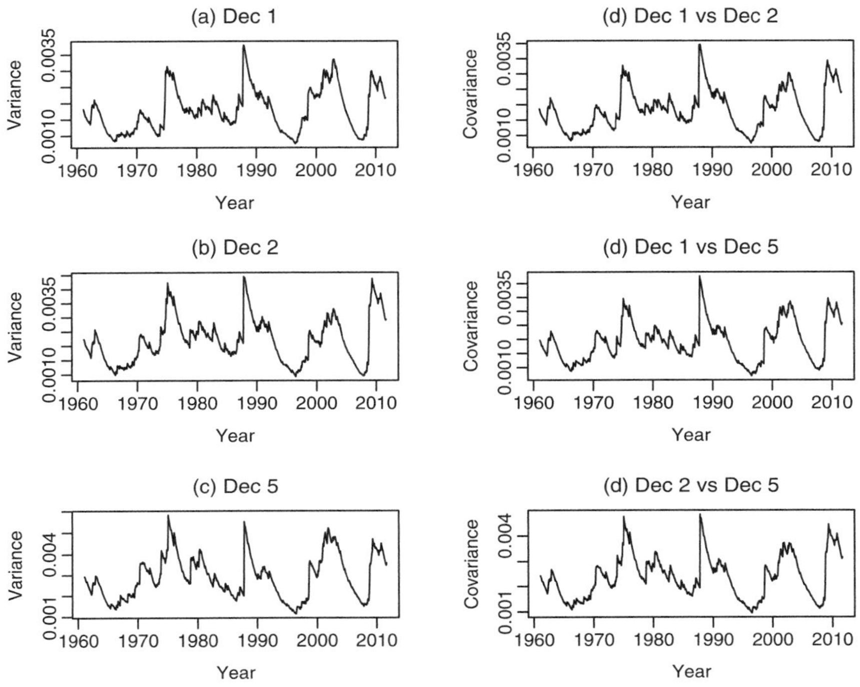

FIGURE 7.3 Time plots of elements of the volatility matrix $\hat{\boldsymbol{\Sigma}}_t$ for the monthly log returns of CRSP decile 1, 2, and 5, portfolios via the EWMA method with $\hat{\lambda} = 0.964$.

of λ is $\hat{\lambda} = 0.964(0.0055)$, where the number in parentheses denotes asymptotic standard error. As expected, all test statistics of Section 7.1 indicate that the standardized residuals of the EWMA model still have conditional heteroscedasticity. All p-values of the test statistics are also close to 0. Figure 7.3 shows the time plots of the fitted conditional variances and covariances of $\boldsymbol{z}_t$ via the EWMA method. The high correlations between the portfolios are understandable. □

Remark: The command for EWMA estimation in the `MTS` package is `EWMAvol`. The command has two options for specifying the smoothing parameter λ. A positive λ denotes a prespecified value and, hence, no estimation is performed. A negative λ denotes estimation. □

R Demonstration: EWMA method.

```
> da=read.table("m-dec125910-6111.txt",header=T)
> rtn=log(da[,2:4]+1)
> m1=VAR(rtn,1)
```

```
Constant term:
Estimates:  0.006376978 0.007034631 0.007342962
Std.Error:  0.001759562 0.001950008 0.002237004
AR coefficient matrix
AR( 1 )-matrix
        [,1]  [,2]     [,3]
[1,] -0.194 0.224  0.00836
[2,] -0.232 0.366 -0.04186
[3,] -0.313 0.452  0.00238
standard error
      [,1]  [,2]  [,3]
[1,] 0.108 0.160 0.101
[2,] 0.120 0.177 0.111
[3,] 0.138 0.204 0.128
at=m1$residuals
> MarchTest(at)
Q(m) of squared scalar series(LM test):
Test statistic:  244.7878  p-value:  0
Rank-based Test:
Test statistic:  215.215  p-value:  0
Q_k(m) of squared series:
Test statistic:  176.9811  p-value:  1.252294e-07
Robust Q_k(m) statistic (5% trimming):
Test statistics:  155.2633  p-value:  2.347499e-05
### Estimation
> m2=EWMAvol(at,lambda=-0.1)
Coefficient(s):
        Estimate  Std. Error  t value Pr(>|t|)
lambda   0.96427     0.00549    175.6   <2e-16 ***
> Sigma.t=m2$Sigma.t
> m3=MCHdiag(at,Sigma.t)  ## Model checking
Test results:
Q(m) of et:
Test and p-value:  59.5436 4.421175e-09
Rank-based test:
Test and p-value:  125.0929 0
Qk(m) of epsilon_t:
Test and p-value:  189.5403 4.518401e-09
Robust Qk(m):
Test and p-value:  228.234 5.57332e-14
```

Remark: It is easy to generalize the EWMA approach to allow for each component series a_{it} has its own smoothing parameter λ_i. The resulting equation is

$$\boldsymbol{\Sigma}_t = \boldsymbol{D}\boldsymbol{\Sigma}_{t-1}\boldsymbol{D} + \boldsymbol{D}_1\boldsymbol{a}_{t-1}\boldsymbol{a}'_{t-1}\boldsymbol{D}_1,$$

where $\boldsymbol{D} = \text{diag}\{\sqrt{\lambda_1}, \ldots, \sqrt{\lambda_k}\}$ and $\boldsymbol{D}_1 = \text{diag}\{\sqrt{1-\lambda_1}, \ldots, \sqrt{1-\lambda_k}\}$. □

7.5 BEKK MODELS

The BEKK model of Engle and Kroner (1995) represents another extreme of multivariate volatility modeling in the sense that it uses too many parameters. For a k-dimensional time series $\boldsymbol{z}_t$, the BEKK(1,1) volatility model assumes the form

$$\boldsymbol{\Sigma}_t = \boldsymbol{A}_0\boldsymbol{A}_0' + \boldsymbol{A}_1\boldsymbol{a}_{t-1}\boldsymbol{a}_{t-1}'\boldsymbol{A}_1' + \boldsymbol{B}_1\boldsymbol{\Sigma}_{t-1}\boldsymbol{B}_1', \quad (7.23)$$

where $\boldsymbol{A}_0$ is a lower triangular matrix such that $\boldsymbol{A}_0\boldsymbol{A}_0'$ is positive-definite and $\boldsymbol{A}_1$ and $\boldsymbol{B}_1$ are $k \times k$ matrices. The model contains $k^2 + [k(k+1)/2]$ parameters and provides positive-definite volatility matrix $\boldsymbol{\Sigma}_t$ for all t. Higher-order BEKK models have also been studied in the literature, but we shall focus on the simple BEKK(1,1) model.

From Equation (7.23), we have

$$\begin{aligned}\text{vec}(\boldsymbol{\Sigma}_t) &= (\boldsymbol{A}_0 \otimes \boldsymbol{A}_0)\text{vec}(\boldsymbol{I}_k) + (\boldsymbol{A}_1 \otimes \boldsymbol{A}_1)\text{vec}(\boldsymbol{a}_{t-1}\boldsymbol{a}_{t-1}') \\ &+ (\boldsymbol{B}_1 \otimes \boldsymbol{B}_1)\text{vec}(\boldsymbol{\Sigma}_{t-1}). \quad (7.24)\end{aligned}$$

Taking expectation and assuming the inverse exists, we obtain the unconditional covariance matrix $\boldsymbol{\Sigma}$ of $\boldsymbol{a}_t$ via

$$\text{vec}(\boldsymbol{\Sigma}) = (\boldsymbol{I}_{k^2} - \boldsymbol{A}_1 \otimes \boldsymbol{A}_1 - \boldsymbol{B}_1 \otimes \boldsymbol{B}_1)^{-1}(\boldsymbol{A}_0 \otimes \boldsymbol{A}_0)\text{vec}(\boldsymbol{I}_k). \quad (7.25)$$

Consequently, for the BEKK(1,1) model in Equation (7.23) to have asymptotic (or unconditional) volatility matrix, all eigenvalues of the matrix $\boldsymbol{A}_1 \otimes \boldsymbol{A}_1 + \boldsymbol{B}_1 \otimes \boldsymbol{B}_1$ must be strictly between 0 and 1 (exclusive). For stationarity conditions of general BEKK models, see Francq and Zakoian (2010, Section 11.3). Next, define $\boldsymbol{\xi}_t = \boldsymbol{a}_t\boldsymbol{a}_t' - \boldsymbol{\Sigma}_t$ as the deviation matrix of $\boldsymbol{a}_t\boldsymbol{a}_t'$ from its conditional covariance matrix $\boldsymbol{\Sigma}_t$. From Equation (7.24), we have

$$\begin{aligned}\text{vec}(\boldsymbol{a}_t\boldsymbol{a}_t') &= (\boldsymbol{A}_0 \otimes \boldsymbol{A}_0)\text{vec}(\boldsymbol{I}_k) + (\boldsymbol{A}_1 \otimes \boldsymbol{A}_1 + \boldsymbol{B}_1 \otimes \boldsymbol{B}_1)\text{vec}(\boldsymbol{a}_{t-1}\boldsymbol{a}_{t-1}') \\ &+ \text{vec}(\boldsymbol{\xi}_t) - (\boldsymbol{B}_1 \otimes \boldsymbol{B}_1)\text{vec}(\boldsymbol{\xi}_{t-1}). \quad (7.26)\end{aligned}$$

It is easy to verify that (a) $E[\text{vec}(\boldsymbol{\xi}_t)] = \boldsymbol{0}$ and (b) $E[\text{vec}(\boldsymbol{\xi}_t)\{\text{vec}(\boldsymbol{\xi}_{t-j})\}'] = \boldsymbol{0}$ for $j > 0$. Therefore, Equation (7.26) is in the form of a VARMA(1,1) model for the $\text{vec}(\boldsymbol{a}_t\boldsymbol{a}_t')$ process. In this sense, BEKK(1,1) model can be regarded as a multivariate generalization of the univariate GARCH(1,1) model. Similarly to the univariate case, $\text{vec}(\boldsymbol{\xi}_t)$ does not form an iid sequence of random vectors. The model representation in Equation (7.26) can be used to study the properties of the BEKK(1,1) model. For instance, it can be used to express $\boldsymbol{a}_t\boldsymbol{a}_t'$ as a linear combination of its lagged values $\boldsymbol{a}_{t-j}\boldsymbol{a}_{t-j}$ with $j > 0$. It can also be used to derive moment equations for the vec($\boldsymbol{a}_t\boldsymbol{a}_t'$) process.

Example 7.2 Consider the monthly log returns of IBM stock and the S&P composite index from January 1961 to December 2011 for 612 observations. The data are shown in Figure 7.2 and obtained from CRSP. Let $\boldsymbol{z}_t$ denote the monthly log return series. As shown in Section 7.1, $\boldsymbol{z}_t$ has no significant serial correlations, but has strong conditional heteroscedasticity. Therefore, we use $\boldsymbol{\mu}_t = \boldsymbol{\mu}$ and apply the BEKK(1,1) model of Equation (7.23) to $\boldsymbol{z}_t$. The QMLE of the mean equation is $\hat{\boldsymbol{\mu}} = (0.00776, 0.00565)'$. The QMLEs and their standard errors of the volatility equation are given in Table 7.3. All estimates but two are statistically significant at the conventional 5% level. The insignificant estimates are the (2,1)th element of $\hat{\boldsymbol{A}}_1$ and $\hat{\boldsymbol{B}}_1$ matrices.

To check the adequacy of the fitted BEKK(1,1) model, we employ the scalar transformed residuals

$$e_t = \hat{\boldsymbol{a}}_t'\hat{\boldsymbol{\Sigma}}_t^{-1}\hat{\boldsymbol{a}}_t - k$$

and apply the portmanteau statistic and the rank-based portmanteau statistic to check the serial correlations in $\hat{e}_t$. Both statistics fail to reject the null hypothesis of no serial correlations. We also apply the multivariate portmanteau statistic and its robust version in Section 7.1 to the standardized residuals $\hat{\boldsymbol{\epsilon}}_t = \hat{\boldsymbol{\Sigma}}_t^{-1/2}\hat{\boldsymbol{a}}_t$. For this particular case, both multivariate portmanteau statistics also fail to reject the null hypothesis that $\hat{\boldsymbol{\epsilon}}_t$ has no serial or cross-sectional correlations. See the R demonstration for detail. In summary, the fitted BEKK(1,1) model in Table 7.3 seems adequate. Figure 7.4 shows the time plots of the fitted volatility series and the conditional covariance between the log returns of IBM stock and the S&P composite index. □

Remark: Estimation of BEKK(1,1) models for k = 2 or 3 in the `MTS` package is via the command `BEKK11`. No higher-order models are available. □

TABLE 7.3 Estimation of BEKK(1,1) Model to the Monthly Log Returns of IBM Stock and S&P Composite Index from January 1961 to December 2011 for 612 Observations

Parameter	Estimates		Standard Errors	
	0.0140	0.0000	0.0041	0.0000
$\hat{\boldsymbol{A}}_0$	0.0084	0.0070	0.0027	0.0019
	0.1565	0.2340	0.0600	0.0858
$\hat{\boldsymbol{A}}_1$	−0.0593	0.4098	0.0325	0.0540
	0.9764	−0.0929	0.0228	0.0323
$\hat{\boldsymbol{B}}_1$	0.0145	0.8908	0.0120	0.0248

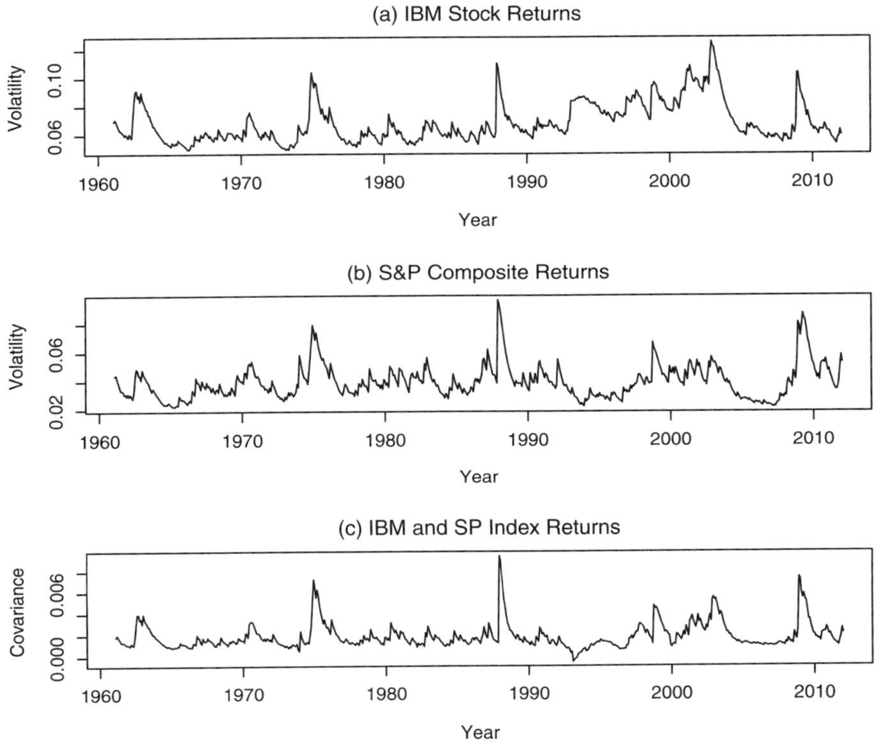

FIGURE 7.4 Time plots of volatilities and conditional covariance of monthly log returns of (a) IBM stock, (b) S&P composite index, and IBM and SP index returns from January 1961 to December 2011 based on a BEKK(1,1) model.

R Demonstration: BEKK(1,1) estimation. Output edited.

```
> da=read.table("m-ibmsp-6111.txt",header=T)
> rtn=log(da[,2:3]+1)
> m1a=BEKK11(rtn)
Initial estimates:  0.00772774 0.005023909 0.06977651 ...
Lower limits:  -0.0772774 -0.05023909 0.0139553 ...
Upper limits:  0.0772774 0.05023909 0.07675416 ...

Coefficient(s):
           Estimate   Std. Error   t value    Pr(>|t|)
mu1.ibm  0.00775929  0.00253971  3.05518 0.00224922 **
mu2.sp   0.00565084  0.00154553  3.65624 0.00025594 ***
A011     0.01395530  0.00408488  3.41633 0.00063472 ***
A021     0.00838972  0.00268086  3.12949 0.00175112 **
A022     0.00700592  0.00193247  3.62537 0.00028855 ***
A11      0.15648877  0.06002824  2.60692 0.00913610 **
```

```
A21     -0.05926387  0.03253895 -1.82132 0.06855810 .
A12      0.23398204  0.08575142  2.72861 0.00636022 **
A22      0.40977179  0.05400961  7.58702 3.2641e-14 ***
B11      0.97639151  0.02283328 42.76178 < 2.22e-16 ***
B21      0.01449633  0.01196030  1.21204 0.22549813
B12     -0.09287696  0.03227225 -2.87792 0.00400306 **
B22      0.89077633  0.02476925 35.96300 < 2.22e-16 ***
> names(m1a)
[1] "estimates"  "HessianMtx" "Sigma.t"
> Sigma.t=m1a$Sigma.t
> at=cbind(rtn[,1]-0.00776,rtn[,2]-0.00565) # Remove
  conditional means
> MCHdiag(at,Sigma.t)
Test results:
Q(m) of et:
Test and p-value:  5.280566 0.8716653
Rank-based test:
Test and p-value:  16.02931 0.0987965
Qk(m) of epsilon_t:
Test and p-value:  29.46281 0.889654
Robust Qk(m):
Test and p-value:  52.13744 0.09462895
```

7.5.1 Discussion

The BEKK models provide a general framework for using the GARCH family of multivariate volatility models. Indeed, the BEKK model enjoys certain nice model representations shown in Equations (7.24) and (7.26). The model also produces positive-definite volatility matrices. However, the model encounters some difficulties in real applications when k is moderate or large. First, the BEKK model, even with order (1,1), contains too many parameters. For $k = 3$, the volatility equation of a BEKK(1,1) model already has 24 parameters. Consequently, it is hard to estimate a BEKK(1,1) model when $k > 3$. Second, limited experience indicates that some of the parameter estimates of a BEKK(1,1) model are statistically insignificant at the traditional 5% level. Yet there exists no simple direct link between the parameters in $\boldsymbol{A}_1$ and $\boldsymbol{B}_1$ and the components of a volatility matrix because the volatility matrix is a nonlinear function of elements of $\boldsymbol{A}_1$ and $\boldsymbol{B}_1$. Furthermore, no methods are currently available to search for simplifying structures embedded in a BEKK model. It seems that an unrestricted BEKK(1,1) model is only applicable in practice when k is small.

7.6 CHOLESKY DECOMPOSITION AND VOLATILITY MODELING

For ease in obtaining positive-definite volatility matrices, Cholesky decomposition has been used in the literature to model multivariate volatility. See Tsay (2010, Chapter 10) and Lopes, McCulloch and Tsay (2013) and the references therein.

In this section, we briefly discuss the framework of multivariate volatility modeling via Cholesky decomposition.

Consider a k-dimensional innovation $\boldsymbol{a}_t$ with volatility matrix $\boldsymbol{\Sigma}_t$. Again, let F_{t-1} denote the information available at time $t-1$. Cholesky decomposition performs linear orthogonal transformations via a system of multiple linear regressions. Let $b_{1t} = a_{1t}$ and consider the simple linear regression

$$a_{2t} = \beta_{21,t} b_{1t} + b_{2t}, \tag{7.27}$$

where $\beta_{21,t} = \text{Cov}(a_{2t}, b_{1t}|F_{t-1})/\text{Var}(b_{1t}|F_{t-1})$. In practice, $\beta_{21,t}$ is estimated by the ordinary least-squares method using the available data in F_{t-1}. Based on the least-squares theory, b_{2t} is orthogonal to $b_{1t} = a_{1t}$ and $\text{Var}(b_{2t}|F_{t-1}) = \text{Var}(a_{2t}|F_{t-1}) - \beta_{21,t}^2 \text{Var}(a_{1t}|F_{t-1})$. Next, consider the multiple linear regression

$$a_{3t} = \gamma_{31,t} b_{1t} + \gamma_{32,t} b_{2t} + b_{3t} = \beta_{31,t} a_{1t} + \beta_{32,t} a_{2t} + b_{3t}, \tag{7.28}$$

where $\beta_{3j,t}$ are linear functions of $\gamma_{3j,t}$ and $\beta_{21,t}$ for $j = 1$ and 2. Again, via the least-squares theory, b_{3t} is orthogonal to both a_{1t} and a_{2t} and, hence, to b_{1t} and b_{2t}. Repeat the prior process of multiple linear regressions until

$$a_{kt} = \beta_{k1,t} a_{1t} + \cdots + \beta_{k,k-1,t} a_{k-1,t} + b_{kt}. \tag{7.29}$$

Applying the least-squares theory, we can obtain $\beta_{kj,t}$ and the conditional variance of b_{kt} given F_{t-1}. In addition, b_{kt} is orthogonal to a_{it} and, hence, b_{it} for $i = 1, \ldots, k-1$. Putting the system of linear regressions from Equation (7.27) to Equation (7.29) together and moving the a_{it} to the left side of the equality, we have

$$\begin{bmatrix} 1 & 0 & 0 & \cdots & 0 & 0 \\ -\beta_{21,t} & 1 & 0 & \cdots & 0 & 0 \\ -\beta_{31,t} & -\beta_{32,t} & 1 & \cdots & 0 & 0 \\ \vdots & \vdots & \vdots & & \vdots & \vdots \\ -\beta_{k1,t} & -\beta_{k2,t} & -\beta_{k3,t} & \cdots & -\beta_{k,k-1,t} & 1 \end{bmatrix} \begin{bmatrix} a_{1t} \\ a_{2t} \\ a_{3t} \\ \vdots \\ a_{kt} \end{bmatrix} = \begin{bmatrix} b_{1t} \\ b_{2t} \\ b_{3t} \\ \vdots \\ b_{kt} \end{bmatrix}, \tag{7.30}$$

where via construction b_{it} are mutually orthogonal so that the volatility matrix of $\boldsymbol{b}_t = (b_{1t}, \ldots, b_{kt})'$ is diagonal. Writing Equation (7.30) in matrix form, we have

$$\boldsymbol{\beta}_t \boldsymbol{a}_t = \boldsymbol{b}_t \quad \text{or} \quad \boldsymbol{a}_t = \boldsymbol{\beta}_t^{-1} \boldsymbol{b}_t. \tag{7.31}$$

Denoting the volatility matrix of $\boldsymbol{b}_t$ by $\boldsymbol{\Sigma}_{b,t}$, we obtain

$$\boldsymbol{\Sigma}_t = \boldsymbol{\beta}_t^{-1} \boldsymbol{\Sigma}_{b,t} (\boldsymbol{\beta}_t^{-1})', \tag{7.32}$$

where Σ_{bt} is diagonal. Equation (7.32) implies that $\Sigma_t^{-1} = \beta_t' \Sigma_{b,t}^{-1} \beta_t$ and $|\Sigma_t| = |\Sigma_{b,t}| = \prod_{i=1}^{k} \sigma_{bi,t}$, where $\sigma_{bi,t}$ is the volatility of b_{it} given F_{t-1}. Consequently, the quasi log likelihood function of $\boldsymbol{a}_t$ is relatively easy to compute via the Cholesky decomposition.

Multivariate volatility modeling via the Choleskey decomposition is essentially making use of Equation (7.31) and the associated conditional covariance matrices. Specifically, the decomposition enables us to simplify the modeling of Σ_t into two steps:

1. Apply univariate volatility models to the components b_{it} of $\boldsymbol{b}_t$, where $i = 1, \ldots, k$.
2. Handle the time evolution of $\beta_{ij,t}$ via time-varying linear regressions, where $i = 2, \ldots, k$ and $j = 1, \ldots, i-1$.

Since the diagonal elements of $\boldsymbol{\beta}_t$ are unity, the volatility matrix Σ_t is positive-definite provided that $\Sigma_{b,t}$ is positive-definite. The latter constraint is easy to handle because it is a diagonal matrix. Consequently, so long as the volatilities of b_{it} are positive for all i and t, Σ_t is positive-definite for all t.

There are many ways to adopt the Cholesky decomposition in multivariate volatility modeling. Lopes, McCulloch and Tsay (2013) apply (a) univariate stochastic volatility models to b_{it} and (b) a simple state-space model for $\beta_{ij,t}$ of the regression coefficients. They demonstrate the approach via analyzing log returns of $k = 94$ assets. Tsay (2010, Chapter 10) illustrates the approach via using GARCH-type of models for the volatilities of b_{it} and for $\beta_{ij,t}$.

7.6.1 Volatility Modeling

In this section, we consider an approach to multivariate volatility modeling using the Cholesky decomposition in Equation (7.30) or Equation (7.31). The approach uses several steps and is designed to simplify the complexity in multivariate volatility modeling. Details are given next.

7.6.1.1 A Procedure for Multivariate Volatility Modeling

1. Apply the recursive least squares (RLS) estimation to Equations (7.27) to (7.29). For the ith linear regression, denote the estimates by $\hat{\boldsymbol{\beta}}_{i,t}$, where $i = 2, \ldots, k$ and $\hat{\boldsymbol{\beta}}_{i,t}$ is an $(i-1)$-dimensional vector.
2. Apply EWMA procedure with $\lambda = 0.96$ to obtain smoothed estimates of $\boldsymbol{\beta}_t$. Specifically, let $\hat{\boldsymbol{\mu}}_i$ be the sample mean of $\{\hat{\boldsymbol{\beta}}_{i,t}\}$ and $\hat{\boldsymbol{\beta}}_{i,t}^* = \hat{\boldsymbol{\beta}}_{i,t} - \hat{\boldsymbol{\mu}}_i$ be the deviation vector. Compute the smoothed estimates of $\boldsymbol{\beta}_{i,t}$ via $\tilde{\boldsymbol{\beta}}_{i,t} = \tilde{\boldsymbol{\beta}}_{i,t}^* + \hat{\boldsymbol{\mu}}_i$, where

$$\tilde{\boldsymbol{\beta}}_{i,t}^* = \lambda \tilde{\boldsymbol{\beta}}_{i,t-1}^* + (1-\lambda)\hat{\boldsymbol{\beta}}_{i,t-1}^*, \quad t = 2, 3, \ldots \tag{7.33}$$

with the initial value of the smoothing being $\tilde{\boldsymbol{\beta}}^*_{i,1} = \hat{\boldsymbol{\beta}}^*_{i,1}$. The sample mean is not used in the smoothing for two reasons. The first reason is to preserve the cases under which some components of $\hat{\boldsymbol{\beta}}_{i,t}$ are close to being a constant. The second reason is to avoid creating drift in the components of $\boldsymbol{\beta}_{i,t}$.

3. Let $\hat{b}_{1t} = a_{1t}$. For $i = 2, \ldots, k$, compute the residual series $\hat{b}_{it} = a_{it} - \boldsymbol{a}'_{i,t}\tilde{\boldsymbol{\beta}}_{i,t}$, where $\boldsymbol{a}_{i,t} = (a_{1t}, \ldots, a_{i-1,t})'$.
4. Fit a univariate GARCH model to each $\hat{b}_{it}$ series and obtain the conditional variance process $\hat{\sigma}^2_{b,i,t}$ for $i = 1, \ldots, k$.
5. Compute the fitted volatility matrix $\hat{\boldsymbol{\Sigma}}_t$ using $\tilde{\boldsymbol{\beta}}_{i,t}$ and the conditional variance processes $\hat{\sigma}^2_{b,i,t}$, where $i = 1, \ldots, k$.

This proposed procedure is easy to implement as it only involves (a) the recursive least-squares method, (b) EWMA procedure, (c) univariate GARCH modeling, and (d) inverses of lower triangular matrices with diagonal elements being 1.

Some discussions of the proposed procedure are in order. First, the choice of smoothing parameter $\lambda = 0.96$ for all regression coefficients is for simplicity. One can use different smoothing parameters for different components a_{it} if necessary. As a matter of fact, one can also estimate the smoothing parameter; see, for instance, Section 7.4. Second, when the dimension k is large, the least-squares estimates $\hat{\boldsymbol{\beta}}_{i,t}$ are not reliable. In this case, some regularization should be used. An obvious choice of regularization is the least absolute shrinkage and selection operator (LASSO) of Tibshirani (1996) and its variants. One can also use sequential LASSO method as given in Chang and Tsay (2010). Third, the choice of univariate volatility model for the transformed series $\boldsymbol{b}_t$ is arbitrary. The GARCH model is used for its ease in estimation. Other models, including stochastic volatility model, can also be used. Finally, the sequential nature of the regression in Equation (7.29) makes the Cholesky approach suitable for parallel computing as one can add a new asset to the portfolio without the need to re-estimate the existing volatility models. See Lopes, McCulloch and Tsay (2013) for demonstration. On the other hand, the sequential nature of the procedure also raises the issue of ordering the asset returns in a portfolio. In theory, ordering of z_{it} in $\boldsymbol{z}_t$ should not be an issue. However, it may make some difference in practice when k is large and T is not.

7.6.2 Application

To demonstrate the Cholesky approach to multivariate volatility modeling, we consider the monthly log returns of three assets. They are IBM stock, S&P composite index, and Coca Cola stock from January 1961 to December 2011 for 612 observations. The data are obtained from CRSP. Figure 7.5 shows the time plot of the data from January 1964 to December 2011. Data of the first three years were omitted from the plots because, as stated later, they are used to initiate a recursive least-squares estimation. Let $\boldsymbol{z}_t$ denote the monthly log returns in the order of IBM, S&P, and Coca Cola. Since the monthly returns have no evidence of serial correlations,

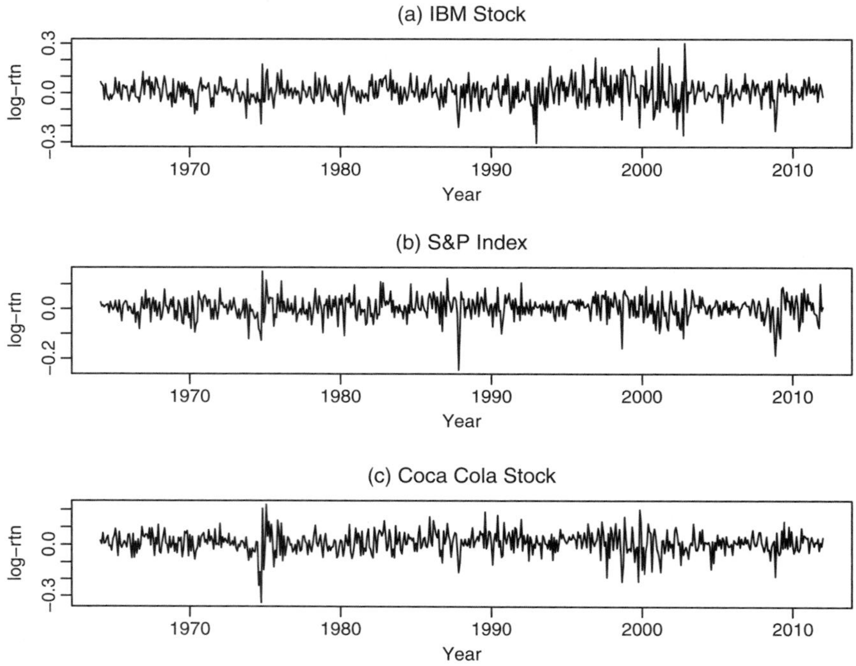

FIGURE 7.5 Time plots of monthly log returns of IBM stock, S&P composite index, and Coca Cola stock from January 1964 to December 2011.

the mean equation consists simply the sample means. We then apply the proposed Cholesky approach of the previous section to estimate the volatility model. For the recursive least-squares method, we use the first 36 observations to compute the initial values so that the volatility estimation uses 576 observations, that is, for years from 1964 to 2011. Univariate GARCH(1,1) models are used for the transformed series $\boldsymbol{b}_t$. Specifically, the three univariate GARCH(1,1) models are

$$\sigma_{1t}^2 = 0.00036 + 0.118 b_{1,t-1}^2 + 0.810 \sigma_{1,t-1}^2$$

$$\sigma_{2t}^2 = 6.4 \times 10^{-5} + 0.099 b_{2,t-1}^2 + 0.858 \sigma_{2,t-1}^2$$

$$\sigma_{3t}^2 = 0.00017 + 0.118 b_{3,t-1}^2 + 0.819 \sigma_{3,t-2}^2,$$

where all estimates are significant at the 5% level. Let $\hat{\boldsymbol{a}}_t = \boldsymbol{z}_t - \hat{\boldsymbol{\mu}}$, where the sample means are $(0.0077, 0.0050, 0.0106)'$. Also, denote fitted volatility matrix as $\hat{\boldsymbol{\Sigma}}_t$. We apply the test statistics of Section 7.1 to check for the conditional heteroscedasticity in the standardized residuals. With 5 lags, all four test statistics fail to detect conditional heteroscedasticity in the residuals at the 5% level. With 10 lags, $Q(10)$ of the transformed scalar standardized residual e_t and the robust $Q_k^r(10)$ of the standardized residuals also fail to detect any conditional heteroscedasticity at the 5% level.

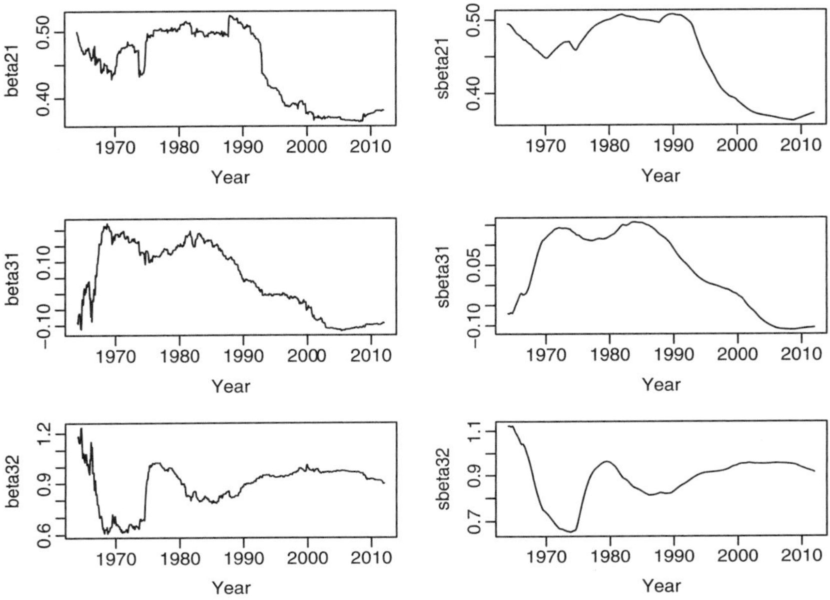

FIGURE 7.6 Time plots of recursive least-squares estimates and their smoothed version used in Cholesky decomposition of monthly log returns of IBM stock, S&P composite index, and Coca Cola stock from January 1964 to December 2011.

However, the rank-based test $Q_R(10)$ of e_t has a p-value of 0.016. In summary, the fitted multivariate volatility model seems adequate.

Figure 7.6 shows the time plots of the $\hat{\beta}_{ij,t}$ of the recursive least-squares method and their smoothed version with $\lambda = 0.96$. For $k = 3$, there are three $\hat{\beta}_{ij,t}$ series corresponding to (i, j) = (2,1), (3,1), and (3,2), respectively. The plots on the left panel are the results of recursive least squares, whereas those on the right panel are the corresponding smoothed estimates. As expected, the smoothed estimates do not have the local fluctuations. Figure 7.7 shows the time plots of the volatilities and the conditional covariances among the three asset returns. The plots on the left panel are volatility series, that is, $\hat{\sigma}_{ii,t}^{1/2}$ for $i = 1, 2, 3$, and the plots on the right panel are conditional covariances. The increase in volatilities during the 2008 financial crisis is clearly seen, especially for individual stocks. Figure 7.8 gives the time plots of the time-varying correlations between the three asset returns. The average correlations are 0.641 and 0.649 between S&P index and the two stocks, respectively. The average correlation between the two stocks is 0.433. The correlation plots, on the other hand, show that correlations between IBM stock and the S&P composite index have decreased markedly after the 2008 financial crisis. The correlations between KO stock and S&P index are less affected by the 2008 financial crisis. Finally, the correlations between the two stocks were deceasing over the past 10 years.

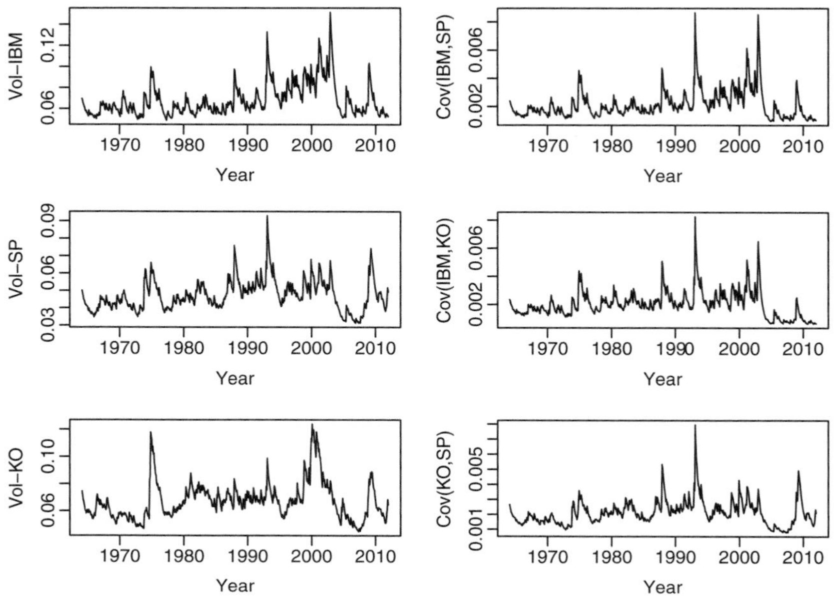

FIGURE 7.7 Time plots of volatilities and conditional covariances for the monthly log returns of IBM stock, S&P composite index, and Coca Cola stock from January 1964 to December 2011.

Remark: Volatility estimation via the Cholesky decomposition is carried out by the command `MCholV` in the `MTS` package. The default starting window for the recursive least squares is 36, and the default exponentially weighted moving-average parameter is 0.96. □

R Demonstration: Multivariate volatility estimation via Cholesky decomposition. Output edited.

```
> da=read.table("m-ibmspko-6111.txt",header=T)
> rtn=log(da[,2:4]+1)
> require(fGarch)
> m3=MCholV(rtn)
Sample means:  0.007728 0.005024 0.010595
Estimation of the first component
Estimate (alpha0, alpha1, beta1):  0.000356 0.117515 0.810288
s.e.                            :  0.000157 0.037004 0.057991
t-value                         :  2.262897 3.175772 13.97261
Component  2  Estimation Results (residual series):
Estimate (alpha0, alpha1, beta1):  6.4e-05 0.099156 0.858354
s.e.                            :  3.1e-05 0.027785 0.037238
t-value                         :  2.034528 3.568616 23.05076
Component  3  Estimation Results (residual series):
Estimate (alpha0, alpha1, beta1):  0.000173 0.117506 0.818722
```

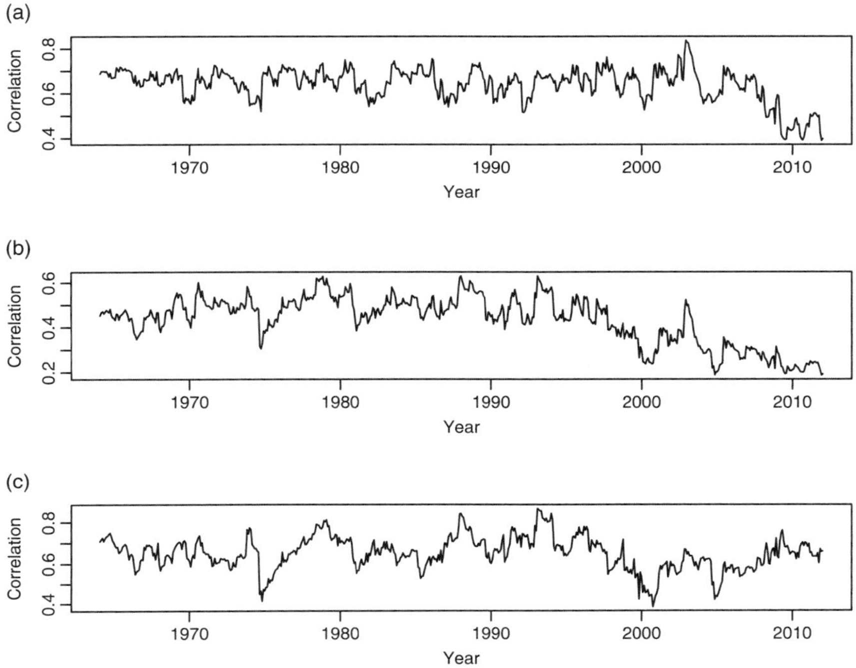

FIGURE 7.8 Time plots of time-varying correlations between the monthly log returns of IBM stock, S&P composite index, and Coca Cola stock from January 1964 to December 2011. (a) IBM versus SP, (b) IBM versus KO, and (c) SP versus KO.

```
s.e.                          :  6.2e-05 0.028651 0.038664
t-value                       :  2.808075 4.101297 21.17521
> names(m3)
[1] "betat"   "bt"      "Vol"     "Sigma.t"
> at=scale(rtn[37:612,],center=T,scale=F)
> Sigma.t= m3$Sigma.t
> MCHdiag(at,Sigma.t)  # use m=10 lags
Test results:
Q(m) of et:
Test and p-value:  15.94173 0.1013126
Rank-based test:
Test and p-value:  21.92159 0.01550909
Qk(m) of epsilon_t:
Test and p-value:  123.7538 0.01059809
Robust Qk(m):
Test and p-value:  95.41881 0.3279669
> MCHdiag(at,Sigma.t,5) # use m=5 lags
Test results:
Q(m) of et:
Test and p-value:  5.71984 0.3344411
```

```
Rank-based test:
Test and p-value:  5.551374 0.3523454
Qk(m) of epsilon_t:
Test and p-value:  59.89413 0.06776611
Robust Qk(m):
Test and p-value:  58.94169 0.07939601
```

7.7 DYNAMIC CONDITIONAL CORRELATION MODELS

Another simple class of models for multivariate volatility modeling is the dynamic conditional correlation (DCC) models. Consider a k-dimensional innovation $\boldsymbol{a}_t$ to the asset return series $\boldsymbol{z}_t$. Let $\boldsymbol{\Sigma}_t = [\sigma_{ij,t}]$ be the volatility matrix of $\boldsymbol{a}_t$ given F_{t-1}, which denotes the information available at time $t-1$. The conditional correlation matrix is then

$$\boldsymbol{\rho}_t = \boldsymbol{D}_t^{-1}\boldsymbol{\Sigma}_t\boldsymbol{D}_t^{-1}, \tag{7.34}$$

where $\boldsymbol{D}_t = \text{diag}\{\sigma_{11,t}^{1/2}, \ldots, \sigma_{kk,t}^{1/2}\}$ is the diagonal matrix of the k volatilities at time t. There are $k(k-1)/2$ elements in $\boldsymbol{\rho}_t$. The DCC models take advantage of the fact that correlation matrices are easier to handle than covariance matrices. As a matter of fact, the idea of DCC models is interesting and appealing. It divides multivariate volatility modeling into two steps. The first step is to obtain the volatility series $\{\sigma_{ii,t}\}$ for $i = 1, \ldots, k$. The second step is to model the dynamic dependence of the correlation matrices $\boldsymbol{\rho}_t$. This is similar to the copula approach to be discussed in a later section.

Let $\boldsymbol{\eta}_t = (\eta_{1t}, \ldots, \eta_{kt})'$ be the marginally standardized innovation vector, where $\eta_{it} = a_{it}/\sqrt{\sigma_{ii,t}}$. Then, $\boldsymbol{\rho}_t$ is the volatility matrix of $\boldsymbol{\eta}_t$. Two types of DCC models have been proposed in the literature. The first type of DCC models is proposed by Engle (2002) and is defined as

$$\boldsymbol{Q}_t = (1-\theta_1-\theta_2)\bar{\boldsymbol{Q}} + \theta_1\boldsymbol{Q}_{t-1} + \theta_2\boldsymbol{\eta}_{t-1}\boldsymbol{\eta}_{t-1}', \tag{7.35}$$

$$\boldsymbol{\rho}_t = \boldsymbol{J}_t\boldsymbol{Q}_t\boldsymbol{J}_t, \tag{7.36}$$

where $\bar{\boldsymbol{Q}}$ is the unconditional covariance matrix of $\boldsymbol{\eta}_t$, θ_i are non-negative real numbers satisfying $0 < \theta_1+\theta_2 < 1$, and $\boldsymbol{J}_t = \text{diag}\{q_{11,t}^{-1/2}, \ldots, q_{kk,t}^{-1/2}\}$, where $q_{ii,t}$ denotes the (i,i)th element of $\boldsymbol{Q}_t$. From the definition, $\boldsymbol{Q}_t$ is a positive-definite matrix and $\boldsymbol{J}_t$ is simply a normalization matrix. The dynamic dependence of the correlations is governed by Equation (7.35) with two parameters θ_1 and θ_2.

The second type of DCC models is proposed by Tse and Tsui (2002) and can be written as

$$\boldsymbol{\rho}_t = (1-\theta_1-\theta_2)\bar{\boldsymbol{\rho}} + \theta_1\boldsymbol{\rho}_{t-1} + \theta_2\boldsymbol{\psi}_{t-1}, \tag{7.37}$$

where $\bar{\rho}$ is the unconditional correlation matrix of $\boldsymbol{\eta}_t$, θ_i are, again, non-negative real numbers satisfying $0 \leq \theta_1 + \theta_2 < 1$, and $\boldsymbol{\psi}_{t-1}$ is a *local* correlation matrix depending on $\{\boldsymbol{\eta}_{t-1}, \ldots, \boldsymbol{\eta}_{t-m}\}$ for some positive integer m. In practice, $\boldsymbol{\psi}_{t-1}$ is estimated by the sample correlation matrix of $\{\hat{\boldsymbol{\eta}}_{t-1}, \ldots, \hat{\boldsymbol{\eta}}_{t-m}\}$ for a prespecified integer $m > 1$.

From Equations (7.35) to (7.37), the two types of DCC models are similar. They all start with the unconditional covariance matrix of $\boldsymbol{\eta}_t$. However, they differ in the way *local* information at time $t-1$ is used. The DCC model of Engle (2002) uses $\boldsymbol{\eta}_{t-1}$ only so that $\boldsymbol{Q}_t$ must be re-normalized at each time index t. On the other hand, the DCC model of Tse and Tsui (2002) uses local correlations to update the conditional correlation matrices. It does not require re-normalization at each time index t, but requires the choice of m in practice. The difference between the two DCC models can be seen by focusing on the correlation $\rho_{12,t}$ of the first two innovations. For the DCC model in Equation (7.37), we have

$$\rho_{12,t} = \theta^* \bar{\rho}_{12} + \theta_1 \rho_{12,t-1} + \theta_2 \frac{\sum_{i=1}^m \eta_{1,t-i}\epsilon_{2,t-i}}{\sqrt{(\sum_{i=1}^m \eta_{1,t-i}^2)(\sum_{i=1}^m \eta_{2,t-i}^2)}}.$$

On the other hand, for the DCC model of Engle (2002) in Equations (7.35) and (7.36), we have

$$\rho_{12,t} = \frac{\theta^* \bar{\rho}_{12} + \theta_1 q_{12,t-1} + \theta_2 \eta_{1,t-1}\eta_{2,t-1}}{\sqrt{(\theta^* + \theta_1 q_{11,t-1} + \theta_2 \eta_{1,t-1}^2)(\theta^* + \theta_1 q_{22,t-1} + \theta_2 \eta_{2,t-1}^2)}},$$

where $\theta^* = 1 - \theta_1 - \theta_2$ and we have used $\bar{\boldsymbol{Q}} = \bar{\boldsymbol{\rho}}$ as they are the unconditional covariance matrix of $\boldsymbol{\eta}_t$. The difference between the two models is easily seen. The choice parameter m of the DCC model of Tse and Tsui (2002) can be regarded as a smoothing parameter. The larger the m is, the smoother the resulting correlations. A choice of $m > k$ ensures that the local correlation matrix $\boldsymbol{\psi}_{t-1}$ is also positive-definite.

From the definitions, DCC models are extremely parsimonious as they only use two parameters θ_1 and θ_2 to govern the time evolution of all conditional correlations regardless of the number of assets k. This simplicity is both an advantage and a weakness of the DCC models. It is an advantage because the resulting models are relatively easy to estimate. It is a weakness of the model because it is hard to justify that all correlations evolve in the same manner regardless of the assets involved. Indeed, limited experience indicates that a fitted DCC model is often rejected by diagnostic checking.

In practical estimation of DCC models, univariate GARCH models are typically used to obtain estimates of the volatility series $\{\sigma_{ii,t}\}$. Let $F_{t-1}^{(i)}$ denote the σ-field generated by the past information of a_{it}. That is, $F_{t-1}^{(i)} = \sigma\{a_{i,t-1}, a_{i,t-2}, \ldots\}$. Univariate GARCH models obtain Var($a_{it}|F_{t-1}^{(i)}$). On the other hand, the multivariate volatility $\sigma_{ii,t}$ is Var($a_{it}|F_{t-1}$). Thus, there is a subtle difference in theory and practice when DCC models are implemented. The practical implication of this difference, if any, is yet to be investigated.

7.7.1 A Procedure for Building DCC Models

A commonly used procedure to build DCC models is given next:

1. Use a VAR(p) model, if necessary, to obtain estimates of the condition mean $\hat{\boldsymbol{\mu}}_t$ for the return series $\boldsymbol{z}_t$ and let $\hat{\boldsymbol{a}}_t = \boldsymbol{z}_t - \hat{\boldsymbol{\mu}}_t$ be the residual series.
2. Apply univariate volatility models, for example, GARCH models, to each component series $\hat{a}_{it}$. Let the volatility series be $\hat{h}_{it}$ and treat $\hat{h}_{it}$ as an estimate of $\sigma_{ii,t}$. That is, $\hat{\sigma}_{ii,t} = \hat{h}_{it}$.
3. Standardize the innovations via $\hat{\eta}_{it} = \hat{a}_{it}/\sqrt{\hat{\sigma}_{ii,t}}$ and fit a DCC model to $\hat{\boldsymbol{\eta}}_t$.

The conditional distribution of $\boldsymbol{\eta}_t$ can either be multivariate standard normal or multivariate standard Student-t with v degrees of freedom. See Equation (7.3).

7.7.2 An Example

To demonstrate the use of DCC models and to compare the two types of DCC model, we consider, again, the monthly log returns of IBM stock, S&P composite index, and Coca Cola stock from January 1961 to December 2011 used in Section 7.6.2. Let $\boldsymbol{z}_t$ denote the monthly log returns in the order of IBM, S&P, and KO. As before, the mean equation for the data is simply $\hat{\boldsymbol{\mu}}_t = \hat{\boldsymbol{\mu}}$, the sample mean, which is $(0.0077, 0.0050, 0.0106)'$. The innovation series is then $\hat{\boldsymbol{a}}_t = \boldsymbol{z}_t - \hat{\boldsymbol{\mu}}_t$. For the DCC modeling, we follow the proposed procedure of the prior section and employ univariate Gaussian GARCH(1,1) models for the individual series. The three fitted Gaussian GARCH(1,1) models for the returns of IBM, S&P index, and KO are, respectively,

$$\begin{aligned}
\sigma_{11,t} &= 4.19 \times 10^{-4} + 0.127a_{1,t-1}^2 + 0.788\sigma_{11,t-1},\\
\sigma_{22,t} &= 4.1 \times 10^{-5} + 0.128a_{2,t-1}^2 + 0.836\sigma_{22,t-1},\\
\sigma_{33,t} &= 2.56 \times 10^{-4} + 0.099a_{3,t-1}^2 + 0.830\sigma_{33,t-1}.
\end{aligned}$$

Using the volatility series of the prior three models, we obtain the marginally standardized series

$$\hat{\boldsymbol{\eta}}_t = (\hat{\eta}_{1t}, \hat{\eta}_{2t}, \hat{\eta}_{3t})', \quad \hat{\eta}_{it} = \hat{a}_{it}/\sqrt{\sigma_{ii,t}},$$

and apply DCC models to $\hat{\boldsymbol{\eta}}_t$.

To handle heavy tails in asset returns, we use multivariate Student-t distribution for the innovations of DCC models. The fitted DCC model of Tse and Tsui (2002) is

$$\boldsymbol{\rho}_t = (1 - 0.8088 - 0.0403)\bar{\boldsymbol{\rho}} + 0.8088\boldsymbol{\rho}_{t-1} + 0.0403\boldsymbol{\psi}_{t-1}, \tag{7.38}$$

where ψ_{t-1} is the sample correlation matrix of $\{\hat{\eta}_{t-1}, \ldots, \hat{\eta}_{t-4}\}$ with $m = k + 1$ and $\bar{\rho}$ is the sample correlation matrix of $\hat{\eta}_t$. The coefficient estimate $\hat{\theta}_1 = 0.8088$ is significant with t-ratio 5.42, but the estimate $\hat{\theta}_2$ is only marginally significant with t-ratio 1.78. The estimated degrees of freedom for the multivariate Student-t innovations is 7.96. The fitted DCC model of Engle (2002) is

$$\begin{aligned} \boldsymbol{Q}_t &= (1 - 0.9127 - 0.0453)\bar{\boldsymbol{\rho}} + 0.9127\boldsymbol{Q}_{t-1} + 0.0453\hat{\boldsymbol{\eta}}_{t-1}\hat{\boldsymbol{\eta}}'_{t-1}, \\ \boldsymbol{\rho}_t &= \boldsymbol{J}_t\boldsymbol{Q}_t\boldsymbol{J}_t, \end{aligned} \tag{7.39}$$

where $\boldsymbol{J}_t = \text{diag}\{q_{11,t}^{-1/2}, q_{22,t}^{-1/2}, q_{33,t}^{-1/2}\}$ with $q_{ii,t}$ being the (i,i)th element of $\boldsymbol{Q}_t$ and $\bar{\boldsymbol{\rho}}$ is defined in Equation (7.38). The two coefficient estimates $\hat{\theta}_i$ of Equation (7.39) are significant with t-ratios 30.96 and 3.56, respectively, and the estimate of the degrees of freedom for the multivariate Student-t innovation is 8.62.

For model checking, we apply the four test statistics of Section 7.1 to $\hat{\eta}_t$ using the fitted time-varying correlation matrices. For both fitted DCC models, only the robust multivariate portmanteau test fails to detect the existence of additional conditional hetroscedasticity in the data. This is not surprising as diagnostic tests typically reject a fitted DCC model. Further results of the test statistics are given in the following R demonstration.

It is interesting to compare the two fitted DCC models for this particular case of three asset returns. First, the parameter estimates of the two DCC models indicate that the DCC model of Engle (2002) appears to have stronger persistence in the time-varying correlations. Second, the two models are qualitatively similar. Figure 7.9 shows the time plots of the fitted time-varying correlations of the two models. The plots on the left panel are correlations of the DCC model in Equation (7.38), whereas those on the right panel belong to the DCC model in Equation (7.39). The overall pattern of the time-varying correlations seems similar, but, as expected, the correlations of the model in Equation (7.39) appear to be more variable. Third, there are differences between the two DCC models, implying that one might want to consider both models in an application. Table 7.4 gives some summary statistics of the time-varying correlations of the two DCC models. The sample means of the correlations are close, but the ranges differ. The summary statistics confirm that the correlations of the DCC model of Engle (2002) are more variable.

Remark: Estimation of DCC models is carried out by the commands `dccPre` and `dccFit`. The command `dccPre` fits univariate GARCH(1,1) models to the component series and obtains the marginally standardized series for DCC estimation. The command `dccFit` estimates a specified DCC model using the marginally standardized series. Its subcommand `type` can be used to specify the type of DCC model. The innovations are either Gaussian or multivariate Student-t. The default option uses multivariate Student-t innovations and fits a Tse and Tsui model. The univariate GARCH model is estimated via the `fGarch` package of Rmetrics. □

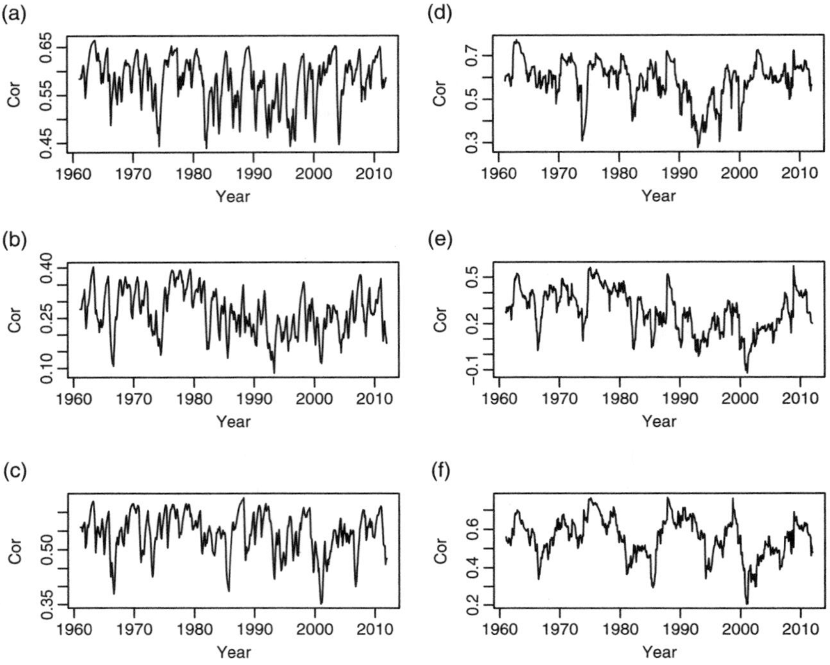

FIGURE 7.9 Time plots of time-varying correlation coefficients for monthly log returns of IBM, S&P index, and KO from January 1961 to December 2011 based on the DCC models in Equations (7.38) and (7.39). The plots on the left panel are for the model in Equation (7.38). (a) IBM versus SP, (b) IBM versus KO, (c) SP versus KO, (d) IBM versus SP, (e) IBM versus KO, and (f) SP versus KO.

TABLE 7.4 Summary Statistics of Time-Varying Correlations of the Monthly Log Returns of IBM, S&P, and KO Based on DCC Models in Equations (7.38) and (7.39)

	Tse and Tsui Model			Engle Model		
Statistics	(ibm,sp)	(ibm,ko)	(sp,ko)	(ibm,sp)	(ibm,ko)	(sp,ko)
Mean	0.580	0.269	0.544	0.585	0.281	0.556
SE	0.050	0.067	0.055	0.095	0.139	0.105
Min	0.439	0.086	0.352	0.277	−0.121	0.205
Max	0.664	0.403	0.638	0.773	0.568	0.764

R Demonstration: Estimation of DCC models.

```
> da=read.table("m-ibmspko-6111.txt",header=T)
> rtn=log(da[,2:4]+1)
> m1=dccPre(rtn,include.mean=T,p=0)
```

```
Sample mean of the returns:  0.00772774 0.005023909 0.01059521
Component:  1
Estimates:  0.000419 0.126739 0.788307
se.coef  :  0.000162 0.035405 0.055645
t-value  :  2.593448 3.57973 14.16662
Component:  2
Estimates:  9e-05 0.127725 0.836053
se.coef  :  4.1e-05 0.03084 0.031723
t-value  :  2.20126 4.141592 26.35486
Component:  3
Estimates:  0.000256 0.098705 0.830358
se.coef  :  8.5e-05 0.022361 0.033441
t-value  :  3.015321 4.414112 24.83088
> names(m1)
[1] "marVol"  "sresi"   "est"     "se.coef"
> rtn1=m1$sresi
> Vol=m1$marVol
> m2=dccFit(rtn1)
Estimates:  0.8088086 0.04027318 7.959013
st.errors:  0.1491655 0.02259863 1.135882
t-values:   5.422222 1.782107 7.006898
> names(m2)
[1] "estimates" "Hessian"   "rho.t"
> S2.t = m2$rho.t
> m3=dccFit(rtn1,type="Engle")
Estimates:  0.9126634 0.04530917 8.623668
st.errors:  0.0294762 0.01273911 1.332381
t-values:   30.96272 3.556697 6.472376
> S3.t=m3$rho.t
> MCHdiag(rtn1,S2.t)
Test results:
Q(m) of et:
Test and p-value:  20.74262 0.02296152
Rank-based test:
Test and p-value:  30.20662 0.0007924436
Qk(m) of epsilon_t:
Test and p-value:  132.423 0.002425885
Robust Qk(m):
Test and p-value:  109.9671 0.0750157
> MCHdiag(rtn1,S3.t)
Test results:
Q(m) of et:
Test and p-value:  20.02958 0.02897411
Rank-based test:
Test and p-value:  27.61638 0.002078829
Qk(m) of epsilon_t:
Test and p-value:  131.982 0.002625755
Robust Qk(m):
Test and p-value:  111.353 0.06307334
```

7.8 ORTHOGONAL TRANSFORMATION

To overcome the difficulty of curse of dimensionality, several dimension-reduction methods have been proposed in the literature for multivariate volatility modeling. In particular, the idea of orthogonal transformation has attracted much attention. For a k-dimensional innovation $\boldsymbol{a}_t$, an orthogonal transformation basically assumes that $\boldsymbol{a}_t$ is driven by k orthogonal latent components b_{it} for $i = 1, \ldots, k$. Specifically, we assume that there exists a nonsingular matrix $\boldsymbol{M}$ such that

$$\boldsymbol{a}_t = \boldsymbol{M}\boldsymbol{b}_t, \tag{7.40}$$

where $\boldsymbol{b}_t = (b_{1t}, \ldots, b_{kt})'$. Ideally, one wishes the components b_{it} are as mutually independent as possible. However, independence is hard to achieve in finite samples and orthogonality is used in lieu of independence. If the transformation matrix $\boldsymbol{M}$ can be found, one can obtain the latent variable $\boldsymbol{b}_t$ by $\boldsymbol{b}_t = \boldsymbol{M}^{-1}\boldsymbol{a}_t$ and apply univariate volatility models to b_{it}. In this section, we briefly summarize some of the existing orthogonal methods and discuss their pros and cons.

7.8.1 The Go-GARCH Model

The most commonly used orthogonal transformation in statistics is perhaps the principal component analysis (PCA). Indeed, PCA has been used in multivariate volatility modeling. See, for instance, Alexander (2001) and the references therein. For non-Gaussian data, the independent component analysis (ICA) is available to perform the transformation. Van der Weide (2002) adopts the concept of ICA to propose a class of generalized orthogonal GARCH (Go-GARCH) models for volatility modeling.

A key assumption of the Go-GARCH model is that the transformation matrix $\boldsymbol{M}$ is time-invariant. Under the assumption, the volatility matrix of $\boldsymbol{a}_t$ becomes

$$\boldsymbol{\Sigma}_t = \boldsymbol{M}\boldsymbol{V}_t\boldsymbol{M}', \tag{7.41}$$

where $\boldsymbol{V}_t$ is the volatility matrix of $\boldsymbol{b}_t$, that is, $\boldsymbol{V}_t = \text{Cov}(\boldsymbol{b}_t|F_{t-1})$ with F_{t-1} denoting the information available at $t-1$. The second key assumption of the Go-GARCH model is that $\boldsymbol{V}_t$ is a diagonal matrix for all t.

Similarly to the ICA, $\boldsymbol{b}_t$ is latent and, without loss of generality, we assume that the unconditional covariance matrix of $\boldsymbol{b}_t$ is the identity matrix, that is, $\text{Cov}(\boldsymbol{b}_t) = \boldsymbol{I}_k$. Consequently, the transformation in Equation (7.40) implies that $\text{Cov}(\boldsymbol{a}_t) = \boldsymbol{M}\boldsymbol{M}'$. This relationship can be exploited to estimate $\boldsymbol{M}$. However, the relationship also implies that the transformation matrix $\boldsymbol{M}$ is not uniquely determined because for any $k \times k$ orthogonal matrix $\boldsymbol{U}$ satisfying $\boldsymbol{U}\boldsymbol{U}' = \boldsymbol{I}_k$ we have $\text{Cov}(\boldsymbol{a}_t) = (\boldsymbol{M}\boldsymbol{U})(\boldsymbol{M}\boldsymbol{U})'$. Further restriction is needed to identify the transformation matrix. Let $\text{Cov}(\boldsymbol{a}_t) = \boldsymbol{P}\boldsymbol{\Lambda}\boldsymbol{P}'$ be the spectral representation of the unconditional covariance matrix of $\boldsymbol{a}_t$. Van der Weide (2002) makes use of the following two lemmas to uniquely determine $\boldsymbol{M}$.

Lemma 7.1 There exists an orthogonal matrix $\boldsymbol{U}$ such that $\boldsymbol{P}\boldsymbol{\Lambda}^{1/2}\boldsymbol{U} = \boldsymbol{M}$.

Lemma 7.2 Every k-dimensional orthogonal matrix $\boldsymbol{W}$ with $|\boldsymbol{W}|=1$ can be represented as a product of $k(k+1)/2$ rotation matrices:

$$\boldsymbol{W} = \prod_{i<j} \boldsymbol{R}_{ij}(\theta_{ij}), \quad -\pi \le \theta_{ij} \le \pi,$$

where $\boldsymbol{R}_{ij}(\theta_{ij})$ performs a rotation in the plane spanned by $\boldsymbol{e}_i$ and $\boldsymbol{e}_j$ over an angle θ_{ij}, where $\boldsymbol{e}_i$ denotes the ith column of $\boldsymbol{I}_k$.

Lemma 7.1 follows directly the singular value decomposition. See also the derivation in the next section. Lemma 7.2 can be found in Vilenkin (1968). The rotation $\boldsymbol{R}_{ij}(\theta_{ij})$ is also known as the *Givens* rotation matrix. A simple way to define the Givens matrix $\boldsymbol{R}_{ij}(\theta_{ij})$ is to start with the $k \times k$ identity matrix, then replace the (i,i) and (j,j) diagonal elements by $\cos(\theta_{ij})$, the (j,i)th element by $\sin(\theta_{ij})$, and the (i,j)th element by $-\sin(\theta_{ij})$. From the two lemmas, we require $|\boldsymbol{U}| = 1$ and can estimate $\boldsymbol{M}$ via the parameterization θ_{ij}. In practice, one can also estimate the transformation matrix $\boldsymbol{M}$ using the methods of estimating ICA discussed in Hyvärinen, Karhunen and Oja (2001).

The theory of orthogonal transformation discussed so far was developed primarily under the assumption that $\{\boldsymbol{a}_t\}$ forms a random sample. For asset returns, the innovation $\{\boldsymbol{a}_t\}$ is serially uncorrelated, but dependent. This gap between theory and practice raises the issue of applicability of Go-GARCH models in analyzing asset returns. In summary, the Go-GARCH models are relatively simple and conceptually appealing. Its main weakness is that the assumptions used are hard to justify in practice.

Remark: Go-GARCH models can be estimated in R using the `gogarch` package of Pfaff (2011). Methods of estimation available are `ica, mm, ml, nls`, standing for independent component analysis, method of moments, maximum likelihood, and nonlinear least squares, respectively. □

Example 7.3 Consider, again, the monthly log returns of IBM stock, S&P composite index, and Coca Cola stock from January 1961 to December 2011. We apply the Go-GARCH(1,1) model to the mean-adjusted returns using the ICA estimation method. In this particular instance, the estimated transformation matrix is

$$\hat{\boldsymbol{M}} = \begin{bmatrix} -0.0087 & 0.0608 & -0.0331 \\ 0.0304 & 0.0307 & -0.0072 \\ 0.0398 & -0.0011 & -0.0471 \end{bmatrix},$$

and the GARCH(1,1) models for the latent variables b_{it} are

$$\sigma_{1t}^2 = 0.0834 + 0.1308 b_{1,t-1}^2 + 0.7676 \sigma_{1,t-1}^2,$$
$$\sigma_{2t}^2 = 1.0888 + 0.1341 b_{2,t-1}^2 + 0.7215 \sigma_{2,t-1}^2,$$
$$\sigma_{3t}^2 = 0.0800 + 0.1027 b_{3,t-1}^2 + 0.8427 \sigma_{3,t-1}^2.$$

Figure 7.10 shows the time plots of the volatility series and the time-varying correlation coefficients between the three log return series. The volatility series are in the left panel of the figure, whereas the time-varying correlations are in the right panel. From the plots, the volatility series of the IBM stock appears to be large and the correlations between the IBM stock and the S&P composite index are relatively high. The remaining two correlation coefficients are relatively low. The sample means of the three correlations are 0.927, 0.126, and 0.206, respectively. These values are rather different from those of the DCC models obtained before. Finally, the four test statistics of Section 7.1 show that the standardized innovations of the fitted Go-GARCH model still have strong conditional heteroscedasticity. This is not surprising because the

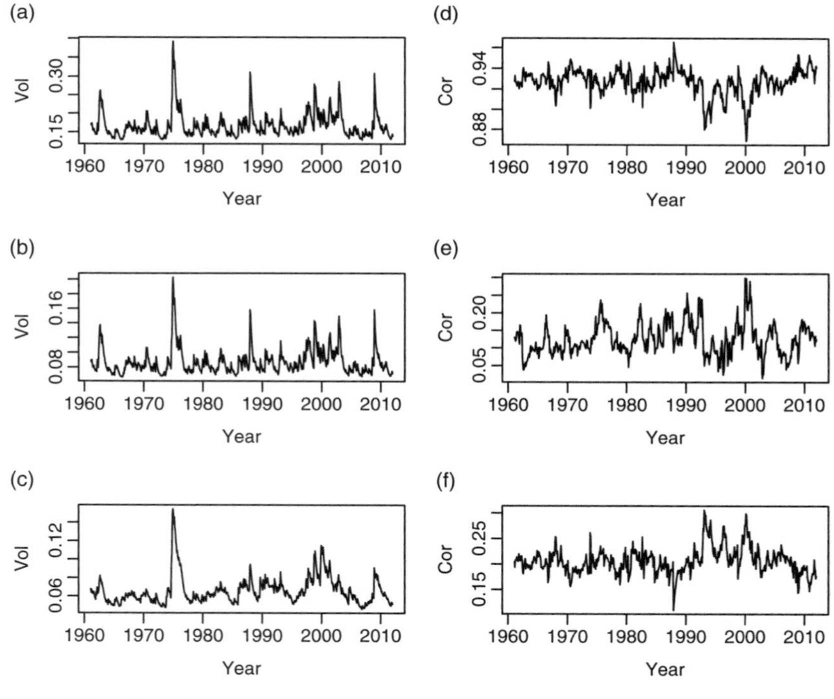

FIGURE 7.10 Time plots of volatilities and time-varying correlations for the monthly log returns of (a) IBM, (b) S&P index, and (c) KO stock from January 1961 to December 2011 via the Go-GARCH(1,1) model estimated by the independent component analysis. (d) IBM versus S&P, (e) IBM versus KO, and (f) S&P versus KO.

squares of the transformed latent variables b_{it} still have significant cross-dependence. For instance, $\text{cor}(b_{2t}^2, b_{3t}^2) = 0.356$, which differs markedly from 0. This issue of nonzero correlations between the squares of fitted latent variables continues to exist for other Go-GARCH estimation methods. □

R Demonstration: Go-GARCH model estimation. Output edited.

```
> crtn=scale(rtn,center=T,scale=F)
> require(gogarch)
> help(gogarch)
> m1=gogarch(crtn,~garch(1,1),estby="ica")
> m1
****************
*** GO-GARCH ***
****************
Components estimated by: fast ICA
Dimension of data matrix: (612 x 3).
Formula for component GARCH models: ~ garch(1, 1)
Orthogonal Matrix U:
          [,1]       [,2]       [,3]
[1,] -0.3723608  0.7915757 -0.4845156
[2,]  0.7625418  0.5585272  0.3264619
[3,]  0.5290344 -0.2479018 -0.8115832
Linear Map Z:   ### Transformation matrix M of the text
            [,1]         [,2]         [,3]
[1,] -0.008736657  0.060758612 -0.033058547
[2,]  0.030419614  0.030713261 -0.007215686
[3,]  0.039780987 -0.001102663 -0.047121947
Estimated GARCH coefficients:
        omega    alpha1     beta1
y1 0.08337071 0.1307602 0.7675916
y2 1.08879832 0.1341139 0.7215066
y3 0.08001260 0.1027129 0.8426915
> sigma.t=NULL   # Obtain the volatility matrices
> for (i in 1:612){
+ sigma.t=rbind(sigma.t,c(m1@H[[i]]))
+ }
> MCHdiag(crtn,sigma.t)   ## Model checking
Test results:
Q(m) of et:
Test and p-value:  23.50608 0.009025087
Rank-based test:
Test and p-value:  26.62406 0.002985593
Qk(m) of epsilon_t:
Test and p-value:  173.8608 2.770358e-07
Robust Qk(m):
Test and p-value:  158.8302 1.044352e-05
> M=m1@Z
```

```
> Minv=solve(M)
> bt=crtn%*%t(Minv)  #Latent variables
> cor(bt^2)
          [,1]      [,2]      [,3]
[1,] 1.0000000 0.1635794 0.1067924
[2,] 0.1635794 1.0000000 0.3563265
[3,] 0.1067924 0.3563265 1.0000000
```

7.8.2 Dynamic Orthogonal Components

Matteson and Tsay (2011) propose a dynamic orthogonal component (DOC) approach to multivariate volatility modeling. The authors also propose a test statistic for checking the existence of dynamic orthogonal components for a given return series $\boldsymbol{z}_t$. The DOC approach is based on the same idea as the Go-GARCH model, but employs a different objective function to estimate the transformation matrix $\boldsymbol{M}$. The new objective function used is to ensure that the transformed latent variables b_{it} are as close to be dynamically uncorrelated as possible.

Consider the transformation of the innovation $\boldsymbol{a}_t$ in Equation (7.40). We call the transformed series $\boldsymbol{b}_{it}$ *dynamic orthogonal components* if

$$\text{Cov}(b_{it}^2, b_{j,t-v}^2) = 0, \quad \text{for} \quad i \neq j; \quad v = 0, 1, 2, \ldots. \tag{7.42}$$

Unlike the Go-GARCH models, the DOC approach does not assume *a priori* the existence of the DOCs b_{it}. Their existence is verified by testing, and it is possible that no DOC exists for a given innovation series $\boldsymbol{a}_t$.

Assume for now that the DOC series b_{it} exist and our goal is to estimate the transformation matrix $\boldsymbol{M}$. To this end, Matteson and Tsay (2011) employ a two-step procedure. In the first step, one performs the spectral decomposition of the covariance matrix $\boldsymbol{\Sigma}$ of $\boldsymbol{a}_t$, say $\boldsymbol{\Sigma} = \boldsymbol{P}\boldsymbol{\Lambda}\boldsymbol{P}'$, where $\boldsymbol{P}$ denotes the matrix of orthonormal eigenvectors and $\boldsymbol{\Lambda}$ is the diagonal matrix of eigenvalues. Define $\boldsymbol{d}_t = \boldsymbol{U}\boldsymbol{a}_t$, where $\boldsymbol{U} = \boldsymbol{\Lambda}^{-1/2}\boldsymbol{P}'$. Then, $\text{Cov}(\boldsymbol{d}_t) = \boldsymbol{U}\boldsymbol{\Sigma}\boldsymbol{U}' = \boldsymbol{\Lambda}^{-1/2}\boldsymbol{P}'\boldsymbol{P}\boldsymbol{\Lambda}\boldsymbol{P}'\boldsymbol{P}\boldsymbol{\Lambda}^{-1/2} = \boldsymbol{I}_k$ and the transformation in Equation (7.40) becomes

$$\boldsymbol{b}_t = \boldsymbol{M}^{-1}\boldsymbol{a}_t = \boldsymbol{M}^{-1}\boldsymbol{U}^{-1}\boldsymbol{d}_t \equiv \boldsymbol{W}\boldsymbol{d}_t, \tag{7.43}$$

where $\boldsymbol{W} = (\boldsymbol{U}\boldsymbol{M})^{-1}$ is referring to as the *separating matrix*. From Equation (7.43), $\boldsymbol{I}_k = \text{Cov}(\boldsymbol{b}_t) = \boldsymbol{W}\text{Cov}(\boldsymbol{d}_t)\boldsymbol{W}' = \boldsymbol{W}\boldsymbol{W}'$. Consequently, the separating matrix $\boldsymbol{W}$ is a $k \times k$ orthonormal matrix with $|\boldsymbol{W}| = 1$. In the second step of estimation, one makes use of the parameterization of Lemma 7.2 to estimate $\boldsymbol{W}$ via minimization a well-defined objective function. Let $\boldsymbol{\theta}$ be the collection of all rotation angles θ_{ij}, where $1 \leq i < j \leq k$. The number of angles is $k(k-1)/2$. For a given $\boldsymbol{\theta}$, we obtain a corresponding $\boldsymbol{W}$, which in turns gives an estimate of $\boldsymbol{b}_t$ via Equation (7.43). Denote the latent vector $\boldsymbol{b}_t$ corresponding to a given $\boldsymbol{\theta}$ by $\boldsymbol{b}_{\theta,t}$.

For volatility modeling, one can use the property in Equation (7.42) to construct an objective function for estimating the separating matrix $\boldsymbol{W}$. Intuitively, one would use the cross-correlation matrices of $\boldsymbol{b}_{\theta,t}^2$ to construct an objective function. However, asset returns tend to contain extreme values because their distributions have heavy tails. These extreme values are likely to have substantial impacts on the estimation of $\boldsymbol{\theta}$. To mitigate the influence of extreme innovations, Matteson and Tsay (2011) propose a robust approach to construct an objective function. Specifically, the authors adopt the Huber's function

$$h_c(s) = \begin{cases} s^2 & \text{if } |s| \leq c, \\ 2|s|c - c^2 & \text{if } |s| > c, \end{cases} \tag{7.44}$$

where c is a positive real number. In applications, $c = 2.25$ is found to be adequate. For a given constant c, apply the Huber's function in Equation (7.44) to each component b_{it} of the latent variable and denote the resulting quantities by $\boldsymbol{h}_c(\boldsymbol{b}_{\theta,t})$. Next, define the lag-$\ell$ covariance matrix of $\{\boldsymbol{h}_c(\boldsymbol{b}_{\theta,t})\}$ as

$$\boldsymbol{\Gamma}_\theta(\ell) = \text{Cov}[\boldsymbol{h}_c(\boldsymbol{b}_{\theta,t}), \boldsymbol{h}_c(\boldsymbol{b}_{\theta,t-\ell})], \quad \ell = 0, 1, \ldots,$$

where, for ease in notation, the dependence of $\boldsymbol{\Gamma}_\theta(\ell)$ on h is omitted. Under the assumption that the latent variables in $\boldsymbol{b}_{\theta,t}$ are dynamically orthogonal, the lag-ℓ covariance matrix $\boldsymbol{\Gamma}_\theta(\ell)$ is a diagonal matrix for all $\ell \geq 0$. In other words, all off-diagonal elements of $\boldsymbol{\Gamma}_\theta(\ell)$ are zero. Matteson and Tsay (2011) make use of this property and the method of moments to construct an objective function for estimating $\boldsymbol{\theta}$ and, hence, $\boldsymbol{W}$.

In finite samples, we can estimate $\boldsymbol{\Gamma}_\theta(\ell)$ by

$$\hat{\boldsymbol{\Gamma}}_\theta(\ell) = \frac{1}{T}\sum_{t=\ell+1}^{T} \boldsymbol{h}_c(\boldsymbol{b}_{\theta,t})\boldsymbol{h}_c'(\boldsymbol{b}_{\theta,t-\ell}) - \left[\frac{1}{T}\sum_{t=\ell+1}^{T} \boldsymbol{h}_c(\boldsymbol{b}_{\theta,t})\right]\left[\frac{1}{T}\sum_{t=\ell+1}^{T} \boldsymbol{h}_c(\boldsymbol{b}_{\theta,t-\ell})\right]'. \tag{7.45}$$

Since $\hat{\boldsymbol{\Gamma}}_\theta(0)$ is symmetric, it has $g = k(k-1)/2$ off-diagonal elements. For $\ell > 0$, $\hat{\boldsymbol{\Gamma}}_\theta(\ell)$ has $2g$ off-diagonal elements. Let $\boldsymbol{f}(\boldsymbol{\theta})$ be the vectorized array of off-diagonal elements of $\{\hat{\boldsymbol{\Gamma}}_\theta(\ell) | \ell = 0, \ldots, m\}$, where m is a positive integer denoting the maximum lag used. Clearly, $\boldsymbol{f}$ is a function of $\boldsymbol{a}_t$, $\boldsymbol{\theta}$, and $h_c(.)$. For simplicity, we use the augment $\boldsymbol{\theta}$ only to emphasize the estimation of $\boldsymbol{\theta}$. The dimension of $\boldsymbol{f}(\boldsymbol{\theta})$ is $p = g(2m+1)$. The objective function used by Matteson and Tsay (2011) is then

$$O(\boldsymbol{\theta}) = \boldsymbol{f}'(\boldsymbol{\theta})\boldsymbol{\Phi}\boldsymbol{f}(\boldsymbol{\theta}), \tag{7.46}$$

where $\boldsymbol{\Phi}$ is a $p \times p$ weighting matrix. In practice, Matteson and Tsay (2011) use a diagonal weighting matrix, that is, $\boldsymbol{\Phi} = \text{diag}\{\phi_{ii,\ell}\}$, where $\phi_{ii,\ell}$ depends on the lag ℓ of $\hat{\boldsymbol{\Gamma}}_\theta(\ell)$. Specifically,

$$\phi_{ii,\ell} = \frac{1 - \ell/(m+1)}{[\sum_{v=0}^{m}(1 - v/(m+1))] \times (g + gI[\ell > 0])}, \quad \ell = 0, 1, \ldots, g,$$

where $I[\ell > 0] = 1$ if $\ell > 0$ and it is 0 if $\ell = 0$. Thus, the weight is smaller for higher-order lags and the quantity $g + gI[\ell > 0]$ is used to account for the number of off-diagonal elements in cross-covariance matrix $\hat{\boldsymbol{\Gamma}}_\theta(\ell)$.

The ideal value of the objective function in Equation (7.46) is 0 so that we estimate the orthonormal matrix $\boldsymbol{W}$ (via $\boldsymbol{\theta}$) by minimizing the objective function $O(\boldsymbol{\theta})$ of Equation (7.46). In other words, we obtain the estimate of $\boldsymbol{W}$ so that the latent variables in $\boldsymbol{b}_{\theta,t}$ are as close to be dynamically orthogonal as possible.

Under some regularity conditions, Matteson and Tsay (2011) prove the consistency and asymptotic normality of the estimate $\hat{\boldsymbol{\theta}}$. The robust transform $h_c(s)$ of Equation (7.44) relaxes certain moment conditions for the innovation $\boldsymbol{a}_t$ in deriving the limiting properties of $\hat{\boldsymbol{\theta}}$. Of course, other robust weighting functions can also be used.

7.8.3 Testing the Existence of DOC

Let $\hat{\boldsymbol{W}}$ be the estimate of the separating matrix $\boldsymbol{W}$ obtained by the method discussed in the prior section and $\hat{\boldsymbol{b}}_t$ be the associated estimate of $\boldsymbol{b}_t$, that is, $\hat{\boldsymbol{b}}_t = \hat{\boldsymbol{W}}\boldsymbol{d}_t$, where $\boldsymbol{d}_t = \boldsymbol{\Lambda}^{1/2}\boldsymbol{P}'$ with $\boldsymbol{\Lambda}$ and $\boldsymbol{P}$ are from the spectral decomposition of the sample covariance matrix of $\boldsymbol{a}_t$. In the presence of k DOCs, the consistency of $\hat{\boldsymbol{W}}$ implies that the estimated series $\hat{\boldsymbol{b}}_t$ is asymptotically a dynamically orthogonal time series. Consequently, to test the existence of DOCs, one can check the hypothesis that all off-diagonal elements of $\boldsymbol{\Gamma}_\theta(\ell)$ are zero for $\ell \geq 0$. In practice, it is easier to use the correlations than the covariances. We define

$$\hat{\boldsymbol{\rho}}_\theta(\ell) = \boldsymbol{J}\hat{\boldsymbol{\Gamma}}_\theta(\ell)\boldsymbol{J}, \quad \ell = 0, 1, \ldots, \tag{7.47}$$

where $\hat{\boldsymbol{\Gamma}}_\theta(\ell)$ is given in Equation (7.45) and $\boldsymbol{J} = \text{diag}\{\Gamma_{11,\theta}^{-1/2}(0), \ldots, \Gamma_{kk,\theta}^{-1/2}(0)\}$. The hypothesis of interest is then $H_0 : \rho_{ij,\theta}(\ell) = 0$ for $\ell = 0, 1, \ldots, m$ and $1 \leq i \neq j \leq k$ versus the alternative hypothesis $H_a : \rho_{ij,\theta}(\ell) \neq 0$ for some ℓ and (i, j), where m is a prespecified positive integer and $\rho_{ij,\theta}(\ell)$ is the (i, j)th element of $\boldsymbol{\rho}_\theta(\ell)$, which is the theoretical counterpart of $\hat{\boldsymbol{\rho}}_\theta(\ell)$.

A natural test statistic to use is a generalization of the Ljung–Box statistics. Specifically, define

$$Q_k^o(m) = T\sum_{i<j}\hat{\rho}_{ij,\theta}^2(0) + T(T+2)\sum_{\ell=1}^{m}\sum_{i\neq j}\hat{\rho}_{ij,\theta}^2(\ell)/(T-\ell), \tag{7.48}$$

where T is the sample size, k is the dimension of $\boldsymbol{a}_t$, and $\hat{\rho}_{ij,\theta}(\ell)$ is the (i, j)th element of $\hat{\boldsymbol{\rho}}_\theta(\ell)$. The superscript o of $Q_k^o(m)$ is used to signify the statistic for testing dynamic orthogonality. For a given m, the number of cross-correlations in the test statistic Q_k^o is $v = k(k-1)/2 + mk(k-1)$. Therefore, under the null hypothesis H_0, $Q_k^o(m)$ is asymptotically distributed as a χ_v^2.

Example 7.4 For comparison purpose, we consider, again, the monthly log returns of the three assets used in Example 7.3. Here, we apply the DOC approach to estimate the multivariate volatility. The mean equation remains $\hat{\boldsymbol{\mu}}_t = \hat{\boldsymbol{\mu}}$, the sample mean. For the innovation series $\hat{\boldsymbol{a}}_t = \boldsymbol{z}_t - \hat{\boldsymbol{\mu}}$, we apply the DOC approach and obtain

$$\hat{\boldsymbol{M}} = \begin{bmatrix} -0.036 & 0.054 & -0.025 \\ -0.027 & -0.000 & -0.035 \\ -0.059 & -0.017 & 0.002 \end{bmatrix}, \quad \hat{\boldsymbol{M}}^{-1} = \begin{bmatrix} -4.85 & 2.51 & -15.07 \\ 16.98 & -12.61 & -4.65 \\ 3.76 & -30.87 & 11.78 \end{bmatrix}.$$

Figure 7.11 shows the time plots of the fitted dynamic orthogonal components $\hat{\boldsymbol{b}}_t = \hat{\boldsymbol{M}}^{-1}\hat{\boldsymbol{a}}_t$. These series are normalized to have unit variance. From the

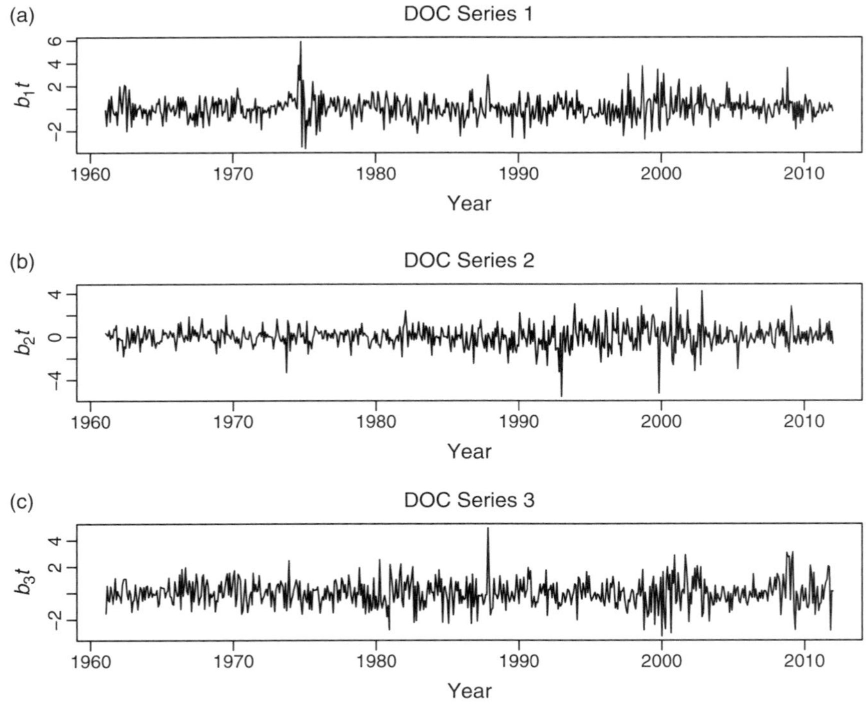

FIGURE 7.11 Time plots of the fitted dynamic orthogonal components for the monthly log returns of IBM, S&P composite index, and KO stock from January 1961 to December 2011. (a) DOC series 1, (b) Doc series 2, and (c) Doc series 3.

transformation matrix $\boldsymbol{M}^{-1}$, the first latent variable $\hat{b}_{1t}$ weights heavily on KO returns, the second variable $\hat{b}_{2t}$ represents a contrast between the three asset returns, and the third variable $\hat{b}_{3t}$ consists mainly the S&P index and KO returns. In this particular instance, the correlation matrix of $\hat{\boldsymbol{b}}_t^2$ is

$$\text{cor}(\hat{\boldsymbol{b}}_t^2) = \begin{bmatrix} 1.000 & -0.005 & 0.149 \\ -0.005 & 1.000 & 0.052 \\ 0.149 & 0.052 & 1.000 \end{bmatrix},$$

which is much close to $\boldsymbol{I}_3$ than that of the transformed series via the Go-GARCH approach shown in Example 7.3.

To estimate the multivariate volatility of the three return series, we apply the univariate GARCH(1,1) models with standardized Student-t innovations to the fitted DOCs and obtain

$$\begin{aligned} \sigma_{1t}^2 &= 0.067 + 0.094\hat{b}_{1,t-1}^2 + 0.837\sigma_{1,t-1}^2, \quad v_1 = 6.63 \\ \sigma_{2t}^2 &= 0.012 + 0.046\hat{b}_{2,t-1}^2 + 0.944\sigma_{2,t-1}^2, \quad v_2 = 5.18 \\ \sigma_{3t}^2 &= 0.052 + 0.104\hat{b}_{3,t-1}^2 + 0.848\sigma_{3,t-1}^2, \quad v_3 = 10.0, \end{aligned}$$

where v_i denotes the estimated degrees of freedom of the Student-t innovations for $\hat{b}_{it}$. Univariate diagnostic checks fail to reject the adequacy of these fitted models. The model for $\hat{b}_{2t}$ is close to being an IGARCH(1,1) model because the constant term 0.012 is not statistically significant with p-value 0.20 and persistence measure of the model is close to 1 at $\hat{\alpha}_1 + \hat{\beta}_1 = 0.99$. The volatility matrices of the three return series are then obtained by $\hat{\Sigma}_t = \hat{\boldsymbol{M}}\hat{\boldsymbol{V}}_t\hat{\boldsymbol{M}}'$, where $\hat{\boldsymbol{V}}_t$ is a diagonal matrix consisting of the conditional variances of $\hat{b}_{it}$. Figure 7.12 shows the time plots of the volatilities and correlations of the three monthly log return series. As expected, the volatilities of the three asset returns were higher during th 2008 financial crisis. The model seems to give more negative correlations between IBM and KO stock returns. Finally, we apply the four test statistics of Section 7.1 to the fitted DOC-GARCH(1,1) model. The portmanteau test of the scalar transformed series and the robust multivariate test fail to reject the adequacy of the model at the 5% level, but the rank test and the multivariate Ljung–Box statistics reject the model.

It is interesting to compare the results of the DOC-GARCH(1,1) model with those of the Go-GARCH(1,1) model shown in Figure 7.10. First, the Go-GARCH model provides larger estimates of volatilities for both IBM stock and S&P composite index. Second, the Go-GARCH model also gives much higher correlations between IBM and S&P returns. The means of the fitted correlations of the Go-GARCH model are 0.927, 0.126, and 0.206, whereas those of the DOC model are 0.606, 0.253, and 0.550. As a matter of fact, the time-varying patterns of the correlations are rather different between the two multivariate volatility models. □

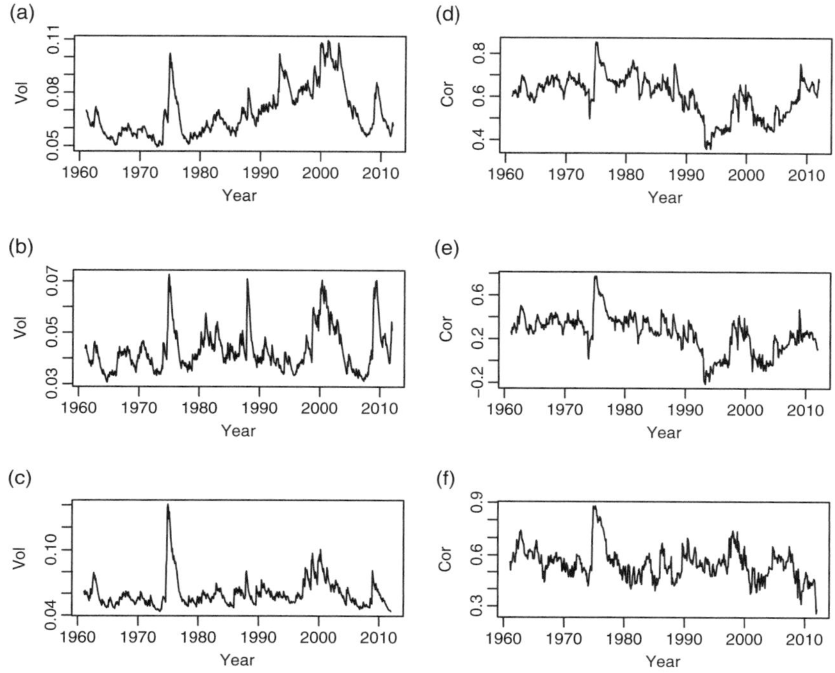

FIGURE 7.12 Time plots of volatilities and time-varying correlations for the monthly log returns of (a) IBM, (b) S&P index, and (c) KO stock from January 1961 to December 2011 via the DOC-GARCH(1,1) model, (d) IBM versus S&P, (e) IBM versus KO, and (f) S&P versus KO.

Finally, other orthogonal transformation methods have also been proposed in the literature. For instance, Fan, Wang, and Yao (2008) propose conditionally uncorrelated components for multivariate volatility modeling.

Remark: The DOC estimation is carried out using an R program developed by Dr. David Matteson. The program is available on his web page. □

7.9 COPULA-BASED MODELS

Turn to copula-based volatility models. Similar to the idea used in the dynamic conditional correlation models, it is intuitively appealing to describe a multivariate random variable by the marginal behavior of individual components and the dependence structure among its components. The copula-based approach provides a way to make use of such a description. In this section, we briefly introduce the concept of copula and discuss some copula-based volatility models. We focus on the case of continuous random variables.

7.9.1 Copulas

Let $[0, 1]$ be the unit interval on the real line. A k-dimensional copula is a distribution function on $[0, 1]^k$ with standard uniform marginal distributions. Denote an element in $[0, 1]^k$ by $\boldsymbol{u} = (u_1, \ldots, u_k)'$ and a k-dimensional copula by $C(\boldsymbol{u}) = C(u_1, \ldots, u_k)$. Then, C is a mapping from $[0, 1]^k$ to $[0, 1]$ with certain properties to be given later.

To gain some insights into copula, consider the case of $k = 2$. Here, a copula C is a mapping from $[0, 1] \times [0, 1]$ to $[0, 1]$ with the following properties:

1. $C(u_1, u_2)$ is nondecreasing in each component u_i,
2. $C(u_1, 1) = u_1$ and $C(1, u_2) = u_2$,
3. For any two two-dimensional points $\boldsymbol{u}_i = (u_{1i}, u_{2i})'$ in $[0, 1] \times [0, 1]$ with $u_{i1} \leq u_{i2}$, where $i = 1$ and 2, we can form a rectangular region in $[0, 1] \times [0, 1]$ with vertexes $(u_{11}, u_{21}), (u_{12}, u_{21}), (u_{11}, u_{22}), (u_{12}, u_{22})$ and have

$$C(u_{12}, u_{22}) - C(u_{11}, u_{22}) - C(u_{12}, u_{21}) + C(u_{11}, u_{21}) \geq 0.$$

Property 3 simply states that the probability of the rectangular region under C is non-negative as we would expect from a two-dimensional distribution function. This probability can be written as

$$\sum_{i_1=1}^{2} \sum_{i_2=1}^{2} (-1)^{i_1+i_2} C(u_{1i_1}, u_{2i_2}) \geq 0.$$

For a k-dimensional copula $C(\boldsymbol{u})$, the prior three properties become

1. $C(u_1, \ldots, u_k)$ is nondecreasing in each component u_i.
2. $C(1, \ldots, 1, u_i, 1, \ldots, 1) = u_i$ for all $i \in \{1, \ldots, k\}$ and $0 \leq u_i \leq 1$.
3. For any two k-dimensional points $\boldsymbol{u}_i = (u_{1i}, \ldots, u_{ki})'$ in $[0, 1]^k$ with $u_{i1} \leq u_{i2}$, where $i = 1, \ldots, k$, we have

$$\sum_{i_1=1}^{2} \cdots \sum_{i_k=1}^{2} (-1)^{i_1+\cdots+i_k} C(u_{1i_1}, u_{2i_2}, \ldots, u_{ki_k}) \geq 0.$$

The first property is required for any multivariate distribution function and the second property is the requirement of uniform marginal distributions. Readers are referred to Nelsen (1999) and Joe (1997) for further information on copulas.

The importance of copulas in multivariate analysis is established by the Sklar's theorem, which shows that all multivariate distribution functions contain copulas and copulas can be used in conjunction with univariate distribution function to construct multivariate distribution function.

Theorem 7.5 (Sklar theorem). Let $F(\boldsymbol{x})$ be a k-dimensional joint distribution function with marginal distributions $F_1(x_1), \ldots, F_k(x_k)$, where $\boldsymbol{x} = (x_1, \ldots, x_k)'$ with $-\infty \leq x_i \leq \infty$. Then, there exists a copula C: $[0,1]^k \to [0,1]$ such that

$$F(x_1, \ldots, x_k) = C(F_1(x_1), \ldots, F_k(x_k)). \tag{7.49}$$

If the marginal distributions are continuous, then C is unique; otherwise, C is uniquely determined on the space $R(F_1) \times R(F_2) \times \cdots \times R(F_k)$, where $R(F_i)$ denotes the range of F_i. Conversely, if C is a copula and $F_1, \ldots, F_k$ are univariate distribution functions, then the function defined in Equation (7.49) is a joint distribution function with margins $F_1, \ldots, F_k$.

A proof of the theorem can be found in Joe (1997), Nelsen (1999), and McNeil, Frey, and Embrechts (2005). These books also discuss many properties of copulas. Here, we only mention a property of copulas that is useful in multivariate volatility modeling.

Theorem 7.6 Let $(X_1, \ldots, X_k)'$ be a random vector with continuous marginal distributions $F_1(x_1), \ldots, F_k(x_k)$ and copula C. If $T_1(X_1), \ldots, T_k(X_k)$ are continuous and increasing functions. Then, $(T_1(X_1), \ldots, T_k(X_k))'$ also has copula C.

Proof. To prove the theorem, let $\tilde{F}_i(x_i)$ be the marginal distribution function of $T_i(X_i)$. Then, $u = \tilde{F}_i(y) = P(T_i(X_i) \leq y) = P(X_i \leq T_i^{-1}(y)) = F_i(T_i^{-1}(y))$ because $T_i^{-1}(.)$ is strictly increasing. Therefore, $y = \tilde{F}_i^{-1}(u)$ and $F_i^{-1}(u) = T_i^{-1}(y)$. The latter implies $T_i(F_i^{-1}(u)) = y$. Consequently, we have $\tilde{F}_i^{-1}(u) = T_i(F_i^{-1}(u))$ for $0 \leq u \leq 1$. Now, by Sklar Theorem,

$$\begin{aligned}
C(u_1, \ldots, u_k) &= Pr(F_1(X_1) \leq u_1, \ldots, F_k(X_k) \leq u_k) \\
&= Pr[X_1 \leq F_1^{-1}(u_1), \ldots, X_k \leq F_k^{-1}(u_k)] \\
&= Pr[T_1(X_1) \leq T_1(F_1^{-1}(u_1)), \ldots, T_k(F_k^{-1}(u+k))] \\
&= Pr[T_1(X_1) \leq \tilde{F}_1^{-1}(u_1), \ldots, \tilde{F}_k^{-1}(u_k)] \\
&= Pr[\tilde{F}_1(T_1(X_1)) \leq u_1, \ldots, \tilde{F}_k(T_k(X_k)) \leq u_k].
\end{aligned}$$

Thus, C is also the copula of $(T_1(X_1), \ldots, T_k(X_k))'$. □

By Theorem 7.6, we can use the correlation matrix, instead of the covariance matrix, in the discussion of commonly used copulas for volatility modeling.

7.9.2 Gaussian and *t*-Copulas

In this section, we discuss Gaussian and t-copulas as they have been used in the literature for multivariate volatility modeling. See, for instance, Creal, Koopman and

Lucas (2011). These two copulas do not have a closed-form expression, but their density functions are available for the maximum likelihood estimation.

Let $\boldsymbol{X} = (X_1, \ldots, X_k)'$ be a k-dimensional Gaussian random vector with mean $\mathbf{0}$ and correlation matrix $\boldsymbol{\rho}$. Denote the cumulative distribution function (CDF) of $\boldsymbol{X}$ by $\Phi_X(.)$. Then, each component X_i follows a standard Gaussian distribution and the copula C of Φ_X or $\boldsymbol{X}$ is the distribution function of $(\Phi(X_1), \ldots, \Phi(X_k))'$, where $\Phi(.)$ is the CDF of $N(0,1)$. We write the Gaussian copula as

$$\begin{aligned} C_\rho^g(\boldsymbol{u}) &= Pr[\Phi(X_1) \le u_1, \ldots, \Phi(X_k) \le u_k] \\ &= \Phi_X(\Phi^{-1}(u_1), \ldots, \Phi^{-1}(u_k)). \end{aligned} \tag{7.50}$$

For instance, consider the bivariate case with correlation coefficient $|\rho| < 1$. The Gaussian copula can be written as

$$C_\rho^g(\boldsymbol{u}) = \int_{-\infty}^{\Phi^{-1}(u_1)} \int_{-\infty}^{\Phi^{-1}(u_2)} \frac{1}{2\pi\sqrt{1-\rho^2}} \exp\left[-\frac{y_1^2 - 2\rho y_1 y_2 + y_2^2}{2(1-\rho^2)}\right] dy_1 dy_2.$$

Figure 7.13 shows the scatter plots of four bivariate Gaussian copulas with $\rho = 0$, 0.9, -0.9, and 0.4, respectively. These plots are based on 1000 random draws from the corresponding Gaussian copulas. The effect of the correlation ρ is clearly seen. Taking the derivatives of the Gaussian copula in Equation (7.50) and using the chain rule, we obtain the density function of a k-dimensional Gaussian copula as

$$c_\rho^g(\boldsymbol{u}) = \frac{f_X(x_1, \ldots, x_k)}{\prod_{i=1}^k f(x_i)}, \quad x_i = \Phi^{-1}(u_i),\ \boldsymbol{u} \in [0,1]^k, \tag{7.51}$$

where $f_X(.)$ and $f(.)$ are the pdf of $\boldsymbol{X}$ and Z, respectively, with $Z \sim N(0,1)$.

Let $\boldsymbol{X}$ be a k-dimensional multivariate Student-t random vector with mean vector $\boldsymbol{\mu}$, positive-definite covariance matrix $\boldsymbol{\rho}$, and v degrees of freedom. The pdf of $\boldsymbol{X}$ is

$$f_{v,\rho}(\boldsymbol{x}) = \frac{\Gamma((v+k)/2)}{\Gamma(v/2)(v\pi)^{k/2}|\boldsymbol{\rho}|^{1/2}} \left[1 + \frac{(\boldsymbol{x}-\boldsymbol{\mu})'\boldsymbol{\rho}^{-1}(\boldsymbol{x}-\boldsymbol{\mu})}{v}\right]^{-(v+k)/2}, \quad v > 2.$$

The covariance matrix of $\boldsymbol{X}$ is $v/(v-2)\boldsymbol{\rho}$. A normalized Student-t distribution with covariance matrix $\boldsymbol{\rho}$ is given in Equation (7.3). By Theorem 7.6, we can use the correlation matrix $\boldsymbol{\rho}$ in our discussion of t-copulas. Also, it does not matter whether we use this conventional pdf or the normalized one in Equation (7.3) in the discussion so long as the proper marginal distributions are used.

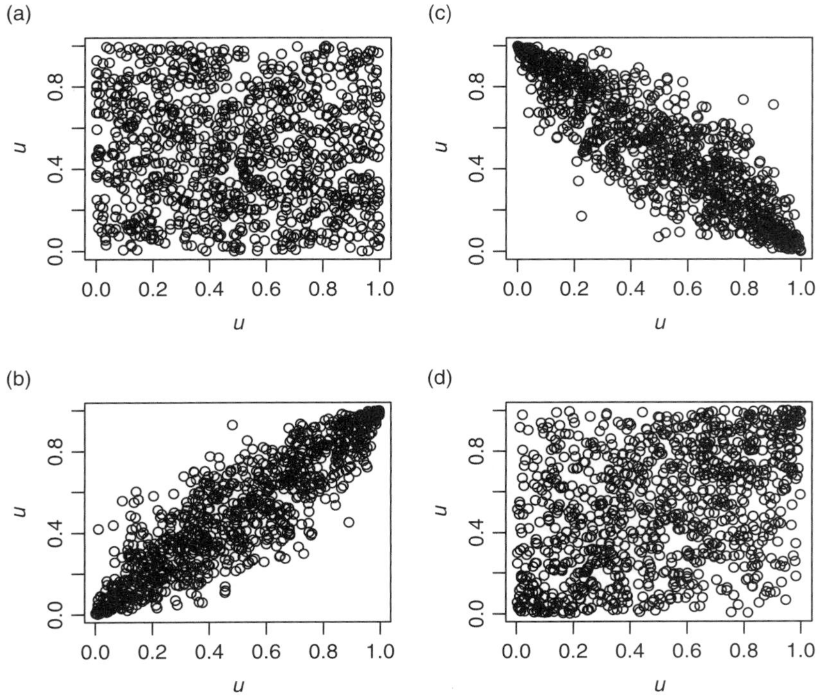

FIGURE 7.13 Scatter plots of bivariate Gaussian copulas with correlation (a) 0, (b) 0.9, (c) −0.9, and (d) 0.4, respectively, based on 1000 realizations.

The copula of $\boldsymbol{X}$ is

$$C^t_{v,\rho}(\boldsymbol{u}) = F_X(t_v^{-1}(u_1), \ldots, t_v^{-1}(u_k))$$
$$= \int_{-\infty}^{t_v^{-1}(u_1)} \cdots \int_{-\infty}^{t_v^{-1}(u_k)} \frac{\Gamma((v+k)/2)}{\Gamma(v/2)(v\pi)^{k/2}|\boldsymbol{\rho}|^{1/2}} \left[1 + \frac{\boldsymbol{x}'\boldsymbol{\rho}^{-1}\boldsymbol{x}}{v}\right]^{-(v+k)/2} d\boldsymbol{x}, \tag{7.52}$$

where $v > 2$ and $t_v^{-1}(.)$ is the quantile function of a univariate Student-t distribution with v degrees of freedom. The pdf of the t-copula is

$$c^t_{v,\rho}(\boldsymbol{u}) = \frac{f_{v,\rho}(x_1, \ldots, x_k)}{\prod_{i=1}^k f_v(x_i)}, \quad x_i = t_v^{-1}(u_i), \ \boldsymbol{u} \in [0,1]^k, \tag{7.53}$$

where $f_v(x)$ denotes a pdf of a univariate Student-t random variable with v degrees of freedom and $f_{v,\rho}(.)$ denotes the integrand in Equation (7.52).

Figure 7.14 shows the scatter plots of 1000 random draws from four bivariate t-copulas. The multivariate Student-t distribution used is a bivariate Student-t with mean zero and covariance matrix

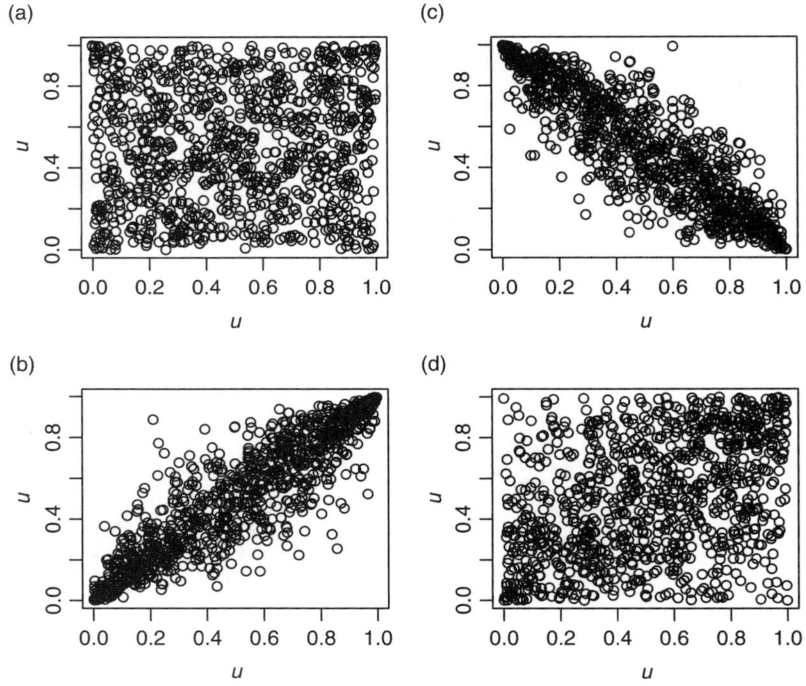

FIGURE 7.14 Scatter plots of bivariate t-copulas with correlation (a) 0, (b) 0.9, (c) -0.9, and (d) 0.4, respectively, based on 1000 realizations. The degrees of freedom used are $v = 6$.

$$\boldsymbol{\rho} = \begin{bmatrix} 1 & \rho \\ \rho & 1 \end{bmatrix},$$

with $\rho = 0.0,\ 0.9,\ -0.9$, and 0.4, respectively. The degrees of freedom are $v = 6$ for all cases. Again, the effect of the correlation ρ on the distribution is clearly seen. It is also interesting to compare these scatter plots with those in Figure 7.13. In particular, the t-copula seems to have data points concentrated around the corners of the plots. This is related to the fact that t-copulas have non-negative tail-dependence, whereas the Gaussian copulas are tail-independent. For further properties of t-copulas, see Demarta and McNeil (2004).

If a k-dimensional random vector $\boldsymbol{X}$ has the t-copula $C^t_{v,\rho}$ and its marginal Student-t distributions have the same degrees of freedom, then $\boldsymbol{X}$ has a multivariate Student-t distribution with v degrees of freedom. If one uses the t-copula in Equation (7.52) to combine any other set of univariate marginal distributions, then one obtains a multivariate distribution function, which is called a meta-t_v distribution. See, for instance, McNeil, Frey, and Embrechts (2005) and the references therein.

7.9.3 Multivariate Volatility Modeling

In this section, we briefly discuss multivariate volatility modeling using Gaussian and t-copulas. We focus on the procedure of using copulas to describe the time-varying volatilities and correlations. For simplicity, we adopt the idea of dynamic conditional correlation (DCC) models of Section 7.7 to govern the time evolution of correlations so that the models employed in this section are highly related to the DCC models. Other approaches can also be used in conjunction with copulas to modeling multivariate volatility.

To introduce another way to describe time-varying correlations, we adopt the parameterization of a positive-definite correlation matrix used in Creal, Koopman, and Lucas (2011) and Van der Weide (2003). Let $\boldsymbol{R}_t$ be a $k \times k$ correlation matrix. Then it can be decomposed as $\boldsymbol{R}_t = \boldsymbol{X}_t'\boldsymbol{X}_t$, where $\boldsymbol{X}_t$ is an upper triangular matrix given by

$$\boldsymbol{X}_t = \begin{bmatrix} 1 & c_{12t} & c_{13t} & \cdots & c_{1kt} \\ 0 & s_{12t} & c_{23t}s_{13t} & \cdots & c_{2kt}s_{1kt} \\ 0 & 0 & s_{23t}s_{13t} & \cdots & c_{3kt}s_{2kt}s_{1kt} \\ \vdots & \vdots & \vdots & \ddots & \vdots \\ 0 & 0 & 0 & \cdots & c_{k-1,kt}\prod_{i=1}^{k-2} s_{ikt} \\ 0 & 0 & 0 & \cdots & \prod_{i=1}^{k-1} s_{ikt} \end{bmatrix}, \tag{7.54}$$

where $c_{ijt} = \cos(\theta_{ijt})$ and $s_{ijt} = \sin(\theta_{ijt})$ with θ_{ijt} being a time-varying angle measured in radians and $1 \le i < j \le k$. Let $\boldsymbol{\theta}_t$ be the $k(k-1)/2$-dimensional vector consisting of all the angles θ_{ijt}. We can model the time evolution of $\boldsymbol{\theta}_t$ in lieu of $\boldsymbol{R}_t$ to describe the time-varying correlations. Creal, Koopman, and Lucas (2011) apply the generalized autoregressive score (GAS) model to $\boldsymbol{\theta}_t$. For simplicity, we postulate that the angle vector $\boldsymbol{\theta}_t$ follows the equation of a DCC model of Section 7.7. For instance, using the DCC model of Tse and Tsui (2002), we employ the model

$$\boldsymbol{\theta}_t = \boldsymbol{\theta}_0 + \lambda_1\boldsymbol{\theta}_{t-1} + \lambda_2\boldsymbol{\theta}^*_{t-1}, \tag{7.55}$$

where $\boldsymbol{\theta}^*_{t-1}$ is a local estimate of the angles using data $\{\boldsymbol{\eta}_{t-1}, \ldots, \boldsymbol{\eta}_{t-m}\}$ for some $m > 1$, where, as before, $\boldsymbol{\eta}_t$ is the marginally standardized innovation, that is, $\eta_{it} = a_{it}/\sqrt{\sigma_{ii,t}}$. In Equation (7.55), λ_i are non-negative real numbers satisfying $0 < \lambda_1 + \lambda_2 < 1$ and $\boldsymbol{\theta}_0$ denotes the initial values of the angles. Unlike the DCC models, here we treat $\boldsymbol{\theta}_0$ as parameters and estimate it jointly with λ_i. Of course, one can fix *a priori* the value of $\boldsymbol{\theta}_0$ using the sample correlations of $\boldsymbol{\eta}_t$ to simplify the estimation. Fixing $\boldsymbol{\theta}_0$ might be needed when the dimension k is large. For the GAS model used in Creal, Koopman, and Lucas, $\boldsymbol{\theta}^*_{t-1}$ is a properly scaled version of the score function of the log-likelihood function of $\boldsymbol{\eta}_{t-1}$.

In what follows, we consider a procedure for modeling time-varying correlations. Again, let $\boldsymbol{a}_t$ be the innovations of a k-dimensional asset return series $\boldsymbol{z}_t$. We use

t-copulas in the discussion because most asset returns are not normally distributed, but one can use Gaussian copulas if desired.

7.9.3.1 Procedure A: Modeling Correlations Using Meta-t Distributions

1. Obtain marginal volatility series σ_{it}^2 of a_{it} via univariate GARCH-type of models and obtain the marginally standardized series $\eta_{it} = a_{it}/\sigma_{it}$.
2. Fit the model in Equation (7.55) with a t-copula for the standardized innovation $\boldsymbol{\eta}_t$ using the pdf of a t-copula with standardized t margins. That is, employ Equation (7.53), but with $f_{v,\rho}(.)$ and $f_v(x)$ being standardized pdfs.

Example 7.5 Consider, again, the monthly log returns of IBM stock, S&P composite index, and Coca Cola stock from January 1961 to December 2011 used in the previous sections. Here, we use univariate GARCH(1,1) models with standardized Student-t innovations for the marginal series. The residual series is $\hat{\boldsymbol{a}}_t = \boldsymbol{z}_t - \hat{\boldsymbol{\mu}}$ with $\hat{\boldsymbol{\mu}} = (0.0077, 0.0050, 0.0106)'$, and the three marginal models for $\hat{\boldsymbol{a}}_t$ are

$$\begin{aligned}
\sigma_{1t}^2 &= 3.88 \times 10^{-4} + 0.116\hat{a}_{1,t-1}^2 + 0.805\sigma_{1,t-2}^2, \ \hat{v}_1 = 9.21 \\
\sigma_{2t}^2 &= 1.2 \times 10^{-4} + 0.131\hat{a}_{2,t-1}^2 + 0.815\sigma_{2,t-1}^2, \ \hat{v}_2 = 7.27 \\
\sigma_{3t}^2 &= 2.16 \times 10^{-4} + 0.105\hat{a}_{3,t-1}^2 + 0.837\sigma_{3,t-1}^2, \ \hat{v}_3 = 7.08,
\end{aligned}$$

where $\hat{v}_i$ denotes the estimated degrees of freedom and all coefficient estimates are statistically significant at the 5% level.

Let $\hat{\eta}_{it} = \hat{a}_{it}/\sigma_{it}$ be the standardized residuals and $\hat{\boldsymbol{\eta}}_t = (\hat{\eta}_{1t}, \hat{\eta}_{2t}, \hat{\eta}_{3t})'$. Following procedure A, we employ the model in Equation (7.55) to describe the time evolution of the angle $\boldsymbol{\theta}_t$ and assume the probability transform $\boldsymbol{u}_t = (u_{1t}, u_{2t}, u_{3t})'$ is from a t-copula with v degrees of freedom, where $u_{it} = t_{v_i}(\hat{\eta}_{it})$ with $t_d(.)$ being the cumulative distribution function of a standardized Student-t distribution with d degrees of freedom. Since t distributions are continuous, the latter assumption implies that the pdf of $\hat{\boldsymbol{\eta}}_t$ is given in Equation (7.53). The maximum likelihood estimates give

$$\boldsymbol{\theta}_t = \hat{\boldsymbol{\theta}}_0 + 0.882\boldsymbol{\theta}_{t-1} + 0.034\boldsymbol{\theta}_{t-1}^*,$$

where $\hat{\boldsymbol{\theta}}_0 = (0.9197, 1.2253, 1.0584)'$, $\hat{v} = 15.38$, and $\boldsymbol{\theta}_{t-1}^*$ denotes the angles corresponding to the correlation matrix of $\{\hat{\boldsymbol{\eta}}_{t-1}, \ldots, \hat{\boldsymbol{\eta}}_{t-4}\}$. Based on the asymptotic standard errors, all estimates are statistically significant at the 5% level, but the standard error of $\hat{v}$ is large. Figure 7.15 shows the time plots of volatilities and correlations of the fitted t-copula model. We apply the test statistics of Section 7.1 to check the validity of the fitted t-copula model to the $\hat{\boldsymbol{\eta}}_t$ series. The four test statistics detect the presence of conditional heteroscedasticity in the standardized residuals. The portmanteau statistics of the scalar transformed series and the robust multivariate test fail to reject the null hypothesis at the 1% level, but the rank statistic indicates that some conditional heteroscedasticity remains in the data.

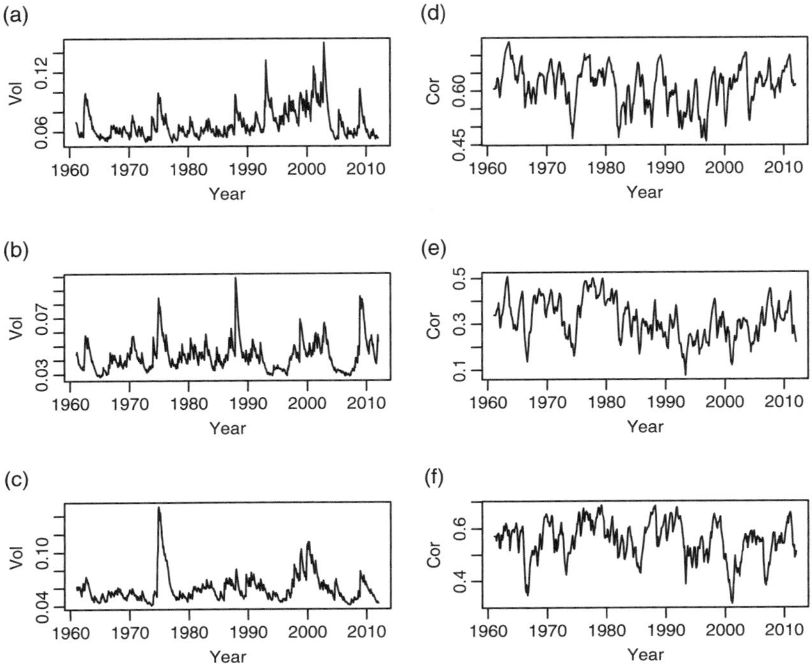

FIGURE 7.15 Time plots of the volatilities and time-varying correlations of the monthly log returns of (a) IBM, (b) S&P index, and (c) Coca Cola stock from January 1961 to December 2011 based on a t-copula with Equation (7.55). The left panel consists of volatility series and the right panel the correlations, (d) IBM versus S&P, (e) IBM versus KO, and (f) S&P versus KO.

Compared with the correlations of the DCC models in Figure 7.9, we see that the correlations of the t-copula model in Figure 7.15 are between those of the DCC models of Tse–Tsui and Engle. The time-evolution patterns of the correlations of t-copula model are closer to those of the DCC model of Tse–Tsui (2002), but with higher magnitudes. The sample means of the three correlation coefficients of the t-copula model are 0.613, 0.319, and 0.555, respectively, whereas the sample standard errors are 0.057, 0.086, and 0.072. The minimum and maximum of the three correlations are 0.460, 0.076, 0.315 and 0.738, 0.509, and 0.687, respectively. These values are similar to those of the DCC models in Table 7.4.

Finally, we also fit the t-copula model to the three monthly log return series, but fix the initial angle $\boldsymbol{\theta}_0$ based on the sample correlation matrix of $\hat{\boldsymbol{\eta}}_t$. In this particular case, $\hat{\boldsymbol{\theta}}_0 = (0.946, 1.289, 1.035)'$, which is close to what we obtained before via joint estimation. The estimates of λ_i become 0.878 and 0.034, respectively, whereas the estimate of the degrees of freedom is $\hat{v} = 14.87$. It seems that, for this particular example, fixing $\hat{\boldsymbol{\theta}}_0$ does not have a major effect on the estimates of other parameters. $\square$

Remark: The estimation of the t-copula models is based on the command `mtCopula` of the `MTS` package. The marginal volatility models of the component series are estimated by the command `dccPre`, which uses the `fGarch` package of Rmetrics. The command `mtCopula` provides two sets of asymptotic standard errors of the estimates based on two different numerical differentiation methods. □

R Demonstration: Estimation of t-copula models.

```
> da=read.table("m-ibmspko-6111.txt",header=T)
> rtn=log(da[,-1]+1)
> m1=dccPre(rtn,cond.dist="std")
Sample mean of the returns:  0.00772774 0.005023909 0.01059521
Component:  1
Estimates:  0.000388 0.115626 0.805129 9.209269
se.coef  :  0.000177 0.036827 0.059471 3.054817
t-value  :  2.195398 3.139719 13.5382 3.014671
Component:  2
Estimates:  0.00012 0.130898 0.814531 7.274928
se.coef  :  5.7e-05 0.037012 0.046044 1.913331
t-value  :  2.102768 3.536655 17.69028 3.802232
Component:  3
Estimates:  0.000216 0.104706 0.837217 7.077138
se.coef  :  8.9e-05 0.028107 0.037157 1.847528
t-value  :  2.437323 3.725341 22.53208 3.830599
> names(m1)
[1] "marVol"  "sresi"   "est"     "se.coef"
> Vol=m1$marVol; eta=m1$sresi
> m2=mtCopula(eta,0.8,0.04)
Lower limits:  5.1 0.2 1e-04 0.7564334 1.031269 0.8276595
Upper limits:  20 0.95 0.04999999 1.040096 1.417994 1.138032
estimates:   15.38215 0.88189 0.034025 0.919724 1.225322 1.058445
std.errors:  8.222771 0.05117 0.011733 0.041357 0.055476 0.051849
t-values:    1.870677 17.2341 2.899996 22.23883 22.08729 20.41412
Alternative numerical estimates of se:
st.errors:  5.477764 0.051033 0.011714 0.041370 0.055293 0.050793
t-values:   2.808107 17.28091 2.904679 22.23173 22.16072 20.83839
> names(m2)
[1] "estimates" "Hessian"   "rho.t"     "theta.t"
> MCHdiag(eta,m2$rho.t)
Test results:
Q(m) of et:
Test and p-value:  19.30177 0.03659304
Rank-based test:
Test and p-value:  27.03262 0.002573576
Qk(m) of epsilon_t:
Test and p-value:  125.9746 0.007387423
Robust Qk(m):
Test and p-value:  107.4675 0.1011374
> m3=mtCopula(eta,0.8,0.04,include.th0=F)  # fix theta_0
Value of angles:
[1] 0.9455418 1.2890858 1.0345744
Lower limits:  5.1 0.2 1e-05
```

```
Upper limits:  20 0.95 0.0499999
estimates:     14.87427 0.8778 0.03365157
std.errors:    7.959968 0.053013 0.011951
t-values:      1.868635 16.55824 2.815811
Alternative numerical estimates of se:
st.errors:   5.49568 0.0529896 0.01191378
t-values:    2.70654 16.56551 2.824592
```

Example 7.6 As a second example, we consider the daily log returns, in percentages, of EXXON-Mobil stock, S&P composite index, and Apple stock from September 4, 2007 to September 28, 2012 for 1280 observations. The simple returns are from CRSP. The left panel of Figure 7.16 shows the time plots of the daily log return series. As expected, the volatilities of the three assets were high during the 2008 financial crisis and the variability of the Apple stock returns appears to be higher than those of the other two series. Let z_t denote the daily log returns, in percentages. Analysis of individual series indicates some minor serial correlations in the Apple stock returns. However, for simplicity, we continue to assume that the mean equation of z_t is just a constant. This provides an estimate of the

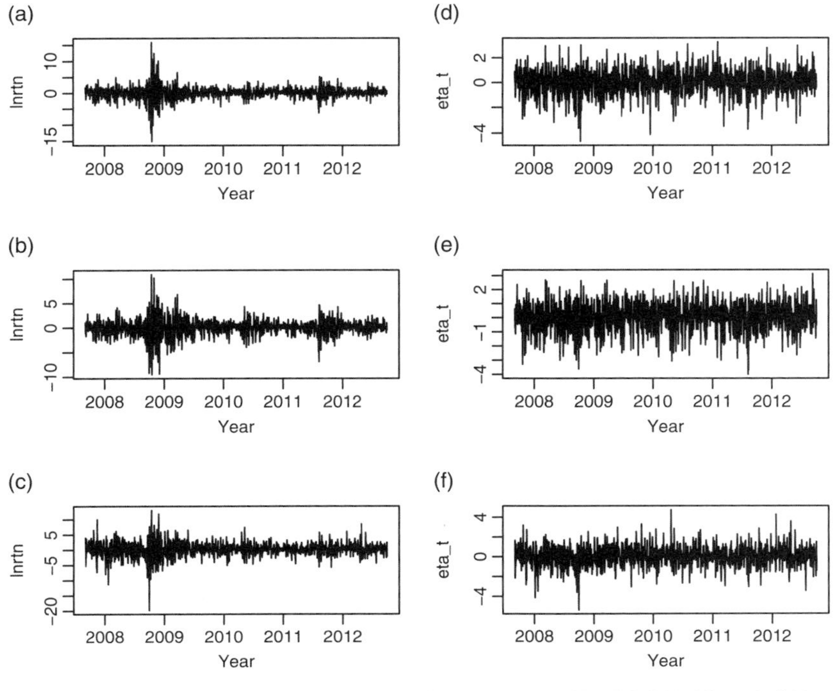

FIGURE 7.16 Time plots of the daily log returns, in percentages, of EXXON-Mobil stock, S&P composite index, and Apple stock from September 4, 2007, to September 28, 2012. The left panel consists of log returns, whereas the right panel the standardized residuals of a univariate GARCH(1,1) model with Student-t innovations: (a) XOM, (b) X&P, (c) AAPL, (d) XOM, (e) S&P and (f) AAPL.

innovation $\hat{\boldsymbol{a}}_t = \boldsymbol{z}_t - \hat{\boldsymbol{\mu}}$, with $\hat{\boldsymbol{\mu}} = (0.014, -0.002, 0.123)'$. Following Procedure A of the section, we start with univariate GARCH(1,1) models with Student-t innovations for the component series. The fitted models are

$$\begin{aligned}
\sigma_{1t}^2 &= 0.044 + 0.104\hat{a}_{1,t-1}^2 + 0.883\sigma_{1,t-1}^2, \quad \hat{v}_1 = 6.99,\\
\sigma_{2t}^2 &= 0.021 + 0.105\hat{a}_{2,t-1}^2 + 0.892\sigma_{2,t-1}^2, \quad \hat{v}_2 = 6.22,\\
\sigma_{3t}^2 &= 0.064 + 0.078\hat{a}_{3,t-1}^2 + 0.911\sigma_{3,t-1}^2, \quad \hat{v}_3 = 6.70,
\end{aligned}$$

where, again, $\hat{v}_i$ denotes the estimate of degrees of freedom, and all coefficient estimates are statistically significant at the 5% level. The right panel of Figure 7.16 shows the time plots of the standardized residuals $\hat{\boldsymbol{\eta}}_t$ with $\hat{\eta}_{it} = a_{it}/\sigma_{it}$ for $i = 1, 2, 3$. The plots of standardized residuals confirm that the variability of Apple stock is indeed higher.

Let $\boldsymbol{u}_t$ be the probability transform of $\hat{\boldsymbol{a}}_t$, where the ith component u_{it} is the probability transform of $\hat{a}_{i,t}$ based on the fitted standardized Student-t distribution with $\hat{v}_i$ degrees of freedom. We apply a t-copula to $\boldsymbol{u}_t$ using Equation (7.55) for the angle vector $\boldsymbol{\theta}_t$, which provides an estimate of the correlation matrix at time t. The fitted model is

$$\boldsymbol{\theta}_t = \boldsymbol{\theta}_0 + 0.94750\boldsymbol{\theta}_{t-1} + 0.02500\boldsymbol{\theta}_{t-1}^*,$$

where, again, $\boldsymbol{\theta}_{t-1}^*$ represents the angles corresponding to the sample correlation matrix of $\{\hat{\boldsymbol{\eta}}_{t-1}, \ldots, \hat{\boldsymbol{\eta}}_{t-4}\}$, and the fitted degrees of freedom is 8.56 with standard error 1.61. The fitted initial angle vector is $\hat{\boldsymbol{\theta}}_0 = (0.6654, 1.1446, 1.0278)'$. All estimates are highly significant based on their asymptotic standard errors. Figure 7.17 shows the time plots of the time-varying correlations and the $\boldsymbol{\theta}_t$ series. The correlations are relatively high with sample means 0.785, 0.427, and 0.635, respectively. From the plot, the angles appear to be negative correlated with correlations. In this particular instance, the degrees of freedom of the t-copula has a smaller standard error compared with the monthly log returns of Example 7.5. This is reasonable as the sample size for daily returns is larger and the daily returns have heavier tails. □

7.10 PRINCIPAL VOLATILITY COMPONENTS

In a recent study, Hu and Tsay (2014) generalize the concept of principal component analysis to principal volatility component (PVC) analysis. Let $\boldsymbol{v}_t = \text{vec}(\boldsymbol{\Sigma}_t)$ be the column stacking vector of the volatility matrix $\boldsymbol{\Sigma}_t$ of a k-dimensional time series $\boldsymbol{z}_t$. The volatility models considered in the chapter imply that

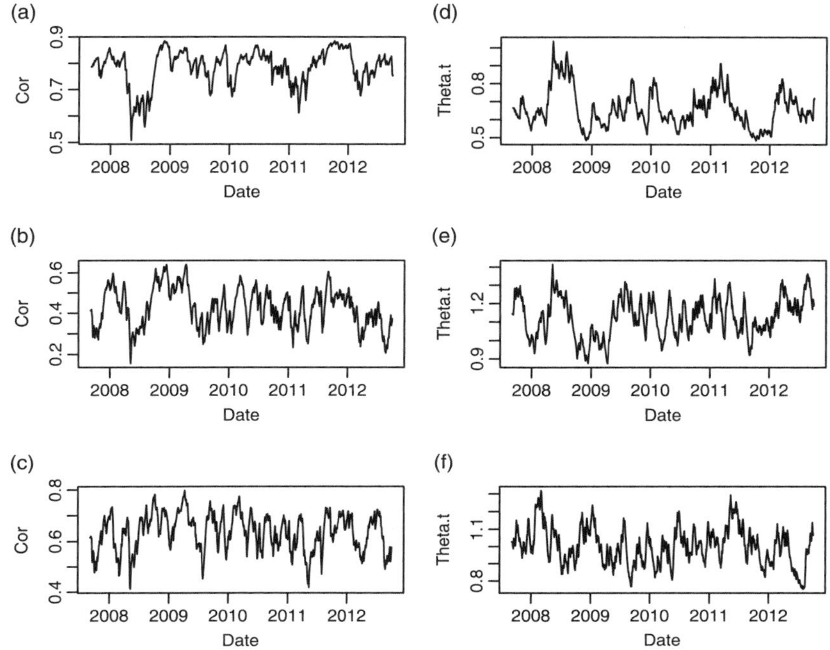

FIGURE 7.17 Time plots of time-varying correlations and angles for the daily log returns, in percentages, of EXXON-Mobil stock, S&P composite index, and Apple stock from September 4, 2007, to September 28, 2012. The results are based on a t-copula with 8.56 degrees of freedom. The left panel consists of correlations and the right panel the angles. (a) XOM versus S&P, (b) XOM versus AAPL, (c) S&P versus AAPL, (d) theta(1,2), (e) theta(1,3), and (f) theta(2,3).

$$\boldsymbol{v}_t = \boldsymbol{c}_0 + \sum_{i=1}^{\infty} \boldsymbol{C}_i \text{vec}(\boldsymbol{a}_{t-i}\boldsymbol{a}_{t-i}'), \tag{7.56}$$

where $\boldsymbol{c}_0$ is a k^2 dimensional constant vector and $\boldsymbol{C}_i$ are $k^2 \times k^2$ real-valued constant matrices. For instance, for the BEKK(1,1) model of Equation (7.23), we have $\boldsymbol{c}_0 = [\boldsymbol{I}_k^2 - (\boldsymbol{B}_1 \otimes \boldsymbol{B}_1)]^{-1}\text{vec}(\boldsymbol{A}_0\boldsymbol{A}_0')$ and $\boldsymbol{C}_i = \boldsymbol{B}_1^i\boldsymbol{A}_1 \otimes \boldsymbol{B}_1^i\boldsymbol{A}_1$. Equation (7.56) says that $\boldsymbol{z}_t$ has conditional heteroscedasticity (or ARCH effect) if and only if $\boldsymbol{C}_i$ is nonzero for some $i > 0$. In other words, the existence of ARCH effects implies that $\boldsymbol{z}_t\boldsymbol{z}_t'$ and, hence, $\boldsymbol{a}_t\boldsymbol{a}_t'$ is correlated with $\boldsymbol{a}_{t-i}\boldsymbol{a}_{t-i}'$ for some $i > 0$. Based on this observation, Hu and Tsay (2014) define the lag-ℓ generalized cross kurtosis matrix as

$$\gamma_\ell = \sum_{i=1}^{k}\sum_{j=1}^{k} \text{Cov}^2(\boldsymbol{a}_t\boldsymbol{a}_t', a_{i,t-\ell}a_{j,t-\ell}) \equiv \sum_{i=1}^{k}\sum_{j=1}^{k} \gamma_{\ell,ij}^2, \tag{7.57}$$

where $\ell \geq 0$ and

$$\gamma_{\ell,ij} = \text{Cov}(\boldsymbol{a}_t\boldsymbol{a}_t', a_{i,t-\ell}a_{j,t-\ell}) = E[(\boldsymbol{a}_t\boldsymbol{a}_t' - \boldsymbol{\Sigma}_a)(a_{1,t-\ell}a_{j,t-\ell} - \sigma_{a,i,j})], \tag{7.58}$$

where $\boldsymbol{\Sigma}_a = [\sigma_{a,ij}]$ is the unconditional covariance matrix of the innovation $\boldsymbol{a}_t$ of $\boldsymbol{z}_t$. The $k \times k$ matrix $\boldsymbol{\gamma}_{\ell,ij}$ is called the *generalized* covariance matrix between $\boldsymbol{a}_t\boldsymbol{a}_t'$ and $a_{i,t-\ell}a_{j,t-\ell}$. See, for instance, Li (1992). A key feature of the generalized covariance matrix is that it remains a $k \times k$ symmetric matrix for the $k \times k$ random matrix $\boldsymbol{a}_t\boldsymbol{a}_t'$.

In Equation (7.57), the square of the generalized covariance matrix $\boldsymbol{\gamma}_{\ell,ij}$ is used because, for the PVC analysis, it is convenient to have a non-negative definite matrix $\boldsymbol{\gamma}_\ell$. Since $\boldsymbol{\gamma}_{\ell,ij}$ of Equation (7.58) is symmetric, $\boldsymbol{\gamma}_{\ell,ij}^2$ is non-negative definite and, hence, the generalized cross kurtosis matrix $\boldsymbol{\gamma}_\ell$ is non-negative definite (or semi-positive definite).

Based on Equations (7.56) and (7.57), Hu and Tsay (2014) define

$$\boldsymbol{G}_m = \sum_{\ell=1}^{m} \boldsymbol{\gamma}_\ell, \tag{7.59}$$

as the mth cumulative generalized cross kurtosis matrix. In particular, $\boldsymbol{G}_\infty = \lim_{m\to\infty} \boldsymbol{G}_m$, provided that the limit exists, is the overall cumulative generalized cross kurtosis matrix. It is interesting to see that for multivariate ARCH(m)-type of volatility models, the cumulative matrix $\boldsymbol{G}_m$ of Equation (7.59) is sufficient to summarize the ARCH effect in $\boldsymbol{a}_t$. The limiting case, $\boldsymbol{G}_\infty$, is mainly for GARCH-type of multivariate volatility models. Note that $\boldsymbol{G}_m$ is symmetric and semi-positive definite by construction.

A nice property of the generalized covariance matrix useful in volatility modeling is as follows.

Lemma 7.3 For a $k \times k$ constant matrix $\boldsymbol{M}$, let $\boldsymbol{b}_t = \boldsymbol{M}'\boldsymbol{a}_t$. Then, $\text{Cov}(\boldsymbol{b}_t\boldsymbol{b}_t', x_{t-1}) = \text{Cov}(\boldsymbol{M}'\boldsymbol{a}_t\boldsymbol{a}_t'\boldsymbol{M}, x_{t-1}) = \boldsymbol{M}'\text{Cov}(\boldsymbol{a}_t\boldsymbol{a}_t', x_{t-1})\boldsymbol{M}$, where x_{t-1} denotes a scalar random variable.

Let $\boldsymbol{m}_s = (m_{1s}, m_{2s}, \ldots, m_{ks})'$ be the sth column of $\boldsymbol{M}$. If $b_{st} = \boldsymbol{m}_s'\boldsymbol{a}_t$ is a linear combination of $\boldsymbol{a}_t$ that has no conditional heteroscedasticity, that is, no ARCH effect, then $E(b_{st}^2|F_{t-1}) = c_s^2$, which is a constant, where F_{t-1} denotes the information available at time $t-1$. Consequently, b_{st}^2 is not correlated with $\boldsymbol{a}_{i,t-\ell}a_{j,t-\ell}$ for all $\ell > 0$ and $1 \leq i, j \leq k$. Therefore, we have $\text{Cov}(b_{st}^2, a_{i,t-\ell}a_{j,t-\ell}) = 0$ for $\ell > 0$ and $1 \leq i, j \leq k$. By Lemma 7.3, we see that $\boldsymbol{\gamma}_{\ell,ij}$ is singular for all $\ell > 0$ and $1 \leq i, j \leq k$. Hence, $\boldsymbol{\gamma}_\ell$ is singular for $\ell > 0$. Consequently, $\boldsymbol{G}_m$ and $\boldsymbol{G}_\infty$ are also singular.

On the other hand, assume that $\boldsymbol{G}_\infty$ is singular and $\boldsymbol{m}_s$ is a k-dimensional vector such that $\boldsymbol{G}_\infty\boldsymbol{m}_s = \boldsymbol{0}$. Clearly, we have $\boldsymbol{m}_s'\boldsymbol{G}_\infty\boldsymbol{m}_s = 0$. Since $\boldsymbol{\gamma}_\ell$ is semi-positive definite, we have $\boldsymbol{m}_s'\boldsymbol{\gamma}_\ell\boldsymbol{m}_s = 0$ for all $\ell > 0$. This in turn implies that

$\boldsymbol{m}_s'\boldsymbol{\gamma}_{\ell,ij}^2\boldsymbol{m}_s = 0$ for all $\ell > 0$ and $1 \leq i,j \leq k$. Consequently, by the symmetry of $\boldsymbol{\gamma}_{\ell,ij}$, we have $(\boldsymbol{\gamma}_{\ell,ij}\boldsymbol{m}_s)'(\boldsymbol{\gamma}_{\ell,ij}\boldsymbol{m}_s) = 0$. This implies that $\boldsymbol{\gamma}_{\ell,ij}\boldsymbol{m}_s = \boldsymbol{0}$ for all $\ell > 0$ and $1 \leq i,j \leq k$. Again, using Lemma 7.3, we see that $b_{st} = \boldsymbol{m}_s'\boldsymbol{a}_t$ is not correlated with $a_{i,t-\ell}a_{j,t-\ell}$ for all $\ell > 0$ and $1 \leq i,j \leq k$. This in turn implies that $E(b_{st}^2|F_{t-1})$ is not time-varying. In other words, the linear combination b_{st} of $\boldsymbol{a}_t$ has no conditional heteroscedasticity.

The prior discussion shows that an eigenvector of $\boldsymbol{G}_\infty$ associated with a zero eigenvalue gives rise to a linear combination of $\boldsymbol{a}_t$ that has no ARCH effect. Hu and Tsay (2014) summarize the result into a theorem.

Theorem 7.7 Consider a weakly stationary k-dimensional vector process $\boldsymbol{z}_t$ with innovation $\boldsymbol{a}_t$. Assume that $\boldsymbol{a}_t$ has finite fourth moments and let $\boldsymbol{G}_\infty$ be the cumulative generalized cross kurtosis matrix of $\boldsymbol{a}_t$ defined in Equation (7.59) with $m \to \infty$. Then, there exist $k-h$ linear combinations of $\boldsymbol{a}_t$ that have no ARCH effect if and only if rank$(\boldsymbol{G}_\infty) = h$.

To make use of Theorem 7.7, Hu and Tsay (2014) consider the spectral decomposition of $\boldsymbol{G}_\infty$. Specifically, suppose that

$$\boldsymbol{G}_\infty\boldsymbol{M} = \boldsymbol{M}\boldsymbol{\Lambda}, \tag{7.60}$$

where $\boldsymbol{\Lambda} = \text{diag}\{\lambda_1^2 \geq \lambda_2^2 \geq \cdots \geq \lambda_k^2\}$ is the diagonal matrix of the ordered eigenvalues of $\boldsymbol{G}_\infty$ and $\boldsymbol{M} = [\boldsymbol{m}_1,\ldots,\boldsymbol{m}_k]$ is the matrix of corresponding eigenvectors with $\|\boldsymbol{m}_s\| = 1$ for $s = 1,\ldots,k$. Then, the sth principal volatility component (PCV) of $\boldsymbol{a}_t$ is $b_{st} = \boldsymbol{m}_s'\boldsymbol{a}_t$.

Consider the sth PVC of $\boldsymbol{a}_t$. From the definition and the spectral decomposition of $\boldsymbol{G}_\infty$, we can premultiply Equation (7.60) by $\boldsymbol{m}_s'$ to obtain

$$\sum_{\ell=1}^{\infty}\sum_{i=1}^{k}\sum_{j=1}^{k}\boldsymbol{m}_s'\boldsymbol{\gamma}_{\ell,ij}^2\boldsymbol{m}_s = \lambda_s^2, \quad s = 1,\ldots,k. \tag{7.61}$$

Let $\boldsymbol{\gamma}_{\ell,ij}\boldsymbol{m}_s = \boldsymbol{\omega}_{\ell,ij,s}$. Then, Equation (7.61) says that

$$\sum_{\ell=1}^{\infty}\sum_{i=1}^{k}\sum_{j=1}^{k}\boldsymbol{\omega}_{\ell,ij,s}'\boldsymbol{\omega}_{\ell,ij,s} = \lambda_s^2.$$

On the other hand, by Lemma 7.3, we have

$$\boldsymbol{m}_s'\boldsymbol{\omega}_{\ell,ij,s} = \boldsymbol{m}_s'\boldsymbol{\gamma}_{\ell,ij}\boldsymbol{m}_s = \text{Cov}(b_{st}^2, a_{i,t-\ell}a_{j,t-\ell}).$$

This equation says that $\boldsymbol{m}_s'\boldsymbol{\gamma}_{\ell,ij}\boldsymbol{m}_s$ can be regarded as a measure of the dependence of the volatility of the portfolio b_{st} on the lagged cross product $a_{i,t-\ell}a_{j,t-\ell}$. This dependence measure may be negative so that we use squared matrices in

Equation (7.61) to obtain a non-negative summation of the dependence measure across lags. In this sense, the eigenvalue λ_s^2 can be considered as a measure of the volatility dependence of the portfolio b_{st}. This provides a justification for the definition of PVC.

7.10.1 Sample Principal Volatility Components

In finite samples, we can estimate the cumulative generalized cross kurtosis matrix $\boldsymbol{G}_\infty$ by a sample counterpart of $\boldsymbol{G}_m$ for a prespecified positive integer m. For instance, Hu and Tsay (2014) estimate the generalized covariance matrix $\gamma_{\ell,ij}$ by its sample version

$$\hat{\text{Cov}}(\boldsymbol{a}_t\boldsymbol{a}_t', a_{i,t-\ell}a_{j,t-\ell}) = \frac{1}{T}\sum_{t=\ell+1}^{T}(\boldsymbol{a}_t\boldsymbol{a}_t' - \bar{\boldsymbol{A}})(a_{i,t-\ell}a_{j,t-\ell} - \bar{A}_{ij}), \qquad (7.62)$$

where T is the sample size and

$$\bar{\boldsymbol{A}} = \frac{1}{T}\sum_{t=1}^{T}\boldsymbol{a}_t\boldsymbol{a}_t' \equiv [\bar{A}_{ij}].$$

The cumulative generalized cross kurtosis matrix $\boldsymbol{G}_m$ is then estimated by

$$\hat{\boldsymbol{G}}_m = \sum_{\ell=1}^{m}\sum_{i=1}^{k}\sum_{j=1}^{k}(1-\frac{\ell}{T})^2\hat{\text{Cov}}^2(\boldsymbol{a}_t\boldsymbol{a}_t', a_{i,t-\ell}a_{j,t-\ell}).$$

The spectral decomposition of $\hat{\boldsymbol{G}}_m$ provides the sample principal volatility components of $\boldsymbol{a}_t$. See Hu and Tsay (2014) for some consistency properties of the sample generalized cross kurtosis matrix under some moment conditions.

An interesting application of the principal volatility component analysis is to detect common volatility factors in a k-dimensional asset returns. See Theorem 7.7. Hu and Tsay (2014) use the approach to detect common volatility factors in the weekly log returns of seven exchange rates shown later.

Example 7.7 Consider the weekly log returns of seven exchanges from March 29, 2000 to October 26, 2011 for 605 observations. The seven currencies are British Pound (GBP), Norwegian Kroner (NOK), Swedish Kroner (SEK), Swiss Franc (VHF), Canadian Dollar (CAD), Singapore Dollar (SGD), and Australian Dollar (AUD). All exchange rates are versus U.S. Dollar. Let $\boldsymbol{z}_t$ be the seven-dimensional log returns of the exchange rates. To remove the serial correlations, a VAR(5) model is used so that we have

$$\boldsymbol{a}_t = \boldsymbol{z}_t - \boldsymbol{\phi}_0 - \sum_{i=1}^{5}\boldsymbol{\phi}_i\boldsymbol{z}_{t-i}.$$

TABLE 7.5 The Eigenvalues and Standardized Eigenvectors of $\hat{\Gamma}_{10}$ for the Residuals of a VAR(5) Model Fitted to the Weekly Log Returns of Seven Exchange Rates

PVC	7th	6th	5th	4th	3rd	2nd	1st
Values	0.449	0.776	1.082	1.200	1.647	1.906	4.076
Prop.	0.040	0.070	0.097	0.108	0.148	0.171	0.366
Vectors	−0.232	0.366	0.165	0.656	−0.002	0.214	−0.197
	−0.187	−0.754	0.177	0.159	−0.102	−0.029	−0.309
	−0.216	0.510	−0.329	−0.313	−0.331	−0.351	−0.235
	−0.219	−0.072	−0.236	−0.016	0.143	0.198	0.631
	0.569	−0.054	−0.399	0.294	0.698	−0.060	−0.038
	0.663	0.173	0.762	0.365	−0.078	−0.850	0.641
	−0.233	0.012	0.191	−0.476	0.605	0.254	0.028

The sample period of the returns is from March 29, 2000 to October 26, 2011, with 605 observations.

Hu and Tsay (2014) focus on the innovation series $\boldsymbol{a}_t$. Univariate ARCH tests show that individually all seven innovations of the exchange rates have conditional heteroscedasticity. Applying the principal volatility component analysis with $m = 10$, we obtain seven principal volatility components. The eigenvalues and their associated eigenvectors are given in Table 7.5. From the table, the smallest eigenvalue is 0.449, which is much smaller than the others and explains only about 4% of the trace of $\hat{\boldsymbol{G}}_{10}$. We applied univariate ARCH tests to the seven PVCs and found that the seventh PVC indeed has no conditional heteroscedasticity, implying the existence of common volatility factors in the exchange rate market. To demonstrate, Figure 7.18 shows (a) the time plots of the first and seventh PVCs and (b) the sample autocorrelations of the squared series of the first and seventh PVCs. From the plots and ACF, it is evident that the first PVC has conditional heteroscedasticity, but the seventh PVC does not. The eigenvector shows that the seventh PVC is approximately

$$b_{7t} \approx 0.2(\text{GBP+NOK+SEK+CHF+AUD})_t - 0.6(\text{CAD+SGD})_t,$$

which suggests a portfolio of exchange rates of European countries and Australian versus that of Canada and Singapore. □

Remark: The sample generalized covariance matrix of Equation (7.62) is likely to be sensitive to extreme innovations in $\{\boldsymbol{a}_t\}$. To reduce the influence of a few aberrant innovations, one can use some robust estimate. See, for instance, the Huber function of Section 7.8. □

Remark: The principal volatility component analysis is carried out by the command `comVol` of the `MTS` package. The program allows for VAR fit to remove serial correlations of the time series before performing the analysis. The default option uses $\hat{\boldsymbol{G}}_{10}$. □

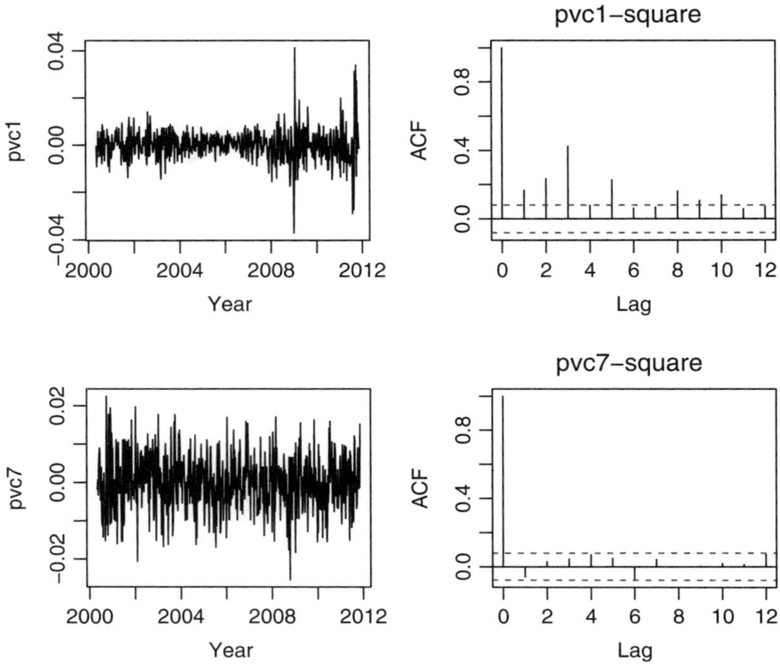

FIGURE 7.18 Time plots of the first and seventh principal volatility components (PVC) of weekly log returns of seven exchange rates and the sample autocorrelations of the squared series of PVCs.

R Demonstration: Principal volatility component analysis.

```
> da=read.table("w-7fx.txt",header=T)
> dim(da)
[1] 606  10
> fx=log(da[,4:10])
> rtn=diffM(fx)
> m1=comVol(rtn,p=5) ## Mean equation is VAR(5).
eigen-values:  4.0759 1.90646 1.64679 1.19989 1.08151 0.776164 0.44903
proportion:    0.36602 0.17120 0.14788 0.10775 0.09712 0.06970 0.04032
Checking:
Results of individual F-test for ARCH effect
Numbers of lags used: 10, 20, 30
Component,(F-ratio P-val) (F-ratio P-val) (F-ratio P-Val)
     [,1]  [,2]     [,3]   [,4]     [,5]  [,6]     [,7]
[1,]    1 21.66 0.00e+00 11.421 0.00e+00 7.725 0.00e+00
[2,]    2  5.78 2.54e-08  4.114 8.66e-09 3.169 7.06e-08
[3,]    3  7.24 7.95e-11  3.804 7.17e-08 2.659 6.66e-06
[4,]    4  4.58 2.76e-06  2.859 3.70e-05 2.279 1.63e-04
[5,]    5  1.49 1.38e-01  0.824 6.85e-01 0.776 7.99e-01
[6,]    6  2.38 9.18e-03  1.872 1.23e-02 1.271 1.55e-01
[7,]    7  1.42 1.68e-01  1.121 3.23e-01 1.126 2.96e-01
```

```
> names(m1)
[1] "residuals" "values"     "vectors"     "M"
> print(round(m1$M,4))
        [,1]    [,2]    [,3]    [,4]    [,5]    [,6]    [,7]
[1,] -0.1965 -0.2139 -0.0020  0.6560  0.1648  0.3660 -0.2319
[2,] -0.3089  0.0292 -0.1021  0.1588  0.1776 -0.7538 -0.1865
[3,] -0.2351  0.3511 -0.3311 -0.3132 -0.3295  0.5094 -0.2157
[4,]  0.6307 -0.1982  0.1428 -0.0160 -0.2357 -0.0725 -0.2187
[5,] -0.0375  0.0600  0.6978  0.2937 -0.3994 -0.0540  0.5686
[6,]  0.6409  0.8503 -0.0782  0.3649  0.7624  0.1732  0.6629
[7,]  0.0279 -0.2534  0.6055 -0.4761  0.1910  0.0124 -0.2332
> rt=m1$residuals%*%m1$M  ### Obtain PVCs.
```

EXERCISES

7.1 Consider the monthly log returns of Fama bond portfolio (6 months), S&P composite index, and Procter & Gamble stock from January 1962 to December 2011 for 600 observations. The simple returns are from CRSP and given in the file `m-bndpgspabt.txt`. The three return series used are in columns 2, 5, and 6. We shall use log returns. Since the bond returns have serial correlations, we employ an ARIMA(1,0,8) model to filter the series. In addition, the scale of bond returns is much smaller so that we use percentage bond returns. The resulting data are in the file `m-bndpgsp-6211.txt`.

- **(a)** Is there conditional heteroscedasticity in the three-dimensional log return series? Why?
- **(b)** Apply the EWMA method with smoothing parameter $\lambda = 0.96$ to obtain time-varying correlations of the data. Obtain the sample means of the correlations.

7.2 Again, consider the three-dimensional monthly log returns of Problem 1. Fit a DCC model of Tse and Tsui (2002) with Student-t innovations to the data and obtain the time-varying correlations. Fit also a DCC model of Engle (2002) with Student-t innovations to the data and obtain the time-varying correlations. Compare the two models. Are the fitted models adequate? Why?

7.3 Again, consider the three-dimensional log returns of Problem 1. Fit a t-copula model with marginal Student-t innovations to the data. White down the fitted model and compute the resulting time-varying correlations.

7.4 Again, consider the three-dimensional log return series of Problem 1. Fit a multivariate volatility model based on the Cholesky decomposition of the data. Is the model adequate? Why?

7.5 Consider the daily log returns of EXXON-Mobil, S&P composite index, and Apple stock used in Example 7.6. Fit a dynamically orthogonal component (DOC) model to the series. Write down the fitted model. Is the model adequate? Why?

7.6 Focus on the daily log returns, in percentages, of EXXON-Mobil stock and the S&P composite index used in Example 7.6. Fit a BEKK(1,1) model to the innovations of log returns. Write down the fitted model and obtain mean, standard error, minimum, and maximum of the correlations.

7.7 The file `m-pgspabt-6211.txt` contains the monthly simple returns of Procter & Gamble stock, S&P composite index, and Abbot Laboratories stock from January 1962, to December 2011. Compute the log returns. Is there conditional heteroscedasticity in the log return series? Fit a Go-GARCH model with ICA estimation method to the log return series. Write down the fitted model. Is the model adequate? Why?

REFERENCES

Alexander, C. (2001). Orthogonal GARCH. In C. Alexander (ed.). *Mastering Risk*: Volume 2: Applications, pp. 21–38. Financial Times Management, London.

Asai, M., McAleer, M., and Yu, J. (2006). Multivariate stochastic volatility: a review. *Econometric Reviews*, **25**: 145–175.

Bauwens, L., Hafner, C., and Laurent, S. (2012). *Handbook of Volatility Models and Their Applications*. John Wiley & Sons, Inc, Hoboken, NJ.

Bauwens, L., Laurent, S., and Rombouts, J. V. K. (2006). Multivariate GARCH models: a survey. *Journal of Applied Econometrics*, **21**: 79–109.

Bollerslev, T. and Wooldridge, J. M. (1992). Quasi-maximum likelihood estimation and inference in dynamic models with time-varying covariance. *Econometric Reviews*, **11**: 143–173.

Chang, C. and Tsay, R. S. (2010). Estimation of covariance matrix via sparse Cholesky factor with lasso. *Journal of Statistical Planning and Inference*, **140**: 3858–3873.

Creal, D., Koopman, S. J., and Lucas, A. (2011). A dynamic multivariate heavy-tailed model for time-varying volatilities and correlations. *Journal of Business & Economic Statistics*, **29**: 552–563.

Demarta, S. and McNeil, A. J. (2004). The t copula and related copulas. Working paper, Department of Mathematics, Federal Institute of Technology.

Dufour, J. M. and Roy, R. (1985). Some robust exact results on sample autocorrelations and test of randomness. *Journal of Econometrics*, **29**: 257–273.

Dufour, J. M. and Roy, R. (1986). Generalized portmanteau statistics and tests of randomness. *Communications in Statistics–Theory and Methods*, **15**: 2953–2972.

Engle, R. F. (1982). Autoregressive conditional heteroscedasticity with estimates of the variance of UK inflation. *Econometrica*, **50**: 987–1008.

Engle, R. F. (2002). Dynamic conditional correlations: a simple class of multivariate GARCH models. *Journal of Business and Economic Statistics*, **20**: 339–350.

Engle, R. F. and Kroner, K. F. (1995). Multivariate simultaneous generalized ARCH. *Econometric Theory* **11**: 122–150.

Fan, J., Wang, M., and Yao, Q. (2008). Modeling multivariate volatility via conditionally uncorrelated components. *Journal of the Royal Statistical Society, Series B*, **70**: 679–702.

Francq, C. and Zakoian, J. M. (2010). *GARCH Models*. John Wiley & Sons, Inc, Hoboken, NJ.

Hu, Y. P. and Tsay, R. S. (2014). Principal volatility component analysis. *Journal of Business & Economic Statistics* (to appear).

Hyvärinen, A., Karhunen, J., and Oja, E. (2001). *Independent Component Analysis*. John Wiley & Sons, Inc, New York.

Joe, H. (1997) *Multivariate Models and Dependence Concepts*. Chapman & Hall/CRC, Boca Raton, FL.

Li, K. C. (1992). On principal Hessian directions for data visualization and dimension reduction: another application of Stein's lemma. *Journal of the American Statistical Association*, **87**:1025–1039.

Li, W. K. (2004). *Diagnostic Checks in Time Series*. Chapman & Hall/CRC, Boca Raton, FL.

Ling, S. and Li, W. K. (1997). Diagnostic checking of nonlinear multivariate time series with multivariate ARCH errors. *Journal of Time Series Analysis*, **18**: 447–464.

Lopes, H., McCulloch, R. E., and Tsay, R. S. (2013). Parsimony-inducing priors for large scale state-space models. Working Paper, Booth School of Business, University of Chicago.

Matteson, D. S. and Tsay, R. S. (2011). Dynamic orthogonal components for multivariate time series. *Journal of the American Statistical Association*, **106**: 1450–1463.

McNeil, A. J., Frey, R., and Embrechts, P. (2005). *Quantitative Risk Management*. Princeton University Press, Princeton, NJ.

Moran, P. A. P. (1948). Some theorems on time series II: the significance of the serial correlation coefficient. *Biometrika*, **35**: 255–260.

Nelsen, R. B. (1999). *An Introduction to Copulas*. Lecture Notes in Statistics, 139. Springer, New York.

Pfaff, B. (2011). The package gogarch. R package available at `http://www.r-project.org` (accessed on July 11, 2013).

Tibshirani, R. (1996). Regression analysis and selection via the Lasso. *Journal of the Royal Statistical Society, Series B*, **58**: 267–288.

Tsay, R. S. (2010). *Analysis of Financial Time Series*. 3rd Edition. John Wiley & Sons, Inc, Hoboken, NJ.

Tse, Y. K. (2002). Residual-based diagnostics for conditional heteroscedasticity models. *Econometric Journal*, **5**: 358–373.

Tse, Y. K. and Tsui, A. K. C. (2002). A multivariate GARCH model with time-varying correlations. *Journal of Business and Economic Statistics*, **20**: 351–362.

Van der Weide, R. (2002). A multivariate generalized orthogonal GARCH model. *Journal of Applied Econometrics*, **17**: 549–564.

Vilenkin, N. (1968). *Special Functions and the Theory of Group Representation*, translations of mathematical monographs. American Mathematical Society, Providence, RI, USA, 22.

APPENDIX A

Review of Mathematics and Statistics

This appendix reviews some mathematics, probability, and statistics useful for studying multivariate time series.

A.1 REVIEW OF VECTORS AND MATRICES

We begin with a brief review of some algebra and properties of vectors and matrices. No proofs are given as they can be found in standard textbooks on matrices; see, for instance, Graybill (1969) and Magnus and Neudecker (1999). Also, a good summary of the results is given in Appendix A of Lütkepohl (2005).

An $m \times n$ real-valued matrix is an m by n array of real numbers. For example,

$$\boldsymbol{A} = \begin{bmatrix} 2 & 5 & 8 \\ -1 & 3 & 4 \end{bmatrix}$$

is a 2×3 matrix. This matrix has two rows and three columns. In general, an $m \times n$ matrix is written as

$$\boldsymbol{A} \equiv [a_{ij}] = \begin{bmatrix} a_{11} & a_{12} & \cdots & a_{1,n-1} & a_{1n} \\ a_{21} & a_{22} & \cdots & a_{2,n-1} & a_{2n} \\ \vdots & \vdots & & \vdots & \vdots \\ a_{m1} & a_{m2} & \cdots & a_{m,n-1} & a_{mn} \end{bmatrix}. \tag{A.1}$$

The positive integers m and n are the *row dimension* and *column dimension* of $\boldsymbol{A}$. The real number a_{ij} is referred to as the (i, j)th element of $\boldsymbol{A}$. In particular, the elements a_{ii} are the *diagonal elements* of the matrix.

Multivariate Time Series Analysis: With R and Financial Applications,
First Edition. Ruey S. Tsay.

An $m \times 1$ matrix forms an m-dimensional column vector, and a $1 \times n$ matrix is an n-dimensional row vector. In the literature, a vector is often meant to be a column vector. If $m = n$, then the matrix is a square matrix. If $a_{ij} = 0$ for $i \neq j$ and $m = n$, then the matrix $\boldsymbol{A}$ is a *diagonal matrix*. If $a_{ij} = 0$ for $i \neq j$ and $a_{ii} = 1$ for all i, then $\boldsymbol{A}$ is the $m \times m$ *identity matrix*, which is commonly denoted by $\boldsymbol{I}_m$ or simply $\boldsymbol{I}$ if the dimension is clear.

The $n \times m$ matrix

$$\boldsymbol{A}' = \begin{bmatrix} a_{11} & a_{21} & \cdots & a_{m-1,1} & a_{m1} \\ a_{12} & a_{22} & \cdots & a_{m-1,2} & a_{m2} \\ \vdots & \vdots & & \vdots & \vdots \\ a_{1n} & a_{2n} & \cdots & a_{m-1,n} & a_{mn} \end{bmatrix}$$

is the *transpose* of the matrix $\boldsymbol{A}$. For example,

$$\begin{bmatrix} 2 & -1 \\ 5 & 3 \\ 8 & 4 \end{bmatrix} \quad \text{is the transpose of} \quad \begin{bmatrix} 2 & 5 & 8 \\ -1 & 3 & 4 \end{bmatrix}.$$

We use the notation $\boldsymbol{A}' = [a'_{ij}]$ to denote the transpose of $\boldsymbol{A}$. From the definition, $a'_{ij} = a_{ji}$ and $(\boldsymbol{A}')' = \boldsymbol{A}$. If $\boldsymbol{A}' = \boldsymbol{A}$, then $\boldsymbol{A}$ is a *symmetric matrix*.

A.1.1 Basic Operations

Suppose that $\boldsymbol{A} = [a_{ij}]_{m \times n}$ and $\boldsymbol{C} = [c_{ij}]_{p \times q}$ are two matrices with dimensions given in the subscript. Let b be a real number. Some basic matrix operations are defined next:

- Addition: $\boldsymbol{A} + \boldsymbol{C} = [a_{ij} + c_{ij}]_{m \times n}$ if $m = p$ and $n = q$.
- Subtraction: $\boldsymbol{A} - \boldsymbol{C} = [a_{ij} - c_{ij}]_{m \times n}$ if $m = p$ and $n = q$.
- Scalar multiplication: $b\boldsymbol{A} = [ba_{ij}]_{m \times n}$.
- Multiplication: $\boldsymbol{AC} = [\sum_{v=1}^{n} a_{iv} c_{vj}]_{m \times q}$ provided that $n = p$.

When the dimensions of matrices satisfy the condition for multiplication to take place, the two matrices are said to be *conformable*. An example of matrix multiplication is

$$\begin{bmatrix} 2 & 1 \\ 1 & 1 \end{bmatrix} \begin{bmatrix} 1 & 2 & 3 \\ -1 & 2 & -4 \end{bmatrix} = \begin{bmatrix} 2 \cdot 1 - 1 \cdot 1 & 2 \cdot 2 + 1 \cdot 2 & 2 \cdot 3 - 1 \cdot 4 \\ 1 \cdot 1 - 1 \cdot 1 & 1 \cdot 2 + 1 \cdot 2 & 1 \cdot 3 - 1 \cdot 4 \end{bmatrix}$$
$$= \begin{bmatrix} 1 & 6 & 2 \\ 0 & 4 & -1 \end{bmatrix}.$$

Important rules of matrix operations include (a) $(\boldsymbol{AC})' = \boldsymbol{C}'\boldsymbol{A}'$, (b) $\boldsymbol{AC} \neq \boldsymbol{CA}$ in general, (c) $(\boldsymbol{AB})\boldsymbol{C} = \boldsymbol{A}(\boldsymbol{BC})$, and (d) $\boldsymbol{A}(\boldsymbol{B} + \boldsymbol{C}) = \boldsymbol{AB} + \boldsymbol{AC}$.

A.1.2 Inverse, Trace, Eigenvalue, and Eigenvector

A square matrix $\boldsymbol{A}_{m\times m}$ is *nonsingular* or *invertible* if there exists a unique matrix $\boldsymbol{C}_{m\times m}$ such that $\boldsymbol{AC} = \boldsymbol{CA} = \boldsymbol{I}_m$, the $m \times m$ identity matrix. In this case, $\boldsymbol{C}$ is called the *inverse* matrix of $\boldsymbol{A}$ and is denoted by $\boldsymbol{C} = \boldsymbol{A}^{-1}$.

The trace of $\boldsymbol{A}_{m\times m}$ is the sum of its diagonal elements (i.e., $tr(\boldsymbol{A}) = \sum_{i=1}^{m} a_{ii}$). It is easy to see that (a) $tr(\boldsymbol{A} + \boldsymbol{C}) = tr(\boldsymbol{A}) + tr(\boldsymbol{C})$, (b) $tr(\boldsymbol{A}) = tr(\boldsymbol{A}')$, and (c) $tr(\boldsymbol{AC}) = tr(\boldsymbol{CA})$ provided that the two matrices are conformable.

A number λ and an $m \times 1$ vector $\boldsymbol{b}$, possibly complex-valued, are a right *eigenvalue* and *eigenvector* pair of the matrix $\boldsymbol{A}$ if $\boldsymbol{Ab} = \lambda\boldsymbol{b}$. There are m possible eigenvalues for the matrix $\boldsymbol{A}$. For a real-valued matrix $\boldsymbol{A}$, complex eigenvalues occur in conjugated pairs. The matrix $\boldsymbol{A}$ is nonsingular if and only if all of its eigenvalues are nonzero. Denote the eigenvalues by $\{\lambda_i | i = 1, \ldots, m\}$: we have $tr(\boldsymbol{A}) = \sum_{i=1}^{m} \lambda_i$. In addition, the *determinant* of the matrix $\boldsymbol{A}$ can be defined as $|\boldsymbol{A}| = \prod_{i=1}^{m} \lambda_i$. For a general definition of determinant of a square matrix, see a standard textbook on matrices (e.g., Graybill, 1969). Let $(\lambda, \boldsymbol{b})$ be an eigenvalue–eigenvector pair of $\boldsymbol{A}$, we have

$$\boldsymbol{A}^2\boldsymbol{b} = \boldsymbol{A}(\boldsymbol{Ab}) = \boldsymbol{A}(\lambda\boldsymbol{b}) = \lambda\boldsymbol{Ab} = \lambda^2\boldsymbol{b}.$$

Thus, $(\lambda^2, \boldsymbol{b})$ is an eigenvalue–eigenvector pair of $\boldsymbol{A}^2$. This property continues to hold for other power of $\boldsymbol{A}$, that is, $(\lambda^n, \boldsymbol{b})$ is an eigenvalue–eigenvector pair of $\boldsymbol{A}^n$ for $n > 1$.

Finally, the rank of the matrix $\boldsymbol{A}_{m\times n}$ is the number of nonzero eigenvalues of the symmetric matrix $\boldsymbol{AA}'$. Also, for a nonsingular matrix $\boldsymbol{A}$, $(\boldsymbol{A}^{-1})' = (\boldsymbol{A}')^{-1}$ and for nonsingular matrices $\boldsymbol{A}$ and $\boldsymbol{B}$, $(\boldsymbol{AB})^{-1} = \boldsymbol{B}^{-1}\boldsymbol{A}^{-1}$ provided that they are conformable. A useful identity of an $m \times n$ matrix $\boldsymbol{A}$ is

$$(\boldsymbol{I}_m + \boldsymbol{AA}')^{-1} = \boldsymbol{I}_m - \boldsymbol{A}(\boldsymbol{I}_n + \boldsymbol{A}'\boldsymbol{A})^{-1}\boldsymbol{A}'.$$

A.1.3 Positive-Definite Matrix

A square matrix $\boldsymbol{A}$ $(m \times m)$ is a *positive-definite* matrix if (a) $\boldsymbol{A}$ is symmetric, and (b) all eigenvalues of $\boldsymbol{A}$ are positive. Alternatively, $\boldsymbol{A}$ is a positive-definite matrix if for any nonzero m-dimensional vector $\boldsymbol{b}$, we have $\boldsymbol{b}'\boldsymbol{Ab} > 0$.

A symmetric $m \times m$ matrix $\boldsymbol{A}$ is nonnegative definite if $\boldsymbol{b}'\boldsymbol{Ab} \geq 0$ for any m-dimensional vector $\boldsymbol{b}$.

Useful properties of a positive-definite matrix $\boldsymbol{A}$ include (a) all eigenvalues of $\boldsymbol{A}$ are real and positive, and (b) the matrix can be decomposed as

$$\boldsymbol{A} = \boldsymbol{P\Lambda P}', \qquad \text{(A.2)}$$

where $\boldsymbol{\Lambda}$ is a diagonal matrix consisting of all eigenvalues of $\boldsymbol{A}$ and $\boldsymbol{P}$ is an $m \times m$ matrix consisting of the m right eigenvectors of $\boldsymbol{A}$. It is common to write the eigenvalues as $\lambda_1 \geq \lambda_2 \geq \cdots \geq \lambda_m$ and the eigenvectors as $\boldsymbol{e}_1, \ldots, \boldsymbol{e}_m$ such that $\boldsymbol{A}\boldsymbol{e}_i = \lambda_i \boldsymbol{e}_i$ and $\boldsymbol{e}_i'\boldsymbol{e}_i = 1$. In addition, these eigenvectors are orthogonal to each other, namely, $\boldsymbol{e}_i'\boldsymbol{e}_j = 0$ if $i \neq j$ provided that the eigenvalues are distinct. The matrix $\boldsymbol{P} = [\boldsymbol{e}_1, \ldots, \boldsymbol{e}_m]$ is an *orthogonal* matrix such that $\boldsymbol{P}\boldsymbol{P}' = \boldsymbol{P}'\boldsymbol{P} = \boldsymbol{I}_m$. For the orthogonal matrix $\boldsymbol{P}$, we have $\boldsymbol{P}^{-1} = \boldsymbol{P}'$. The decomposition in Equation (A.2) can be written as

$$\boldsymbol{A} = \sum_{i=1}^{m} \lambda_i \boldsymbol{e}_i \boldsymbol{e}_i'$$

and is commonly referred to as the *spectral decomposition* of the matrix $\boldsymbol{A}$. It is easy to see that $\boldsymbol{A}^{-1} = \boldsymbol{P}\boldsymbol{\Lambda}^{-1}\boldsymbol{P}' = \sum_{i=1}^{m}(1/\lambda_i)\boldsymbol{e}_i\boldsymbol{e}_i'$.

A useful application of the spectral decomposition of a positive-definite matrix $\boldsymbol{A}$ is to obtain a square-root matrix $\boldsymbol{A}^{1/2} = \boldsymbol{P}\boldsymbol{\Lambda}^{1/2}\boldsymbol{P}'$. Specifically, we have

$$\boldsymbol{A} = \boldsymbol{P}\boldsymbol{\Lambda}\boldsymbol{P}' = \boldsymbol{P}\boldsymbol{\Lambda}^{1/2}\boldsymbol{\Lambda}^{1/2}\boldsymbol{P}' = \boldsymbol{P}\boldsymbol{\Lambda}^{1/2}\boldsymbol{P}'\boldsymbol{P}\boldsymbol{\Lambda}^{1/2}\boldsymbol{P}' \equiv \boldsymbol{A}^{1/2}\boldsymbol{A}^{1/2}.$$

The square-root matrix $\boldsymbol{A}^{1/2}$ has the following properties: (a) $\boldsymbol{A}^{1/2}$ is symmetric, and (b) $(\boldsymbol{A}^{1/2})^{-1} = (\boldsymbol{P}\boldsymbol{\Lambda}^{1/2}\boldsymbol{P}')^{-1} = \boldsymbol{P}\boldsymbol{\Lambda}^{-1/2}\boldsymbol{P}' = \sum_{i=1}^{m}(1/\sqrt{\lambda_i})\boldsymbol{e}_i\boldsymbol{e}_i' = \boldsymbol{A}^{-1/2}$.

Consider, for example, the simple 2×2 matrix

$$\boldsymbol{A} = \begin{bmatrix} 2 & 1 \\ 1 & 2 \end{bmatrix},$$

which is positive-definite. Simple calculations show that

$$\begin{bmatrix} 2 & 1 \\ 1 & 2 \end{bmatrix}\begin{bmatrix} 1 \\ 1 \end{bmatrix} = 3\begin{bmatrix} 1 \\ 1 \end{bmatrix}, \quad \begin{bmatrix} 2 & 1 \\ 1 & 2 \end{bmatrix}\begin{bmatrix} 1 \\ -1 \end{bmatrix} = \begin{bmatrix} 1 \\ -1 \end{bmatrix}.$$

Therefore, 3 and 1 are eigenvalues of $\boldsymbol{A}$ with normalized eigenvectors $\boldsymbol{e}_1 = (1/\sqrt{2}, 1/\sqrt{2})'$ and $\boldsymbol{e}_2 = (1/\sqrt{2}, -1/\sqrt{2})'$, respectively. It is easy to verify that the spectral decomposition holds, that is,

$$\begin{bmatrix} 2 & 1 \\ 1 & 2 \end{bmatrix} = \begin{bmatrix} \frac{1}{\sqrt{2}} & \frac{1}{\sqrt{2}} \\ \frac{1}{\sqrt{2}} & \frac{-1}{\sqrt{2}} \end{bmatrix}\begin{bmatrix} 3 & 0 \\ 0 & 1 \end{bmatrix}\begin{bmatrix} \frac{1}{\sqrt{2}} & \frac{1}{\sqrt{2}} \\ \frac{1}{\sqrt{2}} & \frac{-1}{\sqrt{2}} \end{bmatrix}.$$

A square-root matrix of $\boldsymbol{A}$ is

$$\boldsymbol{A}^{1/2} = \begin{bmatrix} \frac{1}{\sqrt{2}} & \frac{1}{\sqrt{2}} \\ \frac{1}{\sqrt{2}} & \frac{-1}{\sqrt{2}} \end{bmatrix}\begin{bmatrix} \sqrt{3} & 0 \\ 0 & 1 \end{bmatrix}\begin{bmatrix} \frac{1}{\sqrt{2}} & \frac{1}{\sqrt{2}} \\ \frac{1}{\sqrt{2}} & \frac{-1}{\sqrt{2}} \end{bmatrix} \approx \begin{bmatrix} 1.366 & 0.366 \\ 0.366 & 1.366 \end{bmatrix}.$$

If $\boldsymbol{A}$ is not symmetric, then it, in general, cannot be diagonalized. However, it can be written in a *Jordan canonical form*. Details can be found in many textbooks, for example, Graybill (1969) and Lütkepohl (2005, Appendix A).

A.1.4 Comparison of Two Symmetric Matrices

Let $\boldsymbol{A} = [a_{ij}]$ and $\boldsymbol{B} = [b_{ij}]$ be $m \times m$ symmetric matrices. We define $\boldsymbol{A} \geq \boldsymbol{B}$ if $\boldsymbol{A} - \boldsymbol{B}$ is non-negative definite. In other words, $\boldsymbol{A} \geq \boldsymbol{B}$ if $\boldsymbol{b}'(\boldsymbol{A} - \boldsymbol{B})\boldsymbol{b} \geq 0$ for any m-dimensional vector $\boldsymbol{b}$. In particular, we have $a_{ii} \geq b_{ii}$ if $\boldsymbol{A} \geq \boldsymbol{B}$.

A.1.5 Idempotent Matrix

An $m \times m$ matrix $\boldsymbol{A}$ is idempotent if $\boldsymbol{A}\boldsymbol{A} = \boldsymbol{A}^2 = \boldsymbol{A}$. The identity matrix $\boldsymbol{I}_m$ is an example of idempotent matrix. Properties of an idempotent matrix include

1. If $\boldsymbol{A}$ is idempotent, then $\boldsymbol{I}_m - \boldsymbol{A}$ is idempotent.
2. If $\boldsymbol{A}$ is symmetric and idempotent, then rank($\boldsymbol{A}$) = tr($\boldsymbol{A}$).
3. If $\boldsymbol{A}$ is idempotent, then all of its eigenvalues are 0 or 1.
4. If $\boldsymbol{A}$ is idempotent and rank($\boldsymbol{A}$) = m, then $\boldsymbol{A} = \boldsymbol{I}_m$.
5. For an $n \times p$ matrix $\boldsymbol{Z}$ such that $\boldsymbol{Z}'\boldsymbol{Z}$ is a nonsingular $p \times p$ matrix, then $\boldsymbol{Z}(\boldsymbol{Z}'\boldsymbol{Z})^{-1}\boldsymbol{Z}'$ is idempotent.

A.1.6 Cholesky Decomposition

For a symmetric matrix $\boldsymbol{A}$, there exists a lower triangular matrix $\boldsymbol{L}$ with diagonal elements being 1 and a diagonal matrix $\boldsymbol{\Omega}$ such that $\boldsymbol{A} = \boldsymbol{L}\boldsymbol{\Omega}\boldsymbol{L}'$; see Chapter 1 of Strang (1980). If $\boldsymbol{A}$ is positive-definite, then the diagonal elements of $\boldsymbol{\Omega}$ are positive. In this case, we have

$$\boldsymbol{A} = \boldsymbol{L}\sqrt{\boldsymbol{\Omega}}\sqrt{\boldsymbol{\Omega}}\boldsymbol{L}' = (\boldsymbol{L}\sqrt{\boldsymbol{\Omega}})(\boldsymbol{L}\sqrt{\boldsymbol{\Omega}})', \tag{A.3}$$

where $\boldsymbol{L}\sqrt{\boldsymbol{\Omega}}$ is again a lower triangular matrix and the square root is taken element by element. Such a decomposition is called the *Cholesky decomposition* of $\boldsymbol{A}$. This decomposition shows that a positive-definite matrix $\boldsymbol{A}$ can be diagonalized as

$$\boldsymbol{L}^{-1}\boldsymbol{A}(\boldsymbol{L}')^{-1} = \boldsymbol{L}^{-1}\boldsymbol{A}(\boldsymbol{L}^{-1})' = \boldsymbol{\Omega}.$$

Since $\boldsymbol{L}$ is a lower triangular matrix with unit diagonal elements, $\boldsymbol{L}^{-1}$ is also lower triangular matrix with unit diagonal elements. Consider again the prior 2×2 matrix $\boldsymbol{A}$. It is easy to verify that

$$\boldsymbol{L} = \begin{bmatrix} 1.0 & 0.0 \\ 0.5 & 1.0 \end{bmatrix} \quad \text{and} \quad \boldsymbol{\Omega} = \begin{bmatrix} 2.0 & 0.0 \\ 0.0 & 1.5 \end{bmatrix}$$

satisfy $\boldsymbol{A} = \boldsymbol{L\Omega L}'$. In addition,

$$\boldsymbol{L}^{-1} = \begin{bmatrix} 1.0 & 0.0 \\ -0.5 & 1.0 \end{bmatrix} \quad \text{and} \quad \boldsymbol{L}^{-1}\boldsymbol{\Sigma}(\boldsymbol{L}^{-1})' = \boldsymbol{\Omega}.$$

For a positive-definite matrix $\boldsymbol{A}$, the Cholesky decomposition in Equation (A.3) can be written as

$$\boldsymbol{A} = \boldsymbol{U}'\boldsymbol{U},$$

where $\boldsymbol{U} = (\boldsymbol{L}\sqrt{\boldsymbol{\Omega}})'$ is an upper triangular matrix. This decomposition is also known as the LU decomposition.

A.1.7 Partition of a Matrix

In statistical applications, it is often useful to consider the partition of a positive-definite matrix $\boldsymbol{A}$ as

$$\boldsymbol{A} = \begin{bmatrix} \boldsymbol{A}_{11} & \boldsymbol{A}_{12} \\ \boldsymbol{A}_{21} & \boldsymbol{A}_{22} \end{bmatrix},$$

where $\boldsymbol{A}_{ii}$ is $m_i \times m_i$ with $m_1 + m_2 = m$ and $\boldsymbol{A}_{12}$ is an $m_1 \times m_2$ matrix such that $\boldsymbol{A}_{12} = \boldsymbol{A}'_{21}$. Some nice properties of this partition are

1. $(\boldsymbol{A}_{11} - \boldsymbol{A}_{12}\boldsymbol{A}_{22}^{-1}\boldsymbol{A}_{21})^{-1} = \boldsymbol{A}_{11}^{-1} + \boldsymbol{A}_{11}^{-1}\boldsymbol{A}_{12}(\boldsymbol{A}_{22} - \boldsymbol{A}_{21}\boldsymbol{A}_{11}^{-1}\boldsymbol{A}_{12})^{-1}\boldsymbol{A}_{21}\boldsymbol{A}_{11}^{-1}$.
2. $|\boldsymbol{A}| = |\boldsymbol{A}_{11}| \cdot |\boldsymbol{A}_{22} - \boldsymbol{A}_{21}\boldsymbol{A}_{11}^{-1}\boldsymbol{A}_{12}| = |\boldsymbol{A}_{22}| \cdot |\boldsymbol{A}_{11} - \boldsymbol{A}_{12}\boldsymbol{A}_{22}^{-1}\boldsymbol{A}_{21}|$.
3. $\boldsymbol{A}^{-1}$ can be written as

$$\begin{aligned} \boldsymbol{A}^{-1} &= \begin{bmatrix} \boldsymbol{D} & -\boldsymbol{D}\boldsymbol{A}_{12}\boldsymbol{A}_{22}^{-1} \\ -\boldsymbol{A}_{22}^{-1}\boldsymbol{A}_{21}\boldsymbol{D} & \boldsymbol{A}_{22}^{-1} + \boldsymbol{A}_{22}^{-1}\boldsymbol{A}_{21}\boldsymbol{D}\boldsymbol{A}_{12}\boldsymbol{A}_{22}^{-1} \end{bmatrix} \\ &= \begin{bmatrix} \boldsymbol{A}_{11}^{-1} + \boldsymbol{A}_{11}^{-1}\boldsymbol{A}_{12}\boldsymbol{E}\boldsymbol{A}_{21}\boldsymbol{A}_{11}^{-1} & -\boldsymbol{A}_{11}^{-1}\boldsymbol{A}_{12}\boldsymbol{E} \\ -\boldsymbol{E}\boldsymbol{A}_{21}\boldsymbol{A}_{11}^{-1} & \boldsymbol{E} \end{bmatrix}, \end{aligned}$$

where $\boldsymbol{D} = (\boldsymbol{A}_{11} - \boldsymbol{A}_{12}\boldsymbol{A}_{22}^{-1}\boldsymbol{A}_{21})^{-1}$ and $\boldsymbol{E} = (\boldsymbol{A}_{22} - \boldsymbol{A}_{21}\boldsymbol{A}_{11}^{-1}\boldsymbol{A}_{12})^{-1}$.

A.1.8 Vectorization and Kronecker Product

Writing an $m \times n$ matrix $\boldsymbol{A}$ in its columns as $\boldsymbol{A} = [\boldsymbol{a}_1, \ldots, \boldsymbol{a}_n]$, we define the stacking operation as vec($\boldsymbol{A}$) = $(\boldsymbol{a}'_1, \boldsymbol{a}'_2, \ldots, \boldsymbol{a}'_m)'$, which is an $mn \times 1$ vector. For two matrices $\boldsymbol{A}_{m\times n}$ and $\boldsymbol{C}_{p\times q}$, the Kronecker product between $\boldsymbol{A}$ and $\boldsymbol{C}$ is

$$A \otimes C = \begin{bmatrix} a_{11}C & a_{12}C & \cdots & a_{1n}C \\ a_{21}C & a_{22}C & \cdots & a_{2n}C \\ \vdots & \vdots & & \vdots \\ a_{m1}C & a_{m2}C & \cdots & a_{mn}C \end{bmatrix}_{mp \times nq}.$$

For example, assume that

$$A = \begin{bmatrix} 2 & 1 \\ -1 & 3 \end{bmatrix}, \quad C = \begin{bmatrix} 4 & -1 & 3 \\ -2 & 5 & 2 \end{bmatrix}.$$

Then $\text{vec}(A) = (2, -1, 1, 3)'$, $\text{vec}(C) = (4, -2, -1, 5, 3, 2)'$, and

$$A \otimes C = \begin{bmatrix} 8 & -2 & 6 & 4 & -1 & 3 \\ -4 & 10 & 4 & -2 & 5 & 2 \\ -4 & 1 & -3 & 12 & -3 & 9 \\ 2 & -5 & -2 & -6 & 15 & 6 \end{bmatrix}.$$

Assuming that the dimensions are appropriate, we have the following useful properties for the two operators:

1. $A \otimes C \neq C \otimes A$ in general.
2. $(A \otimes C)' = A' \otimes C'$.
3. $A \otimes (C + D) = A \otimes C + A \otimes D$.
4. $(A \otimes C)(F \otimes G) = (AF) \otimes (CG)$.
5. If A and C are invertible, then $(A \otimes C)^{-1} = A^{-1} \otimes C^{-1}$.
6. For square matrices A and C, $tr(A \otimes C) = tr(A)tr(C)$.
7. If A and B are square matrices with (λ_a, e_a) and (λ_b, e_b) being eigenvalue–eigenvector pair of A and B, respectively, then $\lambda_a \lambda_b$ is an eigenvalue of $A \otimes B$ with eigenvector $e_a \otimes e_b$.
8. If A and B are $m \times m$ and $n \times n$ square matrices, respectively, then $|A \otimes B| = |A|^n |B|^m$.
9. $\text{vec}(A + C) = \text{vec}(A) + \text{vec}(C)$.
10. $\text{vec}(ABC) = (C' \otimes A)\ \text{vec}(B)$.
11. $\text{vec}(AB) = (I \otimes A)\text{vec}(B) = (B' \otimes I)\text{vec}(A)$.
12. $\text{vec}(B')'\text{vec}(A) = tr(BA) = tr(AB) = \text{vec}(A')'\text{vec}(B)$.
13. $tr(AC) = \text{vec}(C')'\text{vec}(A) = \text{vec}(A')'\text{vec}(C)$.

14.

$$\begin{aligned} tr(\boldsymbol{ABC}) &= \text{vec}(\boldsymbol{A}')'(\boldsymbol{C}' \otimes \boldsymbol{I})\text{vec}(\boldsymbol{B}) = \text{vec}(\boldsymbol{A}')'(\boldsymbol{I} \otimes \boldsymbol{B})\text{vec}(\boldsymbol{C}) \\ &= \text{vec}(\boldsymbol{B}')'(\boldsymbol{A}' \otimes \boldsymbol{I})\text{vec}(\boldsymbol{C}) = \text{vec}(\boldsymbol{B}')'(\boldsymbol{I} \otimes \boldsymbol{C})\text{vec}(\boldsymbol{A}) \\ &= \text{vec}(\boldsymbol{C}')'(\boldsymbol{B}' \otimes \boldsymbol{I})\text{vec}(\boldsymbol{A}) = \text{vec}(\boldsymbol{C}')'(\boldsymbol{I} \otimes \boldsymbol{A})\text{vec}(\boldsymbol{B}). \end{aligned}$$

15.

$$\begin{aligned} tr(\boldsymbol{AZ'BZC}) &= tr(\boldsymbol{Z'BZCA}) = \text{vec}(\boldsymbol{Z})'\text{vec}(\boldsymbol{BZCA}) \\ &= \text{vec}(\boldsymbol{Z})'(\boldsymbol{A'C'} \otimes \boldsymbol{B})\text{vec}(\boldsymbol{Z}). \end{aligned}$$

In particular, with $\boldsymbol{C} = \boldsymbol{I}$, we have $tr(\boldsymbol{AZ'BZ}) = \text{vec}(\boldsymbol{Z})'(\boldsymbol{A}' \otimes \boldsymbol{B})\text{vec}(\boldsymbol{Z})$.

In multivariate statistical analysis, we often deal with symmetric matrices. It is therefore convenient to generalize the stacking operation to the *half-stacking* operation, which consists of elements on or below the main diagonal. Specifically, for a symmetric square matrix $\boldsymbol{A} = [a_{ij}]_{m \times m}$, define

$$\text{vech}(\boldsymbol{A}) = (\boldsymbol{a}_{1.}', \boldsymbol{a}_{2*}', \ldots, \boldsymbol{a}_{k*}')',$$

where $\boldsymbol{a}_{1.}$ is the first column of $\boldsymbol{A}$, and $\boldsymbol{a}_{i*} = (a_{ii}, a_{i+1,i}, \ldots, a_{mi})'$ is a $(m-i+1)$-dimensional vector. The dimension of vech($\boldsymbol{A}$) is $m(m+1)/2$. For example, suppose that $m = 3$. Then, we have vech($\boldsymbol{A}$) = $(a_{11}, a_{21}, a_{31}, a_{22}, a_{32}, a_{33})'$, which is a six-dimensional vector.

For a $m \times m$ square matrix $\boldsymbol{A}$, the m^2-dimensional vector vec($\boldsymbol{A}$) and the $m(m+1)/2$-dimensional vector vech($\boldsymbol{A}$) are related by the *elimination matrix*, $\boldsymbol{L}_m$, and the *duplication matrix*, $\boldsymbol{D}_m$, as follows:

$$\text{vech}(\boldsymbol{A}) = \boldsymbol{L}_m\text{vec}(\boldsymbol{A}), \quad \text{vec}(\boldsymbol{A}) = \boldsymbol{D}_m\text{vech}(\boldsymbol{A}),$$

where $\boldsymbol{L}_m$ is an $m(m+1)/2 \times m^2$ matrix and $\boldsymbol{D}_m$ is an $m^2 \times m(m+1)/2$ matrix,, both consisting of 0 and 1. Also, for an $m \times n$ matrix $\boldsymbol{A}$, there exists a *commutation matrix*, $\boldsymbol{K}_{nm}$ such that vec($\boldsymbol{A}$) = $\boldsymbol{K}_{nm}$vec($\boldsymbol{A}'$), where $\boldsymbol{K}_{nm} = [\boldsymbol{I}_n \otimes \boldsymbol{e}_1, \boldsymbol{I}_n \otimes \boldsymbol{e}_2, \ldots, \boldsymbol{I}_n \otimes \boldsymbol{e}_m]$ with $\boldsymbol{e}_i$ being the ith unit vector in R^m. Note that $\boldsymbol{K}_{nm} = \boldsymbol{K}_{mn}' = \boldsymbol{K}_{mn}^{-1}$. The commutation matrix can be generalized as follows. Suppose that $\boldsymbol{A}_1, \ldots, \boldsymbol{A}_p$ are $m \times m$ matrices. Define the $m \times mp$ matrices $\boldsymbol{A} = [\boldsymbol{A}_1, \ldots, \boldsymbol{A}_p]$ and $\boldsymbol{A}_* = [\boldsymbol{A}_1', \ldots, \boldsymbol{A}_p']$. Then,

$$\text{vec}(\boldsymbol{A}) = \boldsymbol{K}\text{vec}(\boldsymbol{A}_*) \tag{A.4}$$

where $\boldsymbol{K}$ is an $m^2p \times m^2p$ matrix consisting of 0 and 1 and is given by $\boldsymbol{K} = \boldsymbol{I}_p \otimes \boldsymbol{K}_{mm}$ and $\boldsymbol{K}_{mm}$ is the commutation matrix defined before. That is, $\boldsymbol{K}_{mm} = [\boldsymbol{I}_m \otimes \boldsymbol{e}_1, \boldsymbol{I}_m \otimes \boldsymbol{e}_2, \ldots, \boldsymbol{I}_m \otimes \boldsymbol{e}_m]$ with $\boldsymbol{e}_i$ being the ith unit vector in R^m.

A.1.9 Vector and Matrix Differentiation

Matrix calculus is useful in multivariate statistical analysis. In the literature, matrix calculus is based on *differentials* rather than *derivatives*. See Magnus and Neudecker (1999, Chapters 9 and 10) and Abadir and Magnus (2005, Chapter 13). The first two derivatives, however, can be obtained from the first two differentials. In this review, we use derivatives directly.

In what follows, we assume that all derivatives exist and are continuous. Let $f(\boldsymbol{\beta})$ be a scalar function of the m-dimensional vector $\boldsymbol{\beta} = (\beta_1, \ldots, \beta_m)'$. The first-order partial derivatives of $f(\boldsymbol{\beta})$ with respect to $\boldsymbol{\beta}$ and $\boldsymbol{\beta}'$ are defined as

$$\frac{\partial f}{\partial \boldsymbol{\beta}} = \begin{bmatrix} \frac{\partial f}{\partial \beta_1} \\ \frac{\partial f}{\partial \beta_2} \\ \vdots \\ \frac{\partial f}{\partial \beta_m} \end{bmatrix}, \quad \frac{\partial f}{\partial \boldsymbol{\beta}'} = \left[\frac{\partial f}{\partial \beta_1}, \cdots, \frac{\partial f}{\partial \beta_m}\right],$$

which are $m \times 1$ and $1 \times m$ vector, respectively. The *Hessian matrix* of the second partial derivatives is defined as

$$\frac{\partial^2 f}{\partial \boldsymbol{\beta} \partial \boldsymbol{\beta}'} = \left[\frac{\partial^2 f}{\partial \beta_i \partial \beta_j}\right] = \begin{bmatrix} \frac{\partial^2 f}{\partial \beta_1 \partial \beta_1} & \cdots & \frac{\partial^2 f}{\partial \beta_1 \partial \beta_m} \\ \vdots & & \vdots \\ \frac{\partial^2 f}{\partial \beta_m \partial \beta_1} & \cdots & \frac{\partial^2 f}{\partial \beta_m \partial \beta_m} \end{bmatrix},$$

which is an $m \times m$ matrix. If $f(\boldsymbol{A})$ is a scalar function of the $m \times n$ matrix $\boldsymbol{A} = [a_{ij}]$, then

$$\frac{\partial f}{\partial \boldsymbol{A}} = \left[\frac{\partial f}{\partial a_{ij}}\right],$$

which is an $m \times n$ matrix of partial derivatives. If elements of the $m \times n$ matrix $\boldsymbol{A}$ are functions of a scalar parameter β, then

$$\frac{\partial \boldsymbol{A}}{\partial \beta} = \left[\frac{\partial a_{ij}}{\partial \beta}\right],$$

which is an $m \times n$ matrix of derivatives. Finally, if $\boldsymbol{f}(\boldsymbol{\beta}) = (f_1(\boldsymbol{\beta}), \ldots, f_p(\boldsymbol{\beta}))'$ is a p-dimensional vector of functions of $\boldsymbol{\beta}$, then

$$\frac{\partial \boldsymbol{f}}{\partial \boldsymbol{\beta}'} = \begin{bmatrix} \frac{\partial f_1}{\partial \beta_1} & \cdots & \frac{\partial f_1}{\partial \beta_m} \\ \vdots & & \vdots \\ \frac{\partial f_p}{\partial \beta_1} & \cdots & \frac{\partial f_p}{\partial \beta_m} \end{bmatrix},$$

is an $p \times m$ matrix of partial derivatives and we have

$$\frac{\partial \boldsymbol{f}'}{\partial \boldsymbol{\beta}} \equiv \left(\frac{\partial \boldsymbol{f}}{\partial \boldsymbol{\beta}'}\right)'.$$

Finally, if $\boldsymbol{A} = [a_{ij}(\boldsymbol{B})]$ is an $m \times n$ matrix elements of which are differentiable functions of the $p \times q$ matrix $\boldsymbol{B}$, then the derivative of $\boldsymbol{A}$ with respect to $\boldsymbol{B}$ is defined as $\partial \text{vec}(\boldsymbol{A})/\partial \text{vec}(\boldsymbol{B})$, which is an $mn \times pq$ matrix. Vector and matrix differentiation is often used in multivariate time series analysis to derive parameter estimates.

The chain rule and product rules of vector differentiation are given next:

Result 1 (Chain rule) Let $\boldsymbol{\alpha}$ and $\boldsymbol{\beta}$ are $n \times 1$ and $m \times 1$ vectors, respectively, and suppose $\boldsymbol{\alpha} = \boldsymbol{g}(\boldsymbol{\beta})$ and $\boldsymbol{h}(\boldsymbol{\alpha})$ is a $p \times 1$ vector of functions of $\boldsymbol{\alpha}$. Then,

$$\frac{\partial \boldsymbol{h}[\boldsymbol{g}(\boldsymbol{\beta})]}{\partial \boldsymbol{\beta}'} = \frac{\partial \boldsymbol{h}(\boldsymbol{\alpha})}{\partial \boldsymbol{\alpha}'} \frac{\partial \boldsymbol{g}(\boldsymbol{\beta})}{\partial \boldsymbol{\beta}'}.$$

Result 2 (Products rule) (a) Suppose that $\boldsymbol{\beta}$ is a $m \times 1$ vector, $\boldsymbol{\alpha}(\boldsymbol{\beta}) = (\alpha_1(\boldsymbol{\beta}), \ldots, \alpha_n(\boldsymbol{\beta}))'$ is an $n \times 1$ vector, $\boldsymbol{c}(\boldsymbol{\beta}) = (c_1(\boldsymbol{\beta}), \ldots, c_p(\boldsymbol{\beta}))'$ is a $p \times 1$ vector, and $\boldsymbol{A} = [a_{ij}]$ is an $n \times p$ matrix that does not depend on $\boldsymbol{\beta}$. Then,

$$\frac{\partial[\boldsymbol{\alpha}(\boldsymbol{\beta})' \boldsymbol{A} \boldsymbol{c}(\boldsymbol{\beta})]}{\partial \boldsymbol{\beta}'} = \boldsymbol{c}(\boldsymbol{\beta})' \boldsymbol{A}' \frac{\partial \boldsymbol{\alpha}(\boldsymbol{\beta})}{\partial \boldsymbol{\beta}'} + \boldsymbol{\alpha}(\boldsymbol{\beta})' \boldsymbol{A} \frac{\partial \boldsymbol{c}(\boldsymbol{\beta})}{\partial \boldsymbol{\beta}'}.$$

(b) If $\boldsymbol{A}(\beta)$ and $\boldsymbol{B}(\beta)$ are $m \times n$ and $n \times p$ matrices, respectively, of a scalar β, then

$$\frac{\partial \boldsymbol{A}\boldsymbol{B}}{\partial \beta} = \frac{\partial \boldsymbol{A}}{\partial \beta} \boldsymbol{B} + \boldsymbol{A} \frac{\partial \boldsymbol{B}}{\partial \beta}.$$

(c) If $\boldsymbol{A}(\boldsymbol{\beta})$ and $\boldsymbol{B}(\boldsymbol{\beta})$ are $m \times n$ and $n \times p$ matrices, respectively, consisting of functions of the k-dimensional vector $\boldsymbol{\beta}$, then

$$\frac{\partial \text{vec}(\boldsymbol{A}\boldsymbol{B})}{\partial \boldsymbol{\beta}'} = (\boldsymbol{I}_p \otimes \boldsymbol{A}) \frac{\partial \text{vec}(\boldsymbol{B})}{\partial \boldsymbol{\beta}'} + (\boldsymbol{B}' \otimes \boldsymbol{I}_m) \frac{\partial \text{vec}(\boldsymbol{A})}{\partial \boldsymbol{\beta}'}.$$

Some useful results of vector and matrix differentiation are as follows:

Result 3 (a) For an $m \times n$ matrix $\boldsymbol{A}$ and $n \times 1$ vector $\boldsymbol{\beta}$,

$$\frac{\partial \boldsymbol{A}\boldsymbol{\beta}}{\partial \boldsymbol{\beta}'} = \boldsymbol{A}, \quad \frac{\partial \boldsymbol{\beta}' \boldsymbol{A}'}{\partial \boldsymbol{\beta}} = \boldsymbol{A}'.$$

(b) Let $\boldsymbol{A}$ be $m \times m$ and $\boldsymbol{\beta}$ be $m \times 1$. Then,

$$\frac{\partial \boldsymbol{\beta}'\boldsymbol{A}\boldsymbol{\beta}}{\partial \boldsymbol{\beta}} = (\boldsymbol{A} + \boldsymbol{A}')\boldsymbol{\beta}, \quad \frac{\partial \boldsymbol{\beta}'\boldsymbol{A}\boldsymbol{\beta}}{\partial \boldsymbol{\beta}'} = \boldsymbol{\beta}'(\boldsymbol{A}' + \boldsymbol{A}).$$

Furthermore,

$$\frac{\partial^2 \boldsymbol{\beta}'\boldsymbol{A}\boldsymbol{\beta}}{\partial \boldsymbol{\beta}\partial \boldsymbol{\beta}'} = \boldsymbol{A} + \boldsymbol{A}'.$$

If $\boldsymbol{A}$ is symmetric, then the aforementioned results can be further simplified.

(c) Let $\boldsymbol{\beta}$ be an $m \times 1$ vector, $\boldsymbol{A}$ be a symmetric $n \times n$ matrix, which does not depend on $\boldsymbol{\beta}$, and $\boldsymbol{c}(\boldsymbol{\beta})$ be an $n \times 1$ vector elements of which are functions of $\boldsymbol{\beta}$. Then,

$$\frac{\partial \boldsymbol{c}(\boldsymbol{\beta})\boldsymbol{A}\boldsymbol{c}(\boldsymbol{\beta})}{\partial \boldsymbol{\beta}'} = 2\boldsymbol{c}(\boldsymbol{\beta})\boldsymbol{A}\frac{\partial \boldsymbol{c}(\boldsymbol{\beta})}{\partial \boldsymbol{\beta}'},$$

and

$$\frac{\partial^2 \boldsymbol{c}(\boldsymbol{\beta})\boldsymbol{A}\boldsymbol{c}(\boldsymbol{\beta})}{\partial \boldsymbol{\beta}\partial \boldsymbol{\beta}'} = 2\left[\frac{\partial \boldsymbol{c}(\boldsymbol{\beta})'}{\partial \boldsymbol{\beta}}\boldsymbol{A}\frac{\partial \boldsymbol{c}(\boldsymbol{\beta})}{\partial \boldsymbol{\beta}'} + [\boldsymbol{c}(\boldsymbol{\beta})'\boldsymbol{A} \otimes \boldsymbol{I}_m]\frac{\partial \mathrm{vec}(\partial \boldsymbol{c}(\boldsymbol{\beta})'/\partial \boldsymbol{\beta})}{\partial \boldsymbol{\beta}'}\right].$$

A direct application of this result is in least-squares estimation. Let $\boldsymbol{Y}$ be an $n \times 1$ vector of response variable and $\boldsymbol{X}$ be an $n \times m$ matrix of regressors. Then,

$$\frac{\partial (\boldsymbol{Y} - \boldsymbol{X}\boldsymbol{\beta})'\boldsymbol{A}(\boldsymbol{Y} - \boldsymbol{X}\boldsymbol{\beta})}{\partial \boldsymbol{\beta}'} = -2(\boldsymbol{Y} - \boldsymbol{X}\boldsymbol{\beta})'\boldsymbol{A}\boldsymbol{X}, \tag{A.5}$$

and

$$\frac{\partial^2 (\boldsymbol{Y} - \boldsymbol{X}\boldsymbol{\beta})'\boldsymbol{A}(\boldsymbol{Y} - \boldsymbol{X}\boldsymbol{\beta})}{\partial \boldsymbol{\beta}\partial \boldsymbol{\beta}'} = 2\boldsymbol{X}'\boldsymbol{A}\boldsymbol{X}.$$

Letting Equation (A.5) to 0, we obtain the weighted least-squares estimate of $\boldsymbol{\beta}$ as

$$\widehat{\boldsymbol{\beta}} = (\boldsymbol{X}'\boldsymbol{A}\boldsymbol{X})^{-1}\boldsymbol{X}'\boldsymbol{A}\boldsymbol{Y}.$$

(d) Let $\boldsymbol{\beta}$ be $m \times 1$ and $\boldsymbol{B}(\boldsymbol{\beta})$ be a matrix elements of which are functions of $\boldsymbol{\beta}$. Also, let $\boldsymbol{A}$ and $\boldsymbol{C}$ be matrices independent of $\boldsymbol{\beta}$ such that $\boldsymbol{ABC}$ exists. Then,

$$\frac{\partial \mathrm{vec}(\boldsymbol{ABC})}{\partial \boldsymbol{\beta}'} = (\boldsymbol{C}' \otimes \boldsymbol{A})\frac{\partial \mathrm{vec}(\boldsymbol{B})}{\partial \boldsymbol{\beta}'}.$$

(e) If $\boldsymbol{A}$ is a nonsingular $m \times m$ matrix, then

$$\frac{\partial \text{vec}(\boldsymbol{A}^{-1})}{\partial \text{vec}(\boldsymbol{A})'} = -(\boldsymbol{A}^{-1})' \otimes \boldsymbol{A}^{-1} = -(\boldsymbol{A}' \otimes \boldsymbol{A})^{-1}.$$

(f) For an $m \times m$ matrix $\boldsymbol{A} = [a_{ij}]$, we have

$$\frac{\partial tr(\boldsymbol{A})}{\partial \boldsymbol{A}} = \boldsymbol{I}_m.$$

(g) If $\boldsymbol{A}$ and $\boldsymbol{B}$ are $m \times n$ and $n \times m$ matrices, respectively, then

$$\frac{\partial tr(\boldsymbol{AB})}{\partial \boldsymbol{A}} = \boldsymbol{B}'.$$

(h) For an $m \times m$ matrix,

$$\frac{\partial |\boldsymbol{A}|}{\partial \boldsymbol{A}} = [\text{adj}(\boldsymbol{A})]' = |\boldsymbol{A}|(\boldsymbol{A}')^{-1},$$

where adj$(\boldsymbol{A})$ is the adjoint matrix of $\boldsymbol{A}$. For given i and j $(1 \le i, j \le m)$, by deleting the ith row and jth column from $\boldsymbol{A}$, one obtains an $(m-1) \times (m-1)$ submatrix of $\boldsymbol{A}$, denoted by $\boldsymbol{A}_{(-i,-j)}$. The cofactor of a_{ij} of $\boldsymbol{A}$ is then defined as $c_{ij} = (-1)^{i+j}|\boldsymbol{A}_{(-i,-j)}|$. The matrix $[c_{ij}]$ is called the cofactor matrix of $\boldsymbol{A}$, and its transpose $[c_{ij}]' \equiv \text{adj}(\boldsymbol{A})$ is called the adjoint matrix of $\boldsymbol{A}$. It can be shown that for a nonsingular matrix $\boldsymbol{A}$, $\boldsymbol{A}^{-1} = |\boldsymbol{A}|^{-1}\text{adj}(\boldsymbol{A})$.

(i) If $\boldsymbol{A}$ is a nonsingular matrix, then

$$\frac{\partial \ln |\boldsymbol{A}|}{\partial \boldsymbol{A}} = [\boldsymbol{A}']^{-1}.$$

(j) Let $\boldsymbol{A}, \boldsymbol{B}, \boldsymbol{C}$ are $m \times m$ square matrices and $\boldsymbol{A}$ is nonsingular. Then,

$$\frac{\partial tr(\boldsymbol{B}\boldsymbol{A}^{-1}\boldsymbol{C})}{\partial \boldsymbol{A}} = -(\boldsymbol{A}^{-1}\boldsymbol{C}\boldsymbol{B}\boldsymbol{A}^{-1})^{-1}.$$

(k) Let $\boldsymbol{A}$ be an $m \times m$ matrix elements of which are functions of the vector $\boldsymbol{\beta}$. Then, for any positive integer h,

$$\frac{\partial \text{vec}(\boldsymbol{A}^h)}{\partial \boldsymbol{\beta}'} = \left[\sum_{i=0}^{h-1} (\boldsymbol{A}')^{h-1-i} \otimes \boldsymbol{A}^i\right] \frac{\partial \text{vec}(\boldsymbol{A})}{\partial \boldsymbol{\beta}'}.$$

This result can be shown by mathematical induction.

A.2 LEAST-SQUARES ESTIMATION

The least-squares method is commonly used in statistical estimation. Consider the matrix equation

$$\boldsymbol{Z} = \boldsymbol{X}\boldsymbol{\beta} + \boldsymbol{A}, \tag{A.6}$$

where the dimensions of $\boldsymbol{Z}$, $\boldsymbol{X}$, $\boldsymbol{\beta}$, and $\boldsymbol{A}$ are $n \times m$, $n \times p$, $p \times m$, and $n \times m$, respectively. Often we have $n > m$ and assume that $\boldsymbol{X}'\boldsymbol{X}$ is an invertible $p \times p$ matrix. Here, $\boldsymbol{A}$ denotes the error matrix and $\boldsymbol{\beta}$ the parameter matrix. Let $\boldsymbol{a}_i'$ denote the ith row of $\boldsymbol{A}$. In statistical estimation, we assume further that $\boldsymbol{a}_i$ has zero mean and positive-definite covariance matrix $\boldsymbol{\Sigma}_a$ and that $\boldsymbol{a}_i$ and $\boldsymbol{a}_j$ are uncorrelated if $i \neq j$. The ordinary least-squares estimate of $\boldsymbol{\beta}$ is then

$$\widehat{\boldsymbol{\beta}} = (\boldsymbol{X}'\boldsymbol{X})^{-1}\boldsymbol{X}'\boldsymbol{Z}.$$

A derivation of this estimate can be found in Chapter 2. Let $\widehat{\boldsymbol{A}} = \boldsymbol{Z} - \boldsymbol{X}\widehat{\boldsymbol{\beta}}$ be the residual matrix. Then, the following properties are useful:

(i) $\widehat{\boldsymbol{A}}'\boldsymbol{X} = \boldsymbol{0}$.

(ii) $(\boldsymbol{Z} - \boldsymbol{X}\boldsymbol{\beta})'(\boldsymbol{Z} - \boldsymbol{X}\boldsymbol{\beta}) = \widehat{\boldsymbol{A}}'\widehat{\boldsymbol{A}} + (\widehat{\boldsymbol{\beta}} - \boldsymbol{\beta})'\boldsymbol{X}'\boldsymbol{X}(\widehat{\boldsymbol{\beta}} - \boldsymbol{\beta})$.

These properties can be easily shown. First, $\widehat{\boldsymbol{A}} = [\boldsymbol{I}_n - \boldsymbol{X}(\boldsymbol{X}'\boldsymbol{X})^{-1}\boldsymbol{X}']\boldsymbol{Z}$. Since $[\boldsymbol{I}_n - \boldsymbol{X}(\boldsymbol{X}'\boldsymbol{X})^{-1}\boldsymbol{X}']\boldsymbol{X} = \boldsymbol{0}$, Property (i) follows. Property (ii) follows directly from Property (i), because

$$\boldsymbol{Z} - \boldsymbol{X}\boldsymbol{\beta} = \boldsymbol{Z} - \boldsymbol{X}\widehat{\boldsymbol{\beta}} + \boldsymbol{X}\widehat{\boldsymbol{\beta}} - \boldsymbol{X}\boldsymbol{\beta} = \widehat{\boldsymbol{A}} + \boldsymbol{X}(\widehat{\boldsymbol{\beta}} - \boldsymbol{\beta}).$$

Property (i) says that the residuals are uncorrelated with $\boldsymbol{X}$, and Property (ii) is referred to as the multivariate analysis of variance.

The prior properties of least-squares estimates can be extended to the generalized least squares (GLS) estimates. We focus on the case $m = 1$ so that $\boldsymbol{Z}$ and $\boldsymbol{A}$ are n-dimensional vectors. Suppose that $\text{Cov}(\boldsymbol{A}) = \boldsymbol{\Sigma}$, which is a positive-definite $n \times n$ matrix. Let $\boldsymbol{\Sigma}^{1/2}$ be the positive-definite square root matrix of $\boldsymbol{\Sigma}$. Premultiplying Equation (A.6) by $\boldsymbol{\Sigma}^{-1/2}$, we obtain

$$\tilde{\boldsymbol{Z}} = \tilde{\boldsymbol{X}}\boldsymbol{\beta} + \tilde{\boldsymbol{A}},$$

where $\tilde{\boldsymbol{Z}} = \boldsymbol{\Sigma}^{-1/2}\boldsymbol{Z}$ and $\tilde{\boldsymbol{X}}$ and $\tilde{\boldsymbol{A}}$ are defined similarly. Since $\text{Cov}(\tilde{\boldsymbol{A}}) = \boldsymbol{I}_n$, the prior equation is a homogeneous multiple linear regression. Consequently, the GLS estimate of $\boldsymbol{\beta}$ is

$$\widehat{\boldsymbol{\beta}}_g = (\tilde{\boldsymbol{X}}'\tilde{\boldsymbol{X}})^{-1}(\tilde{\boldsymbol{X}}'\tilde{\boldsymbol{Z}}) = (\boldsymbol{X}'\boldsymbol{\Sigma}^{-1}\boldsymbol{X})^{-1}(\boldsymbol{X}\boldsymbol{\Sigma}^{-1}\boldsymbol{Z}).$$

The residual vector is then

$$\hat{\boldsymbol{A}} = \boldsymbol{Z} - \boldsymbol{X}\hat{\boldsymbol{\beta}}_g = [\boldsymbol{I}_n - \boldsymbol{X}(\boldsymbol{X}'\boldsymbol{\Sigma}^{-1}\boldsymbol{X})^{-1}\boldsymbol{X}'\boldsymbol{\Sigma}^{-1}]\boldsymbol{Z}.$$

Since $[\boldsymbol{I}_n - \boldsymbol{X}(\boldsymbol{X}'\boldsymbol{\Sigma}^{-1}\boldsymbol{X})^{-1}\boldsymbol{X}'\boldsymbol{\Sigma}^{-1}]\boldsymbol{X} = \boldsymbol{0}$, we have (i) $\hat{\boldsymbol{A}}'\boldsymbol{X} = \boldsymbol{0}$. Next, from

$$\boldsymbol{Z} - \boldsymbol{X}\boldsymbol{\beta} = \boldsymbol{Z} - \boldsymbol{X}\hat{\boldsymbol{\beta}}_g + \boldsymbol{X}\hat{\boldsymbol{\beta}}_g - \boldsymbol{X}\boldsymbol{\beta} = \hat{\boldsymbol{A}} + \boldsymbol{X}(\hat{\boldsymbol{\beta}}_g - \boldsymbol{\beta}),$$

we can use Property (i) to obtain

$$(\boldsymbol{Z} - \boldsymbol{X}\boldsymbol{\beta})'(\boldsymbol{Z} - \boldsymbol{X}\boldsymbol{\beta}) = \hat{\boldsymbol{A}}'\hat{\boldsymbol{A}} + (\hat{\boldsymbol{\beta}}_g - \boldsymbol{\beta})\boldsymbol{X}'\boldsymbol{X}(\hat{\boldsymbol{\beta}}_g - \boldsymbol{\beta}),$$

which is the same form as that of the OLS estimate.

A.3 MULTIVARIATE NORMAL DISTRIBUTIONS

A k-dimensional random vector $\boldsymbol{x} = (x_1, \ldots, x_k)'$ follows a multivariate normal distribution with mean $\boldsymbol{\mu} = (\mu_1, \ldots, \mu_k)'$ and positive-definite covariance matrix $\boldsymbol{\Sigma} = [\sigma_{ij}]$ if its probability density function (pdf) is

$$f(\boldsymbol{x}|\boldsymbol{\mu}, \boldsymbol{\Sigma}) = \frac{1}{(2\pi)^{k/2}|\boldsymbol{\Sigma}|^{1/2}} \exp\left[-\frac{1}{2}(\boldsymbol{x} - \boldsymbol{\mu})'\boldsymbol{\Sigma}^{-1}(\boldsymbol{x} - \boldsymbol{\mu})\right]. \quad \text{(A.7)}$$

We use the notation $\boldsymbol{x} \sim N_k(\boldsymbol{\mu}, \boldsymbol{\Sigma})$ to denote that $\boldsymbol{x}$ follows such a distribution. The inverse matrix $\boldsymbol{\Sigma}^{-1}$ is called the *precision matrix* of the multivariate normal distribution. Normal distribution plays an important role in multivariate statistical analysis and it has several nice properties. Here, we consider only those properties that are relevant to our study. Interested readers are referred to Johnson and Wichern (2007) for details.

To gain insight into multivariate normal distributions, consider the bivariate case (i.e., $k = 2$). In this case, we have

$$\boldsymbol{\Sigma} = \begin{bmatrix} \sigma_{11} & \sigma_{12} \\ \sigma_{12} & \sigma_{22} \end{bmatrix}, \quad \boldsymbol{\Sigma}^{-1} = \frac{1}{\sigma_{11}\sigma_{22} - \sigma_{12}^2} \begin{bmatrix} \sigma_{22} & -\sigma_{12} \\ -\sigma_{12} & \sigma_{11} \end{bmatrix}.$$

Using the correlation coefficient $\rho = \sigma_{12}/(\sigma_1\sigma_2)$, where $\sigma_i = \sqrt{\sigma_{ii}}$ is the standard deviation of x_i, we have $\sigma_{12} = \rho\sqrt{\sigma_{11}\sigma_{22}}$ and $|\boldsymbol{\Sigma}| = \sigma_{11}\sigma_{22}(1 - \rho^2)$. The pdf of $\boldsymbol{x}$ then becomes

$$f(x_1, x_2|\boldsymbol{\mu}, \boldsymbol{\Sigma}) = \frac{1}{2\pi\sigma_1\sigma_2\sqrt{1 - \rho^2}} \exp\left(-\frac{1}{2(1 - \rho^2)}[Q(\boldsymbol{x}, \boldsymbol{\mu}, \boldsymbol{\Sigma})]\right),$$

where

$$Q(\boldsymbol{x}, \boldsymbol{\mu}, \boldsymbol{\Sigma}) = \left(\frac{x_1 - \mu_1}{\sigma_1}\right)^2 + \left(\frac{x_2 - \mu_2}{\sigma_2}\right)^2 - 2\rho\left(\frac{x_1 - \mu_1}{\sigma_1}\right)\left(\frac{x_2 - \mu_2}{\sigma_2}\right).$$

Readers can use the package `mvtnorm` in R to obtain density function and random samples of a multivariate normal distribution.

Let $\boldsymbol{c} = (c_1, \ldots, c_k)'$ be a nonzero k-dimensional vector. Partition the random vector as $\boldsymbol{x} = (\boldsymbol{x}_1', \boldsymbol{x}_2')'$, where $\boldsymbol{x}_1 = (x_1, \ldots, x_p)'$ and $\boldsymbol{x}_2 = (x_{p+1}, \ldots, x_k)'$ with $1 \le p < k$. Also, partition $\boldsymbol{\mu}$ and $\boldsymbol{\Sigma}$ accordingly as

$$\begin{bmatrix} \boldsymbol{x}_1 \\ \boldsymbol{x}_2 \end{bmatrix} \sim N\left(\begin{bmatrix} \boldsymbol{\mu}_1 \\ \boldsymbol{\mu}_2 \end{bmatrix}, \begin{bmatrix} \boldsymbol{\Sigma}_{11} & \boldsymbol{\Sigma}_{12} \\ \boldsymbol{\Sigma}_{21} & \boldsymbol{\Sigma}_{22} \end{bmatrix}\right).$$

Some properties of $\boldsymbol{x}$ are as follows:

1. $\boldsymbol{c}'\boldsymbol{x} \sim N(\boldsymbol{c}'\boldsymbol{\mu}, \boldsymbol{c}'\boldsymbol{\Sigma}\boldsymbol{c})$. That is, any nonzero linear combination of $\boldsymbol{x}$ is univariate normal. The inverse of this property also holds. Specifically, if $\boldsymbol{c}'\boldsymbol{x}$ is univariate normal for any nonzero vector $\boldsymbol{c}$, then $\boldsymbol{x}$ is multivariate normal.
2. The marginal distribution of $\boldsymbol{x}_i$ is normal. In fact, $\boldsymbol{x}_i \sim N_{k_i}(\boldsymbol{\mu}_i, \boldsymbol{\Sigma}_{ii})$ for $i = 1$ and 2, where $k_1 = p$ and $k_2 = k - p$.
3. $\boldsymbol{\Sigma}_{12} = \boldsymbol{0}$ if and only if $\boldsymbol{x}_1$ and $\boldsymbol{x}_2$ are independent.
4. The random variable $y = (\boldsymbol{x} - \boldsymbol{\mu})'\boldsymbol{\Sigma}^{-1}(\boldsymbol{x} - \boldsymbol{\mu})$ follows a chi-squared distribution with k degrees of freedom.
5. The conditional distribution of $\boldsymbol{x}_1$ given $\boldsymbol{x}_2 = \boldsymbol{b}$ is also normally distributed as

$$(\boldsymbol{x}_1 | \boldsymbol{x}_2 = \boldsymbol{b}) \sim N_p[\boldsymbol{\mu}_1 + \boldsymbol{\Sigma}_{12}\boldsymbol{\Sigma}_{22}^{-1}(\boldsymbol{b} - \boldsymbol{\mu}_2), \boldsymbol{\Sigma}_{11} - \boldsymbol{\Sigma}_{12}\boldsymbol{\Sigma}_{22}^{-1}\boldsymbol{\Sigma}_{21}].$$

The last property is useful in many scientific areas. For instance, it forms the basis for time series forecasting under the normality assumption, for recursive least-squares estimation, and for Kalman filter. It is also widely used in Bayesian inference.

A.4 MULTIVARIATE STUDENT-*t* DISTRIBUTION

Let $\boldsymbol{y} = (y_1, \ldots, y_k)'$ be a k-dimensional random vector and u scalar random variable. Suppose that $\boldsymbol{y}$ and u are independently distributed according to $N(\boldsymbol{0}, \boldsymbol{\Sigma})$ and the χ^2_n distribution, respectively, where $\boldsymbol{\Sigma}$ is positive-definite. Let $\boldsymbol{x} - \boldsymbol{\mu} = \sqrt{n/u}\boldsymbol{y}$. Then, the pdf of $\boldsymbol{x}$ is

$$f(\boldsymbol{x}|n, \boldsymbol{\mu}, \boldsymbol{\Sigma}) = \frac{\Gamma((n+k)/2)}{\Gamma(\frac{n}{2})(n\pi)^{k/2}|\boldsymbol{\Sigma}|^{1/2}}[1 - \frac{1}{n}(\boldsymbol{x} - \boldsymbol{\mu})'\boldsymbol{\Sigma}^{-1}(\boldsymbol{x} - \boldsymbol{\mu})]^{-(n+k)/2}, \tag{A.8}$$

where $\Gamma(v) = \int_0^\infty e^{-t}t^{v-1}dt$ is the usual Gamma function. We call $\boldsymbol{x}$ follows a multivariate Student-t distribution with degrees of freedom n with parameters $\boldsymbol{\mu}$ and $\boldsymbol{\Sigma}$. See, for instance, Anderson (2003, p. 289). For $n > 1$, we have $E(\boldsymbol{x}) = \boldsymbol{\mu}$ and for $n > 2$, we have $\text{Cov}(\boldsymbol{x}) = n/(n-2)\boldsymbol{\Sigma}$. Multivariate Student-$t$ distribution is useful in modeling time-varying covariance matrix, for example, the volatility matrix of asset returns. This is especially so when $\boldsymbol{\Sigma} = \boldsymbol{I}_k$ and $\boldsymbol{\mu} = \boldsymbol{0}$. In this special case, one can define $\boldsymbol{\epsilon} = \sqrt{(n-2)/n}\boldsymbol{x}$ to obtain the standardized multivariate Student-t distribution. The density function of $\boldsymbol{\epsilon}$ is

$$f(\boldsymbol{\epsilon}|n) = \frac{\Gamma[(n+k)/2]}{(n\pi)^{k/2}\Gamma(n/2)}(1 + n^{-1}\boldsymbol{\epsilon}'\boldsymbol{\epsilon})^{-(n+k)/2}.$$

A.5 WISHART AND INVERTED WISHART DISTRIBUTIONS

Let $\boldsymbol{x}_1, \ldots, \boldsymbol{x}_n$ be independent and identically distributed according to $N(\boldsymbol{0}, \boldsymbol{\Sigma})$, where $\boldsymbol{\Sigma}$ is a $k \times k$ positive-definite matrix. Let $\boldsymbol{A} = \sum_{i=1}^n \boldsymbol{x}_i\boldsymbol{x}_i'$, which is a $k \times k$ random matrix. The pdf of $\boldsymbol{A}$ is

$$f(\boldsymbol{A}|n, \boldsymbol{\Sigma}) = \frac{|\boldsymbol{A}|^{(n-k-1)/2}\exp[-\frac{1}{2}tr(\boldsymbol{\Sigma}^{-1}\boldsymbol{A})]}{2^{(nk)/2}\pi^{k(k-1)/4}|\boldsymbol{\Sigma}|^{n/2}\prod_{i=1}^k \Gamma[(n+1-i)/2]}, \quad \text{(A.9)}$$

for $\boldsymbol{A}$ positive-definite, and 0 otherwise. We refer to $\boldsymbol{A}$ as a random matrix following the Wishart distribution with degrees of freedom n and parameter $\boldsymbol{\Sigma}$ and denote it as $\boldsymbol{A} \sim W(\boldsymbol{\Sigma}, n)$. From the definition, the random matrix $\boldsymbol{A}$ is positive-definite if $n > k$. For the Wishart random matrix, $E(\boldsymbol{A}) = n\boldsymbol{\Sigma}$ and the mode of $\boldsymbol{A}$ is $(n-k-1)\boldsymbol{\Sigma}$ provided that $n > k+1$.

For ease in notation, we define the multivariate Gamma function as

$$\Gamma_k(t) = \pi^{k(k-1)/4}\prod_{i=1}^k \Gamma\left[t - \frac{i-1}{2}\right]. \quad \text{(A.10)}$$

The Wishart density function then simplifies to

$$f(\boldsymbol{A}|n, \boldsymbol{A}) = \frac{|\boldsymbol{A}|^{(n-k-1)/2}\exp[\frac{-1}{2}tr(\boldsymbol{\Sigma}^{-1}\boldsymbol{A})]}{2^{(nk)/2}|\boldsymbol{\Sigma}|^{n/2}\Gamma_k(n/2)}. \quad \text{(A.11)}$$

Some useful properties of Wishart distribution are

(i) If $\boldsymbol{A}_1, \ldots, \boldsymbol{A}_q$ are independently distributed with $A_i \sim W(\boldsymbol{\Sigma}, n_i)$, then $\boldsymbol{A} = \sum_{i=1}^q \boldsymbol{A}_i \sim W(\boldsymbol{\Sigma}, \sum_{i=1}^q n_i)$.

(ii) If $\boldsymbol{C}$ is a nonsingular $k \times k$ matrix such that $\boldsymbol{A} = \boldsymbol{CBC}'$ and $\boldsymbol{A} \sim W(\boldsymbol{\Sigma}, n)$, then $\boldsymbol{B} \sim W[\boldsymbol{C}^{-1}\boldsymbol{\Sigma}(\boldsymbol{C}')^{-1}, n]$.

(iii) Suppose that $\boldsymbol{A}$ and $\boldsymbol{\Sigma}$ are partitioned into q and $k-q$ rows and columns,

$$\boldsymbol{A} = \begin{bmatrix} \boldsymbol{A}_{11} & \boldsymbol{A}_{12} \\ \boldsymbol{A}_{21} & \boldsymbol{A}_{22} \end{bmatrix}, \quad \boldsymbol{\Sigma} = \begin{bmatrix} \boldsymbol{\Sigma}_{11} & \boldsymbol{\Sigma}_{12} \\ \boldsymbol{\Sigma}_{21} & \boldsymbol{\Sigma}_{22} \end{bmatrix}.$$

If $\boldsymbol{A} \sim W(\boldsymbol{\Sigma}, n)$, then $\boldsymbol{A}_{11} \sim W(\boldsymbol{\Sigma}_{11}, n)$.

(iv) Let $\boldsymbol{A}$ and $\boldsymbol{\Sigma}$ be partitioned as in Property (iii). If $\boldsymbol{A} \sim W(\boldsymbol{\Sigma}, n)$, then $\boldsymbol{A}_{1|2} \sim W(\boldsymbol{\Sigma}_{1|2}, n)$, where $\boldsymbol{A}_{1|2} = \boldsymbol{A}_{11} - \boldsymbol{A}_{12}\boldsymbol{A}_{22}^{-1}\boldsymbol{A}_{21}$ and $\boldsymbol{\Sigma}_{1|2} = \boldsymbol{\Sigma}_{11} - \boldsymbol{\Sigma}_{12}\boldsymbol{\Sigma}_{22}^{-1}\boldsymbol{\Sigma}_{21}$.

In multivariate time series analysis, we often need to make inference about the inverse of a covariance matrix. To this end, the random matrix $\boldsymbol{A}^{-1}$ is of interest. If $\boldsymbol{A} \sim W(\boldsymbol{\Sigma}, n)$, the distribution of $\boldsymbol{B} = \boldsymbol{A}^{-1}$ is called the inverted Wishart distribution. The pdf of $\boldsymbol{B} = \boldsymbol{A}^{-1}$ is

$$f(\boldsymbol{B}|\boldsymbol{\Psi}, n) = \frac{|\boldsymbol{\Psi}|^{n/2}|\boldsymbol{B}|^{-(n+k+1)/2}\exp[-\frac{1}{2}tr(\boldsymbol{\Psi}\boldsymbol{B}^{-1})]}{2^{(nk)/2}\Gamma_k(n/2)}, \tag{A.12}$$

for $\boldsymbol{B}$ positive-definite, and 0 otherwise, where $\boldsymbol{\Psi} = \boldsymbol{\Sigma}^{-1}$. We refer to $\boldsymbol{B}$ as the inverted Wishart distribution with n degrees of freedom and parameter $\boldsymbol{\Psi}$ and use the notation $\boldsymbol{B} \sim W^{-1}(\boldsymbol{\Psi}, n)$. The mean of $\boldsymbol{B}$ is $E(\boldsymbol{B}) = 1/(n-k-1)\boldsymbol{\Psi}$ if $n > k+1$ and the mode of $\boldsymbol{B}$ is $1/(n+k+1)\boldsymbol{\Psi}$. A property of Wishart and inverted Wishart distributions useful in Bayesian inference is given next.

Result 4 If $\boldsymbol{A} \sim W(\boldsymbol{\Sigma}, n)$ where $\boldsymbol{\Sigma}$ has a prior distribution $W^{-1}(\boldsymbol{\Psi}, m)$, then the posterior distribution of $\boldsymbol{\Sigma}$ given $\boldsymbol{A}$ is $W^{-1}(\boldsymbol{A} + \boldsymbol{\Psi}, n+m)$.

A proof of this result can be found in Theorem 7.7.2 of Anderson (2003, p. 273).

A.6 VECTOR AND MATRIX DIFFERENTIALS

We have reviewed matrix derivatives in Section A.1.9. However, in some statistical applications, it is more convenient to use matrix *differential*. A useful reference of matrix differential is Schott (2005, Chapter 9). In this section, we briefly summarize some properties of matrix differential that are useful in this book, assuming that the vectors or matrices involved are differentiable.

A.6.1 Scalar Function

If $f(x)$ is a real-valued function of the scalar variable x, then its derivative at x, if it exists, is given by

$$f'(x) = \frac{df(x)}{dx} = \lim_{u \to 0} \frac{f(x+u) - f(x)}{u}.$$

In this case, the first-order Taylor series expansion for $f(x+u)$ is

$$f(x+u) = f(x) + uf'(x) + r_1(u,x),$$

where $r_1(u,x)$ denotes the error term satisfying

$$\lim_{u \to 0} \frac{r_1(u,x)}{u} = 0.$$

We denote the quantity $uf'(x)$ by $d_u f(x)$ and refer to it as the first *differential* of $f(x)$ at x with increment u. The increment u is the differential of x, so that we often write the first differential of $f(x)$ as $d_x f(x) = f'(x)dx$ or simply $df(x) = f'(x)dx$. This idea generalizes to higher-order differentials, namely,

$$f^{(i)}(x) = \frac{d^i f(x)}{dx^i} = \lim_{u \to 0} \frac{f^{(i-1)}(x+u) - f^{(i-1)}(x)}{u}.$$

The vth-order Taylor series expansion is

$$\begin{aligned} f(x+u) &= f(x) + \sum_{i=1}^{v} \frac{u^i f^{(i)}(x)}{i!} + r_v(u,x) \\ &= f(x) + \sum_{i=1}^{v} \frac{d_u^i f(x)}{i!} + r_v(u,x), \end{aligned}$$

where the remainder $r_v(u,x)$ satisfies

$$\lim_{u \to 0} \frac{r_v(u,x)}{u^v} = 0,$$

and $d_u^i f(x) = u^i f^{(i)}(x)$ is the ith differential of $f(x)$ at x with increment u.

If $f(\boldsymbol{x})$ is a re-valued function of the k-dimensional variable $\boldsymbol{x} = (x_1, \ldots, x_k)'$. We have defined the derivative of $f(\boldsymbol{x})$ at $\boldsymbol{x}$ in Section A.1.9) as

$$\frac{\partial f(\boldsymbol{x})}{\partial \boldsymbol{x}'} = \left[\frac{\partial f(\boldsymbol{x})}{\partial x_1}, \ldots, \frac{\partial f(\boldsymbol{x})}{\partial x_k}\right],$$

where

$$\frac{\partial f(\boldsymbol{x})}{\partial x_i} = \lim_{u_i \to 0} \frac{f(\boldsymbol{x} + u_i \boldsymbol{e}_i) - f(\boldsymbol{x})}{u_i},$$

is the partial derivative of $f(\boldsymbol{x})$ with respect to x_i and $\boldsymbol{e}_i$ is the ith column of $\boldsymbol{I}_k$. The first-order Taylor series expansion is given by

$$f(\boldsymbol{x}+\boldsymbol{u}) = f(\boldsymbol{x}) + \sum_{i=1}^{k} \frac{\partial f(\boldsymbol{x})}{\partial x_i} u_i + r_1(\boldsymbol{u}, \boldsymbol{x}), \tag{A.13}$$

where the remainder $r_1(\boldsymbol{u}, \boldsymbol{x})$ satisfies

$$\lim_{\boldsymbol{u}\to 0} \frac{r_1(\boldsymbol{u}, \boldsymbol{x})}{(\boldsymbol{u}'\boldsymbol{u})^{1/2}} = 0.$$

The second term on the right side of Equation (A.13) is the *first differential* of $f(\boldsymbol{x})$ at $\boldsymbol{x}$ with incremental vector $\boldsymbol{u}$ and we write it as

$$df(\boldsymbol{x}) = d_{\boldsymbol{u}} f(\boldsymbol{x}) = \left[\frac{\partial f(\boldsymbol{x})}{\partial \boldsymbol{x}'}\right] \boldsymbol{u}.$$

It is important to note the relationship between the first differential and the first derivative. The first differential of $f(\boldsymbol{x})$ at $\boldsymbol{x}$ with incremental vector $\boldsymbol{u}$ is the first derivative of $f(\boldsymbol{x})$ at $\boldsymbol{x}$ *times* $\boldsymbol{u}$.

The higher-order differentials of $f(\boldsymbol{x})$ at $\boldsymbol{x}$ with the incremental vector $\boldsymbol{u}$ are defined as

$$d^i f(\boldsymbol{x}) = d^i_{\boldsymbol{u}} f(\boldsymbol{x}) = \sum_{j_1=1}^{k} \cdots \sum_{j_i}^{k} u_{j_1} \cdots u_{j_i} \frac{\partial^i f(\boldsymbol{x})}{\partial x_{j_1} \cdots \partial x_{j_i}},$$

and these differentials appear in the vth-order Taylor series expansion,

$$f(\boldsymbol{x}+\boldsymbol{u}) = f(\boldsymbol{x} + \sum_{i=1}^{v} \frac{d^i f(\boldsymbol{x})}{i!} + r_v(\boldsymbol{u}, \boldsymbol{x}),$$

where the remainder $r_k(\boldsymbol{u}, \boldsymbol{x})$ satisfies

$$\lim_{\boldsymbol{u}\to 0} \frac{r_v(\boldsymbol{u}, \boldsymbol{x})}{(\boldsymbol{u}'\boldsymbol{u})^{v/2}} = 0.$$

The second differential $d^2 f(\boldsymbol{x})$ can be written in a quadratic form in the vector $\boldsymbol{u}$ as

$$d^2 f(\boldsymbol{x}) = \boldsymbol{u}'\boldsymbol{H}_f \boldsymbol{u},$$

where $\boldsymbol{H}_f$ is the *Hessian* matrix of the second-order partial derivative defined in Section A.1.9.

A.6.2 Vector Function

The differentials can be generalized to the m-dimensional vector function $\boldsymbol{f}(\boldsymbol{x}) = [f_1(\boldsymbol{x}), \ldots, f_m(\boldsymbol{x})]'$, where $\boldsymbol{x} = (x_1, \ldots, x_k)'$. Here, $\boldsymbol{f}(.)$ is differentiable at $\boldsymbol{x}$ if and only if each component $f_i(.)$ is differentiable at $\boldsymbol{x}$. The Taylor expansion of the previous section can be applied component-wise to $\boldsymbol{f}(\boldsymbol{x})$. For instance, the first-order Taylor series expansion is

$$\begin{aligned}\boldsymbol{f}(\boldsymbol{x}+\boldsymbol{u}) &= \boldsymbol{f}(\boldsymbol{x}) + \left[\frac{\partial \boldsymbol{f}(\boldsymbol{x})}{\partial \boldsymbol{x}'}\right]\boldsymbol{u} + \boldsymbol{r}_1(\boldsymbol{u},\boldsymbol{x}) \\ &= \boldsymbol{f}(\boldsymbol{x}) + d\boldsymbol{f}(\boldsymbol{x}) + \boldsymbol{r}_1(\boldsymbol{u},\boldsymbol{x}),\end{aligned}$$

where the vector remainder $\boldsymbol{r}_1(\boldsymbol{u},\boldsymbol{x})$ satisfies

$$\lim_{\boldsymbol{u}\to\boldsymbol{0}} \frac{\boldsymbol{r}_1(\boldsymbol{u},\boldsymbol{x})}{(\boldsymbol{u}'\boldsymbol{u})^{1/2}} = \boldsymbol{0},$$

and the first *differential* of $\boldsymbol{f}(\boldsymbol{x})$ at $\boldsymbol{x}$ is given by the $m \times k$ matrix

$$\frac{\partial \boldsymbol{f}(\boldsymbol{x})}{\partial \boldsymbol{x}'} = \begin{bmatrix} \frac{\partial f_1(\boldsymbol{x})}{\partial x_1} & \frac{\partial f_1(\boldsymbol{x})}{\partial x_2} & \cdots & \frac{\partial f_1(\boldsymbol{x})}{\partial x_k} \\ \frac{\partial f_2(\boldsymbol{x})}{\partial x_1} & \frac{\partial f_2(\boldsymbol{x})}{\partial x_2} & \cdots & \frac{\partial f_2(\boldsymbol{x})}{\partial x_k} \\ \vdots & \vdots & & \vdots \\ \frac{\partial f_m(\boldsymbol{x})}{\partial x_1} & \frac{\partial f_m(\boldsymbol{x})}{\partial x_2} & \cdots & \frac{\partial f_m(\boldsymbol{x})}{\partial x_k} \end{bmatrix}.$$

This matrix of partial derivatives is referred to as the Jacobian matrix of $\boldsymbol{f}(\boldsymbol{x})$ at $\boldsymbol{x}$. Again, it is important to understand the relationship between the first differential and the first derivative. If we write the first differential as

$$d\boldsymbol{f}(\boldsymbol{x}) = \boldsymbol{D}\boldsymbol{u},$$

then the $m \times k$ matrix $\boldsymbol{D}$ consists of the first derivative of $\boldsymbol{f}(\boldsymbol{x})$ at $\boldsymbol{x}$.

If $y(\boldsymbol{x}) = g(\boldsymbol{f}(\boldsymbol{x}))$ is a real-valued function, then the generalization of chain rule is given by

$$\frac{\partial y(\boldsymbol{x})}{\partial x_i} = \sum_{j=1}^{m} \left[\frac{\partial g(\boldsymbol{f}(\boldsymbol{x}))}{\partial f_j(\boldsymbol{x})}\right]\left[\frac{\partial f_j(\boldsymbol{x})}{\partial x_i}\right] = \left[\frac{\partial g(\boldsymbol{f}(\boldsymbol{x}))}{\partial \boldsymbol{f}(\boldsymbol{x})'}\right]\left[\frac{\partial \boldsymbol{f}(\boldsymbol{x})}{\partial x_i}\right],$$

for $i = 1, \ldots, k$. Putting together, we have

$$\frac{\partial y(\boldsymbol{x})}{\partial \boldsymbol{x}'} = \left[\frac{\partial g(\boldsymbol{f}(\boldsymbol{x}))}{\partial \boldsymbol{f}(\boldsymbol{x})'}\right]\left[\frac{\partial \boldsymbol{f}(\boldsymbol{x})}{\partial \boldsymbol{x}'}\right].$$

A.6.3 Matrix Function

The most general case of functions considered in the book is the $p \times q$ matrix function

$$F(X) = \begin{bmatrix} f_{11}(X) & f_{12}(X) & \cdots & f_{1q}(X) \\ f_{21}(X) & f_{22}(X) & \cdots & f_{2q}(X) \\ \vdots & \vdots & & \vdots \\ f_{p1}(X) & f_{p2}(X) & \cdots & f_{pq}(X) \end{bmatrix},$$

where $X = [x_{ij}]$ is an $m \times k$ matrix. Results of the vector-valued function $f(x)$ of the previous section can be extended to the matrix-valued function $F(X)$ by the vectorization operator vec. Given $F(X)$, let $f(X)$ be the $pq \times 1$ vector-valued function such that $f(X) = \text{vec}(F(X))$. Then, results of the previous section apply. For instance, the Jacobian matrix of $F(X)$ at X is a $pq \times mk$ matrix given by

$$\frac{\partial f(X)}{\partial \text{vec}(X)'} = \frac{\partial \text{vec}(F(X))}{\partial \text{vec}(X)'},$$

where the $(i, j)$the element is the partial derivative of the ith element of vec$(F(X))$ with respect to the jth element of vec(X). This enables us to obtain the first-order Taylor series expansion of vec$(F(X + U))$. The differentials if the matrix $F(X)$ are defined as

$$\text{vec}(d^i F(X)) = \text{vec}[d^i_U F(X)] = d^i f(X) = d^i_{\text{vec}(U)} f(\text{vec}(X)).$$

In other words, the ith-order differential of $F(X)$ at X in the incremental matrix U, denoted by $d^i F(X)$, is defined as the $p \times q$ matrix obtained by unstacking the ith-order differential of f at vec(X) in the incremental vector vec(U).

A.6.4 Properties of Matrix Differential

Basic properties of matrix differentials follow straightforwardly from the corresponding properties of scalar differentials. If X and Y are matrix functions and A is a matrix of constants. Then we have

1. $dA = 0$.
2. $d(\alpha X) = \alpha dX$, where α is a real number.
3. $d(X') = (dX)'$.
4. $d(X + Y) = dX + dY$.
5. $d(XY) = (dX)Y + X(dY)$.
6. $d\text{tr}(X) = tr(dX)$.
7. $d\text{vec}(X) = \text{vec}(dX)$.
8. $d(X \otimes Y) = (dX) \otimes Y + X \otimes (dY)$.
9. $d(X \odot Y) = (dX) \odot Y + X \odot (dY)$.

where $\odot$ denotes the Hadamard product of matrices (element-by-element product). Some other properties of matrix differentials are given next. For proofs of thee results, see Schott (2005, Theorems 9.1 and 9.2).

Result 5 Suppose $\boldsymbol{X}$ is an $m \times m$ matrix and adj($\boldsymbol{X}$) denotes the adjoint matrix of $\boldsymbol{X}$. Then,

1. $d(\text{tr}(\boldsymbol{X}) = \text{vec}(\boldsymbol{I}_m)'d\text{vec}(\boldsymbol{X})$; $\quad \partial\text{tr}(\boldsymbol{X})/\partial\text{vec}(\boldsymbol{X})' = \text{vec}(\boldsymbol{I}_m)'$.
2. $d|\boldsymbol{X}| = \text{tr}[\text{adj}(\boldsymbol{X})d\boldsymbol{X}]$; $\quad \partial|\boldsymbol{X}|/\partial\text{vec}(\boldsymbol{X})' = \text{vec}[\text{adj}(\boldsymbol{X}')]'$.
3. If $\boldsymbol{X}$ is nonsingular,

$$d|\boldsymbol{X}| = |\boldsymbol{X}|\text{tr}(\boldsymbol{X}^{-1}d\boldsymbol{X}); \quad \frac{\partial|\boldsymbol{X}|}{\partial\text{vec}(\boldsymbol{X})'} = |\boldsymbol{X}|\text{vec}(\boldsymbol{X}^{-1'})'.$$

4. $d[\log(|\boldsymbol{X}|)] = \text{tr}(\boldsymbol{X}^{-1}d\boldsymbol{X})$; $\quad \partial\log(|\boldsymbol{X}|)/\partial\text{vec}(\boldsymbol{X})' = \text{vec}(\boldsymbol{X}^{-1'})'$.

Result 6 If $\boldsymbol{X}$ is a nonsingular $m \times m$ matrix, then

$$d\boldsymbol{X}^{-1} = -\boldsymbol{X}^{-1}(d\boldsymbol{X})\boldsymbol{X}^{-1}; \quad \frac{\partial\text{vec}(\boldsymbol{X}^{-1})}{\partial\text{vec}(\boldsymbol{X})'} = -(\boldsymbol{X}^{-1'} \otimes \boldsymbol{X}^{-1}).$$

A.6.5 Application

To demonstrate the application of matrix differentials, we consider the log likelihood function of a random sample $\boldsymbol{x}_1, \ldots, \boldsymbol{x}_T$ from the k-dimensional normal distribution such that the mean vector and covariance matrix of $\boldsymbol{x}_t$ are $\boldsymbol{\mu}_t$ and $\boldsymbol{\Sigma}_t$, respectively. Assume further that $\boldsymbol{\mu}_t$ and $\boldsymbol{\Sigma}_t$ are functions of parameter $\boldsymbol{\theta}$. By ignoring the constant term, the log likelihood function is

$$\ell(\boldsymbol{\theta}) = \sum_{t=1}^{T} \ell_t(\boldsymbol{\theta}), \quad \ell_t(\boldsymbol{\theta}) = -\frac{1}{2}\log(|\boldsymbol{\Sigma}_t|) - \frac{1}{2}tr(\boldsymbol{\Sigma}_t^{-1}\boldsymbol{U}_t),$$

where $\boldsymbol{U}_t = (\boldsymbol{x}_t - \boldsymbol{\mu}_t)(\boldsymbol{x}_t - \boldsymbol{\mu}_t)'$. The first differential of $\ell_t(\boldsymbol{\theta})$ is

$$\begin{aligned} d\ell_t(\boldsymbol{\theta}) &= -\tfrac{1}{2}d[\log(|\boldsymbol{\Sigma}_t|)] - \tfrac{1}{2}\text{tr}[(d\boldsymbol{\Sigma}_t^{-1})\boldsymbol{U}_t] - \tfrac{1}{2}\text{tr}[\boldsymbol{\Sigma}_t^{-1}d\boldsymbol{U}_t] \\ &= -\tfrac{1}{2}\text{tr}(\boldsymbol{\Sigma}_t^{-1}d\boldsymbol{\Sigma}_t) + \tfrac{1}{2}\text{tr}[\boldsymbol{\Sigma}_t^{-1}(d\boldsymbol{\Sigma}_t)\boldsymbol{\Sigma}_t^{-1}\boldsymbol{U}_t] \\ &\quad + \tfrac{1}{2}\text{tr}[\boldsymbol{\Sigma}_t^{-1}\{(d\boldsymbol{\mu}_t)(\boldsymbol{x}_t - \boldsymbol{\mu}_t)' + (\boldsymbol{x}_t - \boldsymbol{\mu}_t)d\boldsymbol{\mu}_t'\}] \\ &= \tfrac{1}{2}\text{tr}[(d\boldsymbol{\Sigma}_t)\boldsymbol{\Sigma}_t^{-1}(\boldsymbol{U}_t - \boldsymbol{\Sigma}_t)\boldsymbol{\Sigma}_t^{-1}] \end{aligned}$$

$$+\tfrac{1}{2}\text{tr}[\boldsymbol{\Sigma}_t^{-1}\{(d\boldsymbol{\mu}_t)(\boldsymbol{x}_t-\boldsymbol{\mu}_t)'+(\boldsymbol{x}_t-\boldsymbol{\mu}_t)d\boldsymbol{\mu}_t'\}]$$
$$=\tfrac{1}{2}\text{tr}[(d\boldsymbol{\Sigma}_t)\boldsymbol{\Sigma}_t^{-1}(\boldsymbol{U}_t-\boldsymbol{\Sigma}_t)\boldsymbol{\Sigma}_t^{-1}]+(\boldsymbol{x}_t-\boldsymbol{\mu}_t)'\boldsymbol{\Sigma}_t^{-1}d\boldsymbol{\mu}_t$$
$$=\tfrac{1}{2}[\text{vec}(d\boldsymbol{\Sigma}_t)]'(\boldsymbol{\Sigma}_t^{-1}\otimes\boldsymbol{\Sigma}_t^{-1})\text{vec}(\boldsymbol{U}_t-\boldsymbol{\Sigma}_t)+(\boldsymbol{x}_t-\boldsymbol{\mu}_t)'\boldsymbol{\Sigma}_t^{-1}d\boldsymbol{\mu}_t,$$

where the second equality holds via Results 5(4) and Result 6, whereas the fifth equality applied the property $\text{tr}(\boldsymbol{ABCD})=[\text{vec}(\boldsymbol{A}')]'(\boldsymbol{D}'\otimes\boldsymbol{B})\text{vec}(\boldsymbol{C})$. Using the prior equality and chain rule, we have

$$\frac{\partial \ell_t(\boldsymbol{\theta})}{\partial \boldsymbol{\theta}'}=(\boldsymbol{x}_t-\boldsymbol{\mu}_t)'\boldsymbol{\Sigma}_t^{-1}\frac{\partial\boldsymbol{\mu}_t}{\partial\boldsymbol{\theta}'}$$
$$+\frac{1}{2}[\text{vec}(\boldsymbol{U}_t-\boldsymbol{\Sigma}_t)]'(\boldsymbol{\Sigma}_t^{-1}\otimes\boldsymbol{\Sigma}_t^{-1})\frac{\partial\text{vec}(\boldsymbol{\Sigma}_t)}{\partial\boldsymbol{\theta}'}.$$

These results are used in multivariate volatility modeling of Chapter 7.

REFERENCES

Abadir, K. M. and Magnus, J. R. (2005). *Matrix Algebra*. Cambridge University Press, New York.

Anderson, T. W. (2003). *An Introduction to Multivariate Statistical Analysis*. 3rd Edition. John Wiley & Sons, Inc, Hoboken, NJ.

Graybill, F. A. (1969). *Introduction to Matrices with Applications in Statistics*. Wadsworth, Belmont, CA.

Johnson, R. A. and Wichern, D. W. (2007). *Applied Multivariate Statistical Analysis*. 6th edition. Prentice Hall, Upper Saddle River, NJ.

Lütkepohl, H. (2005). *New Introduction to Multiple Time Series Analysis*. Springer Verlag, Berlin.

Magnus, J. R. and Neudecker, H. (1999). *Matrix Differential Calculus with Applications in Statistics and Econometrics*. Revised Edition. John Wiley & Sons, Inc, New York.

Schott, J. R. (2005). *Matrix Analysis for Statistics*. 2nd Edition. John Wiley & Sons, Inc, Hoboken, NJ.

Strang, G. (1980). *Linear Algebra and its Applications*. 2nd Edition. Harcourt Brace Jovanovich, Chicago, IL.

Index

Multivariate Time Series Analysis: With R and Financial Applications,
First Edition. Ruey S. Tsay.

WILEY SERIES IN PROBABILITY AND STATISTICS
ESTABLISHED BY WALTER A. SHEWHART AND SAMUEL S. WILKS

Editors: *David J. Balding, Noel A. C. Cressie, Garrett M. Fitzmaurice, Harvey Goldstein, Iain M. Johnstone, Geert Molenberghs, David W. Scott, Adrian F. M. Smith, Ruey S. Tsay, Sanford Weisberg*
Editors Emeriti: *Vic Barnett, J. Stuart Hunter, Joseph B. Kadane, Jozef L. Teugels*

The ***Wiley Series in Probability and Statistics*** is well established and authoritative. It covers many topics of current research interest in both pure and applied statistics and probability theory. Written by leading statisticians and institutions, the titles span both state-of-the-art developments in the field and classical methods.

Reflecting the wide range of current research in statistics, the series encompasses applied, methodological and theoretical statistics, ranging from applications and new techniques made possible by advances in computerized practice to rigorous treatment of theoretical approaches.

This series provides essential and invaluable reading for all statisticians, whether in academia, industry, government, or research.

† ABRAHAM and LEDOLTER · Statistical Methods for Forecasting
AGRESTI · Analysis of Ordinal Categorical Data, *Second Edition*
AGRESTI · An Introduction to Categorical Data Analysis, *Second Edition*
AGRESTI · Categorical Data Analysis, *Third Edition*
ALTMAN, GILL, and McDONALD · Numerical Issues in Statistical Computing for the Social Scientist
AMARATUNGA and CABRERA · Exploration and Analysis of DNA Microarray and Protein Array Data
AMARATUNGA, CABRERA, and SHKEDY . Exploration and Analysis of DNA Microarray and Other High-Dimensional Data, *Second Edition*
ANDĚL · Mathematics of Chance
ANDERSON · An Introduction to Multivariate Statistical Analysis, *Third Edition*
* ANDERSON · The Statistical Analysis of Time Series
ANDERSON, AUQUIER, HAUCK, OAKES, VANDAELE, and WEISBERG · Statistical Methods for Comparative Studies
ANDERSON and LOYNES · The Teaching of Practical Statistics
ARMITAGE and DAVID (editors) · Advances in Biometry
ARNOLD, BALAKRISHNAN, and NAGARAJA · Records
* ARTHANARI and DODGE · Mathematical Programming in Statistics
* BAILEY · The Elements of Stochastic Processes with Applications to the Natural Sciences
BAJORSKI · Statistics for Imaging, Optics, and Photonics
BALAKRISHNAN and KOUTRAS · Runs and Scans with Applications
BALAKRISHNAN and NG · Precedence-Type Tests and Applications
BARNETT · Comparative Statistical Inference, *Third Edition*
BARNETT · Environmental Statistics
BARNETT and LEWIS · Outliers in Statistical Data, *Third Edition*
BARTHOLOMEW, KNOTT, and MOUSTAKI · Latent Variable Models and Factor Analysis: A Unified Approach, *Third Edition*
BARTOSZYNSKI and NIEWIADOMSKA-BUGAJ · Probability and Statistical Inference, *Second Edition*
BASILEVSKY · Statistical Factor Analysis and Related Methods: Theory and Applications

*Now available in a lower priced paperback edition in the Wiley Classics Library.
†Now available in a lower priced paperback edition in the Wiley–Interscience Paperback Series.

BATES and WATTS · Nonlinear Regression Analysis and Its Applications
BECHHOFER, SANTNER, and GOLDSMAN · Design and Analysis of Experiments for Statistical Selection, Screening, and Multiple Comparisons
BEIRLANT, GOEGEBEUR, SEGERS, TEUGELS, and DE WAAL · Statistics of Extremes: Theory and Applications
BELSLEY · Conditioning Diagnostics: Collinearity and Weak Data in Regression
† BELSLEY, KUH, and WELSCH · Regression Diagnostics: Identifying Influential Data and Sources of Collinearity
BENDAT and PIERSOL · Random Data: Analysis and Measurement Procedures, *Fourth Edition*
BERNARDO and SMITH · Bayesian Theory
BHAT and MILLER · Elements of Applied Stochastic Processes, *Third Edition*
BHATTACHARYA and WAYMIRE · Stochastic Processes with Applications
BIEMER, GROVES, LYBERG, MATHIOWETZ, and SUDMAN · Measurement Errors in Surveys
BILLINGSLEY · Convergence of Probability Measures, *Second Edition*
BILLINGSLEY · Probability and Measure, *Anniversary Edition*
BIRKES and DODGE · Alternative Methods of Regression
BISGAARD and KULAHCI · Time Series Analysis and Forecasting by Example
BISWAS, DATTA, FINE, and SEGAL · Statistical Advances in the Biomedical Sciences: Clinical Trials, Epidemiology, Survival Analysis, and Bioinformatics
BLISCHKE and MURTHY (editors) · Case Studies in Reliability and Maintenance
BLISCHKE and MURTHY · Reliability: Modeling, Prediction, and Optimization
BLOOMFIELD · Fourier Analysis of Time Series: An Introduction, *Second Edition*
BOLLEN · Structural Equations with Latent Variables
BOLLEN and CURRAN · Latent Curve Models: A Structural Equation Perspective
BOROVKOV · Ergodicity and Stability of Stochastic Processes
BOSQ and BLANKE · Inference and Prediction in Large Dimensions
BOULEAU · Numerical Methods for Stochastic Processes
* BOX and TIAO · Bayesian Inference in Statistical Analysis
BOX · Improving Almost Anything, *Revised Edition*
* BOX and DRAPER · Evolutionary Operation: A Statistical Method for Process Improvement
BOX and DRAPER · Response Surfaces, Mixtures, and Ridge Analyses, *Second Edition*
BOX, HUNTER, and HUNTER · Statistics for Experimenters: Design, Innovation, and Discovery, *Second Editon*
BOX, JENKINS, and REINSEL · Time Series Analysis: Forcasting and Control, *Fourth Edition*
BOX, LUCEÑO, and PANIAGUA-QUIÑONES · Statistical Control by Monitoring and Adjustment, *Second Edition*
* BROWN and HOLLANDER · Statistics: A Biomedical Introduction
CAIROLI and DALANG · Sequential Stochastic Optimization
CASTILLO, HADI, BALAKRISHNAN, and SARABIA · Extreme Value and Related Models with Applications in Engineering and Science
CHAN · Time Series: Applications to Finance with R and S-Plus®, *Second Edition*
CHARALAMBIDES · Combinatorial Methods in Discrete Distributions
CHATTERJEE and HADI · Regression Analysis by Example, *Fourth Edition*
CHATTERJEE and HADI · Sensitivity Analysis in Linear Regression
CHERNICK · Bootstrap Methods: A Guide for Practitioners and Researchers, *Second Edition*
CHERNICK and FRIIS · Introductory Biostatistics for the Health Sciences
CHILÈS and DELFINER · Geostatistics: Modeling Spatial Uncertainty, *Second Edition*
CHOW and LIU · Design and Analysis of Clinical Trials: Concepts and Methodologies, *Third Edition*

*Now available in a lower priced paperback edition in the Wiley Classics Library.
†Now available in a lower priced paperback edition in the Wiley–Interscience Paperback Series.

CLARKE · Linear Models: The Theory and Application of Analysis of Variance
CLARKE and DISNEY · Probability and Random Processes: A First Course with Applications, *Second Edition*
* COCHRAN and COX · Experimental Designs, *Second Edition*
COLLINS and LANZA · Latent Class and Latent Transition Analysis: With Applications in the Social, Behavioral, and Health Sciences
CONGDON · Applied Bayesian Modelling
CONGDON · Bayesian Models for Categorical Data
CONGDON · Bayesian Statistical Modelling, *Second Edition*
CONOVER · Practical Nonparametric Statistics, *Third Edition*
COOK · Regression Graphics
COOK and WEISBERG · An Introduction to Regression Graphics
COOK and WEISBERG · Applied Regression Including Computing and Graphics
CORNELL · A Primer on Experiments with Mixtures
CORNELL · Experiments with Mixtures, Designs, Models, and the Analysis of Mixture Data, *Third Edition*
COX · A Handbook of Introductory Statistical Methods
CRESSIE · Statistics for Spatial Data, *Revised Edition*
CRESSIE and WIKLE · Statistics for Spatio-Temporal Data
CSÖRGŐ and HORVÁTH · Limit Theorems in Change Point Analysis
DAGPUNAR · Simulation and Monte Carlo: With Applications in Finance and MCMC
DANIEL · Applications of Statistics to Industrial Experimentation
DANIEL · Biostatistics: A Foundation for Analysis in the Health Sciences, *Eighth Edition*
* DANIEL · Fitting Equations to Data: Computer Analysis of Multifactor Data, *Second Edition*
DASU and JOHNSON · Exploratory Data Mining and Data Cleaning
DAVID and NAGARAJA · Order Statistics, *Third Edition*
* DEGROOT, FIENBERG, and KADANE · Statistics and the Law
DEL CASTILLO · Statistical Process Adjustment for Quality Control
DeMARIS · Regression with Social Data: Modeling Continuous and Limited Response Variables
DEMIDENKO · Mixed Models: Theory and Applications with R, *Second Edition*
DENISON, HOLMES, MALLICK and SMITH · Bayesian Methods for Nonlinear Classification and Regression
DETTE and STUDDEN · The Theory of Canonical Moments with Applications in Statistics, Probability, and Analysis
DEY and MUKERJEE · Fractional Factorial Plans
DILLON and GOLDSTEIN · Multivariate Analysis: Methods and Applications
* DODGE and ROMIG · Sampling Inspection Tables, *Second Edition*
* DOOB · Stochastic Processes
DOWDY, WEARDEN, and CHILKO · Statistics for Research, *Third Edition*
DRAPER and SMITH · Applied Regression Analysis, *Third Edition*
DRYDEN and MARDIA · Statistical Shape Analysis
DUDEWICZ and MISHRA · Modern Mathematical Statistics
DUNN and CLARK · Basic Statistics: A Primer for the Biomedical Sciences, *Fourth Edition*
DUPUIS and ELLIS · A Weak Convergence Approach to the Theory of Large Deviations
EDLER and KITSOS · Recent Advances in Quantitative Methods in Cancer and Human Health Risk Assessment
* ELANDT-JOHNSON and JOHNSON · Survival Models and Data Analysis
ENDERS · Applied Econometric Time Series, *Third Edition*
† ETHIER and KURTZ · Markov Processes: Characterization and Convergence

*Now available in a lower priced paperback edition in the Wiley Classics Library.
†Now available in a lower priced paperback edition in the Wiley–Interscience Paperback Series.

EVANS, HASTINGS, and PEACOCK · Statistical Distributions, *Third Edition*
EVERITT, LANDAU, LEESE, and STAHL · Cluster Analysis, *Fifth Edition*
FEDERER and KING · Variations on Split Plot and Split Block Experiment Designs
FELLER · An Introduction to Probability Theory and Its Applications, Volume I, *Third Edition,* Revised; Volume II, *Second Edition*
FITZMAURICE, LAIRD, and WARE · Applied Longitudinal Analysis, *Second Edition*
* FLEISS · The Design and Analysis of Clinical Experiments
FLEISS · Statistical Methods for Rates and Proportions, *Third Edition*
† FLEMING and HARRINGTON · Counting Processes and Survival Analysis
FUJIKOSHI, ULYANOV, and SHIMIZU · Multivariate Statistics: High-Dimensional and Large-Sample Approximations
FULLER · Introduction to Statistical Time Series, *Second Edition*
† FULLER · Measurement Error Models
GALLANT · Nonlinear Statistical Models
GEISSER · Modes of Parametric Statistical Inference
GELMAN and MENG · Applied Bayesian Modeling and Causal Inference from ncomplete-Data Perspectives
GEWEKE · Contemporary Bayesian Econometrics and Statistics
GHOSH, MUKHOPADHYAY, and SEN · Sequential Estimation
GIESBRECHT and GUMPERTZ · Planning, Construction, and Statistical Analysis of Comparative Experiments
GIFI · Nonlinear Multivariate Analysis
GIVENS and HOETING · Computational Statistics
GLASSERMAN and YAO · Monotone Structure in Discrete-Event Systems
GNANADESIKAN · Methods for Statistical Data Analysis of Multivariate Observations, *Second Edition*
GOLDSTEIN · Multilevel Statistical Models, *Fourth Edition*
GOLDSTEIN and LEWIS · Assessment: Problems, Development, and Statistical Issues
GOLDSTEIN and WOOFF · Bayes Linear Statistics
GREENWOOD and NIKULIN · A Guide to Chi-Squared Testing
GROSS, SHORTLE, THOMPSON, and HARRIS · Fundamentals of Queueing Theory, *Fourth Edition*
GROSS, SHORTLE, THOMPSON, and HARRIS · Solutions Manual to Accompany Fundamentals of Queueing Theory, *Fourth Edition*
* HAHN and SHAPIRO · Statistical Models in Engineering
HAHN and MEEKER · Statistical Intervals: A Guide for Practitioners
HALD · A History of Probability and Statistics and their Applications Before 1750
† HAMPEL · Robust Statistics: The Approach Based on Influence Functions
HARTUNG, KNAPP, and SINHA · Statistical Meta-Analysis with Applications
HEIBERGER · Computation for the Analysis of Designed Experiments
HEDAYAT and SINHA · Design and Inference in Finite Population Sampling
HEDEKER and GIBBONS · Longitudinal Data Analysis
HELLER · MACSYMA for Statisticians
HERITIER, CANTONI, COPT, and VICTORIA-FESER · Robust Methods in Biostatistics
HINKELMANN and KEMPTHORNE · Design and Analysis of Experiments, Volume 1: Introduction to Experimental Design, *Second Edition*
HINKELMANN and KEMPTHORNE · Design and Analysis of Experiments, Volume 2: Advanced Experimental Design
HINKELMANN (editor) · Design and Analysis of Experiments, Volume 3: Special Designs and Applications
HOAGLIN, MOSTELLER, and TUKEY · Fundamentals of Exploratory Analysis of Variance

*Now available in a lower priced paperback edition in the Wiley Classics Library.
†Now available in a lower priced paperback edition in the Wiley–Interscience Paperback Series.

* HOAGLIN, MOSTELLER, and TUKEY · Exploring Data Tables, Trends and Shapes
* HOAGLIN, MOSTELLER, and TUKEY · Understanding Robust and Exploratory Data Analysis

HOCHBERG and TAMHANE · Multiple Comparison Procedures

HOCKING · Methods and Applications of Linear Models: Regression and the Analysis of Variance, *Third Edition*

HOEL · Introduction to Mathematical Statistics, *Fifth Edition*

HOGG and KLUGMAN · Loss Distributions

HOLLANDER, WOLFE, and CHICKEN · Nonparametric Statistical Methods, *Third Edition*

HOSMER and LEMESHOW · Applied Logistic Regression, *Second Edition*

HOSMER, LEMESHOW, and MAY · Applied Survival Analysis: Regression Modeling of Time-to-Event Data, *Second Edition*

HUBER · Data Analysis: What Can Be Learned From the Past 50 Years

HUBER · Robust Statistics

† HUBER and RONCHETTI · Robust Statistics, *Second Edition*

HUBERTY · Applied Discriminant Analysis, *Second Edition*

HUBERTY and OLEJNIK · Applied MANOVA and Discriminant Analysis, *Second Edition*

HUITEMA · The Analysis of Covariance and Alternatives: Statistical Methods for Experiments, Quasi-Experiments, and Single-Case Studies, *Second Edition*

HUNT and KENNEDY · Financial Derivatives in Theory and Practice, *Revised Edition*

HURD and MIAMEE · Periodically Correlated Random Sequences: Spectral Theory and Practice

HUSKOVA, BERAN, and DUPAC · Collected Works of Jaroslav Hajek— with Commentary

HUZURBAZAR · Flowgraph Models for Multistate Time-to-Event Data

JACKMAN · Bayesian Analysis for the Social Sciences

† JACKSON · A User's Guide to Principle Components

JOHN · Statistical Methods in Engineering and Quality Assurance

JOHNSON · Multivariate Statistical Simulation

JOHNSON and BALAKRISHNAN · Advances in the Theory and Practice of Statistics: A Volume in Honor of Samuel Kotz

JOHNSON, KEMP, and KOTZ · Univariate Discrete Distributions, *Third Edition*

JOHNSON and KOTZ (editors) · Leading Personalities in Statistical Sciences: From the Seventeenth Century to the Present

JOHNSON, KOTZ, and BALAKRISHNAN · Continuous Univariate Distributions, Volume 1, *Second Edition*

JOHNSON, KOTZ, and BALAKRISHNAN · Continuous Univariate Distributions, Volume 2, *Second Edition*

JOHNSON, KOTZ, and BALAKRISHNAN · Discrete Multivariate Distributions

JUDGE, GRIFFITHS, HILL, LÜTKEPOHL, and LEE · The Theory and Practice of Econometrics, *Second Edition*

JUREK and MASON · Operator-Limit Distributions in Probability Theory

KADANE · Bayesian Methods and Ethics in a Clinical Trial Design

KADANE AND SCHUM · A Probabilistic Analysis of the Sacco and Vanzetti Evidence

KALBFLEISCH and PRENTICE · The Statistical Analysis of Failure Time Data, *Second Edition*

KARIYA and KURATA · Generalized Least Squares

KASS and VOS · Geometrical Foundations of Asymptotic Inference

† KAUFMAN and ROUSSEEUW · Finding Groups in Data: An Introduction to Cluster Analysis

KEDEM and FOKIANOS · Regression Models for Time Series Analysis

KENDALL, BARDEN, CARNE, and LE · Shape and Shape Theory

*Now available in a lower priced paperback edition in the Wiley Classics Library.

†Now available in a lower priced paperback edition in the Wiley–Interscience Paperback Series.

KHURI · Advanced Calculus with Applications in Statistics, *Second Edition*
KHURI, MATHEW, and SINHA · Statistical Tests for Mixed Linear Models
* KISH · Statistical Design for Research
KLEIBER and KOTZ · Statistical Size Distributions in Economics and Actuarial Sciences
KLEMELÄ · Smoothing of Multivariate Data: Density Estimation and Visualization
KLUGMAN, PANJER, and WILLMOT · Loss Models: From Data to Decisions, *Third Edition*
KLUGMAN, PANJER, and WILLMOT · Loss Models: Further Topics
KLUGMAN, PANJER, and WILLMOT · Solutions Manual to Accompany Loss Models: From Data to Decisions, *Third Edition*
KOSKI and NOBLE · Bayesian Networks: An Introduction
KOTZ, BALAKRISHNAN, and JOHNSON · Continuous Multivariate Distributions, Volume 1, *Second Edition*
KOTZ and JOHNSON (editors) · Encyclopedia of Statistical Sciences: Volumes 1 to 9 with Index
KOTZ and JOHNSON (editors) · Encyclopedia of Statistical Sciences: Supplement Volume
KOTZ, READ, and BANKS (editors) · Encyclopedia of Statistical Sciences: Update Volume 1
KOTZ, READ, and BANKS (editors) · Encyclopedia of Statistical Sciences: Update Volume 2
KOWALSKI and TU · Modern Applied U-Statistics
KRISHNAMOORTHY and MATHEW · Statistical Tolerance Regions: Theory, Applications, and Computation
KROESE, TAIMRE, and BOTEV · Handbook of Monte Carlo Methods
KROONENBERG · Applied Multiway Data Analysis
KULINSKAYA, MORGENTHALER, and STAUDTE · Meta Analysis: A Guide to Calibrating and Combining Statistical Evidence
KULKARNI and HARMAN · An Elementary Introduction to Statistical Learning Theory
KUROWICKA and COOKE · Uncertainty Analysis with High Dimensional Dependence Modelling
KVAM and VIDAKOVIC · Nonparametric Statistics with Applications to Science and Engineering
LACHIN · Biostatistical Methods: The Assessment of Relative Risks, *Second Edition*
LAD · Operational Subjective Statistical Methods: A Mathematical, Philosophical, and Historical Introduction
LAMPERTI · Probability: A Survey of the Mathematical Theory, *Second Edition*
LAWLESS · Statistical Models and Methods for Lifetime Data, *Second Edition*
LAWSON · Statistical Methods in Spatial Epidemiology, *Second Edition*
LE · Applied Categorical Data Analysis, *Second Edition*
LE · Applied Survival Analysis
LEE · Structural Equation Modeling: A Bayesian Approach
LEE and WANG · Statistical Methods for Survival Data Analysis, *Fourth Edition*
LEPAGE and BILLARD · Exploring the Limits of Bootstrap
LESSLER and KALSBEEK · Nonsampling Errors in Surveys
LEYLAND and GOLDSTEIN (editors) · Multilevel Modelling of Health Statistics
LIAO · Statistical Group Comparison
LIN · Introductory Stochastic Analysis for Finance and Insurance
LITTLE and RUBIN · Statistical Analysis with Missing Data, *Second Edition*
LLOYD · The Statistical Analysis of Categorical Data
LOWEN and TEICH · Fractal-Based Point Processes
MAGNUS and NEUDECKER · Matrix Differential Calculus with Applications in Statistics and Econometrics, *Revised Edition*

*Now available in a lower priced paperback edition in the Wiley Classics Library.
†Now available in a lower priced paperback edition in the Wiley–Interscience Paperback Series.

MALLER and ZHOU · Survival Analysis with Long Term Survivors
MARCHETTE · Random Graphs for Statistical Pattern Recognition
MARDIA and JUPP · Directional Statistics
MARKOVICH · Nonparametric Analysis of Univariate Heavy-Tailed Data: Research and Practice
MARONNA, MARTIN and YOHAI · Robust Statistics: Theory and Methods
MASON, GUNST, and HESS · Statistical Design and Analysis of Experiments with Applications to Engineering and Science, *Second Edition*
McCULLOCH, SEARLE, and NEUHAUS · Generalized, Linear, and Mixed Models, *Second Edition*
McFADDEN · Management of Data in Clinical Trials, *Second Edition*
* McLACHLAN · Discriminant Analysis and Statistical Pattern Recognition
McLACHLAN, DO, and AMBROISE · Analyzing Microarray Gene Expression Data
McLACHLAN and KRISHNAN · The EM Algorithm and Extensions, *Second Edition*
McLACHLAN and PEEL · Finite Mixture Models
McNEIL · Epidemiological Research Methods
MEEKER and ESCOBAR · Statistical Methods for Reliability Data
MEERSCHAERT and SCHEFFLER · Limit Distributions for Sums of Independent Random Vectors: Heavy Tails in Theory and Practice
MENGERSEN, ROBERT, and TITTERINGTON · Mixtures: Estimation and Applications
MICKEY, DUNN, and CLARK · Applied Statistics: Analysis of Variance and Regression, *Third Edition*
* MILLER · Survival Analysis, *Second Edition*
MONTGOMERY, JENNINGS, and KULAHCI · Introduction to Time Series Analysis and Forecasting
MONTGOMERY, PECK, and VINING · Introduction to Linear Regression Analysis, *Fifth Edition*
MORGENTHALER and TUKEY · Configural Polysampling: A Route to Practical Robustness
MUIRHEAD · Aspects of Multivariate Statistical Theory
MULLER and STOYAN · Comparison Methods for Stochastic Models and Risks
MURTHY, XIE, and JIANG · Weibull Models
MYERS, MONTGOMERY, and ANDERSON-COOK · Response Surface Methodology: Process and Product Optimization Using Designed Experiments, *Third Edition*
MYERS, MONTGOMERY, VINING, and ROBINSON · Generalized Linear Models. With Applications in Engineering and the Sciences, *Second Edition*
NATVIG · Multistate Systems Reliability Theory With Applications
† NELSON · Accelerated Testing, Statistical Models, Test Plans, and Data Analyses
† NELSON · Applied Life Data Analysis
NEWMAN · Biostatistical Methods in Epidemiology
NG, TAIN, and TANG · Dirichlet Theory: Theory, Methods and Applications
OKABE, BOOTS, SUGIHARA, and CHIU · Spatial Tesselations: Concepts and Applications of Voronoi Diagrams, *Second Edition*
OLIVER and SMITH · Influence Diagrams, Belief Nets and Decision Analysis
PALTA · Quantitative Methods in Population Health: Extensions of Ordinary Regressions
PANJER · Operational Risk: Modeling and Analytics
PANKRATZ · Forecasting with Dynamic Regression Models
PANKRATZ · Forecasting with Univariate Box-Jenkins Models: Concepts and Cases
PARDOUX · Markov Processes and Applications: Algorithms, Networks, Genome and Finance
PARMIGIANI and INOUE · Decision Theory: Principles and Approaches
* PARZEN · Modern Probability Theory and Its Applications
PEÑA, TIAO, and TSAY · A Course in Time Series Analysis

*Now available in a lower priced paperback edition in the Wiley Classics Library.
†Now available in a lower priced paperback edition in the Wiley–Interscience Paperback Series.

PESARIN and SALMASO · Permutation Tests for Complex Data: Applications and Software
PIANTADOSI · Clinical Trials: A Methodologic Perspective, *Second Edition*
POURAHMADI · Foundations of Time Series Analysis and Prediction Theory
POURAHMADI · High-Dimensional Covariance Estimation
POWELL · Approximate Dynamic Programming: Solving the Curses of Dimensionality, *Second Edition*
POWELL and RYZHOV · Optimal Learning
PRESS · Subjective and Objective Bayesian Statistics, *Second Edition*
PRESS and TANUR · The Subjectivity of Scientists and the Bayesian Approach
PURI, VILAPLANA, and WERTZ · New Perspectives in Theoretical and Applied Statistics
† PUTERMAN · Markov Decision Processes: Discrete Stochastic Dynamic Programming
QIU · Image Processing and Jump Regression Analysis
* RAO · Linear Statistical Inference and Its Applications, *Second Edition*
RAO · Statistical Inference for Fractional Diffusion Processes
RAUSAND and HØYLAND · System Reliability Theory: Models, Statistical Methods, and Applications, *Second Edition*
RAYNER, THAS, and BEST · Smooth Tests of Goodnes of Fit: Using R, *Second Edition*
RENCHER and SCHAALJE · Linear Models in Statistics, *Second Edition*
RENCHER and CHRISTENSEN · Methods of Multivariate Analysis, *Third Edition*
RENCHER · Multivariate Statistical Inference with Applications
RIGDON and BASU · Statistical Methods for the Reliability of Repairable Systems
* RIPLEY · Spatial Statistics
* RIPLEY · Stochastic Simulation
ROHATGI and SALEH · An Introduction to Probability and Statistics, *Second Edition*
ROLSKI, SCHMIDLI, SCHMIDT, and TEUGELS · Stochastic Processes for Insurance and Finance
ROSENBERGER and LACHIN · Randomization in Clinical Trials: Theory and Practice
ROSSI, ALLENBY, and McCULLOCH · Bayesian Statistics and Marketing
† ROUSSEEUW and LEROY · Robust Regression and Outlier Detection
ROYSTON and SAUERBREI · Multivariate Model Building: A Pragmatic Approach to Regression Analysis Based on Fractional Polynomials for Modeling Continuous Variables
* RUBIN · Multiple Imputation for Nonresponse in Surveys
RUBINSTEIN and KROESE · Simulation and the Monte Carlo Method, *Second Edition*
RUBINSTEIN and MELAMED · Modern Simulation and Modeling
RUBINSTEIN, RIDDER, and VAISMAN · Fast Sequential Monte Carlo Methods for Counting and Optimization
RYAN · Modern Engineering Statistics
RYAN · Modern Experimental Design
RYAN · Modern Regression Methods, *Second Edition*
RYAN · Sample Size Determination and Power
RYAN · Statistical Methods for Quality Improvement, *Third Edition*
SALEH · Theory of Preliminary Test and Stein-Type Estimation with Applications
SALTELLI, CHAN, and SCOTT (editors) · Sensitivity Analysis
SCHERER · Batch Effects and Noise in Microarray Experiments: Sources and Solutions
* SCHEFFE · The Analysis of Variance
SCHIMEK · Smoothing and Regression: Approaches, Computation, and Application
SCHOTT · Matrix Analysis for Statistics, *Second Edition*
SCHOUTENS · Levy Processes in Finance: Pricing Financial Derivatives
SCOTT · Multivariate Density Estimation: Theory, Practice, and Visualization
* SEARLE · Linear Models
† SEARLE · Linear Models for Unbalanced Data

*Now available in a lower priced paperback edition in the Wiley Classics Library.
†Now available in a lower priced paperback edition in the Wiley–Interscience Paperback Series.